D0816615

Fundamentals of Palaeobotany

Fundamentals of Palaeobotany

Sergei V. Meyen
Geological Institute
USSR Academy of Sciences, Moscow

Illustrations by the author

LONDON NEW YORK
Chapman and Hall

First published in 1987 by
Chapman and Hall Ltd
11 New Fetter Lane, London EC4P 4EE
Published in the USA by
Chapman and Hall
29 West 35th Street, New York NY 10001

Printed in Great Britain at the
University Press, Cambridge

ISBN 0 412 27110 9

British Library Cataloguing in Publication Data

Meyen, Sergei V.
Fundamentals of palaeobotany.
1. Palaeobotany
I. Title II. Osnovy paleobotaniki. *English*
561 QE905

ISBN 0-412-27110-9

Library of Congress Cataloging in Publication Data

Meyen, Sergeĭ Viktorovich.
Fundamentals of palaeobotany.

Bibliography: p.
Includes index.
1. Paleobotany. I. Title.
QE904.A1M45 561 86–13000
ISBN 0-412-27110-9

Contents

Foreword

There have been at least ten English-language textbooks of palaeobotany since D. H. Scott published the first edition of *Studies in Fossil Botany* in 1900. Most have been written by scientists who were primarily botanists by training, and were aimed largely at a readership familiar with living plants. They tended to follow a general pattern of an introductory chapter on preservation of plants as fossils, followed by a systematic treatment, group by group. Only Seward in his *Plant Life Through the Ages* departed from this pattern in presenting a chronological sequence.

In the present book, Meyen breaks with this tradition. Although having a basically biological approach, he reaches out into all aspects of the history of plant life and the wider implication of its study. Only half of the present work deals sequentially with fossil plant groups, treated systematically. The remainder then explores those topics which most other textbooks have generally either ignored or have only mentioned rather incidentally – the problems of naming and classifying fragmentary plant fossils, their ecology; biogeography and palaeoclimatic significance and the contribution that they have made to the understanding of living plant morphology, and of the process of evolution.

A further important feature of Meyen's approach to palaeobotany is that he brings palaeopalynology (the study of fossil spores) into the body of the subject, emphasizing the common ground between the study of spores and the plants that produced them. The application of palynology in coal and hydrocarbon exploration, while it brought welcome industrial support to the science, tended to create a gap between those who studied 'whole fossil plants' and those who concentrated on fossil spores. In the western world the former were apt to categorize the latter as upstarts riding on a band-wagon of empirical, applied science and receiving an almost indecent level of financial backing from industry. The result was that many palynologists, recruited from among the geological fraternity, used spores for dating rocks and as palaeoenvironmental indicators, with only a rather cursory glance at their source plants. Equally regrettably many palaeobotanists ignored the findings of palaeopalynologists regarding this as scarcely within the sphere of respectable palaeobiology. Meyen is one of the rather few palaeobotanists who have sought to bridge the gap, and this part of the book highlights some of the messages that palynology can signal to palaeobotanists.

Presenting the fossil evidence for the earliest vascular plants is a challenging area to the textbook author. In the last twenty years, our knowledge in this fast-moving area of palaeobotany has increased enormously, but parts of the emerging picture are tantalizingly incomplete. Meyen abandons the use of the Tracheophyta as a Division (phylum of zoologists), comprising all land plants with vascular tissue, but instead resurrects the Pteriodphyta to cover all free-sporing vascular plants with a well-differentiated sporophyte and free-living gametophyte. The simpler, undifferentiated early vascular plants he places in the Propteridophyta (psilopsids, psilophytes, rhyniophytes *et al.*, of various contemporary authors) – a group name coined over seventy years ago by Newell Arber but dropped by later authors. The remaining vascular plants then belong within the Pinophyta (all gymnosperms) or Magnoliophyta (angiosperms). This apparent reversion to the use of higher taxa which most regard only as evolutionary grades (pteridophytes, gymnosperms), rather than clades, is a piece of somewhat disconcerting nomenclatural empiricism. However, Meyen makes it clear that he regards both groups as embracing a number of independent clades.

The status of the gametophyte in early vascular plants is one of the more awkward areas of the subject to present convincingly in textbook form. Early this century, it was tacitly assumed that they were simply not preserved in the fossil record. Over the last twenty years a number of important discoveries have been made, but they raise more questions than they answer. The Devonian Rhynie chert has figured largely here, in yielding possible vascularized gametophytes resembling axes of *Rhynia*, and puzzling liverwort-like structures with antheridia. Other less well preserved plants, known only in the state of compression fossils such as *Sciadophyton*, contribute to the puzzle. But there are still many gaps; plants with rather convincing archegonia have less persuasive antheridia, and *vice versa*; poor preservation leaves us without any secure link between sporophyte and gametophyte. Meyen documents what we know of this earliest glimpse of the alternation of generations, but the origin of these two fundamental phases in the vascular plant life cycle remains cryptic.

Meyen's guide through the minefield of naming fossil plants is one of the most valuable parts of his book. This is an aspect of palaeobotany which is most daunting to newcomers to the field; one seems to be using the language of names, of genera and species, familiar from living plants, and yet their use with fossils presents paradoxical problems. Modern taxa in the form of genera and species, are seen by Meyen as 'eutaxa', which potentially at least may present all the possible characters shown by that organism. A fossil genus, based on say a detached leaf, offers only a very limited range of characters. This is a typical 'parataxon'. Such parataxa are, in varying degree, artificial concepts, in biological terms. From various kinds of evidence we may piece together several parts of a plant (each with its own generic assignment), and designate it with a rather clumsy string of names in the form of an 'assemblage species', applicable to the 'whole plant'. However, this does not make the generic names of the several parts redundant. Because of the limited characters on which, for example, the parataxa of detached leaves are based, one 'genus' of leaves may have been borne on more than one type of parent plant. The genus *Phyllotheca*, based on cup-shaped whorls of partially fused leaves is common to the Permian flora of Russia and to the southern-hemisphere (Gondwana) flora of the same age. However, it has emerged that these leaf fossils from the two widely separated areas represent two families, with quite different fructifications. In this sense *Phyllotheca* is an artificial (but perfectly valid and useful) generic label for certain types of leaf.

A rather different aspect of the naming problem in palaeobotany is dealt with in a more incisive manner. Some fossil parataxa are (within their limitations) understood quite well, from many well-preserved specimens. Others are less well-founded and of course their systematic assignment (their classification) is less satisfactory. Those parataxa (e.g. genera) which consistently show characters associated with one family may be thought of as approaching a eutaxon in their validity. Others may be less securely assignable, and Meyen relegates these as 'satellite taxa'. Certain genera of conifer shoots can only be reliably attributed to that group as a whole. *Brachyphyllum*, for example, can only be put in the Pinales (Coniferales of other authors), without a family assignment. It is a 'satellite genus of the Pinales'. Other fossil members of the Pinales can be assigned with greater precision, for example the several female cones which form the basis of the extinct conifer family, the Cheirolepidiaceae. The main merit of this segregation of taxa is that our classification can then be based on those fossil parataxa which are regarded as 'respectable'. The boundaries of higher groupings are thus protected from the blurring which results from trying to include these lesser taxa, whose credentials are weakened by poor preservation or lack of biologically distinctive characters.

Some unexpected and disconcerting alliances have emerged from the piecing together of fragmentary plants. The best known was the linking of the supposedly gymnospermous Devonian wood called *Callixylon* with leafy shoots of the spore-bearing plant *Archaeopteris* by Beck. This formed the basis of his new group, the progymnosperms, which combines features (e.g. pycnoxylic secondary wood) of gymnosperms with the free-sporing character of a pteridophyte. This was disturbing to palaeobotanists, who had tended to assume that characters are correlated in a consistent way – what Meyen labels the belief in 'habitual correlation of characters'. If we find an ovate petiolate leaf in a living plant, with net venation and blind vein-endings we would anticipate its being a dicotyledonous plant since such leaves are generally borne on plants of that group. But this correlation is obviously less secure as we go back in time. All of palaeobiology leans upon such correlation, with varying degrees of awareness of its vulnerability. We attribute a chain of cells in a Precambrian chert to the cyanobacteria, since they show the morphology of some members of that group; we attribute a Triassic tooth to the mammals, or a Tertiary jaw to a primate, but just such a process of extrapolating from what we have, to what we suppose the whole organism to have been. We are aware that there are weaknesses in this process; indeed we are dealing with degrees of probability which are not (and probably never will be) susceptible to being quantified.

A similar problem crops up with functional interpretations. Fossil plants with thick cuticle and sunken stomata are too frequently interpreted as xerophytes, although we know that such xeromorphic features may be shown by plants of diverse habitats and their 'adaptive significance' is still only imperfectly understood. A similar frailty in the psychology of palaeobotanists is pilloried by Meyen on the charge of 'reductionism' – the drawing of universal conclusions from a single instance. We do this taxonomically when two species from two different genera (say a leaf and a fructification) are shown to occur in connection. We are apt to conclude that *all* species of *both* genera belong in a single genus, despite the irrefutable evidence that very different selective pressures act on leaves and fructifications, and that these organs have evolved independently. Meyen suggests that the same 'reductionism' shows in our current pre-occupation with the gradualistic/punctuated evolution controversy. A single sequence will be used to argue in favour of the universality of one or the other process, ignoring the possibility that both mechanisms (and possibly others!) may be involved in different phases of an evolutionary pathway.

The diversity of vegetation seen between the northern Tundra and Georgia, between the Baltic and Kamchatka, seems to have acted as a spur to the pursuit of phytogeography in the USSR. In palaeogeographical terms also, the USSR spans two major Palaeozoic floristic provinces within its boundaries, and abuts upon two more. It is therefore not surprising that Meyen and his several Moscow colleagues have been leaders in this field in their 1970 review of the Palaeozoic and Mesozoic plant geography of Eurasia. The impact of plate tectonics on all geological thought has brought palaeobiogeography to the front of the stage, with the added perspective of the palaeoclimatic significance of past floras. Palaeobiogeography is now as important an interface between geology and biology as biostratigraphy has been over the last century. This makes the full treatment of past plant distribution given in the present work especially timely.

Meyen recognizes a hierarchy of phytochoria – floral provinces – rather comparable to the hierarchy of plant nomenclature by which we designate our taxa, the lower ones grouped or nested within higher ones. These phytochoria change with time in response to evolutionary change of the constituent plants, the movement of the continents on which they lived, and the world climatic pattern to which they responded. The diagrammatic representation of this interplay of changing plants and changing ecological associations is a novel and significant feature of this book.

The cuticular covering of plants has a special significance for palaeobotanists. Although it is in the truest sense a very superficial part of the plant body, its composition is such that it is often the only part preserved in a coalified compression fossil which reveals microscopic cellular structure of potential taxonomic value. As such, cuticle studies have played a major role in fossil botany, for they can add a further dimension to the recognition of consistent limits to many taxa (especially of conifers and cycadophytes). Equally important, they may offer a bridge of microscopic detail which serves to link the cuticle of a detached fructification with the vegetative parts of the same species. For these very good reasons the role of cuticle studies in palaeobotany deserves the chapter that they are accorded in this book. However, as the author remarks in a moment of characteristic candour, while fossil cuticle studies are 'very fruitful in certain cases' in others 'they are practically useless'. As with all features shown by living and fossil organisms, we can only seek by trial and error to establish those attributes of plants, living and fossil (and these include cuticle characters) which empirically contribute to consistent and sustainable taxonomic groupings.

In his introduction, Meyen defends the use of line drawings in helping keep down the cost of the book which is certainly a laudable consideration. His drawings are in any event clear and helpful to the reader, and include both representations of original fossils as well as plants restored in varying degree. Some would question his argument that drawings, which are subjective, 'are admissible in a textbook' in contradistinction to an original description. It is in the nature of palaeobotany that the quality of evidence is very varied. What is accepted by one worker as 'xylem' in a putative early vascular plant may fall below the threshold of credibility for another. A drawing is that much more removed from the primary data. The only way a student can have a basis for making up his own mind is to see the original material, or failing that, the photographic evidence. The objection to drawings on these grounds is admittedly a churlish one, but one must still hope that the line illustrations will goad the student to go to original references, to see the published evidence at first hand – or better, to examine the fossils for himself.

Perhaps the best feature of this book is that it makes no concession to the faint-hearted reader who expects to have his ideas pre-packaged and pre-digested, and his attention constantly titivated. The reader has to work. There are no simplistic presentations of encapsulated half-truths to smooth over the bumpy ride through information that we only half understand. This is not a book to read while listening to the radio, or riding on a bus. It expects and commands the full attention of the reader. Many contemporary biology textbooks are rather like fast food produced in large quantities for instant, swift consumption with minimum effort and trouble, and limited enjoyment. Meyen's chapters are like the fish caught by yourself and cooked over a wood fire, contrasted with the fish finger from a microwave. You appreciate the ideas and the concepts all the more from the intellectual effort involved in grasping them.

W. G. Chaloner,
Royal Holloway and Bedford New College,
London University

Introduction

Most palaeobotanical handbooks are written to a traditional pattern. Brief introductory chapters on the preservation of fossil plants and on techniques of study are followed by plant systematics, which occupies most of the text. A much wider representation of palaeobotany was given by Kryshtofovich (1957) and Gothan and Weyland (1973), who considered not only the systematics of fossil plants, but also palaeofloristics. However, these two books are now largely outdated.

In the present book an attempt is made to give the most complete representation yet of modern palaeobotany. In addition to systematics, the book contains chapters on palaeopalynology, epidermal–cuticular studies, the palaeoecology of plants, palaeofloristics, florogenetics, evolutionary problems of palaeobotany, etc. I have tried not just to summarize the information from various palaeobotanical studies, but also to introduce a certain theoretical position, especially concerning the principles of systematics and the theory of evolution. At the same time I have tried to make both the text and the illustrations as compact as possible. For this purpose, numerous alphabetical abbreviations have been introduced into the text, and the figures are tabulated. I have done my best to make the book as cheap as possible and for this reason no half-tone photographs have been included. The drawings are not only cheaper but are also more interpretative, which is admissible in a textbook, as distinct from original descriptions.

The reader will at once notice the uneven distribution of the material by chapters. This is motivated firstly by the availability of the previously published handbooks, and secondly by my competence. The following information is included in very shortened form: (1) the preservation of material and techniques of palaeobotanical studies; these items are well commented on in the books by Taylor (1981) and Stewart (1983); (2) the systematics of the procaryotes and of the eucaryote algae considered in the excellent book by Tappan (1980). For some problems the completeness of the text also varies for other reasons. In Chapter 3 the systematics of plant groups is described differently depending on the contribution of fossil material to the understanding of the bulk and evolution of a particular group. The general notion of certain groups of plants is derived almost entirely from recent material. Though the remains of such plants are known to palaeobotanists, this hardly contributes to the general characterization of taxa. The main significance of fossil remains in this case is that they indicate the time of existence of the taxa. For instance, quite reliable remains of fungi are known since the Devonian and very numerous descriptions of fungi have been published, but all this information little modifies the characteristic of suprageneric fungal taxa derived from recent material. The same is true of many groups of algae and higher plants. All such taxa are considered in this book in brief. However, fully extinct supergeneric taxa of higher plants are described in more detail, especially in those cases where my viewpoint on systematics differs considerably from the most widely adopted viewpoint. Examples are some orders of the classes Ginkgoopsida (Calamopityales, Callistophytales, Peltaspermales, Arberiales) and Pinopsida (Cordaitanthales, Dicranophyllales). The detail in which particular groups of plants are described depends to a considerable extent on my competence. Since I have never studied the systematics of algae, fungi and angiosperms, the corresponding sections are written as compilations of the data and views of other authors. For the same reason I had to request M. A. Akhmetyev to write the section on the Cainozoic floras.

Chapter 4 on palaeopalynology requires special comment. Modern palynology is connected, on the one hand, with stratigraphy and other geological fields, and, on the other hand, with botanical fields (systematics, phylogeny, ecology and plant geography). There was not much point in the cursory consideration of all aspects of palynology and those aspects which stand in the closest relation to the systematics and phylogeny of higher plants were preferred.

As a whole this book lays no claims to the uniform interpretation of the plant world of the geological past. Rather, it has a different function to supply the knowledge of the plant world as a whole which has been obtained from recent material with palaeobotanical data. Accordingly this book may be considered to be a palaeobotanical continuation of handbooks on the morphology, systematics, ecology and geography of recent plants. Data available in these handbooks have been repeated in the present book only in so far as was necesssary, on the one hand, for a general understanding of the text, and, on the other hand, for lecturing on palaeobotany in higher education.

A significant innovation in this book is its chapters on the principles of palaeobotanical systematics and nomenclature, and on the connection of palaeobotany with the theory of evolution, with morphology and with various sections of geology. There is a certain range of general problems considered in palaeobotanical literature; however, the bulk of literature on these problems provides little palaeobotanical material. Here is but one example. The concept of punctuated equilibria and cladistic principles in phylogenetics are intensively discussed in the literature. The concept of punctuated equilibria on the basis of palaeobotanical material has hardly been discussed. Palaeobotany has few articles on cladistics. Thus, all the plant world of the past has been overlooked by the participants in this discussion. I am not sure that palaeobotany has lost something significant as a result of this. At the same time it is hardly advisable to withdraw from discussions which significantly influence the development of theoretical views in general biology.

The 'non-descriptive' chapters of this book pursue no aim of reviewing the literature and various competing views. Rather they contain my general theoretical position, often differing considerably from that of most of my colleagues. The description of my theoretical stance is necessary since it is closely related to many concrete solutions of various problems in morphology and systematics. At present systematics shows a devaluation of suprageneric taxonomical categories. The rank of taxa known for a long time is increasing continuously. Former families acquire the rank of orders and even of classes (e.g. taxads). Numerous suprageneric taxa are outlined by solitary inadequately studied genera (*Cheirostrobales*, *Pseudoborniales*, *Eleutherophyllaceae*, *Scutaceae*, *Glossophyllaceae*, etc.). In problems of suprageneric systematics of higher plants I am conservative, being convinced that suprageneric taxa must be outlined only for plants studied with adequate completeness, while the rank of taxa should be selected to be as low as possible. It is necessary to be much more careful than sometimes occurs, in attributing poorly studied genera to suprageneric taxa of the natural system. Many genera may be related to the suprageneric taxa only as satellites. This purge of the plant system from poorly studied taxa makes the degree of our ignorance more contrasting. It is better to have a less complete system of well-defined taxa, than an extensive system containing numerous poorly understood taxa, which makes just an illusion of knowledge and of understanding of the plant system.

Many extinct suprageneric taxa have formerly been named by genera established for isolated vegetative parts (*Calamitales*, *Phyllothecaceae*, *Lepidodendrales*, *Cordaitales*, etc.). This caused insuperable nomenclatural difficulties for attributing those genera which have been established by fructifications to the suprageneric taxa. Therefore, in this book most of these taxa have been given new names, where possible for formal nomenclatural reasons.

Acknowledgements

The English version of this book has been prepared on the initiative of M. C. Boulter (North-East London Polytechnic), who has contributed much to its publication. I discussed the whole book, certain sections, particular taxonomic decisions, and many other items with very many colleagues, among whom I can mention only Drs M. A. Akhmetyev, I. A. Dobruskina, M. P. Doludenko, M. V. Durante, A. B. Herman, A. V. Gomankov, I. A. Ignatyev, I. Z. Kotova, I. N. Krylov, E. L. Lebedev, V. V. Menner, M. A. Semikhatov, V. A. Vakhrameev, N. A. Volkova, O. P. Yaroshenko (all from the Geological Institute of the USSR Academy of Sciences, Moscow), S. Archangelsky (Argentina), M. Barthel (GDR), Ch. Blanc, J. Broutin (France), W. G. Chaloner (UK), R. Daber (GDR), D. L. Dilcher, W. A. DiMichelle, J. A. Doyle (USA), D. Edwards (UK), V. R. Filin (USSR), E. M. Friis (Denmark), J. Galtier (France), P. G. Gensel (USA), L. V. Glukhova (USSR), L. Grauvogel-Stamm, J. Holmes, Y. Lemoigne, B. Meyer-Berthaud (France), T. L. Phillips, G. J. Retallack (USA), J. B. Richardson (UK), R. Sattler (Canada), H.-J. Schweitzer (FRG), A. C. Scott (UK), T. I. Serebryakova (USSR), C. Sincock (UK), M. S. Solovyova (USSR), D. Storch (GDR), M. Streel (Belgium), A. L. Takhtajan (USSR), T. N. Taylor (USA), B. A. Thomas (UK), A. Traverse (USA), R. H. Wagner (UK), A. L. Iurina, G. A. Zavarzin, S. G. Zhilin (USSR).

I am very grateful to Dr M. A. Akhmetyev for preparing the section of the text dealing with Cainozoic floras, and to Dr I. N. Krylov for assistance in the preparation of the section on stromatolites. Special thanks are extended to Mrs E. H. London and Dr N. N. Smirnov who translated most of the original Russian text into English, and to my wife M. A. Meyen who helped me in so many ways.

Finally, I wish to thank the staff of Chapman and Hall for the highly competent and rapid production of this book.

Abbreviations

The stratigraphic ranges of taxa are usually bracketed together with references to illustrations, and are designated by standard symbols as follows:

O – Ordovician	T – Triassic
S – Silurian	J – Jurassic
D – Devonian	K – Cretaceous
C – Carboniferous	Pg – Palaeogene
P – Permian	N – Neogene

Numerals with the symbols correspond to series (e.g. D_2 – Middle Devonian). A three-fold subdivision of the Carboniferous is adopted, hence C_2 means the Middle (not Upper!) Carboniferous.

Alphabetical abbreviations of taxa in Chapter 3 are derived from Latin or English names. Each abbreviation is valid within a relevant paragraph or section of the text.

Chapter 1

Preservation types and techniques of study of fossil plants

Fossil plants are preserved in three main ways: as impressions, phytoleimmas and petrifactions (for more details see J. M. Schopf, 1975). In the first case the remains of a plant decay completely, leaving its impression on the rock. This is not merely a mechanical impression of the plant on the sediment, which has not quite solidified, but a result of a complex physico-chemical process. The remains liberate decomposition products into the surrounding mineral medium (matrix), producing a peculiar geochemical situation around themselves. Impressions on rough sandstone are often covered by a thin mineral crust rendering the finest detail. Moulds produced by voluminous remains, casts and stone nuclei are often contrasted with the impressions. In reality these are different cases of the same type of preservation when vegetative matter disappears. A leaf impression is an extremely flattened mould and a cast is an impression of a particular inner surface of the remains.

Phytoleimmas ('compression' in literary English) are compressed coalified remains of plants, commonly associated with an impression, if they do not form coal layers. A cuticular bag usually contains structureless coal within the phytoleimma, some tissues sometimes being preserved (vascular, mechanical). Upon leaching of this coal a complete venation may remain in the cuticular bag. During the transformation of organic matter into a phytoleimma and loss of its original structure and chemistry, secondary structures may be formed (e.g. a fringe of thick pinnules, a fold, regular systems of thin cracks) which were sometimes taken for features of the living plant. Phytoleimmas are remains of any size, including spores and pollen.

Petrifactions (true fossils) are remains whose tissues are fully or partially replaced by mineral matter with preservation of their cellular structure (cellular permineralization). These are petrified trunks with persisting wood structure, remains in coal balls (carbonate nodules in coal seams; Scott and Rex, 1985), in volcanics, in calciferous sandstones, etc. The processes of permineralization of cells are still poorly understood physico-chemically, especially if the petrifaction involves fibrillar components of the cell wall (Smoot and

Taylor, 1984), easily decomposing cytoplasmic components (Taylor and Millay, 1977a), including chloroplasts (Niklas and Brown, 1981), and probably chromosomes (Brack-Hanes and Vaughn, 1978). Many biochemical components of plants studied by palaeobiochemistry are preserved by fossilization (Niklas, 1982a).

Different preservation types are sometimes combined in the same remain, where part is just impressed on the matrix, part turned into a phytoleimma, and part petrified. Partial pseudomorphoses are common (known even in miospores), being a replacement of part of the organic matter with mineral matter.

Schopf (1975) suggested that duripartic preservation should also be differentiated (the preservation of solid parts). In this case some tissues of an organism withstand oxidation and other transformations. This includes siliceous and calcareous skeletons of algae if they have not undergone recrystallization. Special cases of preservation are hollows left by roots or by boring algae, traces of bacterial and fungal activity (common in membranes of miospores and acritarchs, in plant cuticles, dinoflagellate cysts, etc.).

Techniques of study depend on the type of preservation. The impressions are studied with reflected light; if a phytoleimma is present, wetting specimens with alcohol or xylene and using crossed polarizing filters, one behind the objective lens of a stereoscopic microscope and one in the illuminator, are very efficient. The microscopic structure of the surface of impressions may also be studied using transparent peels (e.g. of cellulose acetate or replicas) or opaque casts (e.g. of silicone rubber) which may be mounted on stubs for scanning electron microscopy. Replicas are studied with oblique transmitted light. Casts of deep hollows left by plants may be prepared using various materials (paraffin wax, silicone rubber, etc.). Examination of the surface of a specimen may require the removal of the phytoleimma which is burned by heating the specimen in a muffle furnace.

The main techniques of study of phytoleimmas are maceration in oxidizing mixtures and the preparation of transfer preparations. For maceration, Schultze's solution of concentrated nitric acid and potassium chlorate (Berthollet's salt) is most often used. This oxidizes coal to humic acids, which are then removed with an alkali (usually ammonia solution, or potassium hydroxide), leaving cuticular membranes, walls of spores and of pollen grains, resin ducts and bodies and (more rarely) mechanical and vascular tissues. The washed products of maceration are embedded in transparent media (e.g. glycerine-jelly, Canada balsam, various plastics) for examination with transmitted light. They may also be sectioned or fractured using a microtome, or their surfaces may be studied using a scanning electron microscope. If a lump of rock is macerated, yielding plant remains as a result of its destruction, the method is known as bulk maceration (complex maceration).

Transfer preparations are phytoleimmas transferred from the rock on to film, glass or plastic for examination of the side previously obstructed by the rock. To do this, the remains on the rock are first covered with a film or with a transparent resin (plastic) and then the rock on the opposite side of the specimen is dissolved with a suitable acid and its undissolved part removed mechanically. The remains cleared of the rock may then be macerated, cut, etc.

If a phytoleimma is too fragile and it is not possible to prepare a transfer preparation, parts of a plant covered by the rock (e.g. complicated branching axes) have to be exposed by a 'micro-excavating' (degaging) technique. This technique is effective but labour consuming as all stages of the work should be recorded on photographs or drawings. With the same aim a series of sections may be made and the subsequent sections photographed and sketched.

Microfossils are isolated using various separation and enrichment techniques. Thus, to isolate diatoms, carbonate is removed with acetic, formic or hydrochloric acid. To separate palynomorphs (miospores, acritarchs and other organic-walled microfossils), the disintegrated rock is centrifuged in heavy liquids whose specific weight is such as to cause mineral particles to settle and palynomorphs to come to the surface. Making use of the difference in the specific gravity of microfossils it is possible to separate even particular species by centrifugation.

Petrifactions are studied using polished or etched cuts, thin sections and peels. The former are particularly useful for the examination of pyritized and strongly coalified remains. Thin sections are widely used for studies of calcareous algae and wood. Petrifactions in coal balls, tuffaceous and carbonate rock are most often studied at present without thin section making, replacing them with peels. The specimen is sawn into units, their surface is smoothed and etched with acid to dissolve the matrix. After a brief etching the cell walls protrude from the matrix. A solvent for cellulose film is poured over the surface of the saw cut and the film itself is placed on the wet surface. The thin layer of the film facing the specimen is dissolved in the solvent and envelops the protruding cell walls. The dried film is peeled off, bearing a thin layer of the remaining cells adhered, so that the cell structure can be studied with transmitted light in the finest detail. The necessary areas of the film can be dissolved again, liberating the adhered fragments (miospores, tracheids, etc.) for examination by scanning or transmission electron microscopy. If a petrification is on the surface of a lump of rock and it is difficult to prepare sections, the remains may be covered with a transparent resin followed by cutting, polishing, making of peels, etc. For studies of cellular structures of plants in highly metamorphosed coals the method of ion beam etching of polished sections is very efficient (Kizilshtein and Shpitzgluz, 1984).

As different types of preservation may be combined in the same remains, the techniques often have to be combined. Any technique always entails some risk of damaging the specimen. The treatment of specimens which have become nomenclatural types or otherwise historically important or rare is therefore commonly inadmissible.

Chapter 2

Principles of typology and of nomenclature of fossil plants

PARATAXA AND EUTAXA

The possibility of describing and naming fossil plants as if they were living plants was realized even by the founders of palaeobotany, Schlotheim, Sternberg and Adolphe Brongniart. However, even at that time it was clear that special principles were necessary for the typology of fossil plants as well as for their nomenclature. Having introduced the genera *Cladophlebis*, *Pecopteris*, *Neuropteris*, *Odontopteris*, *Taeniopteris* and some others, Brongniart assigned to them the status of form groups with indefinite relation to the natural genera. Later such form groups received the name of parataxa. The term 'parataxon' itself has been introduced in palaeozoology for groups of generic rank, used in the systematics of conodonts, apticha, anapticha, and of other detached parts of animals. It is advisable to use the same term in palaeobotany for form groups of any rank, not only for genera. Thus, such suprageneric groups as anteturmas, turmas, subturmas, and infraturmas in palynology may be considered to be parataxa. The taxa of living plants and the taxa of fossil plants of similar status may then be called eutaxa. The principal difference between eutaxa and parataxa is as follows. The eutaxa may be outlined by any characters of the whole organism. The selection of characters depends on the researcher and his available technique. The parataxa may be outlined using only those characters which are retained on the detached parts in question. The lower the weight of these characters in the system of eutaxa, the more remote will be the resulting parataxa from eutaxa, and the closer to morphological terms. Indeed, in morphology we distinguish simple, pinnate, palmate leaves, leaves with pinnate, palmate, open, and net venation. Such a parataxon as the genus *Taeniopteris* combines any simple leaves having pinnate open venation. It is obvious that this parataxon is closer to a morphological unit than to any eutaxon. Other parataxa because of more complex combination and peculiarity of characters approach eutaxa. This particularly concerns parataxa outlined by isolated fructifications. Such parataxa are often considered to be true eutaxa, which is wrong in principle, as eutaxa cannot be established without sufficient information on associated parts.

The system of eutaxa is often given the name of the natural system as opposed to the formal system of parataxa. Both terms – natural and formal – are not quite valid etymologically and are variously defined in the literature. Originally the notion of a natural system was introduced in botany to counterpoise the Linnean artificial system arranging genera into families by a few characters of the flower. The weight of each character was strictly fixed, i.e. its variability within a certain taxon was not admitted. In the natural system (A. de Jussieu was the pioneer of its creation) the principle was accepted, which was later named by E. S. Smirnov as the congregational principle (Smirnov, 1925. See also: Hennig, 1950; Meyen, 1984a). The weight of characters is not predetermined but is revealed in the course of the taxonomic study. The system is built beginning with the taxa of the lowest rank, which are gradually unified into still larger congregations by all the observed complex of characters. Only after the system of congregations is established is the hierarchy of congregations transformed into the hierarchy of the taxonomic categories of a different rank. It is then possible to conclude what features should be used to characterize taxa and what the weight of a certain feature is in various parts of the resulting system.

The system so derived acquires the status of a privileged (basic) system, underlying any other classification of plants (or of other organisms). Each plant occupies only one place in this system and is given a sole binomial name. Other classifications of plants are possible (e.g. by growth forms, by commercial exploitation, etc.); however, all these classifications use the taxa of the natural (privileged) system as operational units.

The difference between the natural and the artificial system was excellently expounded by Baer early in the last century (1819) (see Baer, 1959). Translated into modern language his explanations mean that the natural system combines organisms into groups by a maximum number of correspondences; it fixes multifaceted relations between organisms and taxa in the general space of characters. At the same time the artificial system may be considered to be an information-retrieval system. It performs the function of a key for the determination of organisms.

Thus, in neobotany we deal with eutaxa distributed in a single natural system by the whole complex of characters. At the same time there are numerous artificial information-retrieval systems (as determination keys). The parataxa are used in botany only for imperfect fungi with their correspondence between imperfect and perfect stages often being unknown. Palaeobotany also comprises eutaxa which are included into the natural system along with eutaxa of the living plants. In addition, there are parataxa partly incorporated in the natural system, and partly staying outside. In the latter case they may exist separately, or may be organized into a system of parataxa of a different rank independent of the natural system. Parataxa are heterogeneous by definition. Some parataxa are outlined by the congregational principle and resemble eutaxa in this respect. Other parataxa are described by a fixed number of characters, each character having a permanent weight. These parataxa are comparable to the taxa of the artificial system of Linnaeus by the principles of their description.

Now for some particular cases. Describing the conifers of the Jurassic flora of Yorkshire, Harris (1979) attempted to establish the connection of different parts as they had been in life. Some of the genera described by him (*Marskea*, *Elatides*, etc.) were given the status of eutaxa and were attributed to families of the natural system. At the same time Harris decided to put the generic classification of vegetative shoots of the Mesozoic conifers in order. There are more than a hundred genera for such remains in the literature. Some of these genera are described by a small number of gross morphological characters and have been outlined irrespective of their position in the system of eutaxa of the conifers (e.g. *Elatocladus*). Other genera were assimilated with certain eutaxa and their separation as distinct genera was approached with caution, which manifested, for example, in the introduction of the suffix '-ites' added to the generic name of a recent genus (e.g. *Pinites*, *Abietites*). The genera of the third kind were described using epidermal characters by which the size and systematic position of the genus were determined. Florin (1958), for example, described the genus *Bartholinodendron* in this way and attributed it to the taxads.

Harris proposed that among Mesozoic conifers only those with known combinations of vegetative shoots and fructifications be attributed to eutaxa. For the vegetative shoots he proposed the use of just a few genera in the system arranged as a determination key (it is reproduced in Chapter 3). This is a typical artificial system. Species of one genus of this system may belong to different families. For example, the species of the genus *Brachyphyllum* may belong to the Araucariaceae and Cheirolepidiaceae.

Describing eutaxa, Harris preserved the nomenclatural independence of parataxa. The description of a eutaxon acquired the following form. At first parataxa introduced for fructifications and approaching the status of eutaxa were described. Different generic names were accepted for male and female fructifications. Then vegetative shoots placed into one of the genera of the artificial system mentioned were described.

This procedure, expressed most clearly in Harris's monographs on the Jurassic flora of Yorkshire, was used more or less consistently by other palaeobotanists beginning with Ad. Brongniart. Adhering to this procedure, Grand'Eury (1877) used different generic names for the associated leaves (*Cordaites*), fructifications (*Cordaianthus*), leafy shoots with leaf scars (*Cordaicladus*), and seeds (*Cordaicarpus*) of the Cordaitanthales.

Establishing the connection in life of various parts often tempted a unification of the corresponding different parataxa. For this reason the generic name *Cordaites* was used not only for the detached leaves but also for the fragments of petrified trunks. The same name *Cordaites* was also used for the whole reconstructed plants. Pollen, at present attributed to the genus *Classopollis*, was originally described by the generic name *Brachyphyllum*, as it was found in microstrobili associated with leafy shoots of the genus *Brachyphyllum*. It was suggested that palaeobotany should seek to build a single system of eutaxa comprising both Recent and fossil plants. So the adoption of independent parataxa for different parts of the same plants was considered to be a forced but provisional procedure.

However, the experience of taxonomic research in palaeobotany testifies to the opposite. The extrapolation of a parataxon name to dispersed parts of different morphological categories, as a rule, creates a confusion entailing phytogeographic, stratigraphic and phylogenetic mistakes. For instance, the generic name *Bennettites* was applied to isolated pollen from the upper Palaeozoic of Siberia (Medvedeva, 1960) implying that the bennettites had appeared in Siberia earlier than anywhere else. The lists of miospores from the Permian north of the Russian platform (Zoricheva and Sedova, 1954) contain the names *Lepidodendron* (a fossil) and *Angiopteris* (a living genus), leading to an erroneous conclusion as to the composition of the ancestral flora.

Palaeobotanists have frequently faced situations where the same parataxon was combined in life with different parataxa of the same morphological category. The example of the genus *Brachyphyllum* has already been mentioned above, whose different species are combined with female fructifications of *Araucarites* (Araucariaceae) and *Hirmeriella* (Cheirolepidiaceae). The miospore parataxon *Calamospora* is combined with strobili of the Calamostachyaceae (*Calamostachys*, *Palaeostachya*, *Pendulostachys*, etc.), Bowmanitales and Noeggerathiales. The leaves of the parataxon *Cordaites* are combined with fructifications *Cordaitanthus*, *Gothania*, *Vojnovskaya*, *Kuznetskia*. The parataxa of *Lepidodendron* and *Sigillaria* trunks correspond to the rhizophores of the parataxon *Stigmaria*. Thus, there is no one-to-one correspondence between parataxa and eutaxa, as well as between the parataxa for parts of different morphological categories (Fig. 1).

Of course more detailed studies may lead to more exact and more persistent correspondences. For instance, the improvement of the generic diagnoses of monosaccate pollen has shown that the genus *Felixipollentites* corresponds to the genus *Gothania* established for fructifications of the Cordaitanthales (Millay and Taylor, 1974). So far the microstrobili *Androstrobus* have been known in combination with polysperms of only one genus, *Beania*. However, even in such cases it is not possible to guarantee that other life-time combinations of the parts will not be found, corresponding to other parataxa. The unexpected combinations of parataxa occur very often. When Zalessky established the genus *Callixylon* in 1911 which had been considered a Devonian member of the Cordaitales for a long time, he had no idea that these pycnoxylic trunks of the gymnospermous type were later to be associated with fern-like fronds of *Archaeopteris* (Beck, 1960). No one thought that trunks of the genus *Poroxylon* (also attributed to the Cordaitales) were associated with the fern-like seed-bearing fronds of *Dicksonites* rather than with leaves of the *Cordaites* type (Rothwell, 1980).

TAXA AND CHARACTERS

Palaeobotany cannot be said to have learnt an adequate lesson from such unexpected discoveries. Palaeobotanists continue believing in (1) the habitual *correlations of characters* and (2) the *persistence* of certain *characters* within the limits of large taxa, though often there are no rationales for either. At present this belief is particularly characteristic of the miospores of various plants and the leaves of gymnosperms and angiosperms. Small quasi-disaccate pollen occurring in the Permian deposits and attributed to the genus *Vitreisporites* have often been called caytonialean pollen, though it has been demonstrated that the same pollen is produced by plants bearing the cladosperms

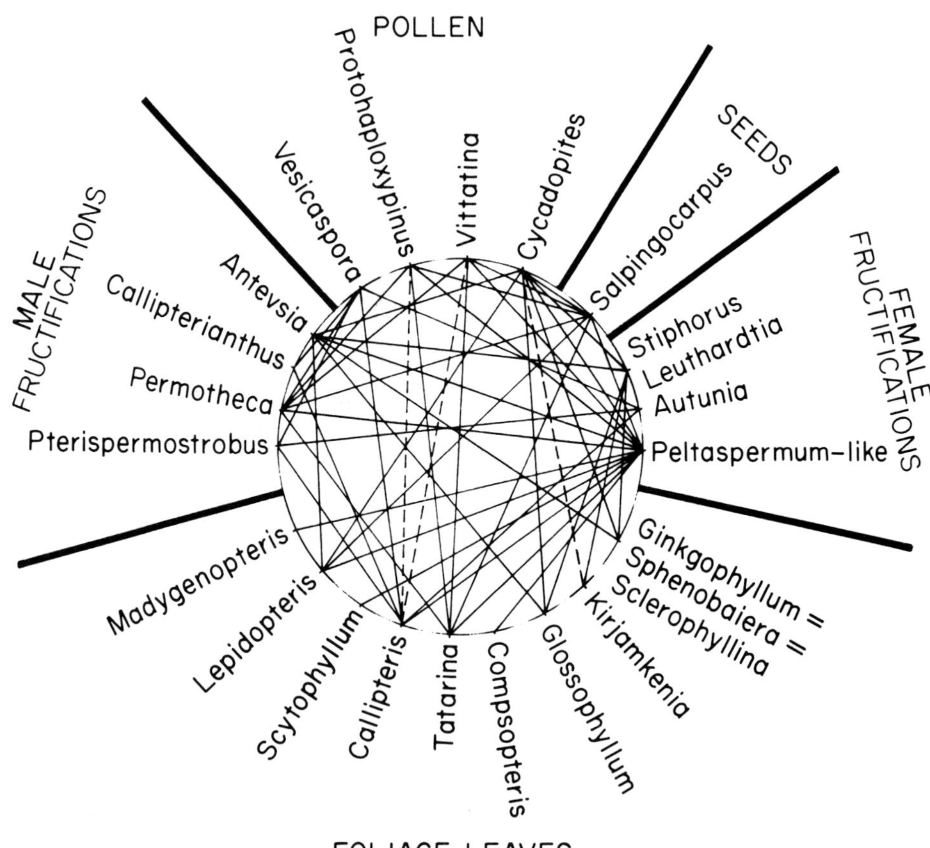

Fig. 1. Multiform correspondence between genera (parataxa) established on different organs (family Peltaspermaceae and its allies).

Sporophyllites, belonging to the Peltaspermales and not to the Caytoniales. The Palaeogene and Upper Cretaceous angiosperm leaves are attributed to Recent genera without due consideration for the probability that the combinations of leaves of a certain type and fructifications in the past could be different from those of the present (see the example of *Fagopsis* in Chapter 3).

Contrary to 'habitual correlations' and to the postulate on persistence of characters within the limits of a taxon the opposite postulates should be suggested stating (1) the absence of firm correlations between characters, (2) the possibility of instability of any character within a taxon (i.e. a potential polymorphism of any character in a taxon), (3) the possibility of parallelism between characters of different taxa. The correctness of these three postulates is obvious from the information on Recent plants. For this very reason, in the time of construction of the classification of plants it was necessary to use the congregational principle, leaving aside the artificial system of Linnaeus. At present a general theoretical foundation may be offered for these three postulates. Namely, in the range of plant characters there are inherent regularities which do not stand in one-to-one correspondence to the distribution of the eutaxa in the natural system.

Applied to plant morphology these regularities were considered in detail by the author in special articles (Meyen, 1971a, 1973a, 1978b,d). The special morphological aspects of the author's views are discussed in Chapter 8; at present only the taxonomic consequences of these regularities are considered. A comparison of both isolated characters and the parts imparted with these characters (mera) in the best studied Recent and fossil plants demonstrates that the polymorphism of each meron is channelled. This means that the modifications of mera in different eutaxa are subordinated to recurrent transformation rules. On the other hand, the range of the polymorphism of mera in eutaxa is not predetermined and may vary depending on a certain representative of the taxon and on the stage of its evolution dealt with.

This distribution of characters and of the corresponding types of mera is best studied for the leaves of vascular plants (Krenke, 1933–1935; Meyen, 1973a; Seybold, 1927). The leaf lamina is dissected according to certain rules repeated in different groups of vascular plants. Therefore, it is not surprising that leaves very similar in their dissection type are present in different parts of the system. If the venation of such leaves is simple, visually indistinguishable forms result. Thus, it is no mere chance that palaeobotanists initially referred to one genus, *Danaeopsis*, the fronds of the Triassic European ferns, of the Gondwana umkomasias (now *Dicroidium*), and of the upper Permian Angara peltasperms (now *Compsopteris*). If all known species of *Dicroidium* are considered, it is possible to see the transition from the once-pinnate to the twice-pinnate leaves (Anderson and Anderson, 1983). The same transition is observed in the Angara *Compsopteris*, *Comia* and *Callipteris* (Meyen, 1971a).

The consequences of the habitual correlation and the postulate of the persistence of characters are obvious in the case of the Arberiales (glossopterids). Most Arberiales have simple leaves with reticulate venation (of the *Glossopteris* and *Gangamopteris* type). The pinnate leaves with open venation found in the Upper Palaeozoic of Gondwana were attributed to the cycadophytes, as there was no idea that the Arberiales might possess such leaves. The habitual correlation was the implicit assumption that the pinnate leaves should correlate with the cycadophyte fructifications, and the postulate on persistence of a character was the assumption of only simple leaves and only reticulate venation in the Arberiales. Now it is known that these pinnate leaves with open venation (*Pteronilssonia*) may be attributed to the Arberiales by epidermal characters. Moreover, almost totally repeated and connected by the same transformation series, forms have been revealed in the orders Arberiales, Cycadales and Bennettitales (Fig. 2; Meyen, 1971a, 1973a, 1978d). It is essential that the relative frequencies of members of the same form series are different in different orders. Thus pinnate leaves are rare among the Arberiales whereas simple ones with reticulate venation are widespread. On the other hand, simple leaves with reticulate venation are rare in the Cycadales, and unknown as yet in the Bennettitales (though their discovery may be expected). However, in both the latter groups pinnate leaves are very common.

Similar relations can easily be presented by alphabetical and numerical symbols, but we will not do so for the moment. It is more important to comprehend the general phenomenology and to try to find a causal explanation for it. The form repetition in different plant groups, and especially the repetition of the transformation rules connecting different forms, obviously indicate that ontogenetic processes are channelled, and there are channels (creodes, after Waddington, 1962) in the character space. It is evident that if creodes characterize ontogeny, they must be manifested both in the taxonomic diversity and the phylogeny of taxa. The regular polymorphism and repetition of forms are fundamental traits of all organ types and of all plant taxa. It is evident that these traits manifest themselves incomparably more commonly in detached parts than in whole plants. It is also obvious that the fewer characters that can be used the more difficult it is to decipher both the affiliation of the different modalities to the same taxon and the affiliation of the independently originated similar forms to the different taxa. The number of characters is directly related to the completeness of preservation of specimens. Therefore, in the

Fig. 2. Parallelism of foliage leaves between Gondwana Upper Palaeozoic Arberiales (*a–c*) and Northern Mesozoic Cycadales (*d–f*): *a*, *Glossopteris indica* Schimp.; *b*, *Pteronilssonia gopalii* Pant et Mehra; *c*, *Rhabdotaenia danaeoides* (Royle) Pant; *d*, *Anthrophyopsis crassinervis* Nath. (Harris initially referred the species directly to *Glossopteris*); *e*, – *Pseudoctenis herriesii* Harris; *f*, *Taeniopteris* sp. Modified from: Meyen, 1984a; figures modified from: Chandra and Surange, 1979 (*a*); Pant and Mehra, 1963a (*b*); Pant and Verma, 1963 (*c*); Harris, 1932 (*d*), 1964 (*e*, *f*).

taxonomy of fossil plant remains, the quality of their preservation should also be taken into consideration.

PECULIARITY OF THE TAXONOMY AND NOMENCLATURE OF FOSSIL PLANTS

Summarizing the above arguments we derive the following main distinctions between the taxonomy (and nomenclature) of living and fossil plants, the eutaxa and parataxa (Meyen and Traverse, 1979):

1. In living plants (Fungi Imperfecti are an exception) each individual, including all parts when found detached, are assignable to a single taxon of any rank and have only a single generic and specific name.

In fossil plants, dispersed parts, even those originally belonging to one individual, when the original connections are not observable, may be referred to several taxa of the same rank and have different generic and/or specific names.

Examples: *Stigmaria* Brongniart, *Lepidodendron* Sternberg, *Lepidostrobus* Brongniart.

2. In living plants, all the individuals of a species belong to the same genus, and all the species of a genus belong to the same family.

In fossil plants, various specimens of a species may or may not belong to the same genus, if the latter were to be established from whole plants. Various species of a genus may be treated as belonging to different families. Examples: *Pecopteris arborescens* (Schlotheim) Brongniart belongs to Marattiaceae, *P. feminaeformis* Schlotheim to Zygopteridales, whereas the affinity of *P. tajmyrensis* Schvedov with higher categories is unknown. Specimens of *Stigmaria ficoides* (Sternberg) Brongniart belong to whole plant genera assigned to Sigillariaceae, Lepidocarpaceae and perhaps other families.

3. Each living plant is ascribable to the whole sequence of taxa mentioned in Article 3 of the Botanical Code (species, genus, family, order, class, division, kingdom).

Fossil plants, if named, should be ascribed to species and genera; higher taxa may not be known. Example: the affinity of *Yavorskyia mungatica* Radczenko is altogether unknown, excepting its attribution to higher plants; it is often mentioned as a problematical cycadophyte, though it could equally well be a peltasperm.

4. In living plants, different degrees of preservation cannot serve as the only basis for establishing separate taxa.

In fossil plants, species and genera may be established for different degrees of preservation. Examples: *Lepidodendron* Sternberg *vs. Aspidiaria* Presl, *Bergeria* Presl, *Knorria* Sternberg; *Taeniopteris* Brongniart *vs. Doratophyllum* Harris (or *Nilssoniopteris* Nathorst).

5. In living plants, different ontogenetic phases of the same life cycle cannot normally serve as the basis for several independent taxa.

In fossil plants, this is quite possible. Examples: the genus *Eddya* Beck may be the sporeling of the adult plants, referred to *Archaeopteris* Dawson. Independent nomenclature for seeds, microspores and megaspores, and for associated shoots also exemplifies this point.

Every fossil genus, even the most thoroughly studied, fits, actually or potentially, at least one of the enumerated points, because in no case does the fossil material yield all the parts of the plant body, and some parts may be unwittingly referred to a separate taxon and hence obtain independent nomenclatural status. This is particularly so with reference to palynological nomenclature, even when the affinities of the miospores are known (see above). In many cases we ascribe a genus to a family or order with confidence, but further investigation may reveal other species which fit the diagnosis of the genus, but which belong to another family or order. It is possible, for example, that the oldest known ovules, referred to *Archaeosperma* Pettitt and Beck, may prove to belong to a plant with *Archaeopteris* fronds, though *Archaeopteris* is usually thought of as progymnospermous. The ovules are in fact associated with *A. hibernica* (Forbes) Dawson, the type species of *Archaeopteris* Dawson.

Lines of demarcation between 'organ-' and

'form-genera', or between fossil and living plants, as discussed in the literature are reflected in various previous editions of the Code of Botanical Nomenclature and correspond to one or more of the five points enumerated above. To reflect all these differences with separate nomenclatural provisions is impossible, because every genus actually or potentially may fulfil all the points or any combination of them. The degree of approximation of fossil plant taxonomy to that of extant plants is a matter of subjective taxonomic assessment.

THE BINARY (DUAL) SYSTEM OF FOSSIL PLANTS

Only a very small number of genera of fossil higher plants is established by a complex of various parts and approaches the status of eutaxa. The overwhelming majority of genera and still more of species can be qualified as parataxa, approaching eutaxa in different degree. The integration of all these taxa into the natural system is evidently only partly possible. For some genera the lowest suprageneric taxon comprising all the species of a particular genus is the higher plants as a whole (*Taeniopteris*, *Pecopteris*, *Sphenopteris*), the seed plants as a whole (*Samaropsis*), etc. Hence it is advisable to construct a general system of fossil plants in the following way. Extinct eutaxa and most completely studied parataxa are included in the natural system (privileged, basic system) along with Recent plants. Those parataxa which can be conditionally assimilated with certain suprageneric eutaxa are included in the latter as satellites (Meyen, 1978e; Thomas and Brack-Hanes, 1984). For instance, the genus *Podozamites* is introduced for the leaves which can belong to different families of the conifers. However, some species of *Podozamites* belong to the family Voltziaceae, so that *Podozamites* may be considered to be a satellite genus of this family. Different species of *Brachyphyllum* belong to the families Araucariaceae and Cheirolepidiaceae. It is not excluded that *Brachyphyllum* contains species belonging to other coniferalean families. Thus *Brachyphyllum* may only be treated as a satellite genus of the order Pinales.

There are genera which cannot be envisaged as satellites in relation to any suprageneric taxon. These are many genera of miospores and some genera established for megafossils (*Taeniopteris*, *Barsassia*, *Glottophyllum*, etc.). For them a general artificial information-retrieval system is to be introduced. However, all parataxa of fossil plants in general may be included in this artificial system.

At present the first version of a general system for genera of fossil higher plants established for megafossils is being prepared by the author. It consists of two parts and an alphabetical index. The first part comprises the usual natural system in which every suprageneric taxon is given in addition to a usual name a letter–figure symbol beginning with N (for Natural). The second part comprises groups of genera corresponding to morphological categories, to their most general varieties, and to preservation types. These groups are arranged in a hierarchy, and in the overall sense correspond to palynological anteturmas, turmas, subturmas and infraturmas. Each group receives a very brief diagnosis, and a letter–figure index beginning with F (for Form-system) instead of a name. All groups of this part of the system are suprageneric parataxa.

Some previously described suprageneric parataxa which have erroneously been considered to be eutaxa may also be attributed to this system. Such are the orders Calamitales and Cordaitales, the families Phyllothecaceae and Anachoropteridaceae, etc. The type material of these parataxa consists of vegetative parts having unknown or vague association with fructifications. Therefore the systematization of these parataxa by productive organs and the selection of names for new taxa established for reproductive organs in accordance with the principles of typification and priority is impossible (Meyen, 1971b, 1976, 1978a). For instance, the genus *Cordaites* which is the type genus for the order Cordaitales is established for vegetative leaves represented by impression–compression remains. With these leaves the fructifications are associated from which at least two independent families may be revealed. According to the nomenclatural rules one genus (as a eutaxon) cannot belong to two suprageneric taxa. So the genus *Cordaites* is to be divided into two genera, leaving within the family Cordaitaceae only those species which associate with the fructifications characteristic of this family. However, we cannot do that. It is not even possible to point out reliably with which fructifications the leaves of the type species *C. borassifolius* (Sternb.) Unger was associated. Therefore it was proposed by the author (Meyen, 1978a) that the taxa with the type genera established for such vegetative parts which may be well characteristic of the various suprageneric eutaxa established with regard to the fructification structure should be excluded from the system of suprageneric eutaxa of fossil plants. The corresponding changes in the names of orders can be easily done, as their names are not protected by rules of priority. The family names retain their priority and cannot be abolished, but there are no obstacles to considering such families as parataxa. For instance, the family Phyllothecaceae may be treated as a parataxon comprising all genera established on vegetative shoots of articulates and having cylindric or conical leaf sheaths (*Phyllotheca*, *Phyllopitys*, *Annulina*, etc.). The name Cordaitaceae may be preserved for the parataxon comprising simple leaves with open fan-shaped or parallel venation (*Cordaites*, *Glottophyllum*, *Poacordaites*, etc.).

Since the aim of this part of the system is purely information retrieval, the main point is the convenience of this system. For this purpose the requirements of the logical strictness may be disregarded. Thus, some genera may be placed in different parataxa at the same time (e.g. if they were established on the basis of vegetative shoots with fructifications). Some parataxa may partially correspond to eutaxa.

These two parts of the system are supplemented with the alphabetical list of all genera, every genus being indexed with letters F- and N-containing symbols.

Using the system described it is easy to select a group of genera to which the studied plant remains may belong, to select the genera with which a newly described or earlier known genus should be compared, and also to produce a list of eutaxa to which plant remains of a certain appearance may belong. This is made possible due to the fact that the parts N and F of this system are connected through the generic alphabetical index.

Designing this two-fold system demonstrated the existence of a great number of poorly described and partly superfluous genera. Thus more than 100 genera are introduced for the vegetative shoots of the conifers, and most of them may be abolished. It is quite possible to abolish many genera for vegetative fern-like leaves. Of more than 4700 genera described so far not more than 1000 are outlined well enough to be used without essential reservations. As a whole the generic system of fossil plants justly deserves the nickname 'Augean stables'. It would be useful to cleanse these stables by the methodical revision of all described genera. However, it is hardly credible that anyone will carry out this tremendous task. Therefore, it is more important at the moment to stop the further emergence of badly described and vague genera.

THE REASONS FOR THE INFLATION OF GENERIC NAMES

There are plants to which several generic names may be attributed, and it is very difficult to solve the problems of interrelations between the relevant genera. Such is the group of genera comprising *Schizostachys*, *Zygopteris*, *Biscalitheca*, *Monoscalitheca*, *Nemejcopteris*. This is a clear example of inflation of generic taxonomy and nomenclature. Several reasons for such inflation may be pointed out. The principal reason is the impossibility of selecting (or unwillingness to select) from the literature all the genera with which a potentially new genus is to be compared. The bibliographic search of the comparable genera is extremely time consuming. The available reference books (Boureau, 1964–1975; Němejc, 1959–1975; Osnovy paleontologii, 1963a,b; Seward, 1898–1919) comprise only part of the described genera, with poorly described uncertain genera being commonly omitted. Moreover, they mostly consider those genera which can be, at least provisionally, connected with the natural system. Numerous genera are described in the floristic regional studies. Even if these genera are included in reference books such as the *Fossilium catalogus* or the *Index of generic names*, a palaeobotanist cannot select exactly those genera which are needed for comparison with a potentially new genus. For this purpose, the above described and adequately detailed dual system of higher plants is necessary. Combining the data of parts F and N of this system it is possible to select the necessary genera. The unavailability of such a system is the main technical reason for generic inflation.

Three other reasons are connected with mistakes in the taxonomic thinking of palaeobotanists. The principal point here is the confusion of the principles of outlining eutaxa and parataxa of different degrees of formalization. Even if the characters available in the hand specimens are inadequate for establishing eutaxa, palaeobotanists conjecture the affiliation of a given plant to a certain eutaxon involving notions relevant only for establishing eutaxa (e.g. the temporal or geographical setting of specimens). A confusion results. For instance, the sole difference between the form-genera *Pecopteris* and *Cladophlebis* is that Palaeozoic fronds are attributed to the former and the Mesozoic fronds to the latter. Although the curved pinnules and the common occurrence of a serrated margin are considered to be characteristic of *Cladophlebis*, these characters are present only in some of its species and are noted among *Pecopteris* species. The distinction between these two genera is quite conventional for the material from the Triassic and the uppermost Permian. Until now the sterile shoots of the Palaeozoic (*Walchia*, *Paranocladus*, *Ullmannia*, etc.) and Mesozoic–Cainozoic (*Brachyphyllum*, *Pagiophyllum*, *Cyparissidium*, *Geinitzia*, etc.) conifers have been attributed to different form-genera, though it is not possible to find any differences in the diagnoses of the parallel Palaeozoic and Mesozoic genera. It is quite impossible to comprehend in what way the Gondwana *Stellotheca* (*Lelstotheca*) differs from the Northern *Annularia*, apart from the geographical setting.

Of course, if one wishes, it is possible to find some minor distinctive characters and to support the independence of the established genus. Reading such substantiation of a genus it is impossible to rid oneself of the impression of a violation of taxonomic logic: first a decision is taken to introduce or to keep a genus and only subsequently is a justification found for this action. Otherwise it is impossible to comprehend why it was necessary to separate the genus *Stellotheca* from *Annularia* or to keep the genus *Bergiopteris* indistinguishable from *Angaropteridium*.

Sometimes the preservation or introduction of typical form-genera is connected with special geographical and stratigraphic confinement of the material. Generally the geographical and stratigraphic criteria are not directly presented when the genera are introduced or retained and are displayed in comparison with the genera only of a certain geographical and stratigraphic confinement.

The next source of inflation of genera is the confusion of two circumstances: (1) a high probability that the given remains belong to a separate genus and (2) the advisability of introducing the genus itself. Thus, it is safe to say that the oldest platyspermic seeds from the upper Devonian, isolated into the special genus *Spermolithus*, belong to a separate eutaxon of generic rank. However, it is impossible to point out the distinction between *Spermolithus* and *Samaropsis*. True, to describe the history of gymnosperms it is convenient to retain the independent genus *Spermolithus*.

Among other sources of superfluous multiplication of the number of genera only one will be mentioned, namely the violation of the congregational principle in systematics and the arbitrary raising of isolated characters to the rank of generic ones without analysis of the general distribution of the character weights in a given group of plants and in the related groups. The situation of those Pleuromeiaceae genera which produced spores of the *Aratrisporites* type is suggestive in this respect. The genera *Annalepis*, *Tomiostrobus*, *Skilliostrobus*, and *Cylostrobus* differ in minor characters of sporophylls, of no generic significance in other lycopsids. It is significant that all these genera are monotypic and the differences indicated for their separation have not even been considered as potential characters of species rank.

THE SPECIES PROBLEM IN PALAEOBOTANY

What the species is and how it should be recognized have been intensively discussed in the neobotanical literature. In palaeobotany the species problem has been discussed very little and mainly for particular plant groups. Meanwhile this problem deserves more thorough consideration as the inflation of species is no less strong than that of genera. Some genera (*Sphenophyllum*, *Pecopteris*, *Sphenopteris*, *Cladophlebis*, etc.) comprise hundreds of species which are very confusing and are obviously too numerous. For example, of more than 100 species of the genus *Phyllotheca* fewer than 20 deserve recognition (Meyen, 1971b).

Indirectly the redundancy of species can be demonstrated by lists of local floras. Sometimes numerous species of one genus are indicated for one layer or a small interval of a section (e.g. *Pecopteris* in the Stephanian of the Euramerian area) which is very suspicious botanically since a high species diversity in ferns of one genus is most uncommon in a community or in a small territory.

The problem of the volume of a species is finally reducible to the evaluation of intraspecific variability and of its admissible limits. Having only detached parts and no possibility of using the biological criteria of species delimitation, a palaeobotanist has to look for other methods to establish the range of intraspecific variability. These methods are:

1. The method of monotopic sampling. If gradual transitions between different morphological varieties of the same parts are observed in one layer and especially on one bedding plane, they represent, most probably, intraspecific variability. Obviously the attribution of these different varieties to one species should be supported by the study of the remains using all available methods, principally with reference to microscopic structure. Using the method of monotopic sampling combined with observations of leaves in attachment (Fig. 3) the author proposed the systematics of species for *Rufloria* and *Cordaites* leaves of the Angaraland (Meyen, 1966). It is found that the leaves attributed earlier to different species and even different genera (*Rufloria* and *Crassinervia*) may be reliably attributed to one species. Figure 4 depicts the intraspecific variability of the fronds of *Rhaphidopteris praecursoria*. The leaves in most pictures of this figure come from a single layer, yield identical cuticle structure, and identical specific resin bodies. A comparable variation is shown by *Dicroidium* (Anderson and Anderson, 1983).

Fig. 3. *Cordaites gorelovae* S. Meyen; variation of leaf outlines within a single rosette; this variation corresponds to differences between several species established on dispersed leaves. Modified from Meyen, 1966.

2. The study of two or more connected organs *in situ* (in organic connection). This is the most reliable method of study of intraspecific variability. This enables one to study the variability of miospores in sporangia, of leaves attached to shoots, of different portions of petrified stems (to evaluate the variability of petrified tissues), etc. A considerable polymorphism of spores and pollen in sporangia of some species has been known for a long time (Harris, 1973; Maheshwari and Meyen, 1975; Pant and Bhatnagar, 1971–1973; Schweitzer, 1960; Sladkov, 1967; etc.). The leaves of *Cordaites* found attached in one case to the same branch (Fig. 3; Meyen, 1966, Fig. 31, Table XXV, Figs 1–6) would have been attributed earlier to six species. Observations of branches attached to stems demonstrated that *Phyllotheca turnaensis* Gorel. and *Koretrophyllites multicostatus* Radcz. probably belong to the same species (Meyen, 1971b). In the same trunk of *Callixylon* the height and width of the medullary rays change much from

the periphery to the centre (Lemoigne *et al.*, 1983), some varieties having been attributed to different species. Very significant changes in the anatomical structure along the same stem are found in *Lepipodendron* (Eggert, 1961), *Calamites* (Eggert, 1962), *Psaronius* (Morgan, 1959), *Cladoxylon* (Bertrand, 1935), *Medullosa* (Delevoryas, 1955) and in other plants.

3. The marker method. The method of monotopic sampling can be supported by the search and observation of separate typical characters, the appearance of which at once in several species of a given genus being unlikely. For instance, in the *Tatarina* flora the leaves of *Phylladoderma* subgen. *Aequistomia* are very common. Leaves with very characteristic large hair bases are found in only one locality. It is highly improbable that this very rare character known from only the one locality would be present in several species of a genus at the same time. If so, then the variation of all other characters of leaves with hair bases (the degree of cutinization of different leaf sides, degree of contrast of papillae on cells, etc.) may be treated as the intraspecific variation.

4. The study of geographically and stratigraphically isolated populations. In the *Tatarina* flora, *Cordaites* leaves are known from only one layer of one locality. They are rather diverse but their variation fully corresponds to that of *C. clercii* Zal., studied by the method of monotopic sampling in the Pechora and Tunguska basins. Moreover, these leaves are accompanied by one species of seeds (*Samaropsis irregularis* Neub.). *C. clercii* is associated with the same seeds in Siberia. The isolated stratigraphic and geographical situation of the layer containing these leaves supports their attribution to one species in spite of their diversity. The same is true of the associated seeds.

5. Reference to the related Recent and fossil taxa, whose intraspecific variability is well studied.

6. The mathematical methods of data treatment (discriminant analysis, etc.).

There are other cases when for one reason or another the remains may be attributed to one species (analysis of the remains in the coprolites of monophagus animals, of the pollen in the micropyle of seeds, the presence of characteristic epiphyllous fungi, etc.). Obviously the more complete the set of the applied methods, the more

Fig. 4. *Raphidopteris praecursoria* S. Meyen; variation of leaves coming mostly from the same bed. Scale bar = 1 cm. Modified from Meyen, 1979b.

reliable is the taxonomic conclusion. Acquaintance with the cases of intended application of the listed methods demonstrates that the range of intraspecific variation may be very wide and greatly underestimated by many palaeobotanists. Palaebotanists widely practise the use of the minutest differences in specimens, even collected in the same layer, for establishing new species, not attempting to analyse the range of intraspecific variation. These 'splitters' are confronted by the 'lumpers' eager to integrate anything into one species. There is no point in discussing in general whether the splitters or the lumpers act more correctly. Both splitting and lumping of species must be supported by the listed or other methods, clearly explained, and well illustrated. Only then might it be worth while to discuss the taxonomic decisions taken. Disregard for these obvious norms has caused the inflation of species in palaeobotany.

THE POLYTYPIC CONCEPT OF THE SPECIES

As early as 1906 De-Vries proposed that we should distinguish 'linneons', being large and sometimes very polymorphic species approximating the species of Linnaeus and the other classics of plant systematics, and 'jordanons', small species, with a minimum variation of characters. De-Vries' proposal is only one of the attempts to overcome the taxonomic difficulties caused by a high variability. In palaeozoology the polytypic species concept was discussed intensively, i.e. the idea of the species as being an assembly of various stable varieties. The fact that the species is not a homogenous assembly of indistinguishable individuals was obvious even to Linnaeus. A study of the polymorphism of natural populations demonstrates that it is not possible in principle to obtain groups homogenous in all characters, for reasons of the uniqueness of individuals. Therefore, the infinite splitting of species to achieve this aim is a senseless occupation. Instead of this it is necessary to introduce species which are most efficient for stratigraphic, biogeographical, and other generalizations, i.e. to take practical requirements into consideration. Subsequently, it is always possible to find methods for describing the intraspecific polymorphism of a given species. In other words, we should learn to describe intraspecific polymorphism rather than exclude polymorphism as a species characteristic. At the same time we can use the categories 'subspecies', 'variety', etc. mentioned in the Code of Nomenclature. True, this is connected with difficulties in typification and priority. Therefore sometimes it is preferable to turn to more simple methods of fixing the diversity of forms, which may be designated by numerical or by alphabetical symbols. This procedure has been elaborated by palynologists in more detail (see Chapter 4).

ASSEMBLAGE-GENERA AND ASSEMBLAGE-SPECIES

Our growing study of fossil plants results in an increased knowledge of the initial associations of dispersed leaves, fructifications, miospores, etc. In some floras, for example the Yorkshire Jurassic flora after Harris's work, many dominating plants are already known in such associations. Lumping together dispersed organs and recognizing their belonging to the same plants always gives rise to the same problem: how such assemblages of the hitherto independent taxa are to be named.

In his well-known monograph on the Palaeozoic conifers (1938–1945), Florin introduced the term 'Kollektiv-Gattung'. Each such Gattung covers 'Organ-Gattungen'. For instance, the Kollektiv-Gattung *Lebachia* established for entire plants, covers Organ-Gattungen *Walchia* (p.p.), *Gomphostrobus*, *Walchiopremnon*, etc. Krassilov (1969) suggested selecting for a reconstructed plant just one of the generic names entered into the reconstruction. The selected name is to be added with a suffix 'rest' (restitution), for example, *Nilssonia* Brongniart 1825 rest. Harris 1941 (for the association of *Nilssonia*, *Beania* and *Androstrobus*).

Both these methods are applicable only when the established association of parts remains valid within the whole selected nominated genus. Unfortunately many genera may enter simultaneously into different life-time associations (Fig. 1), and none of the genera can be selected safely to name the reconstruction. For instance, leaves of the genus *Callipteris* associate with the female fructifications *Autunia* and *Peltaspermum*, as well as with the male fructifications *Pterispermostrobus* and *Callipterianthus*. Also, *Pterispermostrobus*-like synangia associate with the leaf genera *Eremopteris* and *Dicksonites*, and with different pollen belonging to several form-genera. Some of the pollen genera, for example *Vesicaspora*, were found in other microsporangiate genera. It even happens that the foliage ascribed to a single fairly narrow species is associated with different fructifications. Sterile fronds uniformly named *Dicksonites pluckenetii* are associated in the Westphalian D with unmodified seed-bearing fronds, and in the Stephanian with strongly modified fronds deserving separation into an independent genus. (Unfortunately the type material of the species is Lower Permian.)

In the cases of intricate combinations of the constituent genera and species, we need a more sophisticated nomenclature for reconstructed plants than merely one generic or specific name. I believe that the best and easiest way is to call the reconstructed plants by the names of the associated taxa, omitting the author's surname, e.g. *Callipteris conferta–Autunia* sp.–*Pterispermostrobus gimmianus–Vesicaspora* sp. When such a compound name can be repeatedly used in a text some of the constituents can be abbreviated or deleted depending on the context (e.g. '*C. conferta–Autunia–P. gimmianus*' or '*C. conferta–Autunia*'). I propose that such collective taxa be called 'assemblage-genera' (as in *Nilssonia–Beania–Androstrobus*) or 'assemblage-species' (as in *Callipteris conferta* and its associates). Both the terms and the proposed procedure need not be incorporated into the International Code of Botanical Nomenclature. They may be used informally in texts on systematics, phylogeny, ecology, geographical and stratigraphical distributions, etc. Such texts sometimes require this kind of more precise approach to reconstructed plants.

THE CLADISTIC METHODS

Cladistic methods have not had any significant influence on palaeobotanical systematics. Articles where these methods are applied to palaeobotanical material (Doyle *et al.*, 1982; Miller, 1982; Stein *et al.*, 1984) are interesting from the viewpoint of the analysis of phylogenetic relations between some taxa but do not contain taxonomic conclusions. An obvious advantage of cladistics is the organization of the phylogenetic procedure in such a way that consideration of the characters and their analysis is more complete and explicit. It is also important that the alternative phylogenetic schemes are built in this case. However, compared with those former phylogenetic views which were elaborated rather thoroughly (e.g. Banks, 1968; Skog and Banks, 1973) cladistic analysis has not given anything new except for some terms of no evident usefulness and ways of graphic representation of alternative phylogenetic schemes. In general it is possible to agree with the sceptical evaluation of the cladistic boom given by Tatarinov (1984).

Both classical phylogenetic analysis and cladistics in all its varieties allow no possibility of making a final choice from the competing systems of suprageneric taxa. Probably this choice is impossible in principle and we must accept it, as we accepted long ago the impossibility of obtaining perpetual motion. It may be assumed that the diversity of organisms and our cognitive power are controlled by certain limiting laws (similar to thermodynamic laws in the case of perpetual motion) making impossible the final building both of the whole system of organisms and any of its parts.

It is obvious that the complete system is not attainable at least on account of our inability to overcome many aspects of the imperfection of the geological record. Even if we knew the necessary minimum of the extinct forms we would always face the problem of interpreting plant diversity in terms of systematics. The major difficulties here are the weighting of characters and the ranking of taxa. In the final analysis, both depend on the admitted range of variation of particular characters for a given taxon or taxonomic category. There cannot be any a priori assumptions in this respect; everything is solved by the available material, and much depends on the size of a sample as well as on the adopted rules of morphological derivation. The last circumstance is of particular importance. Structures A and B may be regarded as being so different that there is no method for the morphological derivation of B from A or vice versa. The distinctions between the taxa possessing the characters A and B are then considered to be of a high rank. For instance, the fossil leaves of ginkgos and leptostrobans, in spite of their similarity, were attributed to different orders and even to different classes since the morphological distance between the associating female fructifications seemed to be very great. It seemed that no morphological transformations existed which could connect the bivalved capsules of *Leptostrobus* and the seed-bearing organs of *Ginkgo* with their paired seeds supplied with collars. In the same way the degree of morphological distinction of *Taxus* from the other conifers was evaluated. The incorporation of a wider variety of gymnosperms into the analysis has demonstrated that the morphological transformations connecting *Leptostrobus* with *Ginkgo*, or *Taxus* with the other conifers are not so complicated, since they are quite conceivable in closely related gymnosperms (Meyen, 1984a). Accordingly it is possible to bring together the Ginkgoales and Leptostrobales within one class, Ginkgoopsida, and to place *Taxus* in the order Pinales.

A continuous increase in the rank of supra-generic taxa and an increase in the number of orders, classes, and even divisions taking place in systematics can be connected closely with such alleged 'impossibility' of morphological derivation of certain types of organs from each other. Partly it occurs due to canonization of modes of morphological transformations provided by the telome theory (see Chapter 8) that admits only the gradualistic transformations and practically disregards a possible violation of the topology of organs in the course of transformations (in this type of argument, two criteria of homology are canonized, a criterion of position and a criterion of transitional forms). In this connection the morphological hiatuses between taxa are exaggerated artificially and the rank of taxa is accordingly increased without justification.

Chapter 3

Fossil plants systematics

I. PROKARYOTES

Prokaryotes under discussion below involve only bacteria and cyanobacteria (blue-green algae). The classification of living forms of prokaryotes is based on such characters as the configuration of the cells (cylindrical, spiral, spherical, etc.), the mode of their aggregation, mobility, size, mode of feeding, and other features. A character such as the cylindrical form of the cell may occur in combination with any others. Many other characters also combine freely. Because of this the prokaryote system, as was noted by G. A. Zavarzin, assumes the nature of a multidimensional latticework. The nodes of this latticework correspond to taxa, each taxon being defined by a specific combination of characters. Due to the fact that some of the combinations do not occur, missing units occur in the latticework. But the remaining portions of the lattice do not form a structure that may be treated as a phylogenetic tree. In the opinion of G. A. Zavarzin (1974), the construction of a prokaryote system on a phylogenetic basis is, in principle, impossible.

The classification of fossil prokaryotes is fraught with difficulties. Apart from the fact that only the forms and sizes of prokaryotes can be discerned in fossil remains, whereas other characters, most significant in systematics, are, as a rule, unavailable, we lack the means to reconstruct the unpreserved characters according to those preserved. Moreover, the assignment of fossil forms to prokaryotes, in itself, may result in errors, since prokaryotes cannot be reliably distinguished from eukaryotes, that are closely alike in outward aspect. For instance, actinomycetes, cyanobacteria, various algae, fungi and moss rhizoids, may exhibit filaments similar in habit, both septate and non-septate. It is necessary to take into consideration the fact that groups now extinct may have existed during the geological past, and in fossil remains they may resemble prokaryotes in appearance. Inorganic compounds may be easily mistaken for fossil remains of prokaryotes.

DIVISION BACTERIOPHYTA (BACTERIAE)

The treatment of the Bacteria Division is purposely kept rather broad in scope. This division also involves, in addition to bacteria proper, arch-

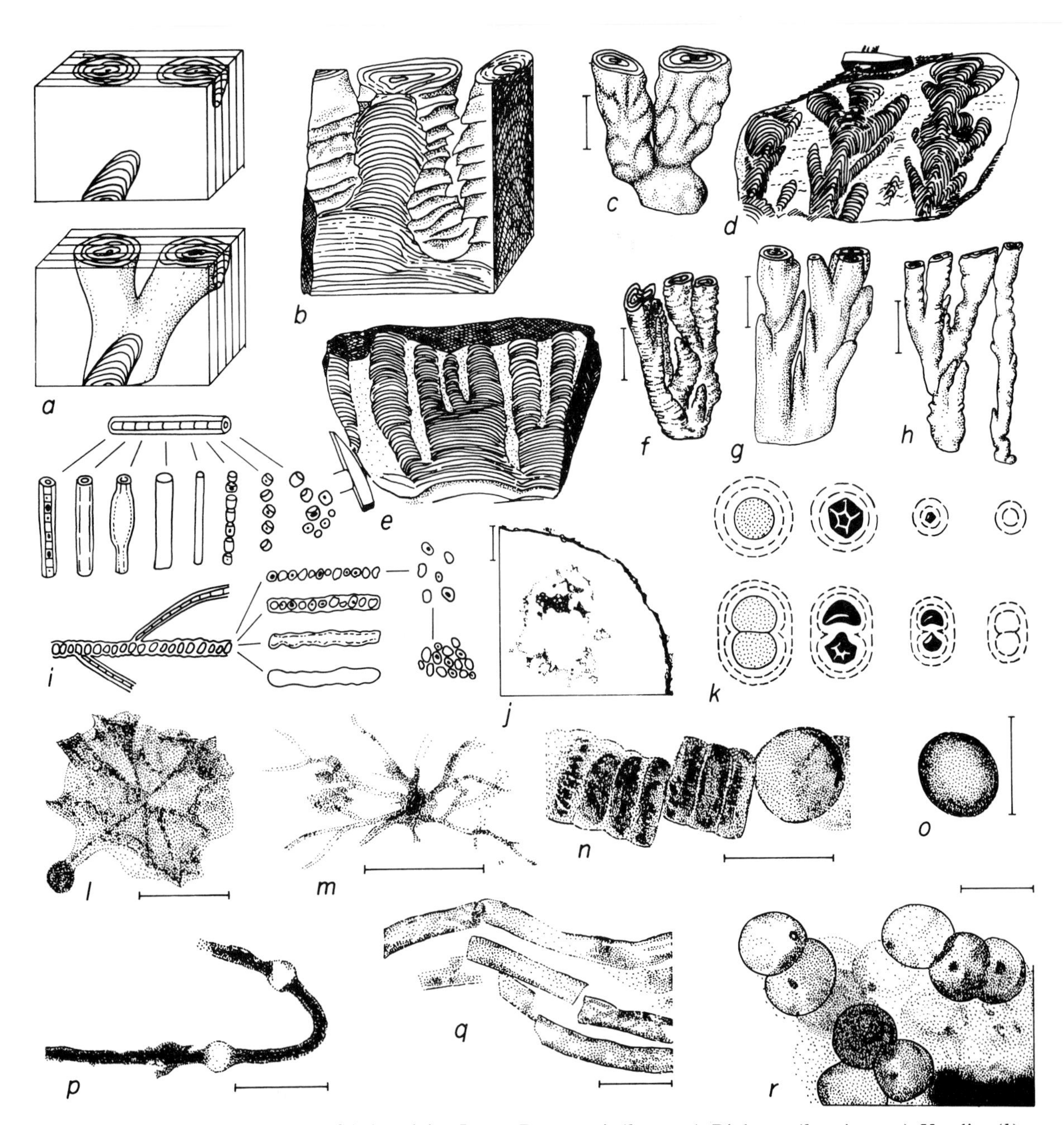

Fig. 5. Prokaryotes and products of their activity; Lower Proterozoic (*l*, *m*, *o*, *p*), Riphaean (*b–g*, *j*, *n*, *q*, *r*), Vendian (*h*); S. Urals (*b–d*, *g*), E. Siberia (*e*), Tien Shan (*f*), Kazakhstan (*h*), Australia (*j*, *n*, *q*, *r*), North America (*l*, *m*, *o*, *p*); *a*, reconstruction of stromatolite columns with the aid of Krylov's method (graphic preparation), scheme of sectioning (above) and reconstructed columns (below); *b*, *c*, reconstructions of *Kussiella kussiensis* (Masl.) Kryl. (*b*) and *Baicalia baicalica* (Masl.) Kryl. (*c*); *d*, *e*, field sketches of *Jacutophyton* (*d*) and *Kussiella* (*e*); *f–h*, reconstructions of *Inzeria toctogulii* Kryl. (*f*), *Gymnosolen ramsayi* Steinm. (*g*) and *Patomia ossica* Kryl. (*h*); *i*, preservation modes of hormogonial (above) and mastigocladian (below) cyanobacteria in algal–bacterial mats; *j*, section of *Glenobotrydion aenigmatis* J. W. Schopf with modified mucilage sheath outside and collapsed cell content inside; *k* decay of cell (stippled or in black) and mucilage sheath (broken line) of coccoid cyanobacteria in unicellular state (above) and during incomplete division ▶

aeobacteria, mycoplasms (bacteria lacking walls), actinomycetes, myxobacteria and spirochaetes. Certain bacteria probably evolved from cyanobacteria similar in type by losing the pigment system.

Bacteria (B.) are microscopic organisms. Their cells commonly do not exceed 10 μm in size. One-celled B. may be rounded (cocci) or elongated; in the latter case they may be rectilinear (rods), curved (vibrio), or spirally twisted (spirillum). Certain bacteria have branching cells. Filamentous forms of B. consist of a single row of cells. Aggregations of B. form sheets and films that are visible with the naked eye. The cell walls of many B. are sufficiently hard, due to which they occur in more or less invariable forms, and may be preserved as microfossils.

The oldest known microfossils assigned to B. were discovered in cherts of the Pre-Cambrian Fig Tree Group (Southern Africa). They are dated at about 3.2×10^9 years b.p. Replicas made from etched cuts of the cherts were studied under the electron microscope. The rod-like bodies found here were organic-walled, less than a micrometre in length, and have been assigned to the genus *Eobacterium*. The same method was applied to studies of cherts of the iron-rich Gunflint Group in Canada, dated at 1.9×10^9 years b.p. Similar bodies have been identified that occurred in disorderly groups and short chains. The thin petrographic sections prepared from silicified stromatolites, recovered from the same group, revealed stellate microfossils that have been described as a particular genus *Eoastrion* (Fig. 5, *m*). However, in outward appearance they cannot be distinguished from the present-day *Metallogenium* mycoplasms, which cause the precipitation of manganese and iron in water bodies. *Eoastrion* may also be related to actinomycetes. Similar microfossils are known from younger sediments. Very thin filaments were recovered from the Pre-Cambrian, which, according to some investigators, may be regarded as filamentous B.

Microfossils, similar in type to cocci, rods, filamentous and spirillum B., are known from rock deposits of different geological periods. Also, various types of B. colonies have been described. The assignment of these microfossils to B. is based not only on their habit. B. have been encountered in bones and plant remains in those very places where the tissue was pathologically afflicted. B. occurring in coprolites are similar to the gut-flora type. Microfossils, similar to iron-depositing bacteria, are restricted to iron ores of the Pre-Cambrian and Phanerozoic. Filaments, resembling extant thiobacterium *Beggiatoa*, have been found in sediments enriched in sulphides.

B. have had a paramount role in the formation and evolution of the biosphere. Studies of recent sediments tend to indicate that the bacterial activity is an important factor in the accumulation of carbonates, iron ores, sulphides, silica, phosphorites, bauxites. At present B. are capable of reworking about 5 million tons of sulphates daily, and can produce about 1.5 million tons of hydrogen sulphide. B. together with blue-green algae participate in the precipitation of carbonates (for instance, in stromatolites). The role of B. in the decomposition of organic remains is widely acknowledged. Traces of B. activities can be discerned in fossil remains most resistant to decay, such as plant cuticles and sporopollenin walls of miospores. Certain portions may be perceived in them that are completely covered by cocci impressions, which, at times, have been wrongly identified as primary microstructural characters of the walls. During early diagenesis of sediments B.

(below); *l–r*, microfossils in silicified stromatolites, *Kakabekia umbellata* Bargh. (*l*), *Eoastrion* (al. *Metallogenium*) *bifurcatum* Bargh. (*m*), *Palaeolyngbya barghoorniana* J. W. Schopf, partly desintegrated filament (*n*), *Huroniospora* (*o*), *Gunflintia minuta* Bargh., filament with heterocyst-like portions (*p*), *Eomycetopsis filiformis* J. W. Schopf (*q*), *Glenobotrydion aenigmatis*, black patches are collapsed cell contents simulating nuclei (*r*; see *j* and *k*). Scale bar = 3 cm (*c*, *f–h*), 10 μm (*l–r*), 1 μm (*j*). Modified from: Krylov, 1975 (*a–h*); Gerasimenko and Krylov, 1983 (*i*); Oehler, 1977 (*j*); Golubič and Barghoorn, 1977 (*k*); Barghoorn and Tyler, 1965 (*l*, *m*); J. W. Schopf, 1968 (*n*, *q*, *r*); Awramik and Barghoorn, 1977 (*o*, *p*).

are the major energy source in the decomposition of organic matter. For detailed information on fossil B. see Tappan (1980).

DIVISION CYANOPHYTA (CYANOBACTERIA)

The name 'Cyanobacteria' (C.) is currently replacing the term 'blue-green algae'. C. (with the exception of cyanelles, that reside within various other algae) have a wall of sufficiently enduring form. They are rounded, elliptical, cylindrical, barrel-like, egg-like, or of other shapes; they may occur in detached forms or colonies, or as multicelled filaments. Commonly C. secrete a mucilage in the form of a thick envelope, that in certain types may be surrounded by a thick wall (sheath). In colonial forms the cells undergo division but remain united by the mucilage in the form of packets. A common mucilage sheath may also unite filamentous types. Certain C. filaments exhibit branching and in some places form multi-layered mats. The filamentous C., in addition to the usual cells, have heterocysts. These are cells with thickened walls and occasionally of larger dimensions. They fix nitrogen and are the source of nitrogen supply to other cells of the filaments. Near the heterocyst the filaments frequently reveal branching or division.

The conditions under which living C. exist may vary widely. The majority of species are freshwater inhabitants, the few species that live in seas occur chiefly within the intertidal zones and areas of low-salinity. Terrestrial forms live in soils, form films on stones, and on the bark of trees. Some C. are capable both of aerobic and anaerobic metabolism. Because of this they dwell in sites that are subjected to continuous changes, alternating between aerobic and anaerobic environments (for instance, in the oozy area of the intertidal zone), where other organisms are incapable of living. C. may exist under conditions of sharp fluctuations of the water salinity, and in a wide temperature range. Certain C. are carbonate secreting (for instance, those types that together with bacteria form stromatolites). Some C. are known that oxidize hydrogen sulphide to elemental sulphur, the latter undergoing deposition on the surface of the cells.

The classification of living C. is based on the morphology of the cells, colonies and filaments, on their pigments, mode of reproduction and other characters. The group arrangement of the fossils is only tentative, although unicellular forms are frequently assigned to Chroococcidiaceae, whereas the filamentous forms are assigned to Oscillatoriaceae. It is probable that during the geological past certain C. groups might have existed that were similar only in outward aspect to the above-mentioned modern taxa. An indirect indication of the probable existence of such groups is found in the recent discovery of green prokaryotes, keyed out in the group Prochlorophyta, that differ from C. in their pigment composition. It is unclear so far whether these organisms should be excluded from C.

C. are represented in the fossil record by remains of three types. They involve, first of all, solitary cells and filaments with preserved organic walls and (or) sheaths. The second type is represented by siliceous pseudomorphs of organic structures. In thin sections these look like three-dimensional cells or their aggregates. The third type includes carbonate-secreting forms. Only calcareous sheaths remain of the latter, from which upon dissolution the organic wall (or sheath) can be liberated. The assignment of many fossils of all three types to C. is still debatable. Calcareous fossils, frequently assigned to C. without any stipulations, are discussed below in the section Calcareous fossils of algal origin (p. 25). Difficulties occur in relation to the well-preserved spheroidal and filamentous microfossils that are represented by siliceous pseudomorphs (Fig. 5, *j*, *n–r*). It has been shown that certain microfossils, described as various C. taxa, are in effect different fossilized stages of posthumous decay of initially uniform cells. The mucilage sheath of the cells, often preserved, simulates their walls; the content of the cell coagulates, and looks like a nucleus (Fig. 5, *k*, *r*). Such microfossils have been erroneously assigned to eukaryotic algae, and on the basis of their first appearance inferences were derived concerning the time of origin of the eukaryotes. Post-mortem alterations of filamentous C. result in an increase in their volume, and the rupturing of certain cells, simulating heterocysts. From detailed investigations of post-mortem alterations in the present-day oscillatorians (Gerasimenko and Krylov, 1983) it became apparent that a single species may yield various fossil types, each being distinguished as a specific microfossil taxon (Fig. 5, *i*). The filaments easily disintegrate into single cells (simulating coccoid C.), that cluster into secondary aggregates. Upon burial the filaments may be fossilized with sheaths or as detached sheaths.

Quite a large number of genera have been established among filamentous and coccoid C. according to the sizes and forms of the cells and filaments, thicknesses of the walls and other characters (Fig. 5, *n–r*).

We have already mentioned Pre-Cambrian C., that were encountered in microbiotas of silicified stromatolites (particularly in microbiotas of the Gunflint Group of Canada, dated at 1.9×10^9 years, and in the Bitter Springs Formation of Central Australia, which is nearly 1×10^9 years old. For an account of current concepts of Pre-Cambrian algal microfossils, refer to J. W. Schopf, (1983). Fossil remains of C. are also known from younger rock formations.

CALCAREOUS FOSSILS OF ALGAL ORIGIN

Carbonate secretion is typical of certain bacteria, cyanobacteria and various types of eukaryotic algae. In some cases carbonates are actively secreted by the organism and deposited into its tissue, on the surface of its body, or in the surrounding mucilage sheath. In other cases carbonates are of passive secretion. It is believed that the plant liberates CO_2 from the water-dissolved calcium carbonates. As a result calcite is formed that precipitates on the surface of the body. The secretion of carbonates may also be elicited by the change in the pH value of the environment caused by the algae. Passive secretion of carbonates may probably occur as a result of the activities of carbonate-secreting bacteria. Fossil remains of the sheaths, or of calcified parts, resulting from these processes, have been recovered from sediments of various ages and are termed 'calcareous algae'. They are typical mostly of marine carbonate sediments, but frequently occur in freshwater deposits, in soils and in the seepage area of springs.

Only those calcareous algae are treated below which are difficult to correlate with extant taxa. Calcification, taken by itself, does not give an indication of the systematic affinity of a fossil. Calcareous sheaths and tissues similar in type may be related to several groups; tubular sheaths, for instance, to filamentous cyanobacteria, green (Conjugatophyceae), red and yellow-green (Xanthophyceae) algae. Other carbonate-secreting filamentous algae might have existed in the geological past. Form-classifications are used in the systematization of such algal fossils. Calcareous structures, where fossil remains of the parent algae are altogether lacking, include most stromatolites and oncolites. Algal remains are discernible in them only in rare instances.

In describing calcareous algae of indeterminate systematic affinities, the palaeobotanists frequently tentatively affiliate these fossils to one or another group of living algae. The variation in the range of assignment is often quite significant. For instance, the genus *Epiphyton*, introduced for the calcareous filaments forming clusters and branching in layers, was affiliated to cyanobacteria, green or red algae. When the disagreement is so wide in scope, it is preferable to have recourse to form-groupings, as adopted in the systematization of acritarchs and dispersed miospores (see Chapter 4), as was suggested earlier by Pia (in Hirmer, 1927). For instance, his group Spongiostromata involves calcareous build-ups, preserving growth zones, but not the organic structure itself (algal sheaths, etc.). Stromatolites and oncolites have been assigned to this group. The group Porostromata (not discussed in this book) includes tubular calcareous sheaths of various algae, that occasionally are aggregated in nodules or layers. No all-inclusive classification has been suggested as yet for the products resulting from the activities of carbonate-secreting algae.

STROMATOLITES (STROMATOLITHI)

The term stromatolites (S.) refers to laminated, frequently shelled, chiefly carbonate organo-sedimentary structures, stratiform, nodular or of columnar shape. These structures were produced as a result of the activities of cyanobacteria, bacteria and other micro-organisms. Comprehensive studies bearing on the case history of S., their structure, origin and classification have been contributed by various authors (Krylov, 1975; Maslov, 1960; Walter, 1976; J. W. Schopf, 1983). The remains of the parent organisms are commonly not preserved in S. Because of this a current tendency is not to assign to S. structures where such remains are easily identifiable (for example, tubular *Girvanella*). In this case they are referred to as algal nodules or algal layers. At the same time, Pre-Cambrian laminated bodies with silicified portions preserving envelopes of cyanobacteria and problematical micro-organisms, are invariably assigned to S. The inconsistency of such an approach is obvious (Krylov, 1975). S. with preserved remains of the parent organism may be termed 'intact' (skeletal – Riding, 1977), whereas those lacking them are termed 'masked'. Accordingly, changes should be made in the group characteristics.

For the first descriptions of S. a common binomial nomenclature was already applied to them. This did not give rise to adverse criticism as long as S. were assigned to animals. When it was discovered that S. are not individual organisms, but merely products of the activities of many organisms, suggestions were advanced that no Latin names should be introduced for them, but descriptive symbols and formulae should be used for their identification. However, the binomial nomenclature is of much wider usage. For the distinction of species, genera and suprageneric taxa (types, groups) attention is focused on the general habit of the S. bodies, the texture and lamination, and, for columnar S., on the forms of the columns, type of branching, their marginal structure and other characters. The evidence available from studies of extant stromatolite-building communities serves to indicate that the S. microstructure results from decomposing bacteria and dead cyanobacteria. S. microstructures may be banded, striated, clotty, etc. They can be studied in thin sections and cuts, and their general pattern may be reconstructed according to the method of graphical preparation, which can be termed 'Krylov's method'. In 1959 I. N. Krylov suggested cutting S. into a series of thin plates and by tracing the changes in the S. contours from one plate to another reconstructing the structure and the changes in lamination within a single build-up (Fig. 5, *a*). The application of Krylov's method allowed variations to be revealed in the textural and structural characters in various places of a single bioherm. The fact was established that S., formerly described as independent genera, may occur in combinations within a single build-up. This provided an additional argument against the application of binomial Latin nomenclature to S. However, as a counter-example was offered the experience gained in palaeobotanic taxonomy and nomenclature, which admits the possibility of distinguishing detached parts as independent form-genera, even when these parts are known to have occurred in combination in living forms (see Chapter 2). Accordingly, the generic name *Jacutophyton* (Fig. 5, *d*) is retained, although different parts of the build-up bearing this name, are ascribed to the genera *Baicalia* (Fig. 5, *c*) and *Conophyton*. Such a genera-set forming one build-up is termed a 'bioherm series'. Build-ups that are similar macroscopically may differ in their microstructure, and vice versa. Because of this a suggestion was made to rely in all cases upon the priority of certain characters, as compared with others (for example, distinction of the genera should always be based on the overall construction of the build-up, whereas distinction of the species should be based on the microstructure). However, the wide diversity of situations resulting from the variability in many characters within the build-up could not be accommodated within such strict bounds. It is necessary to take into consideration the fact that the microstructure may be altered by secondary processes (recrystallization, metamorphism, etc.). In the classification of megafossils palaeobotanists frequently introduce various generic names for the remains, depending upon whether or not their microstructure is known. The same may be applied to S.

The literature reveals over 20 classifications that have been suggested for S. For general acquaintance with S. the following simplified classification may be adopted. According to the form of the build-up, S. may be classified into stratiform, nodular and columnar; also into transitional types between these forms, and combined types (columnar-layered and columnar-nodular S.). Those studied and classified in most detail are columnar S. They are described by a more prolific set of characters and may be more confidently divided into genera and species. They can be classified according to the form of the columns (regularly cylindrical, or irregular in shape, tuberose), style of branching, type of marginal surface of the column (with layers enveloping the columns, or not enveloping them; the surface may be smooth or knobby, with cornices and peaks). The species distinction relies mainly on the microstructure. Some of the genera distinguished in this manner are depicted in Fig. 5, *b–h*.

The current stratigraphic subdivision of the Proterozoic, particularly that of the Riphean, could not have been accomplished without S. assemblages. In the 1960s I. N. Krylov, S. V. Nuzhnov, M. A. Semikhatov, M. E. Raaben and V. A. Komar established that S. assemblages are persistent within vast areas (Siberia, the Urals), and then it was found that the assemblages exhibit the same successive order of occurrence in various regions. Radioisotope datings served to prove that S. assemblages similar in type, distributed over vast territories of the USSR, are practically all of the same age. In later years assemblages nearly the same were traced throughout North America, India, Australia and Africa. The similarity in the stratigraphic succession of S. assemblages in distant areas and continents, and the persistent distribution pattern of these assemblages are empirical generalizations, not quite comprehensible from the biological viewpoint. In fact, it is known that S. are products of the activities of micro-organisms. Studies of extant S. convince us that the form of the build-up depends largely on the hydrodynamical conditions and other ecological factors. Because of this it can be reasonably supposed that S. assemblages might be used for lithofacial reconstructions, rather than for stratigraphic purposes. This is seemingly confirmed by the co-existence in one build-up of forms commonly assigned to different genera. In the course of time it came to be realized that one or another of these genera was of wider stratigraphic range. Nevertheless, although the stratigraphic pattern of S. taxa in the sequence becomes increasingly more complex from year to year, this does not discredit the stratigraphic significance of S. Moreover, studies of Phanerozoic S., intact in the main, made it possible to establish the stratigraphic significance of S. not only for ecological–stratigraphic investigations, but also as guide fossils. To be more exact, Phanerozoic S. are of significance chiefly in lithofacial investigations. Considering that they were produced by photosynthetic organisms, there is reason to believe that they were formed at depths barely exceeding 100 m. It was noted that Phanerozoic S. are restricted to facies that are poor in other organic remains. They occur typically in sediments that were deposited in basins with unstable salinity (see above, various types of metabolism of cyanobacteria). Namely, in such basins they are being formed today. Whether these characteristics of Recent and Phanerozoic S. may extend throughout the Pre-Cambrian, remains as yet unclear. It is thought that the conditions favouring S. formation in the Pre-Cambrian were more diverse, whereas later the stromatolite-forming organisms were pushed aside by others.

ONCOLITES (ONCOLITHI)

Oncolites (O.) in the broad sense of the term were initially defined as nodules with precisely the same type of lamination as in stromatolites, but, by contrast, enveloping the nodule on all sides. These nodules may be small, about a millimetre in cross-section, or larger – fist-size and even greater. At first O. (the same as stromatolites) were regarded as algae, and only later were they found to be products of micro-organism communities (cyanobacteria, bacteria, etc.). Sometimes in O. tubes are encountered of the *Girvanella* type and of other organic remains. The first investigators (W. H. Twenhofel, J. Pia, V. P. Maslov and others) had already attempted the orderly arrangement of O. in taxa, adopting Latin names for them. Later, attention was focused on small O., which, together with some other structures, presumably of algal origin, are termed 'microphytolites' (or 'microproblematics'). The importance attached to these O. increased conspicuously after it came to be realized that they can feasibly be used for the subdivision of the Pre-Cambrian (Zhuravleva, 1964; Radionova, 1972, 1976). The number of identified genera and species of microphytolites increased rapidly. For their distinction such characters as the size of the nodules, thickness of the laminations (if they are present), width of the rays in radial-fibrous forms, structure of the central parts, and others were used. Later on it became apparent that some of these systematic characters are associated with secondary alterations in the nodules, especially due to the recrystallization of calcite; secondary modifications of various initial forms may result in similar end-stages. On the other hand, another problem arose, concerning the distinctions between Pre-Cambrian O. and those Phanerozoic sedimentary structures, which were always described as oolites, studies of which were considered in the field of lithology, not of biostratigraphy. The belief prevailed that oolites were of chemical origin. At the same time organic filaments were discovered in oolites of various ages. Whether oolites are of purely chemical origin, without involving carbonate-secreting organisms, still remains unclear.

II. EUKARYOTES

In this section an outline is given first of the systematics of eukaryotic algae, and then of higher plants. The term 'algae' commonly refers to a certain informal taxon, in the ecological–morphological sense, that involves all the divisions from Pyrrophyta to Charophyta, inclusive, as well as cyanobacteria.

The variation in the systems of classifying eukaryotic algae is very great. For instance, diatoms are occasionally placed in an independent division, named Bacillariophyta or Diatomeae, whereas in other cases they are treated as a class Bacillariophyceae of the division Chromophyta, or as an order Diatomales (Bacillariales) in the class Chrysophyceae. Such a wide disagreement is in part due to the recent wide application of electron microscopy, biochemical and other methods of investigation. The palaeoalgological data offer as yet little for the explication of certain debatable problems bearing on algae systematics. Such characters pertinent to the modern classifications as the chromosome structure, flagellum organization or biochemical properties of pigments cannot be studied in fossil materials. The palaeoalgologist has to rely on less significant characters, but better preserved in fossils, and choose such a system which can be easier adapted to fossils. The system outlined below has been constructed in such a manner as to ensure the least diversion from those systems which are of wide current use in the literature (Denffer *et al.*, 1978; Němejc, 1959; Osnovy paleontologii, 1963a; Stewart, 1983; Tappan, 1980; Taylor, 1981; Gollerbakh, 1977). Following the conventional pattern, but with certain necessary stipulations, the distribution of the genera among supergeneric taxa is shown.

The characters preserved in fossil forms are those according to which a significant parallelism can be observed between divisions in living forms. This parallelism is commonly demonstrated in the levels of morphological differentiation of the algal vegetative body. These levels are as follows: (1) unicellular motile forms with flagella; when the cells are united in colonies then the flagella remain; (2) amoeboid (rhizopodial) structure; detached cells lacking flagella; motile forms issue pseudopodia (as *Amoeba*); (3) palmelloid (gloeomorphic) structure; several cells, (occasionally a large number) submerged in a common mucilage secreted by the cells; (4) coccoid structure; cells lacking flagella and because of this are non-motile; solitary or united in colonies but not filaments; (5) filamentous (trichoid) structure; non-flagellar cells combine into filaments, which may be branching or remain simple; filaments comprising a body may be uniform or differentiated; (6) cladothalloid structure; non-flagellar cells united into a single- or multilayered lamina; upon division of the cells in different planes thalli of cylindrical or complex forms may originate; (7) siphonate structure; very large cells with numerous nuclei within a single cytoplasm; such multinuclear cells (coenocytes) may be tubular or of more complex organization.

Other classifications of the same levels of organization exist. However, no matter which classification is adopted, it can be seen that several levels are present in each division of eukaryotic algae (in some divisions practically all are present), and each level occurs characteristically in several divisions. For example, among both xanthophytaceous and green algae, unicellular flagellar, palmelloid, coccoid, filamentous and siphonate structures are known. Filamentous structures occur in all divisions (in the diatoms, single row colonies may be formed that not everybody regards as filaments). Some of the above-cited levels occur in cyanobacteria and fungi. The palaeobotanist should also take into consideration that each individual of a single species in the course of its life-cycle may pass through several levels. Finally, the very assignment of the fossil to algae may be problematical. For instance, the filamentous stage in the ontogeny of mosses (protonema) may bear a resemblance to a filamentous algae. The calcareous parts of the algae cannot always be easily distinguished from the foraminifer shells (such as filamentous foraminifers *Nubecularia*). The microscopic envelopes enduring maceration, may display affinities to various algae and their cysts, to spores of higher plants, to statoblasts ('resting buds') of bryozoans and eggs of invertebrates. Aggregations of spores of higher plants have been described as algal colonies.

The general principle of construction of a natural system is clearly apparent in relation to the algae. At first it seems difficult to grasp the reason why unicellular forms with flagella, very much alike in outward aspect, have been assigned to different divisions merely because of the difference in their pigments; also, why unicellular, as well as multicellular forms of highly complex organization and highly advanced organs of reproduction have been assigned to one division. The assignment of forms of very different outward aspect to one division and forms similar in appearance to different divisions can be understood only after acquaintance with the entire system. The orderly arrangement of the taxa of lower rank, and their successive grouping into congregations of increasingly higher rank then becomes apparent. Notwithstanding the increasingly greater differences in the forms within these congregations, those characters which persist (pigment composition, flagellar structure, etc.) are of major significance for the designation of taxa of the highest ranks. It is impossible to outline here the entire orderly arrangement of the taxa, from the lowest to the highest ranks. Accordingly, in our descriptions of the superior taxa those characters will be reduced to a minimum that cannot be observed in fossil materials and information on which is available from references on algae systematics (e.g. Flügel, 1977; Johnson, 1961; Maslov, 1956, 1962, 1963, 1973; Osnovy palaeontologii, 1963a; Tappan, 1980; Gollerbakh, 1977).

DIVISION PYRROPHYTA (DINOFLAGELLATES)

Pyrrophyta or dinoflagellates (D.) are characterized by reddish to yellow-green pigments. D. are unicellular, occasionally filamentous forms. Normally they have two long flagella. Inasmuch as some D. lack plastids and are heterotrophic, these organisms have been assigned by certain zoologists to Protozoa (as a class Dinoflagellata). D. are of particular interest due to the great diversity in their chromosome organization and types of mitosis. According to these characters some of the genera (called mesocaryotes) display affinities to procaryotes. Various dinoflagellates (zooxanthelae) are endosymbiotic. They reside in the bodies of foraminifers, radiolarians, other D. (non-photosynthesizing) and in various groups of coelenterates and molluscs. The presence of these

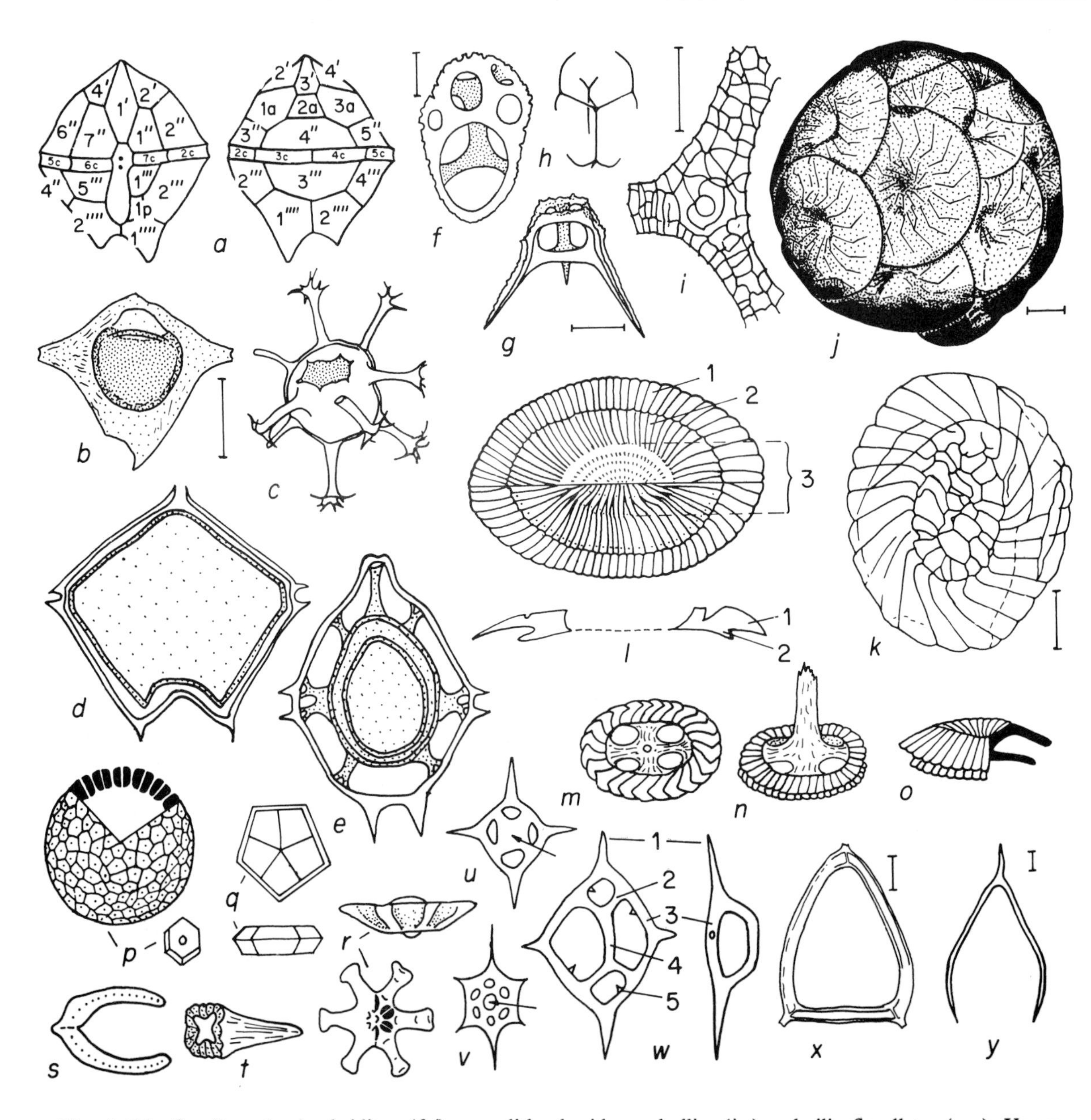

Fig. 6. Dinoflagellates (*a–e*), ebridians (*f–i*), coccolithophorides and allies (*j–t*) and silicoflagellates (*u–y*); Upper Cretaceous (*k*, *y*), Palaeogene (*b*, *f*, *x*), Miocene (*g*, *i*), recent (*a*, *j*); New Zealand (*f*), Bolivia (*g*), USA (*i*, *y*), Ukraine (*k*), Austria (*x*); *a*, arrangement of plates in theca (epitheca: 1′–4′, apical plates; 1″–7″, anterior equatorial, or presingular plates; 1*a*–3*a*, anterior intercalary plates; 1*c*–6*c*, cingulum; hypotheca 1″′–5″′, posterior equatorial, or postsingular plates; 1*p*, posterior intercalary, or postintercalary plates; 1″ ″, 2″ ″, antapical plates with antapical horns; above, apex with apical pore; below, antapex); *b*, cyst of *Rhombodinium glabrum* (Cooks.) Vozzh., peridinioid form; *c*, cyst of *Oligosphaeridium complex* (White) Dav. et Will., chorate form, each projection corresponds to one paraplate; *d*, *e*, cyst formation (stippled) inside of theca, peridinioid (*d*) and histrichosphaeroid, or gonyaulaxoid (*e*) cyst types; *f*, *Craniopsis octo* Hov. et Freng.; *g*, *Parathranium tenuipes* Hov.; *h*, skeleton diagram of *Falsebria*; *i*, *Ebriopsis antiqua* (Schulz) Hov., part of skeleton with scultpure; *j*, coccosphere of *Cyclococcolithus leptoporus* (Murr. et Black.) Kampt.; ▶

endosymbionts is a favourable factor for the survival of the host-organism, as, for instance, in facilitating the rapid growth of reef-building corals.

Of interest to us are only Dinophyceae and Ebriophyceae, treated here as classes. Other D. are unknown among fossil remains. Detailed information on them is available from literature sources (Dodge, 1983; Evitt *et al.*, 1977; Sarjeant, 1974; Tappan, 1980; Vozzhennikova, 1979; Williams, 1978).

Class Dinophyceae (Dinophytes)

Most fossil members of this class (over 200 genera) are affiliated to the order Peridiniales (P.), recovered as fossils from the Silurian onwards. Only single genera of the orders Dinophysiales and Gymnodinales are known in the geological record. Living P. are unicellular, occasionally filamentous forms. The cell is covered by an organic shield – theca, divided into sectors by an equatorial girdle – cingulum, from which a longitudinal groove – sulcus, extends downwards. In the place where the sulcus reaches out from the cingulum the theca has two pores for flagella. The lower flagellum extends along the sulcus and then becomes free. The upper flagellum lies within the sulcus, encircling the entire theca. The theca is made up of many plates that are connected by sharp sutures. The form, number and position of the plates are important systematic characters. For the sake of convenience, the plates are designated by special terms and symbols (Fig. 6, *a*). Different sets of terms have been suggested in application to the theca and its parts (see details in Eaton, 1980). The general form of the theca may vary considerably, mainly due to the formation of protrusions – horns – different in shape and size, which do not exceed five in number. The plates are smooth, or sculptured. They are divided by low ribs into polygonal areas, and are frequently pierced by cavities that are discernible only under the electron microscope.

In the life-cycle of P. and Gymnodiniales there is a resting cyst or dinocyst stage (D.). Beneath the theca a dense D. wall is formed around the protoplast; the wall is impregnated by sporopollenin, at times by calcareous, probably siliceous or other substances. This wall, occasionally composed of several layers, may be well preserved in fossil material. The remains of the walls are also called D. They can be isolated from the rock mass by various palynological methods and are usually studied together with miospores and acritarchs. All these microfossils are often defined by a common term, 'palynomorphs'. Only in the 1960s, through the work of W. R. Evitt, was the fact established that fossil P. are often represented by D. Until then fossil P. were considered to be affiliated to two groups: tabulated and non-tabulated. In the first group plates can be discerned, whereas in the second they are absent, although the cingulum and sulcus are preserved. Studies of the living P. established that tabulated D. bear a reflection of plates unlike non-tabulated ones, although the latter have plates on the theca. In the non-tabulated P. a process or groups of processes can frequently be observed that extend from the outer wall of the D. to the plates. Each plate corresponds to one or a group of processes. Because of this it is possible to reconstruct the theca on the basis of the length and position of the processes (Fig. 6, *c–e*).

---

k, *Arkhangelskiella distincta* Shum., coccolith; *l*, scheme of coccolith structure (1, distal shield; 2, proximal shield; 3, central area with perforations shown by stippling); *m–o*, coccoliths of loxolithid (*m*), podorhabdid (*n*) and coccolithid (*o*) structure; *p–t*, ortholiths (*p*, prismatolith, cell with detached segment and detached prismatolith; *q*, detached pentalith; *r*, asterolith in lateral and surface view; *s*, ceratolith; *t*, sphenolith); *u–w*, scheme of silicoflagellate skeleton (*u*, form with apical plate at arrow; *v*, form with apical window at arrow; *w*, surface and lateral view of skeleton; 1, radial spine; 2, lateral bar; 3, basal ring; 4, apical bar; 5, basal accessory spine); *x*, *Mesocena apiculata* (Schulz) Hanna, corner septae divide skeleton cavity; *y*, *Lyramula furcata* Hanna. Scale bar = 50 μm (*b*), 10 μm (*f*, *g*, *i*, *x*, *y*), 1 μm (*j*, *k*). Modified from: Vozzhennikova, 1979 (*a*, *d*, *e*), Williams, 1978 (*b*, *c*), Tappan, 1980 (*e–i*, *m–t*, *x*, *y*), Haq, 1978 (*j*, *l*, *u–w*), Shumenko, 1968 (*k*).

Forms with processes but with no traces of a sulcus or cingulum had already been found in fossil materials during the last century, and were coined hystrichosphaeres. They were correlated with various algae, also with animals (radiolarians, skeletal elements of sponges, statoblasts of bryozoa, etc.). It was later discovered that in a certain place in the wall the hystrichosphaeres have a split or a definite hole – the archeopyle. After D. were found in extant P. it was established that during germination the cell abandons the cyst through the archeopyle. The discovery of the archeopyle and classification of its types made it possible to determine the correct orientation of D., to correlate extant and fossil D., and also to establish the correspondence between the D. processes and definite theca plates. All of this clarified both the morphology and systematic affinities of D., and urged the need for a special terminology for them. The sets of terms used for the theca elements furnished the basis for the terminology, but to certain terms the prefix 'para' has been added, or the word-base 'theca' has been changed to 'cyst'. For example, the term 'cingulum' has been changed to 'paracingulum', and the term 'epitheca' to 'epicyst'. The suggestion was advanced that instead of 'para' the prefix 'pseudo' should be used, or the D. parts should be referred to as 'reflected elements'. Still greater difficulties arise in relation to systematics. The diagnoses of living P. rely on characters of the theca, not of D., whereas for fossil P. – vice versa. The D. organization has been studied in several tens of living P. forms. However, this is inadequate to establish the relationship between the taxa of extant and fossil P., especially considering the high diversity of both.

All D. have been divided into three types. Proximate D. lack processes and replicate the organization of the theca, including the tabulation (Fig. 6, *d*). Chorate D. reveal a main body bearing strongly developed processes (Fig. 6, *b*, *e*), or other ornamentation. Cavate D. have inserted inner (endocyst) and main (pericyst) bodies, separated by a space (pericoel). Transitional forms (proximochorate) have also been distinguished. D. are further divided according to the types of archeopyle, paratabulation characters, forms and positions of the processes and wall structure. The archeopyle (Fig. 6, *b*, *c*) may be formed as a result of the removal of all apical paraplates, of one precingular plate, the separation of the epicyst from the hypocyst, formation of a meridional fissure, or otherwise. Processes of D. differ in form, often being of complex branching, may become fused basally or apically, form ridges and a reticulate outer wall. At times the processes densely cover the surface, so that the paratabulation cannot be recognized. The inner bodies of the cavate forms differ in their sizes and configuration (they may consist of successively enclosed endocysts). A detailed classification of D. based on a set of characters has been suggested (Darrah *et al.*, 1979; Norris, 1978; Tappan, 1980; Vozzhennikova, 1979; Wilson and Clowes, 1980).

Prior to the establishment of the relationship between hystrichosphaeres and D., the opinion was current that P. first appeared in the Permian, and were abundant only since the Jurassic. D. with archeopyles and residual tabulation, due to which they can be validly correlated with P., have however been identified in the Silurian (genus *Arpulorus*). In older beds palynomorphs (acritarchs) have been encountered with long and occasionally intricate processes. They are similar to D., but their affiliation to P. is as yet problematical. Some of them may belong to Prasinophyceae. P. of a group closely similar to the genus *Gonyaulacysta* are known from the Silurian and lowermost Devonian. Other sections of the Devonian, the Carboniferous and Lower Permian have yielded acritarchs which at times were described as D., but their affinities are indeterminable. The Upper Permian and Triassic have yielded only a small number of indisputable D., but, beginning from the Jurassic, D. occur in abundant quantities, and are of common occurrence further throughout the Cretaceous and Cainozoic. D. are of great stratigraphic significance for the correlation of the Mesozoic and Cainozoic. From the beginning of the Jurassic through to the Quaternary 30 zones of D. have been recognized. In the geological distribution of D. various flourishing phases have been revealed (Silurian–lowermost Devonian, Lower–Middle Jurassic, Palaeocene–Lower

Eocene), divided by periods of decline. P. are abundant in living plankton (surpassed only by diatoms), have a paramount role in trophic chains, and, at times, during periods of 'water bloom' in seas ('red tides'). They reproduce in such great numbers that they provoke mass death of nectonic and benthic organisms. Still, the Recent period is considered to be a phase of decline, according to the taxonomic diversity of P.

Living P. are ecologically of high diversity. Among them are many marine planktonic forms, as well as inhabitants of brackish and freshwaters. Fossil P. with rare exceptions are marine. Firmly determined freshwater P. are scarce and known only from the Upper Cretaceous. The distribution pattern of the P. genera reflects the proximity of the shoreline. On the whole, the thickness of the D. walls increases towards the near-shore zone. D. assemblages reveal a quite distinct climatic restriction, thus allowing a judgement to be made of the warming or cooling of the climate.

Class Ebriophyceae (ebridians)

The systematic affinity of ebridians (E.) is debatable. In some manuals E. are placed among silicoflagellates. E. have an inner siliceous skeleton of the open lattice-work kind, with a triradial or tetraxial symmetry (Fig. 6, *f–i*). Individual branches of the skeletons, in contrast to silicoflagellates (see below), lack an inner cavity. E. differ from typical algae in their heterotrophic feeding (chiefly diatoms), but are similar to some dinoflagellates. They show affinities to the latter in their nucleus structure and pair of flagella. Moreover, forms occur among the dinoflagellates (genus *Actiniscus*) that have an inner skeleton, and are also made of elements lacking an inner cavity. E. are marine plankton organisms. They are shown as fossils since the Palaeocene, but are particularly abundant in the Eocene and Miocene. Living E. are represented by three genera. E. are being studied together with other siliceous organisms – radiolarians, diatoms and silicoflagellates. Due to their restriction to waters of cold and temperate climates, fossil E. may be used for palaeoclimatic reconstructions.

DIVISION CHRYSOPHYTA (CHRYSOPHYCOPHYTA) GOLDEN ALGAE

Golden algae occur in the geological record only as coccolithophores and similar to them carbonate-secreting algae, also as silicoflagellates. Both groups are treated as subclasses of the class Chrysomonadophyceae (Chrysophyceae). Owing to the predominance of flagellate forms and (in part) their phagotrophic nutrition the Chrysomonadophyceae are sometimes included in the Protozoa. This is evident from both autotrophic and phagotrophic nutritions of some coccolithophores. During the motile phase of their life-cycle flagellar forms of *Coccolithus pelagicus* consume bacteria and small algae. Silicoflagellates have pseudopodia, but their employment as a mechanism for feeding has not been recorded. At present, specialists studying coccolithophores tend to distinguish them as an independent class (Haptophyceae or Prymnesiophyceae) and even as a specific division (for details, see Golubev, 1981; Shumenko, 1984). Although coccolithophores are regarded as taxonomically related to silicoflagellates, in practical palaeobotanical investigations they are studied separately. Silicoflagellates are recovered from rocks together with radiolarians, ebridians and diatoms. This requires the dissolution of carbonates, including also the coccolithophore fossil remains. Information on chrysomonad cysts (Chrysostomataceae and Archaeomonadaceae) known since the Upper Cretaceous, may be found in the literature (Cornell, 1970; Tappan, 1980).

Coccolithophores (subclass Coccolithophoridophycidae) and other nannoliths

Coccolithophores rank in various systems from class (Coccolithophyceae) to suborder. Modern coccolithophores are unicellular organisms. Their life-cycle consists of a motile (with flagella) and non-motile (without flagella) phases. During the non-motile phase the cell produces a covering (coccosphere, Fig. 6, *j*) of calcareous plates – coccoliths (C.; Fig. 6, *k–o*). Plates may also develop during the motile phase, but then they are of a different kind. C. are produced within the cell and then forced out on to the surface. Up to 30 C. per cell may be produced. The ontogeny of C. has been studied in detail by Golubev (1981), who considers that coccolithophores can be suitably used as exemplar organisms for studying biomineralization processes, as have bacteria in molecular biology.

Entire coccolithophores are of scarce preservation in fossil materials. Descriptions of the taxa are given according to detached C. Coccolithophores of the same species are characterized by different kinds of C., due to differentiation within the coccosphere, as well as differences in C. produced during various phases of their life-cycle. Different C. of one species may occur in different families of the form-classification. The difficulties in systematics of C. result from their very tiny sizes (from 1 to 15 μm). Modern systematics of C. require the use of an electron microscope. Calcareous microfossils (nannofossils) are known, for which no analogues have been found among oceanic living inhabitants. This refers to discoasterids, thoracosphaeraceans, nannoconids, sphenoliths (Fig. 6, *t*), ceratoliths (Fig. 6, *s*), and other groups, studied together with C. and covered by the term nannoliths. Organisms producing nannoliths comprise the nannoplankton (part of the plankton that passes through the phytoplankton nets). In biostratigraphic studies nannolith assemblages are usually termed 'calcareous nannoplankton', or 'calcareous nannofossils'.

Over 2000 species of nannofossils have been described to date. Among them are many synonyms that result from errors in systematization, and also from the practice of establishing the nannolith taxa irrespective of their life-time relations in coccospheres. Nevertheless, nannoliths are very diverse. At present they are distributed between about 20 families. About half of them are directly included in the subclass Coccolithophoridophycidae. As yet no orderly taxonomic system above the family rank has been proposed, because the system is based on detached elements of the covering, the characters of which do not form sufficiently persistent syndromes and frequently occur in different combinations.

C. can be subdivided into holococcoliths, heterococcoliths and pentaliths, which differ according to the type of calcite crystallization. Crystals comprising holococcoliths (Fig. 6, *p*, *q*) preserve rhombohedral and hexagonal prismatic forms, typical of calcite (both forms may occur in one C.), which are practically unaltered by the cell activity. Most of the C. belong to heterococcoliths (Fig. 6, *j–o*), where the primary crystals are strongly altered and occur in the form specified for the given taxon. For an explanation of the relationship between holo- and heterococcoliths the following comparison has been suggested (S. N. Golubev). A building of complex configuration may be constructed by an immense number of identical bricks, and may be built from a much smaller number of figured ferro-concrete elements. Holococcoliths are analogues of the first type of building, whereas heterococcoliths are analogues of the second type. Pentaliths (Fig. 6, *q*) consist of five identical but not rhombohedral elements. The attribution of pentaliths to coccolithophores has not as yet been proven.

C. differ in behaviour in polarized light, depending upon the orientation of the crystals. Heliolithae exhibit orderly arranged crystals and in cross-polarized light show black cross isogyres. Ortholithae (Fig. 6, *p–t*) have randomly oriented crystals and because of this are of mosaic extinction in the rotating microscope stage. C. are known that differ in the kinds of extinction of their inner and external parts. The behaviour of C. in polarized light is used for the distinction of genera and families. For example, the families Coccolithaceae and Prinsiaceae differ in their extinction and behaviour in the rotating microscopic stage.

Of most widespread occurrence in the Cainozoic are oval or round disc-shaped C. – placoliths (Fig. 6, *j–l*). They consist of two ring-shaped shields, connected by a low cylinder. Another widespread type are discoliths. They are round discs with thickened edges. C. occurring in the form of an inverted short nail with a wide cap are termed rhabdoliths (Fig. 6, *n*). Zygoliths are elliptical forms consisting of a peripheral zone, where the centre is made up of crossbars, exhibiting a cruciform or X-shaped figure (Fig. 6, *m*); frequently parallel bars occur, or only a single bar is present. Other C. and different kinds of nannoliths are known. In the systematics of C. of most importance is the structural plan of the marginal zone, which furnishes the basis for the distinction of families. The genera are distinguished according to the type of structure of the central field and arrangement of the elements of the marginal zone, whereas species are distinguished according to finer details.

Coccolithophores have been firmly established since the Triassic. Nannofossils have been recorded in older formations, from the Cambrian onwards. Their taxonomic position is not quite clear, although the Carboniferous genus *Palaeococcolithus* shows affinities to placoliths. Beginning from the Jurassic nannofossils are highly abundant and are, at times, the major rock-forming constituents. For instance, chalk consists chiefly of C. and other nannofossils, as well as of their loose elements. It probably lacks chemogenic carbonates. Contained in 1 mm^3 of chalk are up to 7×10^{10} C. At depths below 3–4 km C. undergo dissolution. A considerable (probably major) portion of C. is buried in sediments with faecal pellets of plankton coccolithophore-feeders. The organic matter of these pellets preserves C. from dissolution during the long process of their settling at the ocean floor. Due to partial dissolution C. may undergo alterations in configurations, which should be taken into consideration in their classification.

Calcareous nannofossils are highly diversified, exhibiting rapid changes in the stratigraphic sequence, and because of this they are very important stratigraphic indices beginning from the Jurassic. Only in the Jurassic more than 20 nannofossil zones have been identified. The Cainozoic stratigraphic scale, based on C. ('Martini Scale'), according to the number of units in its subdivision, is by no means inferior to that of the foraminifer scale, and is of wide use in correlations based on deep-sea drilling. Considering that nannoliths dwindle in variety in cold waters, according to their distribution it is possible to reconstruct past climatic belts and episodes. It is believed that the cooling of the climate was responsible for the extinction of most of the Cretaceous coccoliths at the Cretaceous/Tertiary Boundary.

Subclass Silicoflagellatophycidae (silicoflagellates)

Silicoflagellates (S.) were first discovered among microfossils and only later among the living plankton (two genera). They are unicellular organisms, between 20 and 100 μm in size, with a single flagellum and thin pseudopodia. The protoplast is enclosed within the siliceous skeleton and only slightly protrudes from it. The skeleton (Fig. 6, *u–y*) is made up of hollow rods with a smooth or finely sculptured surface. The skeleton base forms a rounded, oval or polygonal frame (basal ring). On one side of it are attached supporting bars of the pyramidal central structure, terminating in an apical bar or apical ring. Spines occur in the angles of the basal rings, or at times at the points of attachment of the bars to the central structure. Additional small spines extend from the bars of the basal ring. The skeleton structure, especially its size and the position of the spines, varies considerably within one species, which causes difficulties in their systematics. S. are arranged in groups that derive their names from the typical genus. Dictyochids (from *Dictyocha*) have a square basal ring with spines or nodules in the angles, and an apical bar (Fig. 6, *w*). Distephanids (from *Distephanus*) have a six-angled basal ring with angular spines and a rounded six-angled apical ring (Fig. 6, *v*). Mesocenids (from *Mesocena*; Fig. 6, *x*) have only a round or rounded-polygonal basal ring with small spines, and no central structure. In similar manner other groups and genera, tentatively affiliated to S., have been distinguished (see Haq, 1978; Lipps, 1970; Tappan, 1980).

S. appear in the Lower Cretaceous, where at first they are rare, but quickly become more diverse in the Upper Cretaceous. S. reached peak abundance during the Miocene. Most of the genera disappear at the end of the Pliocene. In the extant plankton only *Dictyocha* and *Distephanus* have been identified. S. have an important stratigraphic significance, particularly for the subdivision and correlation of marine deposits that are poor or lacking fossil remains of carbonate-secreting organisms. S. have been found only in marine waters. The genus *Dictyocha* inhabits warm waters, whereas *Distephanus* inhabits cold waters. According to the proportion of these genera in the sequence inferences may be drawn concerning temperature fluctuations in the environment.

DIVISION BACILLARIOPHYTA (DIATOMEAE)

Diatoms (D.) are occasionally affiliated to Chrysophyta but more frequently they are treated as an independent division. They are unicellular, more seldom colonial organisms with a bipartite siliceous skeleton (frustule). Cells vary in sizes from 0.75 μm to 2 mm. For their systematics of primary importance is the frustule structure. Usually it is compared to a pill-box closed by a lid (Fig. 7, *a–d*). The hypotheca corresponds to the box, while the epitheca corresponds to the lid. Each of them consists of a valve (corresponding either to the bottom of the box, or to the lid) and connecting bands, forming a girdle (which corresponds to the wall). The hypotheca at its edges enters into the epitheca. The hypotheca valve is termed the hypovalve, the epitheca valve is termed the epivalve. There is a valve face and a valve mantle that curves around its edges; at times the mantle is rather high. The growth of the cell may be concomitant with the formation of intercalary bands between the valve and girdle. They originate at the edges of the mantle of the valve and push aside the girdle or the previously created intercalary band. After death the D. skeleton falls apart into valves and bands. Intact frustules are preserved only when all parts are completely united. Certain D. were found that have an organic wall surrounding the frustule. The configuration of the frustule differs according to its orientation. The frustule view from the face or marginal surface of the valve is called accordingly, the valve view or girdle view. In slides some D. with a cylindrical frustule are always shown with orientation for the girdle view.

During vegetative reproduction, which may occur several times during the day, the protoplast increases in growth and shifts aside the hypotheca and epitheca. After division the daughter cells separate, retaining only a hypotheca or epitheca. The frustule part that was inherited by the daughter cell in all cases becomes the epitheca. The hypotheca builds up later. Because of this, during repeated division, part of the frustules being formed becomes ever smaller in size (Fig. 7, *d*). Resting spores of D. have frustules with specific elements (spines, processes, etc.).

D. frustules are pierced by openings – areolae (Fig. 7, *e–i*, *l*, *m*) that are covered around the outer or inner edges by a pellicle (velum), which is also pierced by tiny pores, occasionally with a sculptured margin. The shape and position of the areolae and pores on the velum are important systematic characters. Smooth areas, lacking sculpture and areolae, are called hyaline fields. Septa various in size, that lie in a plane parallel to the valve, may extend inwards from the frustule. Septa (pseudosepta) may also extend from the valves themselves. They are of simultaneous formation with the valve. The inner and outer surfaces are frequently sculptured by chambers, ribs, grooves, thorns, horns, etc.

According to the frustule structure D. are subdivided into two groups that, as viewed from the valve, differ in symmetry. Radially symmetrical ones are termed centric D. (Fig. 7, *b*, *e*, *g–i*), whereas bilaterally symmetrical ones are termed pennate D. (Fig. 7, *c*, *j–o*). Both types are commonly of rectilinear contours in the girdle view. The valves of centric D. are commonly rounded, triangular, quadrilateral, oval and semicircular. Not all of the centric D. are actually radially symmetrical, but according to other characters they are closely related to radially symmetrical D. Their areolae and other ornamentation spread out along radii from a point located in the centre or off-centre. A three-group subdivision has been suggested for centric D. Discoid D. – with round valves and a flat convex or concave face. Gonioid D. – with valves of angular contours. In both types the band is commensurate with the valve diameter. Solenoid D. are similar in their valve configuration to discoid D., but the girdle is very wide, so that

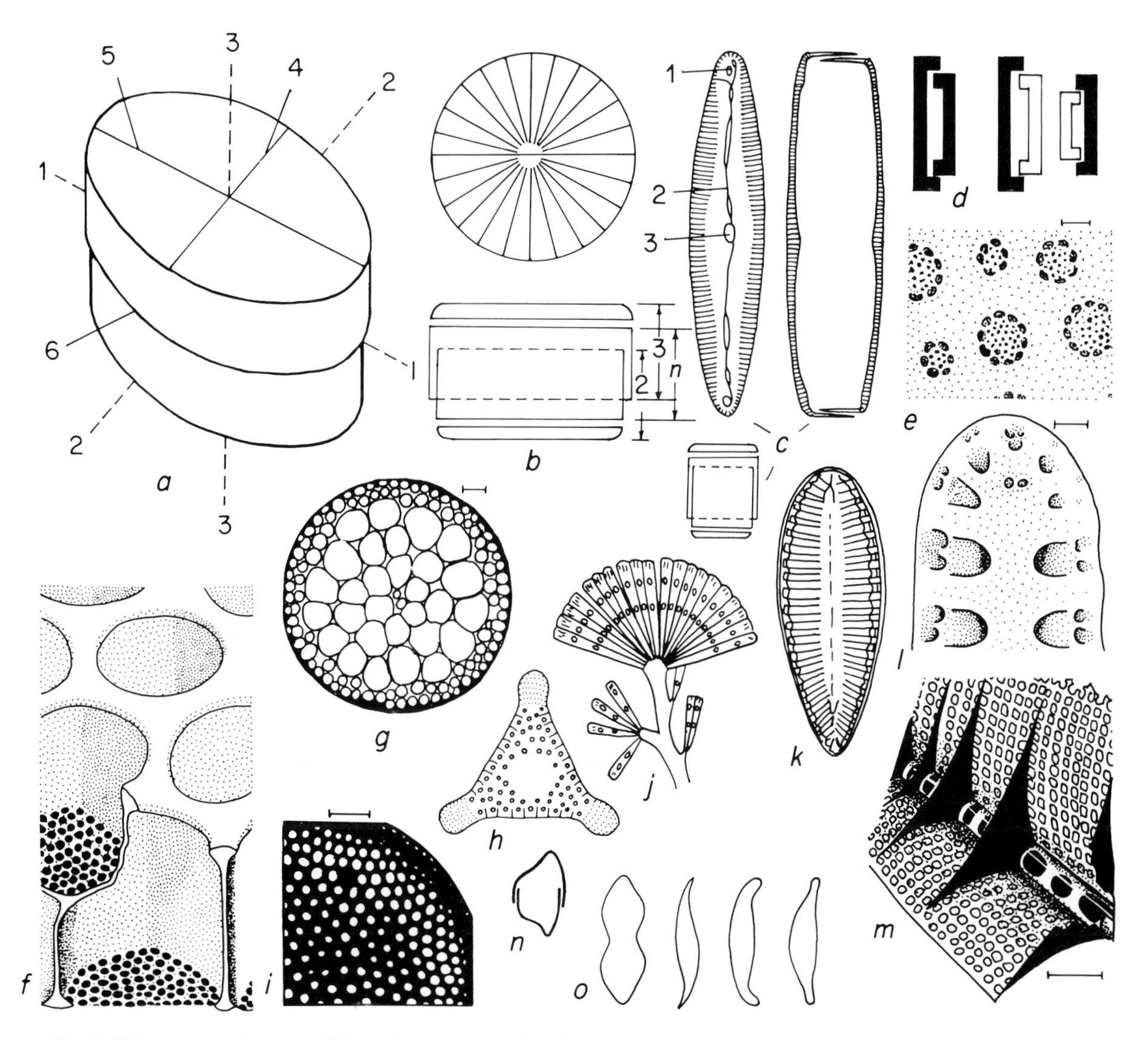

Fig. 7. Diatoms; *a*, elements of frustule symmetry, longitudinal (1), transversal (2) and central (3) axes of symmetry, transversal (4), longitudinal (5) and valvar (6) planes of symmetry; *b*, frustule of centric type, valvar (above) and girdle (below) views (3, epithecal valve; 2, hypothecal valve; *n*, girdle bands); *c*, frustule of pennate type, valvar (upper left) and girdle (right) views, transversal section (below), 1 terminal nodule, 2 raphe, 3 central nodule; *d*, decrease in frustule size during reproduction after Macdonald-Pfitzer's hypothesis (valves of maternal individuals in black, those of daughter individuals in white); *e*, wall perforation in *Triceratium antedeluvianum* (Ehr.) Grun., Miocene, Spain; *f*, structure of areola with velum on frustule inner side (transversal section of frustule wall); *g*, *Rhaphoneis schraderi* Bukry, Miocene, Mexico; *h*, *Triceratium schulzii* Jousé, Upper Cretaceous; *i*, *Coenobiodiscus muriformis* Loebl., Wight et Darley, arrangement of areolae (recent); *j*, *Licmophora flabellata* (Carm.) Ag., recent marine colonial form; *k*, *Surirella capronii* Bréb., recent freshwater carinoid form; *l*, *Delphineis angustata* (Pant.) Andr., Miocene, USA; *m*, *Surirella gemma* Ehr., reconstruction of frustule part with canaliculate raphe; *n*, transversal section of pennate frustule with curved plane of symmetry; *o*, frustule outlines of pennate diatoms. Scale bar = 10 μm (*g*, *i*, *m*), 1 μm (*e*, *l*). Modified from: Osnovy paleontologii, 1963a (*a–c*, *h*, *n*, *o*), Tappan, 1980 (*d*, *e*, *g*, *i*, *l*), Gollerbakh, 1977 (*f*), Fott, 1971 (*j*, *k*, *m*).

the frustule becomes tubular. Occasionally the valves are in the shape of a small cap with a long thorn at the apex. The frustules of pennate D. are more complicated. Three axes and three planes of symmetry can be distinguished (see scheme in Fig. 7, *a*). In centric D. only two types of axes and planes of symmetry can be identified, including a central axis and valvar plane. The frustules of pennate D. (Fig. 7, *c*) have an axial area, frequently with a median slit or raphe. The axial area (pseudoraphe) is a smooth non-structural strip extending along the longitudinal valve axis. The raphe in the simplest case is a slit that cuts the valve lengthwise from the poles, frequently extending to the central part. Such a slit-like raphe occurs on one or both valves. It may differ in length, being occasionally present only within half of the valve. The valve is frequently thickened in the centre (central nodule). Here branches of the raphe may converge. This results in the formation of a central pore. At the poles the slit terminates with a polar pore and polar nodule. The canal raphe is of even more complex organization (Fig. 7, *m*). At the surface it opens with a narrow slit, while inward of the frustule, with numerous openings. Its branches, together with the central nodule to which they extend, are located either on the longitudinal axis of the valve, or shifted on to one of its margins.

Attempts have been made to classify living D. according to the protoplast characters (particularly those of chloroplasts), but most frequently their classification is based on the frustule structure. Centric (Centrophyceae, Centricae) and pennate (Pennatophyceae, Pennatae) D. are now assigned to the rank of classes. Centric D. are divided into orders according to the frustule shape (low-cylindrical, rod-like) and valve configuration (rounded, angular). The orders of pennate D. are distinguished according to the structure of the axial area and raphe (valves lacking a raphe, slit-like raphe on one valve, slit-like raphe on both valves, canal raphe on both valves). Families, genera and species are distinguished according to other characters (for details, see: Osnovy paleontologii, 1963a; Tappan, 1980). Due to the fact that the frustule components may fall apart upon burial, it is not always easy to correlate fossil D. with the living taxa. It is difficult to correlate frustules of resting spores and vegetative individuals. Because of this the systematics of extant D. is also fraught with difficulties. Other difficulties arise in connection with the considerable intraspecific variation of the D. shells.

Since precisely the same characters are used in the systematics of both living and fossil D., it is possible to trace the history of the taxa (Fryxell, 1983; Round and Crawford, 1981). The oldest indisputable D. have been recovered from the Lower Cretaceous (references concerning older finds are in need of checking). The Cretaceous and major part of the Palaeocene yield nearly exclusively centric D., commonly lacking a girdle. These D. do not differ essentially in their valve structure from living forms. At about the boundary between the Cretaceous and Palaeogene two thirds of the composition changes. At the close of the Palaeocene pennate D. appear in conspicuous numbers (they are of singular occurrence from the Upper Cretaceous), but up to the Miocene centric D. are dominant. Pennate D. are at first represented by forms without a raph, (araphid), and much later (during the Eocene) raphid forms appear with a raphe either on one valve, or on both. Beginning from the Miocene pennate D. rapidly become more diverse and more abundant, and at present, both in abundance and diversity they surpass centric D. To be more exact, centric D. dominate the plankton communities at present, whereas pennate D. dominate the benthic communities. Freshwater D. appeared during the Eocene, and ever since up to the present day most predominant among them are pennate D. (all are benthic forms).

D. are of great stratigraphic importance, being most abundant in marine sediments and beginning from the Neogene – also in freshwater deposits. Although there are persistent D. genera (from the Cretaceous to date) and species (from the Miocene to date), it is possible to establish short-living stratigraphic assemblages. For instance, in the Neogene and Quaternary of the Northern Pacific 25 diatomaceous zones have been recognized. D. are of particular stratigraphic importance in those areas where fossil remains of foraminifers and nannoliths are either scarce, or insufficiently characteristic, or altogether absent. This pertains to the high-latitudinal regions. D. may be carried out by currents far away from their original habitats, which is a favourable factor for long-distance stratigraphic correlations between different climatic belts. In recent years D. have been widely used in stratigraphic studies of oceanic sediments, particularly in deep-sea drilling. Commonly they are studied together with silicoflagellates and ebridians.

D. are very sensitive to changes in temperature and water salinity, and also to the presence of nutrients. Because of this the composition of D. may be applied to regional climatic zonation of basins, and as evidence of the salinity in water basins, and can also be used to determine zones of up-wellings in oceans. The changes in D. assemblages are important indicators in climatostratigraphy.

Worthy of note is the rock-building significance of D. In oceans and oligotrophic lakes D. accumulate, forming diatomaceous ooze. The preservation of organic matter of D. together with that of other aquatic organisms results in the formation of decay ooze, which accumulates in lakes. Sediments comprised of D. frustules (diatomites) are of wide occurrence since the Palaeogene.

DIVISION PHAEOPHYTA. BROWN ALGAE

Brown algae (BA.) are one of the most interesting groups of eukaryotic algae. They are multicellular plants with very diverse thalli forms – shoe-string-like, globose, laminar, tubular, pinnate, tree-like, etc. Air bladders are sometimes borne on the thalli. BA. derive their name from the presence of brown and yellow pigments. A characteristic feature of BA. is alternation of generations – of asexual generations (sporophyte), producing zoospores, tetraspores and monospores, and sexual generations (gametophyte), producing gametes. A gametosporophyte may also be formed – a generation resulting from asexual reproduction of gametophytes. The generations may exhibit strict alternations (the same as in higher plants). The sporophytes and gametophytes may differ or be similar in outward appearance. Similarities between BA. and higher plants suggested their phylogenetic links. The living BA. are very diverse and of wide distribution. But the chemical composition of the thalli is unfavourable for their preservation in fossils. In the literature, particularly of the last century, many organic remains have been affiliated to BA. This refers to the genera *Fucoides*, *Caulerpites* and others. Many of them belong to ichnofossils (animal tracks, etc.), others have proved to be leaf remains and shoots of higher plants, and only a few may actually belong to BA. In the Cainozoic thalli imprints with air bladders were encountered that are similar to the living genera *Fucus*, *Cystoseira* and *Ascophyllum* (order Fucales). The genera *Drydenia* (Fig. 8, *a*), *Hungerfordia* and *Enfieldia*, recovered from the Upper Devonian in New York State, resemble BA. thalli. They may be treated as BA. satellite genera. The form-group Vendotaenides from the uppermost Pre-Cambrian (Gnilovskaya, 1984) was tentatively affiliated to BA. The geological record, as a whole, contributes very little to the systematics and phylogenetics of indisputable BA.

Certain Palaeozoic genera, grouped into various orders and families (Nematophytales, Foerstiales, Protosalviniales, Spongiophytaceae, Prototaxitaceae, etc.) show affinities to BA. or red algae. They may also be treated as BA. satellite genera. These genera were first introduced for log fragments (*Prototaxites*) on the one hand, and for leathery compressions with a thick cuticle-like covering (*Protosalvinia*, *Orestovia*, *Spongiophyton*, *Aculeophyton*, *Nematophyton*, *Bitelaria*, etc.) on the other hand.

Remains of *Prototaxites* (S–D; Fig. 8, *b–d*) look like wood fragments. Log fragments reaching 1 m in diameter are known. The logs are made up of a dense mass of thick-walled tubes. Some of them are non-septate, up to 70 μm in diameter, and of longitudinal orientation. Other tubes, up to 7 μm in diameter, fill up the rest of the log space. They are profusely branching near the large tubes and encircle them like a sheath. Slender tubes are septate and bear on the walls minute pores with double bordering (Fig. 8, *c*), similar to the bordered pits in gymnosperm tracheids. The

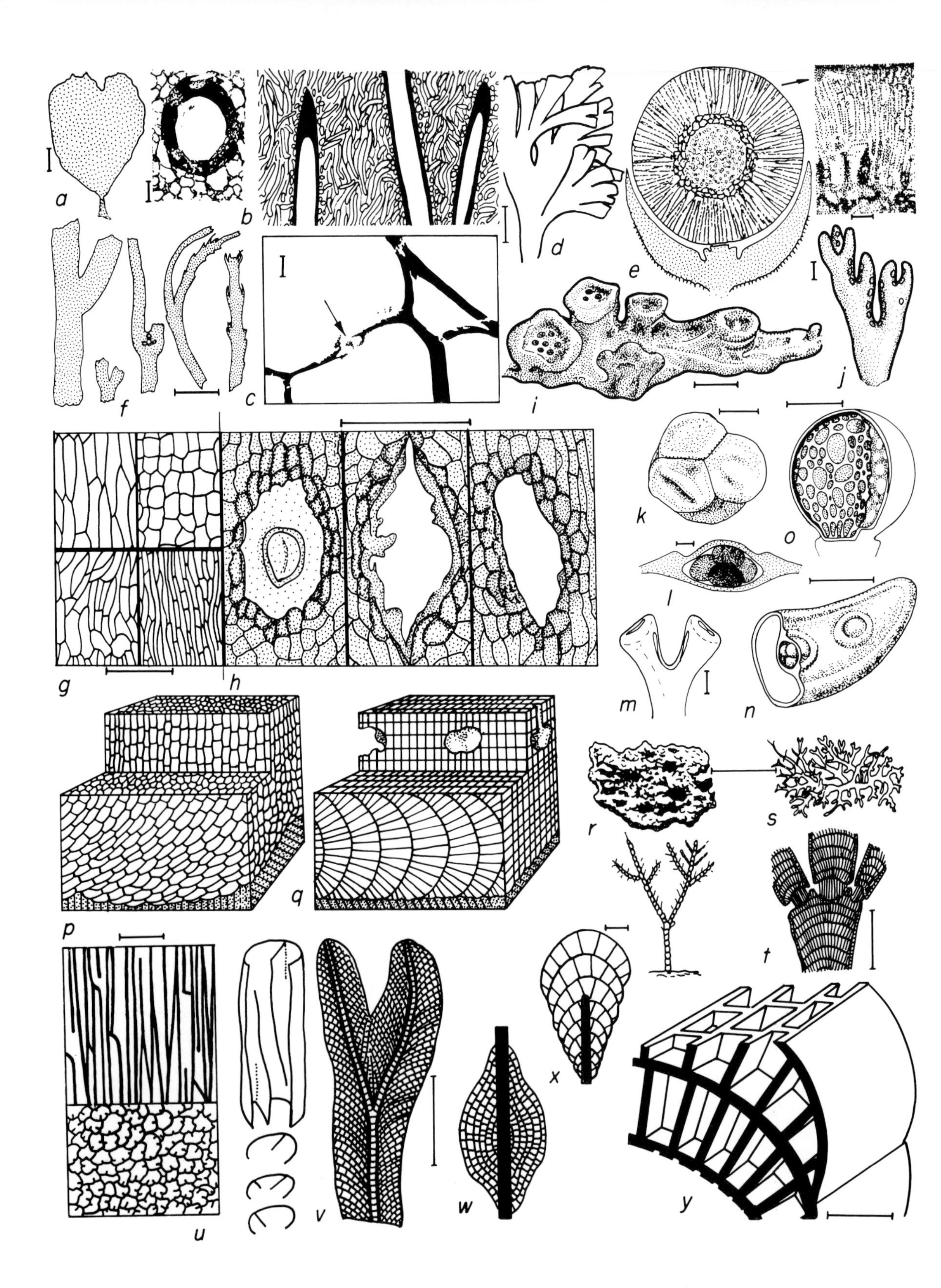
a
b
c
d
e
f
g
h
i
j
k
l
m
n
o
p
q
r
s
t
u
v
w
x
y

Devonian also yielded flattened fragments with outer cuticle-like covering and cellular structure (Fig. 8, *d*). The inner parts of these fossils, which have been assigned to *Nematothallus* and *Nematoplexus*, consist of tubes, either smooth walled or with spiral thickenings. Spore-like bodies have been encountered between the tubes. These remains may be thalli ends, the log parts of which belonged to *Prototaxites*. The chemical composition of *Prototaxites* is closely similar to that of *Nematothallus* (Niklas, 1982a). The probable relationships between both genera were discussed in detail by Schweitzer (1983b). These plants, although affiliated to algae, may have been land-dwelling organisms. Problematical fossils *Pachytheca* (Fig. 8, *e*) frequently occur together with *Prototaxites* (in the Silurian). They are globular multilayered bodies. Their inner parts are either hollow or consist of filaments made up of elongated cells that curve around the inner rim of the outer layer, and spread out into a radial layer. This is followed by a layer of vesicular tissue. In one place the outer layers are interrupted and a swelling is formed here with a structure similar to that of the inner part. The cell walls are composed of an organic substance. In one instance in *Pachytheca* a bowl-shaped subtending structure was found. The *Pachytheca* ontogeny was studied by Schweitzer (1983b), who presented data on the systematic affinity of the genus.

The genus *Protosalvinia* (D_{2-3}; Fig. 8, *i–o*) has been described from fragments of hard leathery compressions. They are forked, with rounded ends, frequently complicated by processes. The ontogeny of the terminal lobes has been reconstructed according to the cell arrangement (Niklas *et al.*, 1976; Niklas and Phillips, 1976). Hypodermal conceptacles with several cavities are located within the rounded and elongated hollows of the terminal lobes. In some cavities tetrahedral tetrads of large (up to 250 μm) spores were encountered that had a trilete proximal slit and a wall composed of maceration-resistant sporopollenin (Fig. 8, *k*–n). In others stalked clavate structures have been found, that are made up of large-celled tissues (Fig. 8, *o*). It is believed that they are plurilocular sporangia or male gametangia.

Other Devonian genera with a cuticle-like covering are known. Some of them have tape-like forked or monopodially dividing thalli (*Orestovia*, *Bitelaria* and others; Fig. 8, *f–h*). Frequently they form accumulations comprising coal seams. In the

Fig. 8. Braun (putative; *a–o*) and red (*p–y*) algae; Devonian (*a–o*), Carboniferous (*v–y*); USA (*a–c*, *i–o*), FRG (*d*), England (*e*), Voronezh area (*g*, *h*); *a*, *Drydenia foliata* Fry et Banks; *b*, *Prototaxites southworthii* Arn. (left) and scheme of thallus structure of *Prototaxites*; *c*, transversal section of pore with double bordering of tube wall in *P. southworthii*; *d*, reconstruction of thallus distal part in *P. psygmophylloides* Kr. et Weyl.; *e*, longitudinal section of *Pachytheca* sp. (left) and detail of wall structure in *P. fasciculata* Kidst. et Lang (right); *e* (from left to right), *Orestovia petzii* Erg. from Kuzbass, *Spongiophyton nanum* Kräus. from Ghana, *Rhytidophyton sulcatum* T. et A. Istch. and *Bitelaria dubjanskii* T. et A. Istch. (two thalli) from Voronezh area; *g* (from left to right, and from top to bottom), cellular structure of *Orestovia voronejiensis* T. et A. Istch., *O. ornata* (Tschirk.-Zal.) T. et A. Istch., *Orestovites fissuratus* T. et A. Istch. and *Rhytidophyton sulcatum* T. et A. Istch.; *h*, capsule and opening in place of detached capsule in *O. voronejiensis*; *i–o*, thalli and reproductive organs of *Protosalvinia* allied to *P. ravenna* White and *P. arnoldii* Bhar. et Venkat. (*i*), *P. furcata* (Daws.) Arn. (*j–n*), and *P. ravenna* (*o*); *i*, *j*, thalli with conceptacles; *k*, spore tetrad; *l*, section of conceptacle with spore tetrad; *m*, thallus with grooves; *n*, detail of groove with conceptacles; *o*, clavate organ; *p*, *q*, thalli of corallinaceous algae with variously ordered cells and sporangial cavities (*q*); *r*, *s*, crustose (*r*) and branched (*s*) thalli of corallinaceous subfamily Melobesiae; *t*, articulated thallus of corallinaceous subfamily Corallineae (right, jointing of calcified segments with the aid of uncalcified nodes); *u*, *Solenopora*, scheme of thallus structure in longitudinal and transversal sections (left) and arrangement of pseudoseptae in cell; *v*, *w*, thallus section in *Ungdarella* (*v*) and *Mametella* (*w*); *x*, *y*, *Stacheia*, thallus section (*x*) and cellular structure (*y*). Scale bar = 10 cm (*d*), 0.5 and 5 cm (*r*, *s*), 1 cm (*a*, *f*, *i*), 1 mm (*j*, *m*, *n*, *t*, *v*), 100 μm (*e*, *g*, *h*, *k*, *l*, *u*, *x*, *y*), 50 μm (*o*), 10 μm (*b*), 1 μm (*c*). Modified from: Banks, 1966 (*a*), Taylor, 1981 (*b*; left), Gothan and Weyland, 1964 (*d*), Hirmer, 1927 (*e*), T. A. and A. A. Istchenko, 1981 (*f–h*), Niklas and Phillips, 1976 (*i*, *k–o*), Phillips *et al.*, 1972 (*k*), Maslov, 1962 (*p*, *q*), Wray, 1978 (*r–t*), Mamet and Roux, 1977 (*u*–y).

Kuznetsk Basin *Orestovia* compressions and closely related plants are the major constituents of coal seams, the coals being of the sapromyxite (barsassite or tomite) type. Coal beds of lesser thickness are known in the Voronezh district. They are composed of the genera *Bitelaria*, *Orestovia*, *Rhytidophyton*, etc. These plants were at first regarded as psilophytes. Their cutin covering has pores of various sizes and structures, which at first were described as stomata. Later Istchenko and Istchenko (1981) established that these pores are of secondary formation taking the place of capsules which rupture in the course of development. Krassilov (1981) found tracheid-like elongated cells with spiral thickenings in the thalli of *Orestovia*. Closely related to these genera is *Spongiophyton* from the Devonian of Poland, Brazil, Ghana and elsewhere. *Spongiophyton* is a very common element of the non-vascular flora of the late Early Devonian (Emsian) in southeastern Canada (Gensel *et al.*, 1984).

The fact that these genera are not higher plants is evident from their chemical composition (Niklas, 1982a), as well as from the presence of a cutin-like substance that impregnates the entire covering tissue (J. M. Schopf, 1978). According to their biochemical characters the genera *Nematothallus*, *Prototaxites*, *Orestovia* and *Spongiophyton* are closely related, whereas *Protosalvinia* occupies an intermediate position between them and higher plants (Niklas, 1982a). According to the organization of the fructifications, thalli forms and multicellular structure, *Protosalvinia* may be affiliated to brown algae and to a lesser degree to red algae. But these algae lack a cutinized covering and spores with a sporopollenin wall. The suggestion was thereby advanced that these genera may belong to a specific algal division, probably of land plants.

DIVISION RHODOPHYTA (RHODOPHYCOPHYTA). RED ALGAE

Red algae (RA) are of widespread occurrence in marine sediments beginning from the Palaeozoic, However, their living species represent only a minor part of their past diversity. This is obvious from the fact that less than 10% of the several thousand living species have calcified thalli (calcite and aragonite are secreted in the mucilage portion of the wall) and may be fossilized.

In contrast to other eukaryotic algae, in RA the motile stage is absent. Their spores and gametes lack flagella. In the course of asexual reproduction monosporangia with a single monospore are formed, or tetrasporangia with four spores (tetraspores), that are arranged decussately, tetrahedrally, or in a row. Groups of sporangia (sori) are inserted on the tips of the thalli branches, at times specialized, or in the surface layer of the thallus. In corallinaceous RA. tetrasporangia are located in conceptacles. Spermatia (male gametes) are produced by spermatangia, that are scattered or occur in groups (sori), whereas in corallinaceans they are situated in conceptacles. A germinating zygote does not produce a sporophyte, but an individual with carposporangia, carpospores of which produce a sporophyte. Such complex alternation of generations has not been observed in other plants.

The most primitive RA are unicellular. Probably many of the Pre-Cambrian one-celled and occasionally budding fossils, assigned to cyanobacteria or green algae, are, in effect, RA (Tappan, 1980). For example, *Eosphaera* (Fig. 9, *a*) and *Huroniospora* (Fig. 5, *o*) are probably of this type. In *Eosphaera* the cells remain within the mucilage sheath during the budding, the same as in the extant *Porphyridium purpureum*. Pre-Cambrian microfossils, occurring in tetrads, may be tetraspores of RA. In the type of their mucilage sheath RA resemble cyanobacteria.

Most of the living carbonate-secreting RA belong to the families Corallinaceae and Squamariaceae. Their thalli are crustose or occur as algal balls (rhodoliths), branching bushes, frequently articulated. The thalli are composed of a large mass of tightly packed filaments of a definite orientation. In section they resemble a multicellular tissue. Usually for studies of RA oriented thin sections are therefore prepared. Several thin sections can be made from a large thallus. More often we have to content ourselves with a single section. An idea of the entire thallus structure can then be gleaned from sections of numerous specimens.

The reproductive organs of fossil RA are either of poor preservation or unknown. In place of the sporangia and conceptacles cavities remain in the thallus (Fig. 8, *q*), occurring either in groups, or solitarily. From the thallus surface canals may extend to them. If the reproductive organs are unknown, then the systematics of RA relies on the form of the thallus microstructure (Fig. 8, *p–y*). Classifications of the microstructure are based on the organization and arrangement of filaments, septae and pores. Zones of different microstructures have been distinguished: hypothallium and perithallium (Fig. 8, *p*, *q*). In crustose forms the first corresponds to the basal part of the thallus (contiguous to the substratum); in the branching forms, to its middle part (median hypothallium). The outer (peripheral) part corresponds to the perithallium zone. The sporangia and conceptacles are restricted to the latter.

According to the form and microstructure of the thalli, and the organization of the reproductive organs (if, at least, their cavities are preserved), fossil RA have been subdivided into the families Solenoporaceae (Cambrian–Miocene), Gymnocodiaceae (Permian–Palaeogene), Squamariaceae (Middle Carboniferous–Recent) and Corallinaceae (Devonian–Recent). The last two families are encompassed in the order Cryptonemiales of the class Florideophyceae. The affinities of the first two families are indeterminate. According to the general structure of the thallus and singular finds of cavities of the reproductive organs, they are commonly affiliated to Cryptonemiales. The problematical Palaeozoic families Epiphytaceae (Cambrian–Devonian) and Ungdarellaceae (Carboniferous–Permian) have also been affiliated to RA. The Ungdarellaceae includes several genera. Some of them (including *Ungdarella*) have been correlated with stromatoporoids, foraminifers and sponges. More detailed studies of Ungdarellaceae serve to indicate that they may be related to RA (Mamet and Roux, 1977).

The best studied family, Corallinaceae, has been divided into several groups. One group (Fig. 8, *p*) includes the living genus *Lithothamnium* (known since the Cretaceous) and other forms with a hypothallium, composed of filaments prostrated along the substratum. The perithallial filaments are not arranged in orderly rows, or the rows are indistinct. In *Lithothamnium* the sori leave cavities that are connected with the thallus surface by pores. In another group transverse septae composed of cells belonging to neighbouring filaments occur on one level, while the filaments themselves are arranged in rows. Because of this in various cross-sections the cell-rows are immediately apparent. In the living genus *Lithophyllum* (known since the Cretaceous) cavities can frequently be observed that remained from conceptacles and have the form of an inverted heart with one apical opening (visible only in a favourable section). Other thalli structures are also known. They occur in crustose, nodular and branching forms. Some coralline algae (Melobesioideae; Fig. 8, *r*, *s*) have continuous thalli, whereas in the Corallinoideae (genus *Corallina*, known since the Cretaceous) they are articulated with uncalcified nodes or genicula (Fig. 8, *t*). These thalli, undergoing burial, fall apart into calcified segments (intergenicula), that form large accumulations. Crustose thalli of the family Squamariaceae are similar to the Melobesioideae.

The family Solenoporaceae (Fig. 8, *u*) is represented by small (up to several centimetres), rounded or crustose thalli, which are closely similar to the first two coralline groups in their anatomical structure. But their cells are much larger and the cellular filaments are thicker. The hypothallium is only rarely discernible and then it consists of one layer of large-celled filaments. Evidently it was not calcified, as in some of the living RA. According to the perithallial organization, the thalli may be with distinct horizontal rows of cells, or without them. The systematics of the Solenoporaceae is also based on the form of the cells, arrangement of filaments, presence and density of the transverse and longitudinal septae in them. In one Silurian *Solenopora* species cavities were observed that probably correspond to conceptacles, and in a Miocene genus, *Neosolenopora*, sporangia were found that are similar to those in Corallinaceae. The independent status of Solenoporaceae from Corallinaceae is not quite evident. It is thought that the genus *Parachaetetes*,

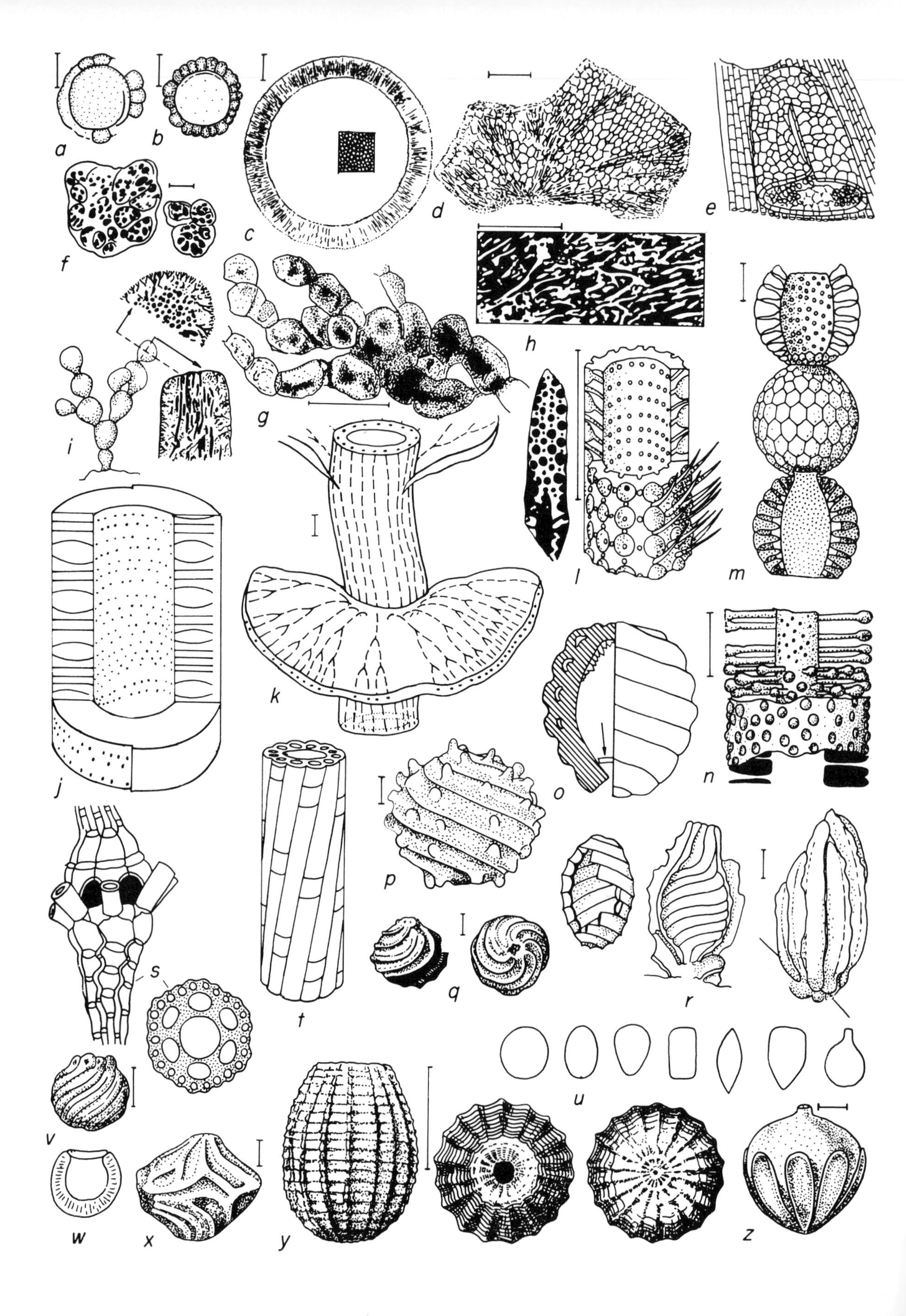
a
b
c
d
e
f
h
i
g
l
m
j
k
o
n
p
q
r
s
t
u
v
w
x
y
z

commonly included in the Solenoporaceae, might belong to cyanobacteria.

The family Ungdarellaceae (Fig. 8, *v–y*) has small rod-like, branching or nodular thalli. As V. P. Maslov (1973) expressed it, they have a 'fountain structure'. The axial zone is composed of a single filament, a bunch of filaments, or a cavity left over by uncalcified filaments, from which filaments or irregular rows of cells slant upwards to the surface. In nodular forms the cells may be arranged in concentric zones. Such forms are similar to foraminifers.

Other fossils are known that have been either directly assigned to RA or tentatively affiliated to them. Some of them belong to foraminifers, sponges, stromatoporoids, green and other algae.

RA are known since the Cambrian. A comprehensive review of the history of Late Palaeozoic RA was presented by Wray (1977), and of Mesozoic RA by Poignant (1977). During the Palaeozoic predominant were Solenoporaceae and putative ancestors of coralline algae (*Archaeolithophyllum* and related forms). During the Mesozoic solenopores were replaced by corallines, that continued to predominate throughout the Cainozoic. Among uncalcified RA the genus *Stenogrammites* (Upper Jurassic of the Volga Region) has been described in the literature. It has a dichotomously branching thallus. This genus has been assigned to the family Phillophoraceae (order Gigartiniales). According to Schweitzer (1983b) the genus *Wahnbachella* (D_1) is conceded to be related to RA. It is represented by narrow dichotomously divided thalli, extending outwards from the flattened basal thallus, and bearing elliptical cystocarpia (?). Such fossils are as yet of uncertain affinity.

RA and calcified fossils closely related to them frequently occur in marine carbonate sediments, and occasionally are the major rock-forming constituents. During the Cainozoic RA were important components of reef-building communities. At times reef massifs and bioherms occur that are made up entirely of these remains. Studies of RA are important for sedimentological reconstructions. Calcified RA live today in the littorals and sublittorals, rarely at depths up to 200 m. The RA belt commonly lies deeper than that of light-requiring green algae. RA readily inhabit shallow water areas, remote from the shore. Branching RA sometimes form nodules, being rolled over the basin floor by the waters. Limestones consisting of such nodules are known in the Tertiary. Lime-

Fig. 9. Green (*a*?, *b*?, *c–n*) and characeous (*o–z*) algae; Pre-Cambrian (*a*), Ordovician (*f*), Silurian (*n*), Devonian (*b*, *d*, *e*, *q*, *y*, *z*), Carboniferous (*h*, *j*, *v*, *w*), Permian (*m*), Jurassic (*r*, *s*), Cretaceous (*l*, *x*), Eocene (*d*, *p*), recent (*i*); North America (*a*, *k*, *v*, *x*), Western Europe (*b–e*, *g*, *h*, *m*, *n*, *p*, *s*), Russian platform (*f*, *j*, *q*, *w*, *y*, *z*), Kopet-Dag (*l*); *a*, *Eosphaera tyleri* Bargh.; *b*, *Eovolvox silesiensis* Kaźm.; *c*, *Tasmanites globulus* (O. Wetz.) Morg., optical section and detail of surface with pores; *d*, *e*, *Parka decipiens* Flem., general view (*d*) and thallus part with sporangium (cellular structure not in a scale); *f*, *Gloeocapsomorpha prisca* Zal., horizontal thin section; *g*, boring alga (Chaetophorales); *h*, *Ortonella furcata* Garw.; *i*, general view and inner structure of erect calcified codiaceous alga (height of specimen 5–20 cm); *j*, *Beresella erecta* Masl. et Kul., scheme of structure of calcareous sheath, canals shown by dots and solid lines; *k*, reconstruction of *Calcifolium*, canals shown in broken line; *l*, *Turkmeniaria adducta* Masl., oblique section and reconstruction; *m*, *Mizzia velebitana* Schub., reconstruction, section of living form (above) and in fossil state (below); *n*, *Rhabdoporella pachyderma* Rothpl., reconstruction, section in fossil state (below); *o*, scheme of gyrogonite structure, basal plate at arrow, secondary calcite deposit in upper part of cavity corresponding to oospore; *p*, *Maedleriella mangenotii* Gramb.; *q*, *Trochiliscus* (al. *Moellerina*) *bulbiformis* Karp.; *r*, *Clavator reidii* Grov., gyrogonite (left), section through calcareous envelopes (middle), outer view of utricle (right); *s*, *C. reidii*, outer view of node with leaf bases, and *C. peckii* Mädl., transversal view of internode; *t*, *Clavator*, structure of central axis with cortical cells; *u*, variety of gyrogonites in vertical section; *v*, *Karpinskya laticostata* (Peck) Peck; *w*, *Quasiumbella pseudorotunda* Brazhn. et Berch.; *x*, *Clypeator corugatus* (Peck) Gramb.; *y*, *Sycidium melo* Sandb.; *z*, *Chovanella maslovii* Iarz. Scale bar = 1 cm (*d*), 1 mm (*h*, *l*, *m*, *y*), 0.5 mm (*v*), 100 μm (*k*, *n*, *p–r*, *x*, *z*), 50 μm (*g*), 40 μm (*b*), 10 μm (*a*, *c*, *f*). Modified from: Kaźmierczak, 1976a (*a*, *b*); Tappan, 1980 (*c*, *v*, *w*); Gothan and Weyland, 1964 (*d*); Niklas, 1975–1976 (*e*); Osnovy paleontologii, 1963a (*f*, *j*, *l–n*); Kaźmierczak, 1976b (*g*); Maslov, 1973 (*h*); Wray, 1978 (*i*); Mamet and Roux, 1977 (*k*); Maslov, 1963 (*o–u*, *x–z*).

stones have also been encountered that consist of small RA fragments. Most of the living calcified RA are inhabitants of warm-water basins of normal salinity.

DIVISION CHLOROPHYTA (CHLOROPHYCOPHYTA). GREEN ALGAE

Green algae (GA), in contrast to the previously described divisions, are represented throughout their fossil record by practically the entire spectrum of living forms – from one-celled to siphonaceous. There are unicellular organic-walled forms (Tasmanitaceae), that are studied together with palynomorphs, multicellular forms preserved as compressions (*Parka*), colonial and siphonaceous calcareous forms observed in petrographic thin sections. The diversity of GA classifications pertains to both the bulk of the division and its constituent taxa. At times charophytes are allied to GA whereas, in contrast, prasinophytes, the same as Conjugales, are excluded from the division. Comparison of the living and fossil forms of GA is fraught with difficulties. The state of affairs is slightly better only in relation to the Siphonophyceae, where the thalli of a complex organization can readily be compared with one of the extant families and even genera. Simpler forms (codiaceous, early dasyclads, receptaculites) may be included in the siphonaceous GA only tentatively.

In the number of species GA outrank most other algae and higher plant groups. Among GA we find unicellular flagellar, coccoid, palmelloid, filamentous, cladothalloid and siphonate forms. They range in length between 1–2 μm and 1 m. The cell wall of many GA is calcified, calcium carbonate combining with pectic substances. The reproductive organs have been recorded only for the siphonaceous forms of fossil GA.

In the system adopted in this book GA are subdivided into six classes. Concise descriptions of them are presented below. The siphonaceous forms are discussed in more detail. Fossil Conjugatophyceae are very rare and are therefore omitted.

The class Prasinophyceae (prasinophytes – PF) has been recognized only recently, whereas formerly it was affiliated to Volvocales. The living PF (up to 10 genera) are mostly unicellular. The cells have no cellulose wall, but are covered by a mucilage or scaly covering. The peculiar characters of PF, including their flagellar organization, place them in a unique position among the GA and eukaryotic algae in general (for details see Tappan, 1980). In the geological record PF are represented first of all by the family Tasmanitaceae. To this family is assigned the genus *Tasmanites* (Upper Pre-Cambrian – N_2; Fig. 9, *c*) and about 30 other genera, occurring from the Pre-Cambrian to Holocene. They were attributed to PF only recently when it was found that they resemble cysts of living PF. The Tasmanitaceae are represented by spheroidal bodies, up to 600 μm in diameter, with a thick (5–20 μm) wall, of yellow, reddish and brown colours. The thickest middle layer is pierced by many radial canals and thinner perforations, that do not completely pierce the wall, but only the upper surface. Under the light microscope the perforations in surface view look like tiny dots whereas the canals look like pores. The arrangement of both is taxonomically significant. Some genera have an ornamented wall, covered, for example, by polygonal thickenings. When the alga emerges from the cyst the wall ruptures, or a round opening is formed (a pylom), the same as in dinoflagellates. Layers of sediments have been encountered in the Devonian that are completely filled with *Tasmanites* walls; in the Permian of Tasmania coal seams occur that are comprised entirely of these remains (tasmanite, or 'white coal'). These tasmanite accumulations are evidence of water blooms. *Tasmanites* are restricted to marine sediments. Owing to the exceptional resistance of their walls, they readily undergo redeposition. PF may also comprise the family Leiosphaeridiaceae, known since the Pre-Cambrian, and including microfossils that might belong to the living family Halosphaeraceae. The leiosphaeridia are represented by thin, smooth envelopes, having a pylom.

The class Volvocophyceae (V.) includes unicellular flagellar forms or palmelloid colonies made up of such cells. The Devonian genus *Eovolvox*

(Fig. 9, *b*) was correlated with V. These are globular hollow colonies (coenobia) made up of round or pear-shaped cells, tightly packed in one layer. Some colonies bear inside another colony, similar, but smaller (daughter?). Pre-Cambrian microfossils *Eosphaera* (Fig. 9, *a*) are similar in aspect, but their cells do not comprise a continuous layer. It was pointed out that similar spheroidal structures may occur in red algae as well as resulting from the decay and subsequent silicification of the cells of cyanobacteria. Certain Palaeozoic microfossils, occurring in carbonates and described as foraminifers (*Archaeosphaera*, *Calcisphaera*, etc.), may be affiliated to V. (Kaźmierczak, 1976a).

In the class Protococcophyceae (PC.) the cells are non-motile (flagella, when present, do not move), solitary or occur in colonies; the thalli are coccoid. Among the fossil PC. best known is the genus *Gloeocapsomorpha* (Fig. 9, *f*). Large accumulations of its colonies comprise the Ordovician combustible shales in Estonia (kuckersites). This genus is commonly assigned to the family Botryococcaceae and occasionally is regarded as synonymous with *Botryococcus*, together with certain Palaeozoic genera that form sapropelic (boghead) coals. Their remains are studied in thin sections or by rock maceration. The genus *Botryococcus*, known since the Early Carboniferous, is made up of colonies of cells that are immersed in a common mucilage. The affiliation of the Carboniferous fossils to this genus was evidenced by chemotaxonomic investigations (Niklas, 1975–1976, 1982a). Certain microfossil genera, beginning from the Pre-Cambrian, have been assigned to PC. Some of them are of quite wide distribution (e.g. *Globochaete* in the Mesozoic of the Mediterranean Region, Mexico, Cuba).

The class Ulotrichophyceae (U.) includes multicellular filamentous and laminated forms. The genus *Parka* (S_2–D_1; Fig. 5, *d*, *e*) is close to the order Chaetophorales. This genus has multilayered and multicellular flattened, rounded or oval, occasionally lobed thalli, up to 7 cm in diameter. Sporangia are located along the entire thallus. At times, when tightly packed, they reveal hexagonal configurations. Sporangia with a unilayered wall are filled with spores on which a trilete fold was noted. The status of *Parka* is in need of more precise specification. A resemblance may be perceived between *Parka* and liverworts of the *Sphaerocarpus* type (Neuber, 1979). It is probable that Chaetophorales may have been ancestors of liverworts. Of affiliation to U. are certain boring algae, that populate carbonate rocks in marine shallows and rework the rocks into a powder (micrite). The oldest boring fossil algae have been recovered from the Ordovician, but their systematic affinity has been definitely established only for the Upper Silurian algae. They are closely related to the living Chaetophorales and are represented by branching filaments of rounded cells (Fig. 9, *g*), where the walls, as depicted in electronic micrographs, are thinly laminated.

The class Siphonophyceae (S.) is remarkable in that the large thalli of complex organization lack cellular structures, but consist of one giant cell (coenocyte) with a single or numerous nuclei. Such a thallus organization is termed siphonate. At times the thallus, divided by septae, is multinucleate. At the base of both the thallus branches and gametangia tight plugs may be formed. Fossil S. are represented by the families Codiaceae (order Siphonales) and Dasycladaceae (order Dasycladales). Their thalli are more or less calcified. Fossil S. lacking a calcareous covering are also known. For example, the genus *Ostrebium*, which resides in empty shells and coral-buildings, is known from the Miocene up to the present.

The order Siphonales. Palaeoalgologists commonly assign to Siphonales and the only family Codiaceae (C.) algal fossils that have calcified thalli of two types: (1) pad-like, or nodular, (2) branching, consisting of rounded or flattened segments. Tappan (1980) considers it more correct to affiliate these algae not to the living C., in which calcified thalli can seldomly be observed, but rather to Caulerpales, among which several families have been recognized. Most of the forms, commonly assigned to C., she includes in the families Udoteaceae and Garwoodiaceae of the order Caulerpales.

Many C. fossils have been described from detached segments. The inner structure of the thalli is rather sophisticated. They consist of intertwined coenocytic threads, the calcified coverings of which become fused into a compact body, that is pierced by canals – coenocyte outgrowths. Several parallel canals extend along the axis of the body forming branches that spread outwards. At their tips are cavities that probably correspond to gametangia, or minute vesicles (utricles). The latter form a system of utricles, that in living C. perform a photosynthetic function (the inner branches of the coenocyte are colourless). In fossil C. the utricles look like tiny holes on the thallus surface. At times the surface is complicated by small projections (papillae), to which the outgrowths from the inner canals extend.

The thallus habit of living C. (Fig. 9, *i*) varies considerably within a single genus (*Codium* includes both pad-shaped and branching forms). In fossil C. great significance is attached to the thallus habit. In addition, important characters are the arrangement, size, mode of calcification of the canals, and the thallus surface structure (the presence of papillae and utricles). Fossil C. may be divided into two groups. One group includes genera that are comparable to living ones (this refers chiefly to articulated branching forms, known since the Ordovician – *Dimorphosiphon*, *Palaeoporella*). Another group, sometimes distinguished as the subfamily Garwoodioideae, includes Silurian–Carboniferous genera, represented by nodules with radial non-septate canals. Distinctions of the genera are based on the size and type of branching of the canals. Best known among them is *Ortonella* (S–C; Fig. 9, *h*). So far the affiliation of these fossils to green algae has not been proven. They might as well be cyanobacteria.

Living and seemingly fossil C. are thermophilic algae. Ever since the Ordovician C. had a paramount role as rock-forming fossils. Fossil C. comprise bioherms and reefs. The living forms may compose calcareous sands and carbonate ooze (when the covering disintegrates into aragonitic crystals). After lithification their biogenic origin cannot be recognized.

In the order Dasycladales (D.) the thallus consists of one gigantic coenocyte, made up of an unbranched or forked central axis with whorls of solitary branches or their bunches. More seldomly the branches are disorderly situated. In contrast to the Codiaceae, their thalli are uninuclear. The main axis and the branches are calcified at the surface, so that in fossil materials it is possible to judge the thallus structure (Fig. 9, *j–n*). Fossil D. in outward aspect look like perforated cylinders, spheres, discs or plates, corresponding to thallus segments. The sporangia are also calcified. The systematics of fossil D. is based on the thallus habitus, type of segmentation, number and arrangement of branches, position and forms of sporangia. The thalli may be twig-like, pear-shaped, globular-stalked, umbellate (stalked disc), articulated. The outgrowths form closed or open tubular cavities (pores) within the covering of the main axis. The open pores may be simple or branching, widening towards the surface. Closed pores narrow at the ends or, contrariwise, terminate with a swelling. Associated with the pores are cavities left by the sporangia. Ultimate outgrowths extend out of the pores and photosynthesize. They may be cake-like, filling the external widened part of the pore, or filamentous in shape.

Many fossil D. genera have been described (Bassoullet *et al.*, 1977; Herak *et al.*, 1977) which commonly are encompassed in one family Dasycladaceae and several tribes. The oldest (Cambrian) D. are weakly articulated; branches are rare. Part of these forms existed up to the end of the Palaeozoic. Two types of thalli have been recovered from the Ordovician: (1) cylindrical, with branching and unbranching pores (traces of branches), at times apically widened, and of disorderly arrangement; (2) spherical-stalked. Affiliated to the first type is *Rhabdoporella* (O–C_1; Fig. 9, *n*). The genus *Vermiporella*, with a cylindrical covering that branches and is pierced by straight or slightly curved pores, was of rock-forming significance in the Ordovician and Silurian. During the Carboniferous an orderly arrangement of the pores (branches) appears, as can be seen in *Beresella* (C; Fig. 9, *j*). These D. are also rock-forming organisms. D. with spherical thalli, have one (*Cyclocrinus*), or several spheres (*Mizzia*; P; Fig. 9, *m*). The reproductive organs of these Palaeozoic D. are unknown. Upwards in the stratigraphic sequence the thallus becomes of increasingly more complex organization. Forms appear with branches gathered in bunches (from the Devonian). Extremely rich and very diverse D. assemblages have been recorded in the Triassic of the Alps and the present-day Mediterranean Region. These D. were reef-builders. Of first appearance in the Triassic is the living genus *Acicularia*, that has an umbel of sporangiophores with spherical cavities for spores. D. are good index fossils, but they are of greater significance for reconstructions of past sedimentary environments. Living D. do not grow at depths below 60 m, and are chiefly thermophilic.

Special mention should be made here of Receptaculites (R.) – puzzling marine organisms (O–C), frequently affiliated to sponges and archaeocyates (for details see Nitecki *et al.*, 1981). They have also been treated as an independent type of animal. Now R. are most commonly included in the division of green algae, being affiliated to siphonaceous, and distinguished as an order Receptaculitales. R. in many respects are affinitive to archaeocyates, which can also be presumably affiliated to algae. R. differ from green algae in that the thallus wall itself, not the mucilage covering, is calcified. The thallus varies in size from 0.5 to 70 cm, and is rounded, pear shaped with an inner cavity, or flattened. It has a double wall, composed of uniform elements (meromes), arranged in rows. The merome consists of an outer plate and inner plate with a ray in between them. The outer plates cover the outer wall, whereas the inner plates cover the inner wall. The ray bears outgrowths that spread out in different directions. The most typical R. are described here, but they may differ in organization (Nitecki and Toomey, 1979; Tappan, 1980). The skeleton evolved through thinning of the walls and eventually through the total disappearance of the inner wall; the heads of the meromes became more simple and the thallus underwent a change, from porous and cone shaped, opening upwards, to entire (without pores), with a closed contour. R. may be treated as a satellite order of green algae, pending detailed studies of well-preserved reproductive organs. R. at times were rock-forming fossils.

DIVISION CHAROPHYTA

Charophytes (Ch.) are frequently included within the division of green algae, owing to the similarity in their pigments and assimilation products (starch). But morphologically Ch. (Fig. 9, *o–z*) are very peculiar. Their thalli resemble shoots of higher plants. At first living Ch. were assigned to horsetails. Ch. reveal distinctly expressed articulation of their bodies and whorled arrangement of the outgrowths. The morphological terminology of higher plants is in current use for descriptions of Ch.

Living Ch. have a multicellular thallus, attached to the substratum by rhizoids. The stems are articulated. Their outgrowths (leaves) are of orderly whorled arrangement in nodes, dividing the stem into internodes. Unicellular processes near the leaf bases are termed stipules. The internode in some Ch. consists of a single tubular cell, whereas in other Ch. the central tube is surrounded by a layer of orderly oriented smaller cells (cortex). The cortex structure varies in different Ch. The cells of the nodes are small, gathered into a low disc, from which leaves and stipules spread out. The leaves are similar in organization to the stems. Leaflets occur at the nodes of articulated leaves. The internodes may bear unicellular processes (spines).

From the peripheral cells of the node antheridia and oogonia are formed. Globular antheridia covered by polygonal shield cells are rarely fossilized. Ellipsoidal and ovate oogonia (Fig. 9, *o–q*, *v–z*) are composed of five spiral cells (primary spirals), enveloping the egg cell from the outside, and wound anti-clockwise (sinistral spiral). At the apex the enveloping cells give off two layers of short cells, that form a coronule with a central pore. The sutures between the enveloping cells leave sharp ridges on the surface of the oospore (fruit). The oospore membrane, formed after fertilization of the egg cell, is impregnated by suberin with silicic acid and becomes hard.

In many Ch. both the vegetative and reproductive organs undergo calcification. Calcified oogonia and fruits of fossil Ch. (gyrogonites) occur most frequently detached (Fig. 9, *p*, *q*, *u–z*). The cells of gyrogonites (partecalcines) correspond to spiral enveloping cells and coronular cells (apical partecalcines), that vary in their forms. In most of the Ch. the gyrogonite consists of five partecalcines. Frequently the partecalcines bear ornamentations. At the ends of the gyrogonites may be pores, apical and basal, the latter being closed by a basal plate (plug). Inside the gyrogonite the organic membrane of the oospore is occasionally preserved. In some fossil Ch. the gyrogonite is enclosed in an outer covering – the utricle (Fig. 9, *r*). It consists of spindle-shaped branches of vegetative origin, that join together at the apex of the gyrogonite (an analogy arises with seed plants: the gyrogonite may be compared to the nucellus, whereas the utricle may be compared to the integument). The utricle branches remain free or their calcified envelopes become fused. Utricles can be easily distinguished from gyrogonites when both of them are preserved. In forms with well-developed utricles gyrogonites may be weakly calcified. When a single calcified envelope is found it is not always easy to decide whether it is an utricle or a gyrogonite. The question as to whether the Palaeozoic Ch. of the genera *Sycidium* and *Chovanella* (Fig. 9, *w–z*) are utricles or gyrogonites is still debatable. Because of this the neutral term 'shell' has been suggested.

The fossil Ch. are studied in reflected light. For the distinction of utricles from gyrogonites, identification of the inner structures, and distinction of Ch. from other organisms (e.g. foraminifers) the preparation of thin sections is necessary. The organic oospore membrane may be liberated by acid-etching of gyrogonites.

The literature contains descriptions of several tens of genera and hundreds of species of fossil Ch. (Conkin and Conkin, 1977; Maslov, 1963, 1973; Tappan, 1980). So far there is no generally accepted suprageneric systematics of them. V. P. Maslov proposed recognizing the classes Trochiliscophyceae (dextral spirals; the only genus *Trochiliscus*; D–C_1; Fig. 9, *q*) and Charophyceae (sinistral spirals) with orders Clavatorales (forms with utricles) and Charales (lacking utricles). The orders have been divided into families, that are differently treated in the literature. The genera *Chovanella* (D; Fig. 9, *z*) and *Sycidium* (D–C_1; Fig. 9, *y*) stand apart, frequently being defined as independent orders. *Chovanella* is treated either as a gyrogonite, or as a gyrogonite and utricle assemblage. The genus *Sycidium* is affiliated to gyrogonites with characteristic polygonal sculpture, corresponding to the sporina relief. The *Sycidium* wall is pierced by thin pores, which have not been observed in other Ch. It was suggested that *Sycidium* should be keyed out as an independent class Sycidiphiaceae. Best studied among the Ch. with utricles is the genus *Clavator* (J_3–K_1; Fig. 9, *r–t*).

The origin of Ch. is unknown (for information on their phylogeny, see Grambast, 1974; Tappan, 1980). They are thought to be phylogenetically related to green algae of the chaetophoralean affinity that have a whorled arrangement of the thalli parts. The oldest known Ch. were recovered from the uppermost Silurian. In the Devonian–Lower Carboniferous of wide occurrence are trochiliscs, chovanellas, sycidia and umbellae (*Umbella* and allied genera are commonly observed in cuts from thin sections). In the Devonian, Charales with sinistral spirals appear (e.g. *Eochara* with 8–13 partecalcines). During the Mesozoic the order Clavatorales, known also in the Palaeogene, played a paramount role. The living *Chara* is known since the Eocene. Neogene Ch. are similar in habit and affinities to extant forms.

Fossil Ch. are successfully used in stratigraphy, particularly in the correlation of Triassic continental beds, where other fossils are sometimes lacking. Living and many fossil Ch. are freshwater, more seldom brackish water plants. In the Miocene and older beds mass accumulations of gyrogonites and vegetative organs of Ch. occur together with marine faunas. It is inconceivable that gyrogonites and extremely fragile vegetative organs could have accumulated at a distance from their living sites. The waters where now extinct Ch. lived were probably of various salinity. It has been noted that in sediments with marine faunas Ch. may be restricted to thin beds lacking marine fauna. Accumulations of fossil Ch. occasionally compose layers termed characites or characeous tuffs.

FUNGI

The term fungi (F.) (Mycetalia) in the present usage refers to an informal group, that covers slime moulds (lower F.) and true (higher) F.

Slime moulds (Myxomyceta) are unknown in the fossil record, although their spores probably occur among palynomorphs, but have been unnoticed by palynologists, or have been assigned to other organisms.

A unified concise description of living F. (over 50 000 extant species) cannot be offered, the same as for all the eukaryotic algae taken together. Among F. are unicellular organisms and complex multicellular bodies. Their reproductive mechanisms are of high diversity, frequently lacking analogues in other organisms. The fragmentary nature of fossil F. does not allow the establishment of reliable extinct suprageneric taxa. Many extinct genera are of uncertain affinities.

A brief outline is presented below of the most pertinent data on those fossil F. which are more frequently encountered by palaeobotanists (see reviews: Baxter, 1975; Graham, 1971; Pirozynski, 1976; Pirozynski and Weresub, 1979; Tiffney and Barghoorn, 1974; Wagner and Taylor, 1982).

1. Septate or non-septate hyphae (Fig. 10, *a*). These fossils are observed on the surface of cuticles or inside petrifactions of higher plants. To hyphae may belong certain tubular microfossils, recovered together with palynomorphs from Pre-Cambrian and younger sediments. Hyphae-like microfossils occur in Pre-Cambrian cherts since the Middle Riphean (about 1 billion years in age).

Attached to the hyphae may be colonies or unicellular vesicles of not quite clear nature, occasionally having small round bodies inside them (spores?). Such vesicles are assigned to sporocytes or sporangia. Identification of such fossils is of particular interest when they are attached to the root system of higher plants and may be ectotrophic mycorrhiza (i.e. symbiotic to plants in which they are found). The hyphae and sporocytes found inside the Devonian *Rhynia* might probably belong to mycorrhiza of oomycetes. Endomycorrhiza has been described from the roots of Carboniferous lepidophytes, cor-

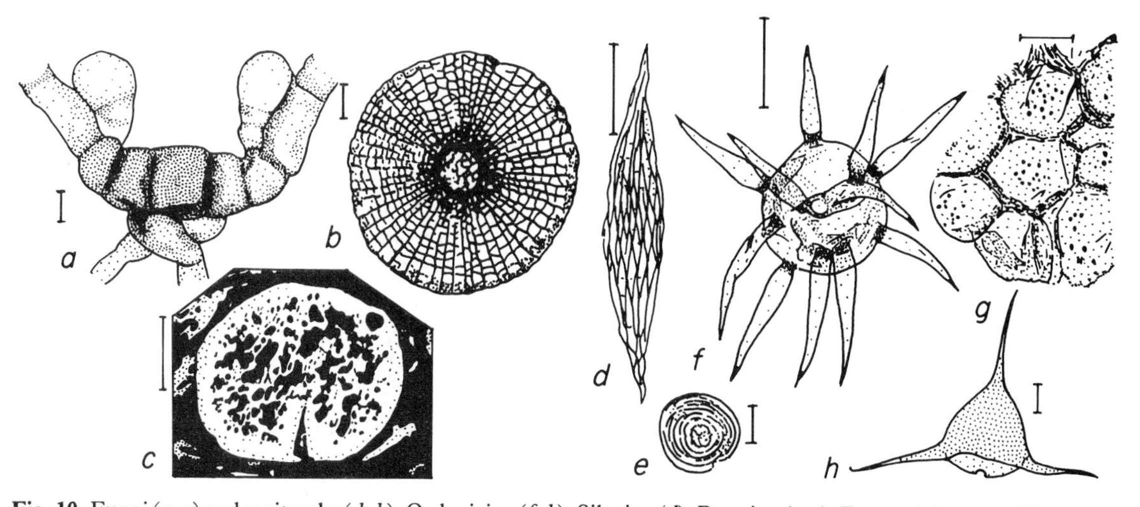

Fig. 10. Fungi (*a–c*) and acritarchs (*d–h*); Ordovician (*f–h*), Silurian (*d*), Permian (*c, e*), Eocene (*a*), recent (*b*); USA (*a, d, f–h*), Australia (*c, e*); *a*, epiphyllous *Meliolinites dilcheri* Daghl., germinating spore with primary hyphopodia and vegetative mycelium; *b*, perithecium of *Microthyrium tunicae* Gol.; *c*, *Coronasclerotes australis* Pickh., sclerotium in coal; *d*, *Leiofusa rhikne* Loebl.; *e*, *Circulisporites parvus* de Jers. with concentric and spiral ribs and grooves; *f*, *Baltisphaeridium perclarum* Loebl. et Tap., projections with basal inner plugs; *g*, *Synsphaeridium* cf. *gotlandicum* Eisen., acritarch aggregate; *h*, *Veryhachium irroratum* Loebl. et Tap. Scale bar = 50 μm (*c, d, f*), 10 μm (*a, b, e, g, h*). Modified from: Taylor, 1981 (*a*); Popov, 1960 (*b*); Stach and Pickardt, 1957 (*c*); Tappan, 1980 (*d, h*).

daitanthaleans and marattiaceous ferns (*Psaronius*). Hyphae have been found with chlamydospores that are nearly like those that occur in living F. of the genus *Glomus* (Wagner and Taylor, 1982). Stubblefield and Taylor (1984) described a wall ultrastructure of the endogonaceous chlamydospores from the aerial axes and roots of Palaeozoic plants. Some non-septate hyphae bear oogonia and antheridia and show affinities to the living parasitic oomycete *Albugo* (Stidd and Cosentino, 1975). These remains have been found in *Nucellangium* seeds. In other cases the plant remains most probably were posthumously infected by F.

2. Sclerotia (densely packed hyphal accumulations that perform vegetative reproduction). They occur in coals since the Carboniferous and are being studied together with other coal constituents both in thin sections and on cut surfaces (Fig. 10, *c*). The relevant fungal component of the coals is termed 'sclerotinite'. Besides sclerotia other F. remains occur in coals.

3. Fruiting bodies of ascomycetes (class Ascomycetes). These fossils belong to fruiting bodies that were situated on leaves of higher plants and have been coined epiphyllous. They are enduring to maceration and can be isolated together with palynomorphs. Peltate fruiting bodies (thyriothecia) together with attached hyphae are encountered on cuticles of gymnosperms (Krassilov, 1967b) and angiosperms (Dilcher, 1975). These fruiting bodies belong to ascomycetes of the order Hemisphaeriales and in outward aspect are flattened, rounded or irregular forms made up of cells of radial arrangement (Fig. 10, *b*). In the middle of the thyriothecium occasionally a 'stoma' can be seen – an opening that serves for shedding of the ripe spores from the asci. The asci and spores are usually not preserved, but at times the successive stages of spore germination have been observed. The systematics of the epiphyllous F. is based on such characters as the presence–absence of a free mycelium, its organization, shape of the fruiting body, stoma organization, cellular pattern. Epiphyllous Hemisphaeriales are known since the Permian, but have been studied in detail in fossil materials from the Cretaceous and Palaeogene.

4. Dispersed spores (Pirozynski, 1976), occurring together with palynomorphs, or adhered to the cuticle of leaves together with fruiting bodies.

5. Bodies resembling chytridia of Chytridiomycetes occur in Carboniferous coal balls. The resemblance to chytridia is particularly apparent in bodies found in the seed tissues, megaspores and pollen (Millay and Taylor, 1978).

The oldest indisputable Basidiomycete has been recorded from the Middle Carboniferous. This is *Palaeancistrus* found in the wood of the Carboniferous fern *Zygopteris* and represented by septate hyphae with clamp connections (Dennis, 1970). Affiliated to ascomycetes are, most likely, the genera *Sporocarpon*, *Dubiocarpon* and *Mycocarpon*, that for a long time were considered problematical, as well as the recently established genus *Coleocarpon*. They represent small bodies, less than a millimetre in size, consisting of 'a spherical central cavity containing scattered spherical asci or ascospores surrounded by an outer wall of pseudoparenchima. Generic and specific distinctions are based principally on the structure and organization of the outer wall. *Mycocarpon* is somewhat different in organization' (Stubblefield *et al.*, 1983, p. 1496). Of other F. only *Palaeosclerotium* is worthy of mention, combining characters both of Ascomycetes and Basidiomycetes (Dennis, 1976). The fruiting bodies are closed, of the cleistothecia types, whereas the spores were produced by asci. These cleistothecia, typical of Ascomycetes, consist of hyphae or have such pores in the septae, the same as in the Basidiomycetes. The vegetative mycelium has clamp connections and pores in the septae of the same type as the Basidiomycetes. This genus was assigned to F. intermediate in type between asco- and basidiomycetes. According to another viewpoint (Pirozynski and Weresub, 1979), spores that were interpreted as ascospores may be tetraspores. The cavities enclosing them can then be correlated with conceptacles of the Silurian–Devonian Nematophytales (see 'Brown Algae'), where the body is also composed of hyphae. Finally, an alternative opinion is that *Palaeosclerotium* unites two different fungi types.

Fossil F. are important not only for studies of the geological history of F. It is thought that F. were the first land inhabitants. The paramount role of F. in ecosystems is evident from the Devonian. Hyphal remains are of common occurrence in Carboniferous and younger sediments. These remains occur mainly in areas with a humid climate, while epiphyllous F. occur mainly in areas with a warm and humid climate. F. fossils, especially their spores, can be used in stratigraphy. For instance spores of the genera *Ctenosporites* and *Pesavis* furnished the basis for the correlation of the Palaeogene of North America, Australia and England (Lange, 1978).

ACRITARCHA AND CRYPTARCHA

By the 1930s, due to the progress in palynological studies, attempts had already been made to isolate spores and pollen from sediments of the Lower Palaeozoic. Microfossils were found that in outward aspect were similar to miospores (works of W. Darrah, A. Reissinger, A. Eisenak). In the 1940s S. N. Naumova reported the discovery of higher plant spores in the Pre-Cambrian. Microfossils of the Lower Palaeozoic and Pre-Cambrian had proved to be more uniform than miospores of younger sediments, so that it was difficult to apply to them conventional systematic characters. This gave rise to much controversy concerning the systematic affinities of these microfossils. The idea that these fossils are spores of the oldest higher plants determined the morphological interpretation of these remains, and, moreover, led to serious observational errors. The drawings depicted a small trilete slit, which seemingly served to prove their affiliation to higher plants.

However, more skilful microscopic examinations together with more circumspect interpretations revealed that compressional folds of the walls occurring by chance were mistaken for the trilete slit, and, at times when the latter was not found on the wall, an imaginary slit was depicted. On the other hand, the simple organization of many walls was a deterrent factor in establishing key-taxa. Because of this for generic and species distinctions any characters were used that were discernible under the microscope, such as the size, wall colour, degree of folding, etc. The peculiarities of preservation of the walls were treated as primary taxonomic characters. Palynomorphs frequently bear traces of corrosion, of bacterial and fungal attack, imprints of the matrix grains, traces of mineral growths, etc. The walls that underwent maceration, frequently exhibit imprints of regular crystals and spherical pyrite grains (framboid forms), that comprise globose accumulations, reticulate cavities, etc. All these preservation features are known in younger miospores, but are not regarded as taxonomic characters. In the oldest microfossils such structures were used for diagnoses of the genera (e.g. *Balvinella*) and species. Random substances were frequently described as microfossils – such as accumulations of organic matter, that underwent alterations in the process of maceration, or were undermacerated, as well as merely contamination. The identity of microfossils in silicified stromatolites was indiscriminately muddled. Dark bodies occurring inside the cells were described as nuclei, while the respective microfossils were described as eucaryotic organisms. This gave rise to much confusion in the systematics of the oldest microfossils.

More thorough investigations (see reviews: Diver and Peat, 1979, Downie, 1973; Volkova, 1965, 1974) established that spores of higher plants are absent from the Ordovician and older beds. It was recommended that the formal inclusion of the oldest microfossils into particular groups of algae or other organisms should be avoided. The suggestion was advanced that they should be united into a specific group Acritarcha (Downie *et al.*, 1963), that encompasses all organic-walled microfossils resistant to palynological processing (acid- and alkali-maceration, separation in heavy liquids). Diver and Peat (1979) proposed that Acritarcha proper and Cryptarcha should be distinguished. The first refers to all microfossils, 5 to 500 μm in size, organic walled and with an inner cavity. They are unicellular forms (sometimes occurring in groups or chains that readily disintegrate, fall apart), have processes, and when the latter are lacking exhibit general conspicuous configurations (polygonal, elongated, spindle shaped, but not spherical). The processes may be in the shape of spines, simple or branching tubes (hollow inside), wings or flanges. The surface may be smooth or microsculptured. Within the wall may be inclusions (inner bodies, grains, dark body). Sometimes the wall may be opened by a rupture (within a considerable part), pylome (as in dinoflagellates) or by such a structure (epityche), that after the formation of the arched fissure the part contoured by the latter folds outwards and downwards, like the flap of an envelope. Thus, those forms remain in the group Acritarcha that can be most easily classified, and with a greater degree of probability may be affiliated with dinoflagellates, prasinophytes, and in rare instances with other eucaryotic algae (green, xanthophytic, chrysophytic).

The group Cryptarcha includes problematical microfossils with less specified characters. These are unicellular, spherical or ellipsoidal bodies, multicellular aggregations (filaments, chains, colonies) consisting of tightly joined cells. Processes are absent, but the wall may be porous or thinly ornamented, opening by a rupture or pylome, with inlain dark bodies, grains or an inner body.

Acritarchs and cryptarchs are divided into genera, that are united into subgroups. The subgroups are distinguished according to the general form, organization and arrangement of the outgrowths, and the fine ornamentation of the wall. The Pre-Cambrian yielded Cryptarcha chiefly of the subgroups Sphaeromorphitae (spherical microfossils with a smooth or weakly oriented unilayered wall), Nematomorphitae (filaments or chains of tightly packed cells) and Synaplomorphitae (aggregates or colonies of tightly packed rounded cells). The genus *Leiosphaeridia* belongs to the first subgroup. It is represented by smooth, spherical, unilayered walls with numerous disorderly arranged compression folds. All three subgroups occur in silicified stromatolites or may be extracted from sediments by maceration methods.

Cryptarcha continues upward into the Cambrian, but the general aspect of the assemblages is determined by Acritarcha proper – forms with spines and outgrowths, occasionally forked (Acanthomorphitae group); spindle-shaped forms with slender ends appear (Netromorphitae group; Fig. 10, *d*), and barrel-or dumbell-shaped forms with processes on the rounded tips (genus *Lophodiacrodium*). Most of the Cambrian acritarchs belong seemingly to dinoflagellates and prasinophytes. In the Ordovician the acritarchs reach peak abundance and maximum diversity. Most typical are the *Baltisphaeridium* and *Micrhystridium* (group Acanthomorphitae; Fig. 10, *f*), that were of first appearance in the Cambrian. They have long outgrowths, simple or with a forked apex. The subgroup Polygonomorphitae (e.g. genus *Veryhachium*; Fig. 10, *h*) includes triangular or polygonal acritarchs, occasionally with angles attenuated into long appendages. Silurian acritarchs are also very diverse. From the Pre-Cambrian throughout the Lower Palaeozoic a general tendency is apparent towards an increase in the complexity of acritarchs and cryptarchs. At first forms appear with a smooth and weakly sculptured wall, then spiny forms, and later, forms with outgrowths, first simple, and then complicated. The wall also becomes more complicated. The occurrence of more complex forms does not lead to the replacement of the simple forms.

By the end of the Silurian the acritarchs became of less diversity, which is due, in part, to the affiliation of microfossils to various algae divisions. The diversity of acritarchs decreases from the Devonian to the Carboniferous, then remains at the same level up to and including the Triassic, and, finally, with the expansion of dinoflagellates drops in the Jurassic. The replacement of acritarchs by dinoflagellates serves as indirect evidence that they are related to different groups of algae (Tappan, 1980). Probably many acritarchs, similar in outward aspect to dinoflagellates, belong to other algae, including prasinophytes, to which *Baltisphaeridium* and several other Palaeozoic genera display affinities in their ultrastructure.

Some acritarch genera continue to the Cainozoic (*Veryhachium* is known from the Palaeogene). In Neogene–Quaternary sediments acritarchs have been recorded in continental freshwater deposits. Some of the older acritarchs probably might have been freshwater organisms, but commonly acritarchs are restricted to marine sediments.

Acritarchs increase in percentage among palynomorphs in the direction away from the shoreline towards the open sea. The differentiation of acritarch assemblages relative to the environment was observed. Assemblages described from the Devonian revealed the occurrence of simple sphaeromorphic acritarchs near reef bodies and their replacement by associations of acanthomorphic acritarchs at a distance from the reef (F. L. Staplin). D. Wall demonstrated that the Jurassic acritarchs with long and complex branching outgrowths exhibit a correlation with calm waters, whereas forms with smooth walls or small outgrowths exhibit a correlation with turbulent environments and coarser sediments. Hence, it is conceivable that the composition of acritarch assemblages may serve as an indicator of past hydrodynamical conditions. The climatic differentiation of Lower Palaeozoic acritarch assemblages has also been noted.

Acritarchs are of great stratigraphic significance, particularly in studies of the Pre-Cambrian and Lower Palaeozoic, and also of younger rock formations with a paucity of fossils or altogether barren of other fossils. Acritarch finds in continental sequences serve as indicators of marine interbeds, which may be used as marker horizons. Acritarchs frequently occur in abundant amounts (up to 900 000 specimens in 1 cm^2), and, evidently, served as the source material in the formation of combustible deposits.

HIGHER PLANTS

The following criteria have been adopted for the distinction of divisions. All the higher plants have been divided into two groups according to the dominance of gametophytes or sporophytes in the life-cycle. Plants with dominating gametophytes constitute one division, Bryophyta. Plants where sporophytes predominate have been subdivided into several divisions as follows:

1. The body is not divided into a root and stem; lateral axes, even though they may differ in position from the main axis, practically do not differ anatomically; the gametophyte is free living. Division Propteridophyta.
2. The body is differentiated into a root (or rhizophore) and stem, and the stem into a main axis and lateral branches, or into an axis and leaves. The gametophyte is free-living. Division Pteridophyta.
3. The female gametophyte develops on a sporophyte; archegonia are well developed. The male gametophyte is at first intracellular, then terminates its development on a sporophyte bearing a female gametophyte. Division Pinophyta (Gymnospermae).
4. Female and male gametophytes are strongly reduced. Archegonia and antheridia are completely reduced. Division Magnoliophyta (Angiospermae).

Upon more detailed consideration of the divisions further in the text other characters, additional to those cited above, will be given. As often happens in systematics, forms occur that are transitional between the divisions, whereas diagnostic features may at times be lacking. For example, in pteridophytes the female gametophyte rarely develops on the parent plant (sporophyte) and in some lycopsids it attains the embryo stage. Even such a character as the sporophyte dominance may not be persistent. Ferns are known where gametophytes are the dominant phase. Because of this the characters of the divisions (the same as the diagnostic features of any other taxa) cannot be regarded as absolute attributes of all individuals, or of all subordinate taxa. They more readily reflect common traits, according to which all members of the division as a whole differ predominantly from members of other divisions. Common traits may be recognized in the process of systematization of the taxa of inferior ranks, upon successive arrangements of them in clumps. We were already faced with this specific nature of biological taxa (the statistic nature of their characters) in considering the algae divisions, and will be confronted with it once again.

The phylogeny of higher plants (without bryophytes) is depicted in Fig. 11.

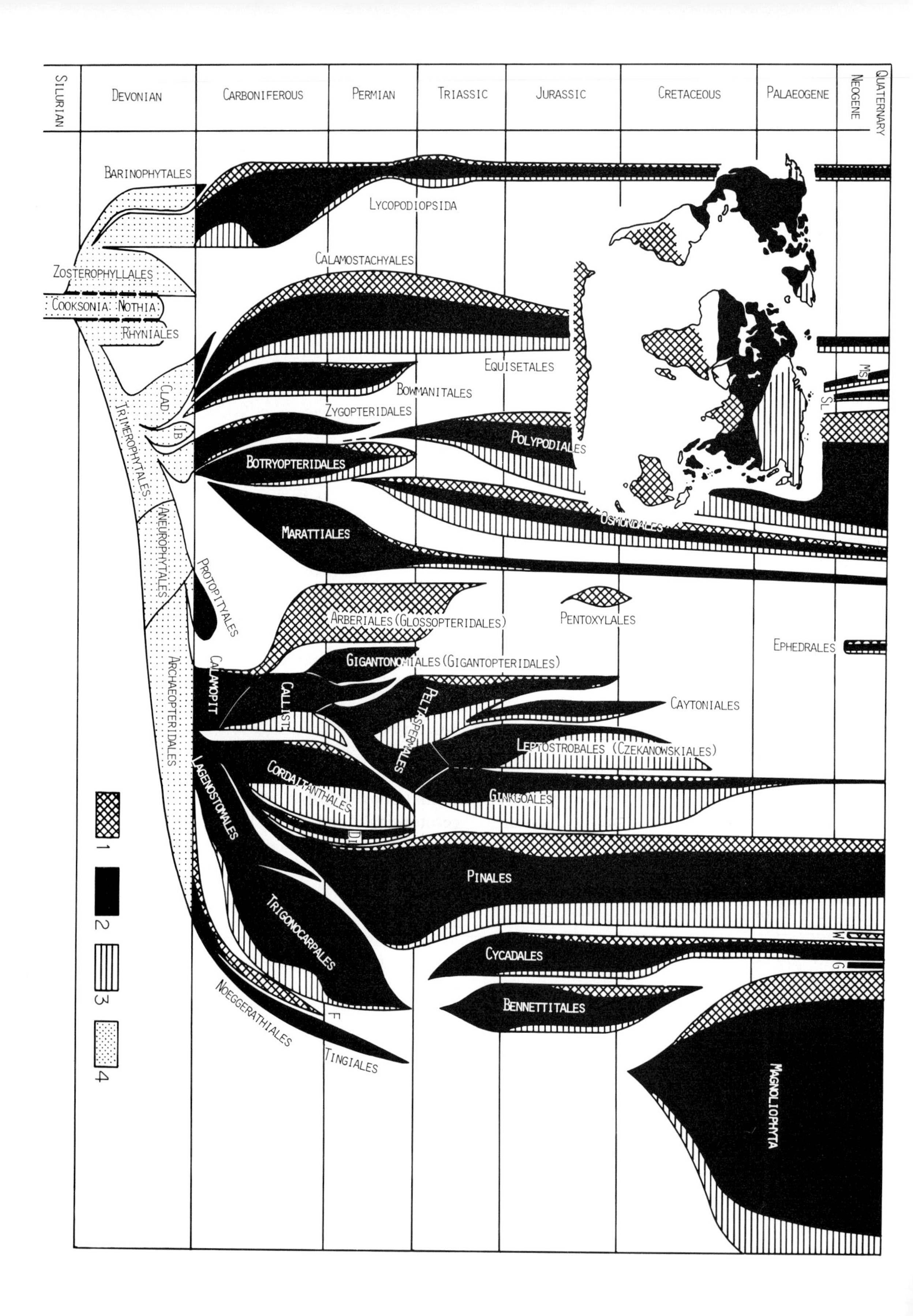

Silurian
Devonian
Carboniferous
Permian
Triassic
Jurassic
Cretaceous
Palaeogene
Neogene
Quaternary
Barinophytales
Lycopodiopsida
Calamostachyales
Zosterophyllales
Cooksonia
Nothia
Rhyniales
Equisetales
Clad
Bowmanitales
Zygopteridales
Trimerophytales
Ib
Polypodiales
Botryopteridales
Aneurophytales
Osmundales
Marattiales
Protopityales
Pentoxylales
Arberiales (Glossopteridales)
Ephedrales
Archaeopteridales
Calamopit
Gigantonomiales (Gigantopteridales)
Callist
Peltaspermales
Caytoniales
Leptostrobales (Czekanowskiales)
Lagenostomales
Cordaitanthales
Ginkgoales
Di
Pinales
Trigonocarpales
Cycadales
Bennettitales
Noeggerathiales
F
Tingiales
Magnoliophyta
MS
SL
W
G
1
2
3
4

DIVISION BRYOPHYTA

Living bryophytes (B.) are represented by about 25 000 species. Fossil B. described in the literature are far inferior in number (no more than 100 species in the Palaeozoic and Mesozoic; Jovet-Ast, 1967; Krassilov and Schuster, 1984; Lacey, 1969). Palaeobotanists frequently fail to recognize B., because parts preserved are chiefly detached leaves and small shoot fragments that may be indistinguishable among the plant debris. The scanty fossil record of B. was wrongly explained by their poor preservation abilities. For the extraction of B. the bulk maceration method is highly expedient (Krassilov, 1967a, 1970). These techniques are comparatively rarely employed in palaeobotanical laboratories.

Gametangia and sporangiophores are of rare occurrence in B. fossils, and dealing only with leaf and vegetative shoot fragments it is not always possible to identify eutaxa, and to trace their geological history. The distribution of fossil B. within the system is usually based on a comparison with living taxa, according to sets of subordinate characters. For example, the Permian genera, assigned to the order Protosphagnales, were compared with extant peat mosses, according to the differentiation of the leaf cells (Fig. 12, *l*, *m*). The assignment of the cells to chlorophyll-bearing was based merely on the greater colour intensity. Thallose forms of liverworts of poor preservation are difficult to distinguish from algae. Some fossils, described as liverworts, may be gametophytes or rooting organs of higher cryptogamic plants. Fossil B., on the whole, have so far contributed little information to the general morphology, systematics and phylogeny of the division. In most manuals B. are divided into the classes Hepaticopsida, Anthocerotopsida and Bryopsida. The data on fossil Anthocerotopsida are as yet unsubstantiated. Many B. described in the literature are of uncertain affinities. In the Upper Carboniferous shoots were encountered that are affinitive to leafy mosses and have been assigned to the genus *Muscites*. From the Devonian the genus *Sporogonites* was described that has a laminar thallus from which unbranching sporangiophores spread out. From the Lower Carboniferous spore tetrads (*Tetrasporites*) have been recorded, that resemble those of sphaerocarpalean liverworts.

Class Hepaticopsida (Marchantiopsida)

Liverworts (L.) may be of thallose or leafy types. Thallose L., which are land creepers, are commonly dorsiventral plants – their dorsal side differs from the ventral to which rhizoids are attached. In leafy L. the leaves, frequently lobed, are borne on the axis. If two rows of leaves are present, they are attached to opposite sides of the axis (which may be upright). The third row of leaves (amphigastria) is located on the ventral side. Gametangia are located on special upright branches.

The classification of fossil L. (Jovet-Ast, 1967; Krassilov and Schuster, 1984) relies chiefly on vegetative parts (Fig. 12, *a–g*). In a simplified form their affiliation to suprageneric taxa may be described as follows. To L. are assigned all thallose forms (although part of them may belong to Anthocerotopsida), and those leafy forms where the leaves lack a midrib and are arranged in two to three rows (shoots with spiral leaf arrangement and/or a leaf midrib are referred to Bryopsida). Leafy L. may be placed in the order Jungermanniales and suborder Jungermaniineae (acrogynous jungermannias). Thallose L. with a lamina consisting of one or only a few layers of similar cells are ascribed to the suborder Metzgeriineae (anacro-

Fig. 11. Phylogeny of vascular plants and geographic pattern of main phyla. Map on inset shows basic floristic zonation in Late Palaeozoic. Taxa with wide transgression of characters are connected without narrowing of phyla. Width of phylogenetic branches mostly reflects role of corresponding taxa in plant assemblages of pertinent time span. 1, Notal (Gondwana) extraequatorial phytochoria; 2, Equatorial and adjacent ecotonal phytochoria; 3, Boreal extraequatorial phytochoria; 4, uncertain phytogeographical situation. Calamopit, Calamopityales; Callist, Callistophytales; Clad, Cladoxylales; Di, Dicranophyllales; F, *Fedekurztia* and allies; G, Gnetales; Ib, Ibykales (= Iridopteridales); Ms, Marsiliales; Sl, Salviniales; W, Welwitschiales.

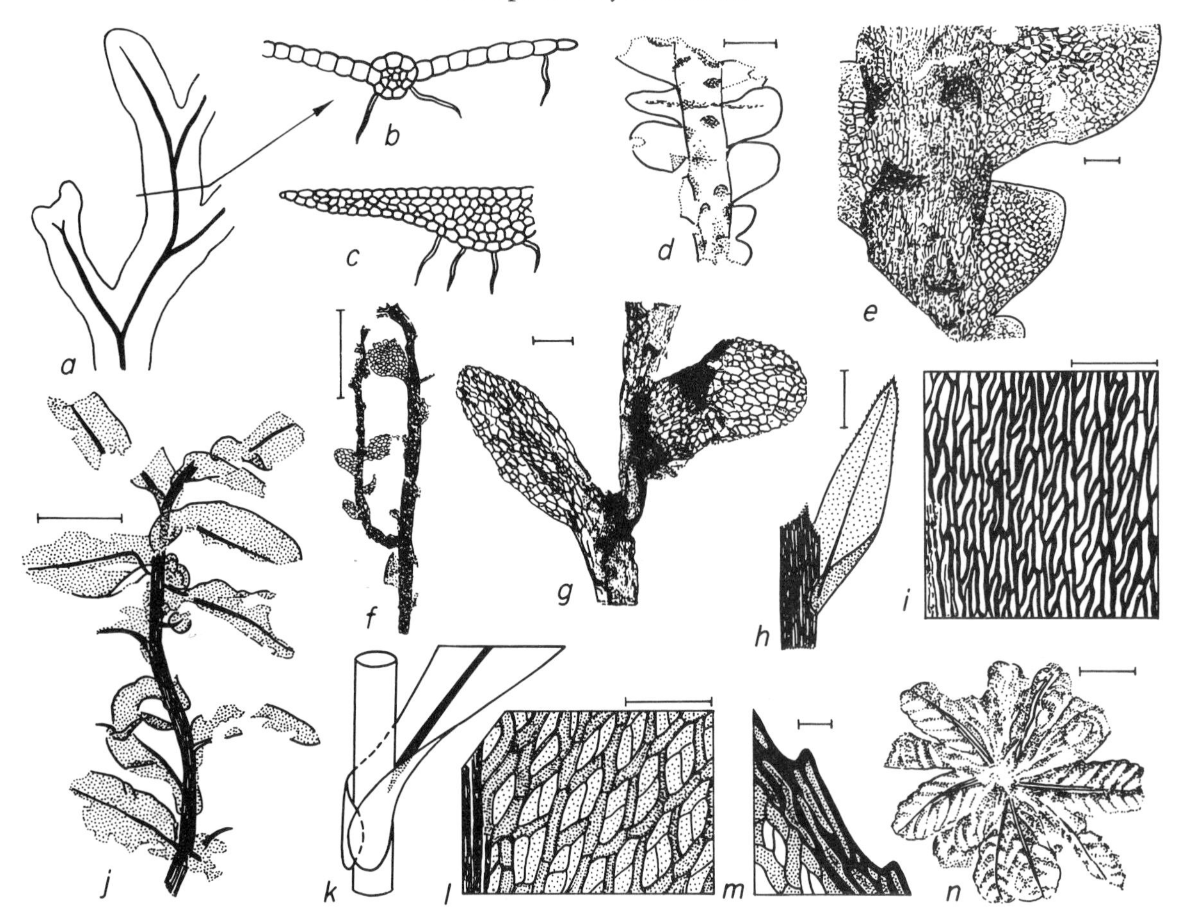

Fig. 12. Bryophytes; Middle Carboniferous (*d, e*), Permian (*h–n*), Upper Jurassic (*f, g*); Western Europe (*d, e*), Bureya (*f, g*) and Pechora (*h, i, l, n*) basins, Russian platform (*j, k, m*); *a, b*, thallose liverwort of *Pallavicinites* type, general view (*a*) and scheme of transversal section showing rhizoids (*b*); *c*, scheme of transversal section of thallose liverworth of *Pelliothallites* type; *d, e*, *Hepaticites kidstonii* Walt., general view (*d*) and cellular structure (*e*); *f, g*, *Cheirorhiza brittae* Krassil., general view (*f*) and part of shoot with two dorsal lobes (*g*); *h, i*, *Intia vermicularis* Neub., leaf attachment (*h*) and cellular structure (*i*); *j–m*, *Protosphagnum nervatum* Neub., shoot (*j*), leaf base (*k*), cells near leaf midrib (*l*), leaf border (*m*); *n*, *Vorcutannularia plicata* Neub. Scale bar = 5 mm (*n*), 1 mm (*f, h*), 0.5 mm (*d*), 100 μm (*e, g, i, l*). Modified from: Hirmer, 1927, after Walton, (*d, e*); Krassilov, 1973a (*f, g*), Neuburg, 1960a (*h, i, l, n*).

gynous L.). L. with a more complex thallus organization, with numerous amphigastria, air chambers and air pores are assigned to Marchantiales. Such an approach is very simplified. For instance, not all leafy L. belong to acrogynous jungermannias and not all thallose jungermannias belong to anacrogynous L. Because of this it is necessary to seek other characters that are typical of the living representatives of one or another family or order.

Most of the Palaeozoic L. are thallose forms, closely related to Metzgeriineae. Their thallus is weakly differentiated, with or without a midrib, and has undivided or lobed wings, with or without rhizoids (Fig. 12, *a–c*). These and other characters serve for the distinction of the genera (*Pallavicinites*, *Metzgeriites*, *Blasiites*, etc.). The genus *Hepaticites* (C; Fig. 12, *d, e*) is externally similar to jungermannias, but, seemingly, still belongs to the Metzgeriineae. Its stems bear two types of unistratose leaves. Large tongue-shaped leaves are attached to the stem in two lateral rows, whereas smaller leaves are ventrally situated in two rows, each leaf being attached to the base of a larger one. *Cheirorhiza* (J_3; Fig. 12, *f*, *g*) is a typical leafy genus with three rows of leaves; the lateral leaves are bilobed. The upper (dorsal) lobe is round, the lower (ventral) is linear, occasionally reduced to a small cell group. Amphigastria, tongue-shaped to hair-like, are tightly pressed to the stem. Rhizoids are multicellular, which is typical of leafy mosses, and attached both to the stem and leaves. Fossil Marchantiales with characteristic air chambers and pores (or without pores) are firmly known since the Triassic. The Devonian *Sciadophytopsis* and *Ricciellopsis*, described as marchantialeans (Istchenko and Shlyakov, 1979), are probably propteridophytic gametophytes. In *Eomarchantites* from the Triassic, the thallus consists of a dorsal and ventral epidermis with interposed air chambers having simple pores. The orders Sphaerocarpales and Calobryales (Haplomitriales) are practically unknown in the geological record. The genus *Najadita* (T_3) may belong to the Sphaerocarpales.

Class Bryopsida (Musci)

This class includes mosses, or leafy mosses (LM.). Their gametophyte consists of a stem covered with leaves, frequently having a midrib. The axis is often supplied with a central conducting tissue. LM. are divided into subclasses: Sphagnidae (peat mosses), Andreaeidae (unknown in the fossil record) and Bryidae. Living peat mosses are represented by the only genus *Sphagnum*, one of the major peat-forming plants on the Earth. Spores and leaves of Sphagnidae are of frequent occurrence in the Quaternary, more seldom in the Tertiary and of singular occurrence in the Cretaceous and Jurassic. Dispersed spores of the genus *Sphagnumsporites* (since T_3) have been tentatively assigned to Sphagnidae.

The order Protosphagnales has been established merely on the basis of vegetative shoots. The branching of the shoots and differentiation of the leaves into cauline and branch (ramal) types has not been observed. Light-coloured cells, that have been interpreted as hyaline, lack pores; they are located in pockets composed of dark cells, occurring in short rows or n-shaped. The leaves of protosphagnalean mosses are similar in the organization of their cell network to young leaves of the living *Sphagnum*, but, in contrast, have a midrib. The genera *Protosphagnum* (P_2; Fig. 12, *j*–*m*) and *Vorcutannularia* (P_2; Fig. 12, *n*) have been studied in more detail.

Bryidean LM. (subclass Bryidae) are distributed between 13–15 orders and are quite common in the Cainozoic. They include either living genera or closely related forms. At times the Cainozoic genera reveal a combination of characters typical of different extant families. For instance, in *Muscites lanceolata* from the Neogene of England, M. Boulter recognized a combination of characters that are typical of the families Hypnodendraceae, Rhizogoniaceae and Mniaceae. Palaeozoic and Mesozoic Bryidae are commonly represented by leaves and leafy shoots. They are assigned to Bryidae, but part of them may belong to Andreaeidae, or to an extinct subclass of LM. Many pre-Cainozoic LM. are included in the collective genus *Muscites*. Typical among the extinct genera is *Intia* (P; Fig. 12, *h*, *i*), having lanceolate or oval leaves with an acute apex. The cells of the leaf lamina are vermiculate, frequently arranged in oblique rows (runners). Certain cell groups can be observed that are similar to those of *Protosphagnum*, but without differentiation of the cells according to their colours. Along the leaf margin lies a border of darker cells, of which the outer ones form teeth (microdenticulation). *Intia*, *Protosphagnum* and certain other Permian mosses are connected by transitional forms and are probably related to a single order of LM.

The origin and time of first appearance of the Bryophyta are unknown. In the Upper Silurian thallose plants occur that were tentatively affiliated to liverworts, but they may equally be treated as gametophytes of associated plants. The Bryophyta were commonly linked in origin with the *Horneophyton*-like plants (see Division Propteridophyta below), having a column of sterile tissue in the sporangia, which is typical of

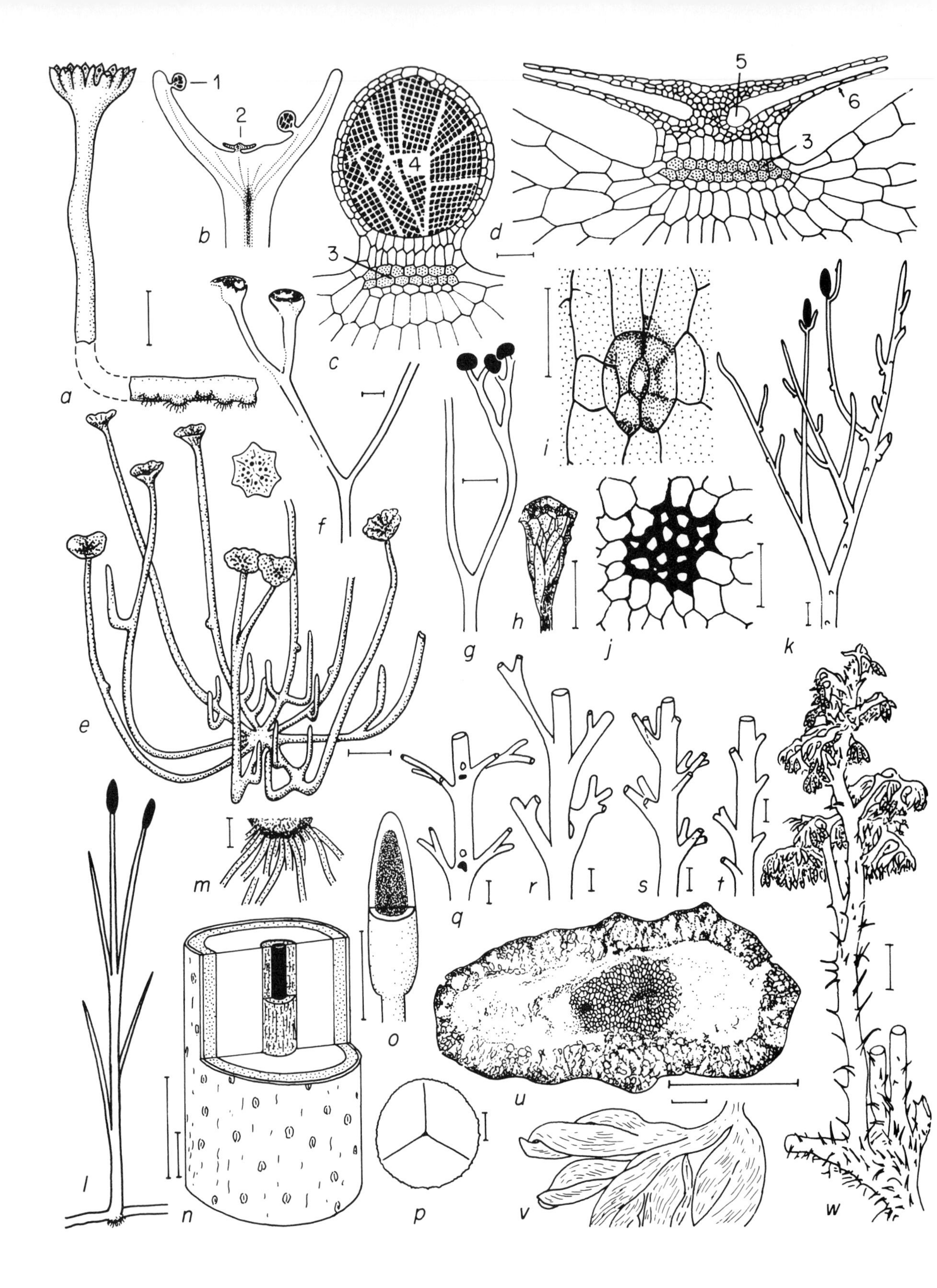

1
2
4
5
6
3
a
b
c
d
e
f
g
h
i
j
k
l
m
n
o
p
q
r
s
t
u
v
w

sporangiophores of mosses. However, no other characters have been observed that give evidence of this phylogenetic linkage. Liverworts are known since the Devonian (only thallose forms in the Palaeozoic, also leafy forms in the Mesozoic). During the Permian leafy mosses predominated, having been restricted to Boreal (Angara) floras (Fefilova, 1978; Meyen, 1982b; Neuburg, 1960a). These mosses could probably have been coal-forming plants. From the Mesozoic liverworts are chiefly known (Anderson, 1976; Anderson and Anderson, 1983; Krassilov, 1970, 1973a; Krassilov and Schuster, 1984). Cainozoic mosses are very similar to those living today.

DIVISION PROPTERIDOPHYTA

Vindication for the recognition of the independent status of the division Propteridophyta (Pt.) was presented above. The name was first suggested by E. A. N. Arber (1921; see also Meyen, 1978c, 1978e). The more customary name 'psilophytes' is invariably and erroneously associated with the genus *Psilophyton*, which belongs to a more advanced group of Pt., that probably deserves to be affiliated to pteridophytes. The widely used term 'rhyniophytes' is also irrelevant, considering that it is associated with the genus *Rhynia*, the anatomical and taxonomic interpretation of which is subject to much controversy. Banks (1968) suggested abstaining from using a taxon uniting all the Pt., and distinguishing rhynias, zosterophylls and trimerophytes as independent categories. However, all three groups share a number of important common features that reflect the initial stage of evolution of higher plants, and according to which all Pt. differ from pteridophytes. Pt. utterly lack secondary tissues, roots, leaves and leaf-like appendages supplied with conducting tissues; their sporangia are always directly attached to axes (not specialized appendages), or occur as their extensions; the axes are always protostelic; the stele is undivided, cylindrical or elliptical. A characteristic feature is the close similarity in the organization of the spores in different groups of Pt. (Allen, 1980; Gensel, 1980; Gensel and White, 1983). Supporting evidence to the unity of Pt. is found in the existence of genera which combine characters of all three above-mentioned groups. This applies to *Renalia*, *Nothia* and *Hsuea*, which are tentatively included below in the class Zosterophyllopsida. The combination of rhynian and zosterophyllean characters can be seen in the Silurian–Lower Devonian genera *Cooksonia* and *Steganotheca*, which, seemingly, stand at the roots of the phylogenetic tree of higher plants, giving an indication of their monophyletic origin.

The absence of reliable information on the organization of Pt. gametophytes is a serious obstacle that prevents the determination of phylogenetic links between the major Pt. groups, as well as between Pt. and other plants. Under the name *Lyonophyton rhyniense* (Fig. 13, *a–d*) bisexual gametangiophores have been described, that con-

Fig. 13. Propteridophytes; Lower Devonian (*a–e*, *i–w*), uppermost Silurian (*f–h*); Western Europe (*a–p*), North America (*q–w*); *a–d*, *Lyonophyton rhyniense* W. et R. Remy (*a*, reconstruction of gametophyte; *b*, gametangiophore; *c*, antheridium; *d*, archegonium; 1, antheridium; 2, archegonium; 3, abscission or nutritive tissue; 4, spermatogenous tissue; 5, egg; 6, archegonial neck); *e*, *Sciadophyton* sp., reconstruction of gametophyte with gametangiophores, and detached gametangiophore (surface view); *f*, *Cooksonia pertonii* Lang; *g*, *C. caledonica* Edw.; *h*, *Steganotheca striata* Edw., sporangium; *i*, stoma of *Rhynia*; *j*, *k*, *R. gwynne-vaughanii* Kidst. et Lang (*j*, vascular bundle in transversal section); *l–p*, *R. major* Kidst. et Lang (*m*, rhizoids; *n*, scheme of anatomical structure, xylem in black; *o*, sporangium; *p*, spore); *q–t*, branching in Trimerophytales (*q*, *Pertica quadrifaria* Kasp. et Andr.; *r*, *P. varia* Granoff et al.; *s*, *Trimerophyton robustius* Hopp.; *t*, *P. dalhousii* Doran et al.); *u*, *Psilophyton dawsonii* Banks et al., cross-section of fertile axis in place of branching; *v*, *w*, *P. crenulatum* Doran, groups of paired sporangia (*v*) and reconstruction of fertile shoot (*w*). Scale bar = 10 cm (1), 1 cm (*e*, *k*, *q–t*, *w*), 5 mm (*a*, *g*, *o*), 1 mm (*f*, *h*, *n*, *u*, *v*), 200 μm (*m*), 100 μm (*c*, *d*), 50 μm (*i*, *j*), 10 μm (*p*). Modified from: W. Remy and R. Remy, 1980 (*a–d*); Remy *et al.*, 1980 (*e*); D. Edwards *et al.*, 1979 (*f*); Gensel, 1976 (*g*); Edwards and Rogerson, 1979 (*h*); Chaloner and McDonald, 1980 (*i*, *n*, *o*); Zimmermann, 1959 (*j*, *m*); D. S. Edwards, 1980 (*k*); Meyen, 1979a (*l*); Doran *et al.*, 1978 (*q–t*); Banks *et al.*, 1975 (*u*); Doran, 1980 (*v*, *w*).

sist of a cylindrical axis and apical bowl-shaped widenings with a lobed margin. Archegonia were attached to elevations in the centre of the bowl and had a long neck. Clavate antheridia with antherozoids were inserted on the inner wall of the bowl. The gametangiophore axis had a vascular bundle of the same organization as in *Rhynia major* and *Horneophyton*. Diclinous gametangiophores have also been described that have a similar bowl-shaped lobed widening and axis, covered by emergences (W. Remy, 1982). The vascular bundle is apically divided and its offshoots spread out funnel-like into a bowl-shaped widening. Round antheridia are inserted on the walls of the bowl below the marginal lobes and slightly immersed, or located among paraphysis-like protuberances of sterile tissue. Archegoniophores are seemingly of the same structure. It was shown that plants of the genus *Sciadophyton* (Fig. 9, *e*), assigned to the family Sciadophytaceae, are, in essence, gametophytes of the same type (W. Remy *et al.*, 1980). To what type of sporophytes these gametophytes correspond is still unclear. Schweitzer (1983a, 1983b) established that certain specimens of *Sciadophyton* bear attached branching axes (young sporophytes) that resemble *Zosterophyllum* in the mode of their branching. On the other hand, the similarity in the anatomical structure between *Lyonophyton*, the above-mentioned diclinous gametangiophores, *Horneophyton* and *Rhynia major*, is indisputable. The above-cited gametangiophores are similar to the upright gametangiophores of liverworts and have nearly the same conducting bundles as bryophytes. These gametophytes might belong to bryophytes, but their affinities with Pt. seems more likely. Their similarity to *Rhynia* and *Horneophyton* and the probable connections between *Sciadophyton* and *Zosterophyllum* seem to suggest that different Pt. had essentially uniform gametophytes, and, moreover, had traits of bryophytic gametophytes. This gives supporting evidence to the unity of all the Pt., and to their kinship with the early bryophytes.

Higher Pt. taxa are based on the structure and arrangement of the sporangia, on the position of the protoxylems in the vascular bundles and on the character of the axes branching.

Class Rhyniopsida

Rhyniopsids have apical, globose or elliptical sporangia. The protoxylem is centrarch. In the more primitive members (Rhyniales) the main axis is of weak expression or indistinguishable, whereas in more advanced plants (Trimerophytales) it is quite distinct.

Order Rhyniales

To this order are assigned plants with dichotomous or trichotomous, isotomous or anisotomous branching of shoots. The axes are smooth. Part of them are crowned by sporangia. They are rounded to elliptical, more seldom kidney shaped. The spores are globose with a distinct trilete slit, often with curvatures and exine thickenings in the place of convergence of the rays; the exine has a fine sculpture.

The genus *Cooksonia* (S–D_1; Fig. 13, *f*, *g*), which has rounded or kidney-shaped sporangia at the ends of the dichotomously branching axes, is the most primitive. The mode of sporangial dehiscence has not been firmly established. At times an indistinct line can be perceived along the distal margin, that probably gives an indication of transversal dehiscence of the sporangia. In this character *Cooksonia* displays an affinity to zosterophylls. In *Steganotheca* (S–D; Fig. 13, *h*) the sporangia are elliptical with a truncated and thickened apex. The mutual arrangement between the proto- and metaxylem in these genera is unclear. Both genera have been assigned to the family Cooksoniaceae, that embraces the most primitive higher plants.

The family Rhyniaceae includes first of all the genus *Rhynia* (D_1; Fig. 13, *i–p*). Two species have been described: *R. major* and *R. gwynne-vaughanii*. To *R. major* (Fig. 13, *l–r*) are assigned naked axes, the branching of which has commonly been described as dichotomous. In effect, they had a thicker axis and extending from it solitary or paired slender axes. The axes bear central vascular bundles with centrally placed small cells, described as protoxylem. Instead of annular and spiral thickenings on the tracheid walls, a microreticulate structure has been observed in *R. major* (D. S. Edwards, 1980; El-Saadawy and Lacey, 1979; Lemoigne and Zdebska, 1980), whereby there is no justification for the assignment of these cells to tracheids, and that of the plant itself, to propteridophytes. The tracheids are surrounded by a thick phloem-like tissue, embraced by a wide cortex. Surficial stomata occur in the epidermis. On the low prostrate part of the axis bunches of rhizoids were borne. At the tips of the axes were elliptical sporangia that dehisced longitudinally with the aid of a special cell layer, similar to the endothecium of seed plants, and interrupted along a narrow band on one side of the sporangium (W. Remy, 1978a). It was thought (Lemoigne, 1970, etc.) that *R. gwynne-vaughanii* is a gametophyte of the same plant. However, the reconstruction of the shoot, performed by serial sectioning, has shown that *R. gwynne-vaughanii* produced sporangia (Fig. 13, *k*). They were situated apically on the axes, from which, below the sporangia, side branches extended upwards, i.e. an overtopping occurred. The shoot was of combined dichotomous and monopodial branching. Some lateral branches ('adventitious') are 'problematical in that their vascular strand is not connected to that of the main axis' (Gensel and Andrews, 1984, p. 64). The sporangia had a thick three-layered wall and a pad consisting of tracheids, and a parenchyma in the lower part of the spore cavity. The sporangia were shed with the aid of a basal abscission layer. Tracheids of the axial bundle had spiral and annular thickenings. These two *Rhynia* species should evidently be assigned to different genera; their affiliation to a single family is doubtful. Other rhynialean genera are poorly known.

Order Trimerophytales (Psilophytales)

In the fragmentary remains the trimerophytes (T.) cannot be validly distinguished from either the Rhyniales or the most primitive pteridophytes. When the material is better preserved one can see that in most of the T., in contrast to the Rhyniales, axes of different orders are present. The lateral axes are widely spaced (Fig. 13, *q–t*; Fig. 14, *a–c*). The sporangia are grouped on special branches, which reveal an anatomical difference from vegetative ones. The latter exhibit a weak tendency towards flattening. It is believed that part of the vegetative lateral axes of T. began to evolve into roots.

In *Psilophyton* (D_{1-2}; Fig. 13, *u–w*; Fig. 14, *c*; Banks, 1980b; Gensel, 1979) no single main axis can be distinguished, but thicker axes of irregular branching can be recognized, from which slender branching axes of no definite arrangement extend. Fertile branches are repeatedly forked, up to seven times, and terminate in spindle-shaped paired sporangia with attenuated apices. In the points of division of the branches of the fertile axes tubercles can be seen – the remains of undeveloped branches (Fig. 14, *c*). The axes are protostelic; the central protoxylem is surrounded by a massive metaxylem, consisting of scalariform tracheids; some tracheids show circular bordered pits. The vascular bundle of the sterile axes is rounded, whereas that of the fertile axes is elliptical, nearly rectangular in cross-section, the protoxylem being strongly flattened. The cortex is two-layered. The outer layer has a collenchymatous tissue. The axes are naked or covered by emergences. The sporangial wall is of similar organization to the outer cortical layer. The sporangia dehisced by a longitudinal slit. The spores are almost identical to those of the Rhyniales. The sporoderm is homogeneous, consisting of two layers, a thinner outer layer representing the ornamentation (Gensel and White, 1983). The spore cavity of the sporangium is covered with a cutinized membrane. In *Trimerophyton* (Fig. 13, *s*), from the nodes of the main axis three branches spread out together in one direction and repeatedly trifurcate. In *Pertica* (Fig. 13, *q*, *r*, *t*; Fig. 14, *a*, *b*) the main axis reaches 3 m in height. The branches are whorled or spirally arranged. From the main axis a practically straight main branch extends, and only its thinner branchlets are dichotomous.

Within the morphological series from plants of the *Rhynia* type to more advanced T. the axes become larger, their branching is increasingly intensified, thicker and more straightened; main axes appear and the branches undergo differentiation into sterile and fertile. Parallel to this the

Fig. 14. Propteridophytes; Lower Devonian; North America (*a–c*, *i–m*, *o*, *p*), Great Britain (*d–h*, *q–y*), W. Siberia (*n*); *a*, *b*, *Pertica quadrifaria*; *c*, *Psilophyton dawsonii*; *d–f*, *Zosterophyllum llanoveranum* Croft et Lang (*d*, folded sporangium in lateral view and viewed from axis; *e*, longitudinal section of sporangium; *f*, same at apex); *g*, *h*, *Z. myretonianum* Penh., stoma and general habit; *i*, *j*, *Renalia hueberi* Gens., reconstruction and spore; *k–m*, *Oricilla bilinearis* Gens., circinately coiled apex (*k*), fertile shoot (*l*), sporangium with transversal dehiscence slit (*m*); *n*, *Sawdonia ornata* Hueb., reconstruction; *o*, *p*, *S. acanthotheca* Gensel et al., reconstruction of fertile shoot, and spore; *q–t*, *Horneophyton lignieri*

metaxylem increases in bulk, while the protoxylem in the fertile branches becomes elongated (as in the axis of primitive progymnosperms and in clepsydroid bundles of ferns).

Class Zosterophyllopsida

Among the zosterophylls (Z.) there are genera (*Zosterophyllum*, *Sawdonia*, *Crenaticaulis*, *Gosslingia*, *Margophyton*) that reveal many important characters in common and form a quite natural grouping. Other genera (*Nothia*, *Renalia*, *Hsuea*, *Koniora*) exhibit certain peculiar features, but it is as yet premature to distinguish them as specific suprageneric taxa. Because of this in this book Z. are described without their subdivision into orders and families. A characteristic feature of Z. are globose or transversely elongated sporangia with a transversal distal slit, occasionally accompanied by a wall thickening. The sporangia dehisced along the slit into two valves. The sporangia are terminal or inserted on short stalks. They are gathered into terminal strobili, or into groups in the middle parts of the axes (fertile zones); more seldomly they are freely scattered over the shoot. Spores are nearly the same as in rhynialeans, frequently with curvatures. The protoxylem is exarch, diffusely scattered over the periphery of the xylem, or less frequently irregularly grouped. The vascular bundle is elliptical in cross-section (*Gosslingia*, *Crenaticaulis*, *Zosterophyllum*, etc.). This character is correlated with a planar branching (Rayner, 1983). In *Gosslingia*, *Crenaticaulis* and *Margophyton* at the point of branching there is a vestigial axillary axis, which is probably homologous to the axillary rhizophore of the living *Selaginella* (note that in *Psilophyton* a reduced axis also occurs at the point of branching of the fertile shoot). The sporangia of *Koniora* were attached slightly below the point of the final or penultimate dichotomy (Zdebska, 1982).

The genus *Zosterophyllum* (D_1; Fig. 14, *d–h*) has slender naked axes, branching dichotomously at right angles, or K-shaped. The intensely and irregularly branching part of the shoot was apparently prostrate on the substratum. The stomata occur mostly on the erect shoots. Sporangia are, in the main, gathered into terminal strobili (up to 60 per strobilus), and attached at the sides of the axis by a thick stalk, which is bent upwards and slightly inwards. In *Z. divaricutum* sporangia descend along the outer side of the axis below its dichotomy. Subgenera have been established according to the arrangement of the sporangia (spiral, two-ranked, on one side of the axis). The distal slit of the sporangium bears a two-lipped transversal thickening. The genus *Sawdonia* (D_{1-3}; Fig. 14, *n–p*) was established for shoots covered by emergences. The sporangia are kidney-shaped with small emergences; the distal slit is accompanied by an elongated swelling. The sporangial stalk is short, with a vascular bundle. The sporangia concentrate in small groups in the middle parts of the axes. The shoots branch irregularly, dichotomously, or exhibit a main axis. Closely related to these two genera are *Crenaticaulis* with two rows of denticular processes, *Margophyton* with lateral rims on the axes, and *Hicklingia*, earlier ascribed to Rhyniales. In *Oricilla* (Fig. 14, *k–m*) the sporangia are arranged in two rows along both sides of the axis and oriented along the axis plane.

The genus *Renalia* (D_1; Fig. 14, *i, j*) was first affiliated with rhynias, owing to the position of the sporangia on the tips of the lateral branches. Many fragments of fertile branches, described in the literature as *Cooksonia*, may belong to *Renalia*. The sporangia in *Nothia* (D_1; Fig. 14, *u–y*) are

(Kidst. et Lang) Darr., fertile apex (*q*), general habit (*r*), forked sporangium in section view (*s*), spore (*t*); *u–y*, *Nothia aphylla* Lyon ex Høeg, section of sporangium in dehiscence zone (*u*), general view of sporangium (*v*), branching fertile shoots (*w*), cross-section of axis (*x*), general view of axis (*y*). Scale bar = 10 cm (*a*, *n*), 1 cm (*b*, *c*, *h*, *i*, *r*), 5 mm (*k*, *l*, *o*, *q*, *w*, *y*), 1 mm (*d*, *e*, *x*), 100 μm (*f*, *u*), 50 μm (*t*), 25 μm (*g*, *j*, *p*). Modified from: Kasper and Andrews, 1972 (*a*, *b*); Banks *et al.*, 1975 (*c*); D. Edwards, 1969 (*d–f*); Lele and Walton, 1961 (*g*); Sporne, 1975 (*h*); Gensel, 1976 (*i*, *j*); Gensel, 1982 (*k–m*); Ananiev and Stepanov, 1968 (*n*); Gensel *et al.*, 1975 (*o*, *p*); W. Remy and R. Remy, 1980 (*q*); Eggert, 1974 (*r–t*); El-Saadawy and Lacey, 1979 (*u–y*).

kidney-shaped, at times fused in pairs, with a transversal slit that is accompanied, not by a thickening of the sporangial wall, but by its attenuation. The sporangia are aggregated into terminal branches, scattered over the shoot, or occur in whorls of three sporangia each. Their stalks are so long that they look like lateral axes. The protoxylem is not exarch, as in other Z., but centrarch. Hence, the genus *Nothia* combines characters of Z. (sporangial organization) and rhynias (sporangia arrangement, centrarch protoxylem). The combination of a centrarch xylem and apical kidney-shaped sporangia with a distal slit is also typical of the genus *Hsuea* (D_1), which bears dichotomous appendages, resembling adventitious roots or rhizophores (Li Cheng-Sen, 1982).

Z. were probably progenitors of lycopsids; the connecting links might have been plants of the *Asteroxylon* type (class Lycopodiopsida).

Class Horneophytopsida

The class Horneophytopsida (H.) was established by Meyen (1978e). The only genus known so far is *Horneophyton* (D; Fig. 14, *q–t*). Its axes were similar in appearance to those in rhynias, but extended from a tuberose body bearing slender rhizoids and lacking conductive tissue. The organization of the axial vascular bundle is about the same as in *Rhynia major* and *Lyonophyton*. Terminal sporangia are tubular with a column of sterile tissue inside. The sporangia are gathered in bundles, having a common spore cavity and a bunch-like branching central column. The sporangia dehisced with an apical pore. The spores are nearly the same as in rhynialeans. In the presence of an axial column H. sporangia resemble sporogonia of the Anthocerotopsida and the sporangia of other bryophytes. H. probably does not give evidence to the origin of bryophytes from propteridophytes, but merely serves to indicate the manner in which the sporangia of higher plants could have evolved from such an axis apex, where the spore mass was initially concentrated only in the cortex. The putative gametophytes of H. are described under the name *Lyonophyton* (see above).

Satellite genera of the division Propteridophyta

Quite a few genera similar to propteridophytes, but poorly studied, are known from the Silurian and Devonian. *Euthursophyton* (D_1) has dichotomously branching axes, densely covered by emergences, and circinately coiled apically. Judging by the exarch protoxylem, this genus may belong to zosterophylls, but its sporangia are unknown. *Taeniocrada* (D) comprises strap-like branching shoots. The sporangia are elliptical (as in the Rhyniales and Timerophytales), gathered in terminal or lateral clusters, whereas in *T. langii* placed by Fairon-Demaret (1985) into a separate genus *Stockmansia* they are seemingly attached at the side of the axes. The apices of young shoots are circinately coiled. Numerous appendages (rhizoids?) are borne on the lower part of the shoot. The affinity of this genus to propteridophytes has not been proven; it might belong to spongiophytes (see Brown Algae).

DIVISION PTERIDOPHYTA

Under the name Pteridophyta (P.) various plant groups were united. Within the P. are grouped the classes Barinophytopsida, Lycopodiopsida, Equisetopsida, Polypodiopsida and Progymnospermopsida. The living psilotes are also included in P. They are usually attributed a high rank of an independent class, or even division, but investigations carried out by Bierhorst (1971) suggest possible links of the psilotes with ferns.

The division of P. into classes does not give rise to any serious problems. Barinophytes have leafless axes and strobiloid aggregations of two-ranked sporangia. Lycopsids (Lycopodiopsida) differ in having microphylly. Their leaves evolved, not as a result of fusion of the axes (as in ferns and, probably, in articulates), but out of enations that were subsequently vascularized. The sporangia are adaxial. Articulates can be distinctly identified by the articulated organization of the stem and its branches, whorled leaves and sporangiophores. Other P. belong to ferns and progymnosperms. The distinctions between these two groups are quite clear (progymnosperms are characterized by secondary wood of the same type as in gymnosperms), but their independence is evident also from their phylogenetic relations; these groups evolved independently from the trimerophytes. The successful application of the phylogenetic principle to the P., as a whole, and to their subdivisions into classes is so far impossible. Plants occur that occupy an intermediate position between lycopsids and articulates (*Eleutherophyllum*, *Estinnophyton*), also between articulates and ferns (*Ibyka*, *Iridopteris*). Bowmanitaleans (sphenophylls) have characters in which they are closer to lycopsids. The Devonian *Enigmophyton* combines characters that are typical of ferns as well as of lycopsids.

Class Barinophytopsida

Barinophytes (B) have been recorded from the Devonian and Lower Carboniferous. Most frequently B. are treated as an independent order of uncertain affinity. B. are characterized by strobiloid sporangial aggregates, inserted on the tips of dichotomous axes (*Protobarinophyton*, *Krithodeophyton*; Fig. 15, *a–d*), or on both sides of the axis (*Barinophyton*; Fig. 15, *e–j*). Sporangia are situated in two rows, transversely elongated and attached to a hooked sporangiophore (Fig. 15, *d*, *h*, *i*); in the genus *Krithodeophyton* they are separated by sterile appendages (Fig. 15, *c*). Similar appendages may be situated under the strobilus. B. includes both homosporous (*Krithodeophyton*, *Protobarinophyton*) and heterosporous (*Barinophyton*; Fig. 15, *f*, *g*, *j*) plants. The mega- and microspores are produced by the same sporangium (Fig. 15, *j*). In the same way as in most heterosporous pteridophytes, the mega- and microspores differ in their sporoderm structure (Cichan *et al.*, 1984). The megaspore sporoderm consists of an unornamented surface layer. The sporoderm of microspores bears irregular gaps in the wall near the outer surface. In the main axis an exarch protostele was found. In the type of the axis branching, form of sporangia, general habit of the strobili and exarch somewhat flattened protostele, B. resemble zosterophylls, but differ from them in a number of advanced characters – thicker axis, pinnate arrangement of the strobili, long sporangiophores, sterile appendages between the sporangia and heterospory. Evidently in other higher plant groups (progymnosperms, articulates) the heterospory also evolved through the differentiation of spores within one sporangium.

Considering that B. probably originated from zosterophylls independent of lycopsids, also the specific characters of B. (they differ markedly from lycopsids in the absence of leaves, quite a different trend in the evolution of the strobili and undissected stele), the B. may be treated as an independent taxon of the same rank as lycopsids, i.e. as a class. In the exarch protostele and structural similarity in the megaspore exine of B. and the Carboniferous lycopsids, B. show certain relationships to lycopsids (Taylor and Brauer, 1983).

Class Lycopodiopsida (Lycopsida)

In some earlier suggested and new systems lycopsids (L.) are divided into two groups. One involves heterosporous forms with a ligule, the other homosporous forms without a ligule. More recently there has been a tendency not to distinguish these groups, but to divide L. directly into orders. The presence of a ligule and the heterospory still remain important taxonomic characters, but now homosporous L. without a ligule are known (order Protolepidodendrales). The presence of a ligule in fossil L. cannot be demonstrated in those cases when it was only present in young shoots, and having withered very early, underwent abscission in the course of ontogeny of the shoot (in the same way as in the living *Selaginella*).

Zosterophylls are regarded as ancestors of L. Interposed between both groups is the order Drepanophycales, where the sporangia are attached either to the stem, or to the axil of an unspecialized leaf. The order Lycopodiales is closely related to these D., but differs in the appearance of strobili and specialized sporophylls with adaxial sporangia. The incoming of ligulate L. is recorded in the Middle Devonian. At first they were homosporous (order Protolepidoden-

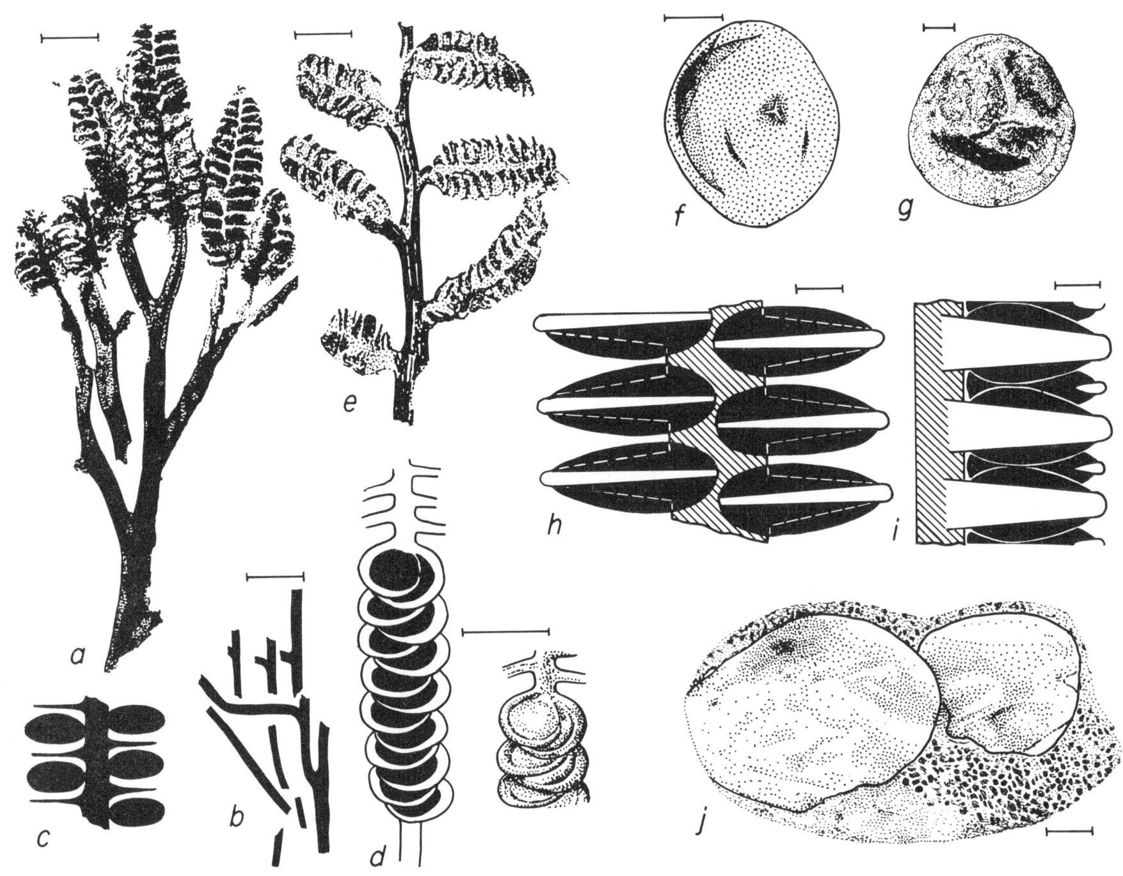

Fig. 15. Barinophytes; Lower Devonian (*c*), upper Lower–lower Middle Devonian (*a*, *b*, *d*), Upper Devonian (*e–j*); W. Siberia (*a*, *b*, *d*), Western Europe (*c*), USA (*e–j*); *a*, *b*, *Protobarinophyton obrutschevii* Anan., fertile shoot (*a*), branching of axes (*b*); *c*, *Krithodeophyton*, scheme of strobilus structure; *d*, *Barinophyton robustius* (Zal.) Brauer, scheme of strobilus structure (left; sporangia in black) and reconstruction of its portion, upper sporangiophores devoid of sporangia; *e*, *B. obscurum* (Dun) White; *f*, *g*, *B. richardsonii* (Daws.) White, megaspore (*f*) and microspore (*g*); *h–j*, *B. citrulliforme* Arnold, scheme of strobilus structure (*h*, *i*; sporangia in black, axis shaded, sporangiophores left white), view from sporangial side (*h*) and lateral (*i*), two megaspores among mass of miospores in sporangium (*j*). Scale bar = 4 cm (*b*), 1 cm (*a*, *e*), 5 mm (*d*), 2 mm (*h*, *i*), 200 μm (*j*), 50 μm (*f*), 10 μm (*g*). Modified from: Ananiev, 1957 (*a*, *b*, *d*); Høeg, 1967, after Dun (*e*); Pettitt, 1965 (*f*, *g*); Brauer, 1980 (*h–j*).

drales). Their subdivided leaves and sporophylls look alike, the sporangia are adaxial and shifted distally from the leaf base. Other ligulate L. are heterosporous. Among them can be distinguished herbaceous plants of indeterminate growth and intensive dichotomous or anisotomous branching of the stem (Selaginellales). The remainder of the ligulate L. include several families which at present can be only tentatively arranged into orders. Usually arborescent L. (Lepidodendraceae, Sigillariaceae and related plants) are contrasted with small-sized L. (Pleuromeiaceae and Isoetaceae). However, analysis of the characters of adequately studied genera makes it immediately apparent that these two groups of families do not show distinctions according to which they may be placed into different orders. Moreover, the families themselves cannot be characterized by any definite set of characters. The Lepidocarpaceae, Sigillariaceae and Pleuromeiaceae are characterized by the formation of strobili, whereas Chaloneriaceae and Isoetaceae lack strobili. However, if this character is accepted as the basis for the distinction of orders, then the Pleuromeiaceae should be separated from the Chaloneriaceae and Isoetaceae, notwithstanding the affinities in other characters, particularly in the rhizophore organization. Recently obtained data indicate that heterosporous L. were characterized by different types of embryogenesis (Pigg and Rothwell, 1983b; Stubblefield and Rothwell, 1981). According to this feature the Chaloneriaceae can be allied to the Isoetaceae and *Bothrodendron-Bothrodendrostrobus*, not to the Lepidocarpaceae. Such a relation of these groups is supported by the organization of the rhizophores. They are cormose in the Chaloneriaceae and Isoetaceae, and stigmarioid in the Lepidocarpaceae. However, in the Pleuromeiaceae strobili crown the unbranched stems, whereas in the Lepidocarpaceae they are laterally attached and inserted on the tips of the branches. Nevertheless, among the Lower Carboniferous L. certain forms probably existed (*Lepidodendropsis*, etc.), that had strobili of the *Lepidostrobus-Flemingites* type, crowning unbranched stems. Iurina (1985) recorded Middle Devonian L. producing compact heterosporous strobili. There is reason to believe that in all these older L. the rhizophores were closer in organization to those of Chaloneriaceae, Pleuromeiaceae and Isoetaceae, rather than to those of the Lepidocarpaceae. Because of this all five families are united below into one order Isoetales. This decision is in good agreement with the proposed phylogeny of lycopsids (Fig. 16). Ligulate and heterosporous genera, which cannot be distributed among these families, are treated as satellite genera to this order.

Thomas and Brack-Hanes (1984) adopted the procedure earlier proposed by the present author (Meyen, 1976, 1978a), according to which the distinction of suprageneric taxa of L. is based on the fructifications, and many genera are treated as satellite taxa. These authors recognize a greater number of families and orders, as compared with those proposed below, but detailed explanations to the suggested system are lacking. Because of this its discussion is deferred.

It should be noted that in badly preserved materials it is not always possible to distinguish L. shoots from those of conifers, mosses and certain other groups. Thereby, it is not surprising that the moss *Polyssaievia* was initially described as *Walchia*, and later as *Lycopodites*. Leafy shoots of L. were occasionally described as *Walchia*, sporophylls were described as seed scales of conifers; cordaitalean bark with leaf cushions (*Lophoderma*) as L. bark, etc. The numerous features of external similarity between L. and conifers still remain a puzzle.

For the distinction of L. from other plant groups, and also for their classification, the epidermal characters may be used (Thomas and Masarati, 1982).

Order Drepanophycales

The Devonian order Drepanophycales (D; Fig. 17, *a–h*) is still closely related to Zosterophyllopsida. The sporangia of *Drepanophycus* and other genera are very similar to those of *Sawdonia* and attached to the stem by a stalk (Schweitzer, 1983b). But in *Baragwanathia* they are inserted on a leaf axil or adaxially on the leaf itself. D. differ from zosterophylls in the stele dissection and appearance of leaf traces. The stele is stellate in cross-section (Rayner, 1984). Near the ends of the stellate xylem are bundles of the protoxylem from which lateral bundles extend. In *Asteroxylon* the stems are covered by long emergences, similar to those of zosterophylls. The lateral bundle extends to the base of the emergence but does not enter it. In *Drepanophycus* and *Baragwanathia* the emergences evolve into spirally arranged, rounded in cross-section or flattened leaves with a vascular bundle in them. Stomata are borne on the leaves and stems. Stubblefield and Banks (1978) described stomata of *D. spinaeformis* Goepp. as paracytic. However, these stomata are more probably actinocytic, whereas the suture, marking the junction of the guard-cells with the underlying mesophyll cells, was mistaken for the boundary

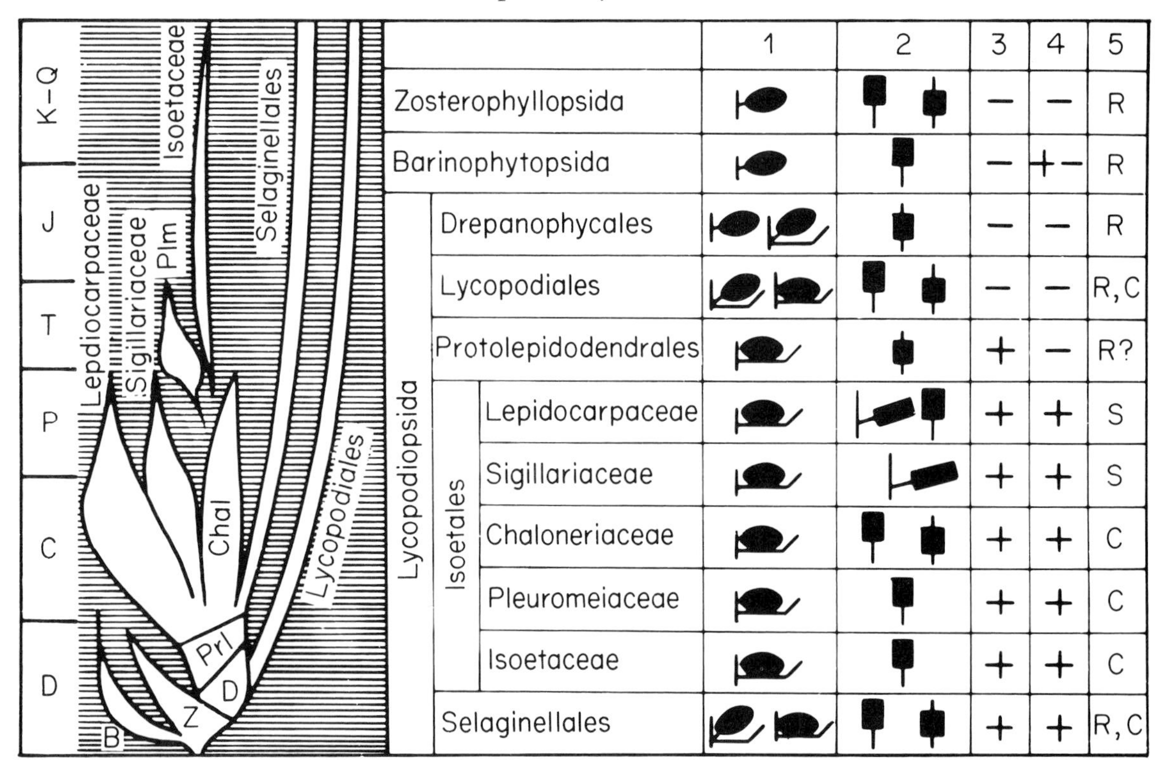

Fig. 16. Phylogeny (left) and distribution of main characters between various Lycopodiopsida and allied taxa. B, Barinophytopsida; Z, Zosterophyllopsida; D, Drepanophycales; Prl, Protolepidodendrales; Chal, Chaloneriaceae; Plm, Pleuromeiaceae; 1, axial, axillary and epiphyllous position of sporangia; 2, arrangement of sporangia into strobili (terminal and lateral) and fertile zones; 3, presence of ligule; 4, presence of heterospory; 5, nature of underground or surficial root-like organs (R, creeping rhizomatous axes similar to aerial shoots; S, *Stigmaria*-like rhizophores; C, cormose rhizophore).

between the subsidiary and guard cells. Gensel and Andrews (1984) unite *Asteroxylon*, *Drepanophycus* and *Kaulangiophyton* into an informal group of 'Prolycopods'.

Order Lycopodiales

This order includes the living genera *Lycopodium* and *Phylloglossum*. They are herbaceous plants with dichotomously and pseudomonopodially branching stems. The sporangia are kidney shaped, dehisce with a transverse slit, and are situated on a short stalk. The sporophylls are gathered in terminal strobili or fertile zones. The genus *Oxroadia* (C_1), that has protostelic stems and homosporous strobili, is tentatively related to the Lycopodiales. Its sporophylls differ from vegetative leaves in that they have a well-developed heel. Fossil shoots, externally similar to *Lycopodium*, are included in the poorly studied collective genus *Lycopodites*, recorded since the Devonian. The Lycopodiales probably include the genus *Carinostrobus* (C_2) comprising permineralized strobili (Baxter, 1971).

Order Protolepidodendrales

The organization of Protolepidodendrales (P.) can be best judged on the genus *Leclercqia* (D_2; Fig. 17, *j–p*), which is closely related to *Protolepidodendron* (Fig. 17, *i*; they are probably taxonomic synonyms). The stele in *Leclercqia* is rounded in cross-section with marginal indentations, from which protoxylem bundles come out. The leaves are dissected into five lobes. The sporophylls are of the same habit, bearing elliptical sporangia, dehisced with a longitudinal slit (in contrast to transversal in the zosterophylls and drepanophycaleans). A small surficial ligule is borne on the leaf distally from the sporangium. The sporophylls are gathered in fertile zones. Spores are trilete with a flange (of the *Aneurospora* type; Streel, 1972). Their heterospory has not been established. The rooting system is not known. It is possible that in P. the axis might have been in part prostrate with adventitious roots, probably corresponding to axillary tubercles and appendages in the zosterophylls (*Crenaticaulis*, *Gosslingia*). These axillary organs could have evolved both into adventitious roots and an axillary ligule.

Closely similar to P. is the genus *Estinnophyton* (its type species was at first placed into *Protolepidodendron*), with twice-forked sporophylls (Fig. 17, *q*). The sporangia were attached by short stalks to the points of the sporophyll dichotomy. Sporophylls of *Estinnophyton* are similar to sporangiophores of *Calamophyton* and *Hyenia* (Cladoxylales), also to *Eviostachya* and *Bowmanites* (Equisetopsida). It was believed that the Bowmanitidae ('sphenophylls') evolved from primitive lycopsids.

Order Isoetales

Below in the Isoetales (I.) are included lycopsids that commonly were referred to different orders (Lepidodendrales, Pleuromeiales, Isoetales, etc.). They probably evolved from the Protolepidodendrales through the acquisition in the vegetative sphere of an upright robust stem, secondary tissues and rhizophores, and, in the reproductive structure, of specialized sporophylls (differing in the form of the laminae from vegetative leaves), strobili and heterosporous reproduction. In some I. the spores were dispersed from dehisced sporangia, the same as in other pteridophytes. However, in different families (Lepidocarpaceae, Sigillariaceae, probably also Pleuromeiaceae) entire sporangia or sporophylls with attached sporangia might have served as units of dispersal (Phillips, 1979; Pigg, 1983). The appearance of secondary tissues furnished the basis for the advent of arborescent forms. However, all (or the majority) of I. were of determinate growth (see Lepidocarpaceae). The existence of a ligule in the vegetative shoots can be recognized by the presence of a ligular pit in the leaf axis or slightly above on the leaf cushion. The ligular pit is preserved as a cast or cutinized sac, with a bottom opening, corresponding to the point of the ligule attachment. When the ligular pit is absent, or the ligule itself is withered (as on vegetative shoots of living *Selaginella*), then such a lepidophyte may be easily mistaken for an eligulate.

A characteristic feature of I. are rhizophores. Their morphological interpretation is debatable; being in the position of roots, the rhizophores differ from them in the endarch xylem and absence of root hairs. Rhizophore appendages, or appendices, also differ from roots in the absence of a cap, in their anatomical structure, and in that they abscissed during the growth of the rhizophores by means of an abscission layer, leaving behind distinct scars. Appendices are of orderly spiral arrangement (the roots are commonly disorderly spaced). It was believed that the appendices are homologous to the leaves. However, the arrangement of the appendices does not follow the Fibonacci series (Charlton and Watson, 1982). The similarity between the rhizophores and stems has embryogenetic grounds. In embryogenesis of lepidocarps (*Lepidocarpon*) the embryo apex is dichotomizing. One of the branches shows structural similarity to *Stigmaria*, whereas the other shows similarity to the stem. Thus, stigmariae did not develop from the embryonal root; ontogenetically they are homologous to the shoot. It is true that in the Chaloneriaceae, Isoetaceae and *Bothrodendrostrobus* the rhizophore developed from the meristeme, appearing between primary roots.

Rothwell and Erwin (1984, 1985) regard all rhizomorphic rooting organs of lycopsids as shoot systems modified for rooting, because of (1) external structural similarities at the apices of *Stigmaria*, *Paurodendron*, *Pleuromeia*, *Cylomeia*, *Protostigmaria* and *Isoetes*; (2) anatomical and

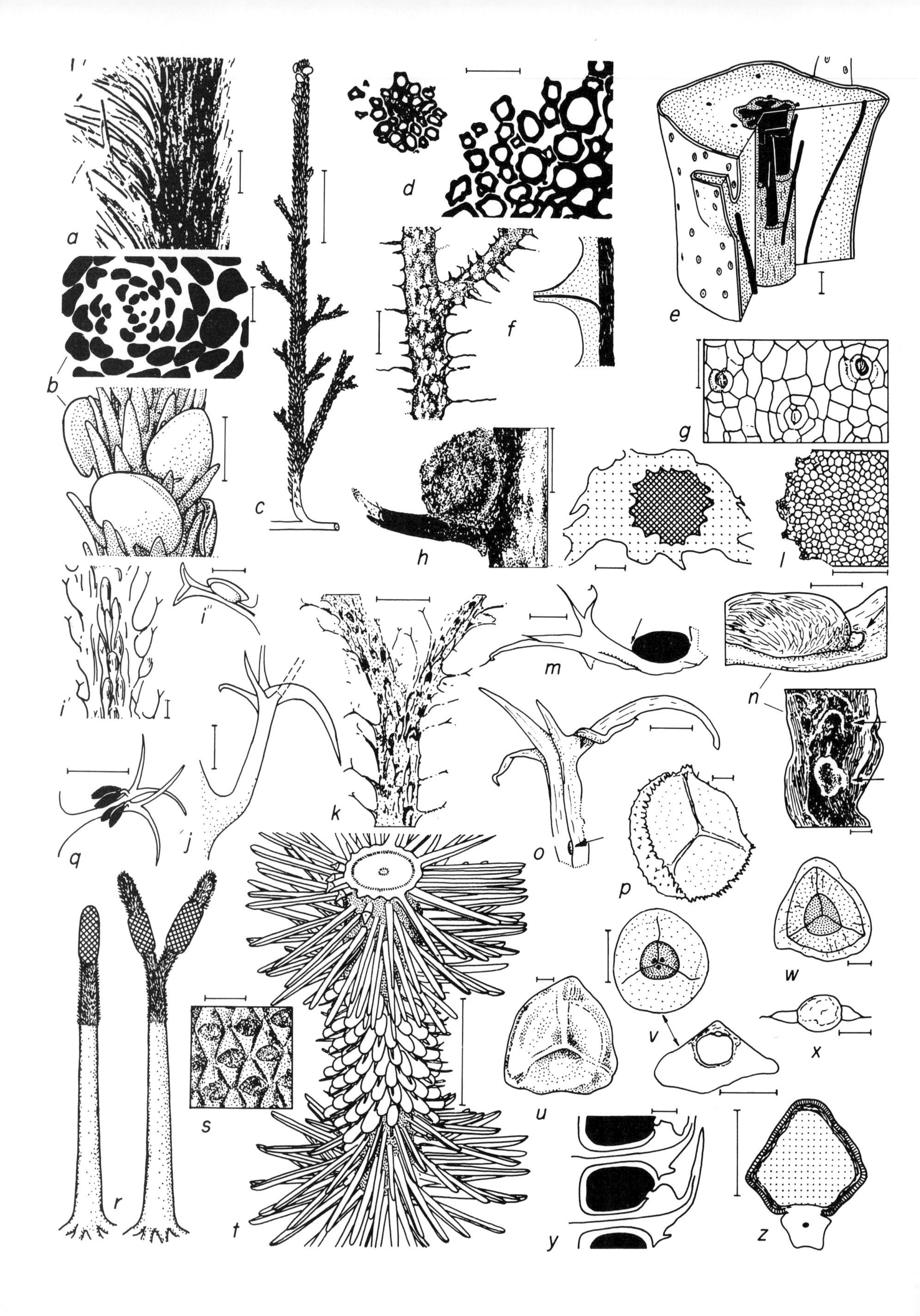
a
b
c
d
e
f
g
h
i
i'
j
k
l
m
n
o
p
q
r
s
t
u
v
w
x
y
z

developmental similarities among *Stigmaria*, *Paurodendron*, *Chaloneria* and *Isoetes*; (3) close correspondence of the vascular tissue in *Lepidocarpon* embryos and in the transition zone of *Paurodendron*. *Stigmaria* is a more primitive type of the rhizomorph. Shorter unbranched (*Paurodendron*) or lobed- and furrowed (*Protostigmaria*, *Isoetes*) forms are treated as derived.

The distinction of families and genera of I. was frequently based on features of low weight, such as the degree of development of the leaf cushions, their size, form and arrangement, structure of the intercushion intervals, etc. The families were delineated according to isolated characters, without taking heed of the syndrome of other characters.

Family Lepidocarpaceae

The plants referred here to the Lepidocarpaceae are commonly included in the order Lepidodenrales (see review: Thomas, 1978). The family name is derived from *Lepidocarpon* – a genus, introduced for strobili. The name *Lepidodendron* does not suit this purpose, because it was proposed for bark imprints. There is no certainty that all the *Lepidodendron* species do belong to the Lepidocarpaceae, considering that the reproductive organs and the anatomical structure of the axes are known only in a small part of the species. The relation between *Lepidocarpon* and allied genera can be described as follows. According to the new strobili nomenclature (Brack-Hanes and Thomas, 1983), the name *Lepidostrobus* (C; Fig. 18, *r*) applies only to homosporous strobili and strobili where the structure of the spores is not known, while those heterosporous strobili that formerly were termed *Lepidostrobus*, are defined by the name *Flemingites* (C; Fig. 18, *o*, *p*, *r*, *s*). Both genera are closely similar morphologically. The strobilus axis is 5–14 cm in length. Sporophylls, borne on the axes, are either spirally arranged or whorled. On their proximal longitudinally convex part sporangia were

Fig. 17. Lycopsids; Lower (*a–f*, *h*, *q*), Middle (*i–p*) and Upper (*g*) Devonian, Middle (*t–z*) and Upper (*r*, *s*) Carboniferous; Australia (*a*), Western Europe (*b–e*, *h*, *j*, *q–s*, *u*), W. Siberia (*f*), USA (*g*, *k–p*, *t*, *v–x*, *z*); *a*, *Baragwanathia longifolia* Lang et Cooks.; *b–e*, *Asteroxylon mackiei* Kidst. et Lang (*b*, arrangement of emergences on cross section of apex, and reconstruction of fertile shoot; *c*, general view; *d*, leaf trace departing from axial vascular cylinder; *e*, block diagram of anatomical structure, xylem in black, phloem densely stippled); *f–h*, *Drepanophycus spinaeformis* Goepp. (*f*, general view and leaf base with vascular bundle departing from stele; *g*, epidermis with stomata; *h*, leaf and sporangium); *i*, *Protolepidodendron scharyanum* Krejči, former representation of leaf structures on leafy shoots (*i′*) and of sporophylls (*i″*); *j*, *Leclercqia*, leaf structure in specimen identified by Kräusel and Weyland as *P. scharyanum*; *k–p*, *L. complexa* Banks et al. (*k*, leafy shoot in fracture plane, leaves look forked; *l*, anatomical structure of axis with leaf bases, and detail of xylem cylinder with protruding protoxylem poles; *m*, sporophyll with sporangium, ligule at arrow; *n*, sporangial structure and ligule at arrow (above), and portion of sporophyll showing scar of detached sporangium (lower arrow) and ligule (upper arrow); *o*, sterile leaf with ligule at arrow; *p*, spore); *q*, *Estinnophyton wahnbachense* (Kr. et Weyl.) Fair.-Dem., reconstruction of sporophyll; *r*, *Sporangiostrobus puertollanensis* W. et R. Remy, reconstruction; *s*, *Bodeodendron hispanicum* Wagn. et Spin., bark with leaf cushions; *t*, *u*, *Chaloneria periodica* Pigg et Rothw. (al. *Polysporia mirabilis* Newb.), reconstruction of fertile zone on leafy shoot (*t*), and megaspore (*u*); *v*, microspores *Endosporites vesiculatus* Kos. from sporangia of *Chaloneria*, polar view and section in polar plane; *w*, spore from sporangium of *Spencerites moorei* (Cridl.) Leism.; *x*, spore *Spencerisporites gracilis* Kos. from sporangium of *Spencerites*, section and polar view; *y*, *Spencerites insignis* (Will.) Scott, scheme of strobilus in longitudinal section, sporangia in black; *z*, *S. moorei*, cross-section of sporophyll in its middle part. Scale bar = 5 cm (*c*, *t*), 1 cm (*a*, *f*, *k*, *s*), 5 mm (*b*, *h*, *q*, *z*), 1 mm (*b*, *e*, *i*, *j*, *m*, *y*), 0.5 mm (*l*, *n*, *o*), 100 μm (*d*, *g*, *n*, *u*, *w*, *x*), 50 μm (*v*), 10 μm (*p*). Modified from: Taylor, 1981 (*a*); Chaloner and McDonald, 1980 (*b–e*); Osnovy paleontologii, 1963a, after Ananiev (*f*, left); Stubblefield and Banks, 1978 (*g*); Gothan and Weyland, 1964 (*h*); Osnovy paleontologii, 1963a (*i*); Fairon-Demaret, 1980 (*j*); Banks *et al.*, 1972 (*k–m*, *p*); Grierson and Bonamo, 1979 (*n*, *o*); Fairon-Demaret, 1979 (*q*); W. Remy and R. Remy, 1975 (*r*); Wagner and Spinner, 1976 (*s*); DiMichele *et al.*, 1979 (*t*); Dràbek, 1977 (*u*); Brack and Taylor, 1972 (*v*); Leisman and Stidd, 1967 (*w*, *z*); Leisman, 1962 (*x*).

3
2
5
1
2
3
4
b
1
a
f
c
d
e
q
g
h
i
j
k
1
2
3
4
1
2
n
o
p
r
l
m
t
u
v
s
w
x
y
z

borne, whereas their distal part was in the form of a triangular lamina of elbow-shaped bending upwards. At the bend the sporophyll formed a heel directed downwards. The sporangia were tightly closed by distal parts of adjacent sporophylls. In bisexual strobili the megasporangia were situated in the lower part of the strobili. Microspores were trilete, rounded-triangular, of the *Lycospora* type (Fig. 18, *m*, *n*). Megaspores were also trilete, up to 3 mm in diameter (Fig. 18, *u*; see also Fig. 78, *s*). It is believed that megaspores with long appendages floated in water, and the appendages held air bubbles. Spermatozoids were freed of micro-gametophytes in the water in abundant quantities. They penetrated distally into the floating megaspore and reached the archegonia. The reproductive biology of these forms is similar to that of aquatic ferns (Stubblefield and Rothwell, 1981).

The genus *Lepidocarpan* had nearly the same anatomical structure of the axis and sporophylls. The strobili were always unisexual. The single megaspore in the sporangium (of the *Cystosporites* type; see Fig. 78; such megaspores have been recovered also from the Upper Devonian) germinated directly in it. In several specimens of megasporangia embryos were found (Phillips, 1979). The sporangia are enclosed in an integument-like covering, formed by the lateral extensions of the sporophyll. A series of co-existing genera have been identified, where at first the number of megaspores in the megasporangia was reduced to one, but the integument-like covering is not yet formed (the genera *Caudatocorpus* and *Achlamydocarpon*). In *Caudatocorpus* (Fig. 18, *u*) aborted megaspores of the tetrad remain quite large. The tetrad is embraced by a granular mass, that forms a caudal wing-like structure of the functioning megaspore. Both the functional and aborted megaspores of *Achlamydocarpon* are

Fig. 18. Lepidocarpaceae (*a–v*), Chaloneriaceae (*w*?) and Sigillariaceae (*x–z*); Carboniferous, including Lower (*t*, *w*), Middle (*c–e*, *m*, *o–q*, *u*, *v*, *y*) and Upper (*x*); USA (*c–e*, *n*, *o*, *u*, *w*, *x*), Western Europe (*d*, *m*, *p*, *q*, *t*, *v*, *y*); *a*, *Lepidodendron*, change of anatomical structure from protostelic in stem base (1) to siphonostelic (medullated protostele) in major stem portion and in larger branches (2), and again protostelic in slender branches (3); *b*, arrangement of tissues in *Lepidodendron* stem (1, phloem; 2, secondary wood; 3, primary wood; 4, pith; 5, cortex with leaf cushions); *c*, *L. scleroticum* Pann.; *d*, *L. vasculare* Binn.; *e*, *L. dicentricum* C. Felix; *f–j*, *Stigmaria ficoides* Brongn. (*f*, general view; *g*, rhizophore cross section in place of appendix departure, wood in black; *h*, cross-section of appendix; *i*, vascular bundle of appendix; *j*, rhizophore with attached appendices); *k*, cross-section of *Lepidodendron* leaf (1, transfusion tissue; 2, phloem; 3, xylem; 4, stomatiferous furrow); *l*, stomatal band in *Lepidodendron* leaf; *m*, microspore *Lycospora* from sporangium of *Lepidostrobus binneyanus* Arber; *n*, microspore of same type, detail, white band in the middle – equatorial flange between proximal (1) and distal (2) surfaces; *o*, sporophyll of *Flemingites* allied to *F. schopfii* (Brack) Brack-Hanes et Thomas; *p*, *Flemingites brownii* (Schimp.) Brack-Hanes et Thomas; *q*, sporophyll *Lepidostrobophyllum maius* (Brongn.) Hirmer; *r*, different degrees of sporophyll expansion (shaded) up to formation of integument-like organ (right), from left to right (cross section view) – *Flemingites* (al. *Lepidostrobus*), *Lepidostrobopsis*, *Lepidocarpopsis* with one functioning and three aborted (shown by broken line) megaspores, *Lepidostrobopsis* with microspores, *Lepidocarpon*; *s*, arrangement of micro- and megaspores in *Flemingites* strobilus; *t*, trichomoid projection on megaspore from *Flemingites allantonensis* (Chal.) Brack-Hanes et Thomas; *u*, content of sporangium in *Caudatocorpus arnoldii* Brack-Hanes, megaspore *Lagenicula* with three aborted megaspores and wing-like projection (stippled) consisting of spongy tissue; *v*, integument-like megasporangium of *Lepidocarpon lomaxii* Scott (distal part of sporophyll abolished); *w*, *Protostigmaria eggertiana* Jenn.; *x*, *Mazocarpon oedipternum* J. M. Schopf, longitudinal section of strobilus (left), cross-section of sporophyll (bottom right), sporangia shown in black, megaspores left in white, and reconstruction of megasporophyll (top right); *y*, megaspore from *Sigillariostrobus czarnockii* Boch.; *z*, reconstruction of *Sigillaria*. Scale bar = 1 m (*c–e*), 5 cm (*w*), 1 cm (*p*, *q*), 5 mm (*o*, *v*), 1 mm (*u*, *x*), 0.5 mm (*y*), 100 μm (*i*), 10 μm (*m*, *t*), 1 μm (*n*). Modified from: Eggert, 1961 (*a*); Snigirevskaya in Grushvitsky and Zhilin, 1978 (*b*); DiMichele, 1981 (*c–e*); Taylor, 1981 (*f*); Eggert, 1972 (*g*); Lemoigne, 1963 (*i*); Kryshtofovich, 1957 (*j*); Hirmer, 1927 (*k*, *l*, *p*, *q*); Thomas, 1970 (*m*); Taylor and Millay, 1969 (*n*); Phillips, 1979 (*o*, *v*, *x*(top right)); Thomas, 1978 (*r*); Andrews, 1961 (*s*, *x*); R. Potonié, 1973 (*t*); Brack-Hanes, 1981 (*u*); Jennings, 1975 (*w*); Bocheński, 1936 (*y*); Němejc, 1963 (*z*).

covered proximally by a spongy sporopollenin mass, resembling underdeveloped massulae of salviniaceous ferns. The megaspore is surrounded by tapetal membranes with orbicules (Ubisch bodies), that are typical of gymnosperms.

It was for long considered that stems of *Lepidodendron* associated with *Lepidostrobus*, but then the association of *Lepidocarpon* with *Lepidophloios* stems was established. Rhizophores of both are referred to the genus *Stigmaria*. A more complex relationship gradually emerged between the genera established on different organs. *Lepidophloios* associated with the megastrobili *Lepidocarpon* and microstrobili *Lepidostrobus oldhamius*. Some *Lepidodendron* associated with the megastrobili *Achlamydocarpon* and microspores *Lycospora* and *Cappasporites* (Di Michele, 1981). Some *Flemingites* species associated with vegetative siphonostelic axes of *Paralycopidites* (C_1–C_2), that had persistent leaves (DiMichele, 1980). These axes correspond, seemingly, to the genus *Ulodendron* (Fig. 19, Ul) introduced for the impression–compression materials.

The genus *Lepidodendron* (C–P; Fig. 18, *a–e*; Fig. 19, Ld) may be treated merely as a satellite genus of the family, since the strobili of most of the species are unknown. Various species were associated with *Lepidostrobus*, *Flemingites*, *Achlamydocarpon*. Their trunks, reaching 40 m in height and 1 m in cross-section, were dichotomously or pseudomonopodially branched. The ends of the branches were leafy. Below the leafy portions the branches were covered by leaf cushions, i.e. the basal widened parts of the leaves. The leaves were lanceolate or linear, up to one metre in length, with a single vein. Stomatal bands, running along both sides of the vein (Fig. 18, *k*, *l*) were frequently immersed in furrows. The leaves left on the cushion a leaf scar with a central minute scar of the vascular bundle, on both sides of which were minute scars of parichnos (strands of the aerenchyme). Above the leaf scar was a ligular pit. Most of the trunk was composed of a multilayered cortex, penetrated by vascular bundles, extending into the leaves, and accompanied by strands of the aerenchyme. A secondary phloem was absent, i.e. the cambium was unifacial. The stele was slender with a certain amount of secondary wood, or without it. The thinnest branches were protostelic, the thicker ones siphonostelic (due to the absence of the phloem inwards from the wood, such a stele is termed a medullated protostele). Judging by the arrangement of the tissue in the axes of different orders including that of the secondary tissue, the trees were of determinate growth (which does not happen in modern trees). The branches became gradually longer and divided, following the division of the apical meristeme, until only several tracheids remained of the stele; the periderm disappeared. Then the growth ceased. Certain species were probably lianas. In the process of ageing the trunks abscissed bark. Various types of preservation of the trunks, different in the degree of decortication, have been distinguished as independent genera. Among these the genus *Knorria* was introduced for the designation of the deep stages of decortication. The anatomical structure of *Lepidophloios* trunks (C_{1-2}; Fig. 19, Lp) is about the same, but the leaf scar is located in the lower part of the transverse leaf cushion.

Rhizophores are assigned to the genus *Stigmaria* (C–P; Fig. 18, *f–j*). The transition from trunk to rhizophores was gradual. The rhizophores were repeatedly forked. Appendages, inserted on the slender branches, reached 1 cm in diameter. The thin pith of the rhizophore was surrounded by a small amount of protoxylem, beyond which was secondary wood, at times without metaxylem, and dissected into radial wedges by medullary rays, accompanying the vascular bundles of the appendages. The vascular bundle of the appendage was suspended in a cavity on a narrow rib of the parenchymatous tissue. Detached appendages of such a structure have been recorded since the Upper Devonian (Meyen, 1982b, text – Fig. 2D). Rothwell (1984) observed the apical structure and growth in *Stigmaria*. The apex tapers, then truncates rapidly, producing a raised rim within which there is a conspicuous groove and a circular central area. Rootlet scars occur on the rim and are absent within the groove. Rothwell suggested that the apical groove may correspond to a radial furrow from which the rootlets emerge during growth, as they do in *Isoetes*. The irregular apical surface is

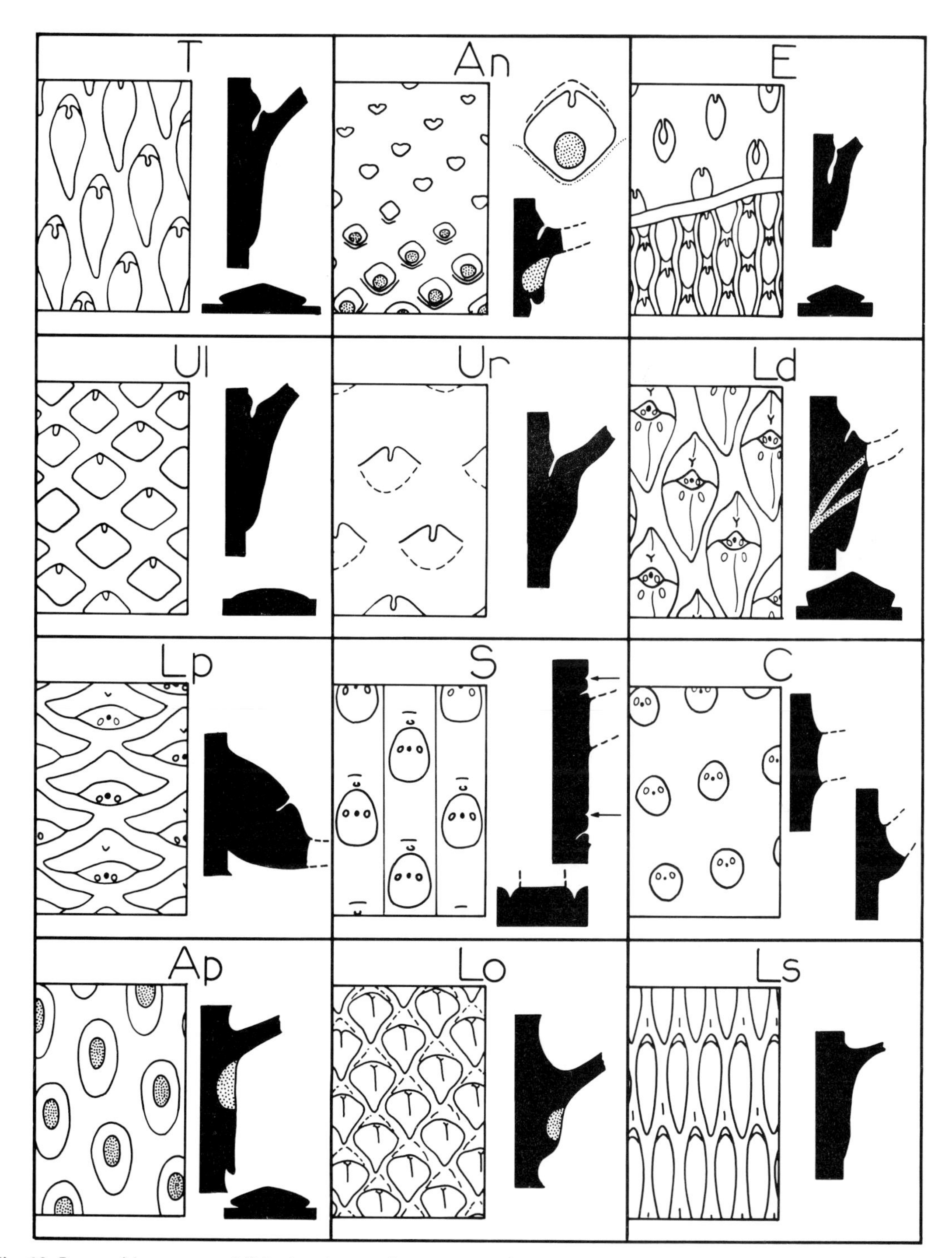

Fig. 19. Lycopsid genera established on impression–compression vegetative shoots; T, *Tomiodendron*; An, *Angarodendron*; E, *Eskdalia*; Ul, *Ulodendron*; Ur, *Ursodendron*; Ld, *Lepidodendron*; Lp, *Lepidophloios*; S, *Sigillaria*; C, *Cyclostigma*; Ap, *Angarophloios*; Lo, *Lophiodendron*; Ls, *Lepidodendropsis*. Modified from Thomas and Meyen, 1984.

interpreted as a plug of parenchymatous tissue that functioned as a root cap of tree roots.

Family Sigillariaceae

The independence of the family Sigillariaceae (S; Fig. 18, *x–z*) is doubtful. S. differ from lepidocarps in their unbranched trunks, and in that the leaf scars do not occur on cushions, but directly on the bark or its longitudinal ridges. However, unbranched forms probably also existed among lepidocarps. As to the ribs on the bark – they might be fused cushions, that belonged to one orthostichy. The S., in the same way as for lepidocarps, are characterized by leaves of two types – with stomatal furrows, and without them. In S. leaves have been reported with two xylem strands in the vascular bundle and with secondary wood. The leaf trace might have been double in the cortex as well. All the well-studied strobili of S. are unisexual. The genus *Mazocarpon* (Fig. 18, *x*; Pigg, 1983) was established for strobili similar to *Flemingites*. Megasporangia were permeated by parenchymatous tissue that left cavities, in each of which was located a single megaspore. S. microspores are assigned to the genus *Crassispora*. The megaspores are smooth walled (= *Laevigatisporites*) or spiny and assignable to *Tuberculatisporites*. Strobili were arranged in horizontal rows and attached by long stalks to the trunks not far from the apex below the leafy portion. The strobili were easily abscissed and disintegrated into the axis and sporophylls. A dispersed unit consisted of a single megaspore associated with a portion of sporangial wall and sterile tissue. Fructifications preserved as impressions–compressions are referred to *Sigillariostrobus*. S. rhizophores are assigned to the genera *Stigmaria* and *Stigmariopsis*. They have fewer medullary rays in the wood and a thinner pith than lepidocarps. The genus *Stigmariopsis*, associated with certain *Subsigillaria*, differs in the branching of its rhizophores. Four main branches at the base of the trunk are further pinnately, not dichotomously, divided.

Genus *Sigillaria* (C–P; Fig. 18, *z*; Fig. 19, S)

Large S. trunks were unbranched or dichotomously divided once or twice near the apex. The trunk or its branches terminated in a bunch of leaves that left leaf scars. The leaf scars are distinct, large with minute scars (of the vascular bundle and two parichnos), and a ligule at the upper edge or slightly above. According to the relief of the cortex, the form, structure and distribution of the scars, S. have been divided into numerous species, that are united into groups (subgenera) and subgroups. The group *Eusigillaria* includes the subgroups *Rhytidolepis* and *Favularia*. A characteristic feature of the first subgroup is the regular longitudinal ribbing of the bark. The ribs are in the shape of flattened cylinders on which leaf scars are borne, that are either narrower than the ribs, or equal in width. A tendency towards the coalescence of the cushions within distinct orthostichies is typical of many lycopsids, but only in *Rhytidolepis* does it culminate in the formation of continuous ribs. The borders between the cushions are marked on the rib by a small transverse fold. As the scars become more closely spaced within the ribs, constrictions of the ribs appear, and then increase. The nature of the ribs as fused leaf cushions then becomes apparent. Forms with closely spaced scars connect the subgroups *Rhytidolepis* and *Favularia*. The latter is characterized by tightly spaced hexagonal scars divided by narrow strips of bark. The trunks of *Rhytidolepis*, from which the outer bark layer abscissed, show swellings of parichnos with an interposed minute scar of the vascular bundle (genus *Syringodendron*). In the group *Subsigillaria* orthostichies are less distinct. Leaf cushions, if they are closely spaced, are lenticular with rounded upper and lower contours and acute side angles, or rhomboid in shape. The leaf scars are of transversal elongation, covering most of the cushion. On the older parts of the trunk the leaf scars spread apart revealing wide areas of smooth bark between them.

Family Chaloneriaceae

The family Chaloneriaceae (Ch.; Pigg and Rothwell, 1983a, 1983b) comprises lycopsids (Fig. 17, *r–z*), earlier assigned to the genera *Polysporia*, *Sporangiostrobus*, *Spencerites* (fertile shoots),

Bodeodendron (bark impressions), *Puertollania* (both fertile and bark remains), *Endosporites* (microspores), *Valvisisporites* (megaspores), etc. Impressions of vegetative shoots of Ch. and lepidocarps may be closely similar. The Ch. stems are apically and basally protostelic, and with a pith in the middle part. The amount of wood increases in the lower part of the stem. At the base the stems grade into a cormose, rounded or subdivided rhizophore, probably covered by appendages. The stems are unbranched or forked at the apex and covered by closely spaced cushions. Parichnos are present. The lamina, both of the leaf and sporophyll, are nearly the same, with a keel and two stomatal bands on the underside (the same as in lepidocarps). The sporangia are clustered in bisexual fertile zones, that may crown the stem in the form of a distal strobilus. Sporangia are attached to the sporophyll by their entire base (*Chaloneria*), or on a distally shifted stalk (*Spencerites*). In *Chaloneria cormosa* the sporangial cavity has longitudinal septa (trabeculae). The microspores have a wide cavum that envelops the entire distal side of the grain. The proximal slit is mono- or trilete. Numerous Lower Carboniferous ligulate heterosporous lycopsids, that have unbranched or sparsely branched stems, not accompanied in the assemblages by stigmariae, seemingly, are sooner affiliated with Ch., rather than with lepidocarps. Some of these forms could have had rhizophores of the *Protostigmaria* type (Fig. 18, *w*). Ch. probably comprise some Upper Palaeozoic lepidophytes of Gondwanaland (*Cyclodendron*; Rayner, 1985) and Angaraland (*Angarodendron*; see Satellite genera of the order Isoetales). In *Cyclodendron* the fertile zones were subterminal, *Angarodendron* appears to have produced fertile zones as well.

Family Pleuromeiaceae

Pleuromeias (P.; Fig. 20, *m–y*), probably evolved from chalonerias, from which they inherited mostly unbranching stems with a distal fertile zone – apical strobilus. The stele is rounded in cross-section with radial projections, that evidently correspond to protoxylem poles (*Pleuromeia sternbergii*). The very existence of projections, similar in habit to those in Protolepidodendrales, but not so numerous, indicates the absence of secondary wood. Rhizophores of P. are subdivided into lobes, and had appendices that after abscission left distinct scars. The leaves are fleshy, narrow-triangular in shape, with a thick midvein. Grauvogel-Stamm (personal communication) observed specimens of *Pleuromeia* bearing persistent leaves in the stem lower part. This suggests that *Pleuromeia* might have been an evergreen plant. In *Pleuromeia rossica* sterile leaves were probably absent (Dobruskina, 1982). In contrast to Chaloneriaceae, the sporophyll lamina differs from that of the vegetative leaves; it is spear-shaped, oval or rounded, with an elongated or rounded sporangium, at times occupying most of the sporophyll lamina, where in this case the distal part is strongly reduced (*Pleuromeia*). Microspores are cavate, monolete (of the *Aratrisporites* type), or trilete (of the *Densoisporites* type), with a granular or spiny exine. Megaspores are smooth or sculptured (from indistinct reticulation to long spines).

Besides the genus *Pleuromeia*, to P. can also be referred the Triassic genera *Tomiostrobus*, *Cylostrobus*, *Skilliostrobus*, *Annalepis*, etc., that differ in the size of their stems (which may be strongly reduced), form of sporophylls, type of spores and other characters. The distinctions between these genera are in need of more precision. It is probable that they may be reduced in number. To P. are probably related siphonostelic stems *Chinlea*, initially regarded as osmundaceous rhizomes. The immediate ancestors of P. might have been the Permian genera of the *Viatcheslavia* and *Signacularia* types.

Family Isoetaceae (quill-worts)

The living quill-worts are represented by the genera *Isoetes* and *Stylites*. Fossil forms resembling *Isoetes*, but as yet inadequately studied, have been assigned to *Isoetites*, recorded since the Triassic. In *Isoetes* the stem is strongly reduced. A current opinion is that the leaves of *Isoetes* are, in essence, underdeveloped sporophylls. The lower part of the cormose axis is a rhizophore. The sporangium is covered by an outgrowth of the sporophyll (velum), which is absent in the families previously described. Sporangia have trabeculae, the same as in some Chaloneriaceae and *Pleuromeia*. The Cretaceous genus *Nathorstiana*, with a short stem and an apical leaf bunch, is commonly affiliated with *Isoetes* (Karrfalt, 1984). On the rhizophore several vertical ridges were borne, from which roots extended. The organization of the sporophylls is unknown. The Permo–Triassic *Takhtajanodoxa* (Snigirevskaya, 1980) probably related to Isoetaceae had short dichotomously branching stems with a simple radially symmetrical rhizophore. The plant was eustelic with secondary wood. The ligule was subtended by a glossopodium. Microsporophylls have not been found. Megaspores are smooth. The position of megasporophylls is uncertain.

Satellite genera of the order Isoetales

Some of the genera described below were formerly treated as members of the families Lepidodendropsidaceae, Caenodendraceae, Leptophloeaceae, etc., which, considering the lack of data on the reproductive organs and major anatomical characters of these plants, is inappropriate. Thomas and Meyen (1984) compiled a key for the identification of lycopsid genera represented by impression–compression remains of vegetative stems.

Genus *Tomiodendron* (C_1; Fig. 19, T; Fig. 20, *d*, *e*)

The trunks, reaching 30 cm in diameter, are unbranched. The leaf cushions are elongate rhomboidal, practically independent in size from the trunk dimensions. They are closely or loosely spaced, at times in widely spaced orthostichies. The leaf with an axillary ligular pit was attached to the upper part of the cushion. Protostelic stems are known. The stele is slender and cylindrical.

Genus *Angarodendron* (C_{1-2}; Fig. 19, An)

The trunks are unbranched, up to 10 cm in diameter, covered by a thick cuticle; stomata are absent. The leaf cushions are transversely rhomboid to lenticular, with a heel, large infrafoliar bladder and ligular pit. Transversal zones of larger more closely spaced cushions alternate with zones of smaller and wider-spaced cushions. Sporophylls are linear with elongate sporangia; megaspores with a small gula are large (up to 9 mm in diameter), of the *Setosisporites pastillus* type; microspores are trilete, of the *Cyclobaculisporites trichacantus* type (data by E. M. Vashchenko and N. G. Verbitskaya, personal communication). Leaves linear, with two compact stomatal bands.

Genus *Ursodendron* (C_1; Fig. 19, Ur; Fig. 20, *f*)

The trunks are up to 6 cm in diameter, occasionally branched. The leaf cushions are transversely elongated, lenticular or rhomboid, at times of indistinct contours, with a large axillary ligular pit. The leaves are persistent. Occasionally the leaves are arranged in orderly horizontal rows. The stele is pentagonal and winged in cross-section.

Genus *Lepidodendropsis* (D?, C_1; Fig. 19, Ls)

The size and diagnostic characters of this genus are both in need of revision. At first to *L.* were assigned Tournaisian lycopsids with slender trunks and elongated spindle-shaped (to hexagonal) leaf cushions, with persistent leaves attached at the upper part. The cushions are arranged in distinct horizontal rows. Later, only on the basis of the configuration and arrangement of the cushions, to L. were assigned Devonian shoots of specific anatomical organization (Iurina and Lemoigne, 1975; Iurina, 1969). However, there is no justification for the assignment of such anatomy to the Carboniferous L. species. Scheckler and Beeler (1984) have found *in situ* upright *Lepidodendropsis* stumps with *Protostigmaria* root bases.

Genus *Cyclostigma* (D_3–C_1?; Fig. 19, C; Fig. 20, *a*–*c*)

This is closely related to *Lepidodendron* according to the type of trunk branching. The juvenile individuals were unbranched (Chaloner, 1984). The plant attained about 8 m in height and 0.3 m in the stem diameter. The leaf cushions are widely spaced, slightly elevated above the bark surface; their rows ascend gently or are practically horizontal. On the leaf scar, that occupies nearly the entire cushion, the minute scar of the vascular bundle and two scars of the parichnos can be seen. The presence of the ligule cannot be excluded. Mega- and microsporangia are elongate. The distal part of the sporophyll is needle shaped with marginal teeth. The terminal strobili externally resemble *Flemingites*. Unlike *Stigmaria* the rhizophores were less divided. The rootlet 'scars' on the rhizophore gradually decrease in dimensions towards the apices of the rhizophore branches (Chaloner, 1984). Appendices were attached to the blunt ends of the branches.

Order Selaginellales

Selaginellales (S.) reliably includes at present only the genus *Selaginella* (known since the Permian), to which Carboniferous plants described as *Paurodendron* may probably also belong. The latter genus comprises rhizomorphs consisting of three successive regions (Erwin, 1984). The stem region shows leaf traces. In the transition zone leaf traces are absent; the centrarch protostele is accompanied by a remnant of the embryonic trace that is similar to the vascular tissue in *Lepidocarpon* embryos. In the rooting zone the primary xylem of a prominent stele is similar to the prismatic tissue of *Isoetes*. Rootlets emerge from a radial furrow of the rhizomorph, developmentally equivalent to the linear furrows of *Isoetes*. Fossil species, similar to *Selaginella*, but inadequately studied for reliable identification, are included in the genus *Selaginellites*. It is probable that to S. belong many dispersed megaspores, known since the Carboniferous and assigned to various genera. S. are herbaceous plants. The branching of the shoots frequently occurs in one plane. In some S., including *Paurodendron* and the Permian *Selaginella harrisiana*, all the leaves are the same and spirally attached. In most species the shoots are heterophyllous, the leaves are four-ranked, but one side of the shoot remains leafless. Sporophylls are clustered in strobili. The number of megaspores in the sporangium may be reduced to one tetrad and even to one megaspore. The gametophytes develop frequently within the sporangia; here the fertilization and development of the embryo may take place (the same as in some *Lepidocarpon*). The stele is stellate with protoxylem strands at the tips of the rays.

To S. may be assigned the peculiar genus *Synlycostrobus* (J) having anisophyllous shoots and strobili attached to the lower part of the scaly leaves. Sporophylls have a strongly developed heel. The microsporangia produced trilete microspores, whereas the megasporangia (not found in attachment) produced only one megaspore tetrad. The herbaceous ligulate heterosporous genus *Miadesmia* (C), frequently placed in a separate family, but as yet not sufficiently studied, can be regarded as a satellite genus of the order Selaginellales.

The literature yields many tens of other lycopsid genera that have been established on impression–compression and petrified trunks, detached leaves, sporophylls, strobili and spores (see review: Thomas and Brack-Hanes, 1984; Chaloner, 1967; Meyen, 1976, 1982b; Thomas and Meyen, 1984). Lycopsids were already highly diversified during the Devonian, when ligulate and heterosporous forms first appeared. The spores *Retispora* (Fig. 20, *z*) from the upper Devonian probably belonged to lycopsids (cyclostigmas?). The Carboniferous lycopsids were most diverse. Among them, besides those described above, many other occasionally quite peculiar forms were encountered, e.g. the genus *Eleutherophyllum* (Fig. 20, *g–l*), with whorled, strongly reduced leaves.

Class Equisetopsida (Articulatae)

Articulates (A.) were of high diversity in the geological past, whereas in the Present plant world they are represented only by the genus *Equisetum*. They appeared in the Devonian and flourished throughout the Carboniferous to the Jurassic. Typical A. can easily be distinguished from other

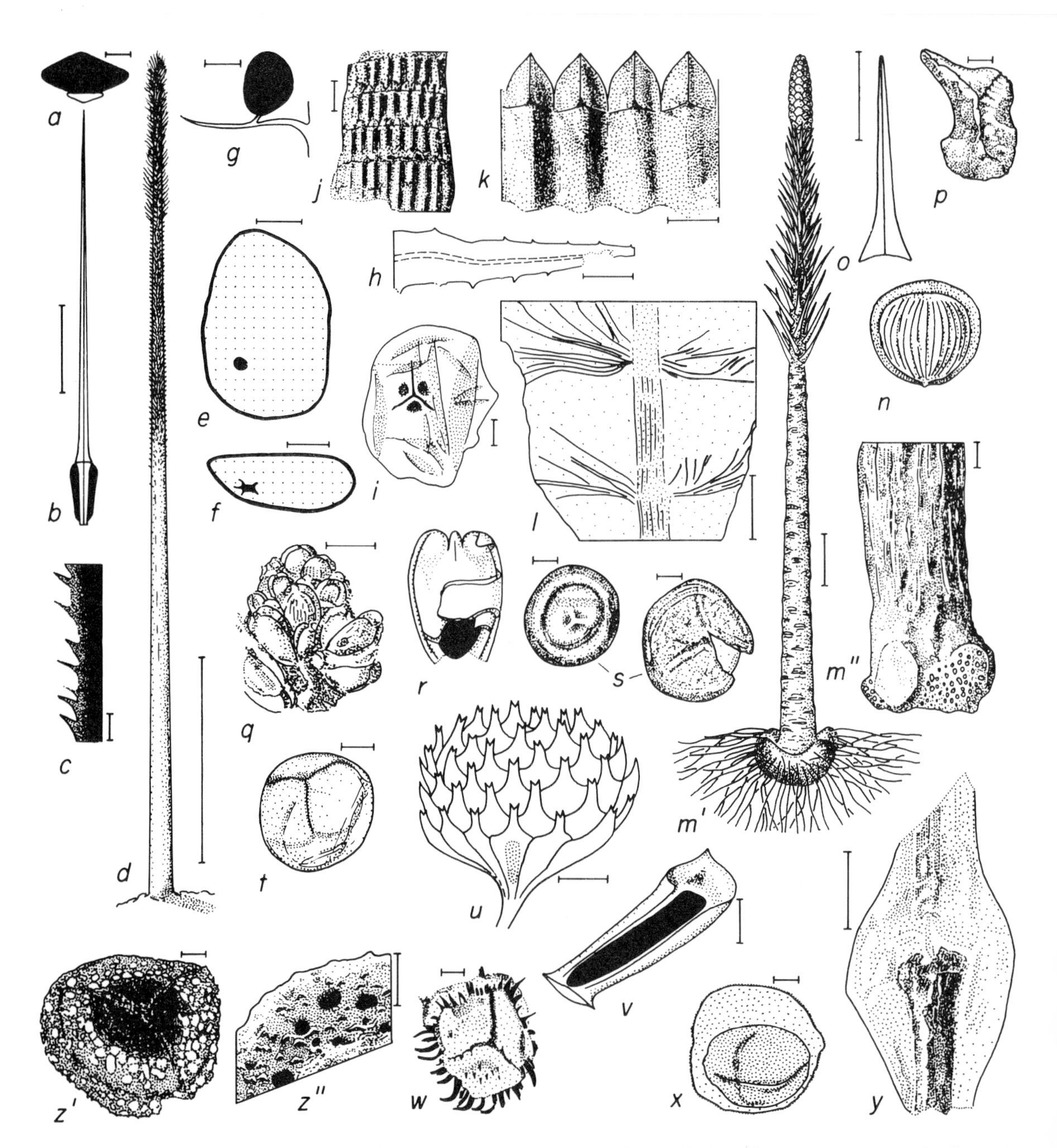

Fig. 20. Lycopsids; Devonian–Carboniferous boundary beds (*a–c*, *z*), Lower Carboniferous (*e–l*), Permotriassic (*x*, *y*), Triassic (*m–w*); Western Europe (*a–c*, *g–o*, *v*, *w*, *z*), Russian platform (*p–t*), Siberia (*d–f*, *x*, *y*); *a–c*, sporophyll of *Cyclostigma kiltorkense* Haught. (*a*, cross section; *b*, abaxial view; *c*, denticulate margin of lamina); *d*, *e*, *Tomiodendron kemeroviense* (Chachl.) Radcz. (*d*, reconstruction; *e*, cross section of stem, stele in black); *f*, *Ursodendron chacassicum* Radcz., stem cross-section with stellate stele; *g–i*, *Eleutherophyllum waldenburgense* (Stur) Zimm. (*g*, sporophyll; *h*, leaf with denticulate margin; *i*, spore); *j–l*, *E. mirabile* Stur (*j*, leafy shoot; *k*, internode crowned by tooth-like leaves; *l*, longitudinal split of shoot with ribbed stele and whorled leaf traces); *m–o*, *Pleuromeia sternbergii* (Münst.) Corda (*m'*, reconstruction; *m''*, stem base with rhizophore; *n*, sporophyll; *o*, reconstruction of leaf); *p–t*, *P. rossica* Neub. (*p*, ▶

pteridophytes, although fossil genera are known that combine characters of A. and other classes (e.g. *Eleutherophyllum*; Fig. 20, *g–l*). In the earlier systems the group of protoarticulates (*Hyenia*, *Protohyenia* and *Calamophyton*) was distinguished, that was regarded as transitional from psilophytes to A. These genera are at present allied to cladoxylalean ferns.

The origin of A. is unknown (see alternative views in Stein *et al.*, 1984). Cladoxylales or Ibykales might have been their ancestors. It was believed that bowmanitaleans (sphenophylls) evolved from lycopsids, owing to their similarity to plants of the *Estinnophyton* type (Protolepidodendrales). The classification of A. given below is based on the following characters (Fig. 21). The subclasses Bownmanitidae and Equisetidae are distinguished mainly according to their stelar organization – exarch or slightly mesarch protostele in the first, arthrostele in the second. Dis-

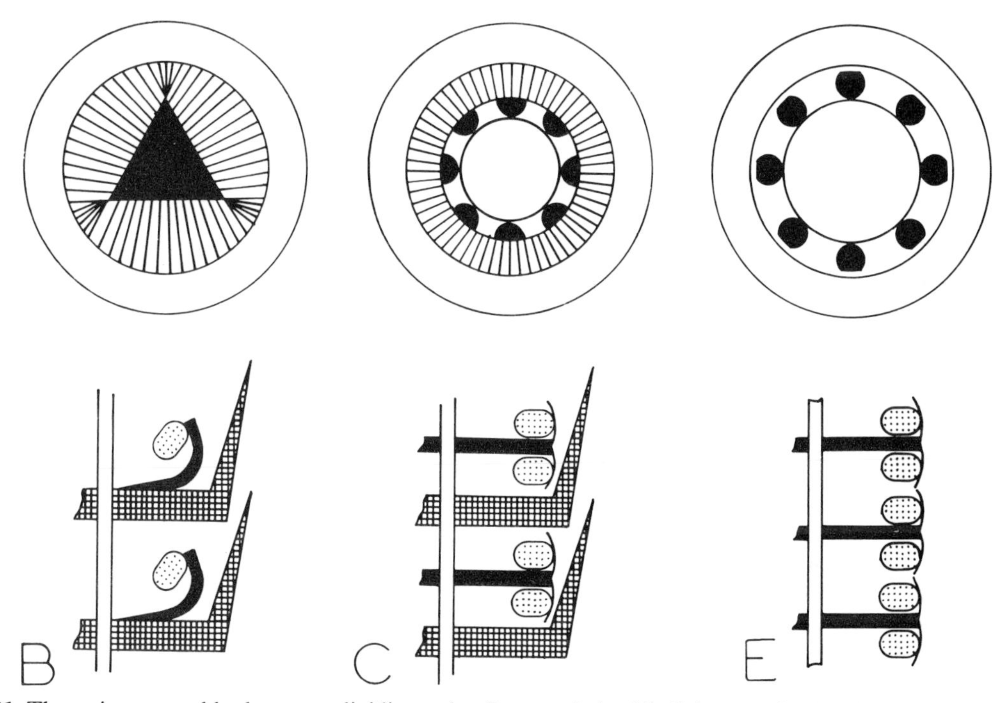

Fig. 21. The main most stable characters dividing orders Bowmanitales (B), Calamostachyales (C) and Equisetales (E). Upper row, stem organization (primary wood in black, secondary – shaded); lower row, organization of strobili or fertile zones (sporangiophores in black, bracts hatched, sporangia stippled).

---

decorticated rhizophore seen from below; *q*, strobilus; *r*, reconstruction of sporophyll, sporangium in black; *s*, microspores of *Densoisporites* type with proximal papillae in left grain, and split exine showing body in right grain; *t*, megaspore); *u*, *Skilliostrobus australis* Ash, reconstruction of strobilus, stippling shows sporangium in one sporophyll; *v*, *w*, *Annalepis zeilleri* Fliche (*v*, reconstruction of sporophyll; *w*, dispersed megaspore associating with sporophylls); *x*, *y*, *Tomiostrobus radiatus* Neub. (*x*, microspore; *y*, microsporophyll); *z*, *Retispora lepidophyta* (Kedo) Playf., general view (*z′*) and detail of flange as seen under SEM (*z″*). Scale bar = 1 m (*d*), 10 cm (*m′*), 5 cm (*o*), 1 cm (*j*, *m″*, *p*, *q*, *u*), 5 mm (*e*, *f*, *k*, *l*, *v*, *y*), 1 mm (*a*, *b*), 0.5 mm (*g*, *h*), 100 μm (*c*, *t*, *w*), 10 μm (*i*, *s*, *x*, *z′*), 5 μm (*z″*). Modified from: Chaloner, 1968a (*a–c*); Meyen, 1981 (*d*) and 1976 (*e*, *f*); R. Remy and W. Remy, 1960 (*g–i*); Sagan, 1980 (*j–l*); Takhtajan, 1956, after Mägdefrau (*m′*, *o*); Dobruskina, 1982 after H. Potonié (*m″*); Takhtajan, 1956, after H. Solms-Laubach (*n*); Neuburg, 1960b, 1961 (*p–t*); Ash, 1979 (*u*); Grauvogel-Stamm and Duringer, 1983 (*v*, *w*); M. Streel in Clayton *et al.*, 1977 (*z*).

tinctions of the orders are based on the presence (Calamostachyales) or absence (Equisetales) of bracts in the strobili. The families differ in the organization and arrangement of sporangiophores. The customary practice of establishing families on the basis of genera introduced for the vegetative parts (Phyllothecaceae, Archaeocalamitaceae, Calamitaceae, Schizoneuraceae, Prynadaeaceae) is rejected here, because similar vegetative shoots (e.g. of the *Phyllotheca* and *Schizoneura* types) associate with essentially different fructifications (Grauvogel-Stamm, 1978; Meyen, 1971b, 1978a).

The monotypic taxa Pseudoborniales and Cheirostrobales are not recognized here as independent A. orders. They are indisputably specific plants, but as yet poorly studied. *Pseudobornia* (D_3; Fig. 22, *a*) are plants with articulated stems, reaching 20 m in height, and half a metre in width. The upper branches terminated in strobili, up to 30 cm, with alternating whorls of bracts and sporangiophores. The latter were forked and upturned, bearing up to 30 sporangia. The leaves are uncommon for A. They were arranged in whorls in fours, and forked up to four times. The end lobes were pinnately divided into thin laciniations. So far it is difficult to correlate *Pseudobornia* with other articulates. Owing to the complex subdivision of the leaf lamina and absence (?) of the pith cavity in the stems, *Pseudobornia* is usually correlated with bowmaniteans (Figs 23, 24). More clearly related to bowmaniteans is *Eviostachya* (D_3; Fig. 22, *b–h*), which had a three-armed (in cross-section) protostele with paired protoxylem poles near the tips of the rays. Strobili were arranged in whorls and consisted of an axis with whorls of branching sporangiophores. Whorls of bracts between the sporangiophores are lacking, but are observed below each strobilus. According to their anatomical structure and organization of the sporangiophores *Eviostachya* is similar to the Middle Devonian Ibykales (see class Polypodiopsida). Leclercq (1957) described and depicted spores of *Eviostachya* as having coni or spines. M. Streel (personal communication) has studied the

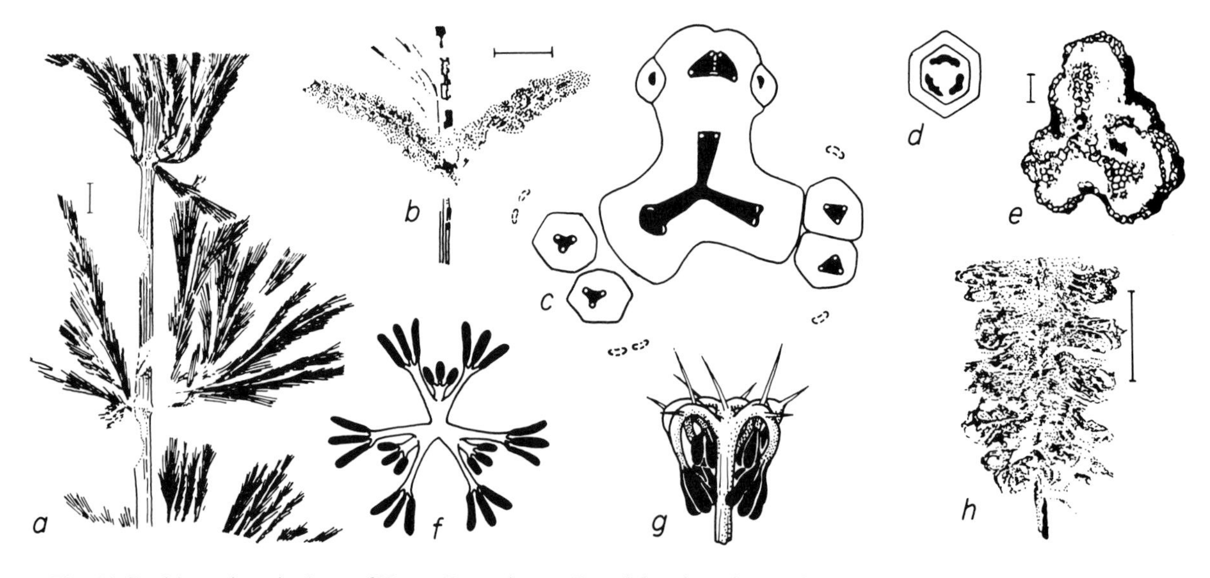

Fig. 22. Problematic articulates of Upper Devonian; *a*, *Pseudobornia ursina* Nath., Bear Island; *b–h*, *Eviostachya hoegii* Lecl., Belgium (*b*, axis with two strobili; *c*, scheme of cross section through lower node of strobilus, pairs of sporangiophores are shown at different stages of departure from node, wood in black, protoxylem left white; *d*, cross-section of sporangiophore; *e*, cross-section of strobilar stalk in its lower part; *f*, scheme of sporangiophore structure with sporangia spread in one plane; *g*, sporangiophore with sporangia in life position; *h*, lower part of strobilus). Scale bar = 1 cm (*a*, *b*), 5 mm (*h*), 100 μm (*e*). Modified from: Nathorst, 1902 (*a*), Leclercq, 1957 (*b–h*).

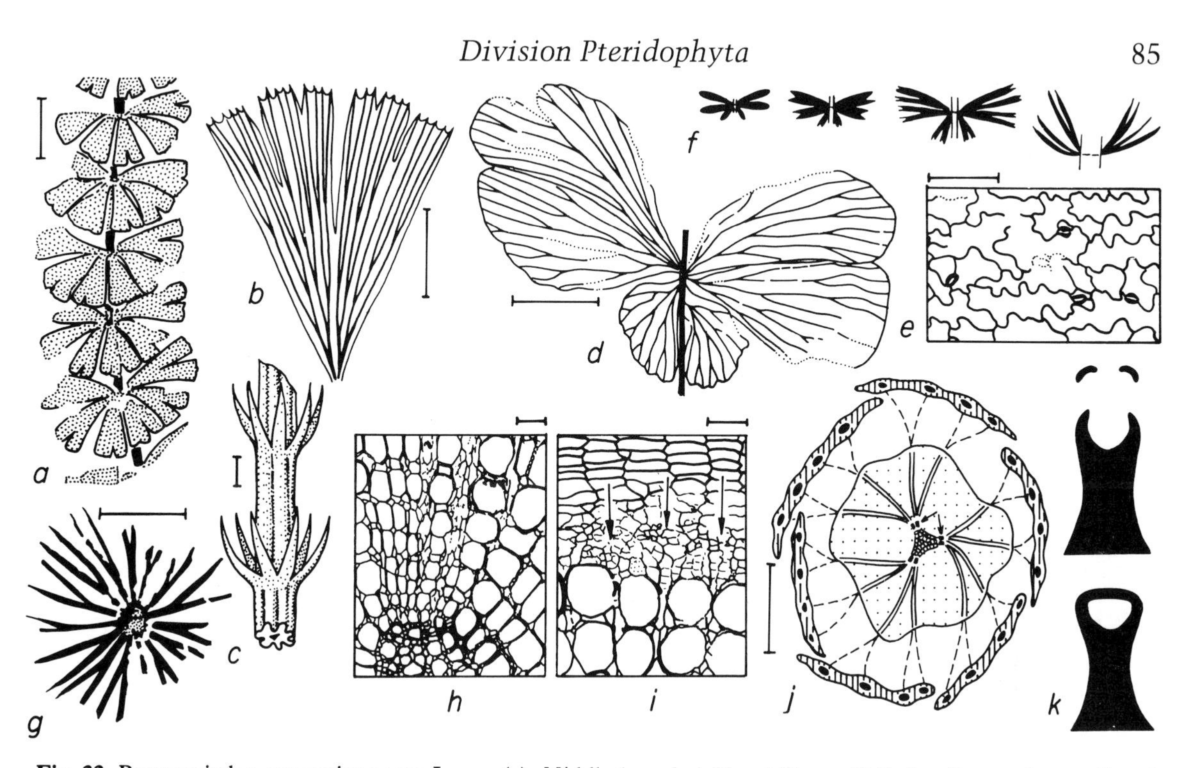

Fig. 23. Bowmanitales, vegetative parts; Lower (*g*), Middle (*a–c*, *h*, *i*, *k*) and Upper (*f*) Carboniferous, Lower (*j*) and Upper (*d*, *e*) Permian; Western Europe (*a*, *b*, *g*, *j*, *k*), India (*d*, *e*,), China (*f*), USA (*c*, *h*, *i*); *a*, *Sphenophyllum saarense* W. Remy; *b*, *S. saxonicum* W. et R. Remy; *c*, *Mesidiophyton* (al. *Sphenostrobus*) *paulus* Leism.; *d*, *e*, *Sphenophyllum* (al. *Trizygia*) *speciosum* (Royle) Zeill. (*e*, epidermis); *f*, *S. oblongifolium* (Germ. et Kaulf.) Unger, leaf polymorphism on branches of different order; *g*, *S. tenerrimum* Ett. ex Helmh.; *h*, *S. plurifoliatum* Will. et Scott, arm end of primary wood, and secondary wood; *i*, same, periphery of secondary wood (bottom), primary phloem at arrows, and secondary phloem (top); *j*, *S. quadrifidum* Ren., scheme of stem cross-section surrounded by sections of leaves, paired protoxylem groups at arrows, double lines show courses of vascular bundles within stem; *k*, *S. insigne* Will., departure of leaf trace. Scale bar = 1 cm (*a*), 5 mm (*b*, *d*, *g*, *j*), 1 mm (*c*), 100 μm (*e*, *h*, *i*). Modified from: Storch, 1980a (*a*, *g*) and 1966 (*b*); Leisman, 1964 (*c*); Pant and Mehra, 1963b (*d*, *e*); Asama, 1966 (*e*); Eggert and Gaunt, 1973 (*h*, *i*); Zimmermann, 1959, after Renault (*j*); Němejc, 1963, after P. Bertrand (*k*).

relevant preparations and concluded that the bodies taken for spores are contamination particles. The genus *Cheirostrobus* (C_1; Fig. 24, *s*, *t*) is known only from a single strobilus fragment. The pith is mixed. Bracts and sporangiophores are arranged in alternating whorls. The bract and sporangiophore above it are supplied by branches of one vascular bundle and divided in the horizontal plane into three segments. Each segment of the sporangiophore terminates with a shield bearing four long sporangia, arranged parallel to the segment stalk. Nothing is known of the vegetative parts.

Subclass Bowmanitidae

Bowmaniteans (B.) are commonly termed 'sphenophylls' in the literature (after the genus *Sphenophyllum* established for vegetative shoots) and treated as an order (Sphenophyllales). The family Sphenophyllaceae has also been introduced. It is probable that plants with essentially different fructifications (including those of the *Cheirostrobus* type) might have had leaves of the *Sphenophyllum* type. Because of this the subclass under consideration is named after the genus *Bowmanites*, introduced for the strobili. Accordingly, the order is

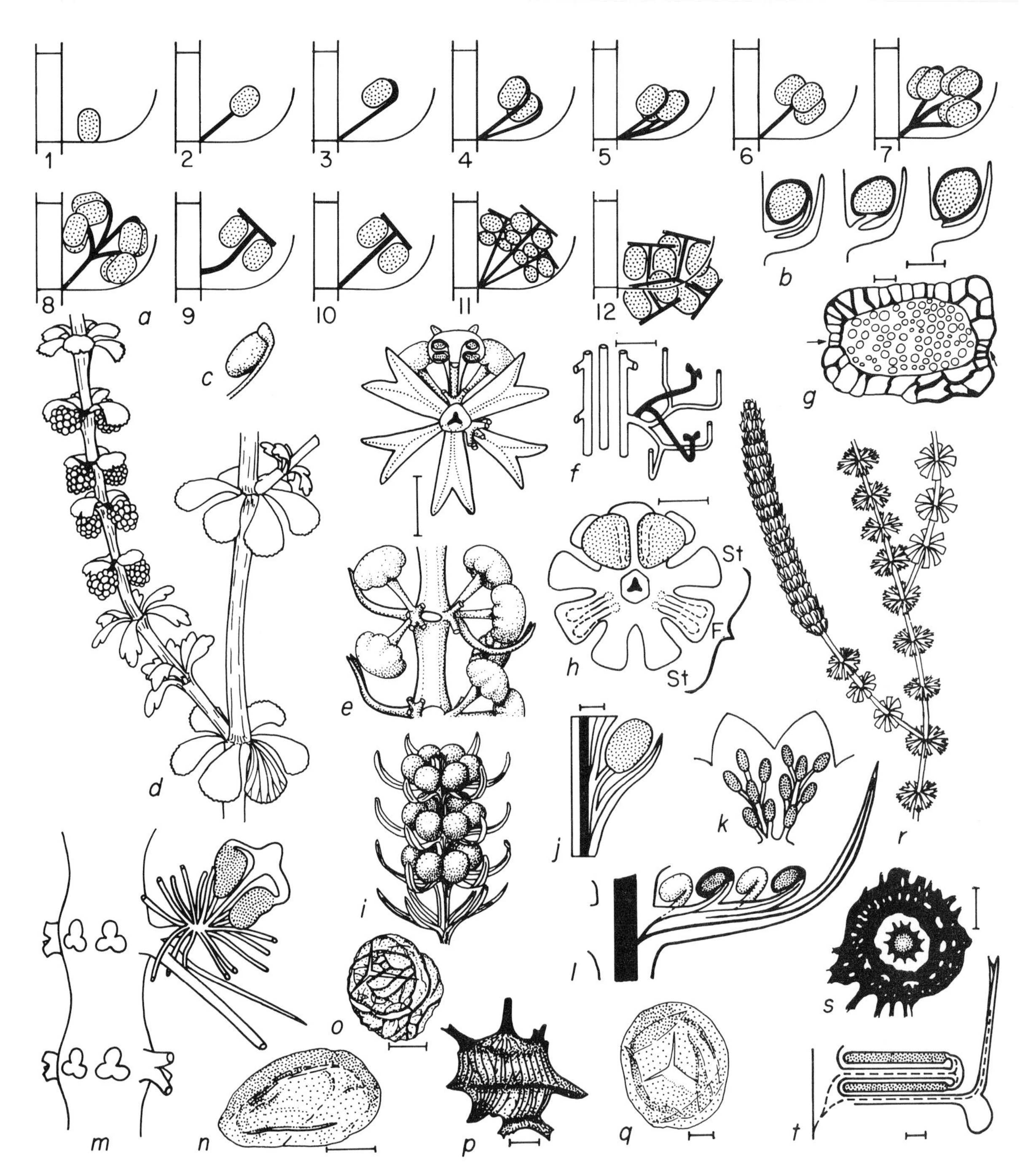

Fig. 24. Bowmanitales, fructifications; Carboniferous (*b*, *c*, *e–u*), Lower Permian (*d*); USA (*b*, *e*, *f*, *h*, *i*), Western Europe (*c*, *d*, *g*, *k–n*, *q*–u), Donbass (*g*, *j*, *o*, *p*); *a*, arrangement of sporangia (in black) and sporangiophores on strobilus metamere, numerals correspond to *Sphenophyllum trichomatosum* Stur (1), *Bowmanites aquensis* W. Remy sp. (2), *B. angustifolius* (3), *B. moorei* Mam. (4), *B. dawsonii* Will. (5), *B. kidstonii* (Heming.) Hosk. et Cross (6), *B. roemeri* Solms-Laub. sp. (7), *B. fertilis* (Scott) Hosk. et Cross (8), *Sphenophyllum majus* after Walton (9), *Aspidiostachys* (10), ▶

termed Bowmanitales and the family Bownmanitaceae (Meyen, 1978a). On the other hand, some B. produced leafy shoots, described in the literature as *Asterophyllites* (this genus is assigned to calamiteans). Such shoots are known in *Mesidiophyton* (Fig. 23, *c*; at times it is regarded as synonymous to *Sphenostrobus* and *Litostrobus*).

B. are known from impression–compression and petrifaction materials. In several instances only was it possible to correlate the genera and species identified according to remains in various states of preservation. For instance, *Sphenophyllum plurifoliatum* in coal balls seems to correlate with the species *S. myriophyllum*, described from impression–compression remains. B. are herbaceous plants. The vegetative shoots are assigned to the genus *Sphenophyllum* (D_3–P; Fig. 23, *a*, *b*, *f*, *g*; Storch, 1980a). The leaves are arranged in whorls usually in numbers divisible by three; oval, wedge-shaped, linear, divided into lobes, or entire; the margin is even, denticulate or lacerated. The leaves in a whorl are of the same or different sizes and shapes. In the latter case the whorl becomes butterfly-shaped. A heterophylly was recorded (the same as in living aquatic and semiaquatic angiosperms). In some species one vein enters the leaf base; it divides dichotomously and all the branch veins reach the distal margin. In other species the veins reach the lateral margins. In this case more than one vein may enter the leaf base. Epidermal cells are elongated along the veins and commonly have sinuous walls. The stomata are concentrated on the lower leaf surface. (Often an independent genus *Trizygia* (P; Fig. 23, *d*, *e*) is distinguished for the shoots with anisophyllous whorls, one vein entering at the leaf base, and its branches reaching the lateral margins; the stomata are of irregular orientation.) The stem is thin, ribbed, occasionally branching. The ribs of the stem pass directly through a node and correspond in position to leaves. The stele is triangular (more seldomly, hexagonal or quadrangular) in cross-section, with exarch or slightly mesarch protoxylem (Fig. 23, *h*, *j*), which may disintegrate leaving a lacuna. Some species produce secondary wood (Fig. 23, *h*, *i*), composed of scalariform and pitted tracheids and a large number of small parenchymatous cells. Bordered pits occur both on tangential and radial tracheid walls. Long tracheids (up to 8 mm) were observed that were probably vessels. The xylem cylinder is surrounded by a phloem and a cortex with distinct ribs, which become smoother as the periderm develops. From the nodes and internodes of the stem adventitious roots may extend. Branch traces come out from the protoxylem poles above the leaf trace and may be regarded as axillary.

Most B. have well-shaped strobili (Fig. 24, *a*–*m*), that belong mainly to the genus *Bowmanites* (= *Sphenophyllostachys*). The strobili are inserted at

---

Peltastrobus (11), *Lilpopia* (12); *b*, variation in sporangial position in *Sphenostrobus* (al. *Litostrobus*) *iowensis* (Mam.) Good; *c*, *Bowmanites roemeri*, sporangiophore; *d*, *Lilpopia raciborskii* (Lil.) Con. et Schaar., reconstruction; *e*, *Peltastrobus reedae* Baxt., reconstruction of part of strobilus; *f*, vascular system of sporangiophores (S) and bracts in *Bowmanites furcatus* Andr. et Mam.; *g*, *B. jablokovii* Snig., section of sporangium, stomium at arrows; *h*, *B. moorei*, scheme of strobilus metamere, brace comprises assemblage of fertile appendage (F) and two sterile bracts (St); *i*, *B. aquensis*, reconstruction of strobilus; *j*, *Sphenostrobus* (al. *Litostrobus*) *novikae* (Snig. et al.) S. Meyen, scheme of vascular system in strobilus; *k*, *l*, *B. dawsonii*, scheme of arrangement of sporangia in surface view (*k*) and in section (*l*); *m*, *B. fertilis*, scheme of strobilus, sporangia are shown in one sporangiophore; *n*, monolete spore from strobilus associating with *Sphenophyllum myriophyllum* Crépin; *o*, *p*, trilete spores showing perisporial relief from sporangia of *B. jablokovii* (*o*) and *B. pterosporus* Snig.; *q*, trilete spore from strobilus attached to shoot of *Sphenophyllum tenerrimum*; *r*, *S. cuneifolium* Sternb., shoot with strobilus; *s*, *t*, *Cheirostrobus petticurensis* Scott, cross-section of strobilus with stellate stele (*s*) and scheme of fertile metamere in longitudinal structure (*t*). Scale bar = 2 mm (*s*, *t*), 1 mm (*b*, *e*, *j*), 0.5 mm (*f*, *h*), 100 μm (*g*), 20 μm (*n*–*q*). Modified from: Good, 1978 (*b*); Hirmer, 1927 (*c*, *l*, *s*); Kerp, 1981 (*d*); Leisman and Graves, 1964 (*e*); Zimmermann, 1959, after Andrews and Mamay (*f*); Snigirevskaya, 1962b (*g*, *o*, *p*); Bierhorst, 1971, after W. and R. Remy (*i*); Snigirevskaya *et al.*, 1977 (*j*); Němejc, 1963 (*k*); Leclercq, 1936 (*m*); R. Remy, 1966–1968 (*n*); Storch, 1980b (*q*); Takhtajan, 1956 (*r*); Sporne, 1975 (*t*).

the tips of the branches which are sometimes shortened. Less commonly the sporangiophores are attached to axils of foliage leaves. The high diversity of the strobili can be arranged in a series where the sporangiophores become increasingly more complicated (Fig. 24, *a*). In the simplest construction the nodes bear whorls of six linear bracts with a short-stalked sporangium in the axil of each bract. In other strobili the bracts coalesce in the lower part into a disc or bowl; the sporangiophores increase in number: they branch, forming peltate structures with two, four or more pendant abaxial sporangia. The sporangiophores accompany each bract, or bracts with sporangiophores alternate with sterile ones. Bracts may be similar to sterile leaves. Occasionally whorls of sporangiophores are attached to the axis somewhat above the whorled bracts. In the most complex strobili the bracts are forked, sporangiophores are also branching, and their widened apices bear a pair of pendant sporangia. In some B. fused sporangia, the formation of a pith in the axis of the strobili was also recorded. In *Peltastrobus* (Fig. 24, *a* (11), *e*) three sterile bracts, alternating with three fertile ones, are attached at a node. In the axil of the fertile bract five sporangiophores are attached, each being crowned by a shield with two rows of sporangia (6–8 sessile sporangia in each row). Only one heterosporous species is known. The spores are mono- or trilete, smooth or sculptured with thin granulation, reticulate, with wing-like extensions (Fig. 24, *n–q*). Considerable variation has been found in the sporoderm ultrastructure (W. A. Taylor, 1984). The genus *Sentistrobus* (Riggs and Rothwell, 1985) comprises unusual strobili lacking bracts and bearing fertile trusses with erect sporangia. This genus resembles *Eviostachya* in the absence of leaf-like bracts. Notwithstanding the high diversity of the strobili, the anatomical structure of the axis is consistent, and nearly the same as in vegetative shoots. So far it is impossible to correlate certain types of the leaves and strobili. The systematics of B. is complicated due to the fact that characters considered important for the distinction of genera exhibit, at times, variations within one species (Fig. 24, *b*).

Usually Lilpopiaceae is defined as an independent family. In the genus *Lilpopia* (P; Fig. 24, *d*) the vegetative shoots are of the *Sphenophyllum* type, but the strobili are situated within one whorl with sterile leaves, alternating with the latter. Strobili consist of a short axis on which are compactly spaced peltate sporangiophores with pendant sporangia. A single strobilus may be compared with a complex-branching axillary sporangiophore of *Bowmanites fertilis* or a group of sporangiophores of *Peltastrobus*. The distinction of a specific family for *Lilpopia* is hardly justified.

These trends of successive complication of the strobili were interpreted as evolutionary lineages (Takhtajan, 1956). However, no directional trend towards the complication or simplification of the strobili can be observed in ascending order of the stratigraphic sequence.

Subclass Equisetidae

The Equisetidae are characterized by the formation of a pith in the stems (Fig. 25, *b*, *e*, *g*). Near the shoot apex the vascular bundles encircle the pith as a ring. Further back the pith is replaced by a cavity in the internodes, with nodal diaphragms. In place of the protoxylem a carinal canal frequently forms (Fig. 25, *h*, *n*). In fossil remains this cavity is infilled with sediment, resulting in the formation of a pith cast with ribs (corresponding to interfascicular medullary rays) and grooves (corresponding to vascular bundles protruding into the cavity), at times complicated by a thin rib (imprint of the carinal canal). Isolated pith casts are assigned to the genera *Calamites*, *Paracalamites*, etc. Occasionally the bark bears longitudinal ribs, strengthened by a mechanical tissue. It is not always easy to decide whether one is dealing with a pith cast or the ribbed outer surface of a shoot. Sutures between the leaves in cylindrical or conical sheaths have been frequently mistaken for surficial ribs.

Order Calamostachyales

The Calamostachyales (C.) are usually termed calamites in the literature (order Calamitales, family Calamitaceae) after the genus *Calamites*, established for pith casts. According to the International Code of Botanical Nomenclature the type species of *Calamites* is *C. radiatus*, including pith casts usually assigned to *Asterocalamites* or *Archaeocalamites*. The assignment of these remains to the order Calamitales in its current definition is therefore doubtful. Because of this it is suggested that the order and family should be named after the genus *Calamostachys*, which has been introduced for fructifications. The family name Calamitaceae, guarded by priority, cannot be abolished, but it may be retained as the name of a form-group including pith casts of articulates.

C. plants varied in dimensions, from small herbaceous forms (including lianas) to arborescent ones with intense branching and thick secondary wood. The genus *Calamostachys* (C–P; Fig. 26, *b–e*) was established for strobili, consisting of whorls of bracts and sporangiophores alternating along the axis. The vascular bundles of the sporangiophores and bracts emerge independently from the axis. The bracts appeared in the ontogenesis earlier than sporangiophores and are anatomically similar to leaves. They are lanceolate or triangular, bent elbow-like upwards, frequently fused in the horizontal part. Sporangiophores are peltate, bearing four sporangia. The spores are rounded, trilete, enveloped from the outside by strap-shaped elaters, which in curled form look like a continuous surficial covering of the spore. During maturation the spore elaters uncoiled. In some species the strobili are heterosporous.

Other genera exhibit a closely similar organization of the strobili axes, bracts and sporangiophores (Gastaldo, 1980). The genera can be distinguished according to the place of attachment of sporangiophores, number of sporangia on a sporangiophore, and other characters. In *Palaeostachya* (C; Fig. 26, *a*, *i*) the sporangiophores are attached at the axil of the bract. The shifting of the sporangiophore is marked by the elbow-shaped bending of its vascular bundle. In *Mazostachys* (C; Fig. 26, *f*) the sporangiophore bears a pair of sporangia and is shifted upwards to the bract whorls. In *Pendulostachys* (C, Fig. 26, *l*) the sporangiophore is attached to the lower surface of the bract, although the vascular system remains independent. The genus *Cingularia* is probably an older synonym of *Pendulostachys*. The second trend in the evolution of the strobili is the appearance of heterospory in the genera *Calamostachys* and *Palaeostachya*. It culminates in the reduction of the number of megaspores to a single one (*Calamocarpon*; C; Fig. 26, *h*, *n*), as can be seen in the Lepidocarpaceae. Of interest is the genus *Weissistachys*, in which the widened distal parts of the sporangiophores are fused in a ring. In *Uralostachya* in each whorl elbow-shaped bracts alternate with peltate sporangiophores. Such an alternation is known in the bowmaniteans. The origin of the C. strobili is unknown. Bracts and sporangiophores of C. may be interpreted as modified sterile and fertile branches.

Notwithstanding the recommendation of the International Code of Botanical Nomenclature (see above), the genus *Calamites* (C–P; Fig. 25, *a*, *b*, *c* (4) *e*, *l*, *m*) is treated below in the traditional delimitation. The typification of the generic name *Calamites* is in need of a change. (Until this is accomplished it is considered convenient to apply this name in its current usage and not to change it to the name *Calamitopsis*, as proposed by W. Remy and R. Remy (1978).) *C.* includes pith casts or stem remains some of which attain 40 cm in diameter. Most of the species are affiliated with Calamostachyaceae, but some species may probably be related to other families (because of this *C.* should be treated as a satellite genus of the order). The pith cast surface is covered by longitudinal ribs (imprints of bundles of primary wood) which alternate, approaching the nodes from both sides. The junction of the grooves in the node forms a transverse zigzag line. The stem remains with preserved surface structures and branch scars are assigned to the genus *Calamitina*. Large scars correspond to the places where the branches were attached, whereas minute scars above the nodal line correspond to leaf traces. Stem remains with preserved anatomical structures are assigned to the genera *Arthropitys*, *Calamodendron*, etc. (Fig. 25, *d*, *f*, *h–k*, *n*). Primary vascular bundles are endarch or mesarch. The secondary wood, occasionally massive, is composed of tracheids with scalariform, reticulate or rounded pitting. The secondary

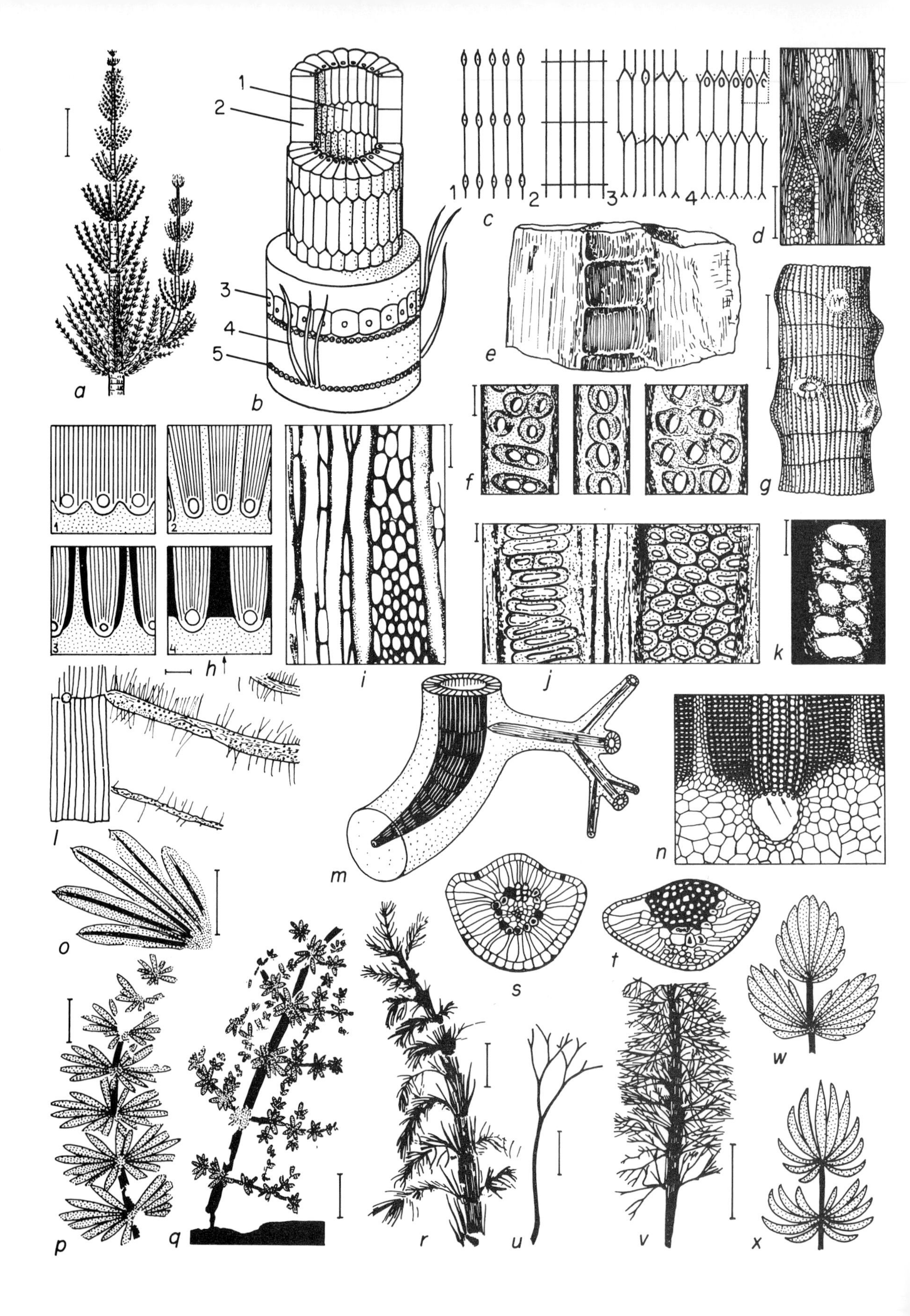

a
b
1
2
3
4
5
c
1
2
3
4
d
e
f
g
h
i
j
k
l
m
n
o
p
q
r
s
t
u
v
w
x

wood massif is dissected into wedges, extending from the primary wood bundles, and subdivided by interfascicular (primary) medullary rays, that sometimes are bordered by a layer of fibrous cells. In the secondary wood narrower secondary (fascicular) rays are present. The tissues outside the xylem cylinder are commonly of poor preservation. When they are well preserved it is impossible to distinguish the usual phloem with sieve tubes. Instead, a peculiar tissue consisting of thin-walled slightly elongated cells can be observed. In *Arthropitys* the formation of a periderm was perceived (Cichan and Taylor, 1983). The rhizomes are similar in their anatomical structure, but are pierced in the cortex by large cavities and have the usual type of phloem with sieve tubes. The genus *Mesocalamites* (Fig. 25, *c* (3)) is closely related to *C.* and includes pith casts, in the nodes of which the alternation of bundles belonging to adjacent internodes is not consistent. Leafy shoots (Fig. 25, *o–r*) are ascribed to the genera *Annularia*, *Asterophyllites* (see satellite genera of Equisetopsida) and *Dicalamophyllum*.

Order Equisetales

The Equisetales include articulates with apical strobili and fertile zones ('staged strobili'), where sterile bracts are absent. The anatomical structure has been studied in detail only in the living genus *Equisetum* and in Lower Carboniferous genera, that sometimes are included in the family Asterocalamitaceae. These genera will be discussed below after the description of other families of the order, which are identified according to the organization and arrangement of the sporangiophores.

Family Tchernoviaceae

To tchernovias belong herbaceous plants with peltate sporangiophores, bearing 10–15 sporangia each, that are closely pressed to the wide base of the shield of the sporangiophore head and cover its entire lower surface (Meyen, 1969b, 1982b; Meyen and Menshikova, 1983). Occasionally the sporangia appear to be laterally fused. Isolated sporangiophores of this type are included in the genus *Tchernovia* (Fig. 27, *f*, *g*; Fig. 28, *f*, *g*). The stalk of the sporangiophore is unbranched. In the

Fig. 25. Vegetative parts of Calamostachyales (*a*, *b*, *c3*, *c4*, d, *e*, *h2*, *h3*, *h4*, *i–t*, *w*, *x*) and archaeocalamiteans (*c1*, *c2*, *f*, *g*, *h1*, *u*, *v*); Carboniferous (*a*, *d*, *e–g*, *i–l*, *n*, *s–v*), Lower (*o–r*) and Upper (*w*, *x*) Permian; Euramerian area (*a*, *d–g*, *i–l*, *n–v*), Cathaysian area (*w*, *x*); *a*, reconstruction of *Calamites* (al. *Calamitina*) *sachsei* Stur; *b*, calamostachyalean stem structure (1, pith cavity with ribs and grooves on inner surface of wood cylinder; 2, secondary wood; 3, branch scars; 4, stem leaves; 5, leaf scars); *c*, course of vascular bundles through node in *Archaeocalamites* (1, 2), *Mesocalamites* (3), *Calamites* (4), leaf traces shown by dots, infranodal canals by ovals, contour marked by dotted line corresponds to *d*; *d*, *Arthropitys communis* Bin., nodal anatomy, leaf trace look as black patch; *e*, *Calamites* sp., secondary wood and pith cavity with nodes; *f*, *Archaeocalamites esnostensis* (Ren.) Leist., pitting of secondary wood tracheids; *g*, *A. radiatus* (Brongn.) Stur, pith cast; *h*, stem cross-section in *Archaeocalamites* (1), *Arthropitys* (2), *Calamodendron* (3) and *Arthroxylon* (4), parenchyme stippled, secondary wood shaded, sclerenchyme in black, carinal canals – oval or round contours; *i*, *Arthropitys communis*, tangential section of secondary wood with medullary rays of different width; *j*, *k*, *A. bistriata* (Cotta) Goepp., pitting of radial walls of tracheids and of cross field (*k*); *l*, *Calamites suckowii* Brongn., underground part with roots; *m*, *Calamites*, scheme of underground part; *n*, *Calamodendron intermedium* Ren., primary vascular bundle with carinal canal showing residual protoxylem tracheids (at arrows), secondary wood above canal is flanged by thick sclerenchyme and interfascicular medullary rays near margins); *o*, *p*, *Annularia carinata* Gutb.; *q*, *A. spicata* (Gutb.) Schimp.; *r*, *Asterophyllites equisetiformis* (Schloth.) Brongn.; *s*, *t*, cross-sections of leaves of *A. charaeformis* (Sternb.) Goepp. (*s*) and *A. grandis* (Sternb.) Gein. (*t*), the latter with thick-walled sclerenchyme cells; *u*, *v*, *Archaeocalamites radiatus*, leaf (*u*) and leafy shoot (*v*); *w*, *Lobatannularia heianensis* (Kod.) Kaw.; *x*, *L. lingulata* Halle. Scale bar = 1 m (*a*), 5 cm (*g*, *v*), 1 cm (*l*, *p–u*), 5 mm (*o*), 1 mm (*d*), 100 μm (*i*), 10 μm (*f–k*). Modified from: Hirmer, 1927 (*a*, *c–1*, *g*, *l*, *s–v*); Boureau, 1964 (*b*) and 1971 (*h*, *w*, *x*); Scott, 1920 (*d*, *n*), Osnovy palaeontologii, 1963a, after Sterzel (*e*); Marguerier, 1970 (*f*, *j*) and 1977 (*k*); Boureau, 1971, after Marguerier (*i*); Eggert, 1962 (*m*); Barthel, 1980 (*o–r*).

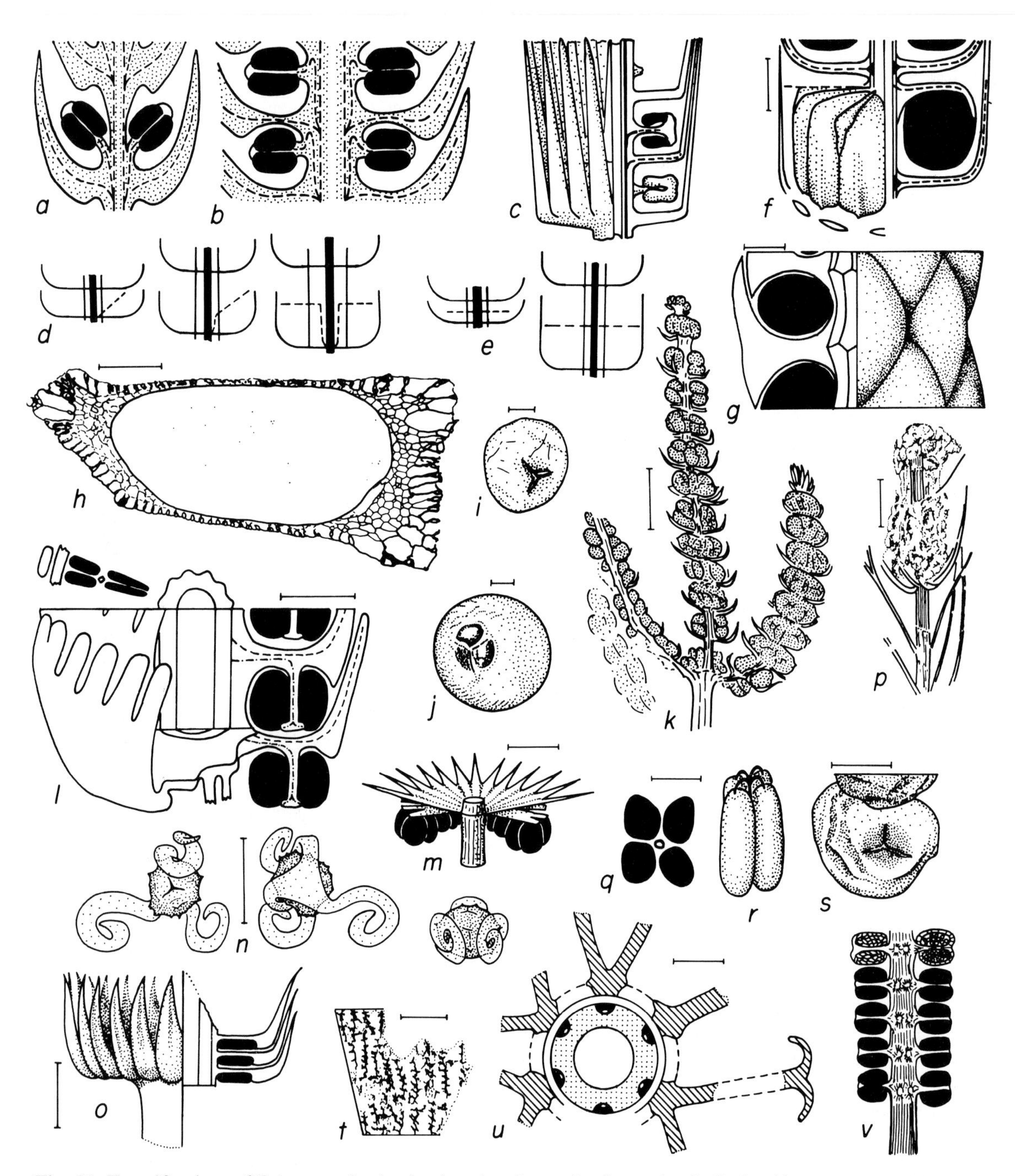

Fig. 26. Fructifications of Calamostachyales (*a–o*) and archaeocalamiteans (*p–v*); Carboniferous (*a–c*, *e*, *f*, *h–v*) and Permian (*d*, *g*); Euramerian area, including North America (*f*, *h–j*, *l*, *n*, *o*) and Europe (*g*, *m*, *p–v*), Angara area (Kuzbass; *k*); *a–e*, vascularization and bract/sporangiophore relation in *Palaeostachya* (*a*) and *Calamostachys* (*b–e*) (*a*, *P. decacnema* Delev.; *b*, *C. americana* Arn.; *c*, *C.* al. *Schimperia* sp.; *d*, scheme of ontogenesis of *C. spicata* var. *roemeri* R. et W. Remy, vascularization of sporangiophore is shown by broken line; *e*, same in *C.* al. *Schimperia*); *f*, *Mazostachys pendulata* Kos.; *g*, strobilus with solitary sporangia between bracts; *h*, section of sporangium of ▶

genera *Equisetinostachys* (C–P; Fig. 27, *b–d*) and *Phyllopitys* (C–P; Fig. 27, *a*) the sporangiophores are gathered in fertile zones, situated either on the lateral branches, or (in some species of *Equisetinostachys*) on the unbranched stem. In the genus *Sendersonia* (P_2; Fig. 27, *e*) the lateral fertile branches are reduced to a single fertile zone, crowned by a bunch of thin leaves. As a consequence, it simulates a lateral strobilus. The leaves become fused in their lower parts into cylindrical or conical sheaths. Vegetative shoots of this type are assigned to the genera *Phyllotheca* (see Satellite genera of the Equisetopsida) and *Phyllopitys*.

Family Gondwanostachyaceae

The genus *Gondwanostachys* (P; Fig. 28, *b*) has fertile shoots with ribbed internodes and spaced

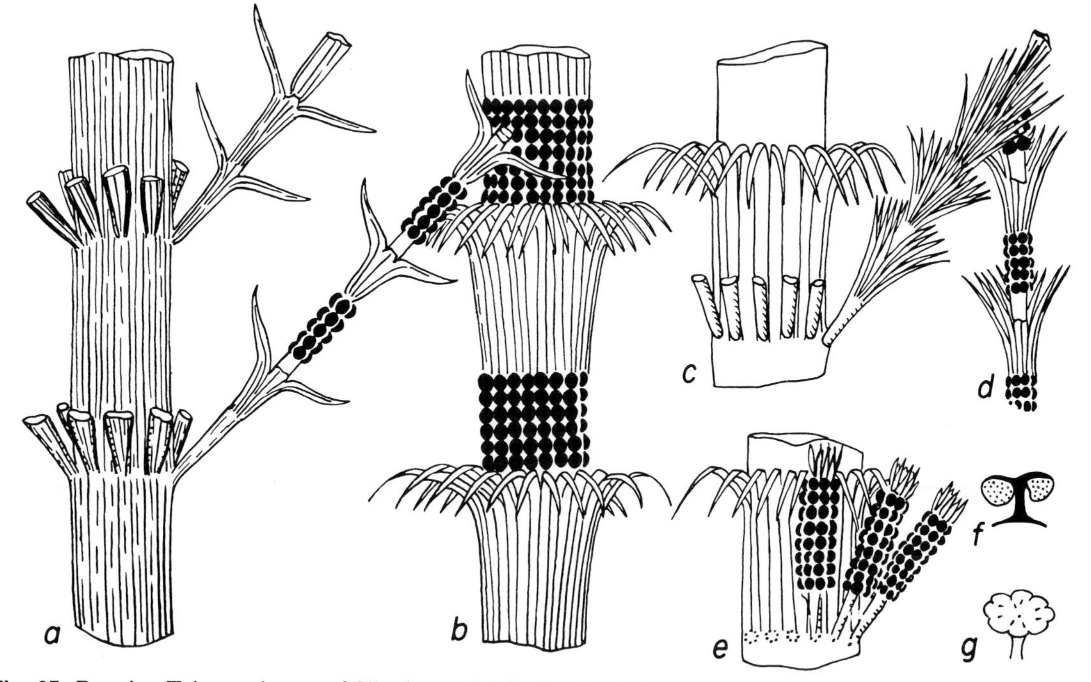

Fig. 27. Permian Tchernoviaceae of Siberia; *a*, *Phyllopitys heeri* (Schmalh.) Zal.; *b*, *Equisetinostachys gorelovae* S. Meyen; *c*, *d*, vegetative shoot (*c*; *Phyllotheca turnaensis* Gorel.) and associating fertile branch (*Equisetinostachys*); *e*, *Sendersonia matura* S. Meyen et Mensh.; *f*, *g*, *Tchernovia*, sporangiophore. Modified from Meyen and Menshikova (1983).

Calamocarpon insignis Baxt., one megaspore occupies entire sporangium; *i*, spore of *Palaeostachya decacnema*; *j*, spore of *Calamostachys binneyi* (Carr.) Schimp.; *k*, *Calamostachys* sp.; *l*, *Pendulostachys cinguláriformis* Good, scheme of strobilus, upper left side – fragment of cross section with one sporangiophore; *m*, *Cingularia typica* Weiss, traditional reconstruction; *n*, microspores *Elaterites triferens* Wils., produced by *Calamocarpon insignis*, elaters are spread or coiled; *o*, *Kallostachys scottii* Brush et Bargh.; *p*, *Pothocites*; *q–s*, *P. grantonii* Pet., cross-section of sporangiophore with 4 sporangia (*q*), reconstruction of sporangiophore (*s*), spore (*s*); *t*, *u*, *Protocalamostachys* (al. *Pothocites*) *arranensis* Walt., sporangial wall (*t*), and scheme of sporangiophore structure together with stem anatomy (*u*; parenchyme stippled, axial vascular bundles in black, protoxylem lacunae within them are left white); *v*, *Pothocites* sp. Scale bar = 1 cm (*p*), 5 mm (*k*, *m*, *o*), 1 mm (*d*, *f*, *g*, *p*, *q*, *u*), 0.5 mm (*h*), 100 μm (*t*), 50 μm (*n*, *s*), 10 μm (*i*, *j*). Modified from: Boureau, 1964 (*a*, *b*); Filin, 1978 (*c*); R. Remy and W. Remy, 1975 (*d*, *e*); Andrews, 1961 (*f*, *m*, *o*); Barthel *et al.*, 1976 (*g*); Good, 1975 (*h–j*, *l*); Baxter and Leisman, 1967 (*n*); Hirmer, 1927 (*p*); Chaphekar, 1965 (*q*, *s*); Walton, 1949 (*t*, *u*); Scott, 1920, after Renault (*v*).

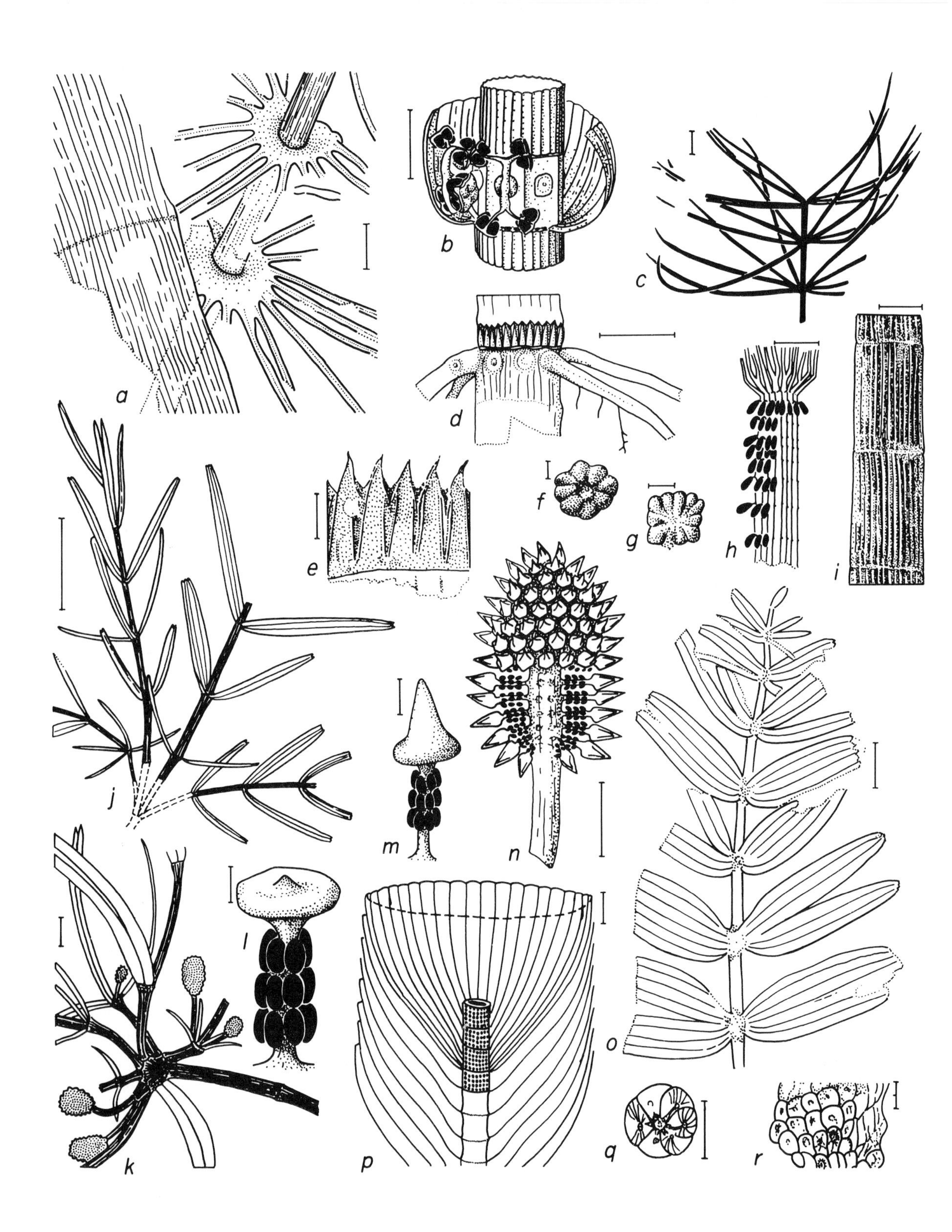
a
b
c
d
e
f
g
h
i
j
k
l
m
n
o
p
q
r

whorls of sterile leaves, that are fused in a bowl-shaped sheath. In the lower part of the internode is one whorl of sporangiophores with a twice-branching stalk. At the tip of each branch is a shield with four pendant sporangia. The vegetative shoots are assigned to the genus *Phyllotheca*.

Family Echinostachyaceae

The genus *Echinostachys* (T; Fig. 28, *j–n*) includes strobili inserted on long stalks into the leaf axils of an intensely branched shoot. The strobilus consisted of numerous whorls of sporangiophores, crowned by a disc-shaped or conical shield. The sporangia were inserted, not on the shield, as in other Equisetales, but in three longitudinal rows on a stalk. The strobili were heterosporous. Spores are of the same type (*Calamospora*) as in many other Equisetales. The vegetative shoots of the plants are assigned to *Schizoneura* (see Satellite genera of Equisetopsida).

Family Equisetaceae

Besides the living genus *Equisetum*, this family includes the genera *Equisetites*, *Neocalamites* and some others. The genus *Equisetum* is known since the Jurassic, and is usually represented by shoot remains with leaf sheaths. Rhizomes have been encountered that exhibit typical root tubercles, as well as strobili, that either crown the stem, or are attached to the tips of the lateral branches.

The genus *Equisetites* (Fig. 28, *d*, *e*) includes forms, that, although similar to the living horsetails, are not sufficiently well preserved as to be validly assigned to *Equisetum*, or they differ somewhat from the extant forms. For instance, in the strobili of the Triassic *Equisetites bracteosus* there are groups of several whorls of sporangiophores with intercalated whorls of sterile leaves. The genus *Equisetostachys* includes detached strobili or fragments of fertile shoots, allied to *Equisetum*. The genus *Neocalamites* (T–K; Fig. 28, *c*) has strobili that are probably similar to equisetaleans, but the leafy shoots are different – with long leaves, occasionally fused in groups. The ribs pass through nodes without alternation, whereas in *Equisetum* they exhibit more or less distinct alternations.

Other members

Other plants are known that may belong to Equisetales. This order is very diverse in foliage structures (leaves are solitary or fused into sheaths, simple or dichotomous), sporangiophore organization and arrangement. These characters probably vary considerably within the genera and families (i.e. as in Bowmanitales and Calamostachyales), so that the number of genera and families should probably be reduced.

The origin of the Equisetales is unknown. Their ancestors evidently belonged to the family Asterocalamitaceae (p. 91). To the genus *Archaeo-*

Fig. 28. Equisetales (*a*, *b*, *d–h*, *j–n*) and satellite genera of Equisetopsida (*c*, *i*, *o–r*); Upper Palaeozoic (*a*), Lower (*h*, *i*) and Upper (*b*, *f*, *g*, *o–r*) Permian, Middle (*j–n*) and Upper (*d*, *e*) Triassic, Rhaetian–Liassic (*c*); India (*a*), Australia (*b*), Greenland (*c*), Western Europe (*d*, *e*, *j–n*), Pechora basin (*f*, *g*), Brazil (*h*), Tunguska basin (*i*), China (*o–r*); *a*, *Phyllotheca indica* Bunb.; *b*, fertile node of *Gondwanostachys australis* S. Meyen; *c*, *Neocalamites hoerensis* (Schimp.) Halle; *d*, *Equisetites conicus* Sternb., part of rhizome with leaf sheath and roots; *e*, *E. glandulosus* Kräus., leaf sheath; *f*, *g*, *Tchernovia striata* Neub., detached sporangiophores with different number of sporangia; *h*, reconstruction of *Notocalamites askosus* Rigby; *i*, *Paracalamites vicinalis* Radcz.; *j*, *k*, *Schizoneura* (al. *Echinostachys*) *paradoxa* Schimp. et Moug., vegetative shoot (*j*), fertile shoot with young strobili (*k*); *l*, *Echinostachys cylindrica* Schimp. et Moug., sporangiophore with megasporangia; *m*, *n*, *E. oblonga* Brongn., sporangiophore with microsporangia (*m*) and microstrobilus (*n*); *o*, *Schizoneura manchuriensis* Kon'no; *p–r*, *Manchurostachys manchuriensis* Kon'no, reconstruction of fertile shoot (*p*), sporangiophore with sporangia (*q*), part of fertile zone (*r*). Scale bar = 5 cm (*d*, *j*), 1 cm (*c*, *e*, *h*, *i*, *k*, *o*, *p*), 5 mm (*a*, *b*, *n*), 2 mm (*r*), 1 mm (*f*, *g*, *l*, *m*, *q*). Modified from: Pant and Kidwai, 1968 (*a*); Townrow, 1955–1956 (*b*); Harris, 1931 (*c*); Kräusel, 1959 (*d*, *e*); Neuburg, 1964 (*f*, *g*); Rigby, 1970–1972 (*h*); Radczenko and Shvedov, 1940 (*i*); Grauvogel-Stamm, 1978 (*j–n*); Kon'no, 1960 (*o–r*).

calamites (= *Asterocalamites*; in strict accordance with the nomenclatural rules these plants should be termed *Calamites*) are referred pith casts and stems including leafy stems with vascular bundles that pass directly, or only occasionally alternating, through the nodes (Fig. 25, *c* (1, 2), *g*, *h* (1)). The leaves are dichotomously divided several times (Fig. 25, *u*, *v*). Fertile shoots consist of fertile zones, the organization and arrangement of which are as yet not clear. In some cases (*Pothocites*; Fig. 26, *p–s*) the axis has several fertile zones divided by whorls of sterile leaves. In other cases (*Protocalamostachys* and fertile shoots, described as *Bornia radiata*) the arrangement of the fertile zone is unclear. In *Protocalamostachys* (Fig. 26, *t*, *u*) the stalk of the sporangiophore was at first branched in the horizontal plane, and then in the vertical plane; on the ends of each branch was an inverted pendant sporangium. In *Pothocites* (Fig. 26, *q*, *r*) the stalk was, seemingly, divided at one point into four branches, at the tips of which were sporangia, also inverted towards the axis. In '*B. radiata*' (Fig. 26, *v*) the stalk was crowned by a shield with pendant sporangia. This series of genera probably illustrates the mode of formation of the peltate sporangiophores in articulates. It should be noted that in peltate sporangiophores of the extant horse-tails the vascular bundles retain the repeated dichotomous branching. More intensive branching of the stalk could have resulted in the formation of sporangiophores of *Gondwanostachys*, whereas the reduction of the branching could have resulted in the formation of the sporangiophores of *Notocalamites* (Fig. 28, *h*).

According to its anatomical structure *Archaeocalamites* is related to *Calamites* (Smoot *et al.*, 1982), but the secondary wood has only fascicular medullary rays, whereas interfascicular rays are absent (Fig. 25, *h* (1)). In *Pothocites* a two-layered spore wall was observed, similar to that known in Calamostachyales when the elaters are still coiled. It is believed that Calamostachyales evolved from asterocalamites. The family Equisetaceae is sometimes derived from the Calamostachyales or both are sometimes even united into one family, according to the presence of elaters on the spores, and certain other characters (Good, 1975). However, one cannot exclude the possibility that the ancestors of the Equisetaceae were herbaceous plants, not related to Calamostachyales, and evolved independently from asterocalamites.

In treatment of the Equisetidae phylogeny attention is usually focused on the structure of the nodes. In *Archaeocalamites*, Tchernoviaceae, Gondwanostachyaceae and, in part, Equisetaceae, the ribs in the nodes are opposite, whereas in the Calamostachyales and in most Equisetaceae, they alternate. According to this character there is a transition from *Archaecalamites* through *Mesocalamites* to *Calamites*. However, much remains still unclear in the evolution of the node and branching of the shoots. In *Archaeocalamites* shoots are known with very indistinct nodes, where grooves, corresponding to primary vascular bundles, are not connected by transverse grooves (Fig. 25, *c* (1)). In other cases such transverse grooves are observed. (In extant horse-tails some of the vascular bundles pass directly through the nodes and are not connected to the adjacent bundles; Bierhorst, 1971.) So far it is still unclear how the first type of nodes evolved into the second type. In *Equisetum* the branching of the shoots is extra-axillary and the branch traces alternate with the leaf traces. Evidently the same type of branching was characteristic of Tchernoviaceae, whereas in Calamostachyales and, probably, in Echinostachyaceae, it was axillary. Extraaxillary branching in extant horse-tails has been explained by the adnation of leaves to the stems, so that the branches should be correlated with the axils of leaves extending from the preceding node and completely adnate to the stem (Filin, 1978). No evidence rejecting (or supporting) this hypothesis has been found in fossil material.

Satellite genera of Equisetopsida

Genus *Paracalamites* (C–P; Fig. 28, *i*)

The pith casts are similar to those of *Archaeocalamites*, but the ribs are more prominent and transverse grooves in nodes are more marked. Branch scars are absent. Such pith casts are known in Tchernoviaceae and Gondwanostachyaceae.

Some of the species of *P*. have been described, not from the pith casts, but from stem remains with a ribbed external surface.

Genus *Annularia* (C–P; Fig. 25, *p–q*)

Leafy shoots with whorls of lanceolate or linear leaves; the whorls are arranged in the plane of the shoot. The leaves have a single midvein; at the base they are fused into a narrow ring. Some of the species belong to the Calamostachyales.

Genus *Asterophyllites* (C–P; Fig. 25, *r–t*)

Leafy shoots with whorls of thin leaves, rectangular in cross-section, attached at an acute angle to the axis. Most of the species belong to the Calamostachyales, but some are, probably, affiliated to the Bowmanitidae.

Genus *Phyllotheca* (C–P; Fig. 27, *c*; Fig. 28, *a*)

Leafy shoots. The leaves are fused over a considerable length into cylindrical, conical or bowl-shaped sheaths. Their form may differ in the stem and branches. The suture between the fused leaves is marked by a groove. The free parts of the leaves are linear. Practically all species, described as *Koretrophyllites* (conical sheaths were mistaken for internodes; see Meyen, 1971b, 1982b), may be assigned to *P*. Different species of *P*. belong to the Tchernoviaceae and Gondwanostachyaceae.

Genus *Schizoneura* (= *Convallarites*; P–T; Fig. 28, *j*, *k*, *o*)

In former reconstructions the type species *S. paradoxa* was depicted with narrow conical strobili, which later were found to belong to conifers *Aethophyllum* (Voltziaceae). The most completely preserved remains of *S. paradoxa* consist of shoots, branching on all sides. The nodes bear branches or leaves, or both. The leaves are free down to the base, or fused in 2–7-s into lanceolate sheaths. In some nodes strobili are attached that have been described as *Echinostachys* (see p. 95). In the Permian species (Gondwana *S. gondwanensis*, Cathaysian *S. striata*, etc.) most frequently two wide lanceolate sheaths have been observed in each node, but species are known with a greater number of sheaths per node. With the Cathaysian *S.* strobili of the *Manchurostachys* type (Fig. 28, *p–r*) are associated. They are totally different from *Echinostachys*, but of indeterminate structure. Fructifications of other S. species are unknown.

Class Polypodiopsida (Pteropsida, Filicopsida). Ferns

In the number of extant genera (about 300) and species (about 10 000) ferns (F.) are inferior only to angiosperms and bryophytes. The number of fossil genera also amounts to several hundreds (Andrews *et al.*, 1970; Boureau and Doubinger, 1975; Němejc, 1963; Osnovy paleontologii, 1963a; Taylor, 1981). In the systematics of F. great difficulties are encountered. F. are characterized by a remarkable diversity of vegetative and reproductive organs. In the distribution of the characters among different taxa a wide variety of combinations is observed, whereas within the taxa the characters themselves exhibit considerable variations. Because of this no definite list of characters can be offered that are persistent within one taxon, particularly within a large one. Attempts to clarify the diagnostics result in continuous reconstructions of the entire system, and recognition of ever smaller genera, families and orders. Owing to this the systematics of ferns is extremely unstable. The disagreements between the classifications already begins at the status of the F., in the whole. Most frequently they are treated as a class, occasionally as a division (Polypodiophyta). F. as a united taxon is not always distinguished. The rank of a group is treated differently even if its limits are the same. For instance, in various systems the osmundaceous F. rank from a family to a subclass. The number of families recognized among leptosporangiate F. varies from 10 to 40. The range of disagreement is evident from the following example. Some authors include the genus *Dennstaedtia* in the order Dicksoniales, others include it in the order Cyatheales; still others distinguish the subfamilies Dicksoniidae and Cyatheoideae of the family Cyatheaceae, whereas they regard dennstaedtias as a separate family including (as subfamilies) numerous groups of ferns that in other systems are assigned to the families Aspleniaceae, Polypodiaceae, etc. Finally, *Dennstaedtia* is at times included in the family Polypodiaceae. Palaeobotanical materials and phylogenetical hypotheses have so far contributed little to the precision and stability of the systematics of F.

In many earlier manuals, as well as in some new ones the logic of the F. system is as follows. All F. are divided into preferns, eusporangiate and leptosporangiate F. The preferns include polystelic and protostelic forms; sporangia are solitary with a multilayered wall, and inserted on the tips of complex-branching axes. Eusporangiate F. also have a multilayered sporangial wall, which develops from several initials; but the sporangia are situated in sori or synangia. In leptosporangiate F. the wall of the sporangia is unilayered; the sporangium develops from one initial cell. The leptosporangiate F. are further divided into isosporous (order Filicales) and heterosporous (orders Marsiliales and Salviniales). The isosporous F. are divided into families chiefly according to the characters of the sporangia (particularly on the basis of the annulus structure) and sori (attention is focused mainly on the presence and organization of the indusium).

The coherency of this system began to break up a long time ago. It came to be realized that, because transitional types occur the division of the sporangia into eu- and leptosporangiate types is merely conventional; in some genera the sporangia differ in their mode of development; the multilayered sporangial wall on maturing may become unilayered. Several Palaeozoic genera exhibited a combination of leptosporangiate characters (unilayered wall in sporangia, distinct sori) with the anatomy of the stems and petioles typical of preferns. It became apparent that a similar annulus structure could have developed independently in different groups. In the Carboniferous ferns *Phillipopteris* (Hamer and Rothwell, 1983) and *Sclerocelyphus* the number of layers in the sporangial walls varied, while no special structure was found for the opening of the sporangia. In the ferns regarded as isosporous leptosporangiates the spores were found to be of two sizes. Whereas earlier the discovery of an oblique annulus in sporangia, exindusiate sori and trilete spores in the genus *Oligocarpia* was sufficient to affiliate it with Gleicheniaceae, now the situation is different. The genus *Sermaya* has the same sori and sporangia, but its petioles are the same as in preferns. The group progymnosperms was singled out from preferns. In short, the criteria for the distinction of higher taxa of F., owing to the overwhelming amount of exceptions, became invalid.

The difficulties encountered in tracing many systematic characters of living forms in fossil F., urge the need to adopt a simplified system for the extinct forms. This is dictated by the necessity of arranging the genera into families and orders at least conventionally. The same considerations compel us to keep a single collective taxa for all F. The basic simplifications in the system adopted below, as compared with the most elaborate classification systems of F., are as follows: (1) Ophioglossales and many families (Davalliaceae, Hymenophyllaceae, Hymenophyllopsidaceae, Platyzomaceae, etc.), not firmly identified in the geological record, are excluded from consideration; (2) the treatment of certain taxa was purposely kept, wherever possible, rather broad in scope; for instance, the families Dicksoniaceae and Cyatheaceae are united; the families Aspleniaceae, Adiantaceae and Grammitidacea, Aspidiaceae, often regarded as independent, are included in the Polypodiaceae; (3) the orders Cladoxylales and Ibykales are not subdivided into families. Notwithstanding these simplifications many genera remain left out of the system, and, in part, are treated as satellite genera to the class Polypodiopsida.

Order Cladoxylales

To Cladoxylales (C.) are referred Devonian and Lower Carboniferous plants with numerous concentrical xylem strands that pass along the stem in various configurations. The strands are situated around the periphery of the stem in a regular ring and radially elongated, or scattered over the entire cross-section of the stem, and then they are irregular in shape, rounded, bent, variously anastomosing and even intersecting the entire diameter of the stem. The diversity depends to a great extent on the level of the cut and the order of branching. The interpretation of the stelar organization of C. (Fig. 29, *d–g*, *k*) depends upon the interpretation of singular xylem strands and the spaces between them. These spaces may correspond to solenostele perforations, leaf-, branch- or root gaps. The general organization, dependent upon the treatment, may be regarded as eustelic, siphonostelic or otherwise. The xylem strands may be rounded or outstretched. The tracheids of the outer zone (usually scalariform) frequently are arranged in orderly rows. If the rows are divided by narrow rays, then the wood is regarded as secondary. A round flattened strand of thin-walled parenchymatous cells, surrounded by protoxylem tracheids, can at times be seen in the centre of a vascular bundle. More frequently such strands are shifted towards both ends of the flattened bundle. This results in something like 'peripheral loops', known in many zygopterid ferns. In another C., in place of the protoxylem, canals are formed (protoxylem lacunae), that are accompanied by protoxylem tracheids of disorderly arrangement. In *Rhymokalon* (D_3; Fig. 29, *f*, *g*) the protoxylem is absent, with the lacunae occurring in axes of the first order, but it accompanies the protoxylem lacunae in axes of higher orders. This genus is important for the understanding of the stelar organization of C. In the cross-section of the axis of the first order an intricately dissected xylem can be observed, the cells of which are mixed in the centre of the axis with parenchymatous ones. Towards the periphery the xylem forms radial plates, interrupted by nests of the parenchyme, so that the plate appears bead-like in cross-section. In the lateral branches the xylem strands are rounded, with protoxylem lacunae in the centre. The strands are anastomosing. The stele of the axis of the first order may be regarded as an actinostele with an incipient pith. The widening of the parenchymatous areas, both in the centre of the axis and within the xylem plates, may lead to the formation of numerous xylem bundles typical of C. This implies that the stelar organization of C. is a strongly subdivided actinostele.

Best studied is the genus *Pseudosporochnus* (D_2–D_3; Fig. 29, *a–d*). Its stem bears a bunch of roots basally and a bunch of branches apically. The branches were either forked, or divided into bunches of three–four branches of the next order, to which the most slender branches were of spiral attachment. The terminal branches were partly sterile, partly fertile, with paired elliptical sporangia on the tips. The vascular system was composed of a large number of radially elongated external bundles and more rounded central bundles. The external bundles showing peripheral loops, did not anastomose. Passing upwards the central rounded bundles become joined to the radial ones. In the branches the external bundles become fused into V- or W-shaped structures, opening outwards. Secondary wood has not been found. The metaxylem consists chiefly of scalariform tracheids, but rounded bordered pitting occurs. From what we saw in *Rhymokalon*, the stelar structure of *P. nodosus* can be derived through separation of singular xylary bundles. But it should be remembered that *Rhymokalon* was recovered from the Upper Devonian, whereas *P. nodosus* was recovered from the Middle Devonian. Moreover, in *Rhymokalon* the protoxylem lacunae are in bundles, whereas in *P. nodosus* there are peripheral loops. The latter are typical of ferns Zygopteridales (see below), while the protoxylem lacunae are typical of articulates, certain ferns, gymnosperms and angiosperms.

To C. also belong the genera *Calamophyton* and *Hyenia*, formerly called protoarticulates and regarded as ancestors of articulates. *Calamophyton* (D_2; Fig. 29, *h–j*) is similar in outward aspect to *Pseudosporochnus*; however the branches on the main axis were at first spirally arranged, later branching resulted in bunches of twigs. Transversal fissures of the axis infilled with sediments have been mistaken for nodes of an articulated shoot. Fructifications on fracture planes looked like a forked stalk with pendant sporangia. The peltate sporangiophores of articulates were derived from this structure. However, uncovering revealed much more complicated branching of the

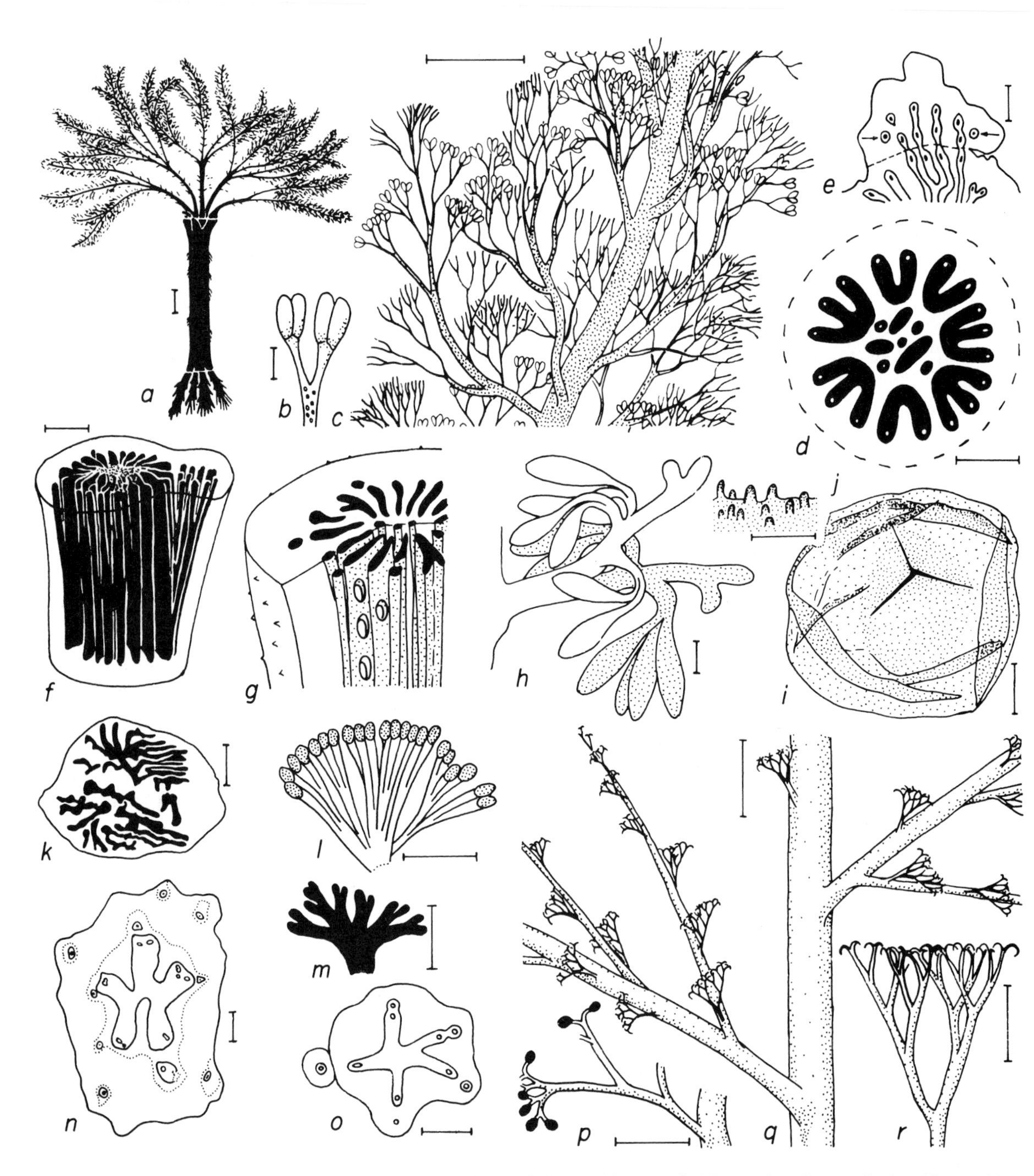

Fig. 29. Cladoxylales (*a–m*) and Ibykales (*n–r*); Middle (*a–d*, *h–r*) and Upper (*f*, *g*) Devonian, Lower Carboniferous (*e*); Western Europe (*a–e*, *h–m*), North America (*f*, *g*, *n–r*); *a–d*, *Pseudosporochnus nodosus* Lecl. et Banks, reconstructions of general habit (*a*), sporangiophore (*b*), lateral branch (*c*), stele of branch of 3rd order (*d*); *e*, *Cladoxylon mirabile* Ung., cross-section of stem in place of branch departure, arrows show vascular bundles of aphlebia; *f*, *g*, *Rhymokalon trichium* Scheckl., stelar organization, in *f*, perforations of plates abolished; *h–m*, *Calamophyton primaevum* Kr. et Weyl., reconstruction of sporangiophore (*h*), spore (*i*) and its ornamentation (*j*), stele (*k*), as well as sporangiophore (*l*) ▶

sporangiophores (Fig. 29, *h*). The anatomical structure of the stem is about the same as in *Pseudosporochnus*, but the bundles with peripheral loops anastomose, forming a strongly dissected arthrostele. *Hyenia* differs from *Calamophyton* in having horizontal rhizomes, not a vertical stem. The structure of the sporangiophores is practically the same. The anatomical structure of the axes is as yet unknown. In the anatomy of their axes *Protohyenia* (D_1–D_2 boundary beds; Zakharova, 1985) and the collective genus *Cladoxylon* are closely related to *Pseudosporochnus* and *Calamophyton*. Usually it is indicated that in *Cladoxylon* the branches were arranged on the stem in bunches or spirally; the sporangiophores were fan shaped, whereas the vegetative appendages were irregularly dichotomous (Fig. 29, *l*, *m*). All these characters are shown for remains that actually belong to *Calamophyton* (Schweitzer and Giesen, 1980). Only the Upper Devonian and Tournaisian petrified stem fragments (Fig. 29, *e*) can be reliably assigned to the genus *Cladoxylon*; other organs of this genus are unknown.

Several other Middle Devonian–Lower Carboniferous genera (*Pietzschia*, *Xenocladia*, etc.) have been described from small fragments of axes having a stelar organization the same as in C. (Gensel and Andrews, 1984; Iurina, 1985; Scheckler, 1974; Stein *et al.*, 1983). The genus *Pietzschia* is of particular interest due to the presence of wedge-shaped sclerenchyma plates interposed between the xylem plates around the periphery of the stem (Daber, 1980).

Order Ibykales (= Iridopteridales)

A group of genera, that may be affiliated both with Cladoxylales (particularly with *Rhymokalon*), and with Zygopteridales, is occasionally united in the order Ibykales (Skog and Banks, 1973), or Iridopteridales (Stein, 1982). The name Ibykales receives priority here because the genus on which it was first based – *Ibyka* – yields both fertile and sterile parts, whereas for *Iridopteris* only stems are known. The xylem of the main stem in cross-section looks like a star with three or more rays. The protoxylem (one–two stranded) is near the ends of the rays and sometimes is replaced by a canal; peripheral loops are absent. Smaller round (with a central protoxylem) or larger elliptical (with protoxylem strands near the narrowed sides) bundles issue from the axial system, either in whorls or alternately. The genera *Iridopteris* (D_2; Fig. 29, *n*) and *Arachnoxylon* (Stein *et al.*, 1984) are known only from fragments of petrified stems. In the genus *Ibyka* (D_2; Fig. 29, *o*–*r*) the branches of the first orders were of spiral attachment, whereas the branches of the last order were nearly whorled with drooping ends. Fertile branches of the last order were dichotomous with solitary sporangia borne on the tips. This genus was affiliated with ancestors of articulates, partly due to the presence of protoxylem lacunae and whorled arrangement of the branches. The whorled arrangement has been noted also in *Iridopteris*. There is reason to believe that articulates might have evolved from Ibykales, especially considering the similarity in the anatomical structure of *Iridopteris* and *Eviostachya* (Fig. 22, *c*).

Order Zygopteridales

The Zygopteridales (Z.) are ferns, where the free sporangia with a two- or multilayered wall and thick stalk are inserted either solitarily, or in groups on the tips of slender branching axes, or on the lower surface of the pinnules. The axes are protostelic, the protostele is solid or mixed; phyllophores and petioles have a C-, V-, W-, or H-shaped xylem. The phyllophores are leafless and

and fertile appendage (*m*) according to initial view when the relevant specimens were included in *Cladoxylon scoparium* Kr. et Weyl.; *n*, *Iridopteris eriensis* Arn., stem cross-section with 5 leaf traces and one branch trace; *o*–*r*, *Ibyka amphikoma* Skog et Banks, section of branching axis (*o*), fertile branch (*p*), general habit (*q*), sterile appendage (*r*). Scale bar = 10 cm (*a*), 5 cm (*q*), 1 cm (*c*, *f*, *p*, *r*), 5 mm (*d*, *l*, *m*), 1 mm (*b*, *e*, *h*, *k*, *n*, *o*), 25 μm (*i*), 10 μm (*j*). Leclercq and Banks, 1962 (*a*–*c*); Leclercq and Lele, 1968 (*d*); Scott, 1923 (*e*); Scheckler, 1975 (*f*, *g*); Leclercq and Andrews, 1960 (*h*); Schweitzer, 1973 (*i*, *j*); Kräusel and Weyland, 1926 (*k*–*m*); Stein, 1982 (*n*); Skog and Banks, 1973 (*o*–*r*).

branch in various planes and may have laminated pinnules. Of interest is the fact that fern-like pinnae and pinnules with a well-developed lamina first appear in Z. only in the Early Carboniferous, i.e. much later than in progymnosperms and gymnosperms. At first all Palaeozoic ferns with the above-described anatomical structure were commonly assigned to Z. However, it was found later that in some Z. the sporangia have unilayered walls and a more slender stalk. According to these characters and the structure of the annulus, these Z. are more closely related to leptosporangiate 'true' ferns, to which they are sometimes assigned (Galtier and Scott, 1985; Taylor, 1981). These genera are united in this book in the order Botryopteridales, which combines characters of Z. (anatomical structure of the stems, phyllophores, and, in part, petioles) and Polypodiales (sporangial structure). On the other hand, *Rhacophyton* (Fig. 30), the systematic position of which has for long been under discussion in the literature, can be assigned to Z. (Andrews and Phillips, 1968; Cornet *et al.*, 1976; Dittrich *et al.*, 1983; Scott and Galtier, 1985).

Many genera of fossil ferns established for fragments of stems, phyllophores and petioles, cannot be classified among Z. and Botryopteridales, whereby they are regarded as satellite to both orders. For instance, the genus *Rhabdoxylon* (C_{2-3}; J. Holmes, 1979) is known, that has protostelic dichotomous axes without differentiated phyllophores. These axes differ from the trimerophytalean ones in the presence of diarch aerial roots. Since the fructifications are unknown, it is impossible to place this genus into any particular order.

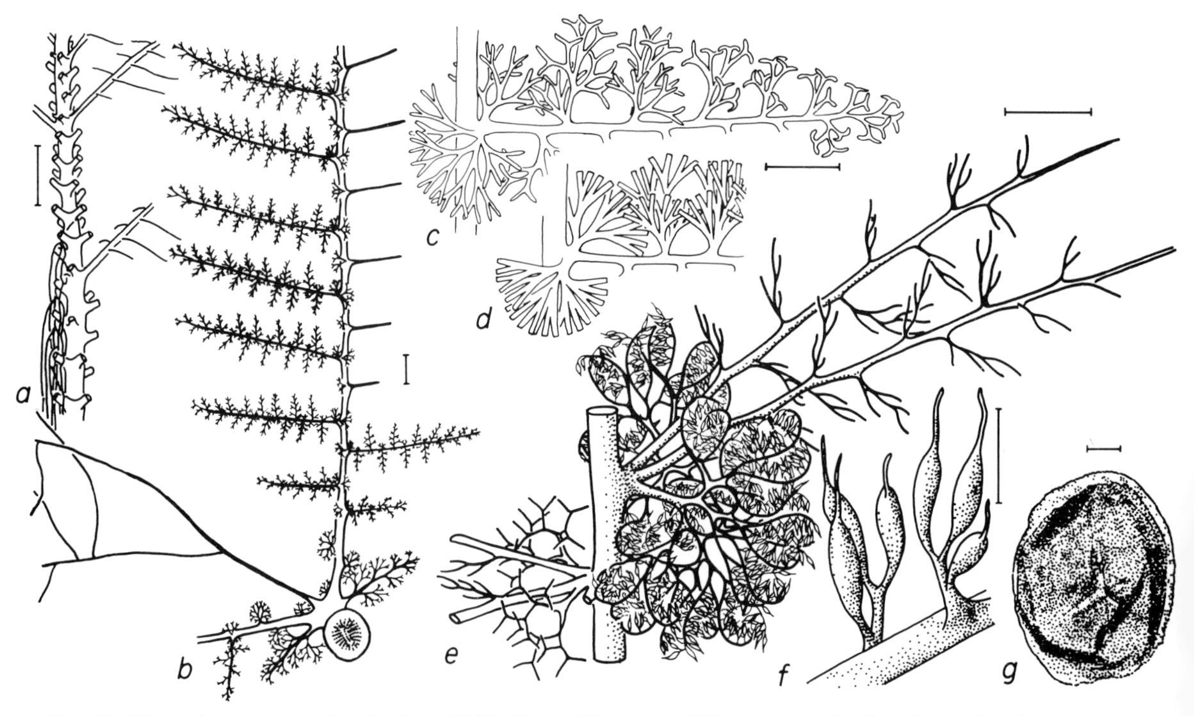

Fig. 30. *Rhacophyton ceratangium* Andr. et Phill.; Upper Devonian; USA; *a*, part of main axis bearing alternating pairs of vegetative rachises (their lower portions are shown) in two rows on opposite sides of axis, at left below – aerial roots; *b*, upper view of vegetative fronds with aphlebia and aerial root (at left), cross-section shows clepsidroid anatomy; *c*, trimerous (slightly flattened) pinna of last order; *d*, strongly flattened pinna of last order from younger beds; *e*, part of fertile frond; *f*, sporangia; *g*, spore. Scale bar = 5 cm (*a*), 1 cm (*b*), 5 mm (*c–e*), 1 mm (*f*), 10 μm (*g*). Modified from: Cornet *et al.*, 1976 (*a–d*); Andrews and Phillips, 1968 (*e–g*).

Apart from this, in the oldest ferns we are faced with an intricate combination of types of organs, which have been described from fragmentary remains under different generic names. For example, petioles, hour-glass-shaped in cross-section, with peripheral loops and narrowed sides (Fig. 31, *k*) are referred to *Clepsydropsis*. Such petioles associate with stems, not only of Z., but also of Cladoxylales. Petioles with a xylem H-shaped in cross-section (Fig. 27) are placed into *Etapteris*. They associate with fertile pinnae of *Corynepteris*, *Biscalitheca* (family Zygoperidaceae) and *Tedelea* (order Botryopteridales). The genus *Ankyropteris* comprises stems with a xylem polygonal in cross-section and a small amount of parenchyma in the centre, from which narrow parenchymatous rays spread out in the form of an irregular star. Such stems associate with petioles of *Etapteris* (in *Tedelea*) and *Clepsydropsis* (in *Austroclepsis* and *Senftenbergia*). On the other hand, Z. are linked by intermediate forms with progymnosperms (via *Rhacophyton*), cladoxylaleans (particularly because of the *Clepsydropsis* petioles), with various botryopterids (according to the anatomical structure of the stems, phyllophores and petioles). In *Rhacophyton* and in one species of *Zygopteris* (Fig. 31, *j*) secondary wood was found, which is not characteristic of ferns. Z. differ from modern 'true' ferns in the thickening of metaxylem tracheids (Galtier and Scott, 1985). In Z. and Botryopteridales the thickening ranges from scalariform to circular bordered pitting, whereas modern 'true' ferns show only the simple scalariform pitting. A comparable evolutionary pattern is known in seed plants, where some younger groups (certain bennettites and more primitive angiosperms) produce only scalariform tracheids, although older seed plants and their ancestors, the progymnosperms, had circular pitting. The similarity between certain Z. and other fern groups may be interpreted in terms of probable phylogenetic linkages and in terms of parallelism. Many Z. had well-developed aphlebiae at the base at the pinnae (Fig. 30, *b*; Fig. 31, *a*, *d*; Fig. 32, *c*, *e*), which is characteristic of many ferns, as well as of some primitive gymnosperms.

Terminological difficulties arise in describing Z., because they do not exhibit distinct division into cauline and foliar parts. Anatomically, radial-symmetrical protostelic (occasionally with a mixed pith) and actinostelic stems can be recognized. Petrified fragments of dorsiventral rachises, attached to the stems, have been described in the literature as petioles. Besides stems and petioles, dorsiventral organs, intermediate in structure – phyllophores – may occur. They are of spiral attachment to the stem and bear either two or four rows of the pinnae. If the entire successive order of branching from the main stem to the branches or pinnules of the last order is unknown, then it is difficult to distinguish the stems, phyllophores and petioles. Frequently any offshoot of the stem, that has a dorsiventral structure, is termed a petiole, although it may probably be a phyllophore. In some Z. a false stem can be formed by the accretion of a compact phyllophore (or petiole) mass and aerial roots around the stem. In the genus *Austroclepsis* (Fig. 33, *h*), that probably was allied to Z., the stem was branched passing through the false stem, in the cross-section of which cuts of several stems can be seen. In other Z. the false stem was lacking due to the sparse stem branching and smaller number of aerial roots. Phyllophores of Z. differed in the form of the wood. The latter can be clepsydroid (as in *Clepsydropsis*; Fig. 31, *k*). The peripheral loops were closed or opened in the places from which petioles issued. In other Z. phyllophore wood was etapteroid in cross-section (Fig. 31, *f*, *g*), or of another type. In petioles the wood cross-section was round or C-shaped (abaxially convex). The most primitive Z. had no leaf laminae; later strongly divided laminae appeared, and finally, increasingly more fused pinnules. In the more advanced forms (*Corynepteris*), sporangia, that were 'transferred' from the leafless sporoclads, were inserted on the pinnules.

Three families can be recognized among Z. The Rhacophytaceae include the most primitive homosporous forms with four-ranked pinnae, incipient dissected pinnules and specialized sporoclads. Stauropteridaceae unite heterosporous forms without leaf laminae. The pinnae in the Zygopteridaceae bear laminated pinnules, at times fertile ones.

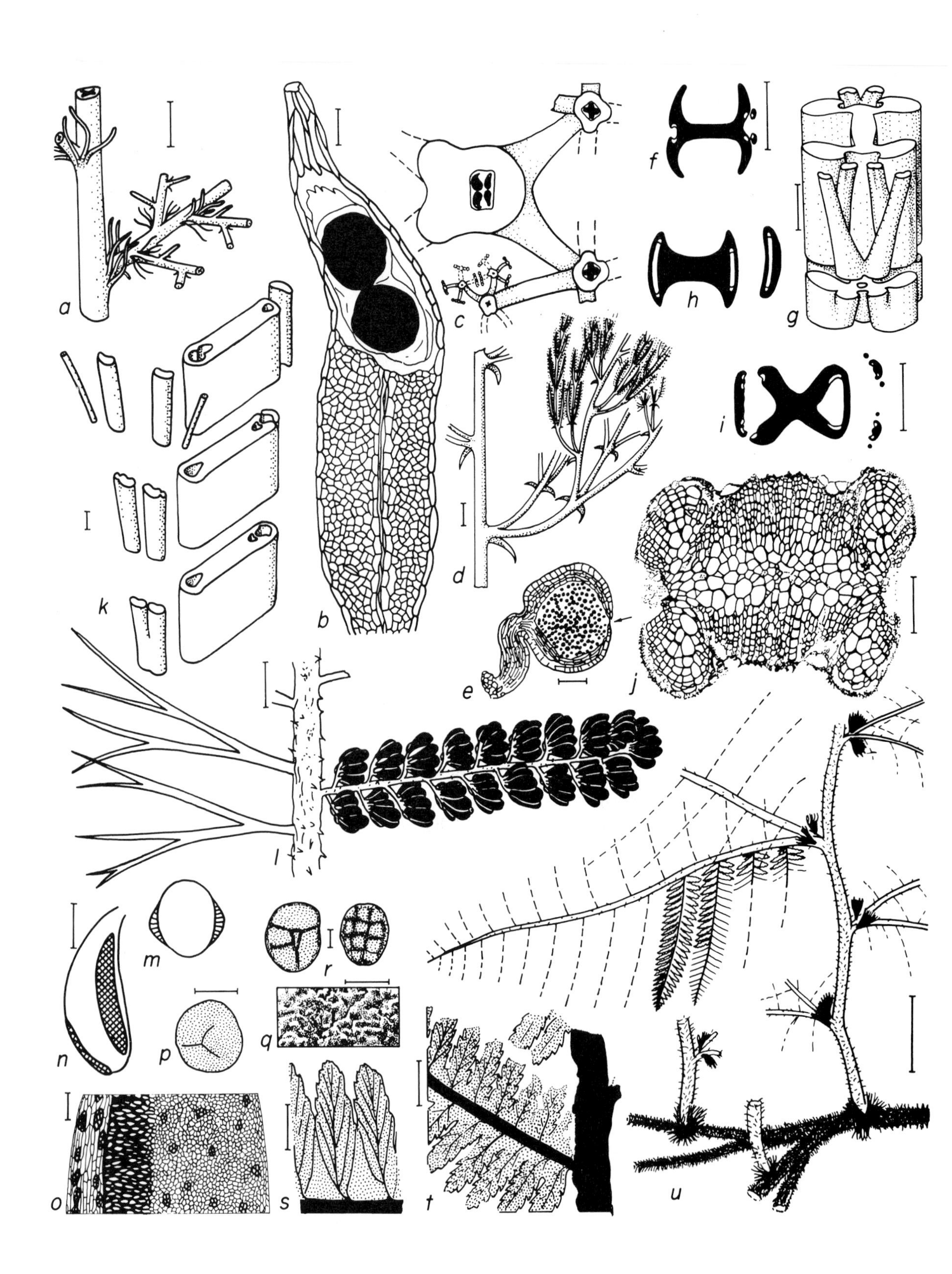
a
b
c
d
e
f
g
h
i
j
k
l
m
n
o
p
q
r
s
t
u

Family Rhacophytaceae

Only the genus *Rhacophyton* (D_3; Fig. 30) has been well studied. The main axis is up to 2 cm in diameter and over 1.5 m in length. Its stele was probably stellate in cross-section. Paired branches of irregular spiral arrangement were accompanied by a pair of aphlebiae and aerial roots. Each branch in the lower part was pinnately divided in one plane. The terminal branches were flattened in various degrees, and approximated strongly dissected pinnules. In the rachises of at least two orders the primary wood was clepsydroid (similar to Fig. 31, *k*) with large peripheral loops (Dittrich *et al.*, 1983). From the narrow ends of the primary bundle daughter bundles issued, at first C-shaped, but soon bifurcating. From the wide sides of the primary bundle secondary wood (tracheids with scalariform and bordered pits) with narrow rays was formed. In thicker rachises secondary wood surrounds primary wood on all sides. Fertile pinnae accompany sterile pinnae and are dichotomously divided. Sporangia inserted on branching sporangiophores are spindle shaped with strongly attenuated apices. Spores have flanges (cavate?).

Family Stauropteridaceae

Only a single genus *Stauropteris* is known here (C; Fig. 31, *a–e*) with complex branching of the leafless axes. Six orders of branching have been traced. At each point of branching the axis issues two daughter axes, at times accompanied by aphlebia-like appendages, so that the pinnae become of four-ranked arrangement. Forms are known with a two-ranked arrangement of the pinnae (Cichan and Taylor, 1982a). Peripheral loops are lacking. The primary wood is cruciform in cross-section. In the thicker axes the rays of the cross separate. The secondary wood is lacking. The plants are isosporous and heterosporous. The sporangia are apical. Some of them, round in shape with a multilayered wall, produce microspores. Others are longitudinally elongated, filled in the bottom part by parenchyme; in the upper part, with a unilayered wall, two to eight megaspores are encased. If only two megaspores functioned, then the other (smaller) megaspores of the same tetrad became aborted. The sporangia dehisced by an apical stomium. *Chacassopteris* (C_1) and *Rowleya* (C_2) may probably be allied to this family.

Family Zygopteridaceae

These plants have well-developed pinnules on vegetative fronds. The fertile axes are either leafless, of the kind of system of axes bearing sporangia bunches, or leafy. The sporangial wall is multilayered with a thick annulus. Best studied is the genus *Nemejcopteris* ($C–P_1$; Fig. 31, *l–u*; Fig. 32, *a* (6)), that has a confused nomenclatural history. Dif-

Fig. 31. Zygopteridales; Lower (*b, i, k*), Middle (*a, c–g*) and Upper (*j, m, n, q–s*) Carboniferous, Lower Permian (*l, o, p, t, u*); USA (*a, j, m, n, r*), Western Europe (*b–g, i, k, l, o, p, q, s–u*); *a*, *Stauropteris biseriata* Cich. et Tayl.; *b*, megasporangium of *S. burntislandica* Bertr., megaspores in black; *c–d*, *S. oldhamia* Bin., scheme of branching and stelar organization (xylem in black), view along main axis (*c*), reconstruction (*d*), section of microsporangium (*e*); *f*, cross-section of wood in phyllophore of *Etapteris diypsilon* (Will.) Bertr., protoxylem left white; *g*, wood in phyllophore of *E. scottii* Bertr. and *E. leclercqae* Smoot et Tayl.; *h, i*, cross-section of wood in *Zygopteris williamsonii* Bertr. (*h*) and *Diplolabis roemeri* Solms-Laub., protoxylem left white; *j*, cross-section of phyllophore of *Zygopteris* sp. with secondary wood; *k*, primary wood in phyllophore and petioles of *Clepsidropsis leclercqae* Galt.; *l–u*, *Nemejcopteris feminaeformis* (Schloth.) Bart., reconstruction of fertile frond (*l*), cross-section of sporangium (*m*) and its general view (*n*) and wall structure (*o*) with multiseriate annulus and sclereid nests, spore (*p*), exine relief (*q*), intracellular gametophyte (*r*), vegetative pinnae (*s, t*), general habit (*u*). Scale bar = 10 cm (*u*), 1 cm (*d*), 5 mm (*a, f, i, l, s, t*), 1 mm (*g, k, n*), 0.5 mm (*j*), 200 μm (*e, o*), 100 μm (*b*), 50 μm (*p*), 10 μm (*q, r*). Modified from: Cichan and Taylor, 1982a (*a*); Andrews, 1961 (*b*); Hirmer, 1927 (*c, d*); Scott, 1920 (*e*); Gothan and Weyland, 1964 (*f, h, i*); Phillips, 1974 (*g*); Smoot and Taylor, 1978 (*g*); Phillips, 1974, after Dennis (*j*); Galtier, 1970 (*k*); Barthel, 1968 (*l, o, s–u*); Mamay, 1957 (*m, n, r*); Galtier and Grambast, 1972 (*p*); Brousmiche, 1976 (*q*).

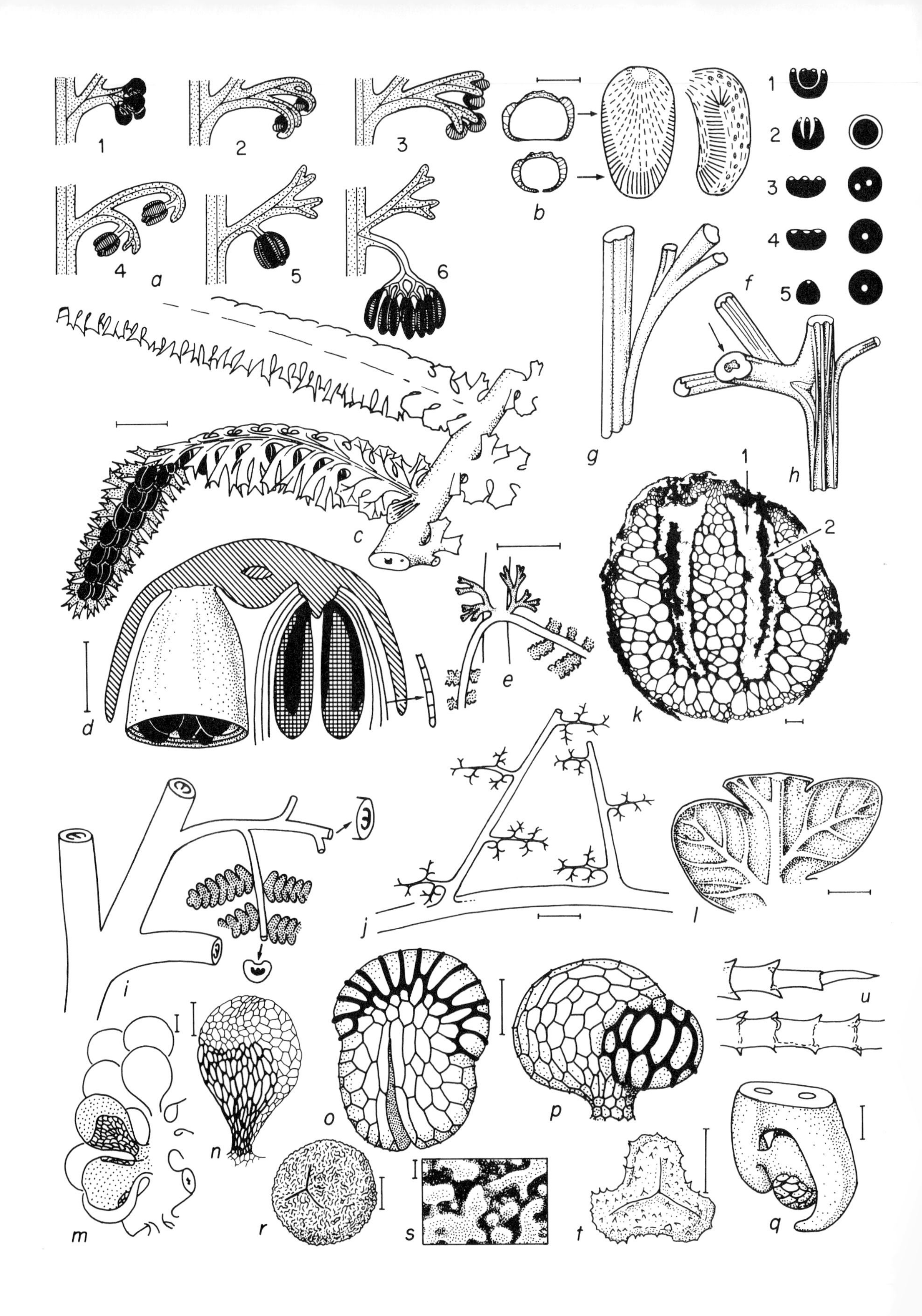

1
2
3
4
a
5
6
b
1
2
3
4
5
f
g
h
c
1
2
d
e
k
i
j
l
u
m
n
o
p
q
r
s
t

ferent parts of it have been described as the genera *Zygopteris*, *Etapteris*, *Biscalitheca*, *Schizostachys*, *Monoscalitheca*, *Pecopteris*, etc. This may be explained to a certain extent by the different state of preservation of the remains (impressions–compressions, petrifications). It seems reasonable to include in the genus *Nemejcopteris* the entire reconstructed plant, when the connections are apparent between the vegetative (of the *Pecopteris feminaeformis* type) and fertile pinnae. Detached fertile pinnae are referred to the genera *Biscalitheca* or *Zygopteris* (their relationship is unclear; see details in Millay and Rothwell, 1983). Permineralized phyllophores of certain species belong to the genus *Etapteris*. The wood in the petioles was C-shaped in cross-section. The anatomical structure of the stem is unknown. In reconstructions (Fig. 31, *u*) *N.* was depicted with a prostrate, densely trimmed stem (rhizome). At times it was branched with upward extending phyllophores, on which were sparsely situated paired pinnae of the penultimate order with numerous scales (the pinnae are in general of four-ranked arrangement). Strongly dissected aphlebiae were attached at the base of the phyllophores and pinnae. Pinnules exhibited a typical denticulate margin. Stalked bunches of banana-shaped sporangia were produced by fertile pinnae. The sporangia had multilayered walls with numerous tiny nests of mechanical cells. The annulus was in the shape of wide lateral bands of cells with strongly thickened walls. Occasionally the spores germinated in the sporangia.

Genus *Corynepteris* (C; Fig. 32, *a* (5), *b–e*)

The pinnae of the penultimate order were situated on axes (of the *Etapteris* type) in four rows. Sterile pinnae have been described in the literature as *Alloiopteris*. Sori, consisting of 5–10 sporangia, were attached by a thick stalk to the lower side of the pinna at the base of each pinnule. At times an involucre was formed around the sori – somewhat like a three-fold indusium (Fig. 32, *d*). The vascular bundle extended through the soral stalk. Sporangia were sessile or with a short stalk, slightly bent inwards of the sori. The annulus was U- or V-shaped, composed of several rows of cells. Nests of mechanical cells were found in the sporangial wall. The spores were trilete, spinose.

Order Botryopteridales

The Botryopteridales (B.) are closely similar to Zygopteridales in the anatomical structure of stems, phyllophores and petioles, but differ from them in their sporangial organization according to which they are more closely related to 'true' ferns

Fig. 32. Zygopteridales (*a*5, *a*6, *b–e*) and Botryopteridales (*a*1, *a*2, *a*3, *a*4, *f–t*); Lower (*a*2, *a*4, *f*5, *g*, *j*, *o–q*), Middle (*a*3, *a*5, *b–e*, *f*3, *f*4, *h*, *r–t*) and Upper (*a*1, *f*1, *f*2, *k–n*, *u*) Carboniferous, Lower Permian (*a*6); Western Europe (*a–c*, *e–h*, *k–q*, *u*), USA (*d*, *r–t*); *a*, position of sporangia on fronds of *Botryopteris forensis* Ren. (1), *B. antiqua* Kidst (2), *Boweria* (3), *Musatea globata* Galt. (4), *Corynepteris* (5), *Nemejcopteris* al. *Biscalitheca* (6); *b*, *Corynepteris scottii* Galt. et Scott, structure of sporangium; *c*, *C. sternbergii* (Ett.) Zeill., reconstruction of fertile pinna; *d*, *C. involucrata* Baxt. et Baxend., scheme of fertile pinna in cross-section, magnified section of involucre at arrow; *e*, *Alloiopteris* (al. *Corynepteris*) *sternbergii* (Ett.) H. Pot., attachment of pinnae and aphlebiae to phyllophore; *f*, cross-section of wood in phyllophores (left row) and stems (right row) in *Botryopteris renaultii* C. Bertr. et Corn. (1), *B. forensis* (2), *B. tridentata* (Fel.) Scott (3), *B. mucilaginosa* Kraent. (4), *B. antiqua* (5); *g*, wood structure in branching phyllophore of *B. antiqua*; *h*, wood of *B. tridentata*, siphonostelic axis (at arrow) departs from branching phyllophore; *i*, *Botryopteris* sp., branching phyllophore bearing pinna with laminated pinnules; *j*, frond of *B. antiqua*; *k*, *B. forensis*, cross-section of vascular bundle (1, phloem zone; 2, sclerenchyme); *l*, pinnules of *B. forensis*; *m*, *n*, sporangia of *B. forensis*; *o*, *p*, sporangia of *B. antiqua*, apical (*o*) and lateral (*p*) views; *q*, subapical sporangial attachment in *B. antiqua*; *r*, *s*, spore of *B. globosa* Darr., general view (*r*) and exine relief (*s*); *t*, spore of *Botryopteris* sp. (species with sporangia abaxially attached to pinnules); *u*, multicellular 'equisetoid' hairs of *Botryopteris* sp. Scale bar = 1 cm (*e*), 5 mm (*c*, *j*), 1 mm (*d*, *l*), 500 μm (*b*), 200 μm (*k*, *m*, *n*), 100 μm (*o*, *p*, *u*), 10 μm (*r*, *t*), 1 μm (*s*). Modified from: Galtier, 1981 (*a*, *q*) and 1970 (*j*, *m–p*); Galtier and Scott, 1979 (*b*, *c*, *e*); Baxter and Baxendale, 1976 (*d*); Snigirevskaya, 1962a (*f*); Phillips, 1974 (*g*, *h*); Delevoryas, 1962 (*i*); Galtier and Phillips, 1977 (*k*, *l*, *u*), Millay and Taylor, 1982 (*r–t*).

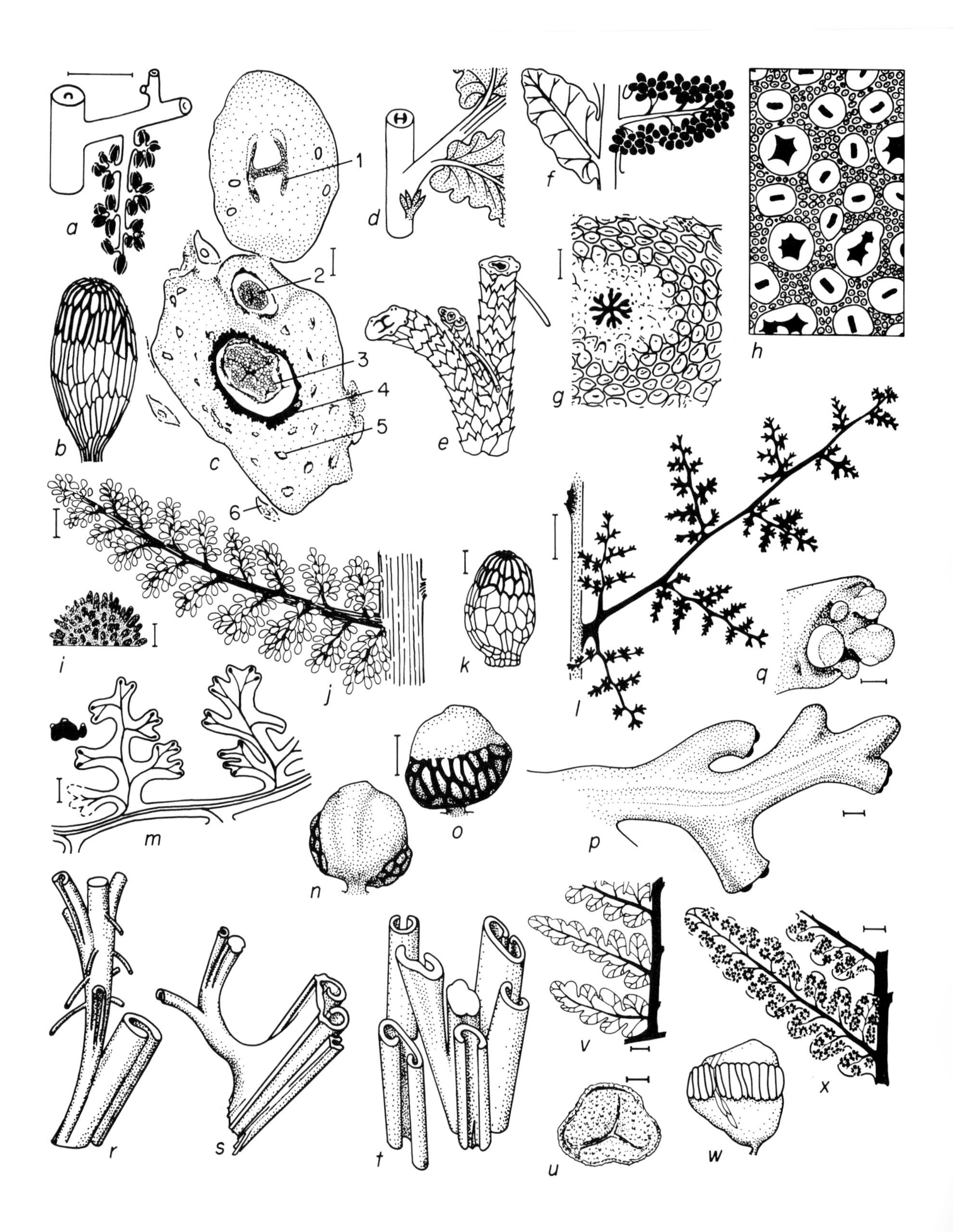

1
2
3
4
5
6
a
b
c
d
e
f
g
h
i
j
k
l
m
n
o
p
q
r
s
t
u
v
w
x

(Osmundales and Polypodiales). In contrast to the Zygopteridales, in B. of sporadic occurrence are siphonostely, axillary branching, and sori composed of a large number of sporangia. A characteristic feature of B. is the absence of a strict order in the interrelations between stems, phyllophores and petioles. Frequently the stems issue from phyllophores, phyllophores from petioles; sometimes only stems and petioles occur. The most primitive B. did not have laminated pinnules, although they are present in most B. Among B. several families can be recognized that differ in the organization and arrangement of the sporangia, also in anatomical features. In the Botryopteridaceae sporangia are either in leafless aggregates, or inserted on pinnules; the annulus is lateral and multiseriate. The sporangia in Tedeleaceae have an apical annulus (the same as in Schizaeaceae), are situated at the margin of modified pinnules, or attached to the terminal branches of the pinnately branching organ. In Sermayaceae the pinnules bear sori or indistinct groups of sporangia on unaltered pinnules; sporangia have an oblique annulus (as in Gleicheniaceae). The sporangia in Psalixochlaenaceae are the same as in Botryopteridaceae, but are situated on the obliquely truncated and flattened apices of the pinnately branching axes. A similarity can be perceived between the above-cited families and various families of Polypodiales, but whether this may be explained in terms of phylogenetic links between the respective families, or in terms of their parallelism, still remains unclear.

Family Botryopteridaceae

Of this family only the genus *Botryopteris* (C–P; Fig. 32, *a* (1, 2), *f–u*) has been well studied. In primitive species the intensely branching axes were 'expinnulate' (Fig. 32, *j*), whereas in the more advanced forms pinnules were produced of the *Pecopteris* (Fig. 32, *i*, *l*) or *Sphenopteris* type. In the most ancient forms to the protostelic axis (it might have issued from the petiole) were attached flattened phyllophores with a wood, elliptical in cross-section and a single protoxylem bundle on the adaxial side. In younger forms the number of protoxylem bundles increased, and the wood gradually acquired an omega-shape in cross-section. In general, no distinctions are apparent between the phyllophores and petioles in the wood structure. In the successive orders of branching it can be seen that from radially symmetrical axes with a cylindrical xylem (stem) extend petioles

Fig. 33. Botryopteridales (*a–f*, *i–x*) and Zygopteridales (*g*, *h*); Lower (*a*, *h*) and Middle–Upper (*b–f*, *i–x*), Carboniferous, Lower Permian (*g*); Western Europe (*a*, *g*, *i–q*, *u–x*), USA (*b–f*, *r–t*); *a*, *Musatea duplex* (Will.) Chaph. et Alv., reconstruction of fertile pinna; *b–f*, *Tedelea* (al. *Ankyropteris*) *glabra* (Baxt.) Egg. et Delev., sporangium (*b*), cross-section of stem with axillary branch and phyllophore (*c*; 1, wood of phyllophore; 2, wood of axillary branch; 3, wood of stem; 4, inner zone of cortex; 5, vascular bundle of scale; 6, scale on stem), sterile pinna with aphlebia (*d*), stem with axillary branch, scales, phyllophore and aerial roots (*e*), sterile and fertile pinnules (*f*); *g*, *Asterochlaena laxa* Stenz., cross-section of stem and numerous petioles; *h*, *Austroclepsis*, cross-section of false stem, stem wood stellate, wood of phyllophores clepsidroid, spaces between stems and phyllophores are filled with aerial roots; *i*, *j*, *Senftenbergia plumosa* (Art.) Radf. var. *ligerensis* Grauv.-St. et Doub. with spores *Raistrickia polymorpha* Grauv.-St. et Doub., exine relief (*i*) and fertile pinna (*j*); *k*, *S. pennaeformis* (Brongn.) Stur, sporangium; *l–q*, *Psalixochlaena cylindrica* (Will.) Hold, pinna (*l*), pinnules on pinna and cross-section of wood in pinna of higher order (*m*), sporangium as seen from stomium (*n*) and annulus (*o*), fertile pinnule (*p*) and its distal part with sporangia (*q*); *r*, *Anachoropteris clavata* Grah., wood of lateral shoot departing from petiole; *s*, *Anachoropteris* sp., wood of adaxial shoot and lateral pinna departing from petiole; *t*, wood of stem *Tubicaulis* bearing petioles *A. involuta* Hosk.; *u*, spore *Granulatisporites parvus* (Ibr.) Pot. et Kremp from sporangium of *Oligocarpia leptophylla* (Bunb.) Grauv.-St. et Doub.; *v–x*, *O. brongniartii* Stur, sterile pinna (*v*), sporangium (*w*), fertile pinna (*x*). Scale bar = 1 cm (*g*, *l*), 5 mm (*a*), 1 mm (*c*, *j*, *m*, *v*, *x*), 200 μm (*k*, *p*), 100 μm (*n*, *o*, *q*), 20 μm (*i*), 10 μm (*u*). Modified from: Chaphekar and Alvin, 1972 (*a*); Taylor, 1981, after Eggert and Taylor (*b*); Andrews, 1961, after Baxter (*c*); Andrews *et al.*, 1970, after Eggert (*d–f*), Radforth (*k*) and Abbott (*v–x*); Hirmer, 1927, after Stenzel (*g*); Grauvogel-Stamm and Doubinger, 1975 (*i*, *j*, *u*); Holmes, 1981 (*l–q*); Phillips, 1974, after Delevoryas and Morgan (*r*); Phillips, 1974 (*s*); Phillips, 1974, after Hall (*t*).

with wood elliptical- to omega-shaped in cross-section; while from the petioles that then become phyllophores, petioles of the next order issue, as well as axes with a cauline wood. The branching may be more intricate and prolific. In addition, it is complicated by aerial roots. As a consequence, in some species a false stem is formed, which may reach 15 cm in diameter. Occasionally the main axis is siphonostelic. Sporangia were inserted on the tips of the branching axes and sometimes formed large aggregations (of several tens of thousands of sporangia). More rarely the sporangia were attached by thick stalks to the lower surface of the pinnule, completely covering it. In the unilayered sporangial wall a lateral annulus with thick-walled cells is conspicuous. The sporangia dehisced along a stomium – a narrow band of thin-walled cells. The spores are trilete, smooth or tuberculate.

B. may prove to be a collective genus. An important character that unites its species is the omega-shaped cross-section of the wood in the petioles. Those *B.* in which the sporangia were situated on the surface of the pinnule, are similar both in the organization and arrangement of the sporangia to Osmundaceae. Some *B.* and a part of the Permian Osmundaceae (*Zalesskya*, *Thamnopteris*) display a similar stem structure (two concentric wood zones). Probably both groups are allied. The close relationship to leptosporangiate ferns is emphasized by the fairly complex structure of the hairs (Fig. 32, *u*). *B.* have a strongly developed secretory system in the cortex, similar to the secretory canals in Marattiales. On the other hand, in *B.* round bordered pits were noted on tracheid walls (besides scalariform and reticulate ones), which is not characteristic of most leptosporangiate ferns.

Genus *Musatea* (C_1; Fig. 32, *a* (4); Fig. 33, *a*) involves fertile pinnae branching in one plane, with slightly flattened terminal branches, on the tips of which were attached groups of 4–6 elongated sporangia, probably slightly fused basally. The sporangial wall had thickened cells, arranged in a kind of elongated annulus. The spores were trilete. Stems, phyllophores and petioles have been described as *Diplolabis* and *Metaclepsydropsis*. *M.* produced narrow laminar pinnules with a specialized mesophyll. Galtier and Scott (1985) and Scott and Galtier (1985) refer *M.* and associated organ genera to the Zygopteridaceae.

Family Tedeleaceae

According to the structure and arrangement of the sporangia the Tedeleaceae are rather similar to Schizaeaceae, but their anatomical structure is, in essence, the same as in typical Zygopteridales. The Tedeleaceae display axillary branching, which is unknown in Schizaeaceae and in Polypodiales, in general.

The genus *Tedelea* (C; Fig. 33, *b–f*) in the structure of its petioles (of the *Etapteris* type) and development of laminate pinnules is close to *Corynepteris* (Zygopteridaceae), but in *T.* the pinnae are two-ranked. The anatomical structure of the axis was of the *Ankyropteris* type. The surface of the axes, phyllophores and petioles in the lower parts is covered by hairs and scaly aphlebiae. *T.* probably had false stems composed of the mass of petioles and aerial roots. Sporangia were grouped along the pinnule margin. Thickened cells of the annulus are gathered at the apex. Closely allied to *T.* is the genus *Senftenbergia* (C–P; Fig. 33, *i–k*), which for a long time was included in the Schizaeaceae, but has axes of the *Ankyropteris* type. Sterile fronds correspond to the form-genus *Pecopteris*, in particular to *P. plumosa*. Fronds of these species were situated on the stems, which in impression–compression materials are described as *Megaphyton* and frequently are tentatively related to the Marattiales. In *Senftenbergia* only sterile fronds have pinnules, whereas sporangia are inserted on the tips of slender flattened branches, whose style of branching imitates the venation in sterile pinnae and pinnules.

Family Psalixochlaenaceae (Fig. 33, l–q)

The sporangial structure in Psalixochlaenaceae (P.) is the same as in Botryopteridaceae, but the sporangia are arranged in sori on somewhat widened and obliquely truncated tips of the pinnules (Fig. 33, *n–q*). The sporangia were not of simultaneous maturation in the sori. The stems were of dichotomous or lateral branching; in the latter case the branching is axillary. The protostele was cylindrical with a mesarch xylem. The leaf vascular bundles were abaxially bent (of the *Anachoropteris* type), both in the main rachis and in the pinnae. *Psalixochlaena* has rachial bundles with two adaxial protoxylem strands. Fronds are pinnate. The genera *Hymenophyllites*, *Boweria*, *Sturia*, etc. described from impressions (J. Holmes, 1981), can probably be affiliated with P. The similarity between P. and extant Hymenophyllaceae has been noted. It should be mentioned that the Hymenophyllaceae are regarded as the most primitive living ferns, according to their exine structure.

Family Sermayaceae

In Sermayaceae the sporangia sit on pinnules, being clustered in indistinct groups or ordered sori. The wall has a distinct oblique or transverse annulus. In *Sermaya* and *Doneggia* the petioles are of the *Anachoropteris* type (Fig. 33, *r–t*). It should be noted that in these petioles an increase in the complexity of the xylem cross-section was perceived; the margins of the xylem plate coil inwards or join in a ring, in the middle of which lies an additional xylem plate. The issue of the *Anachoropteris* petioles from siphonostelic stems has been observed. An independent family Anachoropteridaceae has been recognized but this should rather be treated as a suprageneric form-taxon for fragments of stems and petioles. The genus *Sermaya* (C) was established on petrifactions. Sessile sporangia are situated in sori on the lower surface of the pinnules. The annulus is oblique and two-ranked. The spores are trilete and micropunctate. *Sermaya* and *Oligocarpia* (a genus established for impression–compression specimens of fertile pinnae and known up to the Upper Permian; Fig. 33, *u–x*) are probably closely related (the annulus in *Oligocarpia* is uniseriate though).

Geperapteris (see Satellite genera of Polypodiopsida), *Discopteris* and *Grambastia* can, probably, be affiliated with Sermayaceae. It is thought that Sermayaceae were ancestors of Gleicheniaceae.

Order Marattiales

Some Mesozoic Marattiales (M.) are quite similar to living forms, and are included in the same family Marattiaceae. It was suggested that Palaeozoic M. having stems of the *Psaronius* type should be distinguished as an independent family Psaroniaceae. In the course of investigations the close relationship between the Palaeozoic and recent genera became ever more obvious. M. together with Ophioglossales (not under discussion here) are traditionally assigned to eusporangiate ferns. In M. the sporangia have a thick multilayered wall, occasionally with stomata. They are characterized by synangia that, at times, are hidden in a sporocarpium (bivalved container). In living M. the sporangia may be free, which is regarded as a derived character, notwithstanding the fact that free sporangia are known also in the Carboniferous M. Most likely, the degree of fusion of the sporangia varied in the course of the evolution of M. Fronds of Recent and Mesozoic genera are rarely entire, usually once-, bi- or tripinnate, or palmate. Among the Carboniferous and Permian fronds were at least 4-pinnate forms (of the *Pecopteris* and *Sphenopteris* types). The stele varied from a protostele to a very intricate dictyostele (*Psaronius*). It is believed that M. evolved from Early Carboniferous zygopterids closely related to *Corynepteris*.

The genus *Scolecopteris* (C–P; Fig. 34, *a–e, j*) was established for pinnae of the last order, that were studied from petrifications and well-preserved compressions. The synangia are radial, supplied with a short thick stalk with a vascular bundle, consist of 3–6 (seldom 7) sporangia, fused basally or along the entire length. The sporangial wall was unilayered on the inner upper side and two-layered elsewhere. The sporangia dehisced along the inner upper wall, becoming free. The synangia were arranged in a row along both sides of the pinnule midrib. When the pinnules fused, then the synangia maintained the same position on both sides of the midrib of the fused pinnules. The spores were tri- or monolete, large (up to 120 μm) or small (up to 20 μm), spinose, granulate or verrucose. Various sizes of spores may occur within

a
b
c
d
e
f
g
h
i
j
k
l
m
n
o
p
q
r
s
t'
t''
u
v'
v''
w
x
y'
y''
z

one sporangium. The outer sporoderm layer (perine or sculptine; Millay and Taylor, 1984), characteristic of M. is present. Permineralized stems belong to the genus *Psaronius*, sterile fronds belong to the form-genus *Pecopteris*. Fertile fronds of the *S*. type and allied to it genus *Acaulangium*, when represented by remains in which details of the synangial structure cannot be studied, are assigned to the genus *Asterotheca* (Fig. 34, *k*). A group of genera is known where the sporangia are similar in structure and arrangement to *S*., but differ essentially in details (see Mosbrugger, 1983–1984). For instance, the synangia may be sessile (*Acaulangium*, *Acitheca*, etc.); sporangia – nearly free (*Eoangiopteris*), pinnules – sphenopteroid (*Tetrameridium*, *Radstockia*). In some genera the synangia are elliptical, immersed in the leaf tissue (*Radstockia*; Fig. 34, *l*, *m*; *Millaya*) situated in two rows on both sides of the midrib, covering the entire surface of the pinnule (*Danaeites* = *Orthotheca*), or situated on the apical lobes of the pinnule lamina (*Chorionopteris*). It is believed that M. synangia evolved from radial to bilateral and from surficial to immersed in the leaf tissue. According to these characters the most advanced is the living genus *Danaea*. Although the radial surficial synangia are actually older than the bilateral and immersed, this transformation had already occurred during the Carboniferous (*Eoangiopteris* and *Millaya*) and cannot be regarded as a criterion of the degree of advancement of living forms.

The genus *Psaronius* (C_2–P; Fig. 34, *n–u*) includes petrified stems up to a metre in diameter. The bulk of the stem is composed of numerous tightly packed aerial roots, that are free around the periphery of the stem. Closer to the centre they are joined in places, and then entirely into a complete mass of additional pseudoparenchymatous tissue. (In extant M. the aerial roots are free.) Petioles passing between the roots break abruptly around the periphery of the stem, leaving scars (see genus *Caulopteris* among Satellite genera of Polypodio-

---

Fig. 34. Marattiales (*a–v*) and Osmundales (*w–z*); Middle (*a–m*, *o*, *q*, s–u), and Upper (*n*, t″) Carboniferous, Upper Permian (*w–y*, *z* (at left)), Middle–Upper Triassic (*v*), Lower Cretaceous (*z* (at right)); USA (*a–e*, *j*, *l–o*, *q*, *t*, *u*), Western Europe (*f–i*, *k*, *s*, *v*), Australia (*w*), Russian platform (*x*, *y*, *z* (at left)), South Africa (*z*, at right); *a–c*, *Scolecopteris iowensis* Mam., fertile pinna (*a*), spore (*b*), exine relief (*c*); *d*, *e*, *S. mamayi* Mill., spore (*e*), exine relief (*d*); *f–i*, *Crossotheca crepinii* Zeill., sterile (*f*) and fertile (*g*) pinnae, lateral view and cross section of fertile pinnule (*h*), spore (*i*); *j*, *S. parvifolia* (Mam.) Mill., cross-section of fertile pinna; *k*, *Asterotheca damesii* (Stur) Brousm.; *l*, *m*, *Radstockia kidstonii* Tayl., fertile pinna (*l*) and synangium (*m*); *n*, *Psaronius blickei* Morg., arrangement of vascular bundles (solid lines) and sclerenchyme bands (dotted lines) in cross-section of stem; *o*, *P. simplicicaulis* DiMich. et Phill., change of leaf trace (solid line) and accompanying sclerenchyme (stippling) in course from deep layers of bark (below) to bark surface (above), larger dots mark secretory canals; *p*, reconstruction of *Psaronius* (al. *Megaphyton*) with two-ranked pinnae arrangement; *q*, *P. simplicicaulis*, leaf scar with leaf trace (solid line) and sclerenchyme (stippling); *r*, cross and longitudinal section of *Psaronius* stem; *s*, *P. renaultii* Will., part of leaf trace; *t′*, *Psaronius*, young stem at early stage of development of aerial roots, dense stippling shows secondary parenchyme appearing, at right – departing petiole (simplified); *t″*, more developed stem of *Psaronius* with dicyclic stem (to right) and fully developed secondary parenchyme (stippling) between aerial roots; *u*, *P. simplicicaulis*, vascular system of monocyclic stem; *v*, *Danaeopsis marantaceae* (Presl) Heer, frond (*v′*), venation and position of sporangia (*v″*); *w*, *Palaeosmunda williamsii* Gould, cross-section of stele and inner cortex, stippling marks sclerenchyme and endodermis, black dots in leaf traces and stele show protoxylem, pith shaded; *x*, *y*, *Thamnopteris schlechtendalii* (Eichw.) Brongn., stem cross-section with different wood zones (stippling), leaf traces in cortex and free petioles peripherally (*x*), leaf trace with incipient parenchymatization near protoxylem (*y′*), and change of leaf trace during its departure (*y″*), protoxylem in black; *z*, various stelar types in Osmundales (*Thamnopteris* at right, *Osmundacaulis kolbei* Sew. sp. at left). Scale bar = 50 cm (*p*), 5 cm (*v′*), 1 cm (*g*, *q*, *t*, *u*, *v″*, *x*), 5 mm (*o*), 2 mm (*l*, *w*), 1 mm (*a*, *f*, *h*, *k*, *m*, *s*, *y*), 0.5 mm (*j*), 20 μm (*b*, *c*), 10 μm (*e*), 1 μm (*d*). Modified from: Millay, 1979 (*a–e*, *j*); Brousmiche, 1982 (*f–i*) and 1979 (*k*); Taylor, 1967 (*l*, *m*); Andrews *et al.*, 1970, after Morgan (*n*); DiMichele and Phillips, 1977 (*o*, *q*, *u*); Hirmer, 1927 (*p*, *r*, *y*); Scott, 1920 (*s*); Ehret and Phillips, 1977 (*t*); Hirmer, 1927, after Schimper (*v′*) and Leuthardt (*v″*); Gould, 1970 (*w*); Zalessky, 1927 (*x*); Bierhorst, 1971, after Emberger (*z*).

psida). The stem commonly had an amphiphloic siphonostele at the base; higher up the stem developed a very intricate polycyclic dictyostele with wide gaps. Within the genus groups of species have been distinguished according to the arrangement of the petioles (spiral, two-, four-ranked and multiseriate). Detached permineralized petioles are assigned to *Stewartiopteris* and *Stipitopteris*.

In the genus *Danaeopsis* (T; Fig. 34, *v*) the fronds are once-pinnate with long pinnae decurrent along the rachis. A thick midrib runs through the pinna; the lateral veins are forked, at times joined by marginal anastomoses. Sporangia are situated in double rows (linear sori?). Spores are trilete, smooth or minutely spinose. The anatomical characters of the stems and fronds are unknown. Because of this the assignment of the genus to M. is in need of additional confirmation.

Order Osmundales

The Osmundales (O.) occupy an intermediate position between the Botryopteridales and leptosporangiate ferns, according to the mode of development and multilayered wall of the sporangia, the presence of a group of thickened cells instead of an annulus, the absence of sori, and other characters. O. are sometimes termed protoleptosporangiate ferns. In living O. the stems are short, clothed in aerial roots and have petiole bases from defunct fronds. Fronds are once- and bipinnate. Sporophylls are frequently strongly modified and resemble a brown panicle. The strong frond dimorphism is known since the Jurassic. The sporangia are in groups (not sori) or cover areas on the lower surfaces of the fronds; in some fossil species they cover the entire lower surface of the pinnule. The sporangia are large with a thick short stalk, dehiscing by means of two closely situated groups of cells with thickened walls. In fossil forms one group of such cells may be present. The spores are round, trilete, sometimes germinating in the sporangia. O. are represented by three genera in the living forms, whereas they were more diverse in the geological past. Indisputable O. are known since the Triassic; closely related to them are certain Permian genera. Fossil forms of O. have been identified from petrified stems, the fronds of which are unknown; or from vegetative and fertile fronds, of which the stems are unknown.

According to the stem structure, fossil O. are closely similar to living forms (Fig. 34, *w–z*). In the Upper Permian genera (*Thamnopteris*, *Zalesskya*, *Petcheropteris*, *Iegosigopteris*, *Palaeosmunda*) in the centre of the trunk is a stem surrounded by a compact sheath of petioles and a small number of aerial roots. The stem in *Thamnopteris* (Fig. 34, *x*) and *Zalesskya* is protostelic with a xylem that is distinctly divided into an outer zone of strongly elongated scalariform tracheids, and an inner zone of shorter tracheids that probably functioned as water-storage tissue. The xylem is surrounded by a continuous phloem and endodermis. Leaf traces (Fig. 34, *y*) issued from the outer zone of the xylem but leaf gaps were not formed. The traces passed through the multilayered cortex and entered the petioles, gradually changing in cross-section from rounded to C-shaped. *Palaeosmunda* (Fig. 34, *w*) had a well-developed pith, and the issue of the leaf traces was accompanied by a leaf gap in the xylem ring, that was surrounded on the outside by a phloem. Such a 'dictyoxylic' (not to be confused with dictyostelic) ectophloic siphonostele was termed by Hirmer (1927) a protodictyostele. It is typical of living O. Among the Mesozoic O., of the genus *Osmundacaulis* (sometimes wrongly named *Osmundites*), a true dictyostele appeared in the Lower Triassic, i.e. an amphiphloic siphonostele with leaf gaps. Such a stelar organization is absent in living O. Only sometimes small phloem and endodermis areas develop inside the xylem, which is thus divided into numerous strands.

The systematics of petrified stems of O. is based on the type of stelar organization, structure of the cortex, leaf traces, petioles (in the latter in the Permian secretory cells had already appeared that are typical of O.). Notwithstanding the general similarity between the fossil stems and living O., their affinity to O. is not always evident. Leaves, typical of O., are firmly known only since the Triassic. The genus *Todites* was recorded in the Permian, but it has not been studied in detail. In the Upper Permian of Australia the stems of *Palaeosmunda* accompany leaves, the fructifications of which have not been studied. But in India similar leaves, assigned to *Damudopteris* (Fig. 35, *u–w*), bear sporangia that differ from O. and have an equatorial annulus, a long narrow stomium and unilayered wall. Sporangia occur in indistinct groups. It is probable that among Permian O. were genera that differed in the structure of their sporangia from other O., and constituted an independent family. It should be stressed that according to the structure and arrangement of the sporangia, as well as the morphology of the spores, *Damudopteris* resembles Botryopteridales, particularly *Sermaya* and some species of *Botryopteris*.

Genus *Todites* (P_2?, T–J; Fig. 35, *a–e*). Fronds – bipinnate, pinnules with pinnate venation. Sterile fronds of the *Cladophlebis* type. Fertile pinnules are smaller than sterile pinnules. Numerous solitary sporangia are situated along the veins, occasionally only along the midrib. Multicellular branching hairs were noted between the sporangia. The latter were large, practically sessile, with an apical group of thickened cells, and dehisced with longitudinal slit. The relation between T. and the genera *Acrostichites* and *Acrostichides* is not quite clear.

Krassilov (1978b) includes in the genus *Osmunda* leaves that are of common occurrence in the Jurassic of Siberia and usually have been described as *Raphaelia diamensis*. They are twice-pinnate fronds with lanceolate pinnules having a constricted base and pinnate venation. Sporangia, similar externally to osmundaceous ones, have been described in the Carboniferous genera *Todeopsis* and *Kidstonia*, that are in need of more detailed studies. Some Carboniferous ferns having a bar-shaped or oval foliar xylem (*Grammatopteris*, *Catenopteris*) have been suggested as possible ancestors of the Osmundales (Miller, 1971, cited by Galtier and Scott, 1985). *Anomopteris* (T_{1-2}; Fig. 35, *x*, *y*), with peculiar aphlebiae and numerous sporangia on slightly reduced fertile pinnules, may be regarded as a satellite genus of O.

Order Polypodiales (Filicales)

The order Polypodiales (P.) is treated in this book in the circumscription and content of leptosporangiate ferns (Filicales), as distinguished in earlier systems (Kryshtofovich, 1957; Osnovy palaeonotologii, 1963a; Zimmermann, 1959; etc.), but separately from Osmundaceae. In some recent systems (e.g. Grushvitsky and Zhilin, 1978) leptosporangiate ferns are divided into several orders and the P. comprise only some of their families (e.g. Gleicheniaceae, Matoniaceae, Dipteridaceae, Polypodiaceae, Grammitidaceae). Among the P. only six broad families known from the fossil record are discussed below (Schizaeaceae, Gleicheniaceae, Cyatheaceae, Matoniaceae, Dipteridaceae, Polypodiaceae). They all have sporangia with a unilayered wall and a distinct annulus. Taylor (1981) adopted nearly the same delimitation of the families. All these ferns are homosporous (in the living Platyzomaceae some sporangia have 16 large spores, and others have 32, but smaller ones). The selected six families of P. are distinguished by the following combination of characters. If the sporangia have an oblique annulus and cluster in exindusiate sori, then these ferns are allied with the Gleicheniaceae. In the Schizaeaceae the sporangia are not gathered in sori, have an apical annulus, and the indusium is absent, or instead of it the leaf has an infolded margin. In the Cyatheaceae the sori are embedded in bowl-shaped or bivalved indusia, situated around the margins or on the lower surface of the pinnules. If an indusium is lacking, then other characters become important. In the Matoniaceae the indusium is umbellate (peltate), i.e. it consists of a stalk and shield. The sori are composed of a single ring of sporangia. A characteristic feature is the reticulate venation and dichasial branching of the frond. The Dipteridaceae are similar to the Matoniaceae in the general aspect of the frond, reticulate venation and organization of the sori, but lack an indusium, the same as some fossil Matoniaceae. The sori are situated throughout the leaf in meshes, that are formed by anastomosing veins, whereas in the Matoniaceae the sori are more restricted to the midrib of the pinnule. In the Matoniaceae the annulus is oblique, while in the Dipteridaceae it is practically vertical. In the family Polypodiaceae, in the broad sense, the annulus is distinctly vertical, while the

a
b
c
d
e
f
g
h
i
j
k
l
m
n
o
p
q
r
s
t
u
v
w
x
y

indusium varies considerably (peltate, bowl-shaped, etc.).

Family Schizaeaceae

The extant and, seemingly, fossil Schizaeaceae (S.) are herbaceous plants, lianas. The leaves are commonly differentiated into sporophylls and trophophylls. The former are ribbon-shaped or dichotomous to pinnate, liana-like, with a long rachis of indeterminate growth. The sporophylls frequently have a reduced lamina, so that the sporangia are situated on the branches of the rachis. In other cases the sporangia occur along the pinnule margins. The sporangia are not clustered in sori, and dehisce by a longitudinal slit. The annulus is transversal, apical. The spores are monolete, or trilete. Palynologists frequently assign to S. large dispersed trilete spores with a ribbed exine, but similar spores are also known in other ferns. The genus *Cicatricosisporites* (Fig. 35, *f*), established for dispersed spores, is affiliated with S. It appears near the Jurassic–Cretaceous boundary. In other fossil S. (*Klukia*, *Stachypteris*) the spores have a tuberculate or reticulate exine (Fig. 35, *h*). The fronds of *Klukia* (J–K; Fig. 35, *k–m*) are bi- to quadripinnate. The sporangia are attached laterally on the lower surface of the pinnule, along both sides of the midrib, with a uniseriate apical annulus facing it. Fronds of *Stachypteris* (J–K; Fig. 35, *g–j*) are 3- to 4-pinnate. Sterile pinnules have a lobed margin and pinnate venation, or are narrow, nearly needle-shaped. Fertile pinnae (at times the sporangia occupy only part of the pinnule) with a reduced lamina, resemble a linear spike, consisting of small triangular pinnules, on each of which is situated a large sporangium with an apical annulus. The Mesozoic genera *Ruffordia*, *Pelletixia*, *Schizaeopsis*, *Sellingia*, etc., belong to S.

Family Gleicheniaceae

An important character of the Gleicheniaceae (G.) is the aggregation of sporangia into exindusiate sori. The sporangia are large, dehisce with a longitudinal slit; the annulus is oblique. The stems are chiefly protostelic. The fronds are characterized by their indeterminate growth in length and false dichotomous branching. At the point of branching the growth of the main axis is suppressed, whereas the lateral branches continue to grow. The G. comprise the Mesozoic *Gleichenites* and *Gleichenoides*, probably also the Permian *Chansitheca* and Triassic *Mertensides*. All of them have been poorly studied.

Family Cyatheaceae

Leaves of Cyatheaceae (C.) are repeatedly pinnately dissected. The sori, with few exceptions, are encased in well-developed globose, bowl-shaped or

Fig. 35. Osmundales (*a–e*), Schizaeaceae (*f–m*) and Cyatheaceae (*n–t*), genera *Damudopteris* (*u–w*) and *Anomopteris* (*x*, *y*); Upper Permian (*u–w*), Middle Triassic (*x*, *y*), Middle (*a–e*, *g*, *h*, *k–t*) and Upper (*i*, *j*) Jurassic, Lower Cretaceous (*f*); Western Europe (*a–e*, *g–t*, *x*, *y*), USA (*f*), India (*u–w*); *a*, *Todites denticulatus* (Brongn.) Krass., fertile pinnule, and scheme of attachment of sporangia in cross-section of pinnule (below); *b*, *T.* (al. *Cladophlebis*) *denticulatus*, sterile pinnules; *c–e*, *T.* (al. *Acrostichides*) *williamsonii* (Brongn) Sew., fertile pinnule with crowded sporangia (*c*), sporangium (*d*), spore (*e*); *f*, spores *Cicatricosisporites aralicus* (Bolkh.) Bren. from sporangia of *Pelletixia amelguita* Skog, distal (left) and proximal (right) sides; *g–j*, *Stachypteris spicata* Pom., fertile pinnule (*g*), spore (*h*), fully (*i*) and partly (*j*) fertile pinnae; *k–m*, *Klukia exilis* (Phill.) Racib., frond (*k*; lower pinnae are fertile), sporangium (*l*), fertile pinnule (*m*); *n–t*, *Coniopteris hymenophylloides* (Brongn.) Sew., sterile pinna (*n*) and pinnules (*o*, *p*), fertile pinnae (*q*, *t*), indusium with sporangia (*r*), spore (*s*); *u–w*, *Damudopteris polymorpha* (Feistm.) Pant et Khare, sporangium as viewed from annulus and stomium (*u*), fertile pinna with group of sporangia (*v*), spore (*w*); *x*, *y*, *Anomopteris mougeotii* Brongn., sterile (below) and fertile (above) parts of pinnae viewed from both sides (*x*), reconstruction of pinna with aphlebiae. Scale bar = 1 cm (*k*, *n*, *t*), 5 mm (*y*), 2 mm (*b*, *c*, *v*), 1 mm (*g*, *i*, *j*, *m*, *o–q*, *x*), 200 μm (*a*, *r*), 100 μm (*d*, *l*, *u*), 10 μm (*e*, *f*, *h*, *s*, *w*). Modified from: Harris, 1961 (*a–e*, *g*, *h*, *k–t*); Skog, 1982 (*f*); Barale, 1981 (*i*, *j*); Pant and Khare, 1974 (*u–w*); Grauvogel-Stamm and Grauvogel, 1980 (*x*, *y*).

bivalved indusia, situated along the margin of the pinnules at the tips of the veins, or on the veins at a distance from the margin. The annulus is oblique or nearly vertical. Petrified stems of C. are known with dictyostelic or siphonostelic structures, and petioles (*Protocyathea*, *Ciboticaulis*, *Cyathodendron*, *Thyrsopterorachis*, *Cyathorachis*, etc.). The stems of C. are similar to the marattialean (*Psaronius*) in their thick cover of aerial roots, and, in part, stelar organization. The living genus *Dicksonia* is known since the Jurassic, and *Cyathea* is known since the Palaeogene. Of the extinct forms worthy of mention are the genera *Coniopteris*, *Eboracia*, *Kylikipteris*, *Gonatosorus*, *Disorus* and *Alsophilites*. The fronds in *Coniopteris* (J–K; Fig. 35, *n–t*) are 2- to 3-pinnate, with slender frequently winged rachises and sphenopteroid pinnules, commonly lobed, at times deeply. Fertile pinnules occupy the entire pinnae or its lower part, and are reduced. Along the margin or at the apex of the pinnule are rounded sori (one per pinna or lobe) with a bivalved, bowl-shaped or reniform indusium. At the bottom of it are attached numerous sporangia with an oblique annulus. Spores are trilete, triangular, with a smooth exine.

Family Matoniaceae

The fossil Matoniaceae (M.) (10–12 genera) are known since the Permian–Triassic by their sterile and fertile fronds and petrified rhizomes. The M. fronds are of three main types. In those of the *Matonia* type (Fig. 36, *a*, *b*) the main rachis is apically forked. Each of the branches is bent backwards and issues upward several pinnae (this is a dichopodium, consisting of two unilateral acroscopic sympodia). Fronds of the *Phanerosorus* type (Fig. 36, *h*), externally similar to gleicheniaceous (*Lygodium*), were of indeterminate growth and false dichotomous branching. At the point of branching the rachis trifurcates, after which the growth of the middle axis is suppressed, whereas the lateral axes continue to grow up to the next branching in the same style. Fronds of the *Weichselia* type (Fig. 36, *i*) are large, twice-pinnate, attached at the apex of the stem by a compact rosette. The pinnules of M. are triangular, crescent-shaped or linear, attached with a broad base, and have a midrib. The lateral veins are branching and frequently anastomose. The anastomoses are particularly typical near the attachment of the sori. The latter are arranged in a row along both sides of the midrib, usually not far from it. Short pinnules may bear only one–two sori. The indusia are umbellate. Exindusiate genera (*Phlebopteris*, *Piazopteris* and *Nathorstia*), according to the aspect of their fronds, arrangement of the sori, structure of the sporangia (with a nearly vertical annulus) and spores, are included in M. Permineralized rhizomes of M. are assigned to *Paradoxopteris*, *Yetzopteris*, *Alstaettia* and *Matoniopteris*. They can easily be identified by their polycyclic structure of concentric meristeles. Leaf traces are C-shaped.

Genus *Phlebopteris* (T–K; Fig. 36, *a–g*)

Pinnae issued from backbent branches of the main rachis; bear long linear or short forward bent pinnules. The midrib of the pinnule gives off lateral veins connected by anastomoses. Sori, arranged in a row along both sides of the midrib, consist of 15–16 sporangia, arranged in a circle on a small elevation. The annulus is nearly vertical.

Genus *Weichselia* (J–K; Fig. 36, *i–l*)

Pinnules are thick, xeromorphic, pecopteroid with reticulate venation, attached to upper surface of the rachis, so that the pinna is V-shaped in cross-section, whereas the frond appears plicate. Sori in globose clusters of 12–44, attached to a leafless rachis on one surface. Indusium of adjacent sporangia close up at the margins, so that all the sporangia (about 12 per sori) are hidden inside. Nearly vertical annulus is situated on one side of the sporangium and faces the stalk of the indusium. Spores are subtriangular, trilete; rays are bordered on both sides by exine thickenings. *W.* may belong to a separate family.

Family Dipteridaceae

The Dipteridaceae (D.) are closely similar to the Matoniaceae, so that their independent status is open to doubt. D. fronds resemble matoniaceous ones, but the dichotomous division of the main rachis results in upwardly bent branches (sometimes twisted longitudinally), giving off the pinnae downwards (Fig. 36, *m*). The pinnae are fused to various extents. The venation is reticulate (Fig. 36, *n*). D. fragments may be mistaken for leaves of angiosperms. Exindusiate sori are inserted in meshes, formed by tertiary veins (Fig. 36, *o*). The sporangia are of disorderly arrangement in the sori (in contrast to the ring-like arrangement in the Matoniaceae). The annulus is oblique to nearly vertical (Fig. 36, *p*). In the latter case D. shows affinities to the Polypodiaceae. It has been advocated that the Mesozoic genera, traditionally included in D., should be affiliated with extant plants of *Cheiropleuria* (Polypodiaceae), not with *Dipteris*. Fossil D. are represented by genera of indistinct diagnostics – *Clathropteris* (Fig. 36, *m*), *Dictyophyllum* (Fig. 36, *n–p*), *Thaumatopteris*, *Camptopteris*, *Goeppertella*, *Kenderlykia*, *Apachea*. They are mainly Triassic and Jurassic plants.

In *Hausmannia* fronds (T_3–K; Fig. 36, *q*, *r*) fusion of the pinnae is very deep, so that in some species the fronds are bilobed with entire or incised margins. In other species the fronds are similar in dissection to *Clathropteris*, and in still others the frond lobes are dichotomously divided. The venation is reticulate. The sori are situated in meshes of the venation throughout the lower surface of the frond.

Family Polypodiaceae

The family Polypodiaceae (P.), in this book adopted in a very broad content, includes ferns that sometimes are placed in the families Aspleniaceae, Adiantaceae, Grammitidaceae, Aspidiaceae and Pteridaceae. They have common features: vertical or incomplete annulus, small sporangia; spores are not numerous (less than 100), usually monolete. Quite a lot of living genera have been described from the Cainozoic, and in part from the Mesozoic (*Acrostichum*, *Asplenium*, *Adiantum*, *Onoclea*, *Polypodium*, *Pteris*, *Woodwardia*, etc.). Some fossil genera are closely related to extant ones (*Eorhachis* to *Acrostichum*, *Dennstaedtiopsis* to *Dennstaedtia*). The P. also comprise *Polypodites*, *Aspidites*, *Teilhardia*, *Vargolopteris*, *Onychiopsis*, *Adiantopteris* and other genera, that contribute little to the general characteristics of the family. P. are reliably known since the Jurassic.

Order Marsiliales

Marsiliales (M.) includes the recent genera *Marsilia*, *Pilularia* and *Regnellidium*, recorded from the Tertiary (*Marsilia* was mentioned already in the Upper Cretaceous), and also the Palaeogene genus *Rodeites*. The leaves consist of four (*Marsilia*) or two (*Regnellidium*) leaflets with fan-shaped venation, resembling clover leaflets, or linear, nearly awl-shaped (*Pilularia*). The major distinction of M. are sporocarpia, i.e. closed containers of sori consisting of micro- and megasporangia. A single megaspore (Fig. 36, *s*) develops in each megasporangium. Fossil material does not supplement the characteristics of the order.

Order Salviniales

These ferns float on the water surface. Fertile leaves are water-submerged and bear male and female sori (micro- and megasori), encased in a closed round indusium. Microspores are embedded in a foamy mass (massula). It also accompanies megaspores as a floating apparatus. In each megasporangium one megaspore develops. The families are distinguished on the basis of the structure of the vegetative leaves, indusium (unilayered in Azollaceae and two-layered with a cavity interposed between the layers in Salviniaceae), number of megasporangia in the sorus, and other characters. Sometimes all Salviniales are assigned to one family. In fossil materials detached megaspores with typical massulae, more rarely vegetative and fertile leaves, have been found.

Family Salviniaceae

The only genus *Salvinia* (Fig. 36, *y*) is known since the Upper Cretaceous. Some fossil salvinias exhibit amphisporangiate sori, combining both mega- and microsporangia.

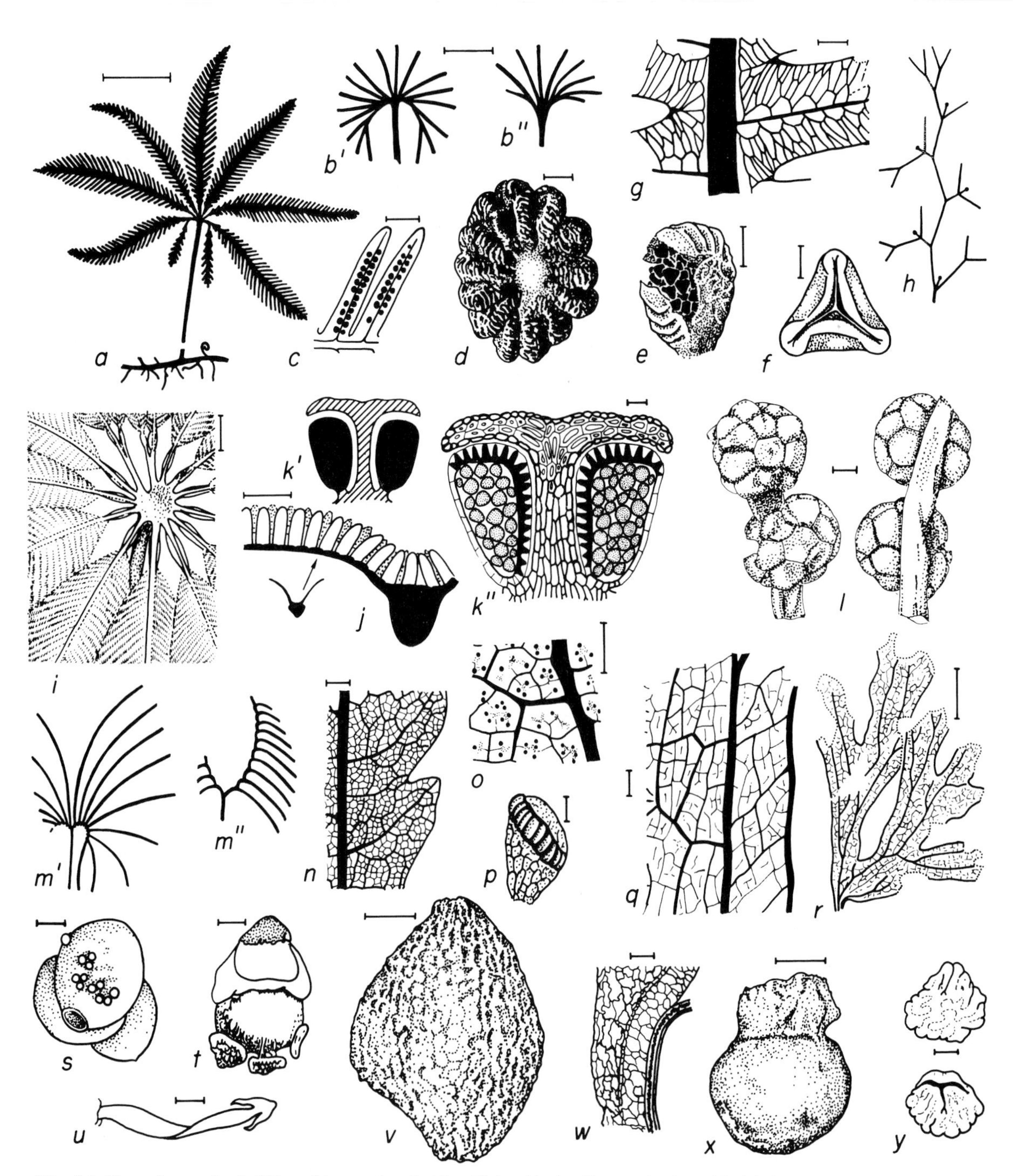

Fig. 36. Matoniaceae (*a–l*), Dipteridaceae (*m–r*), Marsiliales (*s*), Azollaceae (*t–x*) and Salviniaceae (*y*); Upper Triassic (*a–f*), Jurassic (*g, n–r*), Lower (*i–l*) and Upper (*v–x*) Cretaceous, Neogene (*s–u, y*); USA (*a–f*), Western Europe (*g, i–l, o–r*), South of European part of the USSR (*n, s–u, y*), West Siberia (*v–x*); a–*f*, *Phlebopteris smithii* (Daugh.) Arn., frond (*a*), branching of rachis (*b*), fertile pinnae (*c*), sorus (*d*), sporangium (*e*), spore (*f*); *g*, *P. polypodioides* Brongn., venation of pinnules; *h*, stem branching in recent *Phanerosorus*; *i–l*, *Weichselia reticulata* Stokes et Webb., attachment of pinnae ▶

Family Azollaceae

In Azollaceae the microspores are united by a massula, corresponding at times to an entire sporangium. In many species above the surface of the massula jagged hairs – glochodia (Fig. 36, *u*) – can be seen, by which the massulae with the microspores are fastened to the floating megaspores. One megaspore germinates in each megasporangium. Each megaspore is connected with several massulae (Fig. 36, *t*). They surround it on the distal side, whereas others cluster in 3–9 floats on the proximal side. Most common fossils are massulae with glochodia and microspores, megaspores with floats, more rare – megaspores with attached massulae, bearing microspores. The only extant genus is *Azolla*, known since the Lower Cretaceous (for a review of fossils, see Collinson, 1980). Closely related to it are the Upper Cretaceous genera *Parazolla*, *Azollopsis*, *Azinia*, etc., established on megaspores and massulae with microspores. In *Azinia* (Fig. 36, *v–x*) indusia with a hard wall were described as sporocarpia, containing micro- and megasporangia. (Because of this *Azinia* has been classified in a separate family Aziniaceae.) There is indirect evidence indicating that bisexual sori were characteristic of other Upper Cretaceous Azollaceae.

Satellite genera of the class Polypodiopsida

Of the genera described below some (*Caulopteris*, *Pecopteris*, *Lobatopteris*, *Cladophlebis*) are indisputably heterogeneous, including species of different families, orders and even classes. The remaining genera are close to natural ones, but their systematic position cannot be firmly established. The genus *Sphenopteris* is described among the satellite genera of the classes Ginkgoopsida and Cycadopsida.

Genus *Caulopteris* (D_2–P; Fig. 37, *a*)

This genus includes casts and imprints of stems of Palaeozoic arborescent ferns. Their surface shows scars from abscissed fronds, frequently with a distinct leaf trace contour. The scars are spirally arranged in distant orthostichies. Imprints of aerial roots may be preserved between the scars, so that the stem surface appears longitudinally ribbed. Such stems most frequently correspond to the genus *Psaronius* (Marattiales), but some might have belonged to the Zygopteridales and other ferns. Similar stems, but with two vertical rows of scars, have been assigned to *Megaphyton* (Fig. 37, *b*, if the contour of the vascular bundle on the scar is open at the top, or closed and oval), or *Artisophyton* (if the contour is closed with a deep notch at the bottom).

Genus *Pecopteris* (since the Carboniferous; Fig. 37, *c–d*)

This genus has been introduced for sterile compound-pinnate fronds of fossil ferns, that have slightly curved or straight pinnules, attached to the rachis by a wide or weakly constricted base. The margins of the pinnules are entire or lobed, the

---

to stem (*i*), v-shaped attachment of pinnules and pinnae to rachises (*j*), scheme of structure (*k'*) and longitudinal section (*k''*) of sorus, globose aggregates of sori (*l*); *m*, branching of rachis in *Clathropteris* and *Thaumatopteris* (*m'*) and *Dictyophyllum* (*m''*); *n*, venation of *Dictyophyllum acutilobatum* F. Braun; *o*, *p*, *D. rugosum* L. et H., venation of fertile pinna, traces of abscissed sporangia are marked by larger dots (*o*), sporangium (*p*); *q*, *r*, *Hausmannia dichotoma* Dunk., venation (*q*) and general view of leaf (*r*); *s*, *Marsilia* sp., megaspores with stuck microspores; *t*, *u*, *Azolla tomentosa* Nik., megaspore with massulae (*t*) and glochidium (*u*); *v–x*, *Azinia paradoxa* Bal., sporocarpium with floating device covered by sporangial wall (*v*), section of megaspore covered with layer of floating device (to left from band of small cells in spongy tissue; *w*), megaspore without floating device (*x*); *y*, *Salvinia cerebrata* Nik., megaspore, distal (above) and proximal views. Scale bar = 10 cm (*i*), 5 cm (*a*), 1 cm (*b*, *j*, *r*), 5 mm (*n*), 2 mm (*c*, *g*), 1 mm (*o*, *q*), 200 μm (*d*), 100 μm (*e*, *k*, *l*, *p*, *s–v*, *x*, *y*), 10 μm (*f*, *w*). Modified from: Ash *et al.*, 1982 (*a–f*); Taylor, 1981, after Alvin (*i*); Daber, 1968 (*j*); Alvin, 1968 (*k''*, *o*); Hirmer, 1927 (*m*); Osnovy paleontologii, 1963a, after Dorofeev (*n*, *t*, *u*, *y*); Andrews *et al.*, 1970, after Dorofeev (*s*); Balueva, 1964 (*v–x*).

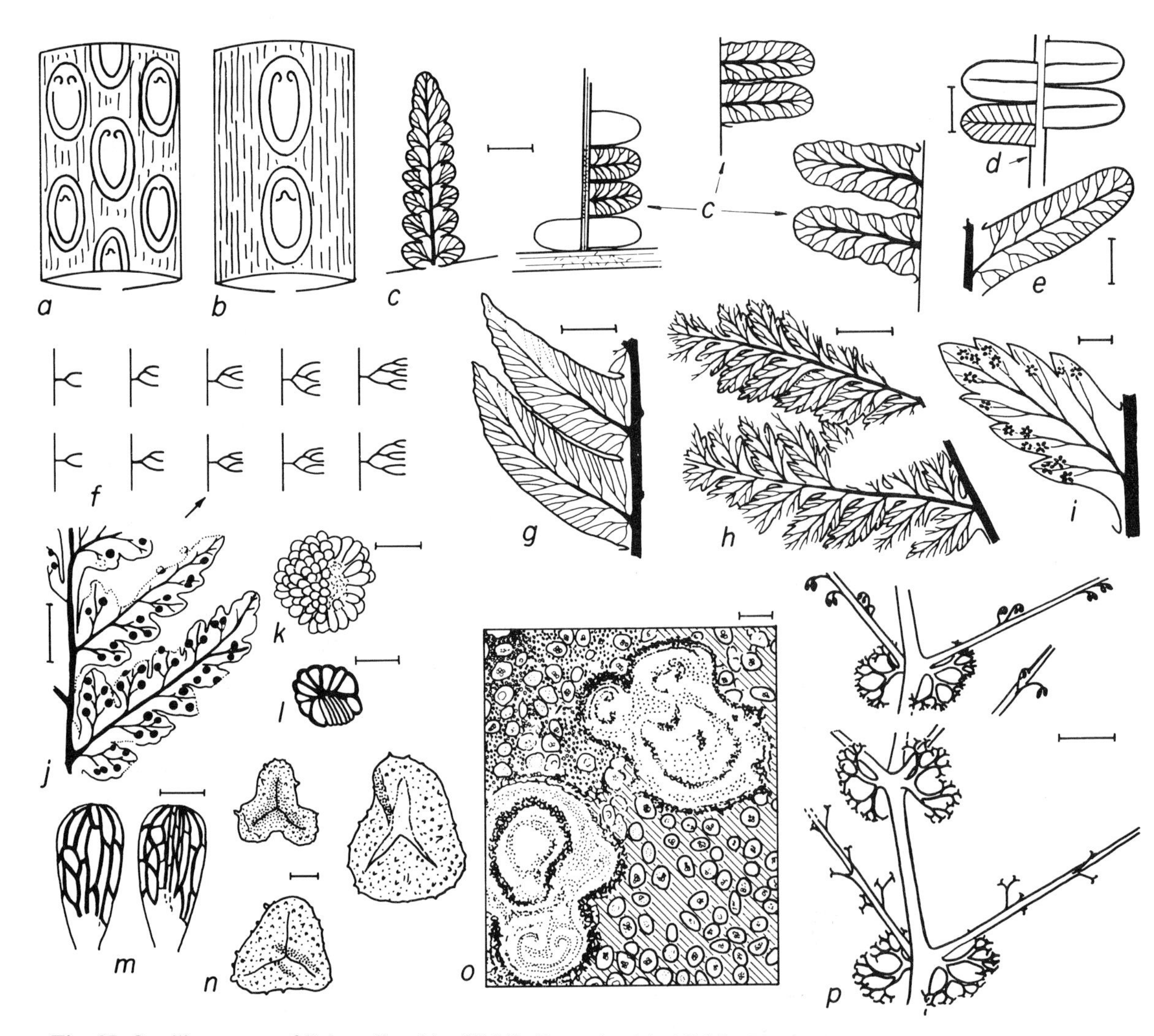

Fig. 37. Satellite genera of Polypodiopsida; Middle Devonian (*p*), Middle–Upper Carboniferous (*a–e*), Lower (*j–n*) and Upper (*h, i*) Permian, Lower Cretaceous (*g, o*); Euramerian palaeofloristic area (*a–e*), Bureja basin (*g*), Kuzbass (*h–n*), Mugodzhary (*o*), Spitzbergen (*p*); *a*, *Caulopteris*; *b*, *Megaphyton*; *c*, *Pecopteris pennaeformis* Brongn.; *d*, *P. arborescens* (Schl.) Brongn.; *e*, *P.* (al. *Lobatopteris*) *corsinii* Wagner; *f*, succession of vein branching in *Pecopteris* (upper row) and *Lobatopteris* (lower row), arrow shows key stage discriminating both kinds of venation; *g*, *Cladophlebis serrulata* Samyl.; *h*, *i*, *Prynadaeopteris karpovii* Radcz., frond (*h*) and fertile pinna (*i*); *j–n*, *Geperapteris imitans* (Neub.) S. Meyen, fertile pinna (*j*), sorus with partly abolished sporangia (*k*), upper (*l*) and lateral (*m*) view of sporangium, different spores from same sporangium (*n*); *o*, *Tempskya rossica* Kidst. et Gw.-Vaugh., part of cross-section of false stem with two branching axes and aerial roots (root hairs are shown in left upper corner, elsewhere space occupied by root hairs is shown by shading); *p*, *Protocephalopteris praecox* (Høeg) Anan. Scale bar = 1 cm (*h*, *p*), 5 mm (*g*, *j*), 2 mm (*c–e*, *i*, *o*), 0.5 mm (*k*), 100 μm (*l*, *m*), 10 μm (*n*). Modified from: Pfefferkorn, 1976 (*a*, *b*); Boureau and Doubinger, 1975, after Corsin (*c*, *d*); Wagner, 1958 (*e*) and 1958–1959 (*f*); Vakhrameev and Doludenko, 1961 (*g*); Radczenko, 1955 (*h*, *i*); Meyen, 1982b (*j–n*); Osnovy paleontologii, 1963a, after Kidston and Gwynne-Vaughan, 1911 (*o*); Andrews *et al.*, 1970, after Schweitzer (*p*).

venation pinnate. The midrib is distinct and can be traced far upward, even to the apex. The lateral veins are simple to repeatedly dichotomizing. After the first division of the lateral vein its branches again divide at the same distance from the first division (in contrast to *Lobatopteris*; Fig. 37, *f*). The suggestion was advanced that only species with simple or once-forked lateral veins should be assigned to *P.*, whereas all other other species (on condition that the genus *Lobatopteris* is included) should be assigned to *Polymorphopteris*. Various species, judging by the associating fertile fronds, belong to Zygopteridales, Botryopteridales, Marattiales, Osmundales, and to various families of Polypodiales. Leaves of the *P.* type are of common occurrence in extant forms. Ferns are known that according to the structure of their pinnules are intermediate between *P.* and other form-genera. Sometimes they are classified as independent form-genera. For instance, the genus *Ovopteris* includes fronds intermediate between *Pecopteris* and *Sphenopteris*. In *Lobatopteris* (since the Carboniferous; Fig. 37, *e*, *f*) the frond is of the same structure as in *Pecopteris*, but the secondary veins differ in their branching. After the first division one branch divides again and then the newly-formed inner branch again divides. This type of 'lobatopteroid' venation may become of more complexity in the course of ontogeny, so that in developed pinnules with strongly branching veins the difference in the venation between *L.* and *Pecopteris* may be obscured.

Genus *Cladophlebis* (known since the Permian; Fig. 37, *g*)

Fronds are compound-pinnate, sterile. Pinnules are crescent-shaped, otherwise similar to *Pecopteris*. Some species exhibit venation as in *Lobatopteris*. It has been suggested that fronds with modified basal pinnules of the pinnae, and other characteristic features, should be referred to different genera. Associated fertile fronds belong to Osmundales, Matoniaceae, Polypodiaceae and other ferns. Many living ferns have fronds of the C. type.

Genus *Prynadaeopteris* (C–P; Fig. 37, *h*, *i*)

Sterile fronds correspond to the form-genera *Pecopteris*, *Sphenopteris* and *Ovopteris*. Pinnules are frequently deeply dissected and assume the aspect of small pinnae. This genus includes sporophylls of poor preservation or inadequately studied. Sori are minute with a small number of sporangia.

Genus *Geperapteris* (P; Fig. 37, *j–n*)

This includes sporophylls of the same type as in *Prynadaeopteris*, but with more complete characteristics of sori, sporangia and spores. Sori are rounded composed of numerous (up to 60 and more) pear-shaped or elliptical, sessile sporangia. The sporangial wall is unilayered, composed in the larger upper part of the sporangium of thick-walled cells. The stomium consists of a few rows of narrow cells. Spores are trilete, rounded-triangular, with small spines or coni. *G.* shows affinities to different genera of the order Botryopteridales (*Doneggia*, *Discopteris*, certain *Botryopteris*, etc).

Genus *Tempskya* (K; Fig. 37, *o*)

This is sometimes keyed out as an independent family (Taylor, 1981). It has a false stem, up to 50 cm in diameter and 6 m in length, consisting of hundreds of stems that pass through a dense mass of adventitious roots with root hairs. In the lower part of the stem only roots were perceived. From siphonostelic stems, up to 4 mm in diameter, thin (1 mm) petioles issue. Seemingly, only the upper part of the stem was covered by leaves. In some localities the stems were found in association with small fronds that have been described as *Anemia fremontii*. Sporangia, stuck in the mass of roots and stems, were of different types, also including those similar to schizaeaceous.

Genus *Protocephalopteris* (D_2; Fig. 37, *p*)

The main axis is zigzag in shape. The branches are attached in pairs alternatively to one or another side of the axis. At the base of the branches from the axis issue dichotomously divided aphlebiae. The branches are straight, bearing paired lateral organs that are also of alternative attachment to one or the opposite side of the branch. Sterile lateral organs are twice-dichotomous, while the fertile organs are once-dichotomous, and terminate in pendant sporangia. In general habit *P.* resembles *Rhacophyton* (Zygopteridales).

Class Progymnospermopsida

The discovery of progymnosperms (P.) is associated with the establishment of organic connections between long known plants, namely between fronds of the genus *Archaeopteris* (they were mistaken for heterosporous ferns) and stems of the genus *Callixylon* (formerly assigned to gymnosperms of the order Cordaitales). The organic connections between these parts, first proven by Beck during the 1960s, were repeatedly confirmed later. In P. we become acquainted with a specific combination of characters: vegetative organs and terminal sporangia as in trimerophytes, or primitive ferns (cladoxylaleans, zygopterids), secondary wood as in gymnosperms, developing in protostelic or eustelic axes (see reviews: Beck, 1976a, 1981: Bonamo, 1975; Gensel and Andrews, 1984). The most primitive genera differ from the most advanced trimerophytes merely in the dissection of the protostele or/and presence of a secondary wood. Sporangia of the trimerophytes (*Psilophyton*) and certain P. (*Tetraxylopteris*, *Protopitys*) are practically of the same organization. The most advanced heterosporous genera of P. attain the gymnosperm level of organization in all parts, excepting the female reproductive organs.

In the lineage from the Early Devonian trimerophytes through the Middle Devonian Protopteridiales to the Archaeopteridales and Protopityales and further to gymnosperms (Calamopityales and Lagenostomales) the following morphological transformations can be perceived: (1) the transition from undivided to a lobed protostele and further to a eustele (Beck *et al.*, 1982–1983) and parallel formation of an increasingly thicker secondary wood; (2) an increase in the number of protoxylem strands in the main axis; (3) the planation of the lateral shoot systems and their fusion in leaf laminae; (4) the development of a regular phyllotaxis; (5) the concentration of sporangia on specialized lateral branches and, at times, their shifting on to pinnules; (6) the formation at first of strobiloid aggregates of fertile pinnules, and then, of well-shaped strobili; (7) the appearance of heterospory and reduction of the number of megaspores in megasporangia; (8) the formation of a distal cavity (cavum) in microspores. These processes were asynchronous, and not all reflected in each lineage. The combination of primitive and advanced features may vary in different genera, which causes serious difficulties in the classification of P. and their distinction from adjacent taxa.

Four groups can be recognized in P. One of them has a persistent protostelic axis (order Protopteridiales), whereas two others have a eustelic axis, with two-ranked (order Protopityales) and spiral (order Archaeopteridales) arrangement of the branches. The genera of the Protopteridiales are homosporous; in the Protopityales the heterospory is indistinct, whereas in Archaeopteridales and Noeggerathiales it is very prominent. The order Noeggerathiales is tentatively assigned to Progymnospermopsida, chiefly on the basis of its putative phylogeneric relationship with the Archaeopteridales (Beck, 1981; Taylor, 1981). Probably there is reason to affiliate with P. the order Ophioglossales, that has a eustele (Schmid, 1982–1983), secondary wood and sporangial aggregates, that may easily be derived from P. sporoclads. Certain genera are known that lack the necessary combination of characters. For instance, the genus *Oocampsa* (the beginning of D_2) is similar externally both to the trimerophytes and early P. Spores have a flange that probably reflects the incipient splitting of the exine and formation of the cavum, which allies *Oocampsa* with *Rellimia*, but whether this genus had secondary wood is as yet unknown. The genus *Chaleuria* (D_2) was heterosporous; spores of two-dimensional classes can be found in the same sporangium. Otherwise from the observed characters it is impossible to decide whether this genus was allied to P., Cladoxylales or Zygopteridales. The anatomy of the axis is unknown. The genera *Cairoa*, *Proteokalon*, *Sphenoxylon*, *Palaeopitys* and some others are tentatively included in P. They have been described from detached permineralized fragments of axes, and the characters of the lateral sterile organs and fructifications are unknown. From the uppermost Lower Devonian Gensel (1984; see also Richardson, 1984) described a plant with a deeply three-lobed mesarch protostele in the main axis, and with a tangentially elongated trace in spirally attached laterals. The protoxylem elements are mostly broken down to form lacunae. The secondary wood is absent. This plant has some characters in common with the Protopteridiales and Ibykales.

Order Protopteridiales (= Aneurophytales)

Protopteridiales (P.; Fig. 38) were for long regarded as the most primitive preferns, from which the zygopterids were supposed to have evolved. However, after more detailed investigations this view was rejected. At present adequately studied are *Rellimia* (al. *Protopteridium*; Fig. 38, *j–m*; Bonamo, 1983; Schweitzer and Matten, 1982), *Tetraxylopteris* and *Aneurophyton* (Schweitzer and Matten, 1982; Serlin and Banks, 1978). *Triloboxylon* has been studied in less detail. The above-mentioned Devonian genera *Cairoa*, *Proteokalon* and *Sphenoxylon* (probably the synonym of *Tetraxylopteris*) display affinities to P. They are probably related to the Lower Carboniferous genera *Stauroxylon* and *Palaeopitys*, that have been described from fragments of permineralized axes and might have belonged to primitive Calamopityales or Lagenostomales. The genus *Eospermatopteris* was established for casts of large stems, that probably had branches of the *Aneurophyton* type (see details in Gensel and Andrews, 1984, pp. 237 ff.).

The axes are protostelic, the primary wood is three-lobed, triangular or cruciate in cross-section (Fig. 38, *a*, *f*). In the more slender axis the xylem is more rounded, the number of lobes being reduced. In the terminal branches the vascular bundles are round or lenticular in cross-section. The protoxylem (Fig. 38, *i*) is endarch, in the shape of singular isometrical nests in the centre of the stele and at the tips of the rays, or elongated discontinuous plates inside the stele (it was thought that these plates are composed of radially aligned cells of the metaxylem). The secondary wood (Fig. 38, *a*, *b*) developed in the thicker axes that also had a periderm, which is typical of plants with secondary wood tissue, but not characteristic of ferns (with the exception of the Ophioglossales). Tracheids of the secondary wood bore round or polygonal bordered pits on all the walls. Medullary rays are uniseriate, high. Secondary wood is lacking in the genus *Reinmannia* (D_2; Fig. 38, *a* (2)), which, according to other characters shows affinities to forms with secondary wood (particularly, *Cairoa*). Sieve cells and parenchyme have been identified in the phloem. Some specimens of *Triloboxylon* produced an outer cortex resembling the *Sparganum*-type of cortex in primitive gymnosperms. The branching of the axes in P. is decussate (*Tetraxylopteris*; Fig. 38, *c*, *d*), spiral (*Rellimia*, *Triloboxylon*, *Reinmannia*, *Aneurophyton*), two-ranked with alternating paired and solitary branches (*Proteokalon*). The terminal branches divide dichotomously in various planes or in one plane, and then become planated (*Triloboxylon*; Fig. 38, *f–i*). When they are fused they can be mistaken for leaves. Such planated structures were called planated branches. The sporangia are elliptical, aggregated, sometimes clustered in large masses or repeatedly branching axes, usually flexed towards the main axis, to which the branches are directed (Fig. 38, *d*, *e*, *l*, *m*). There is a striking resemblance in the sporangial structure of *Psilophyton*, *Tetraxylopteris* and *Protopitys* (Scheckler, 1982). Fertile branches of *Triloboxylon* are small (Fig. 38, *g*) and of spiral arrangement directly on the thicker axes between the branches. Heterospory has not been recorded. Spores are rounded, trilete. The sexine is delimited in the form of a cavum embracing the entire distal side (Fig. 38, *j*).

Order Protopityales

The single genus known here is *Protopitys* (C_1; Fig. 39, *a–c*). These are arborescent plants with a trunk up to half a metre in cross-section. Secondary wood is massive (in the roots non-evident), of typical

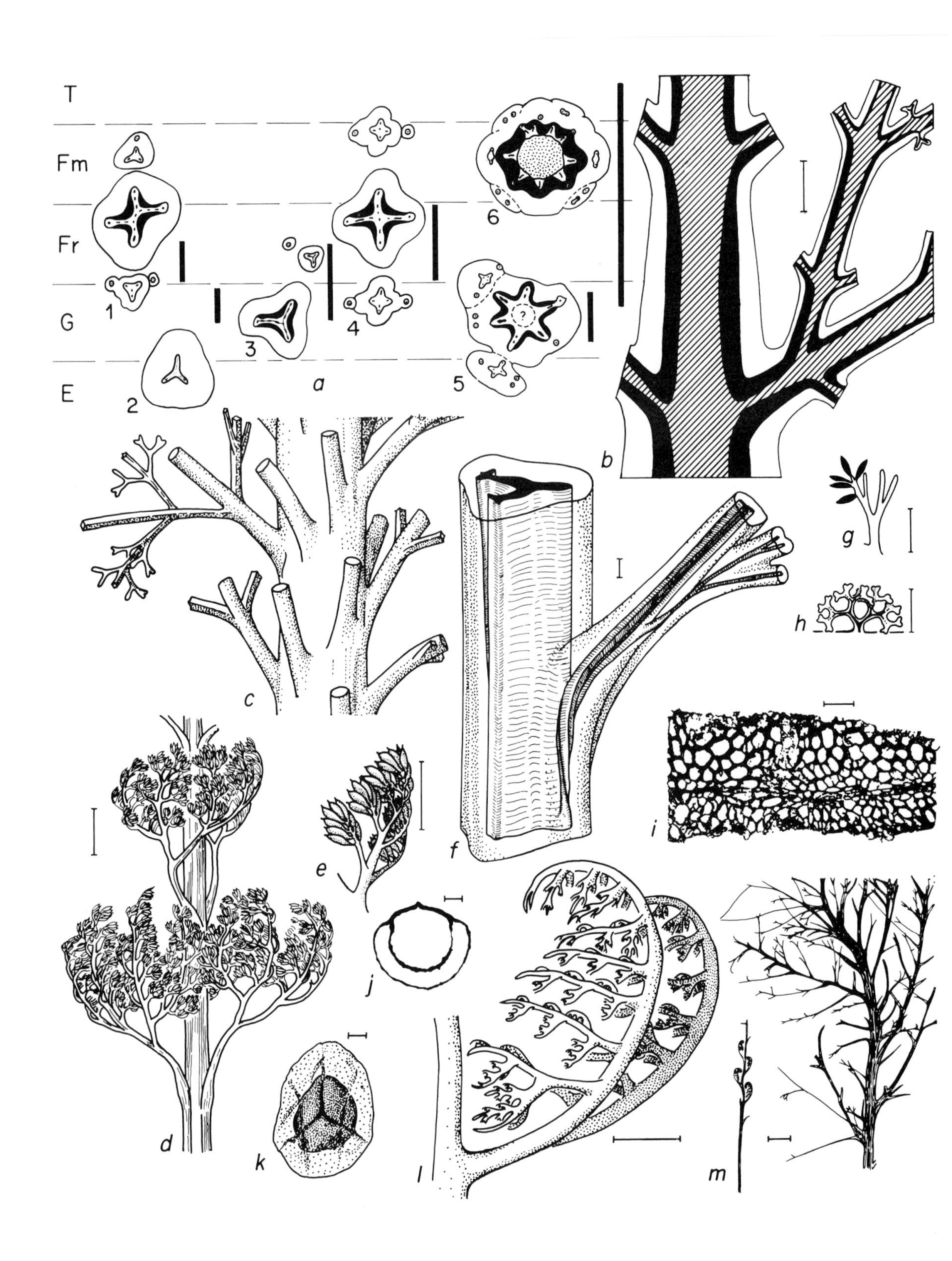

T
Fm
Fr
G
E
1
2
3
4
5
6
?
a
b
c
d
e
f
g
h
i
j
k
l
m

gymnospermous structure. Radial walls of tracheids with one- to multiseriate, round bordered pits. Medullary rays are uniseriate, 2–3 cells high, or biseriate, up to 15 cells high. Pith, elliptical in cross-section, surrounded by metaxylem. Protoxylem strands situated on the tips of paired metaxylem elevations that bulge up into narrowed parts of the pith. The phyllotaxis is alternating, two-ranked. The primary xylem strands first divide, and then the resulting paired strands fuse to form a leaf trace (Gensel and Andrews, 1984). The outward aspect of the photosynthetic organs is unknown. Sporoclads branch pinnately dichotomously. Sporangia are terminal, elliptical with attenuated apex, dehiscing by a longitudinal slit. Spores in some sporangia are on average 80 μm in diameter, in others are about 150 μm, and in a third type are intermediate in size. The largest spores reach 255 μm in diameter. Spores are trilete with a distal cavum, the same as in Protopteridiales. The roots have an oval diarch xylem. This character is clearly of pteridophytic nature.

Order Archaeopteridales

The Archaeopteridales (A.) differs from the Protopteridiales in its eustelic axes, from Protopityales in its chiefly spiral phyllotaxis, and from both groups in its distinctly expressed heterospory. According to the wood structure and organization of the stem and roots, A. reached the level of primitive gymnosperms. Some A. already have well-developed leaf laminae. Only the genus *Archaeopteris* has been well studied. It probably encompasses more than one natural genus (Bonamo, 1975; Gensel and Andrews, 1984). The genera *Siderella*, *Actinopodium* and *Actinoxylon* (Fig. 38, *a* (5)) are known from fragments of petrified axes with two-ranked alternating (in *Siderella* and, evidently, in *Actinopodium*) or spiral (in *Actinoxylon*) branching of the axes. Leaves or leaf-like dissected organs were attached to the branches either spirally, in two rows, or decussately. The genera *Svalbardia* and *Eddya* may be synonymous to *Archaeopteris*. The distinction of families among A. is untimely.

Genus *Archaeopteris* (D_3–C_1; Fig. 39, *e*, *l*, *n–v*) includes large trees with trunks up to a metre and more in diameter. *Archaeopteris* is usually depicted as a tree similar to modern pines. Chaloner (1984) found a large specimen of *A. hibernica* with a group of frond-like branches borne at the apex of a main axis, forming a crown, with a resulting tree-fern-like growth habit. The rooting system was rather primitive, with the predominant dichotomy near the stem base (Snigirevskaya, 1984). Fragments of petrified trunks and roots have been assigned to the genus *Callixylon* (Fig. 39, *f–k*; Lemoigne *et al.*, 1983). The radial tracheid walls are covered by round bordered pits without a torus. The pits are clustered in groups, which on adjacent tracheids form a continuous radial band. The bands with pits on the radial cuts alternate with bands lacking pits (Fig. 39, *f*). In the late wood of the growth layers a smaller number of minor pits occur on the tangen-

Fig. 38. Protopteridiales (*a*(1–4), *b–m*) and Archaeopteridales (*a*(5, 6)); Middle (*j*, *l*, *m*) and Upper (*b–i*, *k*) Devonian; USA (*b–i*, *k*), Western Europe (*j*, *l*, *m*); *a*, schemes of anatomy of axes of different orders, and stratigraphic ranges of genera, secondary wood in black, pith stippled, protoxylem is shown inside primary wood, stratigraphic ranges are shown by vertical lines (T, Tournaisian; Fm, Famennian; Fr, Frasnian; G, Givetian; E, Eifelian; 1, *Proteokalon*; 2, *Reinmannia*; 3, *Triloboxylon*; 4, *Tetraxylopteris*; 5, *Actinoxylon*; 6, *Archaeopteris*); *b–e*, *Tetraxylopteris schmidtii* Beck, scheme of arrangement of primary (shaded) and secondary (in black) wood in axes of different order (*b*), part of vegetative shoot (*c*), fertile shoot (*d*) and its branch of last order (*e*); *f–i*, *Triloboxylon ashlandicum* Matt. et Banks, arrangement of wood in branching axis (*f*), fertile appendage (*g*), leaf-like flattened appendage (*h*), wood in one segment of stele with protoxylem in the middle (*i*); *j–m*, *Rellimia thomsonii* (Daws.) Lecl. et Bon., section (*j*) and polar view (*k*) of spores, fertile branch with abolished sporangia (*l*), fragments of fertile (at left) and sterile (at right) specimens (*m*). Scale bar = 1 cm (*b*, *d*, *m*), 5 mm (*e*, *g*, *h*, *l*), 1 mm (*f*), 100 μm (*i*), 20 μm (*j*, *k*). Modified from: Beck, 1976a (*a*) and 1957 (*c*); Scheckler, 1976 (*b*, *g*, *h*); Bonamo and Banks, 1967 (*d*, *e*); Matten and Banks, 1966 (*f*, *i*); Mustafa, 1975 (*j*); Bonamo, 1977 (*k*); Leclercq and Bonamo, 1971 (*l*, *m*).

a
b
c
d
e
f
g
h
i
j
k
l
m
n
o
p
q
r
s
t
u
v

tial walls (as in some conifers). The roots are protostelic with a quadriradiate metaxylem, each ray terminating in two protoxylem bundles. The trunks have a periderm and wide pith, sometimes transversely septate. Around the periphery of the pith are primary vascular bundles, usually called sympodia. Leaf traces extend outward (Fig. 39, *m*). The points where they issue from the axial traces compose an orderly ascending spiral. In some places the spiral order is violated, due to the difference in the denseness of the leaves on various sides of the axes. The vascular bundles of the branches and leaves display a single phyllotactic system, but the branches and leaves themselves are confined to different orthostichies (Fig. 39, *e*). The apex of the branches of the penultimate order is covered with leaves, the same as the branches of the last order. The latter are arranged in two rows, between which indistinct scars of abscissed leaves can be detected. One bundle entered each branch. It was repeatedly tangentially divided. A cylinder of sympodia was formed, with the outer branches directed into the leaves. One vein entered the leaf base and then repeatedly dichotomized. The leaf lamina was entire or dissected, sometimes in filiform lobes. Although the leaves were arranged spirally or decussately, their laminae were usually oriented parallel to the pinna plane (as in *Metasequoia* and some species of *Abies*). The vascularization in A. (in contrast to gymnosperms) does not give any indication of axillary branching. Although at times pinnules subtend the branches of the last order, this does not imply the axillary origin of the branches. In A. the departure of the leaf traces from the axial sympodia is radial, whereas in gymnosperms (with the exception of the most primitive) it is tangential.

Fertile branches of *Archaeopteris* (Fig. 39, *l*, *o–r*) are either completely fertile, or bear sporangia on the middle pinnae, sometimes only in their middle parts. Completely fertile pinnae terminate in ribbon-shaped sterile appendages – strongly modified leaves. Sporangia are arranged on the acropetal side of the pinna, solitary or in small clusters, on repeatedly dichotomizing branchlets. Occasionally the sporangia pass onto the pinnule lamina, which then becomes modified. Of importance is the fact that the sporangia pass on to the upper surface of the pinnules, not on to the lower one as in ferns. Such fertile pinnules are spirally arranged, so that a strobiloid structure is formed (Fig. 39, *o*). The sporangial stalk is supplied with a vascular bundle, and the wall has stomata. A. were heterosporous plants (Fig. 39, *s–v*). Microspores were round, trilete with a microtuberculate surface. The outer exine layer was composed of a compact mass of fused thick-walled bodies with a small inner cavity. The inner layer was lamellar. The splitting of the exine was absent. The quasisaccate structure of some gymnosperm pollen can easily be derived from the organization of A. microspores (see Chapter 4). The megaspores are similar to the

Fig. 39. Protopityales (*a–c*) and Archaeopteridales (*d–v*); Lower Carboniferous (*a–c*), Middle–Upper (*d*) and Upper (*e–v*) Devonian; Western Europe (*a–c*), Spitzbergen (*d*), North America (*e–v*); *a*, *Protopitys scotica* Walt., fertile shoot; *b*, *P. buchiana* Goepp., scheme of stem anatomy (radial lines – secondary wood, metaxylem stippled, protoxylem – black dots); *c*, *P. scotica*, tracheid of secondary wood with crossed apertures of bordered pits; *d*, *Svalbardia polymorpha* Høeg, vegetative shoot; *e*, *Archaeopteris macilenta* Lesq., branch of penultimate order covered by leaf bases, and leafy branch of last order; *f–i*, *Callixylon newberryi* (Daws.) Elk. et Wiel., radial section of secondary wood with ray and groups of bordered pits arranged into horizontal rows (*f*), surface view of bordered pits (*g*), pits of cross field (*h*), group of bordered pits with crossed apertures (*i*); *j*, uniseriate rays in tangential section of secondary wood of *C. brownii* Hosk. et Cross; *k*, multiseriate rays of *C. newberryi*; *l*, fertile frond of *Archaeopteris macilenta* group, sporangia in black; *m*, arrangement of vascular bundles (sympodia) and their radial branching in axis of *Archaeopteris*; *n*, pinnule of *A. halliana* (Goepp.) Daws.; *o*, strobiloid fertile shoot of *Archaeopteris*; *p–t*, *A. halliana*, partly fertile pinnule (*p*), fertile appendage (*q*), reconstruction of fertile branch with sporangia abolished from lower appendage (*r*), marginal sculpture (*s*) and general view (*t*) of megaspore; *u*, *v*, *Archaeopteris* sp., microspore, sculpture of proximal part and laesura along trilete slit (*u*) and general view (*v*). Scale bar = 1 cm (*a*, *d*, *l*), 5 mm (*e*), 1 mm (*b*, *n*, *p–r*), 100 μm (*f*, *j*, *k*, *t*), 20 μm (*h*, *i*), 10 μm (*g*, *v*), 5 μm (*s*, *u*). Modified from: Smith, 1962 (*a*, *c*); Boureau, 1970, after Solms-Laubach (*b*); Høeg, 1942 (*d*); Beck, 1971 (*e*), 1970 (*f–k*), 1962 (*l*), 1981 (*o*); Phillips *et al.*, 1972 (*n*, *p–v*).

microspores, but much larger. Their exine has the same two layers, but it is covered on the surface by a layer of round bodies with a thick wall and spongy inner part.

The genus *Svalbardia* (Fig. 39, *d*) includes shoots with leaves dissected into narrow dichotomous lobes. The dissection of the leaves in more than one plane was suggested by Matten (1981). In *Eddya* the leaves are larger with fan-shaped venation. It is thought that the plants of this genus represent sporelings of *Archaeopteris*. The genera *Federkurtzia* (Archangelsky, 1981) and *Botrychiopsis* (Middle Carboniferous–Lower Permian of Gondwana) are probably affiliated to A.

It is possible that the reduction of the number of megaspores to a single functional one (the other megaspores of the tetrad are aborted) had already begun among A. After this the formation of the ovule began. During this process the general aspect of the plant and the structure of the microsporoclads remained the same. Those species of *Archaeopteris*, where no megaspores have been found, might have produced seeds (Gensel and Andrews, 1984). It is notable that the oldest ovules *Archaeosperma* have been recovered from the same bed together with leafy shoots of *Archaeopteris hibernica*, no other associating plants being found to which these ovules might have belonged.

Order Noeggerathiales

The Noeggerathiales (N.) were commonly affiliated with articulates and ferns, but most likely they are allied to progymnosperms (Beck, 1981; Taylor, 1981). Best studied is the genus *Noeggerathia* (C; Fig. 40, *a*). The fronds are once-pinnate with

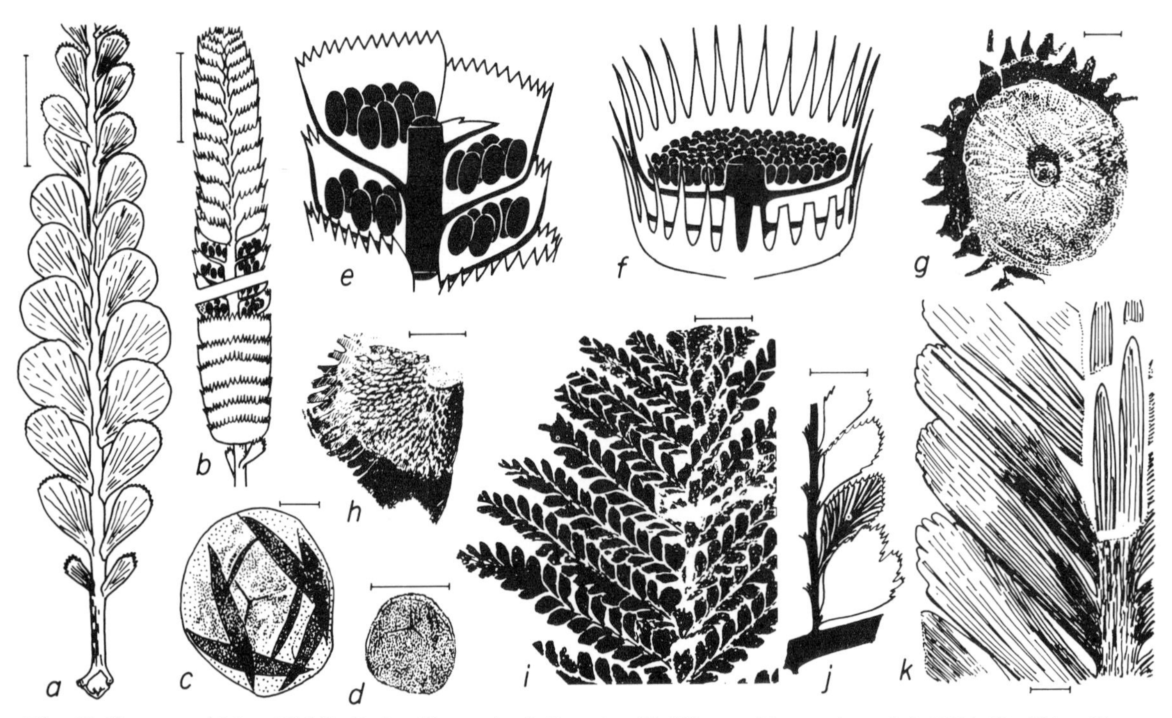

Fig. 40. Noeggerathiales; Middle Carboniferous (*a–j*), Permian (*k*); Western Europe (*a–g*, *i*, *j*), USA (*h*), China (*k*); *a*, *Noeggerathia foliosa* Sternb.; *b*, strobilus of the same plant (*Noeggerathiostrobus bohemicus* Feistm.); *c*, *d*, megaspore (*c*) and microspore (*d*) of *N. bohemicus*; *e*, *f*, schemes of sporophylls of *Noeggerathiostrobus* (*e*) and *Discinites* (*f*); *g*, *D. jongmansii* Hirm., abaxial view of sporophyll; *h*, *Lacoea seriata* Read, adaxial view of sporophyll; *i*, *j*, *Palaeopteridium macrophyllum* Něm., frond (*i*) and pinna (*j*); *k*, *Tingia crassinervis* Halle. Scale bar = 3 cm (*a*, *b*, *i*), 1 cm (*g*, *h*, *j*, *k*), 200 μm (*c*), 100 μm (*d*). Modified from: Boureau, 1964, after Halle (*a*, *b*); Němejc, 1963 (*c*, *d*, *i*, *j*); Beck, 1981, after Leary and Pfefferkorn (*e*, *f*); Zimmermann, 1959, after Hirmer (*g*); Leary, 1978 (*h*); Halle, 1927 (*k*).

obovate or rounded-rhombical pinnules; the venation is fan-shaped; the pinnule margins are denticulate. The strobili (genus *Noeggerathiostrobus*; Fig. 40, *b*, *e*) are composed of bowl-shaped sporophylls, not completely embracing the axes, and bearing mega- or microsporangia. Mega- and microspores (Fig. 40, *c*, *d*) are round, trilete. Bipinnate fronds of the genus *Palaeopteridium* (Fig. 40, *i*, *j*) associate with fructifications of the *Discinites* type (Fig. 40, *f*, *g*), where the sporophylls are bowl-shaped with a denticulate margin, and completely embrace the axis. Unlike *Noeggerathiostrobus*, each megasporangium contains only a single megaspore. Closely related to these genera is *Lacoea* (Fig. 40, *h*). These strobili were, seemingly, associated with entire leaves of the *Lesleya* type (resembling *Taeniopteris*) and once-pinnate leaves of the *Megalopteris* type. Of affiliation to N. is the genus *Tingia* (P; Fig. 40, *k*), where elongate pinnules are situated in four rows, the pinnules in two inner rows being much smaller than the lateral ones. The strobili (*Tingiostachys*) associated with *Tingia* have not as yet been adequately studied. It is believed that their elbow-shaped sporophylls were arranged by fours in whorls and had four sporangia borne in its horizontal part. It is unclear whether these strobili were homo- or heterosporous. The order Tingiales is sometimes recognized.

The similarity between N. and progymnosperms is expressed in the heterospory and form of the pinnules; moreover, the N. sporophylls are similar to the fertile pinnules of strobiloid fructifications associating with some species of *Archaeopteris*.

DIVISION PINOPHYTA (GYMNOSPERMAE)

Gymnosperms (G.) differ from pteridophytes in the appearance of ovules (seeds) – modified megasporangia, embedded in specialized coverings. The ovules are open for the direct access of pollen. Sometimes the pollen germinates outside the ovule, the same as in angiosperms, but these plants (e.g. certain conifers) correspond in all other characters to common G. With the advent of seeds in G. specialized organs develop that bear or accompany the seeds. The male fructifications also undergo modifications.

The morphological terminology, borrowed from pteridophytes (e.g. megasporophyll) or angiosperms (e.g inflorescence), is currently used for the description of G. fructifications. However, the morphological independence of G. fructifications gives justification to the introduction of a specific terminological system (Meyen, 1982a, 1984a) outlined below. The seed with its stalk (seeds without stalks are called sessile) comprise a monosperm, whereas any aggregation of seeds is called polysperm. A complex of sterile organs, differing from foliage leaves and accompanying fructifications, is termed a circasperm. Polysperms may be simple (Fig. 41, *a*–*o*) and compound (Fig. 41, *p*–*r*). Simple polysperms have been divided into four main groups. The first group involves polysperms: (1) racemose (consisting of monosperms arranged around the axis; Fig. 41, *a*); (2) spicate (differ from racemose in the reduction of the stalks; Fig. 41, *b*); (3) umbellate (monosperms on an aborted axis; Fig. 41, *d*); (4) head-like (sessile seeds on a dwarf axis, or monosperms on a fleshy short receptacle; Fig. 41, *c*); (5) pinnate (Fig. 41, *e*); (6) irregularly branching (pleiochasial; Fig. 41, *f*). The second type includes flattened leaf-like polysperms. If the lamina of the polysperm is similar to a vegetative leaf, such a polysperm is called a phyllosperm (Fig. 41, *g*); and if it differs considerably from the leaf, then it is termed a cladosperm (Fig. 41, *h*–*j*). Cladosperms may be with entire or subdivided laminae, having the same types of symmetry as in foliage leaves. By analogy with peltate leaves, there may be peltate polysperms, or peltoids (Fig. 41, *j*). The third type includes semiclosed seed containers – cupules and capsules. The term cupule (Fig. 41, *k*, *l*) refers to an orthotropous or inverted container of variously fused or free circasperm elements. One or several seeds may be attached to the bottom of the cupule. The cupule, which became similar to an integument, where the apex functioned as a pollen capture device is termed an outer integument (or cupuloment). All other seed containers can be termed capsules (Fig. 41, *m*, *n*). They may be bivalved, jug-shaped and inverted, peltate and of other types. The fourth type of polysperms includes fertiligers. In this case polysperms of the former types are attached to a modified or unmodified leaf, and constitute a united complex (Fig. 41, *o*).

No classification of compound polysperms has

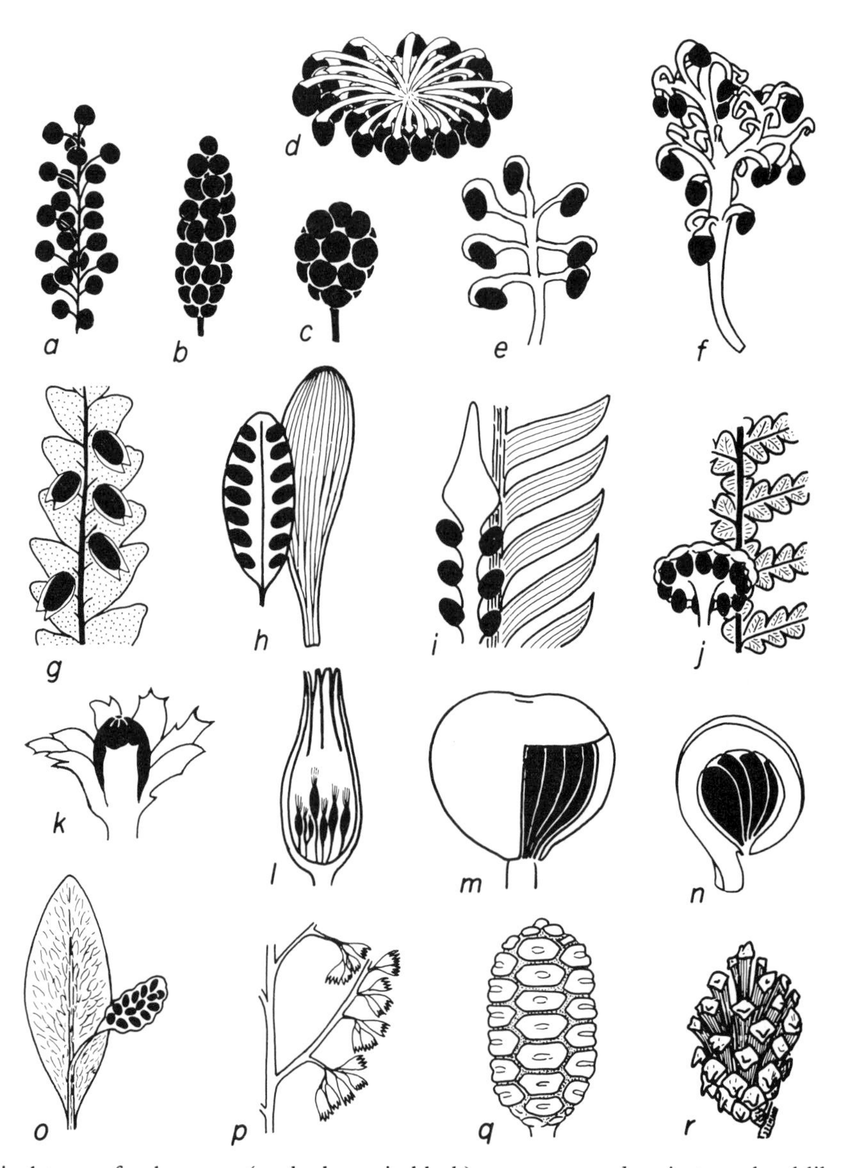

Fig. 41. Principal types of polysperms (seeds shown in black); *a*, racemose; *b*, spicate; *c*, head-like; *d*, umbellate (*Gaussia cristata*); *e*, pinnate (*Trichopitys heteromorpha*); *f*, pleiochasial (*Arberia minasica*); *g*, phyllosperm (*Emplectopteris triangularis*); *h*, cladosperm of *Stiphorus–Leuthardtia* type, and vegetative *Glossophyllum* like leaf; *i*, cladosperm and vegatative leaf of *Cycas*; *j*, peltoid of *Peltaspermum*, and *Lepidopteris*-like frond; *k*, cupule with one seed (*Lagenostoma lomaxii* Scott); *l*, multiseeded cupule (*Calathospermum scoticum*); *m*, multiseeded peltate capsule (*Cardiolepis piniformis*); *n*, multiseeded inverted capsule (*Caytonia*); *o*, fertiliger (*Ottokaria*); *p*, compound polysperm consisting of cupules (*Lagenospermum*); *q*, compound zamioid polysperm (*Zamia*); *r*, compound coniferoid polysperm (*Pinus*). From Meyen, 1984a.

as yet been established. For their designation it is convenient to use the terms of selected taxa. For instance, polysperms similar to the conferalean cone, may be termed coniferoid.

This classification does not imply the common origin of certain types, but merely aims to refine and clarify the usage of descriptive terminology.

In describing male fructifications of G. the terms adopted for pteridophytes may be used. A specific term – microsporoclad – is introduced for those aggregations of microsporangia that branch in different directions, and which cannot be termed either microstrobili, or microsporophylls. Specialized sterile organs, accompanying male fructifications, form a circandr.

The transition from pteridophyte spores to gymnosperm pollen, and problems bearing on the terminology of pollen and prepollen are discussed in Chapter 4. The major trends in the evolution of gymnosperm pollen are depicted in Fig. 42.

In many G. the pollen penetrates into the ovule with the aid of a secretory drop, released from the micropyle and capturing pollen from the air. The secretory drop is drawn into the ovule, carrying in the pollen, that adheres to it. This mechanism had probably already evolved in the earliest G. In the latter the function of capturing the pollen was performed by the nucellar apex, modified into a hollow tubular structure – the salpinx (Fig. 43, *b*). As the integument underwent closure above the nucellus and the micropyle was formed, the function of capturing pollen shifted to the apex of the integument (Fig. 43, *c*), and after the cupule evolved into the outer integument, became shifted to its apex (Fig. 43, *d*). In the trigonocarpalean seeds *Polylophospermum* the outer integument forms an additional air chamber above the attenuated micropyle-like apex, and the pollen was obviously captured by the opening in this chamber (Combourieu and Galtier, 1985). The pollen probably might also have been captured by capsules.

G. are linked by transitional forms with pteridophytes and angiosperms. Some of these forms give evidence of phylogenetic links, such as the progymnosperms that phylogenetically link pteridophytes with gymnosperms. In other cases only certain organs are transitional, where similarities evolved independently in various phylogenetic lineages. This applies to lycopsid megasporophylls that are similar to ovules and seeds with embryos, to pycnoxylic stems in articulates. Probably the 'angiospermization' of various gymnosperms (Krassilov, 1977), can be treated in the same sense, i.e. the acquisition by G. of features typical of angiosperms.

Transitional forms may be observed between the major G. taxa. Soon after pteridosperms were discovered it was suggested that G. should be subdivided into phyllospermids and stachyospermids. The first include G. with seeds on leaf-like organs – megasporophylls. These are pteridosperms and cycads, also bennettites, similar in the foliage aspect and life-forms to the cycads. Stachyospermids involved plants with fructifications on modified shoots, frequently in the shape of cones. This refers to cordaites, conifers and ginkgos. The phyllospermids were associated with manoxylic stems and fern-like leaves, whereas stachyospermids were associated with pycnoxylic stems and simplified leaves – needle-shaped with a simple lamina and fan-shaped or parallel venation. Later the above described syndromes were violated. Among lagenostomalean pteridosperms true pycnoxylic stems were found that produced fern-like foliage. Seed-bearing shoots of lagenostomans can by no means be regarded as modified leaves. They represent stems profusely branching, with cupules borne on the tips of the branchlets. Other pteridosperms were found in association with leaves that were formerly considered typical of conifers (*Phylladoderma*). Leaves of ginkgoid habit were found in bennettites (*Eoginkgoites*) and pteridosperms. In arberians (glossopterids) characters, typical both of phyllospermids and stachyospermids, occurred in various intricate combinations. It was thought that stachyospermids were characterized by platyspermic seeds, whereas phyllospermids were characterized by radiospermic seeds. However, many pteridosperms have platyspermic seeds.

Recent data revealed other alliances between orders of G., and allowed a more natural grouping of orders into classes to be obtained. On the other hand, certain trends in the transformations of

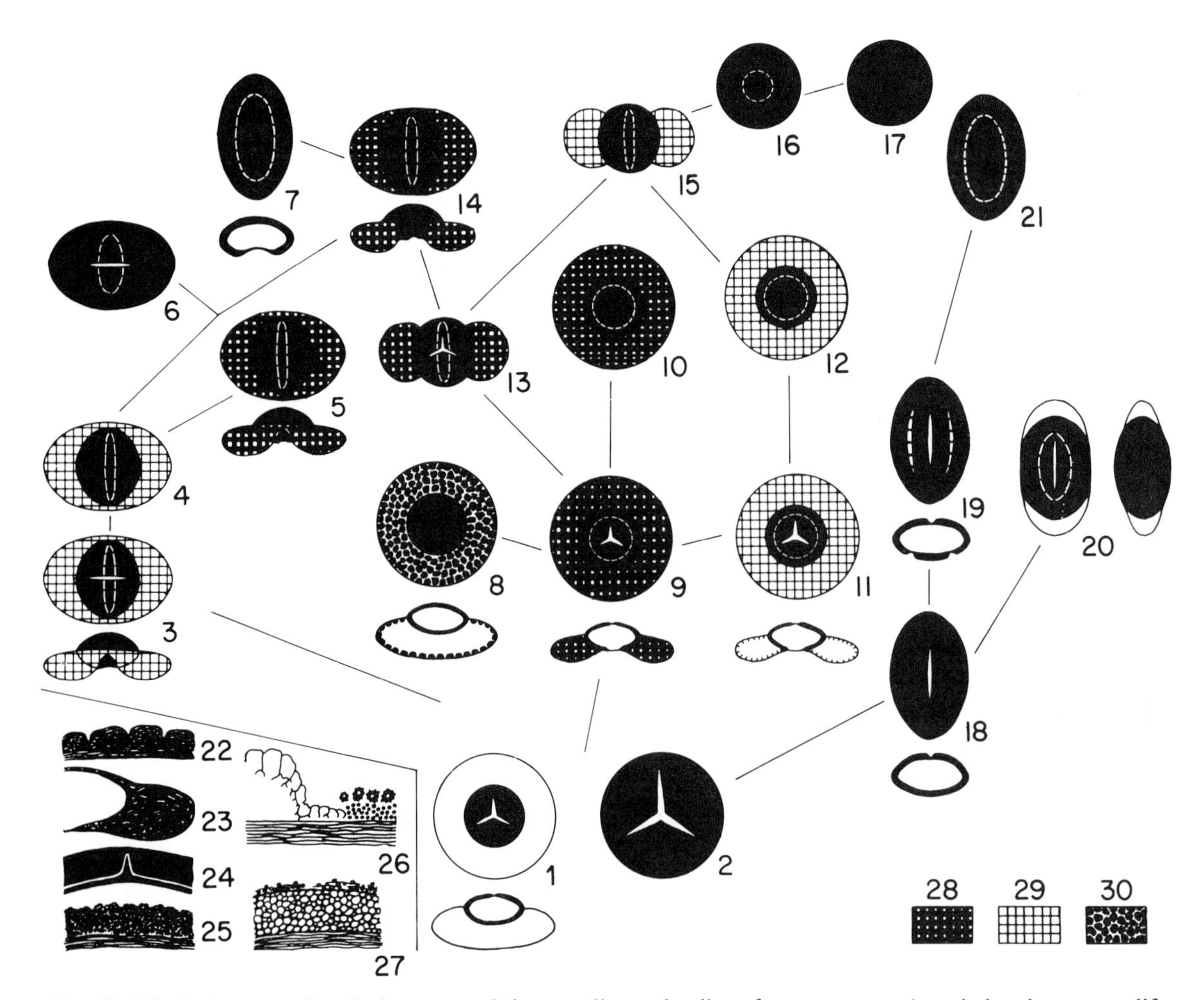

Fig. 42. Principal types and evolutionary trends in prepollen and pollen of gymnosperms (taxa in brackets exemplify certain types); 1, cavate trilete grain (progymnosperms, *Zimmermannitheca*, *Pterispermostrobus*); 2, trilete asaccate grain (progymnosperms, Lagenostomales); 3, monosaccate bilateral grain with monolete proximal slit (older members of Callistophytales); 4, same without slit (younger members of Callistophytales); 5, quasimonosaccate bilateral grain (Cardiolepidaceae); 6, asaccate bilateral grain, occasionally with monolete proximal slit (Arberiales, Peltaspermales); 7, asaccate monocolpate grain (Peltaspermales, Ginkgoales); 8, monosaccate grain with intrareticuloid (*Cladostrobus*); 9, quasimonosaccate grain with proximal slit (Vojnovskyaceae, Rufloriaceae, Arberiales?); 10, same without slit (Vojnovskyaceae, Pinales); 11, monosaccate grain with proximal slit (Cordaitanthaceae); 12, same without slit (Cordaitanthaceae); 13, quasidisaccate grain with proximal slit (Pinales); 14, same without slit (Peltaspermales, Arberiales, Pinales); 15, disaccate grain (Pinales); 16, asaccate grain (Pinales); 17, inaperturate grain (Pinales); 18, monolete grain (Trigonocarpales); 19, grain with three slits or colpi (Trigonocarpales, *Eucommiidites*); 20, monolete grain with saccus-like structure located differently from Nos 2, 4 and 15 (Trigonocarpales); 21, monocolpate grain (Cycadales, Bennettitales); 22, transverse section of striated body (occasionally present in Nos 5–7, 10, 14, 15?, 21); 23, section of quasisaccus; 24, non-laminated exine and proximal slit in grain as in No. 2; 25, granular exine; 26, saccus structure in Pinaceae; 27, alveolar sexine of Trigonocarpales; 28, quasisaccus; 29, saccus; 30, infrareticuloid. From Meyen, 1984a.

fructifications were outlined, which formerly were not taken into account. All this resulted in the re-appraisal of the weight of many characters, and the rearrangement of the G. system (Meyen, 1982a, 1984a). From successive comparison of all the best studied gymnosperm genera and by tracing the characters in the history of the taxa, it became apparent that the major phylogenetic lineages of gymnosperms and the respective taxa of higher ranks are best reflected in the evolution of seeds and cupules.

The oldest seeds can be divided into two types: platyspermic (Fig. 43, *g*; Fig. 44, *a*; Fig. 45, *a*) and radiospermic (Fig. 43, *e*, *f*; Fig. 44, *d*; Fig. 45, *d*). The platyspermic seeds (bilaterally symmetrical) are longitudinally flattened with two vascular bundles passing along the main plane in the wings produced by an integument. More seldomly the bundles divide in the same main plane. Very rarely two additional vestigial bundles pass in the perpendicular plane, quickly disappearing (*Callospermarion*). This can serve to indicate that the platyspermic seeds evolved from radiospermic. The radiospermic (radially symmetrical) seeds are unflattened, and several bundles arranged in a circle pass through the integument. The platyspermic seeds lack a cupule, and at first were situated on leafless axes. Such polysperms characterize the order Calamopityales. The most ancient G. with cupules are united in the order Lagenostomales. Two groups have been recognized here, that so far are of indeterminate taxonomic status. They are tentatively denoted below by the symbols BS and RS. In BS the

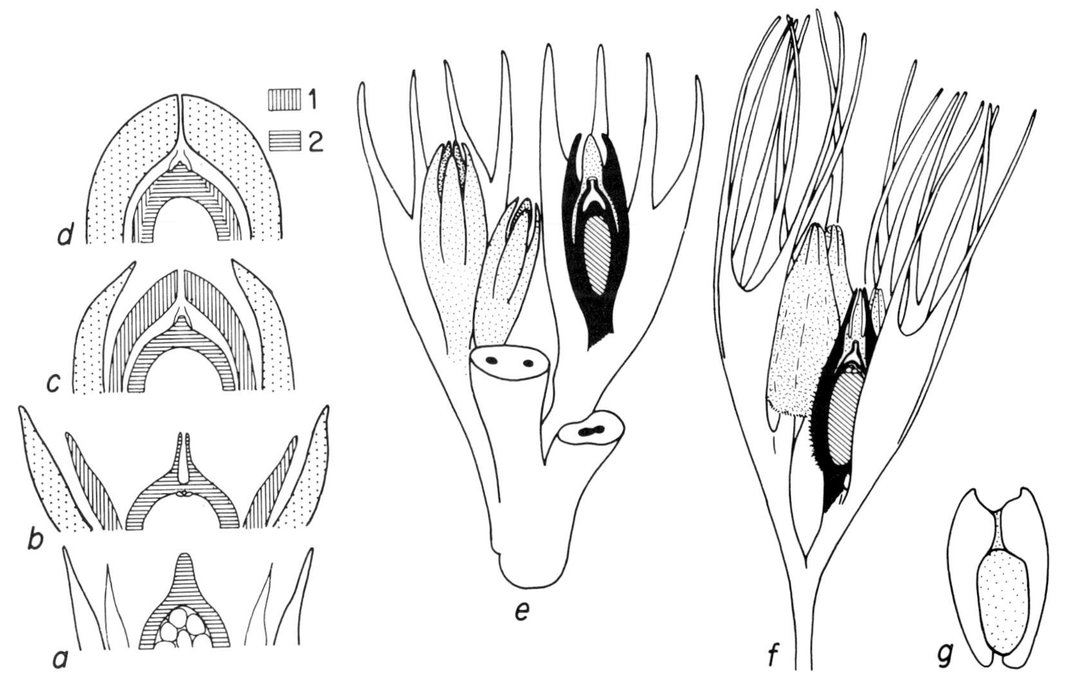

Fig. 43. Origin and transformation of seed envelopes (*a–d*) and oldest seeds (*e–g*); *a*, megasporangium and surrounding sterile axes in hypothetical ancestor of gymnosperms; *b*, oldest seed, megasporangial apex is transformed into salpinx, whereas integumental and cupular lobes are modified; *c*, formation of integumental micropyle, base of salpinx is transformed into lagenostome with central column; *d*, nucellus fused with integument, formation of cupular micropyle; 1, integument; 2, nucellus; cupule lobes and their derivates in *a–d* stippled; *e–g*, oldest (Upper Devonian) seeds of USA (*e*, *f*) and Ireland (*g*); *e*, reconstruction of *Hydrasperma*-like seed bearing cupule, one seed shown in section; *f*, same for *Archaeosperma arnoldii* Pet. et Beck; *g*, platyspermic seed *Spermolithus devonicus* Johns. From Meyen, 1984a (*a–d*); modified from: Gillespie *et al.*, 1981 (*e*); Pettitt, 1970 (*f*); Chaloner *et al.*, 1977, Taylor, 1981 (*g*).

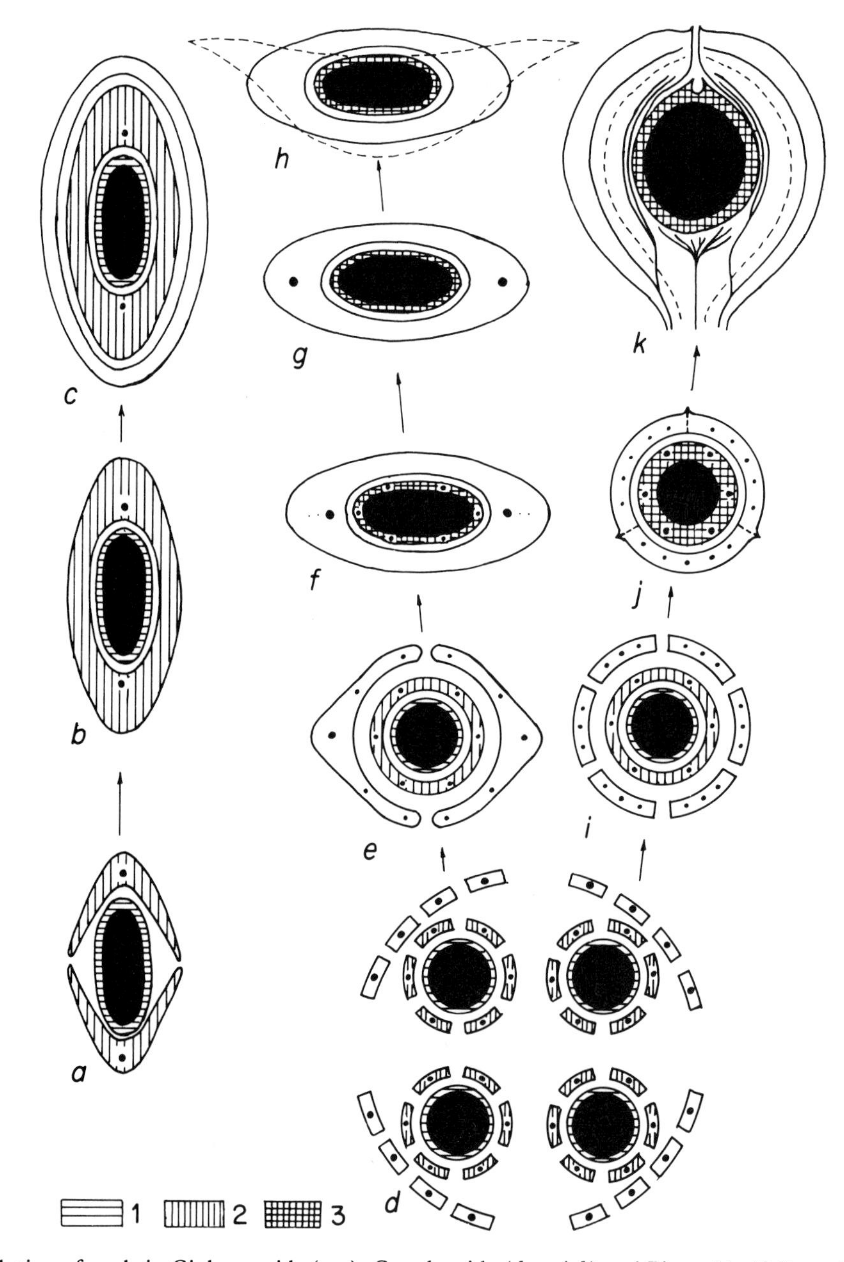

Fig. 44. Evolution of seeds in Ginkgoopsida (*a–c*), Cycadopsida (*d*, *e*, *i–k*) and Pinopsida (*f–h*); *a*, *Lyrasperma*-type with free integumental lobes; *b*, *Callospermarion*-type with integumental micropyle; *c*, *Ginkgo*, outer double contour corresponds to collar which simulates cupule; *d*, *Hydrasperma*-type, cupule with four seeds, cupular and integumental lobes free; *e*, hypothetical stage with bilobed cupule, fused integumental lobes, and non-vascularized nucellus; *f*, *Nucellangium*-type, fusion of integument and nucellus resulted in envelope retaining integumental vascularization ('vascularized nucellus'), external envelope of cupular origin ('external integument' or simply 'integument') with two vascular bundles; *g*, *Mitrospermum*-type, vascularization of nucellus disappeared; *h*, *Pinus*-type, non-vascularized seed, broken line shows bifacial seed contour; *i*, *Lagenostoma*-type, radially symmetrical lobed cupule bears one seed with integumental micropyle; *j*, *Pachytesta*-type, same transformations as in *f*, but cupule and its derivates are radially ▶

cupules are bilaterally symmetrical (Fig. 43, *e*, *f*; Fig. 45, *d*), whereas in RS they are radially symmetrical (Fig. 44, *i*; Fig. 45, *h*). In RS the cupule in some genera remains free, and in others began to close and to undergo modification into an outer integument. The culmination of this process is reflected in the Trigonocarpales (Fig. 44, *j*; Fig. 45, *i*, *j*). The inner integument coalesced with the nucellus, preserving its vascular system. This resulted in the formation of a vascularized nucellus, or inner integument – an envelope of double origin. At the apex of this newly formed organ structures remain that formerly complicated the nucellar apex (Fig. 45, *n*). This is a salpinx with a basal extension (lagenostome), surrounding the elevation of the nucellar apex (central column or nucellar plug). A characteristic feature is the elevation of the megagametophyte inside the nucellus, called a tent-pole.

The same process of fusion of the envelopes was completed in group BS in the order Cordaitanthales (Fig. 44, *f*, *g*; Fig. 45, *e*, *f*). This order is closely related to conifers in a wide complex of characters. In this gymnosperm lineage the vascularization of the nucellus at first varied and then disappeared (Fig. 44, *h*; Fig. 45, *g*). Only two of the vascular bundles of the cupule remained in the outer integument, and, in some instances, additional bundles extending from the main ones and situated in the main plane of the seeds. The resulting seeds cannot be distinguished externally from platyspermic, but are secondary platyspermic.

At the phylogenetic roots of G. stand two orders with a set of multiple characters in common (suggesting their common origin), but having produced different types of seeds – platyspermic (Calamopityales) and radiospermic (Lagenostomales with groups BS and RS). Both orders are linked to progymnosperms by forms transitional in the anatomical structure of the stems. Group BS gave rise to Cordaitanthales, from which conifers evolved (secondary platyspermic G.), while group RS gave rise to Trigonocarpales, which were probably progenitors of cycads and bennettites.

In the most ancient G. the seeds, free microsporangia and synangia, are restricted to leafless axes. At the end of the Early–beginning of the Middle Carboniferous of first appearance in different groups are seeds and synangia borne on the underside of the pinnules, or replacing the pinnules and pinnae of the last order. The descendants of calamopityans, united in the order Callistophytales and Gigantonomiales had platyspermic non-cupular seeds and synangia borne on unmodified (or slightly modified) leaves, so that typical phyllosperms and microsporophylls were formed. In the order Trigonocarpales the seeds and synangia either replace leaf elements (pinnae, pinnules), or are attached to rachises without disturbing the arrangement of the pinnae and pinnules, or remain on the leafless axes. The displacement of the fructifications on earlier formed (already occurring in progymnosperms) vegetative leaves, might, probably have been, in terms of the phylogenetic scale, a very rapid process, similar to that which we noted in zygopterids. Such a rapid displacement of the organs from one place to another is termed homoeotic (homoeosis).

In the descendants of the BS group – cordaitanthaleans, dicranophylls and conifers – no such homoeosis of the fructifications is evident. Instead, already in the Carboniferous, compound, simple racemose and spicate polysperms appear; circasperms and circandrs evolve from sterile scales; microsporangia concentrate on flattened or radially symmetrical structures.

The consequences of the homoeosis of fructifications – posthomoeotic, or postheterotopic modifications – are distinctly expressed in advanced Callistophytales and their descendants, united in the order Peltaspermales. They show the gradual transformation of phyllosperms and microsporophylls into organs that more closely resemble axial organs, so that various cladosperms and microsporoclads are formed, frequently similar in

symmetrical (radial broken lines mark sutures); *k*, *Cycas*, largely reduced vascularization of nucellus. Seeds are shown in transverse (*a–j*) and longitudinal (*k*) sections; black dots denote vascular bundles; 1, nucellus; 2, integument; 3, fused initial nucellus and integument. From Meyen, 1984a.

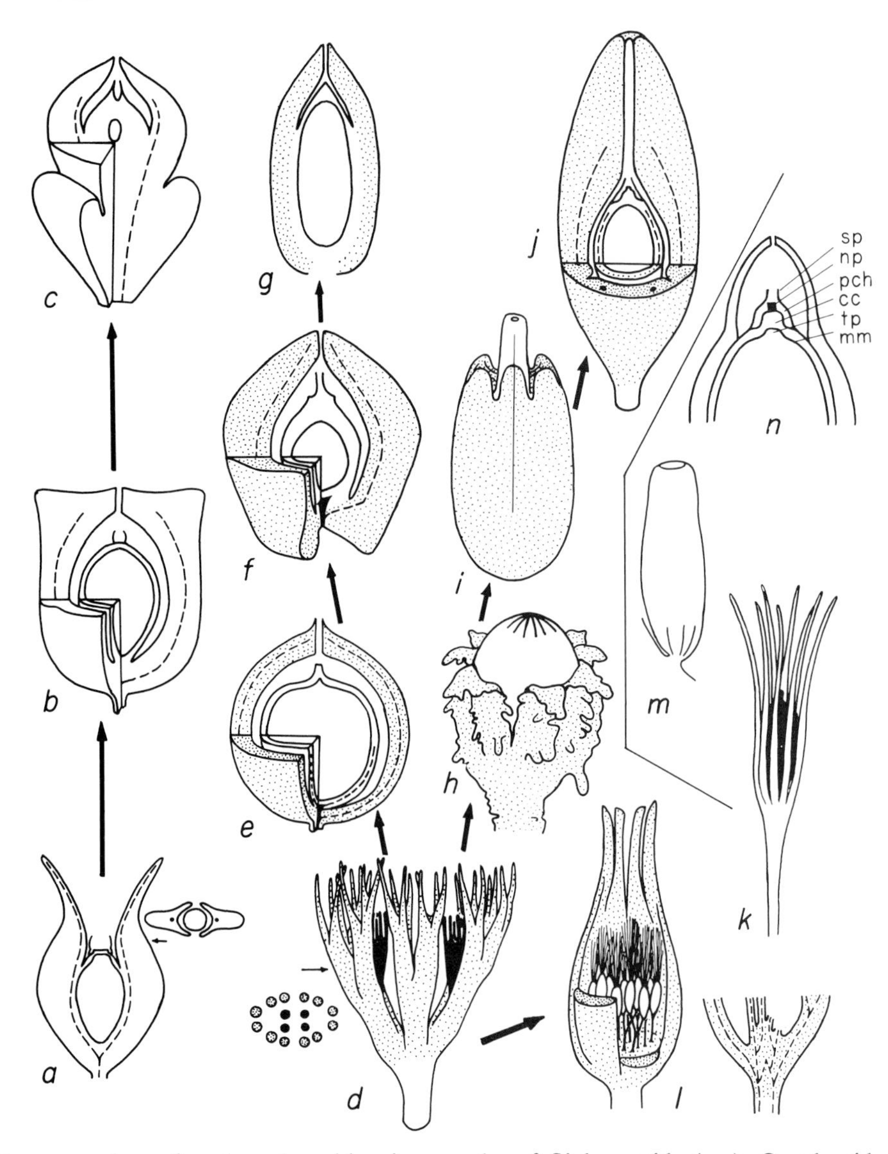

Fig. 45. Reconstructions of seeds and seed-bearing cupules of Ginkgoopsida (*a–c*), Cycadopsida (*d*, *h–n*) and Pinopsida (*e–g*); arrows show same evolutionary trends as in Figs 43, 44; *a*, *Lyrasperma*-type (= Fig. 44, *a*), *L. scotica* (Cald.) Long, Lower Carboniferous; *b*, *Callospermarion*-type (= Fig. 44, *b*), *C. pusillum* Egg. et Delev., Upper Carboniferous; *c*, *Ginkgo biloba* L (= Fig. 44, *c*), seed with collar; *d*, *Hydrasperma*-type (= Fig. 44, *d*), *H. tenuis* Long, cupule with four seeds (in black), to left – scheme of arrangement of seeds and cupule lobes, Lower Carboniferous; *e*, *Nucellangium*-type (= Fig. 44, *f*), *N. glabrum* (Dar.) Andr., Middle Carboniferous; *f*, *Mitrospermum*-type (= Fig. 44, *g*), *M. vinculum* Grove et Rothw., Upper Carboniferous; *g*, *Pinus*-type (= Fig. 44, *h*); *h*, *Lagenostoma* type (= Fig. 44, *i*), *Tyliosperma orbiculatum* Mamay, Middle Carboniferous; *i*, *j*, *Pachytesta*-type (= Fig. 44, *j*), *Stephanospermum elongatum* Hall (*i*) and trigonocarpalean seed (*j*), Middle–Upper Carboniferous; *k*, seed of *Genomosperma kidstonii* Long having free integumental lobes, Lower Carboniferous; *l*, *Calathospermum scoticum* Walt., multiseeded cupule and its vascularization, Lower Carboniferous; *m*, *Stamnostoma huttonense* Long with fused integument, Lower Carboniferous; *n*, seed apex in many Palaeozoic cycadopsid genera (in longitudinal section); mm, megaspore membrane; tp, tent-pole; cc, central column; pch, pollen chamber; np, nucellar plug; sp, salpinx; cupule and its ▶

habit to initially leafless polysperms and microsporoclads. The cladosperms are modified into cupules, capsules, peltoids, or remain flattened (orders Peltaspermales, Caytoniales, Leptostrobales, Ginkgoales). Cladosperms are also formed in the radiospermic G. – Cycadales, that have much in common with the Trigonocarpales. To radiospermic G. are assigned also the Bennettitales, Gnetales and Welwitschiales. The former differ considerably from the Trigonocarpales in the arrangement of their seeds and development of specialized circasperms, but share many common features in the synangial structure – one of the few indications that provides evidence for the origin of the Bennettitales from Trigonocarpales (Stidd, 1980). The Gnetales and Welwitschiales are also regarded as associated in origin with the Trigonocarpales.

The phylogenetic linkages of the Arberiales ('glossopterids') and, particularly, Pentoxylales, are more speculative. They had platyspermic seeds. The platyspermy was, seemingly, primary. The Ephedrales are also platyspermic, but as to whether the platyspermy was primary or secondary is indeterminate. By means of exclusion, it can be inferred that the Ephedrales are related to primary platyspermic G.

If we take into consideration other characters, then proceeding from the modes of transformation of the reproductive organs outlined above, the following general system of gymnosperms can be derived: class Ginkgoopsida includes primary platyspermic G. (Calamopityales, Callistophytales, Gigantonomiales, Peltaspermales, Caytoniales, Leptostrobales, Ginkgoales, Arberiales?, Pentoxylales?, Ephedrales?); class Pinopsida (= Coniferopsida) – secondary platyspermic G. (Cordaitanthales, Dicranophyllales, Pinales), and class Cycadopsida – radiospermic G. (Lagenostomales, Trigonocarpales, Cycadales, Bennettitales, Gnetales, Welwitschiales). The proposed system differs from those earlier suggested in the following salient characteristics:

1. The system is based on fundamental distinctions, established (not postulated) between the major types of G. seeds – platyspermic, radiospermic and secondary platyspermic. Taken together with other pertinent characters, this allows for the precise discrimination of the classes.
2. The group of pteridosperms proves to be heterogeneous, both morphologically and phylogenetically. It can be retained only as an informal unit comprising the genera of uncertain affinities.
3. Ginkgos are again affiliated with leptostrobans (czekanowskias), but can be allied with peltasperms, not conifers (their similarity is chiefly convergent).
4. Ephedrales are assigned to another group of gymnosperms, than Gnetales and Welwitschiales (all three groups are practically unknown in the fossil record, and not discussed below).

The system outlined above reveals the apparent parallelisms between the classes in many characters, especially of the pollen and leaf structure. It makes it easier to explain the diversity of G. However, this parallelism calls, to a greater extent than before, for more circumspect designation and assignment to definite orders and even classes of the form-genera, established on dispersed parts. For reliable distribution of fossil genera in the system of G. it is necessary to know in more detail the organic connections between the parts. This concerns especially the leaves and pollen, produced by plants of various classes, which can be quite similar in type.

Class Ginkgoopsida

In the extant plant kingdom the ginkgoopsids (Gg.) are represented by the genus *Ginkgo*, and probably also by the genus *Ephedra* (which is not discussed here and not taken into account in the characteristics of the class). Primary platyspermic non-cupular seeds of Gg. are situated on leafless axes (Calamopityales, Pentoxylales), phyllosperms (Callistophytales) and their derivatives the

derivates in *d–j*, *l*, stippled. From Meyen, 1984a (figures after Smith, Rothwell, Matten, Lacey, Lucas, Stidd, Cosentino, Grove, Rothwell, Taylor, Millay, Leisman, Roth, Andrews, Schweitzer).

cladosperms (Peltaspermales and their descendants), as well as on fertiligers (Arberiales). Male fructifications are leafless branching microsporoclads, sometimes epiphyllous (Arberiales), or microsporophylls with synangia borne on the underside of unmodified pinnules (Gigantonomiales and some Callistophytales). The pollen varies considerably and evolved from trilete prepollen of the most ancient ferns (Fig. 42 (2)) to saccate (Fig. 42 (3, 4)) and quasisaccate (Fig. 42 (5, 14)), and further to monolete (Fig. 42, (6)) and monocolpate (Fig. 42 (7)) asaccate pollen. Among Gg. were large trees and small prostrate shrubs, perhaps also herbaceous forms and lianas. Growth forms of most of the extinct genera are unknown. The stems were eustelic with one ring of protoxylem strands. Less frequently additional cycles of primary bundles develop. The secondary wood, at times quite thick, is composed of tracheids with round bordered pits on radial walls and pierced by medullary rays. Vascular bundles, extending to the leaf, most frequently depart from two or more primary axial bundles. Leaves are compound-pinnate or dichotomously divided to simple, with fan-shaped or parallel, open or reticulate venation.

Gg. are divided into eight orders (see above), that can be readily recognized according to their polysperm structure. The affinities between certain orders is emphasized by the general similarity in the structure of the foliage and microsporoclads. The orders Callistophytales, Peltaspermales and Ginkgoales are linked by gradual transitions, so that distinctions between them are conventional, although extreme forms in the lineage differ considerably. On the other hand, according to the anatomical structure of the axes and petioles, morphology of the foliage, microsporoclads and prepollen, the order Calamopityales can as yet hardly be distinguished from radiospermic Lagenostomales.

Order Calamopityales

Seed imprints of the Calamopityales (C.; Fig. 47, *a*) have been tentatively assigned to *Samaropsis* (e.g. *S. bicaudata*), whereas petrified remains have been assigned to the genera *Lyrasperma* (C_1; Fig. 45, *a*) *Eosperma*, *Deltosperma*, etc. (Meyen, 1984a). The seeds are elliptical in cross-section. Through each half of the integument one bundle extends, frequently accompanied in the basal part of the seed by additional bundles. The nucellar apex bears a short salpinx with a small basal swelling (lagenostome), at the bottom of which is a central column. The nucellus coalesces with the integument practically throughout its entire length. The habit of the microsporoclads is not sufficiently well known. It was probably the same as in more primitive members of the Lagenostomales. The prepollen is trilete, rounded, found in the pollen chamber of the seeds. Vegetative fronds, seemingly associated with *Lyrasperma*, are compound pinnate, of the *Sphenopteridium* type, with dissected wedge-shaped pinnules. The main rachis was forked, at times had pinnules borne below the point of branching. The venation of the pinnules is fan-shaped (the same as in Archaeopteridales).

Stems were slender, occasionally several centimetres in diameter, eustelic (Fig. 46, *c*, *d*, *g–j*). Around the periphery of the pith, at times mixed (tracheids occur among the parenchyme), several primary bundles are situated. They are radially divided, the same as in Archaeopteridales (Beck *et al.*, 1982–1983), but axillary branching axes can already be observed. Owing to the long passage of the branch trace in the cortex, the axillary and subtending branches correspond to nodes, separated far apart from each other (hetero-axillarity). A quite wide zone of secondary wood, contiguous from the outside to the axial bundles, is composed of tracheids with round pits on radial walls. The rays are multiseriate. Sieve cells occur in the secondary phloem. The cortex is of the *Sparganum* type (Fig. 46, *d*), i.e. short radial sclerenchyma plates are situated around its periphery, run along the stem and, at times, anastomose. In *Calamopitys* several bundles extend towards the frond. They are arranged in a half circle, with the convex part facing downwards (Fig. 46, *c*). Further the bundles fuse. After the first (dichotomous) division of the rachis each of the branches is supplied with a V-shaped bundle, having several protoxylem strands. Further up, the rachis is of pinnate branching. The more slender axes of the same plant are protostelic, and have been assigned to the genus *Stenomyelon* (Fig. 46, *e*, *f*), whereas the petioles have been assigned to *Kalymma*. In the axes and petioles secretory canals can frequently be observed. Of interest is the Lower Carboniferous genus *Diichnia* in which a double leaf trace departs from adjacent axial bundles, the same as in the more advanced genera of the Ginkgoopsida.

It is probable that fronds of the genera *Sphenopteridium*, *Diplotmema* and *Triphyllopteris* should be ascribed to C. Microsporoclads, associating

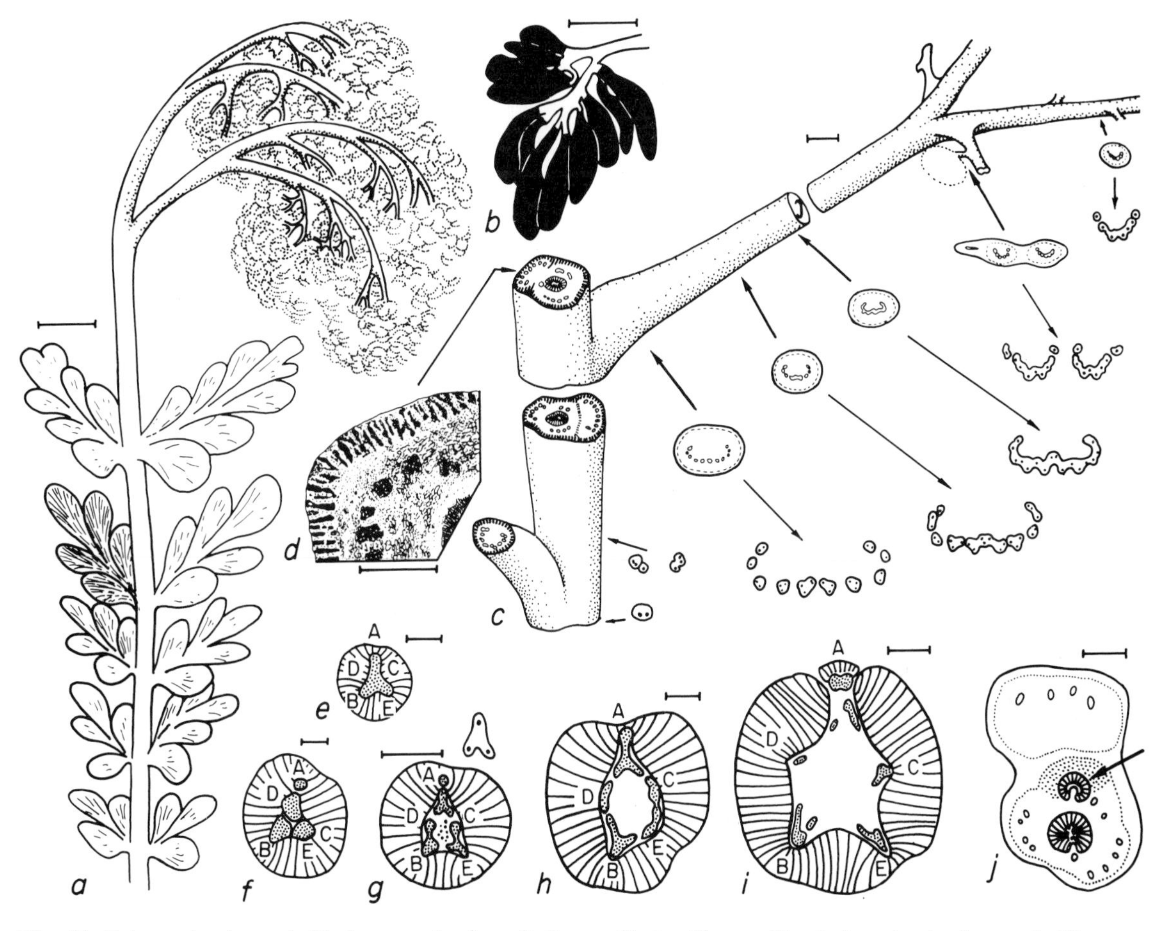

Fig. 46. Calamopityales and allied genera (*a*, *b*, *e*, *f*); Lower Carboniferous; North America (*a*, *b*, *e*, *g–i*), Western Europe (*c*, *d*, *f*, *j*); *a*, *b*, *Triphyllopteris uberis* Skog et Gensel, reconstruction of fertile frond (*a*), bunch of sporangia (*b*); *c*, *d*, *Calamopitys embergeri* Galt., reconstruction of axis with departing pinnae having *Kalymma*-type structure, variation of leaf trace in different places of axis and pinnae is shown (*c*), sector of axis with commencing departure of pinna and *Sparganum* cortex (*d*); *e–i*, formation of eustelic stem, primary wood stippled, secondary wood shown by radial lines, sequence of leaf traces shown by letters A–D; *e*, protostelic axis *Stenomyelon* sp.; *f*, *S. tuedianum* Kidst. et Gw.-Vaugh., primary wood divided into sectors; *g*, *C. americana* Scott et Jeff. with mixed pith (above is arrangement of protoxylem shown by dots in segment of primary wood); *h*, *i*, eustelic stems of *Calamopitys* sp. (*h*) and *C. foersteri* Read; *j*, *Calamopitys* sp., axis with departing petiole of *Kalymma*-type (above) and axillary branch (at arrow). Scale bar = 1 cm (*a*, *c*), 3 mm (*d–j*), 1 mm (*b*). Modified from: Skog and Gensel, 1980 (*a*, *b*); Galtier, 1974 (*c*), 1970 (*d*), 1975 (*j*); Beck, 1970 (*e–i*).

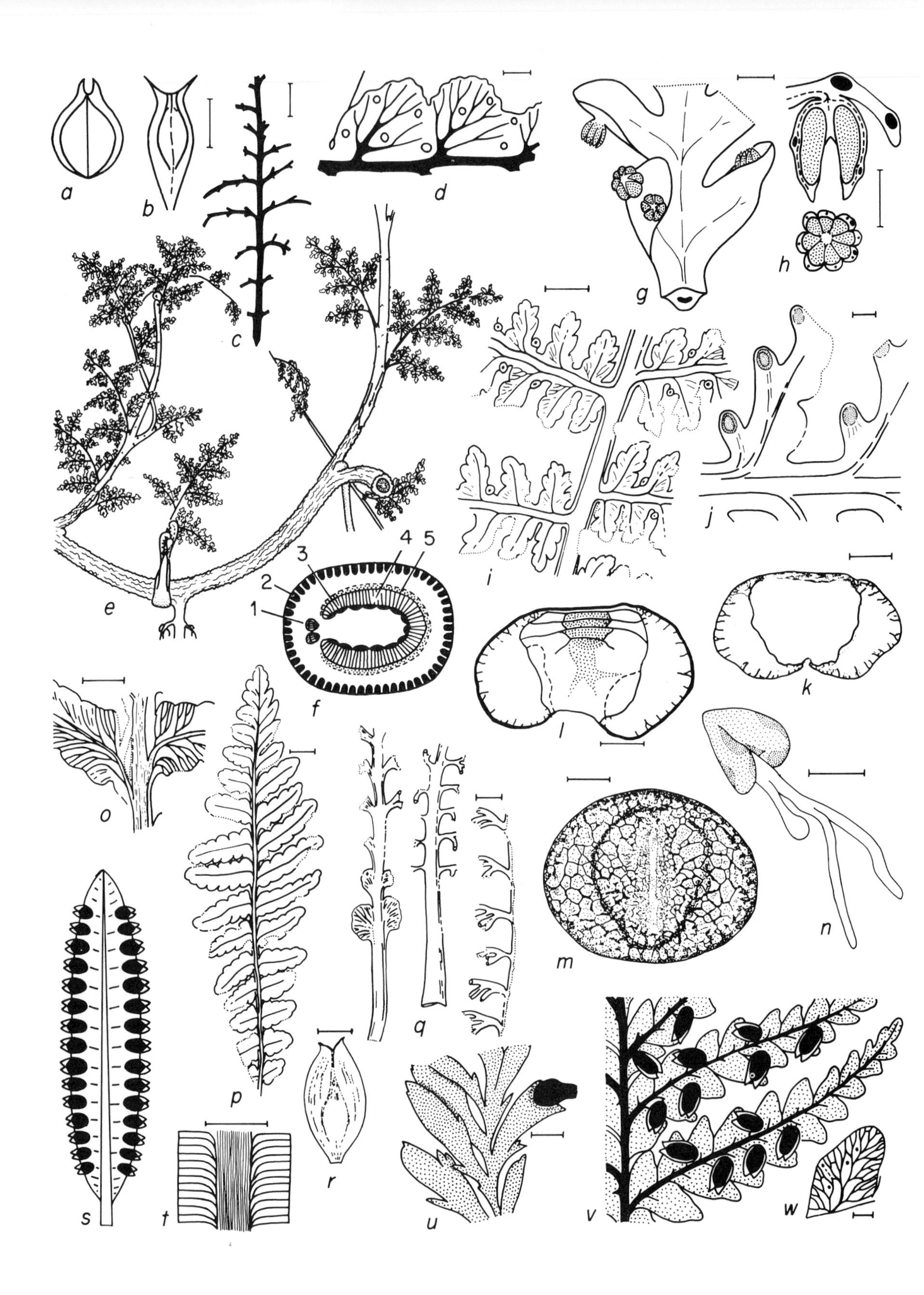

a
b
c
d
e
f
g
h
i
j
k
l
m
n
o
p
q
r
s
t
u
v
w
1
2
3
4
5

with *Triphyllopteris* (Fig. 46, *a*, *b*) are aggregates of profusely branching axes on which free elongated sporangia are borne; spores are trilete.

C. are firmly known from the Upper Devonian and Lower Carboniferous, but might also have existed during the Middle Carboniferous. Some dispersed Upper Devonian petioles and protostelic axes, usually affiliated with progymnosperms, are, perhaps, more closely related to C. On the other hand, the stems and petioles of C. cannot be firmly distinguished from the lagenostomans. It was believed that C. are ancestors of lagenostomans, but it is conceivable that both groups could have evolved independently from the most ancient gymnosperms producing seeds of variable symmetry. The close relationship between C. and the most primitive lagenostomans is quite obvious and emphasized by a wide range of anatomical characters, including such specific features as the *Sparganum* cortex. The oldest known platyspermic seeds *Spermolithus* (Fig. 43, *g*) may belong to C. Their anatomical structure is unknown.

Order Callistophytales

The name of the order Callistophytales (Cp.) is derived from the genus *Callistophyton*, established for petrified stems, which initially have been described under the invalid name *Poroxylon* and assigned the family Poroxylaceae. In old manuals the poroxyls have been erroneously assigned to cordaitaleans. Cp. are similar in their stelar organization to Calamopityales, whereas their seeds (*Callospermarion*) show affinities to *Lyrasperma*, but, by contrast, have a free integument with a well-developed micropyle. According to the structure of the seeds and pollen, Cp. are similar to peltasperms, caytonias, arberias and some conifers. The structure and arrangement of the synangia in Cp. resemble that in the Marattiales. Reconstructions (Fig. 47, *e*) depict a stem prostrate along the ground, up to 3 cm in diameter, with adventitious roots extending from it. The plagiotropic nature of the stem is evident from the distinct bilateral symmetry in the arrangement of the primary xylem, surrounding the flattened pith (Fig. 47, *f*), and frequently in the predominant increment of the secondary wood from two opposite sides. The branching is axillary. The bud is supplied with scale leaves (cataphylls). The vascular bundle, extending to the bud, results from coalesced branches of adjacent axial bundles. The leaf trace has a small amount of abaxial secondary wod. Undergoing separation from an axial bundle in a tangential direction, it continues 2.5 internodes upward, then dichotomizes and

Fig. 47. Callistophytales and allied genera (*a–c*, *o–w*); Middle (*a–e*, *h*, *i*, *n*, *s*, *t*) and Upper (*g*, *j–m*, *o–r*) Carboniferous, Permian (*u–w*); Western Europe (*a*, *i*, *j*), USA (*b–e*, *g*, *h*, *k–n*, *s–u*), Siberia (*o–r*), China (*v*, *w*); *a*, seed of *Cornucarpus actus* (L. et H.) Arber type associating with fronds *Eremopteris artemisiaefolia* (Sternb.) Schimp.; *b*, *c*, seed of *Cornucarpus* type (*b*) and pinnate polysperm with shed seeds (*c*) associating with fronds *E. zamioides* (Bertr.) Kidst.; *d*, pinnules of *Callistophyton boyssetii* (Ren.) Rothw., secretory organs shown between veins; *e*, reconstruction of *Callistophyton*; *f*, *Callistophyton*, scheme of stem anatomy (1, leaf trace; 2, *Sparganum* cortex; 3, phloem; 4, secondary wood; 5, primary wood); *g*, *Callandrium* (al. *Idanothekion*) *callistophytoides* Stidd et Hall; microsporophyll; *h*, *Idanothekion glandulosum* Mill. et Egg., longitudinal (above) and cross section of synangium, secretory canals in black, vascular bundles in broken lines; *i*, *Dicksonites pluckenetii* (Schl.) Sterz., frond with seed scars on unmodified pinnules; *j*, *D. pluckenetii*, pinna with seed scars on modified pinnules; *k–m*, pollen of *Vesicaspora* type from synangia of *Callandrium*, longitudinal section (*k*), intracellular gametophyte (*l*), distal view (*m*); *n*, pollen grain with pollen tube from *Callospermarion* seed; *o*, *p*, *Paragondwanidium sibiricum* (Pet.) S. Meyen, venation (*o*) and general view of frond (*p*); *q*, *Gondwanotheca sibirica* Neub., polysperms associating with fronds of *P. sibiricum*; *r*, *Angarocarpus ungensis* (Zal.) Radcz., seed associating with *P. sibiricum*; *s*, *t*, *Spermopteris coriacea* (Goepp.) Cridl. et Morr., reconstruction of phyllosperm (*s*) and scheme of venation of vegetative leaf (*t*); *u*, *Tinsleya kansana* Mamay, phyllosperm; *v*, *w*, *Emplectopteris triangularis* Halle, phyllosperm (*v*) and venation of vegetative pinnule (*w*). Scale bar = 10 cm (*e*), 1 cm (*c*, *i*, *p*, *u*), 5 mm (*b*, *o*, *q*, *r*, *t*), 1 mm (*d*, *g*, *h*, *j*, *w*), 10 μm (*k*, *n*). Modified from: Seward, 1917 (*a*); Delevoryas and Taylor, 1969 (*b*, *c*), Rothwell, 1981 (*d*, *e*, *h*) and 1972 (*n*); Millay and Taylor, 1979 (*g*) and 1974 (*k–m*); Daber and Helms, 1981 (*i*); Meyen, 1984a (*j*, *p–r*) and 1969b (*o*); Mamay, 1976 (*s*) and 1966 (*u*); Cridland and Morris, 1960 (*t*); Halle, 1927 (*w*).

these branches extend upwards parallel to the stem another 1.5 internodes, then enter the petiole together and somewhat further become fused into a single bundle. The stems, the same as in the Calamopityales, have a cortex of the *Sparganum* type, secretory canals and cavities.

The seeds (*Callospermarion*; Fig. 45, *b*) are small. The integument is three-layered, flattened, with secretory cavities; the axial vascular bundle enters the chalaza and terminates in a bowl-shaped widening. Two lateral branches pass into the sarcotesta and continue through it nearly to the seed apex. Two additional vestigial bundles depart in the perpendicular plane and quickly disappear. The salpinx is formed in the course of the ontogenesis, as a result of lysis of the central apical cells of the nucellus. This is concomitant with the separation of the integument and nucellus. In young ovules the nucellus and integument are fused, as in *Lyrasperma* (Calamopityales). The ovules released a pollen-capturing secretory drop. The seeds *Taxospermum*, where the integument is fused in the lower part with the nucellus, can evidently be affiliated with Cp. Phyllosperm impressions have been described as *Dicksonites pluckenetii*. Specimens coming from the Middle Carboniferous had one seed per lower basal lobe of unmodified pinnules (Fig. 47, *i*). The basal pinnules of the pinna were sterile. The Upper Carboniferous specimens had phyllosperms with modified pinnules bearing a larger number of seeds (Fig. 47, *j*).

The synangia (genus *Idanothekion*; C; Fig. 47, *g*, *h*), the same as the seeds, were situated on unmodified or slightly modified pinnules and consisted of 5–9 sporangia, basally fused and opening inwards. In some *Idanothekion* the vascular bundle passes in the outer wall of the sporangia. The pollen is monosaccate (of the *Vesicaspora* type), but the saccus is divided into two balloons with equatorial interconnections. The inner surface of the saccus bears a reticulum, the same as in conifer pollen (Fig. 47, *k*, *m*). Pollen of earlier Cp. had a proximal slit that disappears in more advanced forms, where the germination was distal (Fig. 47, *n*). In some grains prothallial cells were found proximally and large embryonic cells distally from them (Fig. 47, *l*). The microgametophyte is of the same organization as in living conifers. During a later stage a small generative (antheridial) cell was formed inside a large sporogenous cell. All these stages have been traced in the pollen still encased in the sporangia. The similarity in the structure and ontogeny of the pollen and microgametophyte between Cp. and conifers is one of the most remarkable examples of the parallelism in the evolution of plants.

The genera *Eremopteris* and *Paragondwanidium* occupy an intermediate position between Cp. and Calamopityales. The first comprises fern-like fronds (Fig. 48, *a*), associated with bicornute seeds

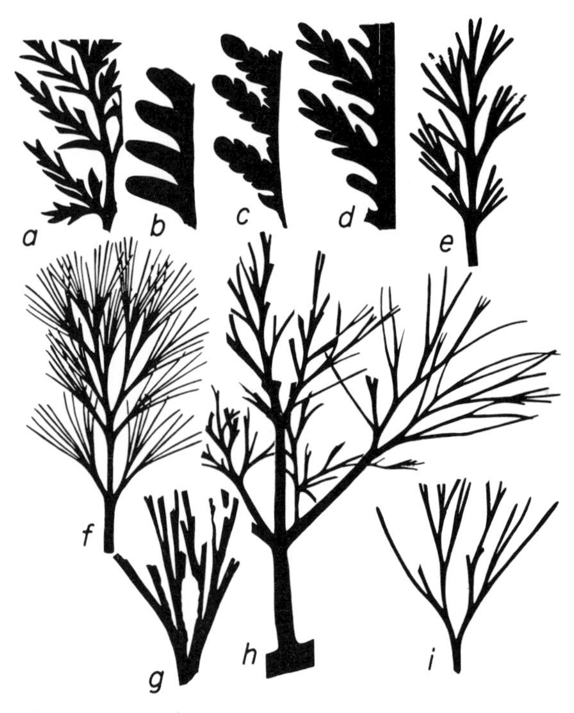

Fig. 48. Transition from pinnate (fern-like) to pinnately-dichotomizing and dichotomizing (ginkgoid) leaves in Ginkgoopsida; Middle (*a*) and Upper (*f*) Carboniferous, Lower (*b–e*, *g*, *h*) and Upper (*i*) Permian; Western Europe (*a–e*, *i*), USA (*f*), Fore-Urals (*g*, *h*); *a*, *Eremopteris artemisiaefolia*; *b*, *Callipteris conferta*; *c*, *C. naumannii*; *d*, *C. lyratifolia* (Goepp.) Gr. 'Eury f. *stricta* Zeill.; *e*, *C. flabellifera*; *f*, *Dichophyllum moorei*; *g*, *h*, *Mauerites*; *i*, *Sphenobaiera digitata* (Brongn.) Florin. From Meyen, 1984a.

(Fig. 47, *a*, *b*), that were borne on pinnate leafless polysperms (Fig. 47, *c*). The seeds resemble *Lyrasperma* in their bicornute apex, but the integument lobes were distinctly closed around a short micropyle. According to this character *Eremopteris* is more closely related to Cp., whereas according to the leafless polysperms it is more closely related to Calamopityales. Fronds of *Paragondwanidium* (Fig. 47, *o*, *p*) associate with pinnate polysperms, either with reduced basal pinnules, or without them (*Gondwanotheca*; Fig. 47, *q*), and seeds with a bicornute apex and long micropyle (*Angarocarpus*; Fig. 47, *r*).

Phyllosperms *Spermopteris* are probably affiliated to Cp. (Fig. 47, *s*, *t*; this genus was regarded as the probable ancestor of cycads, which is unlikely), as well as *Tinsleya* (Fig. 47, *u*) and *Emplectopteris* (Fig. 47, *v*, *w*). The last genus was considered the progenitor of the order Gigantonomiales ('gigantopterids').

Of the form-genera, established on the sterile fronds, it is conceivable that *Mariopteris* (see satellite genera to classes Ginkgoopsida and Cycadopsida), *Karinopteris*, also some *Pecopteris* and *Eusphenopteris*, can be allied to Cp.

Order Peltaspermales

In many classification systems the families Peltaspermaceae and Corystospermaceae (in the foregoing text this family is referred to by the more correct name Umkomasiaceae) together with caytonias are assigned to the informal group 'Mesozoic pteridosperms'. Sometimes peltasperms (P.) are raised to the rank of an order or even subclass (Němejc, 1968). In other cases they are affiliated with medullosans. After detailed comparative studies of many Permian and Mesozoic Ginkgoopsida (Fig. 49) the order Peltaspermales, comprising the families Trichopityaceae, Peltaspermaceae, Cardiolepidaceae and Umkomasiaceae, was keyed out (Meyen, 1984a). P. can be pertinently derived from callistophytans, but differ from the latter in that seed-bearing leaves evolve into cladosperms, that become ever less similar to foliage leaves and increasingly acquire more the habit of specialized fertile shoots. Cladosperms of more primitive P. are still similar to foliage leaves in the type of their dissection and microstructure; in these cases it is difficult to distinguish P. from callistophytans. In more advanced P. the cladosperms are commonly peltate (peltoids) or cupuliform. A seed-bearing disc crowns the stalk and bears abaxial seeds, surrounding the stalk. The seed-bearing lamina may also be pinnately veined with seeds borne in two rows, and palmately veined, when the seeds are borne on the ends of the radii diverging from the base of the cladosperm. The lamina may be reduced to a cupuliform container of a single seed (Umkomasiaceae). The margins of the cladosperm are variously bent downwards. When they are strongly bent down and embrace the stalk, a closed capsule is then formed (Cardiolepidaceae). In all cases the seeds are borne on the underside of the cladosperm. Of importance is the fact that in all P. and its descendants the main plane of symmetry in the seeds is radial, relative to the main axis. In the class Pinopsida the same plane is of tangential orientation relative to the main axis (Fig. 50).

In male fructifications the reduction of the leaf lamina evolved more rapidly and more completely. The synangia in the most primitive P. are of pinnate arrangement, so that the fertile pinna is of the same construction as the vegetative ones (*Callipterianthus*, *Schuetzia*). The Umkomasiaceae retain the flattened lamina bearing sporangia. In all other P. the sporangia are in bunches inserted on the tips of the branchlets, and slightly fused at the base. The P. pollen (Fig. 42 (5–7, 9, 10, 14)) is quasimonosaccate and quasidisaccate (at times bilateral) with a striated or smooth body, to asaccate (smooth or striated).

The foliage of P. is very diverse. Most frequently the leaves are pinnate with open venation. The rachis characteristically is forked. Some genera produced entire leaves with parallel or fan-shaped venation, also palmate leaves. Palmate and pinnate dissections may combine. The fusion of both the pinnae and pinnules can be observed, the venation system being preserved. Leaves formed from such fusion are termed coherent (Asama, 1959). According to the order of the elements that undergo fusion, different leaf types result: unico-

1
2
3
4
5
6
7
8
9
10
11
12
13
14
15
16
17

Fig. 49. Relations between Peltaspermales (Nos 1–14), Ginkgoales (Nos 15, 17) and Leptostrobales (No. 16). Seeds in attachment are shown in black; microsporangia, synangia and pollen stippled. Description of figures within numbered units is given from left to right and from top to bottom. 1. Early Permian callipterids of Euramerian area: fronds of *Callipteris flabellifera* Weiss (see Fig. 48, *e*) and *C. conferta* (Sternb.) Brongn. (see Fig. 48, *b*); pinnate cladosperm *Autunia milleryensis* (Ren.) Kras. (see Fig. 52, *a*) with only one seed preserved at each seed-bearing limb; microsporophyll *Callipterianthus arnhardtii* Ros.; bicornute seed; microsporoclad *Pterispermostrobus gimmianus* W. Remy; seed associating with *Callipteris naumannii* (Gutb.) Sterz.; pollen of *Vesicaspora* type from synangium of *P. gimmianus*; below broken line – pollen of *Protohaploxypinus* and *Vittatina* types probably produced by some older members of Peltaspermales. 2. *Mauerites* frond (see Fig. 48, *g*, *h*) and associating cladosperm *Biarmopteris* (see Fig. 51, *f*, *j*), Lower Permian, Fore-Urals (Subangara area). 3. Frond of *Supaia* and associating pinnate polysperm, Lower Permian, USA. 4. Extreme types of frond variation in *Rhaphidopteris praecursoria* S. Meyen, and associating pollen of *Alisporites* type, Upper Permian, Timan Range (Subangara area). 5. Triassic Umkomasiaceae of Gondwana; forked leaf of *Johnstonia coriacea* var. *coriacea* (John.) Walk.; cupule-like cladosperm of *Pilophorosperma* sp. and axis bearing such cladosperms associating with fronds *Dicroidium odontopteroides* (Morr.) Goth.; bicornute seed detached from cladosperm; microsporoclad *Pteruchus johnstonii* (Feistm.) Town. and its pollen of *Alisporites* type; forked twice-pinnate frond of *Dicroidium zuberi* (Szaj.) Arch. and associating polysperm *Umkomasia* sp.; forked frond of *Xylopteris elongata* var. *rigida* (Dun.) Ret. with linear pinnules. 6. Forked frond of *Ptilozamites nilssonii* Nath., and associating microsporoclad *Harrisiothecium marsilioides* (Harris) Lund., uppermost Triassic, Greenland. 7. Jurassic plants of *Pachypteris–Rhaphidopteris* group; two fronds of *R. dinosaurensis* (Harris) Barale with forked pinnules, Greenland; forked frond of *R. nana* (Harris) Barale, Yorkshire; microsporoclad *Pteroma thomasii* Harris and its pollen of *Alisporites* type, Yorkshire; twice-pinnate frond of *Pachypteris lanceolata* Brongn., Yorkshire; pinna of last order of *R. williamsonis* (Brongn.) Barale, Yorkshire. 8. Family Cardiolepidaceae, Upper Permian, Pechora province (Angara area): leaf of *Phylladoderma* (see Fig. 52, *p*, *q*) with parallel veins and notched apex; peltate capsule *Cardiolepsis piniformis* Neub.; seed *Nucicarpus piniformis* Neub. with long micropylar tube; synangium *Permotheca* and its *Vesicaspora*-like pollen. 9. Same family, uppermost Permian, Western Angaraland (*Tatarina*-flora): linear and rhomboid leaves of *Phylladoderma* subgen. *Aequistomia*; dichotomizing leaf of *Doliostomia*; synangium *Permotheca* and its *Vesicaspora*-like pollen. 10. Fronds of *Callipteris* and *Compsopteris* associating with *Peltaspermum*-like peltoids, Upper Permian, Pechora province. 11. Family Peltaspermaceae from *Tatarina*-flora (below broken line) and various Triassic floras: racemose compound polysperm consisting of two-seeded cladosperms (*Peltaspermum thomasii* Harris) and its isolated cladosperm; pollen of *Cycadopites* type; microsporoclad and cluster of sporangia of *Antevsia*; bipinnate frond of *Lepidopteris*; coherent-leaved *Vittaephyllum*; peltoid of *Peltaspermum rotula* Harris with seeds; racemose polysperm *P. incisum* Pryn. ex Stanisl., its peltoid with seeds concealed inside (compare with peltate capsule of *Cardiolepis*, No. 8), and isolated seed with resin bodies; pollen of *Protohaploxypinus* and *Vittatina* types, and synangium *Permotheca*; head-like compound polysperm *Peltaspermum buevichae* S.M. et Gom.; associating seeds; simple and pinnate leaves of *Tatarina* (note notched apex as in *Phylladoderma*, No. 8). 12. Leaf of *Glossophyllum* type associating with cladosperms *Stiphorus biseriatus* S. Meyen and seeds having bivalved stone and resin bodies, *Tatarina*-flora. 13. Simple (*Glossophyllum*-like) and dichotomizing (*Sphenobaiera*-like) leaves of *Kirjamkenia lobata* Pryn. associating with forked cladosperms *Stiphorus crassus* (Mog.) S. Meyen (cross-section of seed-bearing lamina is shown below) and pollen of *Cycadopites* type; Permo-Triassic of Tunguska basin. 14. Entire leaf of *Glossophyllum florinii* Kräus.; forked leaf of *Sphenobaiera furcata* (Heer) Flor.; pollen of *Cycadopites* type; microsporoclad *Antevsia wettsteinii* (Krass.) S. Meyen; leaf base with two veins, and notched apex of *G. angustifolium* Stanisl.; paired cladosperm *Leuthardtia ovata* Kräus. et Schaar. (associates with *S. furcata*); Triassic of Austria and Donets basin. 15. Two types of leaves of *Sphenobaiera spectabilis* (Nath.) Flor. associating with microsporoclads *Bernettia phialophora* Harris and pollen of *Cycadopites* type; uppermost Triassic of Greenland. 16. Jurassic Leptostrobales; pinnately-dichotomizing leaves of *Czekanowskia microphylla* Harris et Miller, Yorkshire; leaf of *Phoenicopsis* and its base with single vein; dwarf shoot of *Czekanowskia*; compound polysperm and isolated valve of *Leptostrobus laxiflora* Heer. 17. Jurassic and recent Ginkgoales: entire and bilobed (below) leaves of *Ginkgo biloba*; simple linear leaf and its base (with two veins) of *Eretmophyllum*; young ovules of *G. biloba*; seed stone associating with *Baiera furcata* (L. et H.) F. Braun; variously lobed leaves of *G. huttonii* (Sternb.) Heer; dichotomizing leaves of *Sphenobaiera*; star-shaped group of sporangia of *Stachyopitys*; microsporoclad and pollen of *G. biloba*. From Meyen, 1984a.

herent (fusion of the pinnae pinnules), bicoherent (fusion of the unicoherent pinnae of the last order), and tricoherent (fusion of bicoherent pinnae). In the course of this process the venation becomes reticulate at times. In some P. cataphylls can be observed at the shoot base (same as in callistophytans), and the alternation of dwarf and long shoots (as in ginkgos). The anatomical structure of the stem is known only in the genera *Kirjamkenia* (Peltaspermaceae) and *Rhexoxylon* (Umkomasiaceae). In both cases vascular bundles, entering the leaf, depart from different axial primary bundles (as in calamopityalean *Diichnia* and ginkgos). P. are characterized by secretory canals or cavities, in which they display affinities to Calamopityales, Callistophytales and ginkgos.

Four families are distinguished in P.: Trichopityaceae (pinnate or pinnate-dichotomous cladosperms retaining their leaf-like habit; the seeds are of subapical attachment to the terminal lobes), Peltaspermaceae (with peltoids and derived from them multiseeded open cladosperms), Cardiolepidaceae (peltate capsules, nearly closed) and Umkomasiaceae (singular seeds in cupuliform inverted containers).

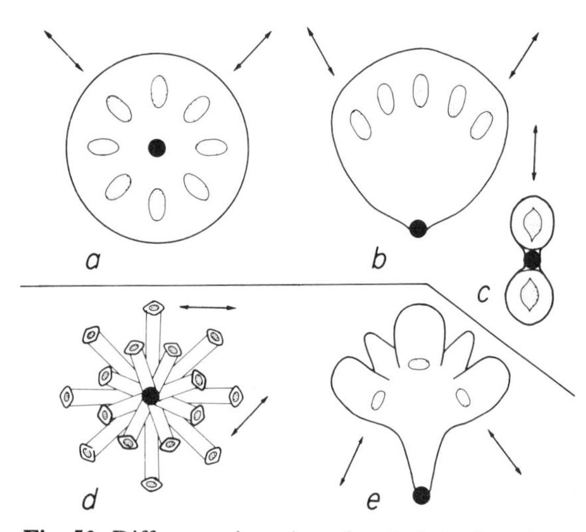

Fig. 50. Different orientation of seeds (stippling shows seed scars only) in relation to polysperm axis (black circle) in Ginkgoopsida (*a–c*) and Pinopsida (*d, e*). The elongation of seed scars admittedly corresponds to seed major plane of symmetry. Abaxial view of *Peltaspermum* peltoid (*a*), inside view of *Leptostrobus* valve (*b*), paired cladosperm of *Ginkgo* (*c*), abaxial view of umbellate polysperm *Gaussia* (*d*), adaxial view of *Pseudovoltzia* seed scale (*e*). From Meyen, 1984a.

The earlier P. are transitional to callistophytans, whereas the more advanced, particularly Triassic, are transitional to ginkgos and leptostrobans (Fig. 49). The genus *Sphenobaiera*, traditionally assigned to the Ginkgoales, is probably partly affiliated with P. It is retained among the former pending new information on female fructifications associating with leaves of the type species. The difficulties encountered in the distribution of P. into families and in the distinction of P. proper from callistophytans, ginkgos and leptostrobans, give confirming evidence to the alliances between these orders.

Family Trichopityaceae

The Trichopityaceae (T.) were formerly included in dicranophylls or ginkgos, in consequence of the misinterpretation of their cladosperms by Florin (1949). He considered that the seeds in *Trichopitys* are inverted and borne on the tips of branchlets, that are spirally inserted on the axillary axis. In Florin's opinion, the seed-bearing axes were spirally situated on the main axis of a compound polysperm. As a matter of fact, the compound polysperm, as well as the simple lateral cladosperms, were pinnate and planated (Fig. 41, *e*; Fig. 51, *e*). The seeds were subapically attached to the lateral branches of the cladosperms (Fig. 51, *b*, *c*). Occasionally a simple flattened seed stalk occurs in the leaf axil instead of the pinnate cladosperm. The laminae of the foliage leaves (including those subtending cladosperms) are dichotomously or pinnate-dichotomously divided. The genus *Mauerites* is closely similar to *Trichopitys* (Fig. 48, *g*, *h*; Fig. 49 (2)). Its cladosperms are pinnate-dichotomous (*Biarmopteris*; Fig. 51, *f–j*); the seeds are of subapical attachment (Fig. 51, *i*) and after being shed leave distinct seed scars (Fig. 51, *h*, *j*). Pinnate cladosperms with subapical seeds were found in association with leaves of the genus *Supaia* (Fig. 45 (3)), which is clearly allied to *Callipteris* (fam. Peltaspermaceae). The genus *Dichophyllum* (Fig. 48, *f*) is interposed between

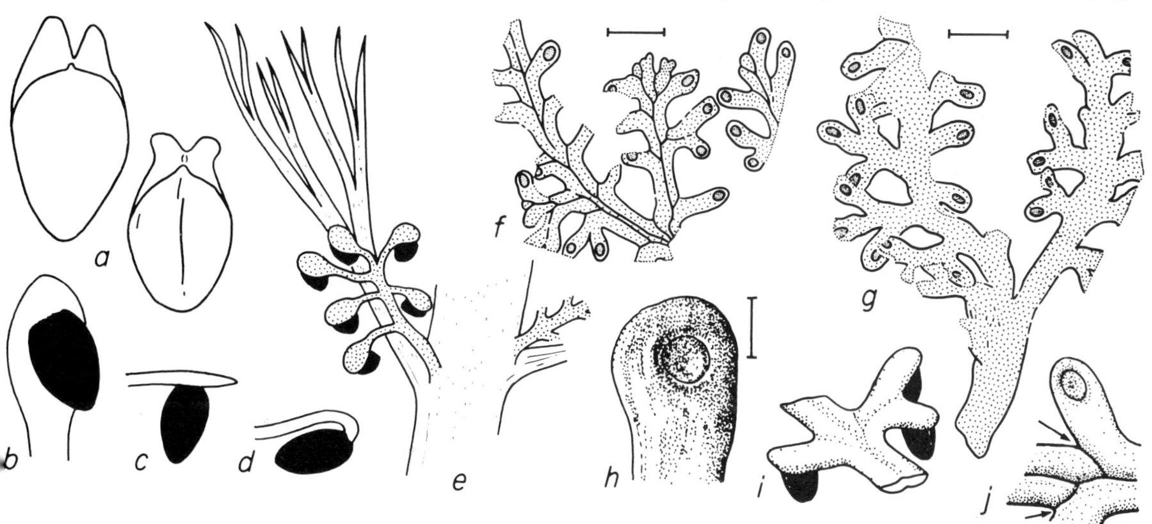

Fig. 51. Trichopityaceae; Upper Carboniferous (*a*), Lower Permian (*b–j*); USA (*a*), France (*b–e*), Fore-Urals (*f–j*); *a*, seeds associating with *Dichophyllum moorei* Elias; *b–e*, *Trichopitys heteromorpha* Sap. (*b*, distal widening of polysperm with abaxially attached seed; *c*, same, lateral view; *d*, same as treated by Florin; *e*, reconstruction of fertile shoot, axillary polysperm stippled); *f–j*, *Biarmopteris pulchra* Zal. emend S. Meyen (*f*, abaxial view of cladosperm bearing subapical seed scars; *g*, nearly full cladosperm with forked lamina; *h*, modified pinnule of cladosperm with abaxial seed scar; *i*, reconstruction of cladosperm, adaxial view; *j*, same in abaxial view showing seed scar and axillary sutures at arrows). Scale bar – 5 mm (*f*, *g*), 1 mm (*h*).

Mauerites and *Callipteris*; associated seeds are bicornute (Fig. 51, *a*), as in many Ginkgoopsida. Male fructifications of the Trichopityaceae are unknown.

Family Peltaspermaceae

The family name Peltaspermaceae (P.) is derived from that of the genus *Peltaspermum* (P–T; Fig. 41, *j*; Fig. 9 (11)), that includes peltoids with a festoon-shaped or deeply incised margin and palmately veined cladosperms. Peltoids are clustered in racemose compound polysperms. The seeds are sessile with a free integument. In the genus *Stiphorus* (Fig. 49 (12, 13)) the cladosperms bear two rows of seeds and are inserted on the tips of a forked stalk. The genera linking the orders Peltaspermales and Callistophytales (*Callipteris*, *Tinsleya*, etc.) can be tentatively assigned to P. *Callipteris* leaves (C_3–P; Fig. 48, *b–e*; Fig. 49 (1, 10); Fig. 52, *b*) associate with cladosperms that have a round-rhombical palmately veined two-sided lamina (*Autunia*; P_1; Fig. 49 (1); Fig. 52, *a*) and racemose aggregations of peltoids (Fig. 49 (10)), typical of younger species.

Microsporoclads of the oldest P. (*Callipterianthus*; P_1; Fig. 45 (1)) are modified pinnae with a completely reduced lamina, that together with the vegetative pinnae are situated on the same rachis of the penultimate order (same as in archaeopterids). Rod-like coprolites (*Thuringia*) were mistaken for synangia, supposedly associated with *Callipteris* leaves. In later P. the sporangia were nearly free (or became so after maturing), arranged in radial bunches on axes of complex branching systems (Fig. 49 (11)). The bunches were probably inserted on the branches, not terminally, but on the underside. Sometimes resin bodies occur in the sporangial wall. The pollen of the oldest P. (Fig. 49 (1, 11 below the dashed line)) was quasisaccate with a ring-shaped or two-winged (with equatorial interconnections) quasisaccus (of the *Vesicaspora* type), with two quasisacci and a striated body (of the *Protohaploxypinus* type) and asaccate striated (of

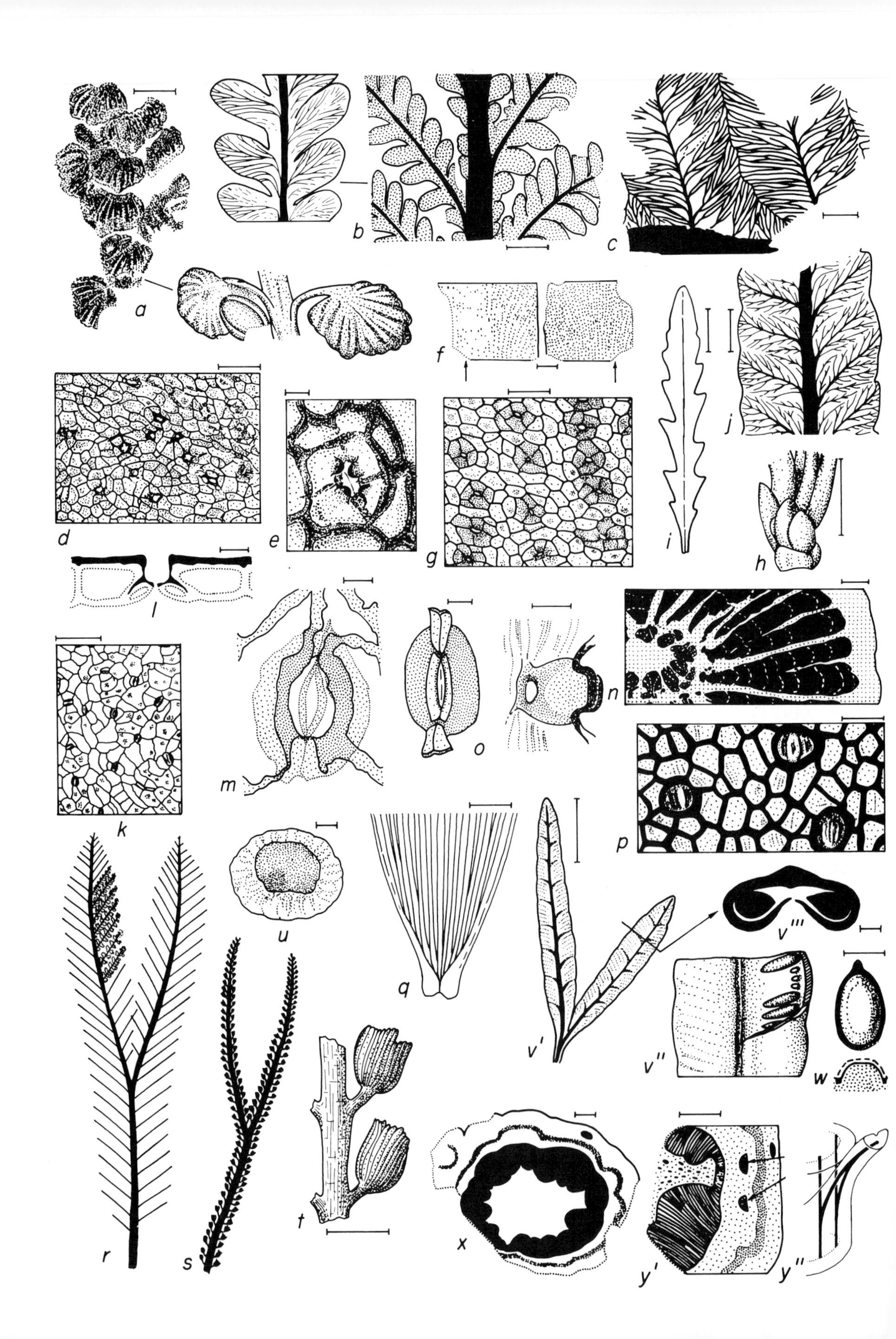
a
b
c
d
e
f
g
h
i
j
k
l
m
n
o
p
q
r
s
t
u
v'
v''
v'''
w
x
y'
y''

the *Vittatina* type). These pollen types are sometimes connected by transitional forms. Pollen of the two latter types were undoubtedly produced by Upper Permian P. with foliage of the *Tatarina* type. At the end of the Permian–beginning of the Triassic some genera appeared among P. that produced asaccate monocolpate pollen (of the *Cycadopites* type; Fig. 49 (11 above the dashed line, 13, 14)).

P. are characterized by practically all types of leaves (Fig. 48, *b–e*; Fig. 49 (1, 10–14); Fig. 52, *b*, *c–j*) known in the order Peltaspermales (see above); many types have been encountered in association with peltoids of the *Peltaspermum* type. These are simple with fan-shaped venation (*Tatarina*), lobed (*Tatarina*), once-pinnate (*Tatarina*, *Compsopteris*) and compound-pinnate (*Lepidopteris*, *Scytophyllum*, *Callipteris*). Besides, leaves have been found that are palmately-dissected (*Kirjamkenia*), pinnate with fused pinnules (*Scytophyllum*) and pinnae (*Vittaephyllum*), in the latter case at times with reticulate venation (*Furcula*, some *Gigantopteris*). Simple leaves *Tatarina* were spirally arranged on dwarf and long shoots; dwarf shoots, as in ginkgos, were accompanied by cataphylls (Fig. 52, *h*). Stomata were situated on both sides of the leaf; radially arranged subsidiary cells, usually bearing papillae, overhang the guard cells (Fig. 52, *d–g*).

The anatomy of the stem is known in *Kirjamkenia* (Fig. 52, *x*, *y*; Sadovnikov, 1983) and resembles that which we see in the Callistophytales. The stems are eustelic; the secondary wood zone is not wide. Large secretory cavities occur in the cortex. Two leaf traces depart from adjacent axial bundles and coalesce higher up, so that only one bundle enters the leaf, further dichotomizing. According to its epidermal characters *Kirjamkenia* shows affinities to *Tatarina* and *Glossophyllum*. The stems and leaves of *Kirjamkenia* associate with paired pinnately-veined cladosperms *Stiphorus* (Fig. 49 (13)). The forked cladosperms *Sporophyllites* (Fig. 52, *v*, *w*), where the seeds are borne in two rows and concealed by the coiled margin of the cladosperm lamina, resemble *Stiphorus*. Their seeds are small, and resemble those associated with *Stiphorus* (Fig. 49 (12)) and Mesozoic ginkgos (Fig. 49 (17)). The associated pollen (of the *Vitreisporites* type) is the

Fig. 52. Peltaspermales; Lower (*a–c*, *r–u*) and Upper (*e–h*, *o–q*, *v*, *w*) Permian, Permo-Triassic (*i*, *x*, *y*), Triassic (*d*, *j–n*); Western Europe (*a*, *b*, *r–u*), USA (*c*), Greenland (*d*), Russian platform (*e–h*, *o* (to right)), Siberian platform (*i*, *x*, *y*), Pechora basin (*j*, *o* (to left), *p*, *q*, *v*, *w*), Gondwana (*k–n*), including Australia (*m*) and South Africa (*n*); *a*, *Autunia milleryensis*, racemose aggregation of polysperms (to left) and cladosperm with attached seed (to right); *b*, *Callipteris conferta*, frond associating with *A. milleryensis*; *c*, *Gigantopteris americana* White, part of frond; *d*, *Lepidopteris ottonis* (Goepp.) Schimp., cuticle; *e*, *Tatarina olferievii* S. Meyen, stoma with papillae on subsidiary cells; *f–h*, *T. conspicua* S. Meyen, arrangement of stomata on lower (to left) and upper (to right) leaf surfaces, leaf axis at arrows (*f*), stomata (*g*), cataphylls of dwarf shoot (*h*); *i*, *T. lobata S. Meyen*; *j*, *Scytophyllum neuburgianum* Dobr., part of pinna; *k*, *l*, *Dicroidium odontopteroides*, leaf cuticle (*k*), cross-section of stoma (*l*; cuticle in black, cell walls shown by dotted line); *m*, stoma with cutinized thickenings of guard cells in *D. superbum* (Shirl.) Town.; *n*, *Rhexoxylon tetrapteroides* Walt., cross-section of stem, wood in black; *o*, *Phylladoderma* subgen. *Phylladoderma*, guard cells with cutin thickenings of guard cells (to left), and *P.* subgen. *Aequistomia*, cuticle of laterally shifted epistomatal chamber of barrel-shaped contour with laterally placed more cutinized guard cells (to right); *p*, *P.* subgen. *P. arberi* Zal., leaf cuticle; *q*, *P.* subgen. *Phylladoderma*, venation at leaf base; *r–u*, frond of *Sphenopteris germanica* Weiss (*r*), associating microsporoclad *Schuetzia anomala* Gein. (*s*), its synangium (*t*) and quasimonosaccate pollen (*u*); *v*, *w*, *Sporophyllites petschorensis* (Schmalh.) Fef. emend. S. Meyen, general view of cladosperm (*v'*), scheme of its cross section (*v''*) and reconstruction of its part with removed margin, seed scars and seeds (*v' ''*), seed (*w*); *x*, *y*, *Kirjamkenia lobata*, stem anatomy, cross-section of stem (*x*; wood in black, zone of cells with dark content stippled, oval black body – secretory organ), double leaf trace (at arrows) in cortex (*y'*), and scheme of leaf trace departure from axial vascular bundles (*y''*). Scale bar = 1 cm (*a*, *b*, *i*, *n*, *q*, *t*), 5 mm (*c*, *h*, *j*, *v*), mm (*f*, *v–y*), 100 μm (*d*, *g*, *k*, *p*), 20 μm (*e*, *o*, *u*), 10 μm (*l*, *m*). Modified from: Kerp, 1982 (*a*, *b*); White, 1912 (*c*); Dobruskina, 1980 (*d*) and 1969 (*j*); Meyen, 1970 (*e–g*); Meyen and Gomankov, 1980 (*h*); Gomankov and Meyen, 1979 (*i*); Townrow, 1957 (*k–m*); Gothan and Weyland, 1964, after Walton (*n*); Neuburg, 1960c (*q*); W. Remy, 1978b (*r*, *s*); Gothan and Weyland, 1964, after W. Remy (*t*, *u*).

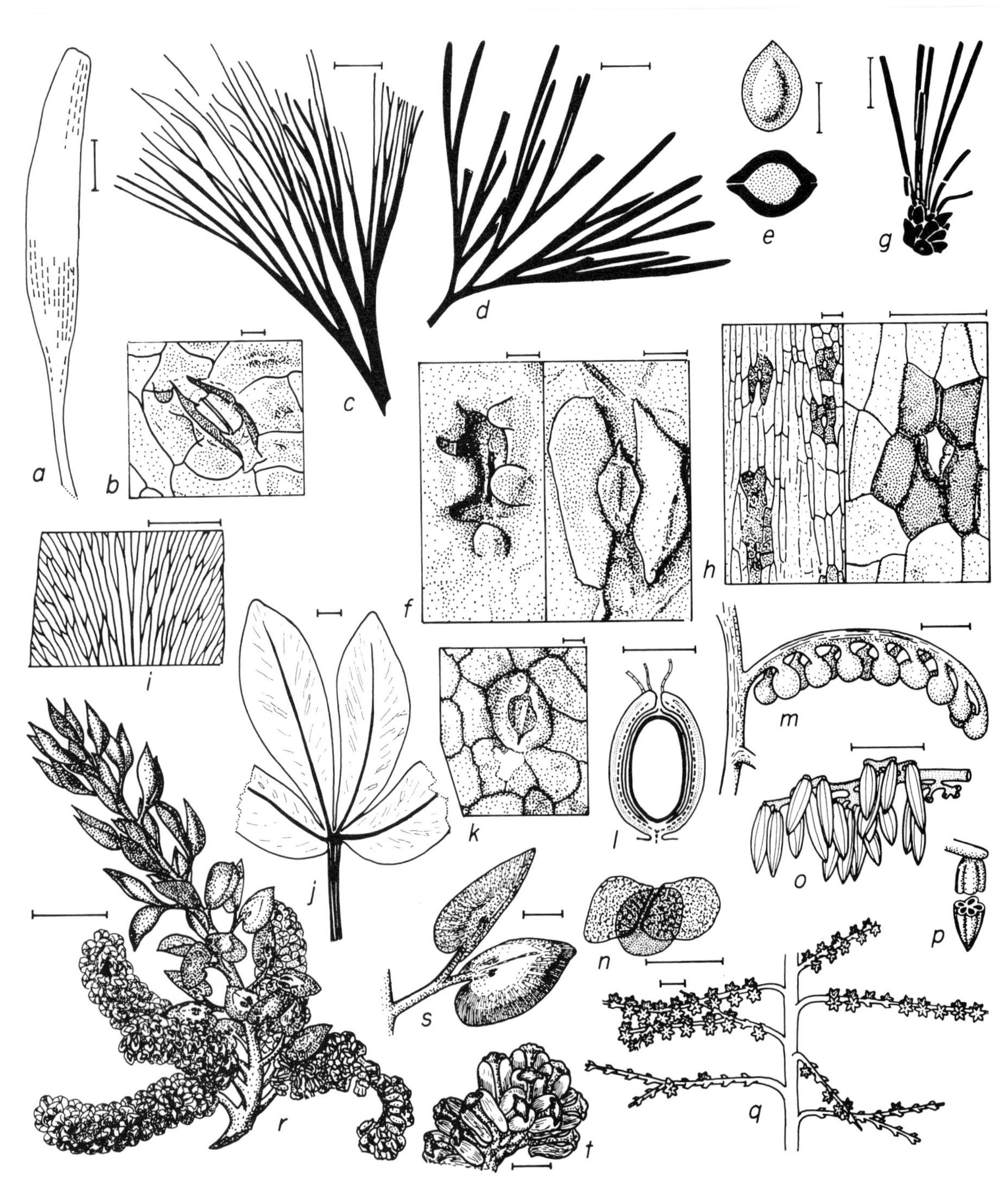

Fig. 53. Ginkgoales (*a–f*), Leptostrobales (*g, h*), Caytoniales (*i–q*), genus *Irania* (*r–t*); Upper Triassic (*j, r–t*), Jurassic (*a–i, k–q*); Western Europe (*a–e, g, j, l–p*), Uzbekistan (*f*), Bureja basin (*h*), Georgia (*i, k*), Mexico (*q*), Iran (*r–t*); *a*, *Eretmophyllum pubescens* Thomas; *b*, *Sphenobaiera longifolia* (Pom.) Flor., stoma with thickenings on guard cells; *c*, *S. pecten* Harris; *d, e*, *Baiera furcata*, leaf (*d*), imprint of half of seed stone, and seed stone in section (*e*); *f*, *Pseudotorellia* sp., surface and inside view of stoma, guard cells with winged thickenings; *g*, *Solenites vimineus* (Phill.) Harris, dwarf

same as in caytonias. These cladosperms were evidently produced by plants with leaves of the genus *Rhipidopsis* (see satellite genera to the division Pinophyta), that have a palmately-dissected lamina and in this respect are similar to leaves in caytonias. Leaves of *Rhipidopsis* are also similar to ginkgoalean leaves, but their veins run into the lateral margins of the lobes, which does not happen in ginkgos.

Probably affiliated with P. are plants with fronds of *Sphenopteris germanica* (P_1; Fig. 52, *r*) and microsporoclads of *Schuetzia* (Fig. 52, *s–u*), where the sporangia were in head-like groups situated in two rows on a thick forked axis.

Family Cardiolepidaceae

Leaves of the Cardiolepidaceae (C.), belonging to the genus *Phylladoderma* (P_2; Fig. 49 (8, 9); Fig. 52, *q*) were wrongly affiliated with cordiates, ginkgos and conifers. The seed-bearing capsules (*Cardiolepis*; P_2; Fig. 41, *j*; Meyen, 1984a) have a peltate organization; the peltoid edges are bent down and inwards, embracing the stalk. The micropylar tubes of the seeds, situated around the stalk, extend to the slit between the stalk and the capsule margin. A thick layer of resinous cells has been observed in the capsule wall. Resin bodies have also been found in the seeds (in dispersed forms they are assigned to the genus *Nucicarpus*; Fig. 49 (8, 9)). Isolated male synangia (some species of *Permotheca*; Fig. 49 (8)) consist of basally fused sporangia with pollen of the *Vesicaspora* type, i.e. with one quasisaccus, divided into two wings (as in early Peltaspermaceae). Leaves of *Phylladoderma* are entire, obovate to linear, with multiple parallel veins coalescing at the apex. One vein enters the leaf at the base. The leaves are amphistomatic. The guard cells bear the same wing-shaped cutinous thickening (Fig. 52, *o*, *p*) as the Mesozoic Umkomasiaceae (Fig. 52, *l*, *m*), ginkgos (Fig. 53, *b*, *f*) and bennettites. Resin strands or resin bodies occur in the mesophyll. A characteristic feature is the thin striation of the cuticle on the underside of the leaf (as in *Mauerites* and the Permian *Rhaphidopteris*). According to the leaf structure C. display affinities to *Kirjamkenia* (Peltaspermaceae), but differ in the organization of the guard cells.

Family Umkomasiaceae (= Corystospermaceae)

The fructifications in the Umkomasiaceae (U.) have been studied in most detail in the genera *Umkomasia* and *Pilophorosperma* (Fig. 49 (5)) that differ little. Their polysperms branched in several planes. The branches issue from the axils of small scaly leaves. At the tips of the branches were situated inverted solitary or paired one-seeded cupules. The micropylar tube was long, bent outwards. The integument was bicornute and free from the nucellus, on which an elongated salpinx was borne. Microsporoclads (*Pteruchus*; T; Fig. 49 (5)) were leafless; to the rachis were attached shields (sometimes strongly reduced) with a lobed margin and numerous singular sporangia on the lower surface. The pollen is quasisaccate, two-winged with a smooth body (Taylor *et al.*, 1984). The stems (*Rhexoxylon*; T; Fig. 52, *n*) are up to 25 cm in diameter. The wood cylinder is divided into wedges by wide medullary rays. Around the

---

shoot; *h*, *Czekanowskia rigida* Heer, leaf cuticle with stomata, and stoma; *i*, *Sagenopteris heterophylla* Dolud. et Svan., venation in upper part of leaflet; *j*, *S. rhoifolia* Presl; *k*, *S. heterophylla*, stoma with thickenings on guard cells; *l*, *m*, *Caytonia nathorstii* (Thomas) Harris, longitudinal section of seed (*l*) and part of pinnate polysperm with capsules (*m*); *n*, pollen of *Vitreisporites* type from *Caytonanthus oncodes* Harris, distal view; *o*, *p*, microsporoclad *Caytonanthus arberi* (Thomas) Harris, reconstruction of microsporoclad (*o*) and of synangium (*p*); *q*, reconstruction of microsporoclad *Perezlaria oaxacensis* Delev. et Gould; *r–t*, *Irania hermaphroditica* Schweitz., reconstruction of fertile shoot (*r*), part of polysperm (*s*) and microsporoclad (*t*). Scale bar = 1 cm (*a*, *c*, *d*, *j*, *m*, *q*), 5 mm (*e*, *g*, *i*, *o*, *r*), 1 mm (*l*, *s*, *t*), 50 μm (*h*), 10 μm (*b*, *f*, *k*, *n*). Modified from: Harris *et al.*, 1974 (*a–e*, *g*); Samylina, 1980 (*f*); Vakhrameev and Doludenko, 1961 (*h*); Doludenko and Svanidze, 1969 (*i*, *k*); Gothan and Weyland, 1964, after Schenk (*j*) and Harris (*o*); Retallack, unpublished (*l*, *m*, *p*); Townrow, 1962 (*n*); Taylor, 1981, after Delevoryas and Gould (*q*); Schweitzer, 1977 (*r–t*).

periphery of the wide pith are additional (perimedullary) wood massifs, each undergoing accretion towards the centre and periphery of the stem. The wood is of the coniferalean type. On the radial walls are 1–2 rows of rounded bordered pits. Several vascular bundles, departing from different axial bundles, enter the main rachis of the pinna, being arrayed in a crescent line (the branch traces are arranged in a circle).

The fronds in U. (genera *Dicroidium*, *Xylopteris*, etc.; T; Fig. 49 (5); Fig. 52, *k–m*) are pinnate, variously dissected, with odontopteroid or pinnate venation. Sometimes only one vein passes through the terminal lobes. The rachis is characteristically forked. According to their epidermal characters some U. display affinities to the Peltaspermaceae and Cardiolepidaceae.

Satellite genera of the family Umkomasiaceae

Genus *Ptilozamites* (T_3–J_1; Fig. 49 (6))

Forked rachis, short pinnules attached by wide bases, parallel venation; epidermal structure similar to that in the Peltaspermaceae. Associated microsporoclads (*Harrisiothecium*) consist of a branching axis system with a bivalved container for the sporangia on the tips. Pollen quasidisaccate.

Genus *Rhaphidopteris* (P_2–J; Fig. 49 (4, 7))

Leaves extremely polymorphic, from once- to thrice-pinnate, frequently with a forked rachis. Leaves amphistomatic. The guard cells bear the same cutinous thickenings as in Cardiolepidaceae. A closely related genus *Pachypteris* (J; Fig. 49 (7)) differs in its hypostomatic leaves that have peculiar extensions ('berets') on the rachis, which can be compared with the scaly outgrowths on the rachis in *Lepidopteris* (Peltaspermaceae). Male fructifications *Pteroma*, similar to those of *Pteruchus*, but with sporangia constricted to the shield, associate with *Pachypteris*. Pollen quasidisaccate (of the *Alisporites* type).

Order Ginkgoales

During the last century it had already become customary to affiliate certain fossil leaves with ginkgos (G.), which frequently was misconceived. For instance, leaves at present included in the genus *Nephropsis* (Cordaitanthales) were initially assigned to ginkgos. Even now the genus *Ginkgophyllum* (C–P) is affiliated with G., although this is based exclusively on the palmate dissection of the leaf lamina. Many genera, hitherto regarded as G., have been transferred into the order Leptostrobales. The genus *Eoginkgoites* was found to be a bennettite, whereas *Phylladoderma* has been included in the family Cardiolepidaceae. In the outline presented below G. are treated in the bulk of a single family Ginkgoaceae (Fig. 49 (15, 17)). The earlier suggested families Sphenobaieraceae, Karkeniaceae, Torelliaceae and Pseudotorelliaceae were defined on the basis of dispersed leaves and inadequately studied fructifications. Only *Ginkgo* and closely related genera of dispersed leaves (Fig. 49 (17); Fig. 53, *a–f*) have been well studied. *Ginkgo* polysperms are forked (seldomly more branched); the stalk has two seeds, one of which frequently remains underdeveloped. The seed-bearing lamina of the cladosperm is reduced to a small asymmetrical collar beneath the seed (Fig. 44, *c*; Fig. 45, *c*; Fig. 49 (17); Fig. 50, *c*). The seeds look apical, but from the evolutionary–morphological point of view, the seed-bearing surface of the collar is abaxial. G. of the genus *Pseudotorellia* had simple ribbon-shaped leaves and the same type of cladosperms, as in *Ginkgo* (Reymanówna, 1980, 1984). Microsporoclads of G. are also strongly reduced. They resemble microstrobili, composed either of peltate microsporophylls (*Bernettia*; Fig. 49 (15)), about the same as in the Umkomasiaceae, or of stalks with a pair of bunch of microsporangia on the tips (*Ginkgo*, *Stachyopitys*; Fig. 49 (17)). Pollen is asaccate (sometimes probably with a strongly reduced ring-shaped quasisaccus), monocolpate, usually curled into a boat. Leaves are rhomboid, bilobed (*Ginkgo*), in other genera they are dichotomously dissected, or simple, elongated (Fig. 49 (17); Fig. 53, *a*, *c*, *d*). Their classification and related problems were discussed by Harris *et al.* (1974), Kirichkova and Samylina (1979) and Krassilov (1972a). Various types of leaves intergrade. The venation is fan-shaped or parallel. Two veins, entering the leaf, issue from adjacent axial primary bundles. Veins dichotomize several times and enter the leaf apex. G. characteristically have secretory canals between the veins or along the veins. Leaves are amphi- and hypostomatic. Stomata are freely situated between the veins and disorderly oriented. Subsidiary cells are cutinized, the same as non-stomatal epidermal cells, and are frequently supplied with papillae (Fig. 53, *f*, left) overhanging the guard cells. The latter frequently bear a winged cutinization (Fig. 53, *b*, *f*, right).

The genus *Karkenia* (J_3–K) has been assigned to an independent family Karkeniaceae. The polysperms are racemose, attached to dwarf shoots and consist of an axis with short seed stalks. The seeds are of the *Allicospermum* type. In contrast to *Ginkgo*, the integument is free to the base. The nucellus is complicated at the apex by a nucellar beak. Frequently only a sclerotesta is perceived, that looks like a nut shell. Whether *Karkenia* seeds have a collar remains unclear. Leaves of *Ginkgoites* and *Sphenobaiera* types associate with the polysperms. The recognition of an independent family Karkeniaceae is still inopportune.

Satellite genera of the order Ginkgoales

Genus *Ginkgoites* (T–K)

This includes species that differ somewhat from *Ginkgo* in their epidermal characters, or utterly lack epidermal characteristics. Such leaves may belong to Ginkgoales, as well as to other gymnosperm orders.

Genus *Baiera* (P?; T–K; Fig. 53, *d*).

Leaves are similar to the most strongly dissected *Ginkgo* leaves; the central sinus reaches nearly to the petiole. More than one vein extends through the terminal lobes (in the closely related genus *Baierella* – one vein). Stomata are chiefly of longitudinal orientation and form indistinct rows (in *Ginkgo* and *Ginkgoites* they are disorderly and scattered). Resin bodies between veins are round or oval. In some species the microstructure of the leaf is unknown. The boundary between *B.* and *Ginkgoites* is vague. Sometimes associated with *B.* leaves are seeds with a thick outer cuticle, stomata on the integument and abundant secretory canals, so that in those features the seeds display affinities with the *Ginkgo* seeds (Fig. 53, *e*). The Permian species of this genus are perhaps affiliated with *Mauerites* and *Trichopitys* (Peltaspermales).

Genus *Sphenobaiera* (P?, T–K; Fig. 44; Fig. 45 (13–15, 17); Fig. 49, *c*)

Leaves attached to dwarf shoots. The lamina, lacking a petiole (in this *S.* differs from *Baiera*), is palmately dissected. The leaf lobes are gathered into two symmetrical groups; they are narrow, wedge-shaped, sometimes apically notched. Two veins enter the base of the leaf. The epidermal structure is known only in some species. The leaves are amphistomatic. The stomatal rows are indistinct. Stomata are chiefly longitudinally oriented. The cuticle of subsidiary cells is thickened on the side of the guard cells. Sometimes papillae of subsidiary cells overhang the latter. Elongated resin bodies occur between the veins. Some Permian and Triassic species probably belong to Peltaspermales. In its gross morphology *S.* is indistinguishable from the Lower Permian *Ginkgophyllum* (Meyen, 1985a) and the Upper Triassic *Sclerophyllina* (Knobloch, 1972).

Order Leptostrobales

This order figures under two names: Leptostrobales and Czekanowskiales. Of these the first is preferable, being derived from the genus *Leptostrobus*, established for fructifications. The peculiarities of leptostrobans (L.) were recognized already during the 1930s. L. polysperms (genus *Leptostrobus*) are racemose with bivalved capsules borne on a slender axis. Most probably the ancestors of L., the same as the ginkgos, were Peltaspermales (Meyen, 1984a). The seed-bearing valve of *Leptostrobus* in its downfolded margin is quite compatible with the bilateral palmately veined *Peltaspermum* (Fig. 49 (11, 16); Fig. 50, *b*), whereas in the paired position of the valves – with paired cladosperms of *Stiphorus* (Fig. 49 (12, 13)) and Umkomasiaceae (Fig. 49 (5)). Tentatively related to L. are microsporoclads of the genus *Ixostrobus*, that are similar to the *Ginkgo* microsporoclads, but, in contrast, have four strongly fused sporangia. Pollen of L., ginkgos and Triassic Peltaspermales cannot be distinguished under the light microscope. In all three groups dwarf shoots with cataphylls are known. Dichotomizing leaves, typical of most L., are also known in the Peltaspermales. The phylogenetic linkage between L. and Peltaspermales may well explain the similarity observed between L. and ginkgos (also descendants of Peltaspermales) in their leaf structure (Harris *et al.*, 1974; Krassilov, 1972a; Kirichkova and Samylina, 1979), pollen, microsporoclads and the presence of cataphylls on dwarf shoots (Fig. 53, *g*). Due to the fact that in L. only one bundle enters the leaf base, they are usually contrasted against ginkgos. However, this dissimilarity within the confines of the class Ginkgoopsida is of secondary significance. One vascular bundle in the petiole (e.g. in *Kirjamkenia*) is formed as a result of fusion of the leaf traces that issue from different axial bundles. The fusion can be restricted to a short interval. Two veins entering the leaf can be easily derived by suppressing the fusion of the bundles, or by shifting the area of fusion inside the cortex. Also, the possibility should be taken into consideration that the adjacent axial bundles, which give rise to the paired leaf trace, might have been tangentially divided branches of a single axial bundle, that continues over a long distance parallel to the axis (as in *Callistophyton*).

Genus *Leptostrobus* (T_3–K_1; Fig. 49 (16); Fig. 50, *b*)

The polysperm axis with small cataphylls at its base bears in a sparse spiral bivalved, practically sessile capsules. The valves are round with a smoothed radial ribbing. Not far from the distal margin several seeds are attached to the valve. After the seeds are shed round scars remain. The valves come in contact along a marginal fringe, frequently covered by a large number of papillae. It is believed that the papillose fringe might have functioned as a stigma, on which the pollen grains germinated (Krassilov, 1973b). In seeds (of the *Amphorispermum* type), associated with one of these species, no pollen was found in the micropyle (a megaspore membrane was, likewise, lacking). In other L. seeds with a megaspore membrane were found in the capsule. The stigma-like fringe has not been observed in more ancient species of this genus. Associated leaves are assigned to the genera *Czekanowskia*, *Solenites*, *Stephenophyllum* and *Windwardia* (?).

Genus *Phoenicopsis* (T–K; Fig. 49 (16))

Linear leaves with a blunt, occasionally denticulate apex and parallel venation, clustered in a bunch at the apex of a dwarf shoot, covered by cataphylls. If the epidermal characteristics are known for leaves of this type, then they are assigned to the genera *Stephenophyllum* (leaves are hypostomatic in the upper part and amphistomatic in the lower part; stomata are arranged in bands between the veins), *Culgoweria* (amphistomatic leaves, stomata in rows, not bands) and *Windwardia* (amphistomatic leaves, stomata on the underside clustered in narrow bands, and on the upper side – in rows). The stomata of all three genera are chiefly of longitudinal orientation; subsidiary cells are more strongly cutinized than non-stomatal cells, and frequently bear papillae, overhanging the guard cells. The genus *Arctobaiera*, where the leaves are sometimes dissected in the upper part, is seemingly closely related to *Stephenophyllum*.

Genus *Czekanowskia* (T_3–K; Fig. 49 (16); Fig. 49, *h*)

This genus differs from *Phoenicopsis* in having very narrow, repeatedly dichotomizing, or pinnately-dichotomous leaves, rectangular in cross-section (Kirichkova and Samylina, 1979; Samylina, 1980). Only one vein enters the leaf, and dichotomizes following the division of the leaf lamina. The genus

Solenites (Fig. 53, *g*) is probably an older taxonomic synonym of *C*. The genus *Hartzia* differs from *C*. in its simple, once-forked leaves. The *Hartzia* leaves associate with polysperms *Staphidiophora*, closely related to *Leptostrobus*.

Genus *Irania* (T_3; Fig. 53, *r–t*)

This genus with bisexual fertile shoots was allied to L. In this genus microstrobili were attached to the lower part of the axis; higher up were arranged short forked appendages with female capsules. The microstrobili were cylindrical, bearing tightly spirally arranged sporangiophores with two pendant sporangia attached at its widened distal part. The female capsules are spindle-shaped, after maturing open (?) and become cordate. No details are known of their structure. It was suggested that a specific order Iraniales should be recognized, but until more information is obtained it seems more appropriate to treat *Irania* as a satellite genus of L.

Order Caytoniales

Caytonias (C.) have for long occupied a secluded place among gymnosperms, owing to the specific inverted capsules with seeds inside them. This feature, together with the reticulate venation of the leaves and elongated tetralocular sporangia, serves to suggest that C. might have been progenitors of angiosperms. Inverted capsules can be easily derived from cladosperms of Peltaspermales, for example, by modification of peltoids of *Peltaspermum* or radial capsules of Cardiolepidaceae into a bilateral organ, or by closing bilateral cladosperms of the *Stiphorus* (Peltaspermaceae) or *Sporophyllites* types. The seeds of C. and *Nucicarpus* (Cardiolepidaceae) are similar in their long micropylar tubes. Microsporoclads and pollen in C. and Peltaspermales are of the same general type. The reticulate venation, typical of C., is also known in the Peltaspermales.

The genus *Caytonia* (J–K; Fig. 53, *l*, *m*) was introduced for female capsules which associate with microsporoclads *Caytonanthus* (Fig. 53, *o*, *p*), pollen *Vitreisporites* (Fig. 53, *n*) and leaves *Sagenopteris* (Fig. 53, *i–k*). Compound polysperms of C. consisted of an axis with pinnately extending ribbed branches, on which inverted round capsules were borne in two closely spaced rows. The manner of attachment of the seeds is not quite clear. Probably they were of U-shaped arrangement. Capsules of different species display dissimilarities. In some species the seeds were probably immersed in the wall of the capsule, whereas in others this was not so. In the seeds (Fig. 53, *l*) apical hairs were noted that probably directed the pollen into the micropyle. The free integument was multilayered with two vascular bundles and a stony layer (a similar layer was perceived in capsules of Cardiolepidaceae and seeds of *Karkenia*); on the apex of the nucellus a short salpinx was borne (as in some Peltaspermales and Ginkgoales). The small quasisaccate pollen penetrated into the capsule with the aid of a secretory drop. Microsporoclads were pinnate. Elongated synangia, consisting of 3–6 completely fused sporangia, were suspended from the terminal branches. It is believed that C. were dioecious plants (microsporoclads and polysperms often occur separately in different assemblages). Leaves are palmate with long petioles, on the ends of which four (less commonly three) leaflets are borne in pairs. Sometimes each pair of leaflets is attached to its own branch of the petiole. A midrib may or may not be present. The venation is reticulate. The stomata are scattered irregularly on the underside of the leaf. Pollen of the *Vitreisporites* type, recorded from the Permian, belonged to the Peltaspermales.

The genus *Perezlaria* (J; Fig. 53, *q*), established for microsporoclads, can also be affiliated with C. Lateral branches with stellate groups of sporangia were borne on the axis (same as in Peltaspermales). Associated leaves (*Mexiglossa*) were initially wrongly assigned to *Glossopteris*.

Order Gigantonomiales (= Gigantopteridales)

Plants, included in the order Gigantonomiales (G.), usually figure in the literature under the name Gigantopteridales (taken from the genus *Gigantopteris*; P; Fig. 54, *d–f*). The leaves are with reticulate venation. The leaves of *Gigantopteris* and of other

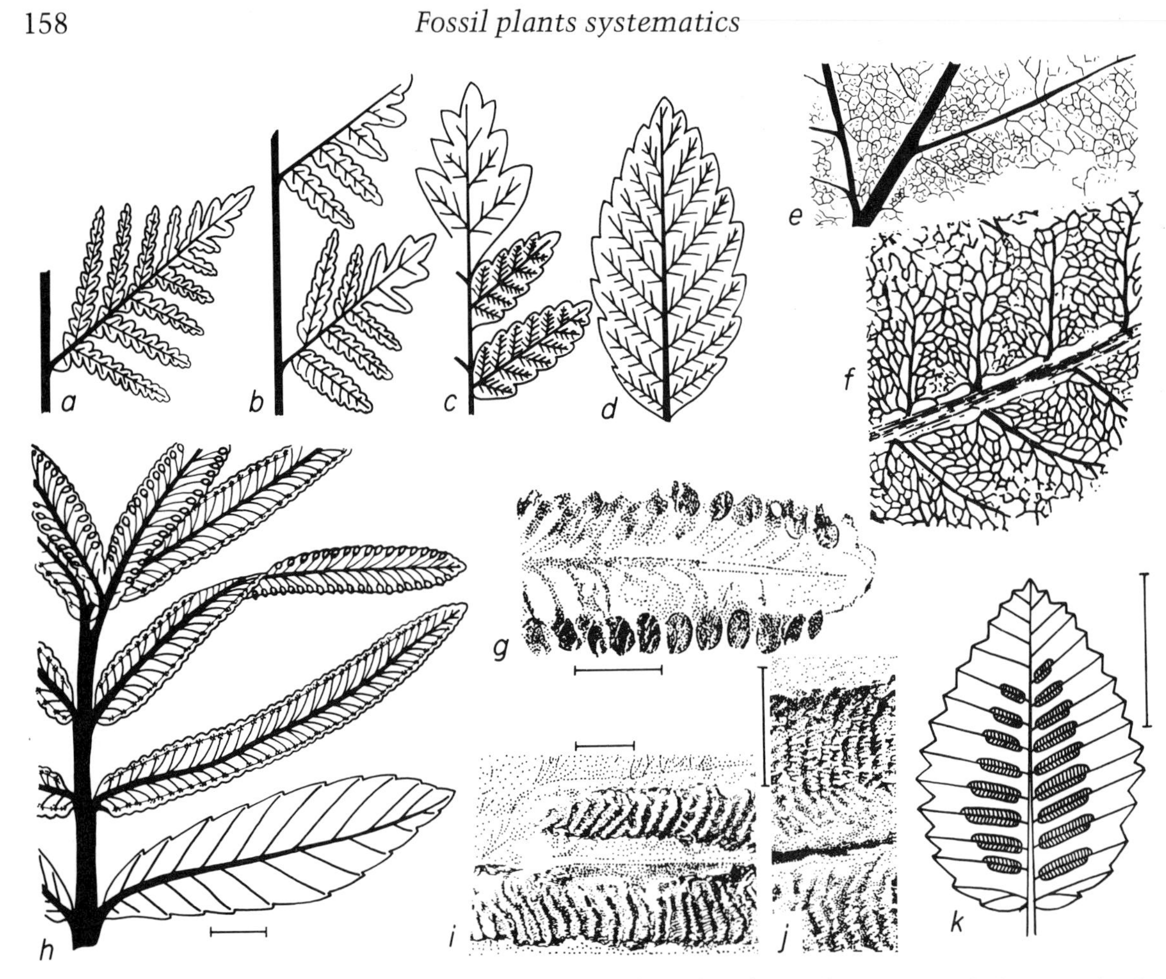

Fig. 54. Gigantonomiales; Permian; China; *a–d*, putative transformation of leaves from compound-pinnate to simple (*a*, *Emplectopteridium*; *b*, *Gigantonoclea lagrelii* (Halle); *c*, *Gigantonoclea* (al. *Bicoemplectopteris*) *hallei* (Asama); *d*, *Gigantopteris* (al. *Tricoemplectopteris*) *taiyuanensis* (Asama)); *e*, *Gigantopteris dictyophylloides* Gu et Zhi; *f*, *Gigantopteris* (al. *Tricoemplectopteris*) *taiyuanensis*; *g*, *h*, *Gigantonomia* (al. *Gigantonoclea*) *fukiensis* (Yabe et Oishi) Li et Yao, seed-bearing pinna (*g*) and reconstruction of seed-bearing frond (*h*); *i–k*, *Gigantotheca paradoxa* Li et Yao, part of pinna with numerous sporangia (*i*, *j*), reconstruction of fertile pinna (*k*). Scale bar = 10 cm (*k*), 2 cm (*h*), 5 mm (*g*), 2 mm (*i*, *j*). Modified from: Asama, 1962 (*a–d*); Gu and Zhi, 1974 (*e*); Halle, 1927 (*f*); Li and Yao, 1983 (*g–k*).

morphologically affiliated genera can be aligned in a sequence, in which we can perceive the successive coalescence of the pinnules of one pinna, and then of the pinnae of different orders (Fig. 54, *a–f*). These elements, undergoing coalescence, retain their former venation. This is similar to the formation of coherent leaves of different orders, as in the family Peltaspermaceae, but G. characteristically reveals reticulate venation. Such a series was constructed for the Permian genera of Cathaysia, where the genus *Emplectopteris*, which already had reticulate venation, was taken as the progenitor (Fig. 47, *v*, *w*), Platyspermic seeds were borne on unmodified pinnules, and there is reason to believe that this genus should be referred to the Callistophytales. Further in the proposed series (*Gigantonoclea lagrelii* – unicoherent leaves; *Gigantonoclea hallei* – bicoherent leaves; *Gigantopteris taiyuanensis* – tricoherent leaves; Fig. 54, *a–d*) the fructifications are known only on bicoherent leaves. The phyllosperms were described as *Gigantonomia* (Fig. 54, *g*, *h*), while the microphylls were described as *Gigantotheca* (Fig. 54, *i–k*). The seeds are borne in a row not far from the pinna margin on its underside. Their structure is unclear, but probably they are platyspermic with the main plane oriented perpendicular to the margin. The indication that the seeds were accompanied by an opening in the leaf lamina near the place of their attachment, is in need of confirmation. The microsporophylls bear numerous sporangia (synangia?), arranged in compact rows along the lateral veins in their lower part. The sporangial structure has not been studied. The inference that the sporangial rows were covered by a common indusium, is not quite convincing.

A tendency is current to affiliate with gigantopterids or even directly refer to *Gigantopteris* all coherent leaf forms of the Upper Palaeozoic and Triassic. As was already stated, some plants with such leaves belong to the family Peltaspermaceae. Obviously, between G. and Peltaspermaceae a far-reaching parallelism can be observed in the formation of coherent leaves of different orders. Taking into consideration the fact that leaves coming from the Lower Permian in the USA, frequently assigned to *Gigantopteris*, belong, most likely, to Peltaspermaceae (Fig. 52, *c*; they associate with cladosperms of the *Autunia* and *Sandrewia* type), it becomes apparent that the informal group 'gigantopterids' cannot be regarded as identical to the order Gigantonomiales. Accordingly, the name Gigantopteridales, proposed by Li and Yao (1983), should be rejected. At present, until more detailed studies have been made of the fructifications, it seems reasonable to treat G. as consisting of the single family Gigantonomiaceae fam. nov. (the type genus – *Gigantonomia* Li and Yao, 1983; the characteristics of the family are the same as the order; the family also includes the genus *Gigantotheca*; the genera *Gigantopteris*, *Gigantonoclea*, etc., established for leaves and commonly assigned to gigantopterids, are to be treated as satellite genera of the family).

If the above described series of the successive increase in the coherency of the leaves reflects the successive order in the phylogeny, and *Emplectopteris triangularis* actually belongs to the Callistophytales, then it is conceivable that G. evolved from callistophytans.

Order Arberiales (= Glossopteridales)

Probably no other group of fossil plants can be named where the morphology, systematics and phylogeny were subject to such tremendous errors and controversy as the Arberiales (A.). On the other hand, extensive investigations carried out in recent years continuously contribute new evidence that highlights the remarkable diversity of the entire group. Because of this the A. systematics is as yet far from stabilization. It is still impossible to divide A. into families.

It is most convenient to begin our acquaintance with A. from the most completely reconstructed plant types. This applies to cladosperms of the genus *Dictyopteridium* (C_3–P_1; Fig. 55, *f*) inserted in the axil of the *Glossopteris* leaves, forming a fertiliger. The margins of the cladosperms were curled downwards and completely concealed the numerous seeds, scattered over the lamina and embedded in a loose network of branching hairs. The integument was free from the nucellus. The nucellar beak is well developed. Quasidisaccate pollen with a striated body (of the *Protohaploxypinus* type) has been found in the pollen chamber. The same pollen type has been found in associated microsporoclads *Eretmonia* (Fig. 55, *l*, *m*), consisting of a modified leaf with reticulate venation. From the axis of the leaf two stalks extend in opposite directions, each stalk revealing repeated branching. The terminal branchlets bear bean-shaped sporangia. These plants were trees with thick pycnoxylic trunks. The wood was of the *Araucarioxylon* type (see satellite genera of Pinophyta). The roots (*Vertebraria*; Fig. 55, *x*, *y*) are composed of an axial massif of primary wood, from which narrow wedges of secondary wood issue, that anastomose by horizontal septae, consisting likewise of tracheids. The gaps between the wedges are infilled with a loose parenchyma (aerenchyma). In thicker trunks the septate and broken into wedges central part is covered by a continuous secondary wood.

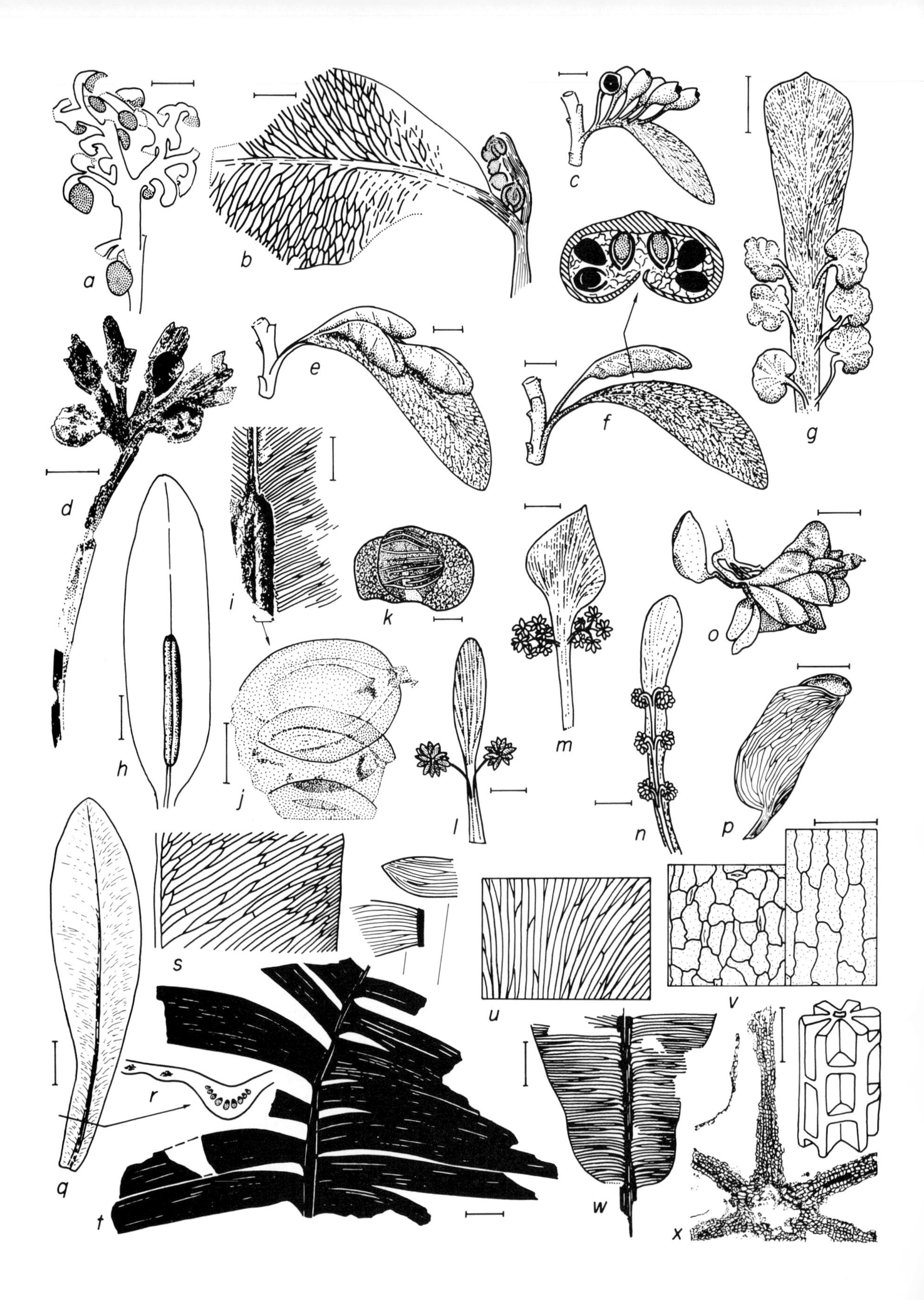
a
b
c
d
e
f
g
h
i
j
k
l
m
n
o
p
q
r
s
t
u
v
w
x

Polysperms of A. are either detached from the subtended leaves, or connected with them. Certain types of polysperms have been placed in different genera, but only a few have been adequately studied and properly interpreted. Several genera have been established for different preservation states of the *Dictyopteridium* cladosperms. The cladosperms vary in the degree of coiling of their margins. Since the fracture plane may pass through different levels of the compression, this simulated an axis surrounded by seeds (one of the former representations of *Dictyopteridium*), a bivalved structure of sterile and seed-bearing valves (another interpretation of the same genus; such fossil remains were assigned to the genera *Scutum*, *Hirsutum*, *Cistella*, *Plumsteadiostrobus*, etc., as well). Several major types of A. polysperms can be recognized. The most primitive were irregularly branching polysperms *Arberia* (Fig. 41, *f*; Fig. 51, *a*). The flattening of the branches and fusion led to the formation of fertiligers of the *Rigbya* (Fig. 55, *d*), *Australoglossa* (Fig. 55, *b*) and further – of *Ottokaria* (Fig. 41, *o*) types. The round or oval cladosperm *Ottokaria* with a dissected margin was attached with the aid of a long stalk to the midrib of the leaf of the *Glossopteris* type, forming a typical fertiliger. The degree of fusion of the cladosperm with the subtended leaf varies from one genus to another. In *Lidgettonia* (Fig. 55, *g*) several cladosperms were attached to the leaf; whereas in *Partha* there were long stalks with apical seed groups. In *Denkania* (Fig. 55, *c*) on each of the stalks one seed was borne, that was surrounded by a cupuliform structure (?). In *Senotheca* the adherence of the bilateral polysperm to the midrib was observed (Fig. 55, *h–j*); in the micropyle of the seeds a pollen of the *Protohaploxypinus* type was found, the same as in *Dictyopteridium*.

Hence, the external diversity in the A. polysperms can be reduced to a few transformations: (1) flattening, fusion and reduction of the seed stalks; (2) formation of the cladosperm and various modifications of it; (3) adherence of the cladosperm to the modified or unmodified leaf and formation of the fertiliger; (4) variations in the number of cladosperms, attached to the leaf. For a comprehensive understanding of the phylogenetic significance of these transformations it is necessary to know more completely in what combinations the polysperms occur with definite microsporoclads and vegetative parts.

Microsporoclads *Eretmonia* have already been described above. In *Glossotheca* a large number of elementary microsporoclads emerge from a modified leaf (Fig. 55, *n*). In *Kendostrobus* stalks with radial sporangial groups on the tips were situated around the axis. From A. microsporangia different types of pollen have been recovered, chiefly of the *Protohaploxypinus* type (Fig. 55, *k*), also asaccate

Fig. 55. Arberiales; Upper Palaeozoic of Gondwana – Brazil (*a*), Australia (*b, f, l*), India (*c, e, g–j, m, n, t, v–y*), South Africa (*d, k, o, p, w*), Antarctica (*x*); *a*, *Arberia minasica* White, branching polysperm; *b*, *Australoglossa walkomii* Holmes, leaf with polysperm attached to petiole; *c*, *Denkania indica* Sur. et Chandra, reconstruction of epiphyllous polysperm; *d*, *Rigbya arberioides* Lacey et al., palmate polysperm; *e–g*, epiphyllous polysperms *Jambadostrobus pretiosus* Chandra et Sur. (*e*), *Dictyopteridium* (*f*; above, cross section) and *Lidgettonia mucronata* Sur. et Chandra (*g*); *h–j* *Senotheca murulidihensis* Ban., seed-bearing leaf with adnate polysperm (*h*), upper part of polysperm and leaf venation (*i*), macerated seeds in polysperm (*j*); *k*, pollen of *Protohaploxypinus* type from sporangium of *Arberiella vulgaris* Pant et Naut.; *l–n*, epiphyllous microsporoclads of *Eretmonia cooyalensis* Holmes (*l*), *E. utkalensis* Sur. et Mahesh. (*m*) and *Glossotheca* (*n*); *o, p*, *Arberiella africana* Pant et Naut., part of microsporoclad (*o*), and sporangium (*p*); *q, r*, *Glossopteris* leaf (*q*) and scheme of its cross section near midrib (*r*); *s*, *G. indica* Schimp., venation; *t*, *Pteronilssonia gopalii* Pant et Mehra, leaf, and details of venation in segment base and apex (above); *u*, venation of *Gangamopteris* leaf in its middle part; *v*, *Glossopteris harrisii* Pant et Gupta, lower (to left) and upper (to right) leaf cuticle; *w*, *Rhabdotaenia danaeoides* (Royle) Pant, leaf base; *x, y*, roots *Vertebraria*, primary wood before secondary wood accretion, cross section (*x*) and general view (*y*). Scale bar = 1 cm (*a–c, e–h, n, q, t, w*), 5 mm (*d, i, l, m*), 1 mm (*j, o*), 0.5 mm (*p, x*), 100 μm (*v*), 10 μm (*k*). Modified from: Rigby, 1972 (*a*); J. M. Schopf, 1976 (*b, h, l–n*) and 1982 (*x*); Retallack and Dilcher, 1981b (*c, e–g*); Lacey *et al.*, 1975 (*d*); Banerjee, 1969 (*i, j*); Pant, 1977 (*k, l*) and 1958 (*o, p, w*); Chandra and Surange, 1979 (*s*); Pant and Mehra, 1963a (*t*); Pant and Gupta, 1968 (*v*).

with a striated body (resembles *Vittatina*). Some A. had, seemingly, quasimonosaccate and quasidisaccate pollen with a smooth body. A striking fact is the parallelism between the pollen of A., Peltaspermales and conifers.

Beds where A. are dominant, yield a great number of pycnoxylic stems with wood of the *Araucarioxylon*, *Dadoxylon*, etc. types, that displayed differences in the pitting of the tracheids, pith structure and other characters (Kräusel *et al.*, 1961–1962; Maheshwari, 1972). So far it has been impossible to establish the correspondence of the above-described genera to any definite genera identified for the fructifications and foliage types. The roots in A. (*Vertebraria*, see above) attain 30 cm in diameter. The evidence available indicates that the structure of the *Vertebraria* type also occurred in leafy branches (Pant, 1977). The A. leaves were in the main, simple with reticulate venation. Such leaves with a midrib, composed of several parallel, sometimes anastomosing vascular bundles, are assigned to the genus *Glossopteris* (Fig. 2, *a*; Fig. 55, *q–s*, *v*). In this genus many species have been recognized that differ in size and configuration of the leaf, form and orientation of the venation meshes and other characters, including the epidermal (Kovács-Endrödy, 1977–1978; Chandra and Surange, 1979). Some species reveal at the leaf base triangular processes directed backwards. When the processes become enlarged and the leaf acquires an arrow-shape, then these forms are assigned to *Belemnopteris*. In *Rhabdotaenia* (Fig. 2, *c*; Fig. 55, *w*) the lateral veins depart from the midrib at nearly right angles; anastomoses are absent. To *Pteronilssonia* (Fig. 2, *b*; Fig. 55, *t*) have been assigned pinnate leaves with a forked rachis and linear pinnules having parallel venation. Leaves of these two types are similar both to cycadaleans and benettitaleans, but according to their epidermal characters cannot be distinguished from typical *Glossopteris*. Evidently, pinnate leaves *Dunedoonia* (W. B. K. Holmes, 1977), where the pinnules have reticulate venation, are merely of convergent similarity with cycads (*Dictyozamites*) and trigonocarpaleans (*Linopteris*, etc.). Simple leaves, which in contrast to *Glossopteris* lack a midrib, have been assigned to the genus *Gangamopteris* (Fig. 55, *u*), including numerous species; whereas leaves with an aborted midrib and open venation are keyed out as *Palaeovittaria*. Probably affiliated to A. are also some leaves with fan-shaped venation, that frequently are regarded as cordaites and ascribed to the genus *Noeggerathiopsis*.

The organic connections between foliage leaves and fructifications have been established only in isolated instances, and at present only for *Glossopteris*, found in organic connection with *Dictyopteridium*, *Senotheca*, *Ottokaria* and certain other fructifications.

Axillary polysperms of early A. have been compared to the axillary compound polysperms of cordaitanthaleans. However, A., most likely, can be allied with those primitive gymnosperms (of the Calamopityales type), that had eustelic stems, platyspermic seeds produced by branching polysperms, and microsporoclads of tridimensional branching. In the lineage from calamopityans to callistophytans solitary seeds were displaced on to the leaves, whereas in A. the entire branching polysperm was shifted on to the leaves. The microsporoclads underwent similar transformations. After these the leaves subtending the polysperms and microsporoclads experienced post-heterotopic transformations. Fronds of the *Sphenopteridium* and *Triphyllopteris* types, associated with Calamopityales, were recorded in Gondwana from the Carboniferous beds underlying plant-bearing formations with A. If A. were descendants of calamopityans, then it is conceivable that the A. seeds were primary platyspermic. The probable descendants of A. were considered Umkomasiaceae, Pentoxylales, Cycadales, Benettitales and angiosperms. However, there is reason to believe that Umkomasiaceae sooner evolved from the Permian Peltaspermaceae. Cycads and bennettites (class Cycadopsida) can be phylogenetically linked with Palaeozoic radiospermic gymnosperms. Angiosperms, evidently, evolved from radiospermic gymnosperms. The origin of the Pentoxylales from A. cannot be excluded.

Order Pentoxylales

The Pentoxylales (P.: Fig. 56) involve several Mesozoic ($J–K_1$) genera established for permineralized stems (*Pentoxylon*, *Nipanioxylon*), leaves (*Nipaniophyllum*), microsporoclads (*Sahnia*) and spicate (or head-like) polysperms (*Carnoconites*). Probably they represent dispersed parts of two plant species (may be genera), that were similar in their foliage and fructifications, but different in the anatomy of their stems (*Pentoxylon*, *Nipanioxylon*). Microsporoclads associated with *Pentoxylon* are assigned to *Sahnia*, whereas those associated with *Nipanioxylon* are as yet unknown. P. in its stem anatomy somewhat resembles trigonocarpaleans (of the *Medullosa* type). In the genus *Pentoxylon*, studied in most detail (the genus *Nipanioxylon* is similar in general features) the wood consisted of 3–9 (most frequently 5) mesarch bundles surrounded by secondary wood, that developed chiefly from the side of the pith (Fig. 56, *b*). These eccentric massives are usually described as meristeles. However, the same as in the case of Trigonocarpales (*Medullosa*), we are confronted with a specific type of eustelic organization (Stewart, 1976, 1983). In *Medullosa* we also find at times forms with a secondary wood increment more developed from the side of the pith (Rothwell, 1976). Tracheids of secondary wood of P. have one-, two-ranked, tightly spaced round or polygonal bordered pits. The medullary rays are uniseriate, 2–7 cells in height.

Dwarf shoots of P. have been encountered (Fig. 56, *a*) with tightly spaced scars of abscissed leaves and scales. In the long shoots leaf scars are rare, the wood is of greater volume and additional peripheral wood bundles appear that alternate with the main bundles. Leaves are entire, ribbon-

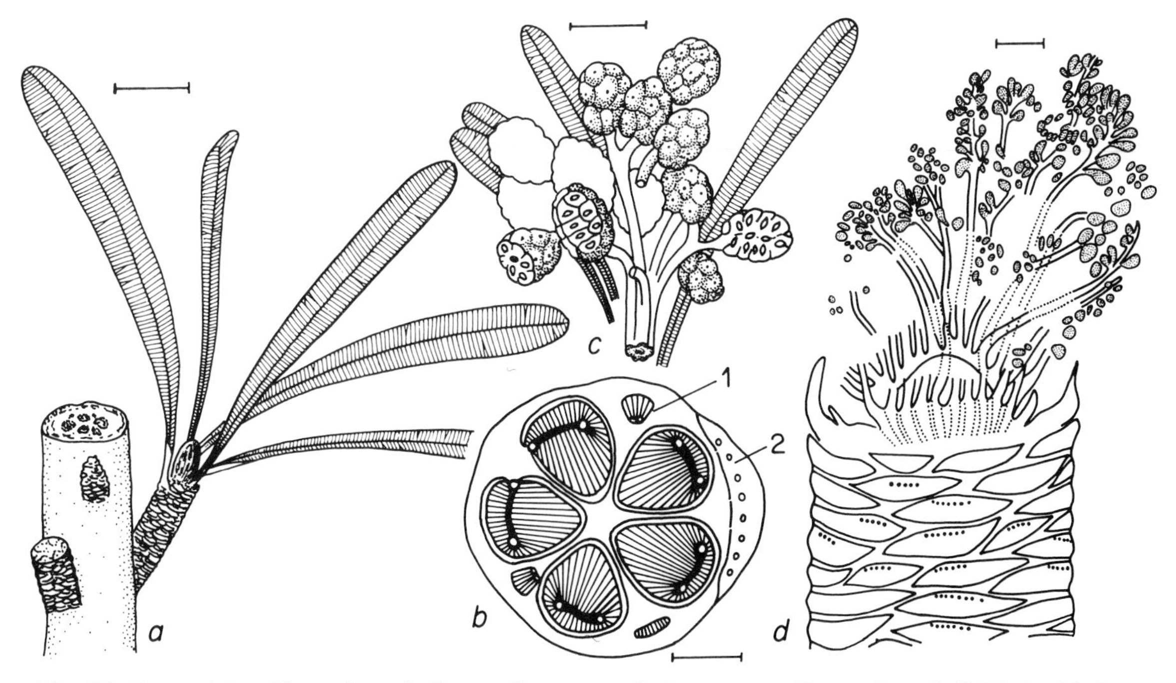

Fig. 56. Pentoxylales; Upper Jurassic–Lower Cretaceous; India; *a*, stem (*Pentoxylon sahnii* Sriv.) with leaves (*Nipaniophyllum raoi* Sahni) on dwarf shoots; *b*, cross section of *P. sahnii*, metaxylem in black with white dots of protoxylem, secondary wood shaded (1, leaf trace; 2, leaf petiole adpressed to stem); *c*, aggregation of polysperms of *Carnoconites compactum* Sriv. and surrounding leaves of *N. raoi* type; *d*, *Sahnia nipaniensis* V.-Mitt., fertile shoot with microsporoclads. Scale bar = 1 cm (*a*, *c*), 2 mm (*d*), 1 mm (*b*). Modified from: Andrews and Mamay, 1955, after Sahni and Vishnu-Mittre (*a*, *c*, *d*); Stewart, 1976 (*b*).

shaped, with a thick midrib consisting of several parallel vascular bundles (as in *Glossopteris*). As in cycads, the vascular bundles in the leaves are diploxylic (upper and lower xylem strands divided by a parenchyme). Six vascular bundles enter the leaf base independently from different axial bundles. The lateral veins depart from the midrib nearly at right angles, occasionally dichotomize and sometimes are jointed by anastomoses. These leaves were first described as having meso-perigenous stomata, the same as in bennettites, and sinuous radial walls of the epidermal cells. Subsequent reports served to indicate that P. are closer to cycads in their epidermal structure.

Polysperms consist of an axis attached to the apex of a dwarf shoot (Fig. 56, *c*). The axis produced up to eight branches, the ends of which served as the receptacle. Platyspermic seeds were borne in a compact spiral, had a two-layered integument free from the nucellus and composed of an outer fleshy layer and an almond-shaped inner stony layer. The fleshy layers of adjacent seeds were tightly adpressed, so that the polysperms resembled a raspberry externally. Microsporoclads (Fig. 56, *d*) were situated on the apex of a dwarf shoot in whorl arrangement, and basally fused. Oval sporangia were inserted on the tips of the forked axis. Pollen is asaccate, monocolpate. The fructifications of P. are known mostly from permineralized specimens. Impression–compression fossils are of less frequent occurrence (White, 1981).

The origin of P. is unknown. The older plants most closely related to them are Arberiales and Peltaspermales. Certain similar features shared by P. with trigonocarpaleans and cycads are, evidently, convergent.

Class Cycadopsida

The class Cycadopsida (C.) includes gymnosperms with radiospermic seeds and cupules that in more advanced orders undergo transformation into an outer integument. These are the orders Lagenostomales, Trigonocarpales, Cycadales and Bennettitales (the still more advanced orders Gnetales and Welwitschiales, not firmly known in the fossil record, are not discussed here). The Lagenostomales (= Lyginopteridales) and Trigonocarpales (= Medullosales) were hitherto placed among pteridosperms. Studies of these plant types led to the discovery of pteridosperms at the beginning of the twentieth century, and resulted in the hypothesis that all plants having seeds and fern-like foliage belong to one large taxon. The fact that they possessed such leaves was regarded as a highly important character and, consequently, the significance of other characters was underrated. This pertains mainly to the structure of male fructifications and seeds, the presence of a cupule. By tracing the gymnosperm phylogeny it becomes evident that the fern-like habit of the leaves was inherited from progymnosperms by different gymnosperms, and this character, taken alone, is not of decisive taxonomic significance. The major trends in the early phylogeny of C. were dealt with above in the general characteristics of gymnosperms.

Order Lagenostomales

The order Lagenostomales (L.) includes the most primitive cycadopsids with branching leafless polysperms, bearing one-seeded or multiseeded cupules. In the most primitive forms (*Eurystoma*, *Hydrasperma*, etc.) the cupule consisted of branching unflattened axes (Fig. 43, *e*, *f*). The flattening and gradual coalescence resulted in a bowl-shaped cover (Fig. 41, *k*). The cupule is distally dissected into lobes. The integument is also frequently dissected into lobes. A series can be constructed that leads from seeds with a deeply dissected integument (*Genomosperma*; Fig. 45, *k*) to seeds with fused integument lobes (*Stamnostoma*; Fig. 45, *m*). It should be noted that in the most ancient seeds from the uppermost Devonian–lowermost Carboniferous (*Hydrasperma*, *Archaeosperma*; Fig. 43, *e*, *f*; Fig. 45, *d*) elements of the integument are rather deeply fused. In the same manner the cupule may have been formed. In the oldest L. the cupule is bilaterally symmetrical. The axis, bearing the cupule, is forked, then the daughter branches divide in the perpendicular plane and to each of the newly-formed branches an ovule is attached from the inside. Above the point of its attachment the branches of the cupule again divide once or twice. The bilaterality of the cupule may be more or less distinct, dependent upon the order of branching and distance between the levels of branching. When the branches of the first order are closely spaced the cupule looks radial. The bipartite subdivision of the oldest cupules is comparable to a forked rachis, that is typical of many primitive gymnosperms. The further branching orders of the cupule can be compared with lateral and terminal pinnae of the fronds.

According to the symmetry of the cupule L. are divided into groups BS and RS, each differing in its evolutionary destiny (see p. 135). In the group BS the bilateral symmetry of the cupule is retained (*Archaeosperma*, *Hydrasperma*, *Stamnostoma*, *Gnetopsis*), whereas in the group RS the cupules became radially symmetrical (*Lagenostoma*, *Tyliosperma*, *Calathospermum*). It is probable that different families correspond to these groups, but as yet they cannot be clearly distinguished, until new, more complete reconstructions of L. with different cupules are available. Probably allied to L. is the genus *Conostoma*, established for radiospermic seeds, that, seemingly, were shed from the cupules.

The megaspores of the oldest L. (*Archaeosperma*) bear apically aborted megaspores of the tetrad. In one case (in *Conostoma*) a megaspore was described with a trilete apical slit under the pollen chamber. This is supporting evidence to the homology between the megaspore membrane of L. and the megaspores of higher cryptogamic plants.

Commonly to L. are related different types of microsporoclads that vary in the branching pattern of the axes, the degree of fusion of sporangia and symmetry of the synangia (Millay and Taylor, 1979). Distinctions between the most primitive microsporoclads of L. and Calamopityales cannot as yet be made. For instance, microsporoclads of the *Telangiopsis* type might have been present in both orders. Some microsporoclads (for example, *Feraxotheca*), probably belonged to L. Leafless microsporoclads of L. sometimes bear single reduced pinnules in the lower parts of the fertile branches. The synangia may be scattered among the pinnules of the leafy fronds. In the most ancient forms free sporangia are in radial groups inserted on the tips of branches, that in the genus *Paracalathiops* (Fig. 57, *a*, *b*) extend out from inner sides of the forked axes (as in some progymnosperms). In other cases the dichotomously divided axis is crowned by radial groups of free sporangia (*Zimmermannitheca*, *Simplotheca*; Fig. 57, *c*). It can be easily conceived that the radial synangia were derived as a result of fusion of the sporangia of one group. Microsporoclads having such radial synangia (Fig. 57, *e*, *f*) are commonly referred to *Telangium*, although the type species of the genus – *T. scottii* – has a different habit and is closer to the impression–compression genus *Pterispermostrobus*. In most of the advanced L. the synangia are situated on the apices of the axes, that are widened in the form of a rounded or arrow-shaped pad (*Feraxotheca*; Fig. 53, *d*). The thick sporangial wall is thinner on the side facing inside of the synangia, where it dehisced. The genus *Crossotheca* with epaulette-shaped synangia, similar to *Feraxotheca*, is usually affiliated with L. It was recently discovered (Brousmiche, 1982), that the type species *C. crepinii* Zeill. belongs to ferns (Fig. 34, *g*, *i*). However, some of the species may be actually allied to L.

The prepollen (Fig. 42 (2); Fig. 57, *a*, *e*, *r*) is round, trilete or monolete, microtuberculate or minutely spinose (of the *Cyclogranisporites* and *Granulatisporites* types). The sporoderm structure has been studied in one specimen of *Crossotheca*, indeterminate to the species. The attribution of this specimen to L. is not fully evident. The exine is two-layered and composed of a thin non-laminated nexine and thick, also non-laminated sexine. The nexine layer bends sharply along the trilete slit and approaches close to the grain surface (Fig. 42 (24)). Such a sporoderm structure can be seen in sporoderms of many homosporous ferns. The Middle Carboniferous genus *Schopfiangium* (Stidd *et al.*, 1985) tentatively affiliated with L., produced globose trilete pollen having lamellate nexine and alveolate sexine similar to that of Trigonocarpales. In some L. the segregation of the exine layer leads to the formation of the cavum ('pseudosaccate pollen'; Fig. 57, *a*).

a
b
c
d
e
f
g
h
15
13
8 5 10 7 4
3
12
11 6 9
14
i
3 8 5 2 7 4 9 1 6
10
13
12
11
15
14
16
I II III IV V
j
k
l
m
1 2
3
4
5
n
o
p
1
2
6
3 4 5
q
r
s
t
u
v
w
x

Distinctions between L. and Calamopityales according to the anatomy of the stem and branches incur difficulties (as in the case of microsporoclads). The systematic status of some of the genera is known, but so far it is impossible to define the characters, typical of the above-mentioned orders, as a whole. The secondary wood consists of tracheids with multiseriate elliptical, rounded, more rarely polygonal bordered pits on radial and (in some genera) tangential walls. The medullary rays vary in width and height in different genera. Among L. both protostelic (Fig. 57, *g*, *p*, *q*, *u*, *v*) and eustelic forms (Fig. 57, *h*, *i*, *k*) have been described. It is unclear as to whether plant types occurred among L. with a protostelic structure of the main axis. L. with cupules *Calathospermum scoticum* (Fig. 41, *l*; Fig. 45, *l*) had axes of different orders; the thickest known axes (of the *Tetrastichia* type; Fig. 57, *g*) were protostelic with a cruciform primary wood surrounded by a thin uniform layer of secondary wood. In another species of the same genus (*C. fimbriatum*) the cupules seemingly associated with axes *Calathopteris*, also protostelic, but already with a mixed pith. The genus *Calathopteris* displayed affinities in many characters of the lateral branches to the genus *Pitus*, that has eustelic pycnoxylic stems, up to 2.5 m in diameter, with a wide pith. Cupules *Stamnostoma* (Fig. 45, *m*; Long, 1979) probably associated with *Pitus*. Protostelic branches of *Pitus* (usually described as *Tristichia*) are similar in organization to the thickest axes of *C. scoticum* and *C. fimbriatum*. Thereby, it is quite possible that the thick lateral branches in the last two species had been mistaken for the main axis. The *Eurystoma* cupules were probably borne on the fronds of *Alcicornopteris* (Long, 1969).

Without knowledge of the axes of all the orders,

Fig. 57. Lagenostomales and allied genera; Lower (*a–c*, *e–g*, *r–t*, *w*, *x*) and Middle (*d*, *h–q*, *u*, *v*) Carboniferous; Western Europe (*a–c*, *g–p*, *w*, *x*), USA (*d–f*, *q–v*); *a*, *b*, *Paracalathiops stachei* W. Remy, pollen (*a*) and microsporoclad (*b*); *c*, *Zimmermannitheca cupulaeformis* W. et R. Remy, part of microsporoclad; *d*, *Feraxotheca culcitaus* Mill. et Tayl.; *e*, pollen of *Telangiopsis arkansanum* Egg. et Tayl.; *f*, *Telangiopsis* sp., part of microsporoclad, and longitudinal section of synangium (wood in black); *g*, *Tetrastichia bupatides* Gord., vascular system, secondary wood shown by radial shading, protoxylem shown by larger dots, to right – *Sparganum* cortex; *h–m*, *Lyginopteris oldhamia* Bin. (*h*, cross section, secondary wood shown by radial lines, primary vascular bundles numbered according to phyllotaxis, *Dictyoxylon* cortex shown by radial dotted lines; *i*, block diagram of stem with petiole, black lines in cortex – sclerenchyme forming *Dictyoxylon* structure; *j*, cross section of cortex with radially aligned sclerenchyme plates; *k*, arrangement of primary bundles in one plane image, Arabic numerals show order of departure of leaf traces, Roman numerals show axial sympodia; *l*, pitting on radial walls of tracheids of secondary wood; *m*, cross section of primary vascular bundle and of secondary wood, 1 ray, 2 tracheid of secondary wood, 3 centrifugal metaxylem, 4 protoxylem, 5 centripetal metaxylem); *n*, *Sphenopteris* (al. *Lyginopteris*) *hoeninghausii* Brongn.; *o*, *S. larischii* Stur; *p*, stem of *Heterangium grievii* Will. (1, petiole base with leaf trace; 2, aerial root; 3, primary wood mixed with parenchyme; 4, secondary wood; 5, cortex with sclerenchyme plates; 6, vascular bundle); *q*, *H. americanum* Andr., vascularization of petiole and pinnae; *r*, *s*, *Telangiopsis* sp. of same type as in *f*, pollen (*r*) and apical synangium (*s*); *t*, *Rhodeopteridium* typ. *trichomanoides* (Brongn.) Zimm., frond associating with synangia and pollen shown in *f*, *r*, *s*; *u*, *Schopfiastrum decussatum* Andr., stem with petioles (to left) and cross-sections of petiole; *v*, *Microspermopteris aphyllum* Baxt., axillary branching, projections on stem and petioles; *w*, seed *Tantallosperma setigera* Barn. et Long, longitudinal and cross sections, megaspore in black, larger dots denote vascular bundles, smaller dots (outside the integument) denote sections of trichomes; *x*, *Buteoxylon gordonianum* Barn. et Long, stem and petiole bases in cross section, secondary wood shown by radial lines, primary wood (in leaf traces) in black, protoxylem denoted by larger black dots around mixed pith in stem centre. Scale bar = 1 cm (*o*, *u*), 5 mm (*f*, *h*, *p*, *x*), 2 mm (*c*, *d*, *g*), 1 mm (*b*, *n*, *s*, *v*, *w*), 0.5 mm (*j*), 100 μm (*a*, *l*, *m*), 10 μm (*r*). Modified from: Karczewska, 1969 (*a*); Millay and Taylor, 1979 (*b–d*, *f*); Eggert and Taylor, 1971 (*e*); Retallack, unpublished (*g*); Scott, 1923, Beck, 1970 (*h*); Osnovy paleontologii, 1963b, after Snigirevskaya (*i*); Hörich, 1906, after Williamson (*j*); Beck, 1970 (*k*); Hörich, 1906, after Williamson and Scott (*l*); Scott, 1923 (*m*, *p*); Taylor and Millay, 1981, after Shadle and Stidd (*q*); Jennings, 1976 (*r–t*); Rothwell and Taylor, 1972 (*u*); Taylor and Stockey, 1976 (*v*); Barnard and Long, 1973 (*w*, *x*).

according to the anatomical structure distinctions between L. and progymnosperms are not warranted, since both protostelic and eustelic forms occur here. The configuration of the protostele may also be repeated – the trilete (*Triloboxylon – Tristichia*) and cruciate (*Tetraxylopteris – Tetrastichia*). However, certain distinctions can be perceived. The L. axes are often characterized by a cortex of the *Sparganum* type. In other L. the outer layer of the cortex is of the *Dictyoxylon* type, i.e. it has radial, frequently anastomosing sclerenchyme plates (Fig. 57, *h–j*). Imprints of these axes simulate shoots of lycopsids with small leaf cushions. Such a cortex structure is unknown in progymnosperms. As a matter of fact in L. a cortex may occur (e.g. in *Stauroxylon*) that does not belong to the above-mentioned types. More significant distinctions between L. and progymnosperms are reflected in the vascularization of the lateral branches. Sympodial strands of the primary xylem in progymnosperms and in more primitive L. are of radial branching, whereas in more advanced L. (as is common in gymnosperms) they are tangential, and only after this become radial passing into the leaf or branch. In L. a flattened V-shaped leaf trace with several protoxylem strands, or several independent leaf traces extended towards the frond; both types have been observed in one species of the genus *Calathopteris*. V-shaped traces extend into the sterile fronds and polysperms, whereas multistranded extend into the microsporoclads. Sometimes, as in ferns, aerial roots emerge from the stems.

Protostelic forms occupy an isolated position among L., the detached stems of which have been described as the genera *Microspermopteris*, *Heterangium* and *Schopfiastrum*. They have an exarch protostele with a metaxylem pierced by a parenchyme, which in *Microspermopteris* (Fig. 57, *v*) and *Heterangium* (Fig. 57, *p*) is grouped in several rays, emerging from the centre. In *Microspermopteris* (as in eustelic *Lyginopteris*) axillary branching was described (Fig. 57, *v*). In *Schopfiastrum* (Fig. 57, *u*) leaf traces emerge alternatively in two rows (fronds of this genus probably belong to *Karinopteris* – DiMichele *et al.*, 1984). Leaf traces in *Heterangium* are paired or solitary at the point where they depart from the stele. Stems of this genus are linked with microsporoclads of the *Telangiopsis* type, producing monolete or trilete rounded spores and foliage *Rhodeopteridium* (Fig. 57, *p–t*). The protostelic genus *Buteoxylon* (Fig. 57, *x*) had leaf traces that are T-shaped in cross-section (in contrast to the V-shaped and multistranded traces in other L.) and accompanied by a small amount of secondary wood. Parenchymatous cells are scattered in the metaxylem. The cortex is of the *Sparganum* type. This genus, which probably associated with radiospermic seeds *Tantallosperma* (Fig. 53, *w*) was keyed out in the family Buteoxylaceae.

L. foliage are of several types. The rachis was characteristically forked not far from its attachment to the axis. Several species of the genera *Rhodeopteridium* (Fig. 57, *t*), *Diplotmena* and *Sphenopteris* (Fig. 57, *n*, *o*) definitely belong to L.

The genus *Calathospermum* (C_1; Fig. 41, *l*; Fig. 45, *l*) was described from jug-shaped, deeply lobed cupules with a large number (up to 70) of markedly elongated seeds (*Salpingostoma*). The seed stalks are simple or forked. The integument consisted of lobes, fused along half the length. The salpinx is long with a basal widening (lagenostome). The cupules were of pinnate arrangement on the axes. Impressions of compound polysperms are referred to the genus *Megatheca*. Microsporophylls (*Telangium? affine*) are profusely branched in various planes. Synangia, consisting of 4–5 basally fused sporangia, are situated on the tips of the branchlets. The prepollen was round, trilete. The axes (*Tetrastichia*, *Calathopteris*) associated with C. have been discussed above (p. 167).

The genus *Lagenostoma* (C_2; Fig. 41, *k*) has been established for detached one-seeded cupules with a lobed margin. The outer wall of the cupule is sparsely covered by clavate trichomes. The polysperm axes are leafless, of pinnate branching; in some species the axes of the last order are of dichotomous branching. The integument of the seeds is fused with the nucellus. A short salpinx crowned the lagenostome, having a strongly developed central column. Microsporoclads (*Telangium scottii*) are leafless, pinnate; the synangia consist of 7–8 basally fused sporangia. The prepollen is round, trilete with a reticulate relief. The stems (*Lyginopteris*; Fig. 57, *h–m*) is eustelic, with a thick cortex of the *Dictyoxylon* type. The axial vascular bundles divide tangentially. The branch traces are crescent-shaped, abaxially convex. Fronds are of the *Sphenopteris* type (Fig. 57, *n*, *o*). The main rachis is forked at a distance from the stem. The rachis is of pinnate branching.

Order Trigonocarpales

The name of the order Trigonocarpales (T.) is derived from the genus *Trigonocarpus*, established from seed casts of indeterminate inner structure. Seeds of the same type, but with preserved inner structures, have been assigned to the genus *Pachytesta*. T. represents a quite natural plant group. It differs from lagenostomans in the vascularized nucellus (inner integument) that is free from the integument (outer one, or former cupule), synangia consisting of completely fused sporangia, and a specific stem organization, simulating a polystely. The pollen is predominantly monolete with a characteristic alveolar sexine. Some T. genera display in some characters more affinities to lagenostomans, than to T., which tends to emphasize the kinship between both orders.

The T. seeds are radiospermic. Best studied among them are *Pachytesta* seeds (C_2–P_1; Fig. 45, *j*). They attain 11 cm in length and 6 cm in cross-section. The thin inner integument is pierced by vascular bundles arranged in a circle, and crowned by a small lagenostome with a central column. The outer integument is thick, three-layered with a lignified middle layer (sclerotesta), that has three large ribs and various numbers of additional ones. the outer layer is fleshy (sarcotesta). The inner integument has three to several tens of vascular bundles. Three radial layers of cells in the outer integument have been interpreted as the place of fusion (sutures) of hitherto free cupular lobes. In the closely allied genus *Stephanospermum* the apical lobing of the outer integument was observed, which gives indication to the incomplete fusion of the cupular lobes. Other T. seeds differ in the number of main ribs and other secondary features. Transitions between seeds, typical of T. and lagenostomans have not as yet been observed. The oldest T. seeds are known from the Serpukhovian. They belong to the genus *Rhychogonium*, which is externally similar to the Visean *Boroviczia* seeds, but their anatomical structure has not been studied. The manner of attachment of T. seeds is not quite clear. Frequently large synangia have been mistaken for seeds. Judging by the few specimens of fairly good perservation, T. seeds replaced pinnules on the fronds, or were attached to the rachis of the pinna of the last order, i.e. T. had phyllosperms.

Male fructifications of T. are represented by synangia (Millay and Taylor, 1979; Stidd, 1981). Only in a few genera did the tips of the sporangia remain free. The T. synangia are very diverse (Fig. 58, *a–j*, *n*, *o*). Some have three or four sporangia surrounding an axial column of parenchymatous tissue (*Rhetinotheca*), or axial cavity (*Aulacotheca*, *Halletheca*; Fig. 58, *a*, *b*). In other cases the sporangia are fused into very intricate aggregates, the interpretation of which is debatable. In *Dolerotheca* (C_{2-3}; Fig. 58, *g–j*) laminae with sporangia formed a campanulate structure. According to one interpretation the laminae were radially aligned and dichotomously divided; according to another they were aggregated into three segments. In *Sullitheca* (Fig. 58, *e*, *f*) the laminae system was much simpler. The laminated synangia of the genus *Parasporotheca* (Fig. 58, *c*, *d*) are triangular, infolded inwards. The adjacent synangia partially cover each other within a pinnate aggregate. The synangia of *Potoniea* (Fig. 58, *n–q*) are close to those of *Dolerotheca*, *Sullitheca* and *Stewartiotheca*, but the sporangia were aggregated within the sporangium in groups of four–six. These groups appear to correspond to initial synangia, the fusion of which resulted in the compound campanulate structure. This is evidenced by the location of the vascular bundles along each sporangium on the outer side of the group. Such an outer arrangement of the vascular bundles was noted in the synangia of *Halletheca* (Fig. 58, *a*), consisting only of one row of sporangia, and in the permineralized synangia of other T. The T. synangia can be easily derived from the

a
b
c
d
e
f
g
h
i
j
k
l
m
n
o
p
q
r
s
t
u
v
w
x
y

lagenostomalean synangia, and more complex forms from those of less complexity. The oldest synangia of the trigonocarpalean type were recorded in the Upper Tournaisian (J. Galtier and B. Meyer-Berthaud, personal communication, 1984).

In most T., in contrast to lagenostomans, the prepollen was elliptical with a longitudinal proximal slit, and occasionally two additional distal slits (Fig. 58, *k–m*). These grains can be easily mistaken externally for mono- and tricolpate pollen. The exine is three-layered. The thinly laminated nexine is covered by a thick layer of alveolar sexine, which is overlain by a tectum layer, covered by a more or less amount of hollow grains (orbicules, or Ubisch bodies) of tapetal origin (Fig. 58, *m*). The genus *Potoniea* stands apart from others in that it has a trilete prepollen, the same as in lagenostomans (Fig. 58, *p*, *q*). In *Parasporotheca* the sexine is exfoliated, so that two sacci are formed exhibiting a sophisticated system of outer folds. Unlike the Ginkgoopsida and Pinopsida, the narrow ends of the distal sulcus (not its sides) are facing the sacci, whereas the proximal slit is parallel (not perpendicular) to the sulcus. In all cases the proximal slit is of the same structure as in pteridophytes. Because of this one can maintain that T. and lagenostomans had a prepollen and not pollen proper. In some T. the onotogenetic development of the sporoderm has been studied (Fig. 58, *q*).

Little is known concerning the attachment of T. synangia to their fronds. The synangia *Aulacotheca* were found aggregated on a leafless branchlet (Fig. 58, *b*). The relationship is apparent between the *Potoniea* and paripinnate neuropterids; also it was shown that these synangia are arranged on branching leafless axes. The synangia in *Dolerotheca* were attached to the rachis, replacing pinnae of the last order on the fronds of *Alethopteris* (Fig. 58, *w*). The laminated synangia *Whittleseya* have been found in attachment to imparipinnate fronds of *Neuropteris*, where they replace the lateral pinnules in the

Fig. 58, Trigonocarpales, male fructifications and anatomy; Lower (*r*). Middle (*a*, *b*, *e–h*, *m–q*, *s*), Middle–Upper (*k*, *l*, *v–x*), Upper (*c*, *d*, *i*, *j*, *t*, *y*) Carboniferous, Lower Permian (*u*); USA (*a–j*, *m–o*, *r–t*, *w–y*), Western Europe (*p*, *u*), Euramerian area in general (*k*, *l*, *q*); *a*, *Halletheca reticulata* Tayl., synangium, vascular bundles denoted by peripheral black dots, sections of tubular sporangia stippled; *b*, *Aulacotheca iowensis* Egg. et Kryd., leafless microsporoclad; *c*, *d*, *Parasporotheca leismanii* Den. et Egg., microsporoclad (*c*) and section of synangium (*d*), section of tubular sporangia stippled; *e*, *f*, *Sullitheca dactylifera* Stidd et al., general view of synangium (*e*), section of lower part of synangium (*f*), stippling shows hollow area in which sporangia dehisced, vascular bundles shown by dots; *g*, *h*, *Dolerotheca formosa* J. M. Schopf, bottom view of synangium (*g*) and its magnified fragment (*h*), tubular synangia shown dehisced; *i*, *j*, *Dolerotheca*-like synangium, cross sections at level of sporangia and narrow lisigenous lacunae (*i*; branching line shows sclerenchyme plates) and at level of vascular system (*j*; branching line shows main bundles, short lines show departure of distal bundles); *k*, *l*, pollen of *Monoletes* type, proximal view and cross section (*k*), distal view (*l*); *m*, *Aulacotheca iowensis*, section of exine near proximal slit (laminated nexine below, alveolar sexine above, in centre – aperture accompanied by hollow area and nexine swelling; at right above sexine, Ubisch bodies and tapetal membranes); *n*, *o*, *Potoniea illinoense* Stidd, synangium (*n*), vertical lines – sections of tubular synangia arranged in groups star-shaped in cross section (*o*); *p*, prepollen from *Potoniea* synangium associating with *Paripteris linguaefolia* Bertr.; *q*, exine of *Potoniea illinoense* in immature (above) and mature (below) states, nexine is laminated, sexine is granular; *r*, *Quaestora amplecta* Mapes et Rothw., cross-section of stem and adpressed petioles, metaxylem in black, protoxylem shown by white dots, secondary xylem shaded, petiolar leaf traces shown by white circles; *s*, *Medullosa primaeva* Baxt., same designations; *t*, *M. noei* Steidtm., cross-section of stem and petioles (numerals mark sequence of phyllotaxis); *u*, *M. solmsii* Schenk var. *lignosa* Web. et Sterz., cross-section of part of stem with two extrafascicular layers of secondary wood; *v*, primary vascular system of *Medullosa*; *w*, frond of *Medullosa* type with *Myeloxylon* petioles, *Alethopteris* fronds and *Dolerotheca* synangium (at arrow to right); *x*, *Alethopteris* pinnule and its section; *y*, *A.* typ. *grandinii* (Brongn.) Goepp., vascularization of pinnule in its upper part. Scale bar = 1 cm (*b*, *c*, *g*, *r*, *s*), 5 mm (*d*, *e*, *t*, *x*), 2 mm (*a*, *f*, *i*, *j*, *n*), 1 mm (*h*, *o*), 100 μm (*k*, *l*), 20 μm (*p*), 5 μm (*m*). Modified from: Taylor and Millay, 1981 (*a*); Eggert and Kryder, 1969 (*b*); Millay and Taylor, 1979 (*c*, *d*, *g*, *h*, *n*); Stidd *et al.*, 1977 (*e*, *f*); Rothwell and Eggert, 1982 (*i*, *j*); Taylor, 1978 (*m*) and 1982 (*q*); Stidd, 1978 (*o*); Laveine, 1971–1972 (*p*); Mapes and Rothwell, 1980 (*r*); Basinger *et al.*, 1974 (*s*, *t*); Seward, 1917 (*u*); Rothwell, 1976 (*v*); Ramanujam *et al.*, 1974 (*w*, *x*); Mickle and Rothwell, 1982 (*y*).

pinnae of the last order. In T., as in the class Ginkgoopsida, the fructifications thereby underwent displacement from specialized leafless shoots on to leafy fronds.

The T. stems were described as genera *Medullosa* and *Sutcliffia*, and multistranded petioles as the genus *Myeloxylon*. It was for long considered that the stelar organization in *Medullosa* (Fig. 58, *s–v*) differs essentially from that in lagenostomans. A striking feature of the cross-section of the stem is the separation of the primary vascular bundles, each of which is surrounded by its own secondary wood massive. Because of this these stems were considered polystelic, consisting of several meristeles embedded in a parenchyme. A closer examination of the course of the primary bundles in *Medullosa* (Beck *et al.*, 1982–1983; Stewart, 1983) has shown that this was a misconception. Each bundle consists of a tangentially outstretched metaxylem massif with protoxylem nests, arranged on the ends and along the outer margin of the bundle. The branching bundle undergoes subdivision and the protoxylem strands occur in the distal part of the daughter bundle. By tracing the pattern of separate bundles and their protoxylem strands in a series of consecutive sections, it becomes apparent that they constitute the same system of sympodia as in eustelic progymnosperms and lagenostomans. Due to the fact that the bundles are divided at acute angles to the stem axis, the branching of various levels occurs together in one cross-section, giving the impression of multiple independent bundles. Their attribution to a united system of sympodia is masked also by the increment of secondary wood, often endocentric, as a consequence of which each bundle looks like an independent meristele. Stems with one, two bundles of the same type as in *Medullosa*, were assigned to the genus *Sutcliffia*. The independence of *Sutcliffia* was subject to doubt. Petioles *Myeloxylon* (Fig. 58, *w*) are pierced by numerous vascular bundles and commonly have a *Sparganum*-type cortex.

The origin of the anatomical structure of stems of the *Medullosa* type is unknown. The segmentation of the xylem cylinder with the accretion of secondary wood on all sides in each segment can be discerned already in the Lower Carboniferous (*Bostonia*). The Lower Carboniferous genus *Quaestora* (Fig. 58, *r*) has the same type of petioles as in *Medullosa*, but the stems are protostelic with a cruciform in cross-section stele, whereas the phyllotaxis is decussate. Evidently the segmentation of the wood in *Medullosa* evolved through the fragmentation of the protostele, that was either stellate or cruciform in cross-section.

T. roots with an exarch actinostele and three-, five protoxylem rays have been described.

The T. metaxylem consists of tracheids with scalariform and circular pitting. In secondary wood the tracheid pitting is circular; some pits occur on tangential walls. Secondary wood massif is pierced by parenchymatous rays.

The relationships between the axes of *Medullosa* and the foliage leaves of *Neuropteris* (Fig. 59, *a*, *b*, *e*, *f*) and *Alethopteris* (Fig. 58, *w*, *y*), also, between the axes of *Sutcliffia* and foliage of *Linopteris* (Fig. 59, *g*, *h*), are indisputable. These genera and others, established for leaves and associated with T. are discussed in the section Satellite genera of the classes Ginkgoopsida and Cycadopsida.

Order Cycadales

Cycads (C.) are closely related to trigonocarpaleans according to their seed structure (Fig. 44, *k*), and, to a certain extent, the anatomy of their stems, but differ markedly in the arrangement of the seeds on cladosperms (Fig. 60, *d*, *e*, *i*). Only in a few forms (*Cycas*) do the cladosperms retain the leaf-like lamina with seeds attached to its margin. Such cladosperms occur in loose aggregates on the stem apex. In most of the living genera the cladosperms look like lignified scales with two pendant seeds. These cladosperms are aggregated in compact cones that resemble externally coniferalean cones. Hence, in the phylogenetic lineage from trigonocarpaleans to C. the phyllosperms undergo the same transformations into more specialized cladosperms as in the class Ginkgoopsida.

Fossil C. are known mostly from leaf remains. Fructifications and the anatomical structure of the stems have been described only in single genera. Judging by the leaves, fossil C. were probably

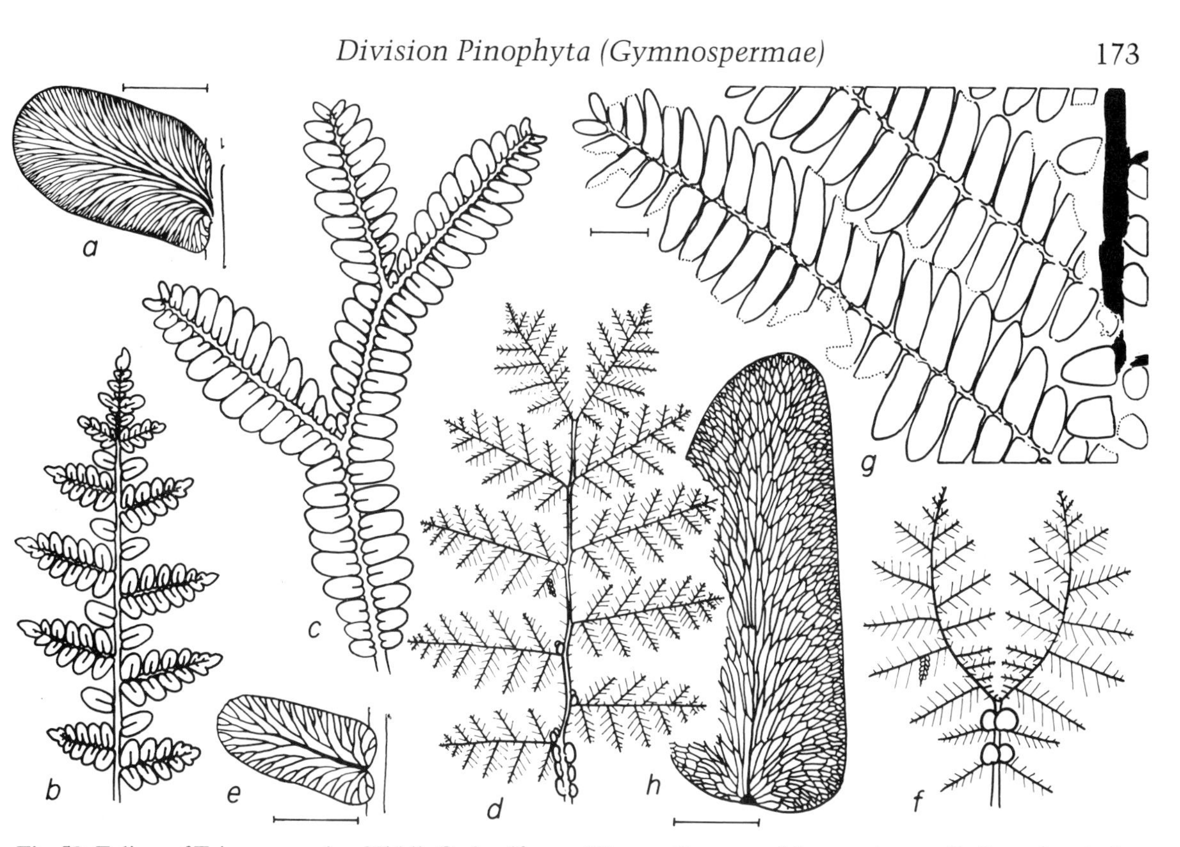

Fig. 59. Foliage of Trigonocarpales; Middle Carboniferous; Western Europe; *a*, *Neuropteris ovata* Hoffm., pinnule; *b*, *Neuropteris*, imparipinnate frond; *c*, *Paripteris*, paripinnate frond; *d*, *Neuropteris*, construction of frond with basal aphlebiae; *e*, *f*, *N. rarinervis* Bunb., pinnule (*e*) and construction of frond (*f*); *g*, *Linopteris neuropteroides* (Gutb.) H. Pot., frond with subsidiary (intermediate) pinnules on main rachis; *h*, *L. subbrongniartii* Gr. 'Eury, pinnule venation. Scale bar = 1 cm (*g*), 5 mm (*a*, *e*, *h*). Modified from: Laveine, 1967 (*a*, *d–f*, *h*); Laveine *et al.*, 1977 (*b*, *c*); H. Potonié, 1904 (*g*).

more diversified as compared with the extant forms. Living C. have thick cormose or columnar-shaped unbranching (pachycauline) stems, also known in some extinct C. For instance, the Triassic genera *Michelilloa* and *Lyssoxylon* described from stem fragments, display affinities in their anatomical characters to the living *Dioon*. Among the fossil C. were also leptocauline forms, i.e. having a crown of leaves borne on slender repeatedly divided branches. This applies to the Triassic *Antarcticycas* (Smoot *et al.*, 1985), Jurassic nilssonias (genus *Nilssoniocladus*; Fig. 60, *a*) and Triassic *Leptocycas*. The anatomical structure in leptocauline forms is known only in *Antarcticycas*. Both lepto- and pachycaulic C. have a manoxylic stem, i.e. the main bulk was occupied by a pith and cortex. The primary vascular bundles with an inward facing xylem are arranged in a ring or in several concentric rings. The secondary wood is not large in amount. It is composed of tracheids with round or scalariform bordered pits. The leaf traces are paired, emerging from an axial bundle and extending further out on various sides; encircling the stem they enter the leaf from opposite sides of the stem. Passing through the cortex they divide, so that several bundles enter the leaf. In *Antarcticycas* only some leaf traces are girdling and the extent of the girdling appears to be limited. The cycadalean affinity of this genus has been established on the basis of the stem anatomy, whereas accompanying fructifications and leaves are still to be discovered.

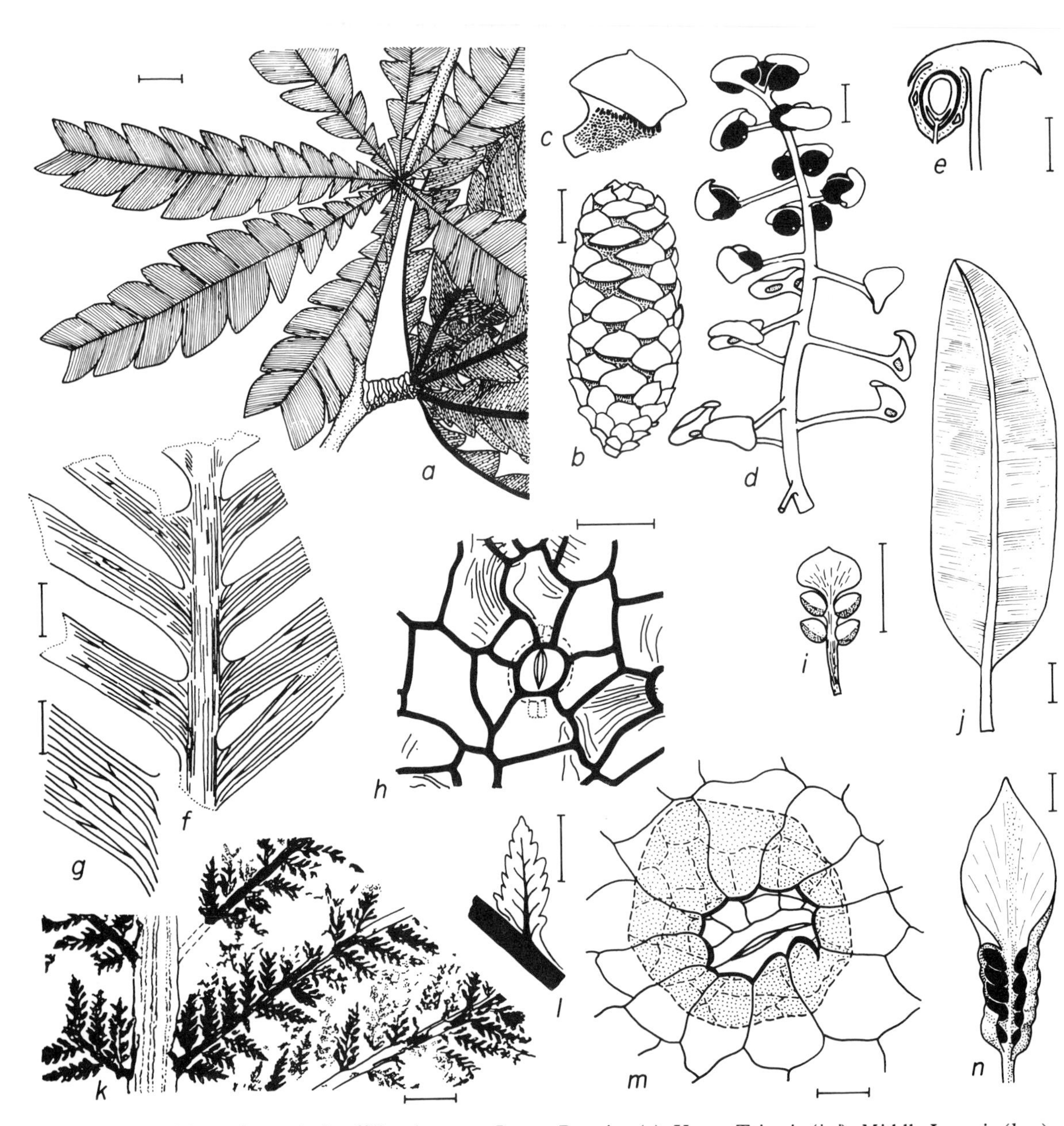

Fig. 60. Cycadales and putatively affiliated genera; Lower Permian (*n*), Upper Triassic (*i*, *j*), Middle Jurassic (*b*, *g*), Lower Cretaceous (*a*, *k–m*); Japan (*a*), Western Europe (*b–j*), Argentina (*k–m*), USA (*n*); *a*, *Nilssoniocladus* (al. *Nilssonia*) *nipponensis* Kim et Sek., reconstruction of leafy shoot; *b*, *c*, *Androstrobus manis* Harris, microstrobilus (*b*) and detached microsporophyll with sporangia (*c*); *d*, *e*, *Beania gracilis* Carr., polysperm, seeds in black, seed scars stippled (*d*), seed-bearing appendage and seed shown in longitudinal section (*e*); *f*, *g*, *Ctenis falcata* Sew., leaf fragment (*f*) and detail of venation (*g*); *h*, *C. nathorstii* Flor., stoma; *i*, *j*, cladosperm *Palaeocycas integer* (Nath.) Flor. (*i*) and associating leaf *Bjuvia simplex* Flor. (*j*); *k*, *l*, *Ticoa harrisii* Arch., frond (*k*) and pinnule (*l*); *m*, *T. magnipinnulata* Arch., stoma with sunken guard and neighbouring (stippled) cells; *n*, *Archaeocycas whitei* Mamay, reconstruction of cladosperm. Scale bar = 10 cm (*i*, *j*), 1 cm (*a*, *b*, *d*, *f*, *k*), 5 mm (*e*, *g*, *l*, *n*), 50 μm (*h*), 20 μm (*m*). Modified from: Kimura and Sekido, 1975 (*a*); Andrews, 1961, after Harris (*b–e*); Gothan and Weyland, 1964, after Florin (*f*, *g*); Florin, 1933 (*h–j*); Archangelsky, 1963 (*k–m*); Mamay, 1976 (*n*).

The classification of fossil C. leaves relies on the epidermal characters and their attribution to different genera may depend upon whether or not these characters are known. For instance, simple leaves with a thick midrib and numerous lateral veins, emerging at an open angle, are assigned to the form-genus *Taeniopteris*, if their epidermal structure is unknown. The same leaves with an epidermis of the cycadalean type are assigned to the genus *Doratophyllum*, whereas leaves with a bennettitalean epidermis are assigned to the genus *Nilssoniopteris*. Other plants also had a similar foliage. C. leaves are sometimes simple (Fig. 60, *j*), but more frequently pinnate (Fig. 60, *a*, *f*, *g*, *k*, *l*) with linear, lanceolate, oval, wedge-shaped or other pinnules (segments). The venation is open or reticulate. The margin of the leaf is smooth or denticulate. Radial walls of the epidermal cells are straight, curved, occasionally sinuous. The stomata are perigenous (haplocheilic). The guard cells are commonly sunken, their walls thickened, usually due to the accretion of lignin, more seldomly of cutin (Fig. 60, *h*, *m*).

Flattened cladosperms of the *Cycas* type (Fig. 60, *i*) have been reported in fossil C., but attached seeds have not been encountered. Strongly specialized cladosperms, clustered in compact cones, have not as yet been described in extinct C. Compound polysperms of the genus *Beania* (J; Fig. 60, *d*, *e*) stand apart from others. A slender axis bears spirally arranged thin cladosperms crowned by upward and outwardly bent shields with paired seeds. Judging by the cutinized membranes, the integument was strongly fused with the nucellus. The stony middle layer of the integument was surrounded by outer and inner fleshy layers. The inner layer was penetrated by vascular bundles. The pollen was found in the micropyle, and is the same as in the microsporangia of *Androstrobus*. Polysperms *Beania* associated with leaves of the *Nilssonia* type and scales of the *Deltolepis* type. The genus *Dirhopalostachys* (J_3–K_1) where the seeds were evidently concealed in expanded and closed cladosperms, was compared with *Beania*. It was suggested that the family Dirhopalostachyaceae should be keyed out, which does not seem advisable, in view of the present knowledge of the genus *Dirhopalostachys*. Doludenko and Kostina (personal communication) studied the coniferalean polysperms from the Jurassic of Middle Asia, which in outward aspect resembled *Dirhopalostachys*. The paired cladosperms of this genus may prove to be strongly modified bipartite scales of the *Schizolepis* type.

The C. microstrobili differ strikingly from trigonocarpalean synangia. They represent compact organs with fleshy sporophylls, bearing sori of microsporangia on the underside. Such strobili are known in Mesozoic C. (*Androstrobus*; J; Fig. 60, *b*, *c*; *Leptocycas*). The pollen is asaccate, monocolpate, frequently boat-shaped. The outer layer of the sexine is continuous with a fine sculpture, whereas the inner layer is spongy. The nexine is lamellar (as in most gymnosperms).

C. probably evolved from plants related to trigonocarpaleans, but no firmly established transitional forms are known. Usually as such are regarded the Lower Permian genera *Archaeocycas* (Fig. 60, *n*) and *Phasmatocycas* with leaves of the *Taeniopteris* type and seeds on the leaf petiole. These genera can be compared with the Carboniferous *Spermopteris* (see order Callistophytales; Fig. 47, *s*, *t*). So far it is unclear whether the seeds of the above-mentioned Lower Permian genera were radio- or platyspermic. *Phasmatocycas* seeds, judging by the structure of the cuticular membranes, did not have at the nucellar apex a pollen chamber surrounded by a circular swelling, and in this respect they are closer to *Beania* than to the Cycadaceae genera, that have a pollen chamber.

It is currently considered that cladosperms of the *Cycas* type are the most primitive representatives of living C. However, leaves corresponding to *Cycas* in their epidermal characters, first appear only in the middle of the Upper Cretaceous.

Several families have been distinguished among C. All living and closely related fossil genera are commonly united into one family Cycadaceae (sometimes the genus *Cycas* is placed in an independent family). The family Nilssoniaceae (at times raised to the rank of an order) includes plants with polysperms *Beania*, microstrobili *Androstrobus* and foliage *Nilssonia*.

The genera identified from dispersed leaves are

regarded below as satellites of C., in the whole. The genus *Nilssonia* has been assigned to this status, owing to the association of these leaves with vaious fructifications.

Satellite genera of the order Cycadales

Genus *Ctenis* (T–K; Fig. 60, *f*, *g*)

Leaves are simple-pinnate; pinnules elongated with an obtuse or slightly acute apex, attached to a rachis at an open angle. Veins are nearly parallel, dichotomize and anastomose (reticulate venation). The midrib is absent. Stomata are perigenous, arranged on underside of leaf, irregularly oriented. Radial walls of the epidermal cells are straight. Fructifications unknown. The leaf outline in the genus *Pseudoctenis* (T_3–K_1) is nearly the same as in *Ctenis*, but the venation is open. Several parallel veins, at times dichotomizing, enter the pinnule base. Stomata are perigenous, occur on the leaf underside, oriented perpendicular to the veins. The radial walls of the epidermal cells are straight. The species *P. lanei* is linked with the microstrobili *Androstrobus prisma*.

Genus *Nilssonia* (T–K)

Leaves lanceolate to nearly linear, entire or variously, at times irregularly segmented; margin smooth or denticulate. Leaves on lateral dwarf shoots (genus *Nilssoniocladus*; Fig. 60, *a*). Veins parallel, emerge at right angles from rachis and very seldomly dichotomous. The radial walls of the epidermal cells are straight. Stomata are perigenous of irregular orientation with 5–6 subsidiary cells, bearing papillae overhanging a stomatal pit.

Other genera have been established on leaves with linear pinnules and parallel venation (*Paracycas*), entire leaves with pinnate open venation, straight, rarely dichotomizing veins (*Doratophyllum*, *Bjuvia*; Fig. 60, *j*). Judging by the epidermal characters certain fern-like plants may be assigned to cycads, e.g. the Cretaceous genera *Ticoa* (Fig. 60, *k–m*), *Ruflorinia*, *Mesodescolea*, *Almargemia* and *Mesosingeria*.

Order Bennettitales

The discovery of bisexual fructifications in bennettites (B.) at the end of the last century led to the hypothesis on the origin of angiosperms from B., this group being the only known extinct gymnosperms with bisexual fructifications. At present this hypothesis is no longer popular. However, in textbooks certain obsolete conceptions of B. are still current, namely the belief that all B. had bisexual fructifications. The first erroneous reconstruction of the *Cycadeoidea* 'flower' is still being reproduced.

Among the radiospermic gymnosperms no other plants are known besides trigonocarpaleans with which the origin of B. may be, at least indirectly, linked. The similarity in the organization of the synangia in both groups is striking. The B. seeds are radiospermic, but in contrast to the trigonocarpaleans, they sometimes have a free cupule. The vascularization of the seeds is strongly reduced, as can frequently be discerned in gymnosperms, where the seeds are concealed by various envelopes. The B. pollen is monocolpate, not monolete. Other important distinctions are apparent between both orders, so that the distance between them is of quite a wide range.

B. are commonly subdivided into families Williamsoniaceae and Bennettitaceae (= Cycadeoideaceae). The stems in williamsonias were slender and profusely branching (leptocaulic organization; Fig. 61, *g*). These plants were evidently shrubs. The stems in Bennettitaceas were thick, columnar- or barrel-shaped (pachycaulic organization; Fig. 61, *a*). Probably some williamsonias were also pachycaulic. The distinctions between the families according to the life forms are in need of more precision. Although the life forms of B. repeat to a great extent that seen in cycads, the fructifications differ essentially. B. microsporophylls (Fig. 61, *b–d*, *h*, *k*) are leaf-like, sometimes pinnate, palmate or cup-shaped, bearing microsporangia or synangia on the inner side or along the margin. Seeds are not situated on cladosperms, but on the expanded apex of the lateral shoot (receptacle), intermixed with interseminal scales. The expanded apices of the scales

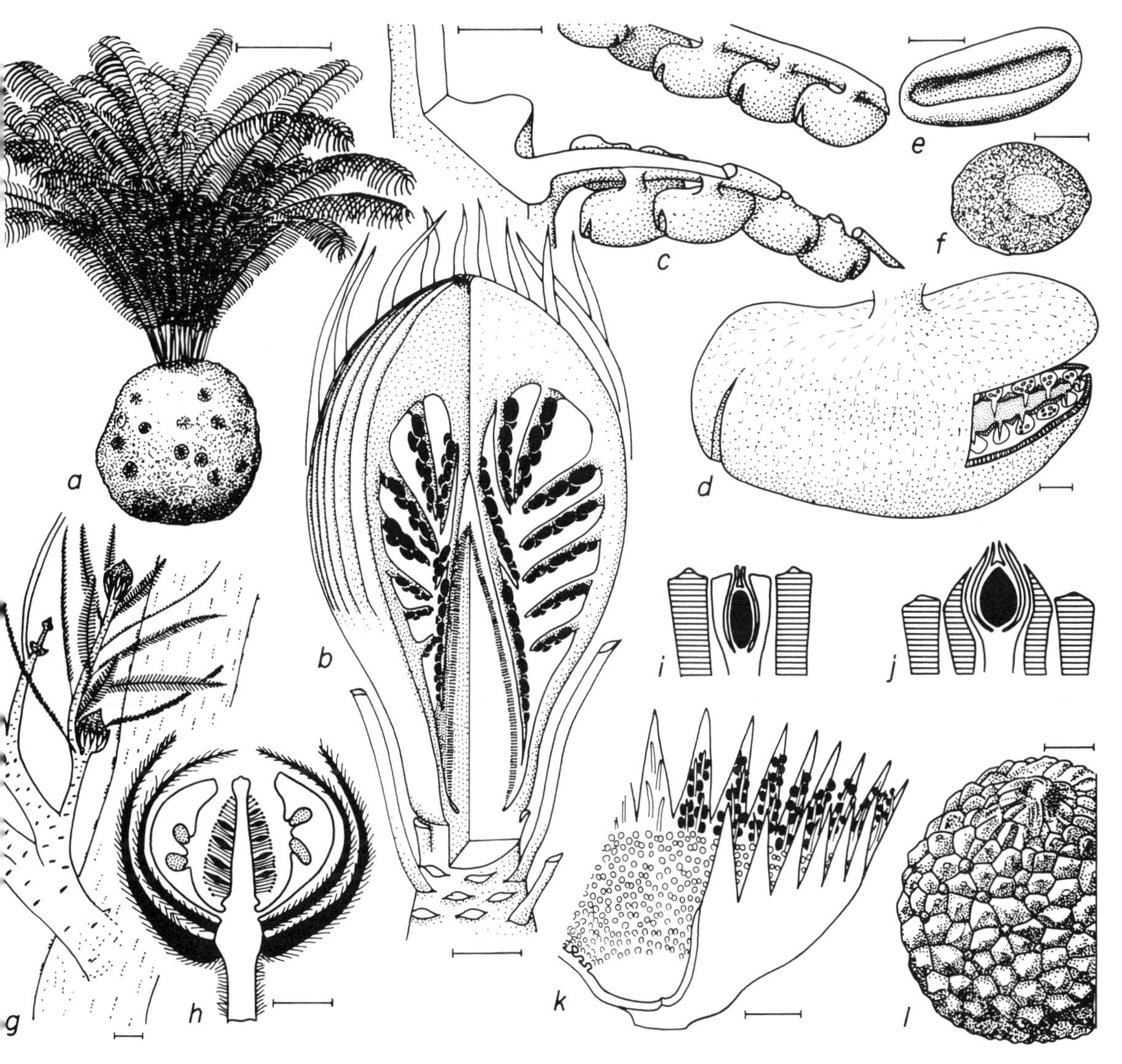

Fig. 61. Fructifications of Bennettitales; uppermost Triassic–lowermost Jurassic (*i, j*), Jurassic (*a–h, k, l*); USA (*a–e*), England (*f–h, k*), Greenland (*i, j*), Mexico (*l*); *a–e*, *Cycadeoidea dacotensis* (McBride) Ward, reconstruction (*a*), fructification (*b*; synangia in black), microsporophyll (*c*), synangium (*d*), pollen (*e*); *f*, *Williamsoniella lignieri* Nath.; *g*, reconstruction of organic connection between axes *Bucklandia pustulosa* Harris, fructifications *Williamsonia leckenbyi* Nath. and leaves *Ptilophyllum pecten* (Phil.) Nath.; *h*, *Williamsoniella coronata* Thomas, fructification structure, seeds and bracts in black, sporangia stippled; *i, j*, scheme of relations between seeds and interseminal scales (shaded) in *Bennetticarpus crossospermus* Harris (*i*) and *Vardekloeftia conica* Harris (*j*); *k*, male fructification *Weltrichia sol* Harris, synangia in black, resiniferous organs denoted by circles; *l*, *Williamsonia netzahualcoyotlii* Wiel., reconstruction of female fructification. Scale bar = 10 cm (*a, g*), 5 mm (*b, h, k, l*), 1 mm (*c*), 100 μm (*d*), 10 μm (*e, f*). Modified from: Taylor, 1981, after Delevoryas (*a*); Crepet, 1974 (*b*); Stidd, 1981 (*c, d*); Taylor, 1973 (*e*); Harris, 1974 (*f*), 1969b (*g, h, k*), 1932 (*i, j*); Delevoryas and Gould, 1973 (*l*).

are adjusted one to another in such a manner that they form a continuous smooth surface with pits in the corners of the converging scales (Fig. 61, *l*). The micropylar tubes of the seeds extend to these pits. B. leaves are morphologically similar to cycadalean, but differ in their epidermal structure.

Valid reconstructions of the overall habit have been presented only for a few B., so that it is difficult to judge the connections between various types of stems, leaves and fructifications. The characteristics for families are therefore given below based on fructifications, whereas the leaf genera are treated as satellite genera of the whole order

Family Bennettitaceae

This family is known chiefly from the genus *Cycadeoidea* (J–K; Fig. 61, *a–e*). This genus is perhaps the older synonym of the genus *Bennettites*, which has been poorly studied. The stems in *Cycadeoidea* are thick, barrel-shaped, conical or cylindrical, up to 60 cm in diameter. When the stem is branched the plant resembles in appearance several fused pineapples. In the centre of the stem is a wide pith surrounded by a ring of mesarch strands of the primary wood. The secondary wood is thin, facing the periphery of the stem, and composed of scalariform tracheids. In addition to numerous wide medullary rays conjugated with leaf traces, narrow uni- and biseriate rays are present. The leaf trace, at first C-shaped, soon divides into several mesarch vascular bundles arranged (in cross-section) in a horseshoe shape. The entire group extends directly into the leaf and is not broken into two girdling bundles as in cycads. The surface of the stem is covered by a dense sheath of leaf bases. In the upper region of the stem buds with unfolded pinnate leaves have been encountered.

Bisexual fructifications, scattered along the entire stem, were frequently depicted as open flowers. A perianth was represented as consisting of pinnate bracts attached to the base of a receptacle, and followed upwards of the axis first by stretched pinnate microsporophylls, and then by numerous seeds divided by interseminal scales. Later it was noted that microsporophylls do not occur in an unfolded state, and that the synangia on the folded microsporophylls are mature or even free of pollen. It is now admitted that the microsporophylls were bent inwards. Their lower part bore lateral appendages that reached to the upper part of the microsporophyll, but did not become fused to it (Fig. 61, *b*). The synangia (Fig. 61, *c–e*) consisted of 20–30 elongated sporangia, that were arranged in two rows on both sides of the slit-like cavity. Mature synangia dehisced into two valves. The pollen was monocolpate, with a two-layered exine. The exine is frequently crushed into a system of folds, mistaken for cells of the gametophyte. The female part of the bisexual fructification consisted of a conical receptacle, covered by hundreds of radiospermic seeds and interseminal scales. A rudimentary cupule was probably present at the seed base. The integument was fused with the nucellus. The fructification was subtended by an involucre of pinnate bracts.

Family Williamsoniaceae

Some of the genera included in this family are sometimes placed in the family Wielandiellaceae which is characterized by slender branching shoots. The williamsonias *sensu stricto* are conceded to be pachycauls, which was inferred from the reconstruction of *Williamsonia sewardiana*. Later it was demonstrated that this reconstruction was incorrect and that the shoots in *W. sewardiana* were quite richly branched and not thick. Nevertheless, some fructifications of *Williamsonia* were seemingly produced by pachycaulous plants. The anatomical structure of the stems is practically unknown. The fructifications were borne at the forks of the dichasially branching shoots or attached to the lateral branches. The fructifications were unisexual, as well as bisexual. Polysperms belong to the genera *Williamsonia* (Fig. 61, *g*, *l*) *Bennetticarpus* (Fig. 61, *i*), *Vardekloeftia* (Fig. 61, *j*) and *Wiellandiella*; male fructifications to *Weltrichia* (Fig. 61, *k*) and *Bennettistemon*; bisexual to *Williamsoniella* (Fig. 61, *f*, *h*) and *Sturiella*. Other genera can also be named. The organization of both the polysperm and 'gynoecium' of bisexual fructifications are in principle the same. In *Vardekloeftia* a free cupule was reported, and the integument is free from the nucellus. In *Wielandiella* the cupule was not found. It may seem strange that genera, both with and without a cupule are included in one family. However, it is quite possible that in the non-cupulate seeds the cupule was transformed into an outer integument (as in trigonocarpaleans), whereas in cupulate seeds it again became free (as in lagenostomans). Comparable variations in the number of seed envelopes can be observed in angiosperms, where this character is of secondary taxonomic significance and occasionally varies within one genus. The homology of interseminal scales remains unclear. They might probably have been monosperms that underwent sterilization. As in *Cycadeoidea*, the fructifications were subtended by one or several rows of bracts, often variously fused. Isolated fructifications with closed parts have been encountered that resemble externally flower buds of angiosperms.

Microstrobili (Fig. 61, *k*) consist of whorls of microsporophylls fused at the base or at a long distance, forming a cup-shaped organ. The microsporophylls are simple or pinnate covered by rows of synangia, which have not been studied in detail. The pollen is of the same type as in *Cycadeoidea*.

The foliage of williamsonias is of the *Nilssoniopteris* (Fig. 58), *Ptilophyllum* (Fig. 58, *a–d*), *Anomozamites*, and other types. The leafy shoots were of lateral and dichasial branching.

Descriptions of some form-genera of B. leaves are presented below. There is a striking parallelism in the variation between the B. and cycadalean leaves. The dispersed leaves can be properly discriminated only on the basis of epidermal characters. It is believed that the bennettitalean stomata are mesoperigenous, although analysis of underdeveloped stomata shows that they could have developed according to the mesogenous type. In any event, in describing the B. stomata it is preferable to use terms referring to the structure of the mature stomatal apparatus, rather than those defining the type of ontogeny (see Chapter 5). Accordingly, the B. stomata can be termed paracytic, whereas the stomata of cycads are actino- or petalocytic. In B. the guard cells have one lateral subsidiary cell on each side, so that they are both encased in a rectangular frame of radial walls (Fig. 62, *g*, *p*). If encircling lateral cells are present then they are also encased in this frame. The guard cells exhibit a specific organization (Fig. 62, *c*, *d*, *g*, *l*, *p*). They have attenuated poles and an extended middle part, bearing cutin thickenings on the dorsal side (faced to the leaf surface). The latter are well resistant to maceration, so that from the inner side of the cuticle the pair of guard cells looks like a butterfly with outspread wings, its body corresponding to thickenings along the stomatal aperture, and the aperture itself. The cutin 'wings' are connected on both ends of the stomatal aperture by a cutin bridge. The thickening along the stomatal aperture blends with the cutin border of the stomatal pit. As a consequence, if the guard cells are sunken, then the butterfly-shaped thickening is suspended on a cutin funnel, corresponding to the proximal (faced to the stomatal aperture) walls of the subsidiary cells. Sometimes the butterfly-shaped thickening is less regular and transformed into a rather broad frame with polar projections. Cycadalean stomata have a similar system of guard cell thickenings, but usually resulting from the lignification – not cutinization – of the walls.

The radial walls of the B. epidermal cells are most frequently sinuous on one or both sides of the leaf (Fig. 62, *c*, *j*, *l*). The sinuosity may reach in amplitude to half the cross-section of the cell. Frequently the radial walls seem jagged because of the system of lateral cutin thickenings on both sides of the cutin rib, that passes along the radial wall (Fig. 62, *j*). It was thought that sinuous walls are typical of all B., whereas cycads characteristically have straight radial walls. However, later straight walls were discovered in B. and sinuous walls in cycads. Sinuous walls may also occur in ferns, including those that have leaves similar to bennettites. But in ferns the cuticle is either very thin or absent, whereas the guard cells do not bear cutin thickenings.

Descriptions are given below of satellite genera of B. leaves.

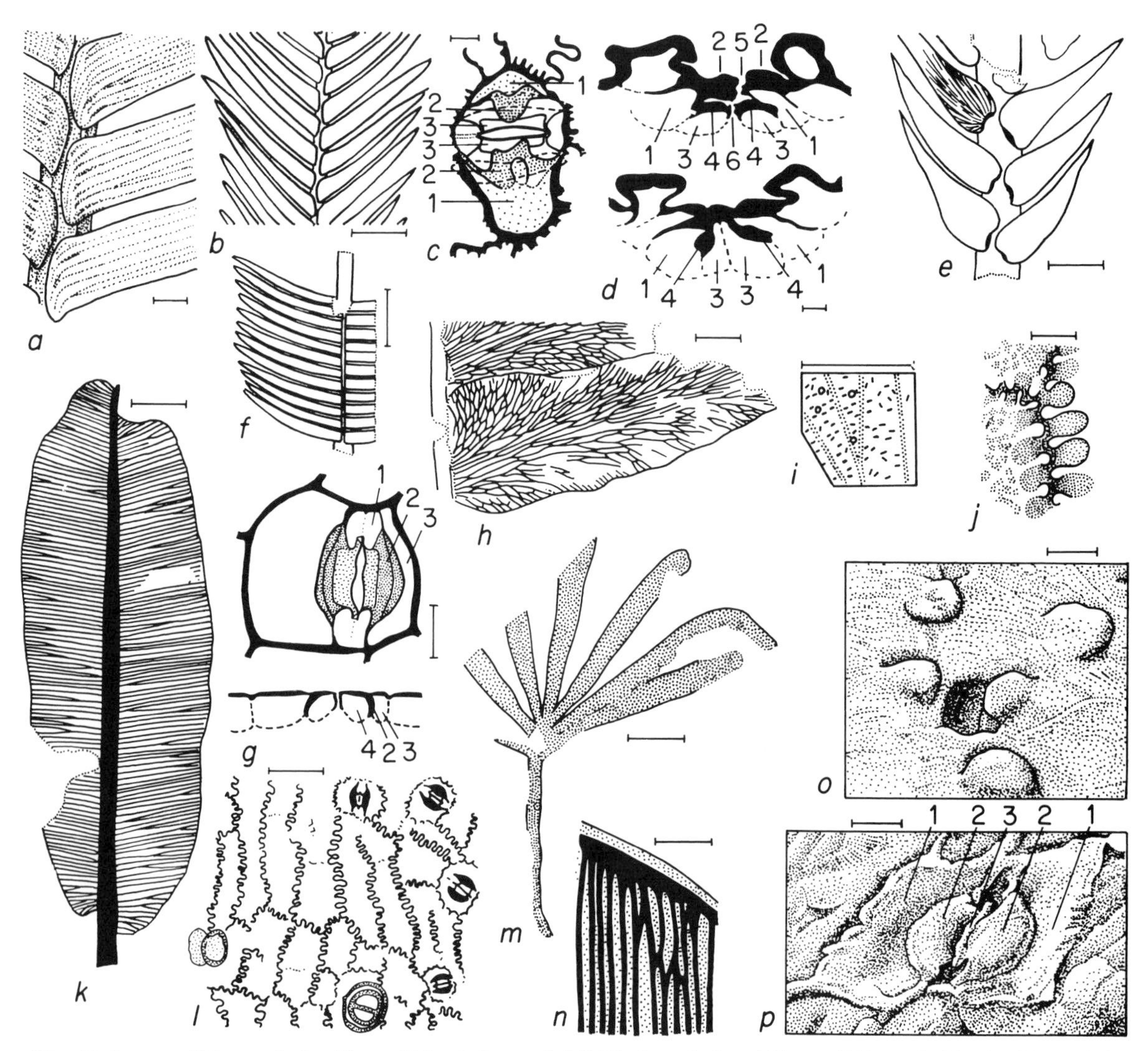

Fig. 62. Leaves of Bennettitales; Upper Triassic (*g*, *m–p*), Middle (*a–e*, *h–l*) and Upper (*f*) Jurassic; England (*a–e*, *h–l*), Georgia (*f*), Greenland (*g*), USA (*m–p*); *a–c*, *Ptilophyllum pectinoides* (Phil.) Mor., attachment of pinnules to upper surface of rachis (*a*), part of pinna (*b*), stoma (*c*), subsidiary cells (1) bearing papillae (2), guard cells (3); *d*, *P. pecten*, cross section of cuticle of stoma in its middle (above) and polar (below) parts (cell contours shown by dashed line), subsidiary cells (1) and their papillae (2), guard cells (3) and their dorsal cutin thickenings (4), stomatal pit (5) and aperture with cutin bordering (6); *e*, *Zamites quiniae* Harris; *f*, *Pterophyllum paradoxum* Dolud.; *g*, *P. rosenkrantzii* Harris, poles (1) and wing-like cutinization (2) of guard cells, subsidary cell (3), guard cell lumina (4); *h*, *i*, *Dictyozamites hawellii* Sew., reticulate venation of pinnules (*h*) and epidermis topography (*i*; circles denote hair bases; dashes denote orientation of stomata; dotted lines denote veins); *j*, jagged-sinuous suture above radial walls in *Anomozamites nilssonii* (Phil.) Sew.; *k*, *l*, *Nilssoniopteris major* (L. et H.) Flor., leaf (*k*) and its epidermal structure (*l*) with four stomata and two hair bases; *m*, *Eoginkgoites sectoralis Bock*, leaf; *n–p*, *E. davidsonii* Ash, venation near leaf margin and marginal vein (*n*), stoma as seen outside (*o*) and inside the cuticle (*p*), subsidiary cells (1), wing-like thickenings of guard cells (2), cutin border along aperture (3). Scale bar = 1 cm (*b*, *e*, *f*, *k*, *m*), 2 mm (*a*, *h*, *n*), 1 mm (*i*), 50 μm (*l*), 10 μm (*c*, *d*, *j*, *o*, *p*). Modified from: Harris 1969b (*a–e*, *h–l*) and 1932 (*g*); Doludenko and Svanidze, 1969 (*f*); Ash, 1976, 1977 (*m–p*).

Genus *Nilssoniopteris* (T_3–K_1; Fig. 62, *k*, *l*)

Leaves are simple, oval, ribbon-shaped, petiolate. The leaf lamina is attached to the midrib in such a manner that it (the lamina) is equally open on both sides of the leaf. The lateral veins are simple or forked extending out at an open angle. The stomata are arranged on the underside of the leaf in the bands between the veins and have an irregular orientation. The upper epidermis is composed of rectangular cells. Usually the leaf surface bears hairs.

Genus *Pterophyllum* (T_2–K; Fig. 62, *f*, *g*)

Leaves are pinnate with linear pinnules attached to the wide rachis by the whole width of their bases, which are sometimes slightly expanded. The pinnules are not of strict lateral attachment to the rachis, but shifted to its upper surface. The veins are parallel, at times dichotomizing at the exit from the rachis. The stomata occur only on the underside. They are either aggregated in distinct bands and oriented perpendicular to the veins, or uniformly scattered over the leaf and of irregular orientation. *Tyrmia* and *Bureja* are considered synonyms of *P*.

Genus *Ptilophyllum* (T_3–K_1; Fig. 62, *a–d*)

Leaves are lanceolate to nearly linear, pinnate, with a long petiole expanded at the base. Pinnules are alternating, inclined towards the leaf apex; their bases overlap the rachis upper surface. The pinnules are basally constricted, sometimes ovate or with parallel margins, always with a rounded asymmetrical apex. The veins slightly diverge and are sometimes dichotomous. The stomata are gathered in bands between the veins of the leaf underside, and of irregular orientation.

Genus *Dictyozamites* (J–K_1; Fig. 58, *h*, *i*)

Leaves similar to *Ptilophyllum*, but with rounded pinnules; veins frequently dichotomizing and anastomosing. The epidermal structure is nearly identical to that of *Ptilophyllum*.

Of other satellite genera mention should be made of *Otozamites* (differs from *Dictyozamites* in its venation, and from *Ptilophyllum* in rounded pinnules, more often dichotomizing and more strongly diverging veins), *Zamites* (Fig. 62, *e*; this genus is closely similar to *Ptilophyllum*, but the pinnules are more widely spaced and often the attachment to the rachis is accompanied by a swelling; stomatal bands are absent), *Zamiophyllum*, *Neozamites*, *Sphenozamites*, *Anomozamites*, *Pseudocycas*, *Cycadolepis* (scale leaves). In the genus *Eoginkgoites* (Fig. 62, *m–p*) the leaves resemble ginkgoalean ones in the dissection of the leaf lamina.

Satellite genera of the classes Ginkgoopsida and Cycadopsida

The genera described below are commonly assigned to pteridosperms. Various species of some of these genera may probably belong to different families, orders and even divisions (*Sphenopteris*).

Genus *Sphenopteris* (known since the Middle Devonian; Fig. 63, *a*)

The fronds are compound-pinnate, at times with a forked main rachis. The pinnules are entire or lobed, with a constricted base. The venation of the pinnules ranges from anisotomous to pinnate. The rachis is often winged. This genus is treated in the literature in a different bulk. Sometimes it includes all the fronds fitting the above diagnosis. Other taxonomists ascribe to *S*. only pteridosperms and narrow the diagnosis of the genus. Some species of the genus are frequently placed in an independent genus *Eusphenopteris*.

Genus *Mariopteris* (C_2; Fig. 63, *b*, *c*)

The fronds are compound-pinnate, the pinnules are attached by the wide base, or sphenopteroid (as in *Sphenopteris*). Basal pinnules of the pinnae are larger than the others, due to the expansion of the catadromous part, which is variously dissected. Fertile fronds have been placed in the genus *Fortopteris*, assigned to ferns, although its attribu-

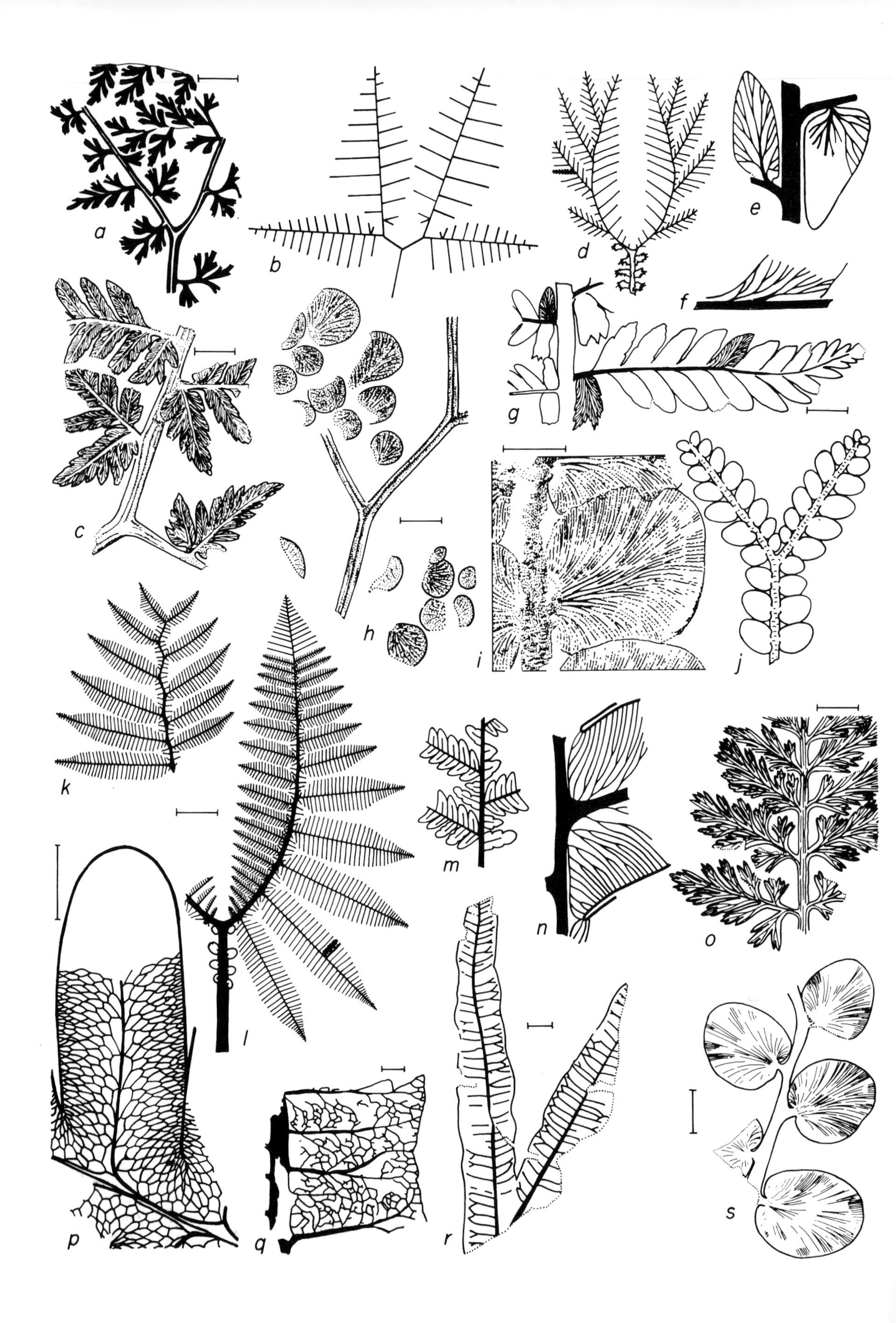
a
b
d
e
f
c
g
h
i
j
k
l
m
n
o
p
q
r
s

tion to Callistophytales is more likely. It has been suggested that according to the type of branching of the rachis and frond outline some of the species should be placed in the genus *Karinopteris* (Boersma, 1972; DiMichele *et al.*, 1984).

Genus *Odontopteris* (C_2–P; Fig. 63, *d–g*)

Fronds, diverse in structure, bearing entire pinnules with 'odontopteroid' venation, i.e. from the rachis several veins, variously diverging, enter the pinnule. The midrib is absent. The veins are strongly decurrent and dichotomize near the rachis. Because of this it is difficult to apprise how many veins entered the pinnule. At times the dichotomous branching beginning from one strongly decurrent vein was interpreted as odontopteroid venation (*O. brardii*; Fig. 63, *f*, *g*; *O. minor*; Fig. 63, *d*, *e*; etc.). In other cases (*O. subcrenulata* and related species) it is evident that several veins entered the pinnule. The genus is in need of revision.

Genus *Alethopteris* (C–P_1; Fig. 58, *w–y*)

Fronds are compound-pinnate. Pinnules, varying from subtriangular to nearly linear, are connected by a wing of the rachis. The venation is pinnate; the midrib extends to the pinnule apex. Lateral veins are simple or forked. Short veins enter the wing of the rachis. Most of *A.* species probably belong to the Trigonocarpales.

Genus *Lonchopteris* (C_2; Fig. 63, *p*)

Fronds are similar to *Alethopteris*, but have reticulate venation. A series of forms between these two genera can be constructed showing the progressive increase in the number of anastomoses, among which the lateral veins gradually become imperceptible. Species retaining a proportion of free lateral veins are placed in the genus *Lonchopteridium*.

Genus *Callipteridium* (C_2–P_1; Fig. 63, *k–n*)

The fronds are similar to *Alethopteris*, but in contrast, have subsidiary pinnae and pinnules. (The subsidiary pinnules are attached between pinnae directly to the rachis of a higher order. A similar shift in the branching order occurs also in the pinnae. Pinnae situated directly on the pinna rachis of two orders higher than they themselves are termed subsidiary.) The pinnules are tongue-shaped or subtriangular. The venation is pinnate. The midrib is strong, extending to the pinnule apex. Probably the typical representatives of *C.* belong to the Trigonocarpales.

Genus *Neuropteris* (C_1–P; Fig. 59, *a*, *b*, *d–f*)

Fronds are compound-pinnate. The largest fronds attain 2–3 m in length, but smaller forms also occur. There are subsidiary pinnules and pinnae. The pinnules are fused at the pinna apex that terminates in an unpaired pinnule. The pinnules are basally constricted, at times auriculate, sessile

Fig. 63. 'Pteridosperms' (satellite genera of Ginkgoopsida and Cycadopsida); Lower (*h–j*) and Middle–Upper (*a–g*, *k–p*, *s*) Carboniferous, Upper Triassic (*q*, *r*); Western Europe (*a–n*, *p*, *s*), Kuzbass (*o*, *s*), Greenland (*q*, *r*); *a*, *Sphenopteris elegans* Brongn.; *b*, scheme of *Mariopteris* frond; *c*, *M. muricata* (Schloth.) Zeill.; *d*, *e*, *Odontopteris minor* Brongn., scheme of frond (*d*), basal pinnules of pinnae (*e*); *f*, *g*, *O. brardii* Brongn., venation of pinnule base in middle part of pinna (*f*), pinna (*g*); *h*, *Cardiopteridium spetsbergense* Nath.; *i*, *Fryopsis* (al. *Cardiopteris*) *polymorpha* (Goepp.) Wolfe; *j*, *Fryopsis*, scheme of frond; *k*, *l*, fronds of *Callipterdium pteridium* (Schloth.) Zeill., schemes of modifications B (*k*) and A (*l*); *m*, subsidiary (intermediate) pinnules in *Callipteridium*; *n*, *C. gigas* (Gutb.) Weiss, venation of pinnule base; *o*, *Angaridium potaninii* (Schm.) Zal.; *p*, *Lonchopteris rugosa* Brongn., pinnule venation; *q*, *r*, *Furcula granulifera* Harris, venation (*q*) and general view of leaf (*r*); *s*, *Angaropteridium cardiopteroides* (Schm.) Zal. Scale bar = 30 cm (*k*, *l*), 1 cm (*a*, *c*, *g*, *h*, *o*, *r*, *s*), 5 mm (*i*, *p*), 1 mm (*q*). Modified from: Brongniart, 1828 (*a*); Boersma, 1972 (*b*); Huth, 1912 (*c*); Laveine, 1967 (*d*); H. Potonié, 1903 (*g*); Gothan and Weyland, 1964 (*h*); W. Remy and R. Remy, 1977 (*i*); Laveine *et al.*, 1977 (*j*, *m*); Wendel, 1980 (*k*, *l*); Wagner, 1965 (*n*); Zalessky and Čirkova, 1935 (*o*); Bocheński, 1960 (*p*); Harris, 1932 (*q*, *r*); Neuburg, 1948 (*s*).

or with a small petiole. The midrib or a bundle of parallel dichotomizing veins extend along the pinnule axis. The lateral veins are bent backwards and repeatedly dichotomize. Some species had round aphlebiae situated at the base of the frond. The aphlebiae occurring detached are assigned to the genus *Cyclopteris*. Numerous *N.* species differ in the form, sizes and venation of the pinnules, as well as in their epidermal characters. Transitional species between *N.* and other genera are sometimes keyed out into independent genera (*Neuralethopteris*, *Neurodontopteris*, etc.). Transitions are observed from *N.* to large fronds with small, broad, tongue-shaped pinnules (genus *Margaritopteris*). Species without a midrib are sometimes placed in the genus *Cardioneura*. Most *N.* species belong to the Trigonocarpales. Fronds of the genus *Reticulopteris* (C_2–P_1) exhibit the same organization as those of *N.*, but the venation is reticulate. Both genera are linked by transitional forms that have a small number of anastomoses between the veins (Josten, 1962).

Genus *Paripteris* (C_{1-2}; Fig. 59, *c*)

Fronds are compound-pinnate, but the pinnate dissection is combined with the sympodial. The pinnae of the last order are crowned by a pair of pinnules. The pinnule structure is the same as in *Neuropteris*. Distinctions between both genera are possible merely on the basis of the structure of the pinnae apices. Because of this the independent status of *P.* is not unanimously admitted (for instance, all Angara paripinnate fronds are included in *Neuropteris*). Some P. species associate with synangia of *Potoniea* (Trigonocarpales).

Genus *Linopteris* (C_2; Fig. 59, *h*)

Fronds are similar to those in *Paripteris*, but of reticulate venation. If the paripinnate construction is rejected as a generic character, then species of the genus *Reticulopteris* are assigned to *L. L.* pinnae have been found in attachment to axes of *Sutcliffia* (Trigonocarpales).

Genus *Fryopsis* (= *Cardiopteris*; C_1; Fig. 63, *i*, *j*)

Fronds are simple-pinnate with a forked rachis. Pinnules are rounded or ovate, sessile. A bundle of closely spaced veins enters the pinnule base; the veins are dichotomizing and bending backwards (cardiopteroid venation).

Genus *Cardiopteridium* (C_1; Fig. 63, *h*)

Fronds are at least twice-pinnate, often with a forked rachis. Pinnules are rounded or asymmetrical, sometimes slightly dissected, sessile or with a short petiole. Venation is cardiopteroid, but a single (?) vein enters the pinnule.

Genus *Angaropteridium* (C–P; Fig. 63, *s*)

Fronds are simple pinnate or with a forked rachis (species with a forked rachis have been placed in the genus *Abacanidium*). The pinnules are rounded or more elongated, often markedly auriculate, at times asymmetrical, attached to rachis by a thick petiole. Venation is cardiopteroid, but one wide vein enters the pinnule. In those species where the epidermal structure has been studied, the stomata occur in compact bands on one surface of the leaf and are sometimes sunken into furrows interposed between the veins. Detached dissected aphlebioid pinnules are assigned to *Tchirkoviella*. The relationship between *A.* and the genus *Bergiopteris* is unclear.

Genus *Angaridium* (C_2–P_1; Fig. 63, *o*)

Fronds are simple or twice-pinnate. Pinnules are wedge-shaped with a rounded distal margin and fan-shaped venation (veins end in small marginal teeth) to strongly dissected into linear lobes. Terminal fusion of twice-pinnate forms gives rise to forms referred to *Paragondwanidium* (Callistophytales; Fig. 47, *o*, *p*).

Genus *Furcula* (T; Fig. 63, *g*, *r*)

Leaves are simple with ribbon-shaped forked lamina. Lateral veins, departing at open angles from the midrib, branch out in a crowded reticulum. *F.* probably belongs to the order Peltaspermales, in which similar coherent leaves are known that have fused pinnules and pinnae of various orders (genera *Scytophyllum* and *Vittaephyllum*; Fig. 49 (11); Fig. 52, *j*). In some of these leaves reticulate venation was perceived.

Class Pinopsida (Coniferopsida)

In the class Pinopsida (P.) three orders are here recognized, namely, the Cordaitanthales, Dicranophyllales and Pinales. All three have in common platyspermic seeds. In P., in contrast to Ginkgoopsida, the platyspermy is secondary (see the description of the division Pinophyta p. 131). Some Cordaitanthales retain the nucellar vascularization, that disappears in all other P. The vascular system in the seeds of most conifers is totally reduced and retained only in a few genera where the seeds are exposed, not being concealed among seed scales (Taxaceae, *Cephalotaxus*, some Podocarpaceae). P. are characterized by pycnoxylic stems and leaves simple in structure: needle-shaped, scaly, or larger with fan-shaped or parallel venation, more seldom phyllodial. Polysperms are usually compound, less frequently the seeds are not in polysperms, but isolated. Male fructifications are so diversified that no uniform characteristics can be offered for them. Judging by the seed structure and stem anatomy, it is conceivable that P. evolved from lagenostomans of the BS group (see p. 135). The modification of the bilaterally symmetrical cupule led to the formation of the outer integument, just as the modification of the radial cupule in lagenostomans of the RS group resulted in the formation of the outer integument in trigonocarpaleans.

The origin of other organs, typical of P., is as yet speculative. The most primitive representatives of P. – *Cordaitanthus* and *Dicranophyllum* of the Lower Carboniferous (Serpukhovian) – are inadequately studied. Indirect inferences concerning the structure of their fructifications can be drawn from the data on Middle Carboniferous and even Upper Carboniferous species. Their seeds are aggregated in strobiloid polysperms, the origin of which is unknown. The branching polysperms in lagenostomans can be compared to the branching seed stalks in the Middle Carboniferous species *Cordaitanthus pseudofluitans* (Fig. 64, *a*), whereas the scales of the axillary complex of the same species can be compared to cataphylls known at the base of the fronds in lagenostomans (*Lyginopteris*). Microsporoclads of some cordaitanthaleans (*Kuznetskia*) are branched and comparable to those of lagenostomans, but, by contrast, are strongly reduced, branching in one plane and with free sporangia. The strobiloid organization of P. male fructifications appears to have evolved by means of a gamoheterotopic process, i.e. through the transfer of characters of one sex to another; in this case through the transfer of the strobiloid architectonics from the female to the male sex (Meyen, 1984b; see also Chapter 8). P. leaves differ considerably from fern-like fronds of lagenostomans, but it is conceivable that the leaves of the cordaitanthaleans evolved via phyllodization, i.e. complete reduction of the pinnae and pinnules of fronds and transformation of the petiole into a normal foliage leaf. Scaly leaves of P. could have been derived from cataphylls of lagenostomans. The phylogenetic relations between the orders are discussed below.

Order Cordaitanthales

The name of this order, current in the literature (Cordaitales), was derived from the generic name *Cordaites*. This genus was established for leaves accompanying essentially different fructifications of at least two families (Cordaitanthaceae, Vojnovskyaceae). The Gondwana species of *Cordaites* (commonly not credibly assigned to *Noeggerathiopsis*) probably belonged to Arberiales. Because of this the order was given the name Cordaitanthales (Meyen, 1978a).

In Cordaitanthales the seeds are in simple polysperms, consisting of spirally arranged sterile scales and monosperms, the mutual arrangement of both along the polysperm axis varying from one family to another. The simple polysperms are subtended by bracts and grouped into compound polysperms, attached to leafy shoots. The com-

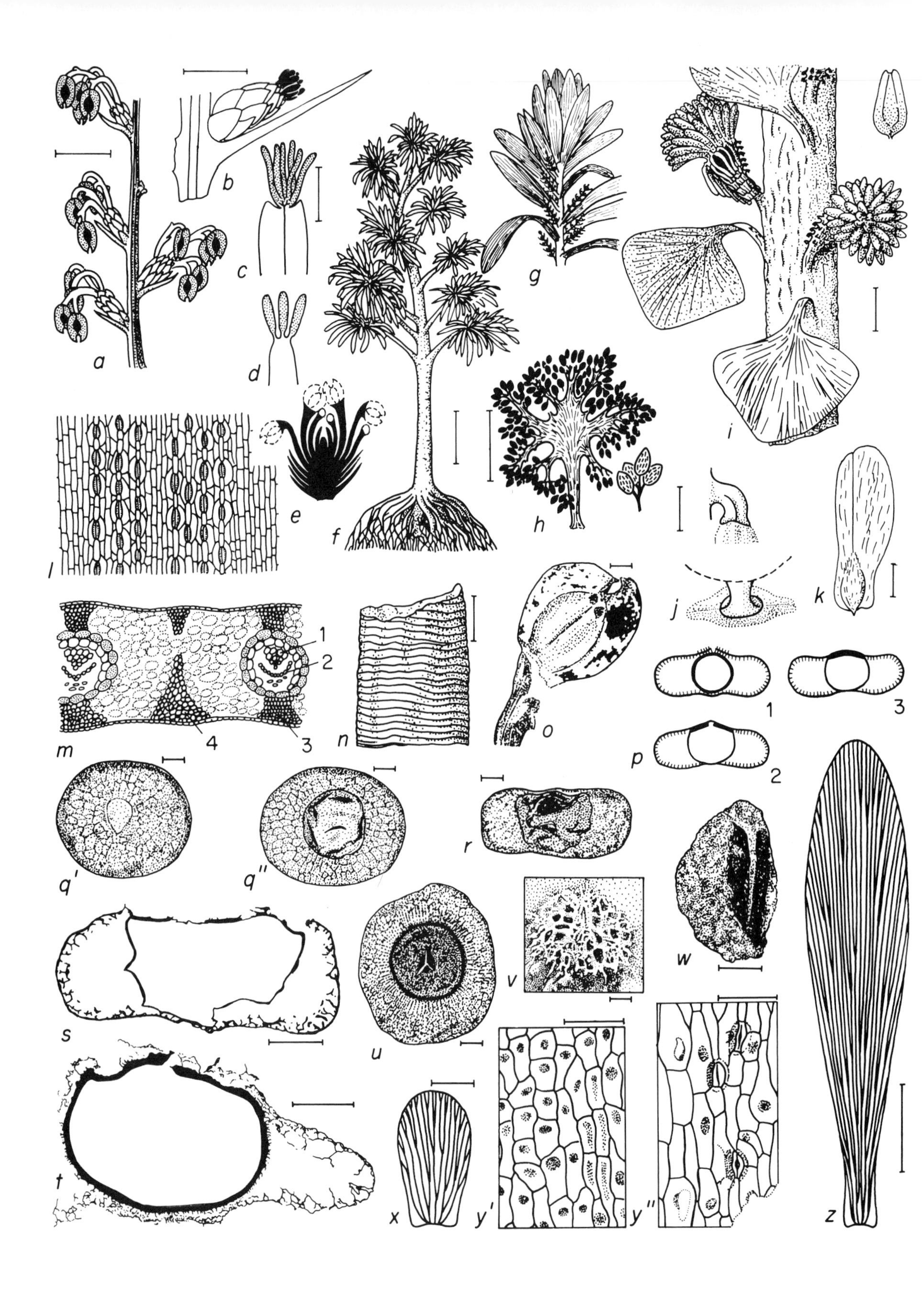

a
b
c
d
e
f
g
h
i
j
k
l
m
1
2
3
4
n
o
p
1
2
3
q'
q''
r
s
t
u
v
w
x
y'
y''
z

pound polysperms were arranged in the same orthostichies as the subtending leaves, and attached at some distance above the leaf axil. In some genera, only isolated simple polysperms were encountered, and whether they were united into compound polysperms is unknown. The leaves are simple with fan-shaped or nearly parallel venation. In the family Rufloriaceae the stomata occur in dorsal (i.e. on the lower leaf side) furrows. A typical feature of cordaitanthalean leaves is 'false veins' (Fig. 64, *m*) – hypodermal sclerechymatous strands, occurring in various numbers between and along the veins. If, as stated above, we admit that cordaitanthalean leaves evolved from lagenostoman petioles, then the 'false vein' system can be feasibly compared to the *Sparganum*-type cortex, typical of such petioles. In the cordaitanthalean leaves the division of the xylem in the vascular bundles into centrifugal and centripetal (Fig. 64, *m* (1 and 2)) can be compared to the similar division of the xylem in the leaf trace of lagenostomans (Fig. 57, *m* (3 and 5)).

According to the structure of the fructifications and, in part, of the leaves, the cordaitanthaleans can be divided into families Cordaitanthaceae, Vojnovskyaceae and Rufloriaceae, of which only the first has been adequately studied.

Family Cordaitanthaceae (C.)

The external habit of C. is commonly reconstructed by analogy with that of living arborescent conifers. Recently Brauer (1984) found a branched stem of the *Dadoxylon* (*Araucarioxylon*). The lower 9.5 m of the stem are unbranched. The specimen allows the restoration of a tree approximately 12 m high with a crown some 6 m across. However, among C. were trees of lower heights with aerial roots; these trees were probably mangrove dwellers (Fig. 64, *f*). Evidence confirming the suggestion that some of the roots were aerial is provided by the traces of borings caused by insects (Cichan and Taylor, 1982b). Perhaps some C. were small trees and shrubs (Constanza, 1983, cited in Brauer, 1984; Rothwell and Warner, 1984).

The characteristics of C. given below are based on the fructifications. Although they associate with leaves of *Cordaites* (Fig. 64, *g*, *l*, *m*), stems of *Mesoxylon* and *Araucarioxylon*, roots of *Amyelon* and *Premnoxylon*, pith casts of *Artisia* (Fig. 64, *n*) and pollen of *Florinites*, these genera should preferably be treated as satellites of the order as a whole, and partly even of gymnosperms, since some of these genera associated with fructifications different from those in C. This applies particularly to

Fig. 64. Cordaitanthaceae (*a–g*, *l–v*) and Vojnovskyaceae (*h–k*, *w–z*); Middle–Upper Carboniferous (*a–g*, *l–v*), Permian (*h–k*, *w–z*); Western Europe (*a*, *c*, *g*, *n*, *o*), USA (*b*, *d–f*, *l*, *m*, *p–v*), Kuznetsk (*h*, *k*, *w*), Pechora (*i*, *j*) and Tunguska (*x–z*) basins; *a*, compound polysperm of *Cordaitanthus pseudofluitans* (Kidst.) Crook. type; *b*, *C. concinnus* Delev. sp., male axillary complex; *c*, *d*, microsporphylls of *C. penjonii* Ren. sp. (*c*) and *Gothania* (*d*); *e*, *C. duquesnensis* Rothw. sp., axillary polysperm; *f*, *g*, reconstruction of tree (*f*) and fertile branch (*g*) of *Cordaites*; *h*, microsporoclad *Kuznetskia planiuscula* S. Meyen; *i*, *Vojnovskya paradoxa* Neub., compound polysperm and isolated seed; *j*, *V. paradoxa*, seed stalk in lateral and axial view, seed stippled; *k*, *Sylvella alata* Zal.; *l*, *m*, *Cordaites crassus* Ren., lower epidermis (*l*) and cross-section of leaf (*m*), centrifugal (1) and centripetal (2) xylem, hypodermal tissue accompanying vein (3) and corresponding to false vein (4); *n*, pith cast *Artisia*; *o*, *Cordaitanthus pseudofluitans*, seed with seed stalk; *p*, pollen transformation from forms with proximal slit and outer sculpture around it (1; *Felixipollenites*) to forms with proximal slit and only inner structure (2; *Sullisaccites*) and further to forms without proximal slit (3; *Florinites*); *q*, *r*, *Cordaitanthus concinnus*, pollen, distal side with leptoma (*q'*), proximal side with slit (*q''*), lateral view (*r*); *s*, *Sullisaccites kentuckiensis* Mill. et Tayl., longitudinal section of pollen grain; *t–v*, *Felixipollenites macroreticulatus* Mill. et Tayl., pollen in section (*t*), proximal view with slit (*u*), sculpture around proximal slit (*v*); *w*, pollen of *Kuznetskia tomiensis* Gorel. et S. Meyen; *x*, *Crassinervia tunguscana* Schved.; *y*, upper (*y'*) and lower (*y''*) epidermis of *Cordaites* (al. *Sparsistomites*) *gorelovae* S. Meyen; *z*, scheme of venation of *Cordaites* (al. *Sparsistomites*) typ. *gracilentus* (Gorel.) S. Meyen – *gorelovae*, veins are shown rarefied. Scale bar = 50 cm (*f*), 1 cm (*a*, *h*, *i*, *n*, *z*), 5 mm (*b*, *k*, *x*), 1 mm (*c*, *j*, *o*), 100 μm (*y*), 20 μm (*u*, *v*), 10 μm (*q–t*, *w*). Modified from: Meyen, 1984a (*a*, *d*, *e*, *h*, *x*, *z*), 1982b (*w*), 1966 (*y*); Andrews, 1961, after Delevoryas (*b*); Florin, 1944a (*c*); Cridland, 1964 (*f*); Grand'Eury, 1877 (*g*); Harms and Leisman, 1961 (*l*, *m*); Gothan and Weyland, 1964 (*n*); Florin, 1944b (*o*); Millay and Taylor, 1976 (*p*) and 1974 (*q–v*).

the genera established on stem and root fragments with preserved anatomical structures. C. are known to have had stems with secondary wood of the *Araucarioxylon* type, and a pith (genus *Artisia*) with transversal cavities. Similar stems with a septate pith were seemingly characteristic of various gymnosperms, including arberians and conifers. The genus *Mesoxylon* is linked with microsporoclads of *Gothania*, seeds *Mitrospermum*, leaves of *Cordaites felicis* and pollen of *Felixipollenites* or *Sullisaccites* (Trivett and Rothwell, 1985), and has not as yet been found in association with other parts. The genus *Pennsylvanioxylon*, established for permineralized woods, also belong to C.

Genus *Cordaitanthus* (C_2–P_1; Fig. 64, *a–e, g*) includes both female and male fructifications, although they probably should be assigned to different form-genera. C. fructifications are spicate with a bilaterally symmetrical axis. Its stele was elliptical in cross-section and composed of numerous contiguous mesarch bundles, at times with a small secondary wood tissue. From the narrow sides of the stele alternating vascular bundles emerge. The bundles on each side of the axis formed two paired rows. Thereby, although the bracts were four-ranked, their arrangement in impression–compression fossils appears two-ranked (Fig. 64, *a*). Lateral polysperms or microsporoclads were inserted in the axils of the bracts, and both had a secondary axis, where the vascular bundle is derived from fused branches of two vascular bundles of the main axis. Above the point of fusion a pith appears in the bundle. Both the polysperms and microsporoclads bore spirally arranged sterile scales at the base. Some of the scales in the polysperms were replaced by branching seed stalks each with an apical seed or by a solitary seed borne apically on a simple shorter stalk. In some cases all the sterile scales were situated in the proximal region of the lateral polysperm, and all the seed stalks were situated in its distal part (Fig. 64, *e*). Dispersed seeds are referred to the genera *Nucellangium* (Fig. 44, *f*; Fig. 45, *e*), *Mitrospermum* (Fig. 45, *f*), *Samaropsis* and, probably, *Rhabdocarpus*, *Cordaicarpus*, etc. They have characteristically a free integument with two vascular bundles. In *Mitrospermum compressum* the integumental bundles divide in the main plane of the seed. In some species vascular bundles are present in the nucellus and they are then arranged in a ring (which implies a primary radially symmetrical seed structure).

The microsporoclads produced bunches of microsporangia attached to the margin of the scales (Fig. 64, *b–d*), clustered at the end of the axis, or (more rarely) situated among sterile scales. The vascular bundle penetrates the scales and in the fertile ones reaches the apex, becoming dichotomously divided here. The terminal branches of the bundle enter the base of each sporangium. In some species the bundle branches in various planes, so that the sporangia are not arranged in one plane. The pollen is monosaccate (Fig. 64, *p–v*) of the *Florinites*, *Sullisaccites* and *Felixipollenites* types. The true saccus has a large cavity and inner reticulum (remnant of the alveolar sexine). More primitive forms have a well-developed proximal trilete slit, which in more advanced forms becomes reduced, with parallel formation of the distal sulcus. Obviously, the germination changed from proximal (prepollen stage) to bipolar and then to distal (pollen stage). The generic name *Gothania* is used for microsporoclads of nearly the same habit, but the sporangia were arranged in one row along the scale distal margin; and the pollen was of the *Felixipollenites* type (Fig. 64, *p* (1), *t–v*); such pollen has been found in *Mitrospermum* seeds.

Considering that vojnovskyalean microsporoclads were not strobiloid, there is reason to believe that microsporoclads of the same habit were produced by primitive Cordaitanthaceae or by common ancestors of this family and vojnovskyans. It is thereby conceivable that the strobiloid habit, typical of female fructifications of all cordaitanthaleans, was of secondary appearance in male fructifications of the Cordaitanthaceae. It is possible that this character was transferred from fructifications of one sex to another (gamoheterotopy).

Family Vojnovskyaceae

Vojnovskyans (V.) were initially introduced as a separate order, since it was considered that they had bisexual strobili. More detailed investigations have shown that this was a misconception and that B. can be compared to Cordaitanthaceae according to their general polysperm organization, although V. microsporoclads are altogether different. Polysperms are known only in *Vojnovskya* (C_3–P; Fig. 64, *i, j*). On the thick axis in a loose spiral are scaly leaves of the *Nephropsis* type. Somewhat above the point of attachment of such a leaf to the axis an obconical polysperm is attached. The basal part of its axis is naked, and probably bore a vestigial sterile scale, whereas the upper expanded part is covered by backward bent seed stalks with widened apices. Evidently the seeds were attached to the stalk in such a manner that the main plane of symmetry was of tangential orientation to the polysperm axis. Short appendages, recognized at present as seed stalks were earlier mistaken for sterile scales. The expanded polysperm apex was covered by linear interseminal scales, with several interposed seeds. The seeds usually abscissed before burial of the polysperm. Leaves of the *Nephropsis* type can be correlated with bracts, whereas polysperms can be correlated with lateral polysperms (axillary complexes) of *Cordaitanthus*. *Vojnovskya* differs from *Cordaitanthus* in: (1) the spiral arrangement of the bracts and lateral polysperms; (2) the attachment of the polysperms not directly in the axil of the bract, but higher along the axis; (3) leaf-like habit of the bracts; (4) reverse arrangement of the seed stalks and sterile scales on the axis. The seeds found in association with *Vojnovskya* are turret shaped, of the *Samaropsis* type, with rather thick wings and a notched apex. Affiliated to V. are also certain seeds of the *Samaropsis* type with a thin samara, seeds of the genus *Sylvella* (P; Fig. 64, *k*) with an elongated asymmetrical samara and a very long micropyle, and also some *Tungussocarpus* species with a fleshy integument.

In the Cordaitanthaceae the male fructifications are of the same construction as the compound polysperms, whereas in V. the male and female fructifications differ in their structure. The microsporoclads (genus *Kuznetskia*; P; Fig. 64, *h, w*) look like flattened profusely branching axes systems, occasionally fused in the central region. The terminal branchlets bear solitary sporangia with quasimonosaccate pollen (of the *Cordaitina* type). Longitudinal rod-like thickenings are scattered over the sporangial wall.

V. fructifications are linked with leaves of the *Cordaites* type (Fig. 64, *y, z*), at times indistinguishable in their gross morphological characters from leaves associated with *Cordaitanthus*, but different in the epidermal characters. For instance, in V. such regular stomata files as in most Cordaitanthaceae leaves are not known. The anatomical structure of V. stems is unknown. In burials V. is commonly associated with ruflorias. Because of this it is impossible to determine to which family should be related the associated woods of the genera *Araucarioxylon*, *Mesopitys*, *Eristophyton*, *Metacaenoxylon*, *Septomedullopitys*, *Amyelon*, *Dadoxylon* (Lepekhina, 1972). Some species of *Araucarioxylon*, *Eristophyton* and *Mesopitys*, also the genera *Septomedullopitys* and *Metacaenoxylon* have been found, so far, only in beds from which ruflorias are absent.

The phylogenetic relations between V. and Cordaitanthaceae will be discussed after the description of the family Rufloriaceae.

Family Rufloriaceae

The family name Rufloriaceae (R.) is unsuitable, because it is derived from *Rufloria* (C_2–P; Fig. 65, *a–e*) – a genus established for leaves. The evidence so far available does not indicate, however, that such leaves were present in different families. R. leaves are characterized by stomatiferous furrows, running along the veins on the lower side (dorsal furrows). According to other characters *Rufloria* leaves are similar to *Cordaites*. Both genera exhibit complete parallelism in the species according to the configuration and size of the leaf, structure of its base (wide and bordered, narrow at times attenuated) and apex (smooth or minutely denticulate).

R. polysperms are mostly racemose, but the axis is at times strongly aborted, and the seed stalk arrangement is in the form of an umbrella (*Gaussia*; C_2–P; Fig. 65, *i, j*). The degree of shortening of the axis varies in the genus *Krylovia* (C_{2-3}; Fig. 65, *f, g*). Otherwise the axis remains long, whereas the

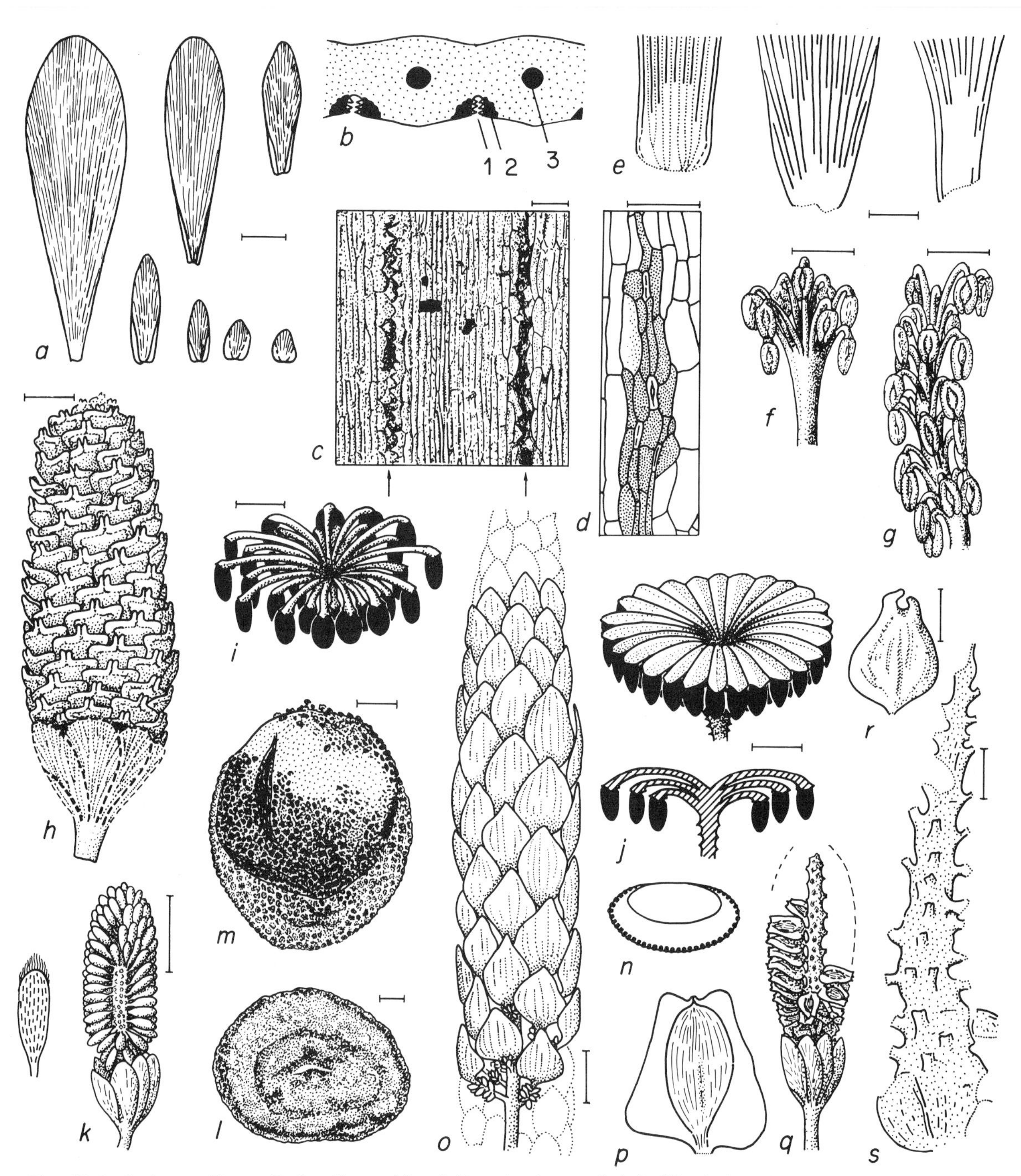

Fig. 65. Rufloriaceae; Upper Carboniferous (*f*, *g*, *i*), Permian (*a*, *c–e*, *h*, *j–s*); Siberia (*a*, *c–j*, *m–o*), Pechora basin (*k*, *l*, *p–s*); *a*, transition from normal leaves to cataphylls in *Rufloria brevifolia* (Gorel.) S. Meyen; *b*, *Rufloria* leaf, scheme of cross section (1, dorsal furrow; 2, hypodermal strands; 3, vein); *c*, dorsal furrows (at arrows) on lower leaf side of *Rufloria* (drawn after cellulose peel); *d*, stomatal band in dorsal furrow of *Rufloria*; *e*, arrangement of dorsal furrows (solid lines) and veins (dotted lines) in leaf base of various species of *Rufloria*; *f*, *g*, *Krylovia sibirica* Chachl. with dwarf (*f*) and elongated (*g*) axis; *h*, reconstruction of *Bardocarpus depressus* (Schmalh.) Zal. (basal scales are shown ▶

seed stalks are reduced, so that the polysperm becomes spicate (*Bardocarpus*; P; Fig. 65, *h*). Polysperms with a long axis and aborted seed stalks were mostly typical of Permian R. Of this type is the genus *Suchoviella* (P; Fig. 65, *p*, *q*), where the polysperm axis bears a rosette of scales (of the *Lepeophyllum* type) below the seed-bearing part. Sterile scales were probably borne in the lower part of the axis in other R. In this case the R. polysperms can be correlated with those of other Cordaitanthales, but differ from them in the leaf-like habit of the scales and their displacement down to the base of the polysperm, and the shift of the seeds to its upper part. In certain Permian R. polysperms might probably have been associated with leaf-like bracts. Scaly leaves are known of the *Nephropsis* type (subgenus *Sulcinephropsis*) with dorsal furrows, the same as in *Rufloria*. R. seeds belong to the genera *Bardocarpus*, *Samaropsis* and *Tungussocarpus*.

Microsporoclads are known only in the Upper Permian R. and are assigned to the genera *Pechorostrobus* (Fig. 65, *k*, *l*) and *Cladostrobus* (Fig. 65, *m–o*). Initially the genus *Pechorostrobus* was wrongly included in the family Vojnovskyaceae (for details see Meyen 1984a). In this genus the microsporangia are spirally arranged along the axis and the lower part of the axis bears a circandr of scaly leaves (of the *Lepeophyllum* type). Occasionally thin distal hairs occur on the sporangia. The pollen is quasimonosaccate (of the *Cordaitina* type) with a proximal slit. In *Cladostrobus* the axis of the microstrobilus bears, instead of sporangia, microsporophylls that consist of a thin stalk and distal rhomboid lamina. The sporangia are clustered in the middle part of the stalks and, seemingly, arranged around it. The pollen (of the *Cladaitina* type) had a saccus embracing the body from the equatorial and distal sides (Fig. 42 (8); Fig. 65, *m*, *n*; initially it was erroneously thought that the saccus embraced the body proximally – Maheshwari and Meyen, 1975). The degree of the saccus offlap from the body varies considerably. In some grains the saccus is unrecognizable. The saccus bears inner polygonal projections with narrow slits in between, so that a negative reticulum (intrareticuloid) appears. The proximal aperture is absent. Pollen of older R. is closer to vojnovskyalean pollen. *Bardocarpus* seeds are accompanied by pollen of the *Potoniesporites* type with an equatorial quasisaccus. This pollen resembles externally *Cordaitina* and *Florinites*, and even more the pollen of the oldest conifers (Lebachiaceae). Of interest is the fact that *Cladostrobus* is also somewhat similar in its general organization to conifers of the families Lebachiaceae and Voltziaceae.

The phylogenetic relations between R. and the families Vojnovskyaceae and Cordaitanthaceae are unknown. The Cordaitanthaceae appeared earlier than the other two families, but the well-studied representatives of Cordaitanthaceae come from the Middle Carboniferous or younger deposits, and obviously could not have been ancestors of R. and vojnovskyans. Primitive characters in the order Cordaitanthales were probably the branching seed stalks and branching (not strobiloid) microsporoclads. No genera are known where the combination of both these characters occurs. In the Cordaitanthaceae only strobiloid male fructifications have been recorded. It is inconceivable how such a structure could have evolved from branching axes. At the same time, the male and female fructifications in cordaitanthaleans reveal a striking similarity in their general organization. It is probable that the ontogenetic programme of strobiloid male fructifications was borrowed from compound polysperms. (The transfer of characters from one sex to another is not unusual in organisms.) A similar explanation is applicable to the

---

hypothetically); *i*, *Gaussia cristata* Neub.; *j*, *G. scutellata* Neub.; *k*, *Pechorostrobus bogovii* S. Meyen, sporangium with hairs and thickenings in wall (to left), and general view of microstrobilus (to right); *l*, *P. bogovii*, pollen with proximal slit; *m–o*, *Cladostrobus lutuginii* Zal., pollen (*m*) and scheme of its meridional section (*n*), microstrobilus (*o*); *p*, *q*, *Suchoviella synensis* Ignat. et S. Meyen, seed (*p*), polysperm (*q*); *r*, *s*, *Suchoviella* sp., seed (*t*) and polysperm axis (*u*). Scale bar = 1 cm (*a*, *h*), 5 mm (*e–g*, *i–k*, *o*, *r*, *s*), 100 μm (*c*, *d*), 10 μm (*l*, *m*). Modified from: Meyen, 1966 (*a*, *d*), 1964 (*c*, *e*), 1982b (*f*, *g*, *i–l*), 1984a (*p–s*); Maheshwari and Meyen, 1975 (*m*, *o*).

remarkable similarity (unusual for gymnosperms) between *Pechorostrobus* and *Suchoviella* (Fig. 65, *k*, *g*) – microstrobili and polysperms linked with leaves of *Rufloria synensis*. Such a general succession in the evolution of Cordaitanthalen fructifications seems quite possible. The most primitive and as yet unknown forms could have been compound polysperms with spirally arranged lateral polysperms, where sterile scales were intermingled with branching seed stalks. Microsporoclads of these plants were nearly the same as in lagenostomans. Later microsporoclad modifications in the Cordaitanthaceae resulted in the same general construction as in compound polysperms, whereas in vojnovskyans the microsporoclads retained their primitive organization, but became strongly reduced. In the lateral polysperms of the vojnovskyans and Cordaitanthaceae the seed stalks become unbranched, a character which we find in the ancient ruflorias. What type of microsporoclads was characteristic of the ancient ruflorias is not precisely known, but indirect evidence suggests that they were of the strobiloid type. The strobiloid organization in them probably evolved via the same genetical mechanism as in Cordaitanthaceae. The suggested scheme fits well the available data on the structure of fructifications in Cordaitanthales and their distribution in the stratigraphic section. However, as will be shown below, the suggested sequence of forms should be in agreement with what is known of the history of the order Dicranophyllales.

Satellite genera of the order Cordaitanthales

Genus *Cordaites* (C–P; Fig. 64, *g*, *l*, *m*, *y*, *z*)

Although the genus was introduced for isolated leaves, restored plants with fructifications of *Cordaitanthus* were also assigned to it. Since leaves of the *C.* type associated with other fructifications it seems reasonable to treat this genus in its original content and circumscription. *C.* leaves are strap-shaped, oval, lanceolate with an expanded (at times bordered) or attenuated base; two or more veins entering the base dichotomize following the expansion of the lamina, and continue nearly parallel into the strap-shaped laminae. Frequently between and along the veins run strands of a hypodermal mechanical tissue, the imprints of which form false veins varying in number. The leaf epidermal structure is of several types: with regular stomatal files in narrow or wide bands, or stomata scattered over the lower surface; variations are apparent in the stomatal structure. These differences in the epidermal types are sufficient for the distinction of separate genera. Considering that in the type material of the genus the epidermis is unknown, it has been suggested that the name *C.* should be retained for leaves with unknown characters of the epidermis. Leaves in which the epidermal characters have been studied can be placed in the genera *Sparsistomites*, *Angophyllites*, *Europhyllites*, etc. Developed leaves of C. may be transitional to scaly leaves assigned to the genera *Crassinervia* (Fig. 64, *x*), *Lepeophyllum* and *Nephropsis*. Branches with scars of abscissed leaves are referred to the genus *Cordaicladus*.

Genus *Artisia* (C–P; Fig. 64, *n*)

Pith casts with traces of transverse diaphragms corresponding to parenchymatous plates. The habit of the casts varies depending on the thickness and configuration of the diaphragms and the interposed areas. A septate pith is characteristic of the genera *Cordaioxylon*, *Mesoxylon*, *Solenoxylon*, *Septomedullopitys*, also of stems of some other gymnosperms, progymnosperms and certain angiosperms.

Order Dicranophyllales

In most of the gymnosperm classification systems dicranophylls (D.) are either omitted or treated as a separate order. Němejc (1968) recognized the class Dicranophyllopsida with the orders Dicranophyllales and Trichopityales. Trichopityans belong to the order Peltaspermales (see above). In determining the systematic affinity of D. it is better to rely on characters of those genera that constitute a natural set of forms indisputably involving *Dicranophyllum* (C–P; Fig. 66, *a–e*, *f* (1)). This genus was established for leafy shoots and detached leaves, where the lamina may be linear, simple or dichotomizing up to four times. In some species (including the type species) along the leaf margins run stomatiferous (dorsal) furrows, that are of the same habit as in the older representatives of *Rufloria* (see above). Sometimes the stomatal band, remaining compact and distinctly divided from the rest of the epidermis, is not immersed (Fig. 66, *e*). A typical feature is the marginal microdenticulation. The genus *Mostotchkia* (C–P; Fig. 66, *f* (1), *n*, *o*) is close to *Dicranophyllum*. Its leaves are always simple and marginal teeth are of very rare occurrence. Further in the D. leaf sequence is the genus *Entsovia* (C–P). In one of its species coupled furrows run along the margins of the linear leaf (Fig. 66, *f*(2)); in all other species the paired furrows are distributed over the entire width of the leaf (Fig. 66, *f*(3), *g–j*). Numerous, at times coupled, furrows are typical of the genus *Slivkovia* (Fig. 66, *f*(4), *k–m*), but in this case the leaves are scaly and the leafy shoots resemble coniferalean. In *Slivkovia*, as in *Dicranophyllum*, the leaf margin is microdenticulate (Fig. 66, *m*). An important fact is that the structure of the furrows and stomata are the same in *Mostotchkia*, *Entsovia* and *Slivkovia*. *Mostotchkia* have characteristically axillary buds (Fig. 66, *o*), that evidently correspond to the same organs in *Dicranophyllum* (Fig. 66, *a*), which have been misinterpreted as microstrobili. Hence, the vegetative parts in the above-cited genera serve to indicate that they belong to a single group (Meyen and Smoller, in press). To this group may be related the Lower Permian *Lesleya delafondii* (the type species *L. grandis* is probably affiliated to Noeggerathiales) established for simple broad leaves with a midrib and numerous dichotomizing lateral veins. According to the structure of the stomatal bands and stomata, also according to the marginal microdenticulation, this species displays affinities to *Dicranophyllum*. *Zamiopteris* (see satellite genera of the Pinophyta) reveals the same stomatal bands and stomata. D. probably also comprise the poorly known genus *Acanthophyllites*.

It is not incidental that attention is called here to the marginal microdenticulation in D. This character has never been reported in Ginkgopsida, but is very typical of conifers and has also been noted in *Cordaites*.

Polysperms are known in *Dicranophyllum*. On the lateral forked appendages below the point of furcation, seeds are borne, but whether they were helically arranged, or in two rows, is unclear. Isolated racemose polysperms with seeds spirally attached by short seed stalks can probably also be related to D. According to the structure of these polysperms D. can be affiliated only with Pinopsida and allied to Cordaitanthales and primitive conifers (Lebachiaceae).

Order Pinales (conifers)

The systematics of fossil conifers (C.) is in a confused state. The main reason for this is that without the application of special palaeobotanical techniques (principally epidermal analysis) it is frequently impossible to systematically arrange the vegetative shoots and to gain an insight into the structure of the fructifications, particularly female ones. Proceeding on the basis of external characters of fructifications, pollen and leafy shoots, palaeobotanists have often boldly affiliated poorly studied fossil remains with living genera and families. Many genera and species have been established on poorly studied material.

Sometimes parts of other plants have been mistaken for C. remains. The Articulate *Phyllopitys* was initially allied with Taxodiaceae. The moss *Polyssaievia* was at first described as *Walchia*. Strobili and dispersed micro- and megasporophylls of lycopsids (particularly of the family Pleuromeiaceae) were misinterpreted as cones and seed scales of Araucariaceae. Leafy shoots of C. were indiscriminately mixed with those of lycopsids. Wood fragments of cordaitanthaleans, progymnosperms and lagenostomans were regarded as coniferalean. The peltaspermaceous leaves *Pachypteris* were described as a C. genus (*Retinosporites*). Peltaspermalean leaves of *Phylladoderma* were also wrongly assigned to C. These errors are not fortuitous. They reflect the profound and versatile parallelism in the morphology and anatomy of various groups of higher plants. It seemed natural to assume that lanceolate leaves of *Phylladoderma* with one

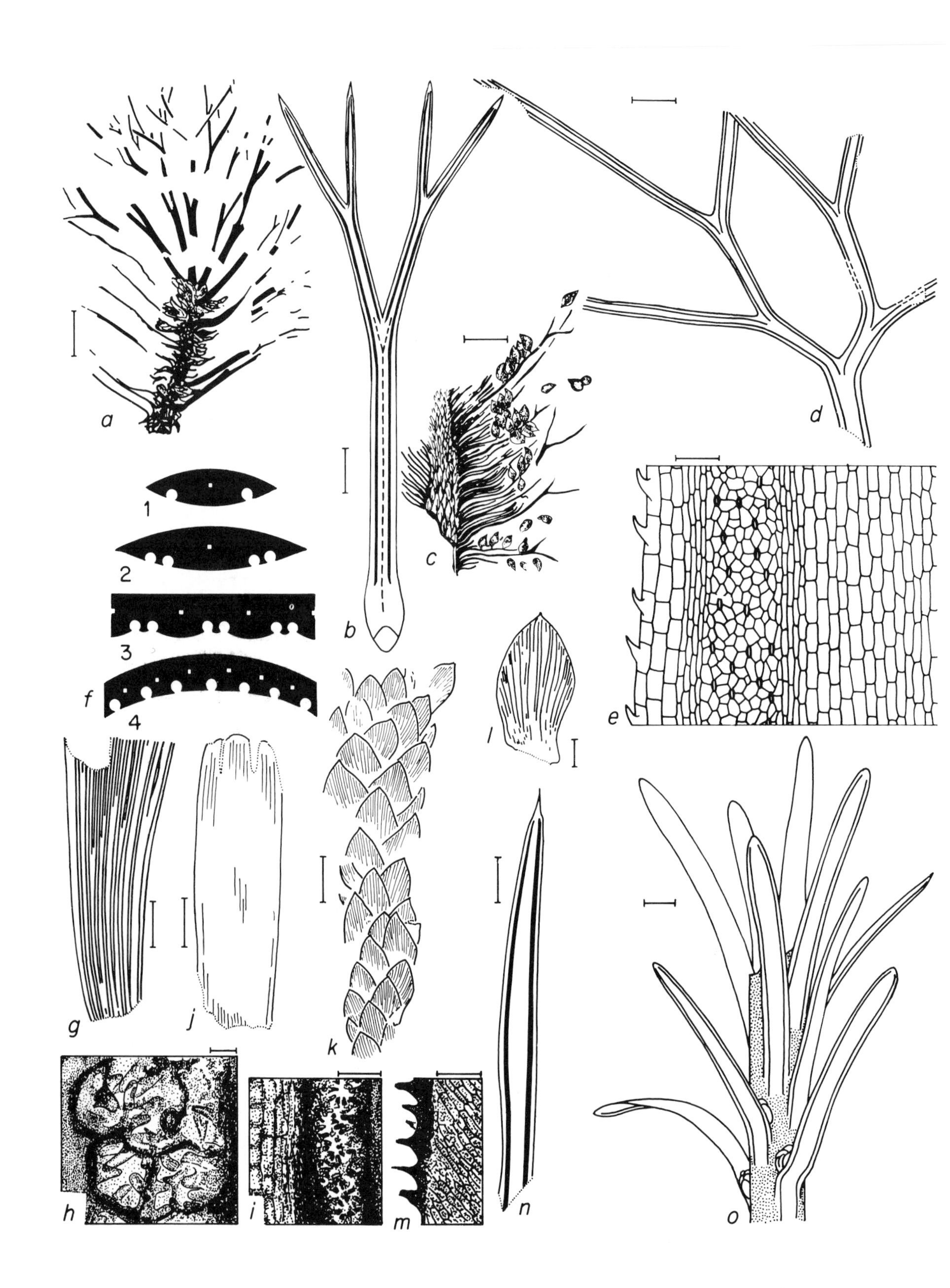

a
b
c
d
e
1
2
3
f
4
g
j
k
l
h
i
m
n
o

vascular bundle at the base and parallel veins, converging apically, belonged to C. Just as striking is the similarity between the microstrobili in Voltziaceae and the genus *Cladostrobus* (Rufloriaceae). Of wide repetition are the types of saccate and quasisaccate pollen. The similarity between different parts of C. and lycopsids furnished the basis for the hypothesis of direct phylogenetic linkage between them. Both groups produced similar types of foliage and display affinities even in the microstructural characters (stomata arranged in compact bands, scattered, or in furrows; mucronate apex, marginal microdenticulation, etc.). Sporophylls of lycopsids resemble one-seeded scales in C. Many other examples illustrating the similarities can be given.

Other problems arose, bearing on the controversy regarding the interpretation of the C. morphology, primarily of their female cones. Some morphologists regarded the seed scale as a modified seed-bearing leaf (i.e. a cladosperm), whereas others regard it as a strongly modified shoot. Following Florin's classical work the second viewpoint triumphed. Adherents of the first viewpoint still persist, appealing to observations on terats in living conifers. Ovule envelopes have also for long been a subject of animated debates, particularly the aril in Taxaceae and epimatium in Podocarpaceae. The morphological interpretation of organs is complicated by numerous parallelisms, to which is added the despecialization of the organs. For example, compound polysperms of the Upper Permian genus *Sashinia* look more primitive than in all other Palaeozoic Pinopsida, because the bracts and circasperm parts differ from foliage leaves only in size; moreover, seed stalks also resemble them. However, according to the structure of the microsporophylls and pollen the same C. cannot be regarded as so primitive. The similarity between various parts of a compound polysperm and foliage leaves, is, undoubtedly, of a secondary nature, and can be accounted for by the distribution of one ontogenetic programme on to all the above-cited organs. This fact should be taken into consideration in reconstructions of the phylogeny of C.

Now let us turn to the general characteristics of C. According to the growth forms C. are close to cordaitanthaleans and certain Ginkgoopsida. Among C. are very large trees as well as small shrubs. Conifers of the genus *Aethophyllum* (Voltziaceae) were herbaceous. C. stems are in general of the same organization as in other pycnoxylic progymnosperms and gymnosperms. They are eustelic with thick secondary wood and a thin cortex. The Triassic *Yuccites* seems to have produced unbranched succulent stems devoid of secondary wood. Primary wood consists of bundles ('sympodia') arranged in a ring, with leaf traces emerging from it tangentially and then extending outwards (Fig. 67, *a–c*). In the most advanced C. (Taxodiaceae, Cupressaceae, some Podocarpaceae) the branches of adjacent sympodia coalesce in pairs (Fig. 67, *b*) and from the point of coalescence

Fig. 66. Dicranophyllales; Upper Carboniferous (*a–d, n*), Lower (*e, j*) and Upper (*g–i, k–m, o*) Permian; Western Europe (*a–c, e*), Kuzbass (*d*), Russian platform (*g–i, o*), Fore-Urals (*j*), Pechora (*k–m*) and Tunguska (*n*) basins; *a–c*, *Dicranophyllum gallicum* Gr. 'Eury, vegetative shoot with axillary buds (*a*), scheme of leaf structure, thick lines – stomatiferous furrows, veins shown by dashed line only in lower part of leaf (*b*), female shoot, bark with leaf cushions (*c*); *d*, *D. effusum* Chachl., marginal stomatiferous furrows shown by solid lines; *e*, *Dicranophyllum* sp., lower epidermis with stomatal band; *f*, schemes of cross sections of leaves in *Dicranophyllum* and *Mostotchkia* (1), *Entsovia lorata* Fef. (2), *E. rarisulcata* S. Meyen (3), *Slivkovia* (4), white squares denote veins, round notches denote stomatiferous furrows; *g–i*, *E. rarisulcata*, arrangement of paired stomatiferous furrows on leaf, each furrow shown by one line (*g*), stomata with papillae on subsidiary cells (*h*), stomatiferous furrow (*i*); *j*, *E. kungurica* S. Meyen, leaf outline; *k–m*, *Slivkovia petschorensis* S. Meyen, leafy shoot (*k*), leaf with stomatiferous furrows (*l*), cuticular marginal teeth (*m*); *n*, leaf of *Mostotchkia longifolia* Chachl., stomatiferous furrows shown by thick lines; *o*, *M. gomankovii* S. M. et Smol., shoot with axillary buds, stomatiferous furrows shown by solid lines decurrent down stem. Scale bar = 2 cm (*j*), 1 cm (*a, c*), 5 mm (*b, d, g, k*), 2 mm (*l, n, o*), 200 μm (*e*), 100 μm (*i, m*), 20 μm (*h*). Modified from: Osnovy paleontologii, 1963b, after Renault and Zeiller (*a, c*); Barthel, 1977 (*b, e*); Meyen, 1969c (*g–j, l, m*) and 1976–1978 (*k*); Meyen and Smoller, in press (*d, f, n, o*).

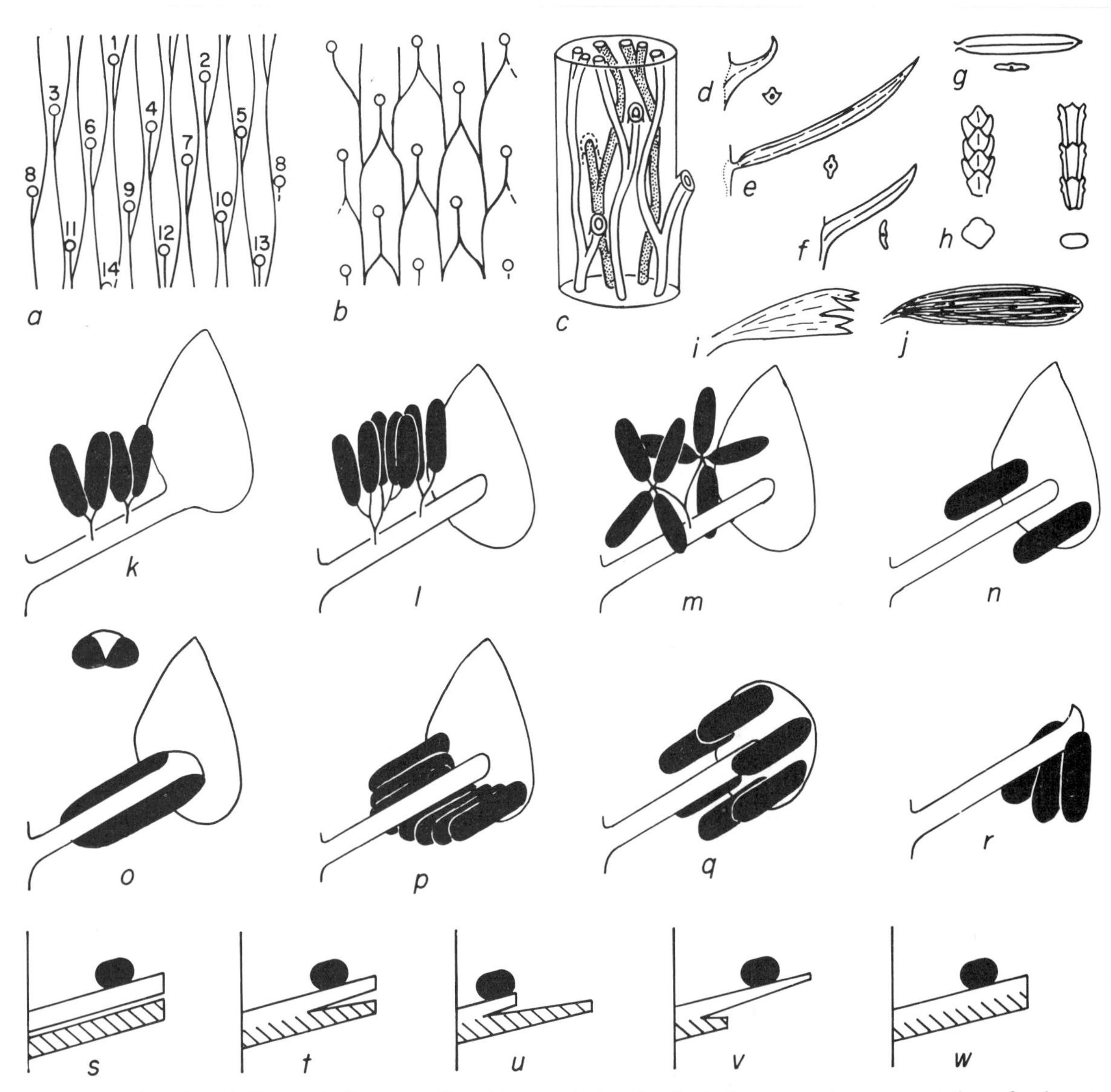

Fig. 67. Diversity of diagnostic features of conifers; *a*, *b*, longitudinal diagrammatic representation of primary vascular system as illustrated in one plane (seen from inside of stem), numerals indicate sequence of phyllotaxis (*a*, undulating free bundles of *Abies* type; *b*, anastomosing bundles of whorled phyllotaxis of cupressoid type); *c*, arrangement of undulating and tangentially divided primary bundles; *d–j*, leaves, shoots and their cross-sections (*d*, type I, *Dacrydium*; *e*, type I, *Picea*; *f*, type I, *Dacrydium*; *g*, type II, *Sequoia*; *h*, type III, *Cupressus*, *Tetraclinis*; *i*, dissected leaf of *Buriadia*; *j*, type IV, *Podozamites*); *k–r*, microsporophylls (*k*, episporangiate with distal lamina, *Dvinostrobus*, simplified; *l*, *m*, episporangiate peltate, *Darneya* (*l*), *Sertostrobus* (*m*); *n*, perisporangiate peltate, *Lebachia*; *o*, hyposporangiate peltate with sporangia fused with stalk, *Pityanthus*; *p*, perisporangiate peltate, *Willsiostrobus*; *q*, perisporangiate peltate, *Taxus*; *r*, hyposporangiate without distal lamina, *Cephalotaxus*); *s–w*, relations between bracts (shaded) and seed scales in mature polysperms (*s*, bract free, *Walchiostrobus*; *t*, bract partly fused with seed scale, *Pseudovoltzia*; *u*, seed scale reduced partly fused with bract, Araucariaceae; *v*, bract reduced and partly fused with seed scale, *Swedenborgia*; *w*, full fusion of seed scale and bract, *Pinus*). Modified from: Namboodiri and Beck, 1968 (*a–c*), Laubenfels, 1953 (*d–h*).

a leaf trace emerges. The primary vascularization system in such C. resembles externally a fern dictyostele, but is altogether different in origin (Beck *et al.*, 1982–1983). Secondary wood consists of tracheids (with circular bordered pitting on radial walls) and a ray parenchyme. In the Lebachiaceae the pitting in the tracheids is multiserial, the pits are closely spaced and of polygonal outline. Such pitting (araucarioid) is characteristic of wood in the Cordaitanthales and Araucariaceae. However, Palaeozoic C. had tracheids with less crowded circular pits, as in most living C. At the intersection of the ray cells and tracheids, i.e. on the cross-fields, the pits differ from other pits in the tracheids and may vary in type; this feature is widely used in systematics of C. Resin canals in the wood, unknown in Palaeozoic C., appear only in Mesozoic genera, but resin canals in C. leaves are known ever since the Permian. Stewart (1983) presented a review on the C. genera, described from permineralized stem remains.

C. leaves (Laubenfels, 1953) are of four major types (Fig. 67, *d–j*). The most ancient leaves that persist in some living C. are awl-shaped with a wide base, helically arranged and not adpressed to the stems. In the Permian a second type appears (Fig. 67, *h*): leaves are scale-like, adpressed to the stem, helically arranged or decussate. Since the Triassic flattened leaves are known that have a somewhat attenuated base (third type; Fig. 67, *g*), also broader leaves with multiple parallel veins (fourth type; Fig. 67, *j*). At times leaves can be found that have a forked apex (*Buriadia*). In the living genus *Phyllocladus* (Podocarpaceae), and in the Upper Cretaceous genus *Protophyllocladus* instead of leaves phyllodia appear. Deciduous leaves with an abscission layer, that leave a decurrent leaf cushion on the stem, are known since the Upper Permian (*Quadrocladus*). The stomata are perigenous (haplocheilic), scattered all over the leaf or restricted to one side; they occur in files, bands or groups, and are often irregularly spaced. Most frequently the guard cells are sunken and oriented along the leaf, but may be oblique or transversal to the leaf axis. The leaf margin often bears microdenticulation.

According to the structure of female fructifications the living C. are subdivided into two groups, the first including taxads, and the second including all others. In taxads solitary seeds are borne on the ends of the branchlets in a fleshy envelope (arillus). C. are known from the Palaeozoic, where solitary seeds without an arillus are borne on leafy shoots (Buriadiaceae). All other living C. have seeds attached to various seed scales subtended by a bract, and, in most cases, aggregated in compound polysperms (cones). Sometimes the cones are reduced to a single seed, inserted on the widened fleshy receptacle in the axil of the reduced bract. In the Carboniferous and Permian lebachians instead of a seed scale a dwarf shoot (axillary complex) is inserted in the axil of the bract, bearing in the upper part long seed stalks (monosperms) with terminal or abaxially attached seeds, and in the proximal part sterile scales (circaperm). The position of the seed stalks and sterile scales can be reversed. In the primitive C. (Fig. 68) the axillary complex may be very similar to the lateral polysperms in Rufloriaceae and Vojnovskyaceae. The seed scale was initially free from the subtending bract (Fig. 67, *s*; Fig. 68, *f*, *i–k*). In different lineages both organs then became increasingly more fused (Fig. 67, *t*, *w*; Fig. 68, *l–n*), and, in addition, either the seed scale (Fig. 67, *u*) or the bracts (Fig. 67, *v*) underwent reduction.

The C. male fructifications are presented by microstrobili, which in more advanced C. occur in compact groups, where each strobilus is inserted in the axil of the bract. Microsporophylls are helically arranged on the axis. Several major types of microsporophylls are distinguished (Fig. 67, *k–r*). Of most rare occurrence are radial microsporophylls, where the stalk is crowned by a shield on which are inserted sporangia surrounding the stalk (peltate microsporophylls of taxads). In most C. the microsporophylls are bilateral. The stalk is crowned by a distal shield of triangular, rhomboid or other outline. The sporangia are attached to it at the level of the stalk or below the place of its attachment. Finally, the sporangia can be attached to the stalk itself. They are either directly situated on it or occur on the tips of slender branchlets, extending out from the stalk.

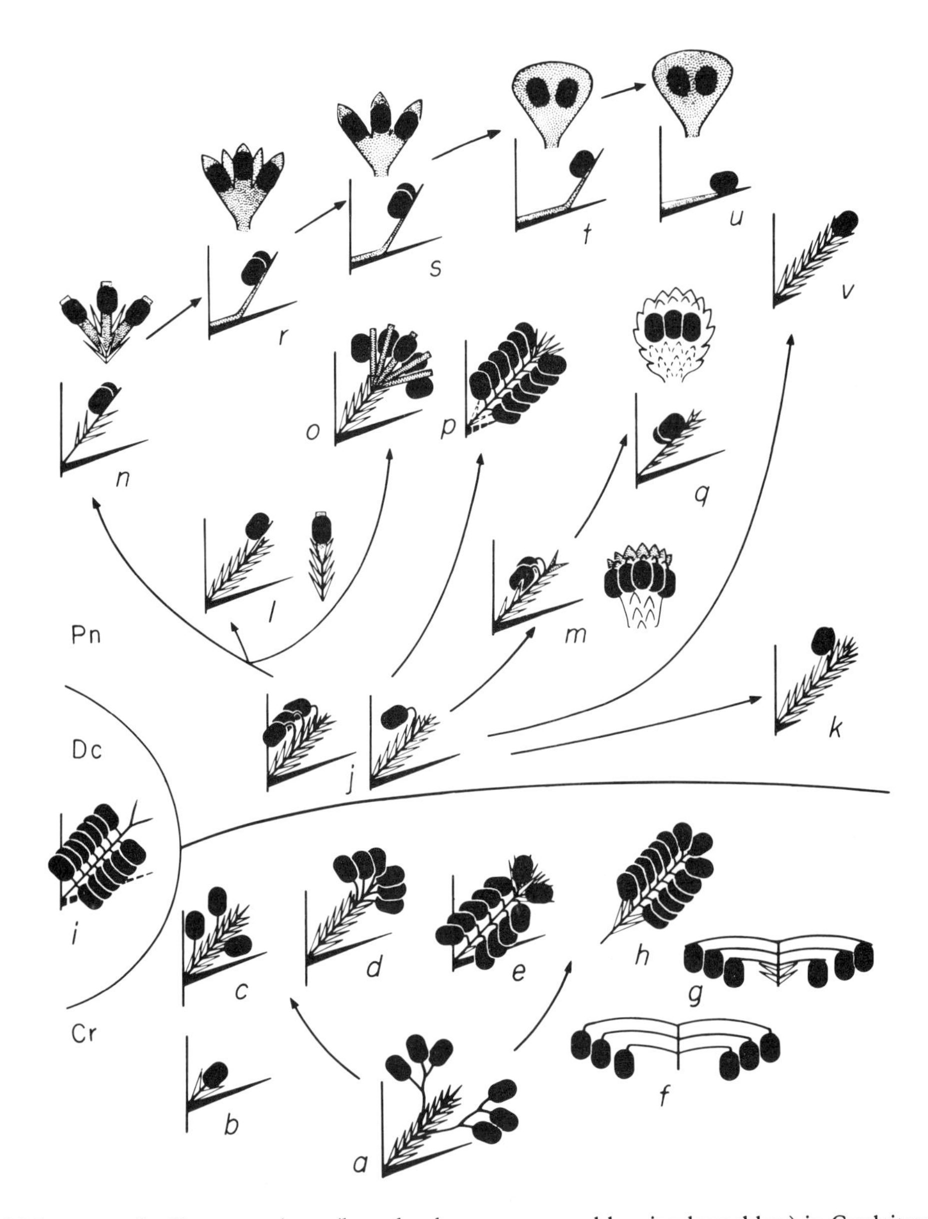

Fig. 68. Main types of axillary complexes (lateral polysperms or seed-bearing branchlets) in Cordaitanthales (Cr), Dicranophyllales (Dc) and Pinales (Pn), and probable ways of their transformations; bracts and seeds in black, seed scales (*n*, *r*–*u*), leaf-like seed stalks (*l*, *o*) and sterilized seed stalks (*m*) stippled, sterile scales (members of circasperm) shown by narrow white triangles, homologous organs shown identical, spirally arranged sterile scales and monosperms shown two ranked; *a*, *Cordaitanthus pseudofluitans* (Kidst.) Crook.; *b*, *C. diversiflorus* Crook. (sterile scales problematical); *c*, *C. zeilleri* Ren. sp.; *d*, *C. duquesnensis* Rothw. sp.; *e*, *Vojnovskya*; *f*, *Gaussia cristata* Neub.; *g*, *G. scutellata* Neub.; *h*, *Suchoviella*; *i*, *Dicranophyllum*; *j*, '*Lebachia*' *lockardii* Mapes et Rothw.; *k*, *Buriadia*; *l*, *Lebachia* (al. *Walchia*) *piniformis* (Schloth.) Flor.; *m*, *Kungurodendron*; *n*, *Walchiostrobus*; *o*, *Sashinia*; *p*, *Timanostrobus*; *q*, *Concholepis*; *r*, *Pseudovoltzia*, *Voltzia*; *s*, *Swedenborgia*, *Aethophyllum* and allied Triassic genera; *t*, *u*, more advanced Mesozoic and Cainozoic conifers with partly (*t*) and completely (*u*) fused seed scale and bract; *v*, taxads.

The C. pollen (and prepollen) is quasisaccate (in most Palaeozoic and Triassic C.), saccate and asaccate. The number of sacci varies from one to three (rarely up to five). Analogues of mono- and disaccate forms are present among quasisaccate pollen. In primitive C. only proximal and bipolar apertures are observed, whereas in more advanced forms only the distal aperture is retained or the pollen is inaperturate.

In the classification adopted in this book C. are subdivided into 12 families, 5 of which are extinct. The most significant characters that provided the basis for the distinction of the families in fossil C. are: the presence of specialized seed scales, the degree of fusion of the scale with the bract, the number of seeds and their position on the scale, the position of the seeds when the seed scale is lacking, the presence and structure of additional envelopes around the seeds, the seed structure, microsporophyll habit, foliage type, and its epidermal characters. In recent classifications a tendency is apparent towards the grouping of C. families into orders and the elevation of C. to the rank of a subclass or even class. Most frequently taxads are placed in a distinct taxon of higher rank (up to a class). The classification adopted below for the 12 families is in better agreement with the current conceptions on the phylogeny and diversity of C.

It is considered that C. evolved from Cordaitanthales. This can be inferred from the multiple similitudes displayed between lebachians and cordaitanthaleans. It is a fact that the lateral polysperms in vojnovskyans and certain lebachians (*Timanostrobus*) are nearly the same in organization. A close similarity can be observed between pollen of lebachians and ancient ruflorias, between the wood in C. and Cordaitanthaceae. In the course of investigations the affinities between C. and Cordaitanthales become more and more apparent (Meyen, in press). However, C. also display affinities to Dicranophyllales, particularly in some fine and peculiar features, which are difficult to explain in the light of their independent origin. These features are under discussion below in the description of Lebachiaceae. The Voltziaceae evolved from lebachians and gave rise to other families with axillary polysperms (Taxodiaceae, Pinaceae, Araucariaceae, etc.). The Buriadiaceae and Taxaceae probably also evolved from lebachians.

As the investigations continue the roots of the extant genera and families descend deeper down in the geochronological scale. Nearly all families known today are already reported in the Jurassic and Lower Cretaceous. However, more detailed studies of specific extinct taxa lead to the conclusion that inferences concerning the time of origin of the living genera and families can be drawn only on the basis of restored organic connections between different organs. The point is that finds of isolated organs, indistinguishable from the same organs in a living genus, are still insufficient for the identification of this genus. For instance, the fact is known that the Eocene leafy shoots, the same as those in *Taxodium*, combine with female cones and microstrobili that are typical of the living genus *Metasequoia*. These Eocene conifers were 'synthetic types'; they combine characters of different extant taxa, in this particular case characters of genera of one family. 'Synthetic types' encountered in older deposits combined characters of different extant families. In this respect of interest is the Jurassic genus *Pararaucaria* established on petrified cones. It combines characters of the Pinaceae and Taxodiaceae, whereas certain characters are typical of the Araucariaceae and Cupressaceae. This probably supports the opinion on the phylogenetic linkages between the above-cited families, but as yet it is impossible to define their interrelations more precisely.

Family Lebachiaceae

The names Lebachiaceae and *Lebachia* are nomenclaturally illegitimate (and should be substituted by Walchiaceae and *Walchia*; Clement-Westerhoff, 1984). Nevertheless, following long-standing tradition, they are retained here.*

* The taxonomy and nomenclature of these oldest conifers need substantial revision. Since the specimen which is the nomenclatural type of both *Walchia* and *Lebachia* is sterile and even its cuticular structure cannot be studied, the present author (Meyen, in press) suggested

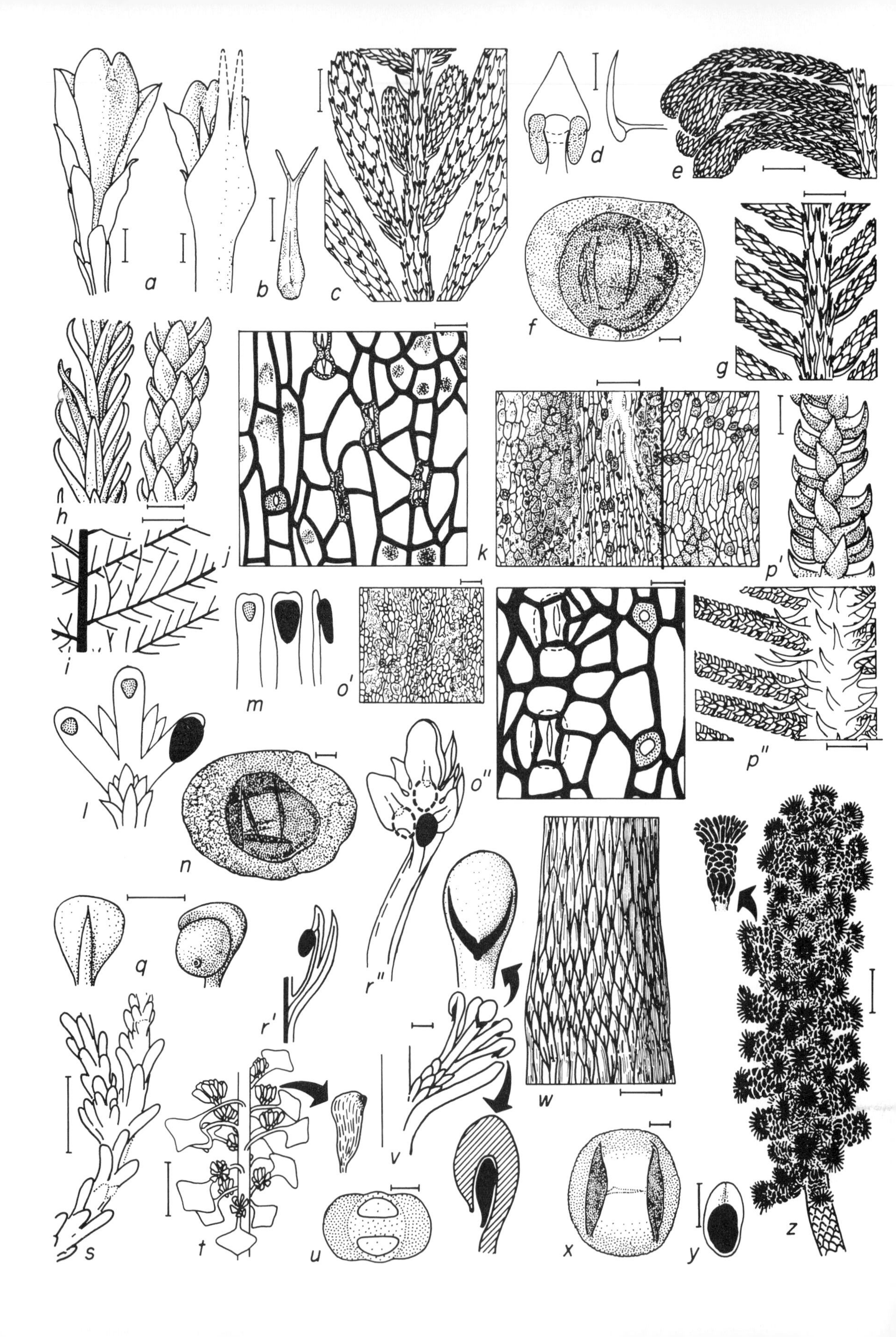
a
b
c
d
e
f
g
h
i
j
k
l
m
n
o'
o''
p'
p''
q
r'
r''
s
t
u
v
w
x
y
z

Lebachians (L). differ from cordaitanthaleans in their needle-shaped, sometimes forked leaves, the beginning of the flattening of the axillary complex (not in all genera) and simultaneous shifting of the seed stalks (monosperms) on to its adaxial side. The number of seed stalks may be reduced, so that eventually merely one seed stalk remains. The seeds are borne on bent back apices of the seed stalk or attached to the lower side of the stalk under the apex (this side of the seed stalk which shifted on to the adaxial side of the axillary complex, is faced towards the apex of the compound polysperm). The L. bracts are similar to foliage leaves, but usually larger (in cordaitanthaleans the bracts and leaves differ considerably, even when the bract is leaf-like). Sterile scales of the axillary complex are also similar to the foliage leaves, but much smaller. In the axillary complex of some L. a secondary derived gradual transition between sterile scales and seed stalks can be observed, that is not evident in cordaitanthaleans; variously underdeveloped seeds were borne on the intermediate organs. Such intermediate organs may be termed sterilized seed stalks. The structure of L. microsporophylls has been inadequately studied. The pollen (or pre-pollen) is monosaccate or quasimonosaccate. Compound polysperms ('cones') of L. and microstrobili crown ordinary leafy shoots and have no specialized stalk, as in more advanced conifers. Another primitive feature of L. is the decrease in the leaf sizes from the thicker to thinner branches.

Genus *Lebachia* (C_2–P_1; Fig. 69, *a*, *c–k*; to this genus probably belong also the vegetative shoots from the Westphalian B, that were treated as a specific genus *Swillingtonia* – Scott and Chaloner, 1983). The better studied species were small trees. The secondary wood is similar to that in cordaitanthaleans (of the *Dadoxylon* type). The pith was seemingly at times septate, but the pith casts (of the *Tylodendron* type; Fig. 69, *w*) were altogether different from those in cordaitanthaleans (*Artisia*).

---

the introduction of separate generic names for sterile shoots and female fructifications of these plants. The genus *Walchia* is suggested to comprise only sterile shoots. The female fructifications are referred to as a new genus (*Lebachiella*). The family name (Lebachiellaceae) is derived from the name of this new genus. *Walchia* is treated as a satellite genus of this family, and the family Walchiaceae is treated as a form-group of sterile coniferalean shoots.

Fig. 69. Lebachiaceae (*a–k*, *s–z*) and early Voltziaceae (*l–r*); Upper Carboniferous (*g*), Lower (*a–f*, *h–p*, *w*) and Upper (*q–v*, *x–z*) Permian; Western Europe (*a–r*, *w*), north of European part of the USSR (*s–v*, *x–z*); *a*, axillary polysperm of *Lebachia piniformis* (Schloth.) Flor. as viewed from side of clavate seed stalk (to left; seed abscissed) and forked bract (to right); *b*, forked leaf *Gomphostrobus bifidus* (Gein.) Zeill.; *c*, compound polysperms of *L. piniformis*; *d*, microsporophyll of *L. hypnoides* (Brongn.) Zeill, ; *e*, *f*, *L. piniformis*, young microstrobili (*e*) and pollen grain (*f*); *g*, *L. hirmeri* Flor., branch of penultimate order; *h*, branches of ultimate order of *L. parvifolia* Flor. and *L. hypnoides*; *i–k*, *L. hypnoides*, scheme of branching of vegetative shoot (*i*), stomata and hair base (*j*), upper epidermis with stomatal bands (to left) and lower epidermis (to right) with stoma and hair bases (*k*); *l*, *m*, *Walchiostrobus gothanii* Flor., scheme of seed scale structure, numerous sterile scales accompany three-lobed seed scale showing one seed (in black) and two seed scars (stippled) (*l*), seed scale lobe (seed stalk) with seed scar (to left) and with seed still attached (*m*); *n–p*, *Ernestiodendron filiciforme* (Schloth.) Flor., pollen grain (*n*), leaf cuticle (*o'*), stomata and hair bases (*o''*), branch of last order (*p'*), branch of penultimate order (*p''*); *q*, *Ullmannia frumentaria* (Schloth.) Goepp., seed scale viewed from side of bract (to left) and seed; *r*, *Pseudovoltzia liebeana* (Gein.) Flor., axillary complex in longitudinal section (*r'*) and in adaxial view (*r''*), one seed shown in black, remainder shown by dashed line; *s*, *Quadrocladus dvinensis* S. Meyen; *t*, *Dvinostrobus sagittalis* Gom. et S. M., general view of microstrobilus and a sporangium (to right); *u*, *Scutasporites*-like pollen from sporangium of *D. sagittalis*; *v*, *Sashinia*, axillary polysperm and seed stalk, seeds in black, above – abaxial view of seed stalk, below – longitudinal section of seed stalk; *w*, pith cast *Tylodendron speciosum* Weiss; *x–z*, *Timanostrobus muravievii* S. Meyen, pollen grain in proximal view, quasisaccus stippled (*x*), seed, megaspore membrane in black (*y*), reconstruction of compound polysperm and simple polysperm (to left) (*z*). Scale bar = 1 cm (*b*, *c*, *e*, *g*, *i*, *p''*, *q*, *s*, *w*, *z*), 5 mm (*t*), 2 mm (*h*, *p'*, *v*), 1 mm (*a*, *d*, *y*), 100 μm (*k*, *o'*), 20 μm (*f*, *j*, *n*, *o''*, *u*), 10 μm (*x*). Modified from: Florin, 1944 (*a*, *c–e*, *g–j*, *o''–p''*, *w*), 1940 (*b*), 1938 (*f*), 1939a (*k*), 1939b (*n*, *o'*); Schweitzer, 1963 (*q*, *r*); Meyen, 1976–1978 (*s*) and 1984a (*t–v*).

They were covered by spindle-shaped cushions with acute ends. The grooves between the cushions correspond to imprints of primary vascular bundles, protruding towards the pith. The branches of the last order were of pinnate arrangement. The leaves borne on them were awl-shaped with microscopic teeth or hairs along the margin. The stomata are aggregated in short and wide bands, one on each leaf face. The leaves on the thicker branches of the penultimate order have a forked apex (genus *Gomphostrobus*; Fig. 69, *b*). Compound polysperms are elongate-elliptical, bracts are forked (of the *Gomphostrobus* type). The lateral polysperms are flattened, the axis bears in the lower part spirally attached sterile scales, and distally a single clavate seed stalk with a notched apex (Fig. 69, *a*). The seeds were attached under the notch. The structure of the axillary complex was wrongly interpreted by Florin (for details see Clement-Westerhoff, 1984). He mistook the seed stalk for the seed itself, and this misinterpretation was widely current in the literature. The seeds correspond to the genus *Samaropsis*. In contrast to Cordaitanthaceae, the microstrobili differ in their organization from that of the compound polysperms. They lack bracts and axillary shoots; instead, microsporophylls with a distal shield are borne on the axis, and a pair of sporangia are inserted on both sides of the point of attachment of the stalk. The species '*L*'. *lockardii*, where the seed stalks were backward bent and seeds were borne apically, has been arbitrarily assigned to *Lebachia* (Fig. 70, *a*). The genera *Kungurodendron* and *Ortiseia* are close to *Lebachia*. In *Kungurodendron* (P; Fig. 70, *b*) several backward bent seed stalks were adaxially inserted in the axillary complex; the abaxially attached sterilized seed stalks were replaced more proximally by sterile scales, similar to the foliage leaves in their marginal microdenticulation. Prepollen was quasimonosaccate with a thin columellar-like quasimonosaccus, that completely closed the distal side leaving the proximal side free, at times with a monolete slit. In *Ortiseia* (P; Fig. 70, *c–f*) the axillary complex had only one wide seed stalk with abaxially attached seeds. The chalaza of the seed (Fig. 70, *f*) was

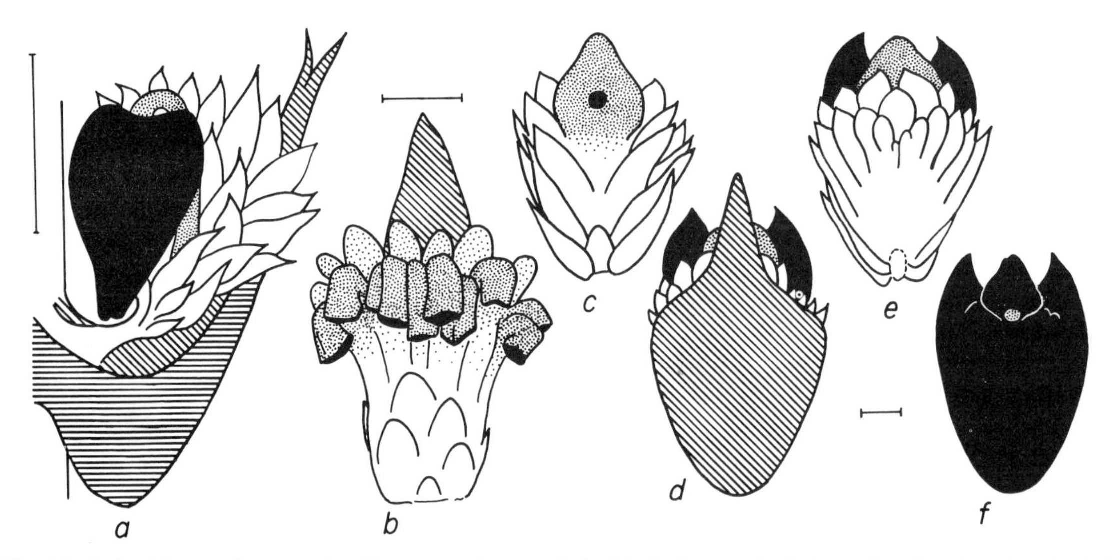

Fig. 70. Lebachiaceae, bract and axillary complex; seeds in black, bracts shaded, seed stalks densely stippled, sterilized seed stalks sparsely stippled, seed scar in black (*c*) or stippled (*f*); Upper Carboniferous–lowermost Permian (*a*), Lower (*b*) and Upper (*c–f*) Permian; USA (*a*), Fore-Urals (*b*), Western Europe (*c–f*); *a*, '*Lebachia*' *lockardii* Map. et Rothw.; *b*, *Kungurodendron*, adaxial view; *c–f*, *Ortiseia leonardii* Flor. emend. Clem.-Westerh., adaxial view with abolished bract (*c*), abaxial view with bract (*d*) and without it (*e*), seed with chalasa at top (*f*). Scale bar = 2 mm. Modified from: Mapes and Rothwell, 1984 (*a*); Clement-Westerhoff, 1984 (*c–f*).

complicated by projections. Prepollen was of the *Nuskoisporites* type with an annular saccus and trilete proximal aperture.

The genus *Timanostrobus* (P; Fig. 69, *x–z*) had more primitive lateral polysperms. They were not flattened and had a large amount of seeds that were replaced near the apex by sterile scales. It is interesting to note that bracts were not found below the lateral polysperms. The prepollen was about the same as in *Kungurodendron*.

The genus *Sashinia* (P_2; Fig. 69, *s–v*) is very peculiar. It displays affinities to other L. in its leaf-like bracts and primitive lateral polysperms, but its circasperm is composed of ordinary foliage leaves. The seed stalks, situated in a bunch at the apex of the axis, have the same epidermal structure and type of apex as the leaves, but the apex was flexed and the seeds were borne beneath the flexion on the abaxial side (Fig. 69, *v*). According to the structure of the pollen, microsporophylls and epidermis, the *S.* leaves are closer to Voltziaceae. The pollen was quasidisaccate with equatorial and proximal taeniae also of quasisaccate structure. The microstrobili (*Dvinostrobus*) were composed of microsporophylls that had a slender stalk with sporangia inserted on it and a rhomboid distal shield. *S.* associated with leafy shoots of the genus *Quadrocladus* (Fig. 69, *s*). The shoots are of irregular branching. Lateral dwarf shoots with smaller leaves are also present. The leaves are short, subtriangular and spathulate to long, linear. The lamina is lenticular to round in cross-section. The stomata occur all over the leaf, at times arranged in short files. The margins are sometimes microdenticulate. The shoots of *Quadrocladus* probably associated with fructifications that are referred to different genera.

At times the genera *Carpentieria* and *Lecrosia* with forked leaves are also assigned to L. But these genera can in equal measure be affiliated to Dicranophyllales.

From the description given above it becomes immediately apparent that L. display several features common to Dicranophyllales: furcation of some of the leaves, marginal microdenticulation, compact stomatal bands (in some L.). If the seeds in dicranophyllans were of spiral arrangement in the lateral polysperms, as in *Timanostrobus*, then this could be another essential common feature.

Family Buriadiaceae

Genus *Buriadia* (C_3?–P; Fig. 71, *a–f*).

The B. shoots were of irregular branching and had wood with circular bordered pits. The leaves were either needle-shaped or with a forked apex, more seldomly wedge-shaped with a denticulate apex. The leaves had marginal microdenticulation. The stomata were gathered on the lower surface (or on the lower faces of the leaves, triangular in cross-section) into loose stomatal bands. The seeds were scattered over the shoot without any apparent order. They were inverted; the long micropylar tube was adpressed to the slender seed stalk. The seeds were flattened, covered along the margin by numerous hairs. Inside the seeds a megaspore membrane was not observed. In *Walkomiella* the leaves are not forked; they are wider and more tightly pressed to the axis, whereas the stomata are more compactly arranged in bands. The seeds, in contrast to *Buriadia*, are orthotropous. Microstrobili of buriadians are not known. So far nothing can be said of the phylogenetic relations of this family.

Family Voltziaceae

Voltzians (V.) differ from lebachians in the formation of a seed scale, that was derived through the fusion of monosperms (seed stalks) on the adaxial side of the lateral polysperm. The seeds were first abaxially attached near the apex of the seed stalk and then shifted proximally. In more primitive V. the seed scale was accompanied by sterile scales, some of which probably corresponded to sterilized seed stalks. The sterile scales then disappeared and the bract underwent partial fusion with the seed scale. Microsporophylls were peltate with a slender stalk and distal shield, that had a strongly developed heel. Sporangia were attached directly to the shield under the stalk, or attached to the stalk by thin branching sporangiophores. The pollen is quasisaccate.

a
b
c
d
e
f
g
h
1
2
3
4
i
j
♀
♂
k
l
m
n
o
p
q
r
s
t
u
v
w
x
y

To V. can be assigned conifers that are separated into the family Cycadocarpidiaceae (= Podozamitaceae). The major distinction between them was formerly perceived in the structure of the leaves and bracts, that have numerous parallel veins converging at the apex. This character is not sufficient for the distinction of the family, inasmuch as in certain Triassic conifers, clearly related to each other in other characters, the bracts may be either narrow with one vein, or broader with numerous veins. Leaves of both types can occur together in one family, which is exemplified in the living Araucariaceae and Podocarpaceae.

The most primitive among V. is the genus *Walchiostrobus* (C_3–P_1; Fig. 69, *l–p*), which also comprises leafy shoots, usually included in the genus *Ernestiodendron* (Fig. 69, *o, p*). Florin (1938–1945, 1951) allied *Ernestiodendron* with *Lebachia* and, due to the misinterpretation of the axillary complex (Clement-Westerhoff, 1983; Meyen, 1984a and in press), included it in the family Lebachiaceae. He considered that these plants had free seed stalks that were situated on a shortened axis, and that the seeds were borne on the tips of the backward bent apices. It is suggested in the present book that the name *Walchiostrobus* should be applied to isolated compound polysperms, also for restored plants. The compound polysperms are known only from impressions. Bracts are linear. The seed scale is composed of three–five flattened and basally fused monosperms (Fig. 69, *l, m*) with subapical seeds, that upon abscission leave large scars. Monosperms are accompanied by sterile scales. The structure of the microstrobili is unknown. Pollen (Fig. 69, *n*) is about the same as in *Lebachia*. The shoots were of pinnate branching, leaves awl-shaped, stomata arranged in files all over the leaf. A characteristic feature is the presence of hair bases, the same as in *Lebachia* (Fig. 69, *j, o*). *Walchiostrobus* occupies an intermediate position between the Lebachiaceae and Voltziaceae.

The lateral polysperms in more advanced V. lose their external similarity to lebachian polysperms, as a result of the formation of a lobed seed scale with more proximal seeds, also the reduction in the number of sterile scales (or sterilized seed stalks) and partial fusion of the bracts with the seed scale. This evolutionary stage is represented in *Pseudovoltzia* (P_2; Fig. 69, *r*) and *Voltzia*. In *Ullmannia* (P_2; Fig. 69, *q*) the seed scale is not dissected, bears only one seed and probably corresponds to the monosperm in *Ortiseia*, or to one lobe of the seed scale in *Pseudovoltzia*. According to their epidermal characters *Ullmannia* and *Pseudovoltzia* are closely related. *Ullmannia* associates with microstrobili consisting of peltate microsporophylls with numerous sporangia

Fig. 71. Buriadiaceae (*a–f*), Voltziaceae (*g–w*), Palissyaceae (*x, y*); Upper Palaeozoic (*a–f*), Middle (*g–l, n, o, w*) and Upper (*m, p–v, x, y*) Triassic; India (*a–f*), Western Europe (*g–l, n–p, v–y*), Ukraine (*m, q–t*), Japan (*u*); *a–f*, *Buriadia heterophylla* (Feistm.) Sew. et Sahni, shoot with seed (in black) and variously dissected leaves (*a*), seed in attachment to shoot (*b*), forked leaf (*c*), leaf epidermis near margin (stippled), stomata shown by circles (*d*), stoma (*e*), pollen from seed micropyle (*f*); *g–j*, *Aethophyllum stipulare* Brongn., axillary complex (*g*; seeds in black, bract shaded), pollen (*h*; 1, laesura of proximal side; 2, subequatorial thinning of exine; 3, quasisaccus; 4, sulcus), microsporophyll, sporangia in black (*i*), general view of monoecious shoot (*j*); *k*, *Voltzia heterophylla* Brongn.; *l*, *Sertostrobus laxus* Grauv.-St., microsporophyll; *m*, *Cycadocarpidium* (al. *Podozamites*) *toretziensis* Stanisl., shoot with polysperm and axillary complex with leaf; *n, o*, *Darneya peltata* Schaar. et Maub., microsporophyll (*n*) and microstrobilus (*o*); *p*, *Swedenborgia cryptomeroides* Nath., axillary complex as seen from side of seeds (in black) and bract (at arrow); *q*, compound polysperm of *S. tyttosperma* Stanisl.; *r–t*, *Borysthenia fasciculata* Stanisl., shoot with compound polysperm (*s*), axillary complexes with three (*r*) and two (*t*) seed stalks; *u*, *Cycadocarpidium osawae* Kon'no; *v*, *C. schwabii* Nath.; *w*, *Yuccites vogesiacus* Schimp. et Moug.; *x, y*, *Palissya sphenolepis* Braun, part of compound polysperm and reconstruction of lateral polysperm (*x*), compound polysperm (*y*). Scale bar = 10 cm (*j, w*), 1 cm (*k, o, r, y*), 5 mm (*a, m, n, q, s–v*), 2 mm (*g, p, x*), 1 mm (*b, d, i, l*), 0.5 mm (*c*), 50 μm (*e*), 25 μm (*f, h*). Modified from: Pant and Nautiyal, 1967 (*a–f*); Grauvogel-Stamm, 1978 (*g–j, l, n, o, u, v*); Mägdefrau, 1968, after Schimper and Mougeot (*k*) and Schlüter and Schmidt (*w*); Stanislavsky, 1976 (*m, q–t*); Florin, 1951 (*p*) and 1944b (*x* (to left), *y*); Gothan and Weyland, 1964, after Schenk and Nathorst (*x*, to right).

attached to the lower part of the distal shield. Pollen is quasidisaccate and operculate (of the *Jugasporites* type; Foster, 1983).

Genus *Voltzia* (Fig. 71, *k*) has been established for vegetative heterophyllous shoots. The structure of its seed scales has been insufficiently studied. Evidently they were of the same type as in *Pseudovoltzia*. Best studied among the more advanced V. is the genus *Aethophyllum* (T; Fig. 71, *g–j*). These were herbaceous plants with a creeping root (rhizome?), producing secondary wood, and aerial dioecious shoots without secondary wood. Grauvogel-Stamm (1978) described entire plants only 25 cm tall, bearing microstrobili. Leaves are long and very narrow with several parallel veins. Loose female cones were situated on the apices of the upper branches. They were long with helically arranged spear-shaped bracts fused in the lower part with a five-lobed seed scale. Microstrobili (*Willsiostrobus*) are composed of peltate microsporophylls; the sporangia are attached to the sides and heel of the shield and outstretched along the stalk. The pollen is quasidisaccate.

Genus *Cycadocarpidium* (T–J_1?; Fig. 71, *m*, *u*, *v*)

Most frequently occurs as isolated bracts with an attached axillary polysperm. Bracts are from leaf-like, similar to the *Podozamites* (see below), to reduced, not extending beyond the lobes of the seed scale, but with characteristic parallel venation. Attached to the lobes are inverted seeds, which have not been studied in detail. Most frequently there are two seeds, but sometimes, up to four. The seeds are attached to the middle part of the lobes. *C.* polysperms associate with leaves of *Podozamites*. Microstrobili *Willsiostrobus* were found associated with one *P.* species. Various forms can be arranged into a sequence in which the bracts undergo gradual reduction to a small appendage (*Swedenborgia*; T; Fig. 71, *p*, *q*).

Genus *Borysthenia* (T; Fig. 71, *r–t*)

Leaves are long with one vein, gathered in bunches on the tips of dwarf shoots. Compound polysperms are similar to those in *Aethophyllum*, but shorter; bracts are very long, linear with one vein.

Genus *Yuccites* (T_1; Fig. 71, *w*)

Leaves are large (up to 20 cm in width and 1.20 m in length; Grauvogel-Stamm, personal communication), strap-shaped with a broad base. Several veins entering the base, soon dichotomize and then continue further nearly parallel to the leaf apex and lateral margins. *Y.* was long regarded as a Mesozoic representative of cordaitaleans. Microstrobili *Willsiostrobus* accompany *Y.* The stems were unbranched and devoid of secondary wood (i.e. succulent – Meyen, 1985b).

Genus *Podozamites* (P_2?; T–K; Fig. 71, *m*) can be assigned to V. as a satellite genus

Leaves are lanceolate with an acute or obtuse apex. One vein enters the narrow base, soon undergoes dichotomous division several times. Further the veins run parallel and converge at the apex. The epidermal structure is of two types. *P.* comprises leaves with distinct stomatal and stomata-free zones on the lower side; on the upper side stomata are lacking. The stomata are in regular files and oriented mainly across the leaf. Two rectangular subsidiary cells embrace the stomatal aperture along the whole length. The guard cells are nearly surficial. Leaves with longitudinally oriented stomata, gathered in files between the veinlets, are assigned to the genus *Lindleycladus*. It still remains unclear with what type of fructifications both types of leaves associate. Triassic *P.* species associate with female cones of *Cycadocarpidium* and *Swedenborgia*, which are unknown in younger deposits. Fructifications, associated with Jurassic and Cretaceous species, are unknown. However, Krassilov (1982) reported the discovery of polysperms of the *Swedenborgia* type and microstrobili of the *Darneya* type associated with the Lower Cretaceous *P.*

Family Cheirolepidiaceae

Fossil remains of cheirolepidians (Ch.) were frequently included in the family Araucariaceae.

The relationship between both families is to a large degree unclear. If we admit the recent interpretation of the genus *Hirmeriella* and closely related genus *Tomaxellia*, then Ch. would differ from the Araucariaceae in their strongly developed lobed seed scale (in Araucariaceae it is reduced to a small ligule), also in their free bracts, so that upon disintegration of the cone the bracts and seed scales undergo separate burial. Probably both families will be eventually united into one. Microstrobili (*Classostrobus*; Fig. 72, *f*, *g*) are attached singly to leafy shoots. Microsporophylls are peltate with 2–8 sporangia attached to the shield under the stalk. The pollen is peculiar (genus *Classopollis*; T–Pg; Fig. 72, *h*, *i*). It is round, asaccate, in dispersed state often gathered in tetrads. Around the grain extends a groove (rimula), shifted distally from the equator, and accompanied by a circular girdle with a striated structure. On the distal pole is a round area lacking sexine (pseudopore, cryptopore), on the proximal is a poorly defined trilete slit, sometimes absent altogether. The sexine is composed of columellae, covered by a tectum with micropores and orbicules (Ubisch bodies) bearing the same sculpture as the grain itself. Taylor and Alvin (1984) studied in detail the ultrastructure and ontogenesis of *Classopollis* from sporangia of *Classostrobus comptonensis* (Alvin *et al.*, 1978). In their opinion the complicated stratification of the exine and the presence of tapetal membranes and sculptured orbicules is the result of adaptation to the pollen-receiving organ. Microstrobili with such pollen were assigned to the genus *Masculostrobus*, but its type-species has a different pollen. It was suggested that to *Masculostrobus* should be assigned microstrobili with only two free sporangia attached to the shield (Grauvogel-Stamm and Schaarschmidt, 1979).

Leafy shoots are of different types. They may be *Brachyphyllum*-, *Pagiophyllum*-, *Geinitzia*- or *Cupressinocladus*- like (see Form-genera of vegetative shoots). The whorled leaves of *Frenelopsis* (K; Fig. 72, *b–e*) are basally fused into a cylindrical sheath with short free portions of the leaves. In *Pseudofrenelopsis* only one leaf per node is usually present. Its basal portion forms a short cylindrical sheath and the free leaf lamina is much reduced. The genus *Suturovagina* produced both *Brachyphyllum*-like and sheath-like leaves, the latter embracing most or all of the stem, leaving either a gap or a suture between two lateral edges (Zhou Zhiyun, 1983).

A common feature shared by all Ch. leaves is the arrangement of the stomata, which can be either freely scattered or in files, but not in bands. The guard cells are of different orientation and deeply sunken. Papillae of subsidiary cells frequently extend forth into the round stomatal pits (Fig. 72, *e*). Some Ch. were large trees with secondary wood of the *Protocupressinoxylon* type and leaves of the *Cupressinocladus* type (Fig. 72, *a*). Ch. are of first appearance in the upper Triassic and continued up into the Cretaceous (the pollen of *Classopollis* continued up to the Palaeogene). The origin of Ch. is unknown. Their seed scale can be derived from the voltzialean.

Genus *Hirmeriella* (= *Cheirolepidium*; T_3–J; Fig. 72, *j*, *k*)

This genus was established for female cones, usually represented by isolated seed scales, more rarely by isolated bracts. On the proximal side lies a single (seldom two) inverted seed, covered by a winged outgrowth of the seed scale. (The outgrowth may be homologous to the flexed seed stalk apex of *Sashinia*.) The distal part of the scale is divided into several lobes. The isolated bracts are larger than the seed scales and can remain on the axis after abscission of the seed scales. Associated shoots are of the *Brachyphyllum*, *Pagiophyllum* (Fig. 72, *m*) and *Geinitzia* (?) types. Closely related to *H.* is the genus *Tomaxellia* (Fig. 72, *l*), that has two seeds in the scale and associates with leaves of the *Elatocladus* type, and microstrobili where the microsporophylls lack a distal shield.

Family Palissyaceae

The Palissyaceae (P.) is a poorly studied family. It is generally considered (following Florin, 1951), that the female cone in P. consists of long spear-shaped bracts, spirally arranged on the axis, under the axillary polysperm. The latter in *Palissya* (T_3)

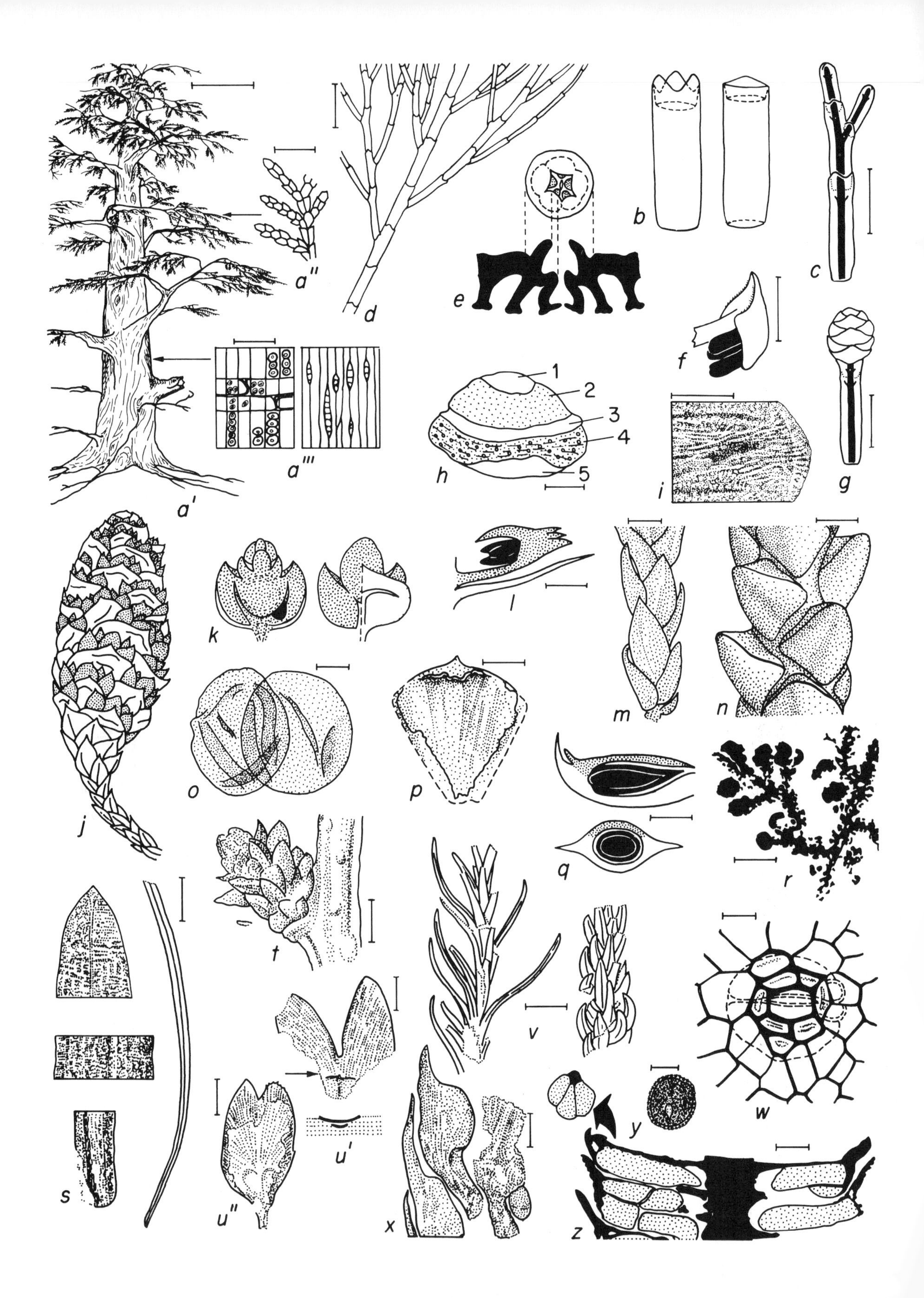

a''
d
b
c
e
f
a'''
1
2
3
4
5
h
i
g
a'
j
k
l
m
n
o
p
q
r
t
s
v
w
y
u''
u'
x
z

is of sympodial branching several times (the upper branches are sterile), whereas in *Stachyotaxus* (J_1) it is once-forked. Seeds, borne on the tips of the branches, are partly closed by a fleshy arillus. However, the illustrations available do not support the suggested interpretation of the lateral polysperms. They merely depict lateral polysperms with something like seeds borne on the adaxial (?) side, attached to the axis of the compound polysperm (Fig. 71, *x*, *y*).

P. leaves are helically arranged. They are narrow-linear or linear-lanceolate, with one vein, hypostomatic. On both sides of the vein is a stomatal band with regular files of stomata. Microstrobili are unknown.

Probably closely related to *Palissya* is the genus *Metridiostrobus* (T_3; Delevoryas and Hope, 1981), the polysperm structure of which has not been adequately studied. The systematic affinity of P. remains problematical. P. might probably belong, not to conifers, but more readily to Ginkgoopsida, and their polysperms could have been depicted in an inverted position.

Family Araucariaceae

Fossil remains belonging to different families, including all leafy shoots of the genera *Brachyphyllum* and *Pagiophyllum* were assigned in the literature to araucarians (A.). Seed scales of A. and other conifers, also lycopsid sporophylls, were described under the name *Araucarites*. To *Dammarites* were referred casts that externally resemble A. cones, but actually belonged to vegetative shoots with persistent bases of fleshy leaves (Hluštik, 1974). Wood fragments, assigned to the genus *Araucarioxylon*, partly do belong to A. (Mesozoic species), but many of them are of different affinities. Genera that were only superficially studied (*Agathopsis*, *Albertia*, etc.) were also included in A. On the other hand, more detailed examinations have shown that certain Jurassic and Cretaceous fossils can be directly assigned to the genus *Araucaria* (Stockey, 1982).

A. wood is practically the same as in the cordaitanthaleans. Bordered pits on the radial walls of the tracheids are often crowded and then hexagonal (araucarioid pitting). Resin canals are absent. Leaves are helically arranged. They are

Fig. 72. Cheirolepidiaceae (*a–m*), Araucariaceae (*n–r*), Pinaceae (*s–u*), Taxodiaceae (*v–z*); uppermost Triassic–lowermost Jurassic (*j*, *k*), Middle (*m–z*) and Upper (*a*) Jurassic, Cretaceous (*b–i*); Western Europe (*a–k*, *m–z*), Argentina (*l*); *a*, conifer with shoots *Cupressinocladus valdensis* Sew. (*a″*), wood *Protocupressinoxylon* (?) *purbeckense* Francis (*a″ ′*) and strobili *Classostrobus*; *b*, scheme of stem segment of *Frenelopsis* with tooth-like fused (to left) and cylindrical (to right) leaves; *c–e*, *F. alata* (K. Feistm.) Knobl., scheme of shoot branching, wood in black (*c*), general view of shoot (*d*), stomatal structure in section of cuticle and in surface view (*e*); *f*, *Classostrobus comptonensis* Alvin et al., microsporophyll associating with *Pseudofrenelopsis*, sporangia in black; *g*, *F. alata* with microstrobilus of *Classostrobus* type; *h*, scheme of pollen *Classopollis* associating with shoots *F. oligostomata* Rom. (1, distal 'pore'; 2, distal band; 3, rimula; 4, equatorial band; 5, proximal pole); *i*, structure of equatorial band of *Classopollis* from microsporangium of *C. comptonensis* (see *f*); *j*, *k*, *Hirmeriella muensteri* (Schenk) Jung, compound polysperm before abscission of seed scales (stippled; *j*), seed scale (stippled) and bract from both sides, seed in black (*k*); *l*, scheme of lateral polysperm of *Tomaxellia teguistoi* Arch., seed scale stippled, seed in black; *m*, *Pagiophyllum maculosum* Kend., leafy shoots associating with *H. kendalliae* Harris; *n–r*, araucariaceous plant with vegetative shoots *Brachyphyllum mammillare* L. et H. (*n*, *r*), polysperms *Araucarites phillipsii* Carr. (*p*, *q*; ligule, i.e. seed scale stippled; seed in black) and microstrobili (*r*); *s–u*, pinaceous plant with leaves *Pityocladus scarburgiensis* Harris (*s*), brachyblasts (*t*), seed scales and bracts *Schizolepis liasokeuperianus* Braun (*u′*, strongly dissected seed scale and reduced bract; *u″*, seed scale with imprints of two seeds); *v–z*, *Elatides williamsonii* (L. et H.) Nath., vegetative shoots of *Geinitzia* type (*v*), stomata on seed scale (*w*), part of polysperm (*x*), pollen (*y*), longitudinal section of microstrobilus and cross-section of microsporphyll, sporangia stippled (*z*). Scale bar = 1 m (*a′*), 1 cm (*a″*, *d*, *f*, *g*, *l*, *r*), 5 mm (*c*, *p*, *q*, *s*, *v*), 2 mm (*m*, *n*, *u*, *x*), 1 mm (*t*), 200 μm (*z*), 100 μm (*a‴*), 20 μm (*o*, *w*, *y*), 10 μm (*h*, *i*). Modified from: Francis, 1983 (*a*); Watson, 1977 (*b*); Alvin and Hluštik, 1979 (*c*, *g*); Hluštik, 1978 (*d*); Alvin, 1977 (*e*); Alvin *et al.*, 1978) (*f*, *i*); Pons and Broutin, 1978 (*h*); Stockey, 1981 (*j*) and Alvin, 1982 (*k*), after Jung; Archangelsky, 1968a (*l*); Harris, 1979 (*m–z*).

awl-shaped, scaly or with a lanceolate leaf lamina and numerous veins, that seldomly dichotomize, and extend to the leaf apex, where they converge. Stomata are in regular files, less frequently in indistinct bands. Leafy shoots are assigned to the genera *Brachyphyllum* (Fig. 72, *n*), *Pagiophyllum* (Fig. 72, *m*), *Allocladus*, *Araucariodendron*, *Ussuriocladus* (?), etc.

Female cones are compact. Woody bracts are fused with the seed scale that bears a single inverted seed. The integument is fused on one side with the seed scale (Fig. 72, *q*). In *Araucaria* the seed scale protrudes in the shape of a small appendage – ligule. The fertile unit, abscissed from the cone axis, includes the bract, seed scale and seed. Only in a few A. do the seeds separate from the scales upon burial. Fossil parts of A. cones are referred either to *Araucarites* (Fig. 72, *p*, *q*) or directly to the genus *Araucaria*, known since the Jurassic. Permineralized cones of the Jurassic *Araucaria mirabilis* allowed the species to be assigned to the modern section *Bunya* (Stockey, 1975). Various developmental stages of the gametophyte are preserved in it. Embryos with two cotyledons were found in the seeds, which formerly was regarded as a very advanced character within the genus. In the Jurassic A. one-seeded scales were found, that are also considered an advanced character.

Microstrobili (Fig. 72, *r*) are larger than in other conifers and reach 25 cm in length. Microsporophylls are peltate. Twenty long sporangia are attached to the shield under the stalk. Isolated microsporophylls of A., Cheirolepidiaceae and Voltziaceae can be identified only on the basis of the epidermal characters of the shield and the pollen structure. A. pollen (Fig. 72, *o*) is asaccate, round, lacking pores and a true colpus. Dispersed pollen is assigned to the genus *Araucariacites* known since the Triassic.

Family Pinaceae

The Pinaceae (P.) is the most widespread family of living conifers, where the genus *Pinus* is the most preponderant tree genus in the world. P. has played a paramount role in vegetation, at least from the end of the Palaeogene. Other plant remains have frequently been allied with P. For instance, when no evidence was available on the cuticle structure, linear leaves with one vein of the *Pityophyllum* type were then assigned to P., although leaves of the same type were found in other conifers, as well as in the Ginkgoales (*Toretzia*). The Jurassic genus *Pseudoaraucaria*, erroneously placed into Araucariaceae, has been referred to P., because of its close affinities to *Pseudolarix*.

The most ancient P. have been recorded from the Middle Jurassic (seed scales *Schizolepis* and associated shoots *Pityocladus*). P. are known from remains of various parts. Secondary wood fragments are ascribed to the genera *Pinoxylon*, *Piceoxylon*, *Keteleerioxylon*, *Pseudolaricioxylon*, etc. Three- and five-leaved dwarf shoots, characteristic of *Pinus*, are known in the Upper Cretaceous. Female cones, typical of *Pinus*, were recorded since the Lower Cretaceous. Seed scales *Schizolepis* (T_3–K; Fig. 72, *u*) are bilobed; the sinus between the lobes may be very deep. The fossil pollen of P. is disaccate, monosaccate or asaccate.

The P. from the Jurassic and Cretaceous variously combine characters of living genera belonging to one or different tribas, whereas *Pararaucaria* combines characters both of P. and Taxodiaceae, as well as certain characters of Araucariaceae. These synthetic types, regrettably, are of little aid in phylogenetic reconstructions. Moreover, the similarities to other families can be convergent. The same applies to the genera. Characters of different living P. genera are combined in the Cretaceous species of *Pityostrobus*. *Nellostrobus* (J_1), *Pseudoaraucaria* and *Nansenia* (K_1) also belong to synthetic types. The latter genus was established on leafy shoots and resemble *Keteleeria* and *Tsuga*, and according to certain characters resemble *Abies* and *Picea*. The Lower Cretaceous genus *Keteleerioxylon* was established for petrified wood. According to the structure of the resin canals it is closely similar to *Pinus*, *Picea*, and *Larix*, whereas in other characters it is similar to *Keteleeria*. The Neogene *Colchidia* combines characters of *Pinus* (long and narrow needles in bunches) and *Cathaya* (stomatal structure).

Owing to the fact that in addition to synthetic types, typical *Pinus* cones have been recovered from the Lower Cretaceous, it is conceivable that this genus, or one closely related to it, stands at the roots of the phylogenetic tree of P. The genus *Schizolepis* displays affinities in some characters (non-winged seeds, free bract, absence of regular stomatal files on associated leaves of *Pityocladus*) to Voltziaceae, which are also known to have differentiated short and long shoots (the inception of such a differentiation has already been noted in *Sashinia* – Lebachiaceae). The Upper Triassic genus *Compsostrobus* displays affinities to P. Its two-seeded scales are in the axil of the bract, microsporophylls bear two sporangia under the stalk, pollen is quasidisaccate or disaccate (of the *Alisporites* type). Associated leaves are close to the genus *Thomasiocladus* (J). They have regular files of longitudinally oriented stomata, arranged in two bands on either side of the midrib. However, *Thomasiocladus* in its epidermal characters is more closely allied to Cephalotaxaceae, and differs markedly from Voltziaceae.

The above-mentioned genus *Schizolepis*, also the following two genera below, can be treated as satellite genera of P.

Genus *Pityostrobus* (J–Tr.).

P. species have not been affiliated to any living genera, either because of their poor preservation, or because they combine characters of different genera. For example. *P. palmeri* combines characters of *Pinus* and *Cedrus*; *P. hallii* those of *Pinus* and *Picea*; *P. hueberi* those of *Picea*, *Keteleeria*, *Pseudolarix* and *Tsuga*; *P. virginiana* those of *Picea*, *Pinus* and *Cedrus*. Other combinations are known. Because of this the diagnosis for the genus P. can be given only in a generalized form. It is represented by compact cones with two-seeded scales. The inverted seeds occur on the proximal part of the scale. The bract, if it can be observed, is variously fused with the seed scale. The distinction of species and comparison with living genera is based on the shape of the cones and scales, seed characters, arrangement of the resin canals, vascularization of the seed scale and bracts, etc.

Genus *Pityocladus* (T_3–K; Fig. 72, *s*, *t*)

This includes long vegetative shoots bearing deciduous leaves, and short persistent shoots covered by tightly spaced scales and linear, easily abscissing leaves of the *Pityophyllum* type, with a single vein. The epidermal characters are about the same as in the genera of Pinaceae. In *Pityocladus scarburgensis*, associated with female cones of *Schizolepis*, the stomata are scattered, which is not typical of the Pinaceae. The affiliation to the Pinaceae of species where the epidermal characters are unknown is uncertain. The genus *Archaeolarix* (K_1) displays affinities to *Pityocladus*.

Family Taxodiaceae

T. remains have been described in the literature as *Abietites* (Pinaceae), *Podocarpoxylon* (Podocarpaceae), *Cephalotaxopsis* (the type species belongs to Taxaceae, whereas certain species belong to T), etc. In systematization of leafy shoots the intraspecific polymorphism was frequently neglected and, as a consequence, specimens of one species were assigned to different species and even genera. Extinct T., the same as Pinaceae, were characterized by a combination of characters that differed from those in the living genera, i.e. among T. synthetic types were also present. For example, *Parataxodium* combines characters of *Taxodium* and *Metasequoia*, whereas *Elatides* (*sensu* Harris) combines characters of *Cryptomeria*, *Cunninghamia* and *Taxodium* in the structure of the cones and foliage, and of *Cryptomeria* and *Sequoia* in the wood structure. *Parasequoia* combines characters of *Sequoia* and *Sequoiadendron*. *Sequoia coutsiae* (Palaeogene) comprises shoots and cones with characters of *Sequoiadendron*, *Sequoia*, *Athrotaxis* and even of *Cupressus*. In contrast to the living *Metasequoia*, which can be rather well distinguished from *Taxodium* and *Sequoia*, being as it were interposed between them, the fossil species can only tentatively be included in this genus. In *M. occidentalis* the shoots with juxtaposed leaves (as in *Metasequoia*) bear cones with spirally inserted scales (as in *Sequoia*). The foliage in *M. milleri* is the same as in *Taxodium*. Because of this when isolated parts are found they can be assigned to extant genera only if the connections between different parts are known (cones, leaves, etc.).

Modern T. are trees, but in the geological past among them were evidently shrubs. T. display affinities to Pinaceae in many characters. In some T. the primary axial vascular bundles anastomose (Fig. 67, *b*), which is not known in Pinaceae. Wood fragments of T. are referred to *Taxodioxylon*, *Cupressinoxylon*, *Podocarpoxylon*, *Glyptostroboxylon*, *Taiwanioxylon*, etc.

Leaves are of various types: (1) linear-lanceolate with a petiolate base; (2) awl-shaped, rhomboid in cross-section, at times laterally flattened; (3) scaly. These three types correspond to the form-genera *Elatocladus*, *Geinitzia* and *Pagiophyllum*. According to the epidermal characters the genera *Sewardiodendron*, *Haiburnia* and *Farndalea* display affinities to T. The first two correspond morphologically to *Elatocladus*, whereas the third corresponds to *Pagiophyllum*. They exhibit essential epidermal distinctions: in *Sewardiodendron* the leaves are hypostomatic with two bands of transversely oriented stomata; in *Haiburnia* they are amphistomatic with scattered longitudinally oriented stomata; in *Farndalea* they are epistomatic with two stomatal bands, chiefly of transverse orientations (at times singular stomata or a short stomatal band is present also on the underside). Epidermal structures of these and other types can be seen in younger T. as well.

T. cones are small, round, composed of peltate or flattened seed scales, nearly completely fused with the bracts. The fusion develops in the course of the cone ontogenesis. In *Cryptomeria* the distal part of the seed scale is lobed (as in Voltziaceae). Seeds of the Cretaceous T. are placed into the genera *Taxodiastrum*, *Alapaja*, etc. (Dorofeev, 1979).

T. microstrobili are small, situated on the tips of the branches or in the axil of the leaves, solitary, gathered in groups or racemose aggregates. In *Elatides* (*sensu* Harris) the microsporangia are fused and coalesced with the stalks (as in Pinaceae). The pollen is round, asaccate and acolpate; the pore is usually defined by a conical or hook-like elevation. More seldomly the pore is simple or absent. The pollen frequently splits into two halves. The pollen genera *Perinopollenites*, *Taxodiaceaepollenites*, *Sequoiapollenites*, etc. are commonly affiliated with T.

The similarity between certain T. and the Voltziaceae gives implication to their phylogenetic relationship. On the other hand, the previously mentioned *Pararaucaria* combines characters of both T. and Pinaceae. The most ancient T. (*Elatides sensu* Harris) are known from the Middle Jurassic. During the Jurassic the Sciadopityaceae evidently underwent segregation. Ancient synthetic types provide little evidence of more exact phylogenetic relationships between the genera. It can be assumed that *Sequoia*, *Metasequoia*, *Sequoiadendron* and probably *Taxodium*, were of common origin. In this respect it is of interest that the Eocene species *Metasequoia milleri*, established for microstrobili, associates with branches, cones and wood of the same genus. However, the associated leaves are practically indistinguishable from those of *Taxodium*.

Genus *Elatides sensu* Harris (J_2–K; Fig. 68)

Trees with abscissed shoots. Leaves persistent, rhomboid in cross-section, helically arranged on decurrent leaf cushions. The vascular bundle is accompanied by a resin canal. The stomata occur in bands, of transverse or irregular orientation. Subsidiary cells are papillose. The wood (of *Cupressinoxylon* type) has a large amount of xylem parenchyma and taxodioid pits in the cross-field. Resin canals pass through the cortex. Female cones are terminal, ovoid, up to 6 cm in length, composed of persistent helically arranged scales. The scales are flattened with a long stalk and acute apex. The part free of the bract (ligule) is weakly lobed. Situated on the proximal side of the ligule are 3–5 orthotropous seeds. Microstrobili are in apical clusters. Three fused microsporangia are attached to the shield of the microsporophyll and extend along the stalk, becoming fused to it. The pollen is round with a poorly visible pore evidently situated on a shallow papilla. Microstrobili *Stenomischus* with free sporangia probably associated with some of the species.

Besides the genera mentioned above, *Parasequoia*, *Cunninghamiostrobus*, *Parataxodium*, *Sphenolepis*, *Protosequoia* can also be assigned to T.

Outstanding among T. is the living genus *Sciadopitys*, which is frequently distinguished as a specific family of its own. Its leaves are of two types. One type comprises brown scaly leaves that are helically arranged on long shoots. On the apices of the shoots they are arranged in false, closely spaced whorls and bear in their axils small dwarf shoots, crowned by a single linear leaf with a stomatiferous furrow on the upper side. On both sides of the furrow is a vein. It is believed that these leaves evolved through fusion of two leaves, the furrow corresponding to their lower surfaces. Such leaves, known since the Jurassic, are assigned to *Sciadopitytes*, or to *Sciadopitys*. Leafy shoots with such leaves are assigned to *Sciadopitophyllum*. The pollen in *Sciadopitys* is elliptical, with a short sulcus, and with spinose granules on their surface. A similar pollen associates with the Jurassic *Sciadopitytes*.

Family Cupressaceae

Although the Cupressaceae (C.) produce such female fructifications as 'cone-berries' of junipers, in all other characters of the fructifications most C. display affinities to Taxodiaceae. The similitude is evident in the wood structure, leaf epidermis, pollen. The major distinctions between the families are revealed in the phyllotaxis (chiefly decussate in C.) and other characters of the vegetative shoots. Fossil C. are known from wood remains, leafy shoots sometimes bearing fructifications, also from seeds and pollen. Commonly fossil C. are directly included in the extant genera, or in a few fossil genera, such as *Thuites*, *Cupressinocladus*, that were introduced for vegetative shoots. The affiliation of the latter to C. is at times questionable. For instance, Jurassic species belonging to Cheirolepidiaceae (Fig. 72, *a*) have been included in the genus *Cupressinocladus*, although the Upper Cretaceous species of the same genus can most likely be assigned to C.

Without knowledge of the epidermal characters the discrimination of the leafy shoots of various C. is frequently impossible (for example, leaves of *Cupressus*, *Juniperus* sect. *Sabina*, *Diselma* and *Widdringtonia* are very much alike). Very similar flattened shoots with downfolded margins of the lateral leaves can be observed in *Microbiota* and *Biota*, also in some species of *Cupressus*, *Chamaecyparis*, *Thuja*, *Libocedrus*, etc. It should be remembered that decussate and whorled phyllotaxis, as well as flattened shoots are also characteristic of the Cheirolepidaceae. Among members of C. synthetic types are known. In *Cupressinocladus interruptus* (Newb.) Schweitzer the leafy shoots resemble those of *Thuja*, the branching pattern is opposite as in *Libocedrus*, and the female cones are comparable to those of *Chamaecyparis* (Vincent and Basinger, 1984).

From the increasing fusion of the seed scales with the bracts it can be inferred that C. evolved from Taxodiaceae (not vice versa). Leafy shoots, wood fragments of the *Cupressinoxylon* or *Juniperoxylon* types, and female cones with decussate scales, found in the Upper Triassic, are sometimes affiliated with C. However, these fossil remains have been inadequately studied. Indisputable C. have been recorded since the Upper Cretaceous. Most of the living genera are known since the Palaeogene. According to the epidermal characters the phyllodes of the genus *Protophyllocladus* (K_2) were compared with C.; usually they are assigned to the Podocarpaceae, despite the absence of stomatal files (Krassilov, 1979).

Family Podocarpaceae

General geographical considerations had a strong bearing on the systematics of fossil Podocarpaceae (P.). Taking into account the fact that the living P. are nearly exclusively restricted to the Southern Hemisphere, Florin (1963) advanced the hypothesis that in the geological past their distribution pattern was the same. As a consequence, the fossil remains of Northern P. were affiliated with other families.

P. are much more diverse in the structure of their foliage and female fructifications, as well as in the arrangement of the microstrobili, in comparison with other coniferalean families. Some P. produce cones of the ordinary habit, whereas in others only a fleshy receptacle remains of the cone with a single seed. Taken separately, dispersed parts of fossil P. can be easily confused with parts of other plants. The two-seeded scales have been mistaken for a cycadalean cladosperm of the *Beania* type. The saccate pollen of P. can easily be confused with the pollen of Pinaceae and Peltaspermales. It is significant that the pollen of some of the Mesozoic P. were termed *Tsugaepollenites*. The Jurassic *Retinosporites* which later was found to belong to Peltaspermales (as a younger synonym of *Pachypteris*; Bose and Roy, 1967–1968) was initially affiliated to the living genus *Acmopyle*. Shoots of the Jurassic P. of the genus *Cyparissidium* were wrongly assigned to Cupressaceae. The genus *Tritaenia* (K_1) that cannot be clearly distinguished from *Podocarpus*, was described from leaves that at first were assigned to *Abietites* and affiliated with Pinaceae.

P. are represented both by trees and shrubs. At present the diagnostic characters of the wood, typical of the family as a whole, cannot be defined. Wood fragments are assigned to the genera *Podocarpoxylon* and *Circoporoxylon*. In general of most frequent occurrence in P. are tracheids with uniseriate bordered pits, but species have been encountered where araucarioid pitting predominates. Resin canals are absent.

P. produce all the major leaf types known in conifers. Of these two–three leaf types may successively appear in the course of ontogenesis from seedling to the mature plant. In several genera the leaves are flattened, not dorsiventrally, but laterally (bilateral leaves; Fig. 67, *f*). *Phyllocladus*, in addition to needle-shaped leaves, develops phylloclads with pinnate venation. Isolated leaves or leafy shoots of P. are included in the genera *Elatocladus*, *Brachyphyllum*, *Pagiophyllum*, *Cyparissidium*. Seemingly all P. are characterized by stomata of longitudinal orientation. In narrow leaves with one vein the stomata are arranged in two bands on both sides of the vein. In broad leaves with multiple veins and phylloclades the stomata are scattered over the surface of the lamina.

P. microstrobili are small, resembling the pinaceous. They consist of an axis and peltate microsporophylls. Two sporangia are attached to the shield under the stalk and sometimes are fused to the lower keel of the stalk. Microstrobili are solitary, inserted in the leaf axil, or occur in groups of various patterns. At times the group of microstrobili is situated in the axil of the leaf. The group may consist of a simple or branching axis with several bracts, each bract bearing a microstrobilus in its axil. The evolution of the microstrobili is ordinarily viewed as proceeding from a group of microstrobili to solitary ones, which is doubtful, inasmuch as the microstrobili in all primitive conifers are solitary.

The pollen is chiefly disaccate, more seldom with three or more sacci. In the living *Saxegothaea* sacci are absent. The diversity in the size, inclination and outline of the sacci in the disaccate grains is comparable to that observed in Pinaceae. Sometimes a trilete structure on the proximal pole can be observed.

The female cones in P. can be reduced to a single seed scale, situated in the bract axil on a receptacle (fleshy shoot apex). The seed scale nearly completely embraces the seed or only its lower part and is termed an epimatium. In some P. the seeds are orthotropous and the epimatium is cup-shaped. The degree of fusion of the integument with the nucellus, also the invertedness of the seeds may vary even within one genus. The degree of fusion of the epimatium with the integument and bract also varies. According to a certain belief, the epimatium may correspond to the aril in the Palissyaceae and the lagenostomalean cupule, but studies of the Triassic and Jurassic P. do not support this viewpoint. In *Rissikia* the axil of the three-lobed bract bears three seed scales with inverted seeds. In *Mataia* the distal region of the seed scale is infolded and embraces the lower parts of the seeds. In living P. the degree of protection of the seeds is even more pronounced, so that only an opening in the jar-like epimatium extends to the micropyle. The structure of *Rissikia* is comparable to that seen in other conifers. Hence, this provides corroborative evidence for the homology between the epimatium and the seed scale.

The female fructifications have been well studied in *Rissikia*, *Mataia*, *Scarburgia* (see below), *Trisacocladus* (associate with microstrobili that produced pollen with three quasisacci), *Nipaniostrobus*, *Nipanioruha*, *Mehtaia*, and *Sitholeya*. Cones, tentatively assigned to *Podocarpus*, have been reported since the Lower Cretaceous.

The modification of the seed scale into a seed envelope is known in the Araucariaceae and Cheirolepidaceae, which first appear in the geological record in the uppermost Triassic–Jurassic, i.e. they co-existed with the first P. The Triassic *Rissikia* is similar to Voltziaceae and *Cryptomeria* of the Taxodiaceae. It has been suggested that the genus *Tricranolepis*, commonly assigned to the Cycadocarpidiaceae, should be included in P. *Rissikia* differs from the above-cited families in their strongly dissected bracts. According to the structure of the pollen and microstrobili, P. show a propensity towards Pinaceae and those Voltziaceae in which the microsporophylls are hyposporangiate. Actually the sporangia are more numerous in Voltziaceae. It is conceivable that P. might have evolved from the Voltziaceae.

Genus *Rissikia* (T; Fig. 73, *d–k*)

The branches of the last order bear small scale leaves at the base, and, more distally, helically arranged linear, laterally flattened leaves (as in *Podocarpus* of the *Dacrycarpus* section), or leaves rhomboid in cross-section. Along each leaf face is an indistinct stomatal band consisting of irregular files of longitudinally oriented stomata. Four subsidiary cells bear small papillae or are proximally thickened. Microstrobili are round. Microsporophylls are peltate, with two sporangia attached to the lower part of the shield and to the stalk. The pollen is disaccate. The exine is striated on the proximal side, in which *R.* differs from all known P. Female cones are situated on the tips of leafy shoots. The bracts are divided into three deep lobes; in turn, the margins of each are notched into three shallow lobes. One lobe of the three-lobed seed scale corresponds to each of the main lobes. Each of the seed scale lobes bears one or two inverted seeds, attached by a long stalk to the lower part of the seed scale.

Genus *Mataia* (J; Fig. 73, *l*, *m*)

Leaves are dorsiventrally flattened with a distinct midrib. On either side of the latter is a stomatal band composed of longitudinal stomatal files. The cone is loose; its axis is leafy at the bottom, then naked, and further up bears bracts and seed scales. The bracts are small, triangular. The seed scale has a thick stalk and a broad sagittate lamina, which is infolded in the distal part and conceals the lower parts of the inverted seeds.

Genus *Scarburgia* (J; Fig. 73, *a–c*)

Leafy shoots (of the *Cyparissidium* type) are slender with tightly spaced and slightly deflected lanceolate amphistomatic leaves. The stomata are scattered, most frequently of longitudinal orientation. The leaf margin is microdenticulate. The female cones are loose with a slender axis. The seed scale has a short stalk and triangular lamina. The seed is round, concealed by the epimatium. The nucellus has an apical beak. The microstrobili (of the *Pityanthus* type) are cylindrical with a small number of basal sterile scales. The sporophyll stalk is long, triangular in cross-section, crowned by a shield with a heel. Two sporangia are attached to the lower part of the shield and to the keel of the stalk. Pollen is disaccate.

Family Taxaceae

Living T. are shrubs and small trees. Most T. have supplementary spiral thickenings on the tracheid walls of secondary wood. On the basis of this character even Palaeozoic wood fragments (genera *Taxopitys*, *Prototaxoxylon*, *Parataxopitys*, *Platyspiroxylon*) have been tentatively affiliated with T. However, such thickenings are also known in other conifers (for example, Pinaceae), and could probably have been present in other gymnosperm orders.

T. leaves are linear, with a narrowed or petiolate base, usually expanded in the shoot plane. Leaves

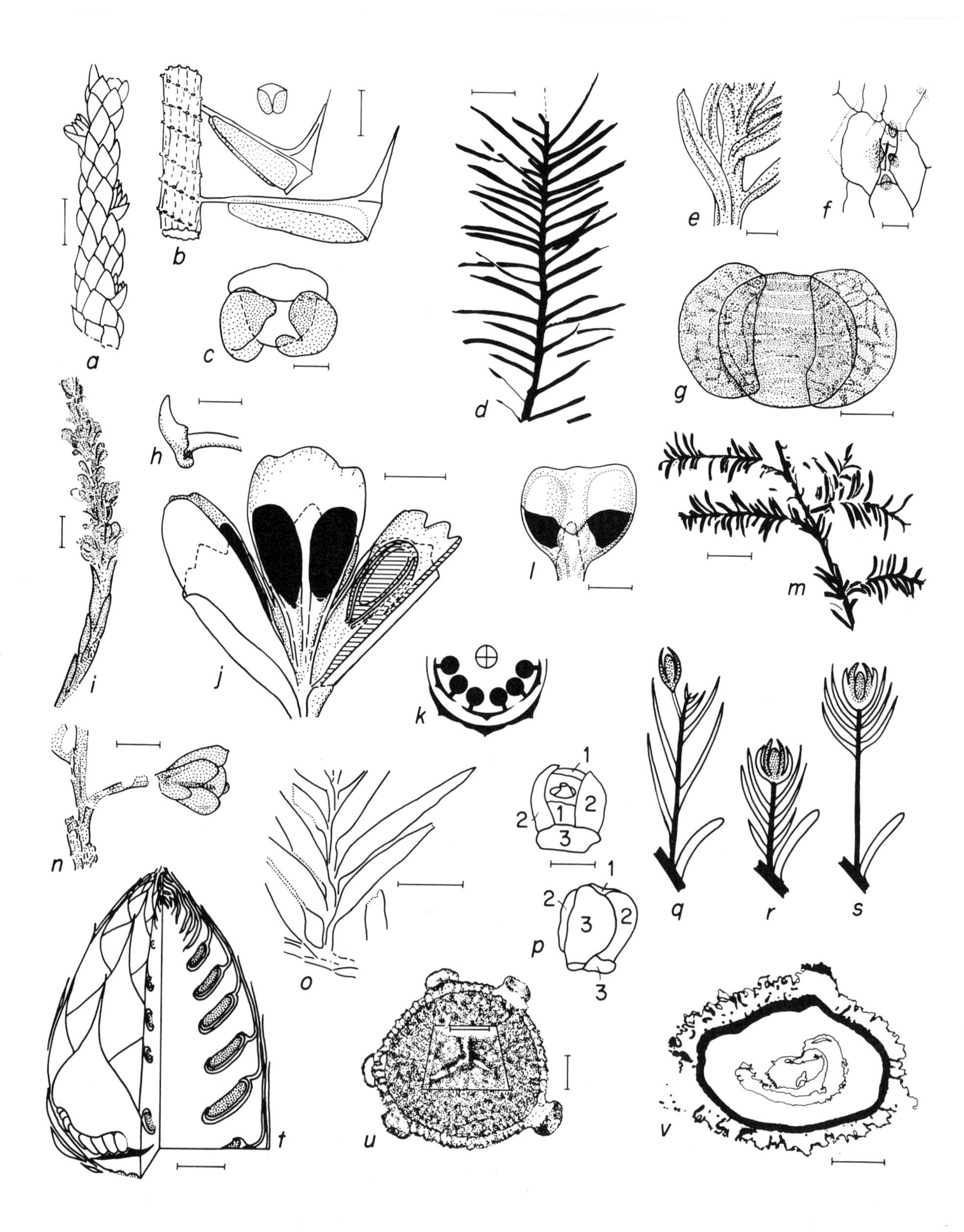

a
b
c
d
e
f
g
h
i
j
k
l
m
n
o
1
2
1
2
3
1
2
3
2
3
p
q
r
s
t
u
v

are hypostomatic, stomata are longitudinally oriented, arranged in two bands along both sides of the midrib, and within the bands in files.

Microstrobili in T. are solitary (more seldom in small clusters). Microsporophylls are radial with a distal shield in the centre of which is attached a stalk, and around it, sporangia (*Taxus*). In *Torreya* and *Marskea* the sporangia are abaxially attached to the distal expansion of the stalk. In this case the shield may be reduced. In *Austrotaxus* the microstrobili consist of sporangiophores, each being subtended by a bract and crowned by synangia of three–four sporangia. Owing to the peculiar traits of the microstrobili it was suggested that this genus should be separated in an independent family Austrotaxaceae. T. pollen is round, asaccate, at times with a vestigial sulcus.

Female fructifications resemble externally cones of certain Podocarpaceae, but are, in effect, dissimilar in their organization. On the short apex, or on a small lateral branchlet one seed is borne, that is surrounded by a fleshy envelope – the aril, leaving the seed apex exposed (Fig. 73, *r*, *s*). In the lower part the arillus is closed by scales. The arillus may be fused with the integument. The vascularization of the integument and arillus differs in different T. The integument has two (very rarely three or four) strands or altogether lacks them. The aril also can be non-vascularized. If vascularization is present, it consists of several bundles arranged in a ring. In the course of ontogenesis, as in all conifers, the integument originates in the form of two tubercles, that develop into a ring-shaped swelling which later expands upwards. Frequently, the aril develops from the ring-shaped swelling. There can be no doubt concerning the homology between the T. integument and that in other conifers. The aril is frequently regarded as comparable to the podocarpaceous epimatium, i.e. to a seed scale. Accordingly, the true apical position of the T. seed is contested, since it supposedly develops on one side of the apex. According to another viewpoint, the seed scale, being a modified axillary shoot, is lacking in T. The aril in T. is sometimes homologized with the palyssiaceous aril and regarded as an innovation. It is also believed that the T. aril evolved via fusion of scales situated below the seed. Fossil material does not unambiguously support one or another viewpoint. Arillate T. seeds are known since the Triassic (genus *Palaeotaxus*).

The apical position of T. seeds is confirmed by their vascular system, in which no traces can be found which indicate that the seed was situated in the axil of any one of the scales surrounding the aril. The vascular system of the aril is a direct continuation of the vascular system of the axis, but it has been studied only in living forms. The scales, aril and seeds, studied in most detail in the Jurassic *Marskea*, reveal the same arrangement as in living T. Solitary seeds, not aggregated in cones,

Fig. 73. Podocarpaceae (*a–m*); Taxaceae (*n–p*, *r*, *s*) and their comparison with Lebachiaceae (*q*); genus *Lasiostrobus* (*t–v*); Upper Carboniferous (*t–v*), Upper Triassic (*d–k*), Middle Jurassic (*a–c*, *l–p*), recent (*r*, *s*); Western Europe (*a–c*, *n–p*), South Africa (*d–k*), New Zealand (*l*, *m*), North America (*t–v*); *a–c*, *Scarburgia blackii* (Harris) Harris, leafy shoot of *Cyparissidium blackii* (Harris) Harris type (*a*), microstrobilus *Pityanthus scalbiensis* van Kon. van Citt. and cross section of microsporophyll, microsporangia stippled (*b*), pollen (*c*); *d–k*, *Rissikia media* (Ten.-Woods) Town., leafy shoot (*d*) and its part (*e*), stoma (*f*), pollen (*g*), microsporophyll with abscissed microsporangia, lateral view (*h*), compound polysperm (*i*), assemblage of three seed scales with subtending bracts, seeds in black, left seed shown in section and shaded (*j*), diagram of the same assemblage (*k*); *l*, *m*, *Mataia podocarpoides* (Ett.) Townr., seed scale with bract (shown by dashed line), seed in black (*l*), leafy shoot (*m*); *n–p*, *Marskea jurassica* (Flor.) Harris, monosperm in attachment (*n*), leafy shoot of *Elatocladus* type (*o*), monosperm and surrounding scales, numerals mark successive pairs (*p*); *q–s*, homologization between dwarf fertile shoots of Lebachiaceae (*q*), *Taxus* (*r*) and *Amentotaxus* (*s*) as suggested by Harris; *t–v*, *Lasiostrobus polysacci* Tayl., reconstruction of microstrobilus (*t*), pollen with five sacci, inset shows somewhat magnified trilete slit (*u*), cross section of pollen (*v*; in centre, collapsed pollen content). Scale bar = 1 cm (*d*, *h*, *m*), 5 mm (*i*, *t*), 2 mm (*a*, *e*, *i*, *j*, *n*, *p*), 1 mm (*l*), 0.5 mm (*b*), 20 μm (*c*, *f*, *g*), 5 μm (*u*, *v*). Modified from: Harris, 1979 (*a–c*, *n*, *o*) and 1976 (*p–s*); Townrow, 1967 (*d–m*); Taylor, 1970 (*t*, *u*); Taylor and Millay, 1977b (*u*, inset; *v*).

are known in the Palaeozoic Buriadiaceae, but here the seeds are not apical. The opinion has been stated that T. cannot be phylogenetically linked with all the above discussed conifers, and comprises an independent phylogenetic lineage of gymnosperms. However, T. can be derived from Lebachiaceae (Fig. 73, *q–s*), by comparing the scaly dwarf seed-bearing shoot of T. to the axillary shoot in Lebachiaceae. In this case it has to be admitted that the seed shifted from the lateral position to the apical, which is quite probable considering the high 'mobility' of the seeds in primitive Pinopsida. There is then no need for the complete phylogenetic segregation of T. from all other conifers. Apart from this, it is necessary to take into account the common features between T. and other coniferalean families. It is relevant that the genus *Austrotaxus* occupies an intermediate position between T. and Podocarpaceae, and, as a consequence, has been placed in one or other of these families by different taxonomists. The Podocarpaceae are known to have a cup-shaped radially symmetrical epimatium, similar to the aril in T. The pollen, including the fine exine structure, is similar to the asaccate pollen in other conifers, particularly Taxodiaceae. T. display affinities to certain Pinaceae in the additional spiral thickenings of the tracheids. T. also have many traits in common with Cephalotaxaceae. In those classification systems where Taxales is considered an independent order some taxonomists also include Cephalotaxaceae in this order.

Well-developed forms of T. appear in the geological record at the same time as other surviving coniferalean families (Pinaceae, Araucariaceae, Taxodiaceae, Podocarpaceae). In Mesozoic T. we again encounter synthetic types that combine characters of different extant genera. For instance, in the decussate arrangement of the scales beneath the seeds *Marskea* displays affinities to *Torreya* and *Amentotaxus*, and in its epidermal structure (ridge-like bordering of stomatal bands and concentration of papillae in the bands) to *Taxus*. *Marskea* and *Amentotaxus* both have a basally naked fertile dwarf shoot, and according to the microsporophylls *Marskea* is closer to *Torreya*. Evidently the Mesozoic genera *Tomharrisia*, *Bartholinodendron*, *Florinia*, etc. described from leafy shoots, belong to T.

Genus *Marskea* (J; Fig. 73, *n–p*)

Branches of the last order are attached nearly in juxtaposition in one plane. Leaves are juxtaposed, decussate in adjacent pairs, but due to the twisting of their bases spread out in one shoot plane. The leaf lamina is flat with two stomatal bands consisting of poorly defined files of longitudinally oriented monocyclic stomata. The subsidiary and all non-stomatal cells of the stomatal band are papillose. The seeds are borne on the tips of the lateral dwarf shoots that extend out singly. The lower part of the dwarf shoot is naked; crowded, decussately attached scales appear near the seeds. The aril is thin, but with a thick cuticle and has a small opening for the seed apex; it is probably to a small extent free from the integument. Microstrobili are solitary, the distal ends of microsporophylls are acuminate and bent upwards. Microsporophylls bear abaxially 2–3 microsporangia. Pollen is globose, smooth with an area of a thinner sporoderm.

Family Cephalotaxaceae

Cephalotaxads (C.) are at present represented by a single genus *Cephalotaxus*. Sterile shoots recovered from the Cretaceous were assigned to this genus according to their epidermal and morphological characters. The genus *Thomasiocladus* (J–K_1?) shows affinities in its epidermal characters and phyllotaxis to *Cephalotaxus*. The Cretaceous genus *Cephalotaxopsis*, at times included in C., belongs in part to Podocarpaceae, and in part to Taxodiaceae.

C. are highly significant for the understanding of the coniferalean systematics. According to the wood structure (tracheids bear spiral thickenings) and structure of the leafy shoots C. are close to taxads. Particularly great is the similarity between C. and *Torreya*. Their leaves are juxtaposed. The paired leaves are arranged in a biseriate helix. C. leaves are hypostomatic, whereas in seedlings they are amphistomatic, i.e. the same as in taxads. The stomatal bands and stomata are of similar organization. Microstrobili are clustered on head-like structures, subtended by scales. Each microstrobilus is subtended by a bract borne on its stalk. Microsporophylls are few, and consist of a stalk with an apical mucro, under which 3–8 sporangia are attached. Sometimes the bract of the microstrobilus is compared to the distal lamina of the microsporophyll in other conifers, while the microstrobilus itself is compared to a system of branching sporangiophores in voltziaceous (?) *Darneya* and *Sertostrobus*. According to the structure of their microspores C. are related to taxads.

The female cone consists of an axis with several decussate bracts, each bearing a pair of axillary ovules (later only one seed develops for the whole cone). The seed stalks are short with a transverse swelling frequently interpreted as a reduced epimatium. This swelling more likely corresponds to a distal expansion of the seed stalk, as is typical of the Cordaitanthales (Fig. 64, *e*, *j*). It can be assumed that the axis of the axillary shoot on which the two seeds were borne was nearly completely fused with the bract. The pair of seeds with their vascular bundles are then comparable to the whole axillary complex in lebachians. Evidence in favour of this viewpoint is found in teratological forms of C. where the seeds (their number in the axil of the bract may increase to four) are accompanied by scales surrounding a group of seeds. These forms are, in fact, of the same organization as the polysperm in Lebachiaceae. In C. two vascular bundles extend all along the integument – an ancient character typical of the cordaitanthaleans.

C. thereby display many affinities to Podocarpaceae (it is no wonder that part of the Cretaceous *Cephalotaxopsis* was transferred to Podocarpaceae), lebachians and taxads. This gives additional support to the affiliation of the taxads with other conifers.

Satellite genera of the order Pinales

Among the genera described below are *Lasiostrobus* – a genus of uncertain systematic affinities, and a group of form-genera for vegetative shoots (for which an identification key is given).

Genus *Lasiostrobus* (C_3; Fig. 73, *t–v*)

Only microstrobili are known. They are oval in shape with an acuminate apex. The thick axis bears tightly packed in a spiral fleshy, hypostomatic and hairy microsporophylls; their distal part is bent upwards and ends in a long mucro. At the point of flexion is a small fleshy heel. To the lower surface of the microsporophylls all along their length to the heel are attached 7–10 elongate thick-walled microsporangia, arranged in a single transverse row and dehiscing by a longitudinal slit. The pollen is round with 3–8 small saccus-like swellings on the distal side not far from the equator. At times a trilete slit can be perceived. The exine is composed of a homogeneous nexine and three-layered sexine, where the middle layer is columellate and remotely resembles the columellar layer in angiosperms. The surface of the sexine is densely covered by tubercles and rugae.

Form-genera of vegetative shoots

The key and diagnoses of the genera presented below are given after Harris (1969a, 1979), with certain amendments. Many genera included in the key are currently identified on the basis of epidermal characters. If the epidermal characters are known, then they are used either for the distinction of species, or for the segregation of certain genera. For example, the species of *Elatocladus* can be distinguished according to epidermal characters, or morphologically identical leaves, are distributed according to epidermal characters among various genera *Thomasiocladus* (Cephalotaxaceae), *Tomharrisia* and *Bartholinodendron* (Taxaceae), *Farndalea* and *Sewardiodendron* (Taxodiaceae). Decisions concerning the feasibility of keying out the leaf genera on the basis of epidermal characters alone should be made individually in each case (see also Stewart, 1983).

1. Leaves undivided. 2
   Leaves forked . . . . *Carpentieria*
   (see also: Lebachiaceae; *Buriadia*, Buriadiaceae)
2. Leaves contracting to base and apex 3

Leaves contracting only to apex 5
3. Leaves with one vein 4
Leaves with many veins. . . . *Podozamites*, *Lindleycladus*, *Yuccites*, *Aethophyllum* (see family Voltziaceae and its satellite genera), *Araucariodendron* (Araucariaceae?)
4. Leaves very long (the length-to-width ratio exceeds 20:1, frequently up to 100:1), usually borne on dwarf shoots. . . . *Pityocladus* (satellite genus of Pinaceae) *Borysthenia* (Voltziaceae)
Leaves shorter (length-to-width ratio less than 20:1) borne on elongate shoots *Elatocladus*
5. Phyllotaxis helical 6
Phyllotaxis decussate or whorled . . . . *Cupressinocladus* (see also *Frenelopsis*, Cheirolepidiaceae)
6. Leaves elongate, the length-to-width ratio not less than 5:1 7
Leaves shorter 8
7. Leaves flattened in cross-section *Elatocladus*
Leaves round or rhomboid in cross-section. . *Geinitzia*
8. Free part of leaf longer than width of leaf cushion 9
Free part of leaf shorter than width of leaf cushion . . . . *Brachyphyllum*
9. Free part of leaf adpressed to stem . . . . *Cyparissidium*
Free part of leaf spreading 10
10. Free part of leaf thin, flattened, spreads in shoot plane . . . . *Elatocladus*
Free part of leaf thick, oriented upwards, not spread in shoot plane . . . . . *Pagiophyllum* (see also: *Haiburnia*, Taxodiaceae)

The diagnoses for some genera given in the key are outlined below. These diagnoses may be applicable to conifers of any age.

Genus *Brachyphyllum* (Fig. 72, *n*)

Leaf arrangement helical. Leaf consists of a basal cushion, tapering into a free part, the length of which (i.e. upper surface) or the whole length of the leaf is less than the width of the cushion.

Genus *Cupressinocladus* (Fig. 72, *a*)

Phyllotaxis decussate, or in alternating whorls. Leaves are small, scale-like, or longer, dorsiventrally flattened, spreading, but not constricted basally into a petiole.

Genus *Cyparissidium* (Fig. 73, *a*)

Phyllotaxis helical. Free part of leaf contracting gradually from cushion; somewhat flattened, adpressed to stem, elongated. Length of leaf exceeds width of cushion.

Genus *Elatocladus* (Fig. 73, *o*)

Phyllotaxis helical, more seldom leaves are opposite, but then arranged in one plane. Leaves elongated, dorsiventrally flattened, diverging from stem. The leaf has one vein. In some species the leaf base is petiolar.

Genus *Geinitzia* (Fig. 72, *v*)

Phyllotaxis helical. Free part of leaf crescent-shaped or turned back, needle-like, of same thickness in vertical and lateral dimensions; merges with cushion without contracting.

Genus *Pagiophyllum* (Fig. 72, *m*)

Phyllotaxis helical. Leaf width nearly the same as the cushion, length exceeds width of cushion. Leaf lamina at times slightly narrowed at base, width exceeds thickness in cross-section.

Satellite genera of the division Pinophyta

The genera described below were tentatively affiliated with definite orders. For instance, leaves with palmate dissection of the lamina (*Rhipidopsis*, *Psygmophyllum*, *Ginkgophyllum*) were tentatively included in the Ginkgoales, whereas pinnate leaves (*Yavorskyia*) were included in cycadophytes (Cycadales and Bennettitales). Such a distribution of the genera between orders, although tentative, deceives the reader in that it leads to a false belief concerning the possible appearance of ginkgos in the Carboniferous, and implies the presence of cycadophytes in the centre of Angaraland in the Permian, etc. It is more advisable to place such genera in parataxa as adopted in palaeopalynology (anteturma, turma, etc.). This was already accomplished in relation to leaves with an entire lamina and parallel or fan-shaped venation (Maheshwari and Meyen, 1975). For leaves that presumably belonged to pteridosperms the groups suggested are Gondwanides, Compsopterides, Cardiopterides, Syniopterides (Osnovy paleontologii, 1963b). A more detailed classification for gymnosperm leaves was proposed by Němejc (1968). The genera established from permineralized wood fragments and impressions–compressions of seeds are separately discussed.

Genus *Ginkgophyllum* (C–P)

Leafy shoots with helically arranged leaves. Usually only isolated leaves are found. They are wedge-shaped, repeatedly dichotomously divided, with a small number of veinlets in the end linear lobes. The distinctions of *G.* from *Sphenobaiera* are unclear.

Genus *Psygmophyllum* (P; Fig. 74, *f*, *g*)

The bulk of the genus is variously treated in the literature. Treating it in a strict sense, and taking *P. expansum* as the type species, to this genus can be assigned large leaves (or phyllodes?), dissected into lobes of irregular outline. The dissection is mixed pinnate, palmate and forked. From the constricted base into the leaf and its lobes extends a thick midvein, that, seemingly, is composed of several strands. The lateral veins curve outwards. The margin is dissected into large lobes. In smaller leaves the lamina is less dissected; in the smallest ones it is undivided with slanting marginal lobes and a midrib only near the petiole. The stomata are not in bands and files. Sometimes the genera *Syniopteris*, *Iniopteris* and *Comipteridium* are included in *P.* (Burago, 1982). From *P.* transitions can be observed to *Mauerites*, *Rhaphidopteris* and *Comia*, i.e. to genera belonging to the order Peltaspermales.

Genus *Rhipidopsis* (P–T_1?; Fig. 74, *h*)

Leaves are palmate, petiolate. The lobes are wedge-shaped with a straight or convex margin. One vein enters the lobe base, dichotomizing several times. Its branches begin to emerge at the margin in the lower third of the lobe. The lateral lobes are smaller than the others, at times much reduced. *R.* differs from ginkgoalean leaves in the more proximal emergence of the veinlets at the margin. *R.* leaves were probably produced by plants with fructifications *Sporophyllites* (Peltaspermaceae, Fig. 52, *v*, *w*). This refers only to the type species *R. ginkgoides*. No information is available as yet on the fructifications linked with other species of the genus.

Genus *Ruehleostachys* (T)

Isolated candle-like microstrobili consist of a thick axis with bunches of 6–8 long sporangia tightly arranged in a shallow spiral. The pollen is quasidisaccate. To the long naked stalk of the microstrobilus was attached a rather long leaf with a petiolate base and indistinct longitudinal venation. The leaf is amphistomatic with a large number of randomly scattered longitudinally oriented stomata. The guard cells are superficial. The subsidiary cells are not conspicuous among the non-stomatal epidermal cells. *R.* is at times affiliated with the conifers (Grauvogel-Stamm and Schaarschmidt, 1979).

Genus *Zamiopteris* (P; Fig. 74, *a–e*)

Leaves resembling those of *Cordaites*, but, in contrast, the lateral veins curved outwards, whereas the axial veins are clustered in a bundle. On the lower surface compact stomatal bands are interposed between the veins. The stomata are oriented along the veins or irregularly. In the stomatal bands structure *Z.* is close to *Lesleya delafondii* and to certain species of *Dicranophyllum*. But leaves have been encountered which according to their

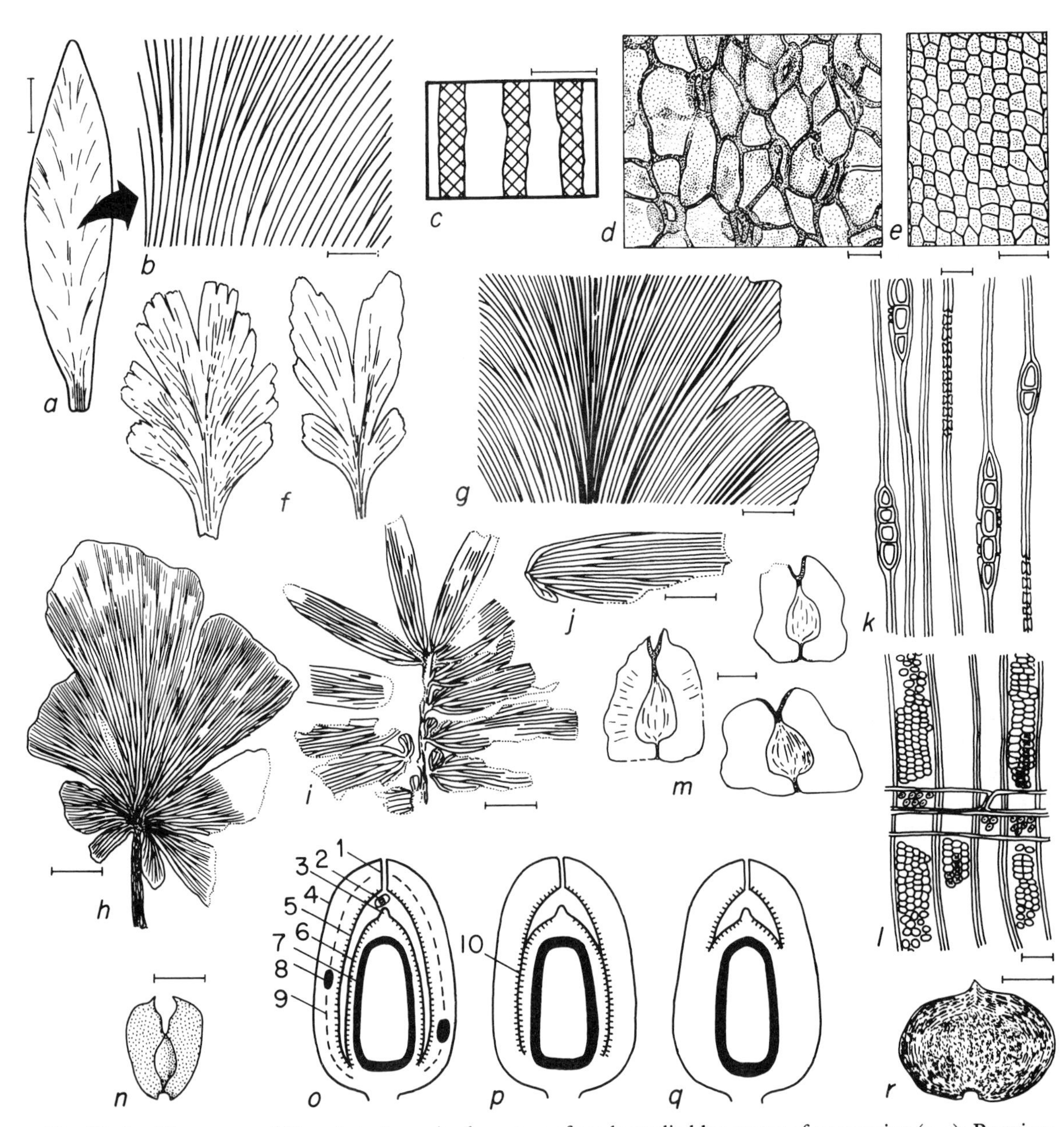

Fig. 74. Satellite genera of Pinophyta (*a–n*, *r*); characters of seeds studied by means of maceration (*o–q*); Permian; Siberia (*a*, *b*, *i–m*, *r*), Pechora basin (*c–e*, *h*), Fore-Urals (*f*, *g*), Western Europe (*n*); *a*, *b*, *Zamiopteris glossopteroides* Schmalh.; *c*, *d*, *Zamiopteris* sp., arrangement of stomatal (shaded) and non-stomatal bands (*c*), stomata (*d*); *e*, *Z. neuburgiana* S. Meyen, upper epidermis; *f*, *g*, *Psygmophyllum expansum* (Brongn.) Schimp.; *h*, *Rhipidopsis ginkgoides* Schmalh.; *i*, *j*, *Yavorskyia mungatica* Radcz., leaf (*i*) and pinnule (*j*); *k*, *l*, *Araucarioxylon rangeiforme* Lep., uniseriate rays in tangential section (*k*), pitting of tracheids and cross fields (*l*); *m*, *Samaropsis irregularis* Neub.; *n*, *S. ulmiformis* Goepp.; *o–q*, maceration resistant seed membranes (*o*, seed with free integument; *p*, seed with integumental and nucellar cuticles stuck together; *q*, seed with fused integument and nucellus), inner cuticle of integument and nucellar cuticle shown by short dashes corresponding to sutures between cells (1, micropyle; 2, pollen; 3, nucellar beak or salpinx; 4, external cuticle of integument; 5, inner cuticle of integument; 6, nucellar cuticle; 7, megaspore membrane;

morphological and epidermal characters are transitional from *Z.* to *Cordaites*. Obviously *Z.* are allied either to Cordaitanthales, or Dicranophyllales (Meyen and Smoller, in press).

Genus *Yavorskyia* (P_2–T_1; Fig. 74, *i, j*)

Leaves pinnate; pinnules are nearly uniform, linear or lanceolate, attached by a constricted base into which one vein enters. It is forked and its branches extend to the margins, emerging in them in the middle of the pinnule. Each branch issues more veinlets extending upward towards the leaf apex. In some species the pinnule is accompanied on the acroscopic side (facing the leaf apex) of the base by a small lobe into which one vein enters, further dichotomizing.

Genera for wood fragments

The fossils described below may include not only gymnosperms but also progymnosperms, especially when the Lower Carboniferous and Devonian specimens are included. The classification of wood fragments is based mainly on secondary wood characters. The following characters are taken into account: (1) the pitting of the tracheids (araucarioid or mixed, presence of a torus; the group arrangement of pits); (2) types of thickening on the tracheid walls (trabeculae, crassulae, i.e. transverse ridges between pits or groups of pits; thin or flat ribbon-like spiral thickenings); (3) the presence of a wood parenchyma; (4) wall thickness in ray cells, their pitting; (5) the presence of ray tracheids; (6) the type of pitting on the cross-field (cupressoid, piceoid, taxodioid, pinoid, fenestrate, podocarpoid); (7) the width and height of rays; (8) the presence and structure of resin canals. When primary wood, pith and cortex are preserved, of importance are then the mesarch, exarch or endarch type of maturation of primary wood, the pith septation, and presence of tracheids and secretory cavities in the pith, also various characters of the bark (for details see: Kräusel *et al.*, 1961–1962; Lepekhina, 1972; Lepekhina and Oleinikov, 1978; Lepekhina and Yatsenko-Khmelevsky, 1966; Maheshwari, 1972). A survey of the major genera (chiefly of conifers) established on the wood structure was presented by Stewart (1983).

Genus *Araucarioxylon* (recorded since the Carboniferous; Fig. 74, *k, l*)

Pitting on radial walls of tracheids is araucarioid, pits on cross-fields are cupressoid. Spiral thickenings and woody parenchyme are absent. Tangential and horizontal walls of ray cells, smooth, lacking pores. Width of rays is up to 5 cells. It was suggested that only specimens with uniseriate rays should be included in *A.*, whereas those with multiseriate rays should be included in *Dadoxylon*. No matter how we treat these two genera, in any case they include species of cordaitanthaleans, coniferaleans and evidently other gymnosperms.

Genus *Dadoxylon* (recorded since the Carboniferous)

Stems with the same wood type as in *Araucarioxylon*, but in contrast, primary wood is preserved, which in *D.* is endarch. The pith is aseptate, homo- or heterocellular. According to another interpretation *D.* differs from *Araucarioxylon* in the ray width (see above).

Dispersed seeds

Many dispersed seeds (and ovules; for reasons of space both are treated together below) exhibit salient characters, enough to identify certain orders. This refers to *Pachytesta* (Trigonocarpales), *Lagenostoma* (Lagenostomales), *Mitrospermum* and *Nucellangium* (Cordaitan-

8, secretory organ; 9, trace of vascular bundle; 10, stuck cuticles of integument and nucellus, sutures faced to opposite sides); *r*, *Tungussocarpus tychtensis* (Zal.) Such. Scale bar = 2 cm (*a, h*), 1 cm (*g, i*), 5 mm (*j, n, r*), 2 mm (*b, m*), 0.5 mm (*c*), 100 μm (*e, k, l*), 20 μm (*d*). Modified from: Meyen, 1970 (*e*) and 1982b (*m*); Zalessky, 1937 (*g*) and 1934 (*h*); Radczenko, 1936 (*i, j*); Lepekhina, 1972 (*k, l*); Goeppert, 1864–1865 (*n*); Neuburg, 1948, after Zalessky (*r*).

thales), etc. Other genera (e.g. *Samaropsis*) are obviously collective. The classification of seeds depends upon the state of their preservation. Some genera are established on characters, that are visible in imprints, others require maceration of compressions, still others were introduced for petrified remains. Important characters in studies of imprints and non-macerated compressions are the seed outlines, structure of their base (notched, round, attenuated) and apex (flattened, round, beaked, auriculate, notches of various forms, etc.), presence of a border or samara around the central part (nucleus), their form and proportions, surface characters in various parts of the seed (smooth, wrinkled, ribbed, with imprints of hairs, etc.), other supplementary characters. Seeds of such preservation are exemplified below in the genera *Samaropsis* and *Tungussocarpus*.

Maceration of impressions–compressions serves to widen the range of applicable characters (Fig. 74, *o–q*). According to the number of cutinized membranes it is possible to establish whether the seeds had an outer integument (cupule); the adherence or extension of the membranes towards the chalaza allow judgement to be made of the degree of fusion of the integument with the nucellus. From the outlines of the membranes it is possible to reconstruct the micropyle, special structures on the nucellar apex (beak, lagenostome, salpinx). The seed apex can bear a funnel, hook-shaped ears, hairs and other projections. Maceration helps to reveal resin canals and bodies, the stony layer, at times the course of the vascular bundles. Frequently, but not always, the megaspore membrane can be liberated. At times archegonia are visible. Pollen may be found in the micropyle, from which can be judged the organic connections between seeds, pollen and male fructifications. An alien pollen not belonging to this plant may occur in the micropyle. However, the repeated occurrence of a specific pollen type in the micropyle of particular seeds provides reliable evidence of the organic connections between these parts.

If the macerated seeds exhibit peculiar characters, then they can be assigned to genera of limited circumscription. This applies to seeds with an apical funnel, which have been separated into the genus *Stephanostoma*. Distinctive features convenient for classifications are however, lacking in a large number of seeds. In this case the genera are understood in a very broad sense and are, in part, of a collective nature. The classification then relies on such features as the degree of fusion of the integument with the nucellus, degree of cutinization of both, the presence of a megaspore membrane, stony layer, resin canals and cavities (genera *Amphorispermum*, *Bysmatospermum*, *Allicospermum*, *Chitospermum*).

Genus *Samaropsis* (recorded since the Upper Devonian; Fig. 74, *m*, *n*)

Seeds are platyspermic (bilaterally symmetrical and flattened), varied in shape (round, subtriangular, trapezoid, cordate, etc.). Distinctions are clearly apparent between the nucleus and surrounding wide border of a more delicate tissue, frequently membranaceous, interrupted at the base and/or apex. This genus probably also includes certain winged angiosperm fruits.

Genus *Tungussocarpus* (P; Fig. 74, *r*)

Seeds are bilaterally symmetrical, fleshy, with an ill-defined central body. Seed outline is from round to kidney-shaped or transversely oval. Apex slightly acuminate, base notched. Various species probably belong to the families Vojnovskyaceae and Rufloriaceae.

DIVISION MAGNOLIOPHYTA (ANGIOSPERMAE)

The systematics of fossil angiosperms (A.) compared with that of other higher plants, is much more strongly influenced by the systematics of extant forms. It is notable that in other divisions surviving to date, extinct families, orders and classes have been established, whereas among the extinct A. practically no such taxa ranking higher than genera have been established; apart from this some fossil genera have been separated, not so much because of their peculiar characters, as because the data available on the fossil remains are insufficient for their comparison with extant genera. It is not fortuitous that the names of many fossil A. genera are derived from names of living genera. This state of affairs developed under the influence of earlier investigations, when imprints of the oldest A., including the Lower Cretaceous, were habitually attributed to living families and genera, standing far apart in the system. The belief was current that angiosperms underwent a long evolutionary development prior to the beginning of the Cretaceous. Fossil remains assigned to different orders were reported in pre-Cretaceous sediments. As a consequence, it was thought that A. originated during the Triassic, and even during the Permian. It was assumed that A. were for long inhabitants of such sites (e.g. uplands), which were unfavourable for the preservation of their remains. Since, supposedly, the palaeobotanical material did not provide evidence of the origin of A. themselves, nor of their major sub-groups, the development of A. phylogeny was based exclusively on comparative analysis of modern taxa, including their geographical distribution.

Nevertheless, evidence accumulated over a long period that served to indicate that the significance of palaeobotanical materials for A. phylogeny was seriously underrated. The widespread palynological investigations performed during the 1950s–1960s have shown that Palaeogene pollen and even more so the Cretaceous pollen of A., because of their peculiarities, cannot always be feasibly attributed to modern genera or even families. Palynology did not give confirmation to the conjectured sudden expansion of modern A. types during the middle Cretaceous, and did not provide evidence justifying the postulated pre-Cretaceous long-term evolution of A. Moreover, cuticular studies frequently revealed serious errors in the identification of leaf remains based solely on their macroscopic characters. From revision of the Palaeogene floras by means of the cuticular method it became evident that up to 60% of the genera were earlier erroneously determined.

Studies of Silurian, Devonian and Early Carboniferous plants during the 1960s–1970s convincingly showed that palaeobotanical materials, when competently treated, are of great significance in deciphering the origin of major taxa of higher plants. Probably due to the success of palaeobotanical studies in regard to the Palaeozoic, the records on the oldest A. were subjected to detailed examinations and revision. This was conducive to the rejection of many ideas that held sway over many years. New materials continue to turn up. So far it is impossible to suggest a classification system for A. which can encompass even the most interesting fossil forms of the oldest A. Because of this an outline is given here only of the major data available on those fossil A. which are of paramount importance to the knowledge of the evolution of A. as a whole. Information on the systematics of A. based on the modern members, is not presented here.

The advent of A. can be viewed as the next important step in the evolution of reproduction. In gymnosperms, in contrast to spore-producing plants, the megaspore is not free living any more and is encased in the ovule which remains open for the pollen, that enters directly into the micropyle; whereas in A. the ovule itself is isolated from direct contact with the pollen and enclosed in a specialized container (carpel, ovary). The morphological tendency towards the encasement of the ovule was apparent in gymnosperms. For instance, in Cardiolepidaceae and Caytoniales the ovules are completely concealed in the cupules. But the pollen penetrated into the micropyle with the aid of an exudate mechanism (secretory pollen-receiving drop). In these plants the ovule is just as well isolated from the physical environment as in A., that do not have completely closed carpels; but functionally, according to the pollination strategy, these plants remain typical gymnosperms. It is believed that in some Leptostrobales the pollen germinated on the distal surface of the capsule, so that pollen tubes – not the pollen – entered the ovule (Krassilov, 1973b, 1977b). In conifers the pollen at times germinates on the seed scale or on the bract, fused with the scale. These plants, displaying functional affinities to A., are morphologically indistinguishable from plants which retain the pollination strategy typical of gymnosperms. A. combine both the encasement of the seed in special containers, inaccessible for the pollen, together with the germination of the pollen on a special receptive part of these containers – stigmas. The pollen tube grows in the direction of the ovule and upon reaching it produces double fertilization; then follows the development of the seed and of the fruit enclosing it. Pollen capture by the stigma and double fertilization are the major diagnostic features of A.

A. differ from gymnosperms in other characters. A. display a multi-aperturate pollen, vessels in wood and bisexual fructifications; leaves have reticulate venation composed of veins of several orders, developing in the course of the intercalary growth of the leaf lamina. Many A. are herbaceous plants, this fact having been so far validly established only for one gymnosperm – *Aethophyllum* (Voltziaceae). Not only the growth, but also the entire structure of the herbaceous plants may be highly determinate. In gymnosperms such a degree of determination of the structure is characteristic only of individual organs. Of wide development among A. are various modes of asexual reproduction (vegetative, apomictic, etc.), which are very common in spore-producing plants and not characteristic of gymnosperms. A., as compared with gymnosperms, are much more diverse in their growth forms, in many physiological and ecological features.

In the systems currently accepted, A. are divided into two classes: dicots (Magnoliopsida, Dicotyledons) and monocots (Liliopsida, Monocotyledons). No persistent characters are known for the distinction of classes. A specified character may be only a dominant one within a definite class. For example dicots are characterized by two cotyledons, reticulate venation of the leaves and the number of perianth members divisible by five, less frequently by four. Monocots usually produce a single cotyledon, leaves with parallel venation and the number of perianth members divisible by three. However, many exceptions in these and other characters can be observed in both classes. Because of this the assignment of fossil genera to monocots or dicots relies, as a rule, not on such generalized characters, but on comparison with one or another living genus or family.

Extant dicots are commonly divided into seven classes, their names being derived from living genera. (In earlier systems other names, shown below in parentheses, were used, reflecting some characteristic feature. It should be remembered that the introduction of new names was concomitant with the change in delimitation of the subclasses.) They include: Magnoliidae (Polycarpicae), Ranunculidae, Hamamelidae (Amentiferae; Monochlamydeae), Caryophyllidae (Centrospermae), Dilleniidae, Rosidae and Asteridae (Compositae). The subclass Ranunculidae is often included in the Magnoliidae. Monocots (class Liliopsida) are derived from dicots and are divided into four subclasses – Alismatidae, Liliidae, Commelinidae and Arecidae. These subclasses, in the same way as the classes, cannot be strictly distinguished by any persistent characters, but merely by the trends in the distribution of certain characters. The Magnoliidae include the so-called 'ranalian complex' and certain other families. The term 'ranalian complex' applies to an A. group with a well-developed perianth, which may or may not be differentiated into petals and sepals. These A. have numerous stamens that develop (during ontogenesis) from the periphery of the flower towards its centre (centripetal stamens). The gynoecium is apocarpous, the ovules are bitegmic and crassinucellate (with a massive nucellus). In various families semiclosed carpels, vessel-less wood and monoaperturate pollen occur. The above-cited characters are all regarded as primitive for A. in the whole. A widely accepted hypothesis maintains that the first A. belonged to the ranalian complex and produced large solitary bisexual flowers with multiple perianth members, numerous stamens and free follicle-like carpels. All these parts were helically arranged. Pollination was performed by insects. Frequently the flowers of *Magnolia* are taken as a model of this primitive flower-type. Further evolution of such a flower is viewed as involving the reduction of the number of constituent members, their fusion, transition to a whorled arrangement, development of unisexual flowers, formation of different inflorescences, reduction of the perianth, etc.

The subclasses of extant A. are further divided

into a large number of orders and families (Takhtajan, 1980). Characteristics of all these taxa are not given below. The systematics of extant A. may be adapted to fossil A. only if the fossil forms can be directly compared to living A. according to detached parts, without employing the whole syndrome of characters defining any specific suprageneric taxon. For example, suppose that we find leaves similar to oak leaves, and associated acorns. Then we can reasonably claim that the given flora yields the genus *Quercus* which belongs to the order Fagales of the subclass Hamamelidae. In many cases such direct comparisons of fossil remains with extant forms are precarious or utterly impossible due to certain peculiarities of the hand specimens. In these cases, palaeobotanists dealing with other higher plants usually establish suprageneric taxa, regarding them as extinct. The same procedure may be principally applied to the systematics of fossil A., but for unknown reasons it is not employed by palaeobotanists. Instead they merely establish new genera, which according to certain indirect indications are included in suprageneric taxa of living forms or remain outside of them.

Nevertheless, acquaintance with the most ancient A., particularly with their fructifications and pollen, clearly reveals their singularity, which at times is sufficient for the distinction of suprageneric taxa of a very high rank, perhaps even of independent classes. Most Cretaceous A. seem to belong to extinct orders and families pending identification. Many specialists have arrived at this conclusion.

Considering that all Cretaceous and many Cainozoic A. can be only tentatively compared to living families, with a high degree of uncertainty, it seems reasonable to arrange the fossil A. among special suprageneric parataxa of dispersed parts. Such a parataxa system has been successfully employed in the classification of A. fossil pollen. Recently a similar system was proposed for dispersed leaves by Krassilov (1979).

Classification of dispersed leaves

A typical feature of A. leaves is their reticulate venation. It is no wonder that leaves of certain Triassic and Jurassic gymnosperms because of their reticulate venation were mistaken for A. leaves, and this upheld the opinion that A. appeared already in the pre-Cretaceous. However, the reticulate venation in gymnosperms differs essentially from that in A. In A. it is composed of veins of several orders, i.e. the anastomoses connect thicker and thinner veins. The ends of the thinnest veins, if they do not emerge in the margins and do not coalesce with adjacent veins, blindly terminate in the mesophyll of the venation meshes. In gymnosperms veins of one order anastomose. As a result, each mesh of the reticulum, if it is not located along the midrib, is encircled by veins of the same thickness, or similarly narrowing towards the margin. This type of venation is known in trigonocarpaleans, arberias, caytonias, bennettites, cycads and other gymnosperms (Fig. 47, *w*; Fig. 49 (11); Fig. 52, *c*; Fig. 53, *i*; Fig. 54, *f*; Fig. 55, *b*, *s*, *u*; Fig. 59, *h*; Fig. 60, *g*; Fig. 62, *h*, *n*; Fig. 63, *p*). Only in the Triassic genus *Furcula* (Fig. 63, *q*, *r*), allied to peltasperms, in some gigantonomias (Fig. 54, *e*) and living *Gnetum* is the reticulate venation similar to that of A., particularly of older ones: veins of various orders coalesce.

The difference between gymnosperms and A. in the venation of mature leaves is connected with the different ontogeneses of the leaf lamina (Hickey and Doyle, 1977). Although the architectonics of the gymnosperm leaves in relation to the sequence of ontogenesis have not been studied, they are believed to be the same as in ferns, with similar venation including reticulate. In ferns two phases of meristematic activities are recognized in the leaf ontogenesis. The apical phase led to the formation of the petiole and rachis from the apical meristem. During the second – marginal – phase the meristematic cells along both sides of the embryonic rachis produce pinnules and their veins. In simple leaves of dicots with pinnate venation three overlapping phases are observed. The apical phase is of short duration and leads to the formation of a small tubercle and a procambium, which subsequently gives rise to the midrib. This is followed by a short-term marginal phase, when two marginal meristems produce the embryonic leaf lamina and secondary veins, that are of pinnate arrangement and differentiated from the embryonic axis on to the margin. After this a long-term phase of diffuse intercalary meristematic activities sets in. The meristem corresponding to this stage is termed the plate meristem. Due to the latter the leaf surface becomes larger and tertiary and thinner veins are formed in spaces between the primary and secondary veins. Different orders of the veins and interposed meshes correspond to synchronized discrete phases of activities of the plate meristem in various, but orderly arranged, leaf parts. Hence, the reticulate venation in A., in contrast to that in ferns and probably gymnosperms, develops from the plate meristem, not from the marginal one. It is conceivable that the observed gradual increase in the complexity of the venation in A. leaves along with the progressive increase in its orderly arrangement, occurring upwards of the geological section (see below), reflects the gradual rearrangement of the ontogenetic processes, including the formation of the intercalary plate meristem.

In the classification and identification of fossil A. leaves palaeobotanists are confronted with serious difficulties in finding extant forms with leaves similar to the fossil hand specimens. Moreover, leaves of A. standing systematically far apart may be closely similar and reveal merely subtle distinctions, which draw the attention only of palaeobotanists. This calls for a special classification system for fossil and modern A. leaves, which should reflect the necessary taxonomic characters. The arrangement of leaf types in this system may not necessarily depend upon the systematic affinity of the parent plant, its assignment to any definite family, order or subclass, since this system aims chiefly to assist the palaeobotanist in selecting plant leaves with which the fossil remains can be feasibly compared. It is clear that such a system should facilitate the comparison of fossil A. with definite groups of extant A. Otherwise one may overlook a group of A. with a leaf type that is of special interest to the palaeobotanist.

The afore-mentioned system, proposed by V. A. Krassilov (1979) (p. 227), meets these requirements, although as yet it only includes leaves of some dicots. In this system 10 groups of leaves have been recognized, each named after the type of living A.: Laurofolia, Nymphaefolia, Ficofolia, Platanofolia, Viburnifolia, Betulifolia, Fagofolia, Proteaefolia, Rosifolia and Legumifolia.

In the classification of A. leaves a wide range of characters is used. Great significance is attached to the venation. According to the number and arrangement of the primary veins the venation can be (Krassilov, 1979): (a) pinnate – one primary vein (midrib) extends above the middle of the leaf; (b) plumage (Krassilov's term) – the midrib does not extend to the centre of the leaf; (c) palmate – two or more primary veins extend above the middle of the leaf far from the margins; (d) dichasial – the primary veins pass along the margin and give off branches upward (acroscopic); (e) fan-shaped – primary veins are of fan-shaped arrangement; (f) plumage – fan-shaped – the middle vein of the fan-shaped system is of plumage branching; (g) psilobasal – the primary veins do not branch at the base (naked). The vein branching (if it exists) may be: (a) uniform – successive veins are of the same branching; (b) decreasing – each next vein is of less intensive branching than the previous one; (c) palmately-pinnate – the basal veins extend to the middle of the leaf; (subtypes: basal, suprabasal, mixed (basal veins do not emerge from one point)); (d) subpalmately-pinnate – basal veins do not extend to the middle of the leaf; subtypes are the same.

Important characters are the form of the veins (straight, oblique, curved) and their density. Several types of venation can be recognized, dependent upon the behaviour of the vein along the margin: craspedodromous, semicraspedodromous, etc. (Fig. 71, *h*, *i*; for descriptions of other types see: Dilcher, 1974, Hickey, 1973; Krassilov, 1979). In addition, other types are distinguished according to the fine marginal venation (Fig. 75, *c*), the nature of the tertiary venation (Fig. 75, *d*), the behaviour of veins of higher orders and the structure of the areoles (Fig. 75, *e–g*), structural details of the marginal denticulation (Fig. 75, *a*) and other characters.

Examples of several genera established for leaves of the oldest A. are presented below. Illustrations of some other genera are given in Fig. 76, *e*.

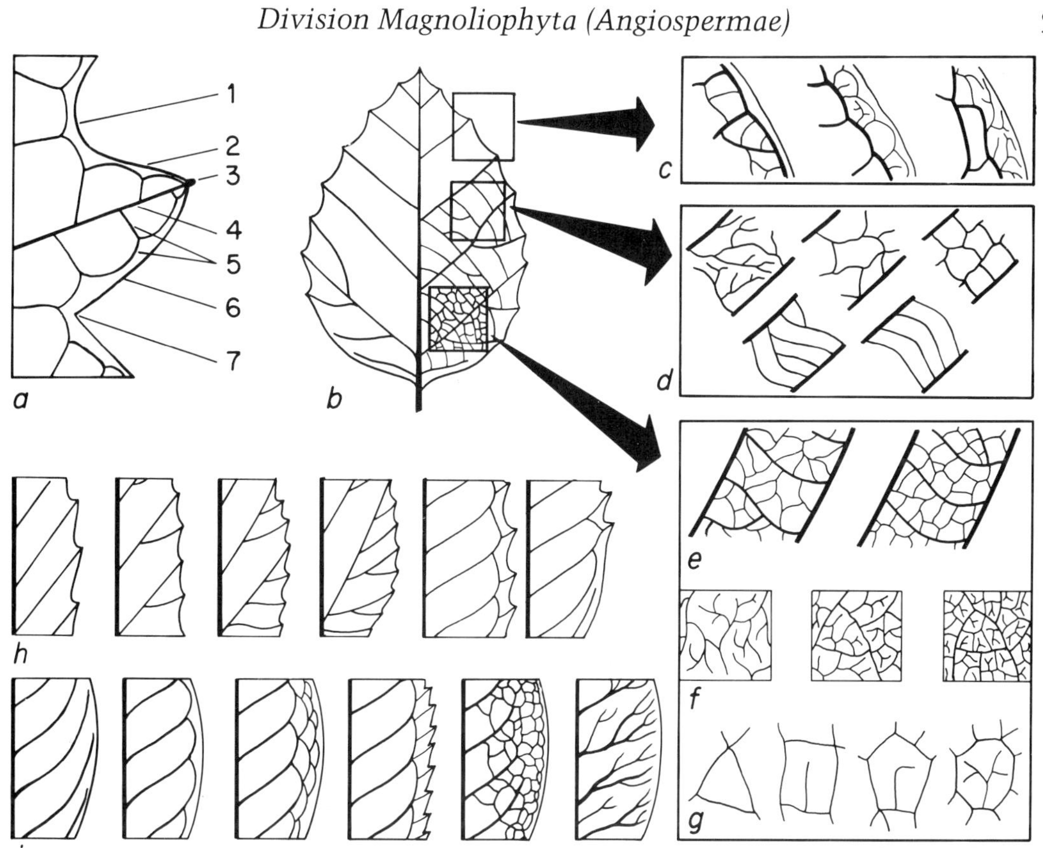

Fig. 75. Elements of angiosperm leaf morphology exploited in diagnostics of taxa; *a*, structure of marginal tooth (1, rounded notch; 2, apical side of tooth; 3, gland; 4, main vein; 5, subsidiary veins; 6, basal side of tooth; 7, acute notch); *b*, scheme of arrangement of elements; *c*, fine venation near margin (from left to right: fringed, meshed, imperfect); *d*, tertiary venation (upper row, branching, irregularly reticulate, orthogonal reticulate; lower row, branching scalariform, scalariform); *e*, irregular reticulate (to left) and orthogonal venation of higher order; *f*, areolae (from left to right) incompletely closed, imperfect, well developed; *g*, areolae (from left to right) non-filled triangular, quadrangular with straight veinlet, pentagonal with flexed veinlet, polygonal with branching veinlet; *h*, craspedodromous proper (four figures to left), semicraspedodromous and mixed venation; *i*, camptodromous venation (eucamptodromous, brochidodromous, festoonal-brochidodromous, brushed-brochidodromous, reticulodromous, cladodromous). Courtesy of A. B. Herman.

Genus *Araliaephyllum* (recorded since the Lower Cretaceous; Fig. 76, *e* (8))

Leaves are ternate with or without supplementary lateral lobes, at times smooth-lobed, in the extreme variety entire, with entire margins. The lobes are elliptical, sinuses are rounded. Venation is flexed – palmate – pinnate, basal or suprabasal, with a marginal vein along the sinuses. The tertiary veins are scalariform, branching.

Genus *Menispermites* (recorded since the Lower Cretaceous; Fig. 76, *e* (5))

Leaves are medium-sized and large, ternate with incipient lobation; the leaf base is peltate, the margin is sinuous, the venation is dichasial or flexed-palmate-pinnate, basal. Veins radiate from the petiole attachment point. Branches of secondary veins are reticulodromous or eucamptodromous. Tertiary venation is irregularly reticulate or branching-scalariform.

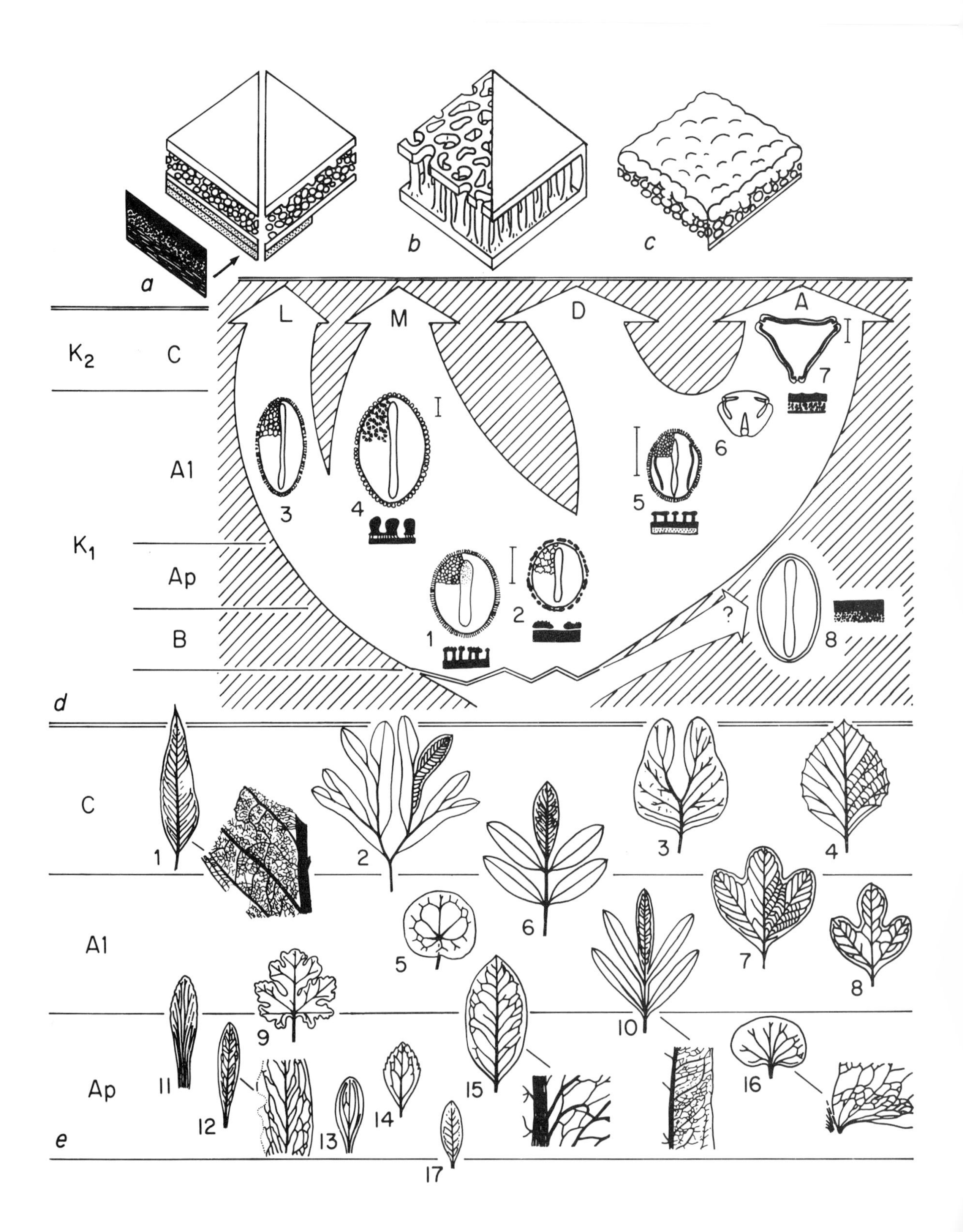

a
b
c
L
M
D
A
K2
C
A1
K1
Ap
B
1
2
3
4
5
6
7
8
?
d
C
A1
Ap
1
2
3
4
5
6
7
8
9
10
11
12
13
14
15
16
17
e

Genus *Sapindopsis* (recorded since the Lower Cretaceous; Fig. 76, *e* (6, 10))

Leaves are compound pinnate, leaflets opposite or coupled, upper ones frequently coalesced, decurrent, lower with a short petiole. Leaflets are lanceolate, with attenuated apex, entire. Venation is flexed-pinnate, camptodromous. Tertiary venation is irregularly pinnate.

It has been observed that the foliage of a definite subclass or order shows a tendency towards a particular type. For instance, Rosidae are the only large dicot group with compound pinnate leaves. Certain taxa occur among the angiosperms, where the leaves can easily and unmistakably be identified (e.g. palms). But, as a whole, merely on the basis of gross morphological characters of the leaves (shape, venation), without any knowledge of their epidermal characters and associated organs, errors are nearly inevitable in the identification of living taxa, because the particular leaf-type might be a synthetic-type component (see below: The oldest Angiosperms).

Probable ancestors of angiosperms

The literature available on the A. origin is voluminous, but is concerned in the main with extant forms. Through comparisons of the latter attempts are made to derive the most primitive families, and according to them to conceive an idea of the habit of A. progenitors; then from this to reconstruct the probable phylogenetic lineages connecting this progenitor with living taxa. An amazing variety of phylogenetic schemes has been proposed. Some of them hold angiosperms to be monophyletic, stemming from the ranalian complex of the A. Other schemes are polyphyletic. In this case the ranalian complex plays the role of only one of its roots. Phylogenetic schemes for A., based chiefly on the analysis of living forms, are not discussed below. The presently available palaeobotanical materials are insufficient for the construction of any definite phylogeny of A. taxa, and do not spotlight any particular group of gymnosperms in the rank of an order that might have been ancestral to A. Because of this we will merely content ourselves with the acquaintance with the oldest known A., and discuss the problem of A. ancestry in general form, in terms of the gymnosperms that could have been in principle progenitors of A. The palaeobotanical data can shed light on the origin of the carpel, stamens, pollen, the flower as a whole, the fruits, seeds, leaves and wood. Little information is available on the evolution of certain biochemical characters of A.

The problem bearing on the origin of the carpel and gynoecium in the whole is subject to much controversy. A popular hypothesis is that of a conduplicate origin of the carpel, derived from the ovule-bearing leaf (phyllosperm or cladosperm). According to this hypothesis, during the ontogenesis of the progenitor this leaf was at first encased in the bud in folded form with the ovules

Fig. 76. Pollen (*a–d*) and leaves (*e*) of oldest angiosperms; *a*, exine structure of gnetaleans, *Eucommiidites* and putative ancestors of angiosperms (left side; to left, scheme of exine structure as seen under transmission electron microscope), some Amentiferae, Rosidae, Asteridae and many other angiosperms (right side); *b*, exine of *Clavatipollenites* and other angiosperms with perforated (left side) and entire (right side) tectum; *c*, some Magnoliidae and Asteridae; *d*, chronological distribution of oldest angiosperm pollen types (1, *Clavatipollenites*; 2, *Retimonocolpites*; 3, *Liliacidites*; 4, *Stellatopollis*; 5, *Tricolpites*; 6, tricolporate pollen; 7, *Normapolles*; scale bar = 10 μm; 8, recent Magnoliidae with granular exine, K_1 Lower Cretaceous; B, Barremian; Ap, Aptian; Al, Albian; C, Cenomanian; L, lineage towards monocots; M, lineage towards Magnoliidae; D, lineage towards more advanced dicots; A, lineage towards Amentiferae); *e*, chronological distribution of oldest angiosperm leaves (1, leaf of *Magnoliaephyllum* type associating with infructescence *Prisca reynoldsii* Ret. et Dilch.; 2, dichotomously-compound leaf; 3, *Liriophyllum*; 4, platanoid leaf; 5, peltate leaf *Menispermites*; 6, *Sapindopsis*, compound pinnate leaf; 7, *Araliopsoides*; 8, *Araliaephyllum*; 9, *Vitiphyllum*; 10, *Sapindopsis*, pinnate leaf; 11, *Plantaginopsis*; 12, *Rogersia*; 13, *Acaciaephyllum*; 14, *Quercophyllum*; 15, *Ficophyllum*; 16, *Proteaephyllum*; 17, leaf from Lower Cretaceous of Transbaikalia). Modified from: Doyle *et al.*, 1975 (*a–c*); Doyle, 1977 (*d*) and 1978 (*e*); Hickey and Doyle, 1977, Retallack and Dilcher, 1981c (*e*).

facing inwards. This early ontogenetic stage of the progenitor was retained in the adult forms of the first A. The ovules matured, whereas the leaves remained folded along the midrib. The leaf margin gradually evolved into the stigmatic surface for pollen reception. In this case, pteridosperms with seed-bearing fronds are specified as the probable progenitors. The Permian genus *Archaeocycas* (Fig. 60, *n*) was offered as an example (it was also regarded as the probable progenitor of cycads). It can be envisioned how the leaf lamina in *Archaeocycas* expanded, whereas the ovules remained hidden in something like a follicle (Mamay, 1976).

Adherents of other hypotheses bearing on the origin of the carpel, choose other fossil forms as A. prototypes. If the carpel is regarded as the homologue of a cupule, then attention is focused on the cupulate lagenostomans, particularly on the cupule of *Hydrasperma* (evidently, teratological), in which, besides seeds, microsporangia were inserted (Long, 1977a, 1977b). A. were linked in origin also with Arberiales (glossopterids), and the gonophyll – a seed-bearing organ fused with a sterile leaf-like bract – was regarded as the ancestral fructification type. Arberias were seemingly suited to the role of A. progenitors, considering that, in contrast to other 'pteridosperms' the seed-bearing cupule-like organs were attached to the upper surface of the main leaf (Fig. 55, *c*, *e–g*). Many angiosperms are also characterized by the adaxial arrangement of ovules on the carpels. Gymnosperms with bisexual fructifications, namely bennettites and *Irania* (Fig. 53, *r–t*) were also considered the progenitors of A. Special attention was devoted to certain leptostrobans with something like a stigma along the distal margin of the capsule, to plants of the *Dirhopalostachys* type with seeds encased in capsules, to caytonias and certain other fossil gymnosperms (Krassilov, 1977b).

It still remains unknown from what type of gymnosperms A. could have evolved. Each alternative hypothesis, explaining the origin of a definite group of A. characters, cannot explain the origin of others. For example, leptostrobans give evidence for the probable mode of formation of the stigma, but from their ovules it is too difficult to derive the A. ovules. Krassilov (1977b) suggested that the typical syndrome of A. characters was formed at the expense of several groups, as a result of asexual reciprocal transfer of genes between them, e.g. as a result of virus transduction. If this process did actually occur, then it should be manifested not only in the evolution of the flower, but also in other parts of A. However, as will be shown below, a gradual differentiation in the morphological characters can be perceived in the successive assemblages of pollen and leaves of the oldest A.

In discussions bearing on the probable A. ancestry the following circumstances should be taken into account. First of all, certain gymnosperm characters are quite persistent in the course of the long evolution of the major taxa. For example, among Pinopsida during their entire past history leaves with reticulate venation never appeared. In the class Ginkgoopsida the seeds were only abaxially attached to a flattened seed-bearing organ. In the class Cycadopsida a strong tendency can be observed towards the repeated formation of reticulate venation, and practically no tendency towards the exfoliation of the exine with the formation of the sacci. On the other hand, the gymnosperms of different classes exhibit similar modifications, resulting in very similar organs of independent origin. This applies, for example, to the remarkable similarity between the pollen (including the intracellular gametophyte) in Callistophytales and conifers, leaves in the family Cardiolepidaceae and various conifers. If the above-described gymnosperm phylogeny is in general correctly outlined, then the numerous similarities between ginkgos and conifers should be interpreted as having been independently derived.

Hence, we should not seek simply gymnosperm forms similar to A. according to certain chosen characters, even striking ones, since they could have evolved independently at different times. It is necessary to find such characters in A. that are well persistent in gymnosperms over the entire period of their evolution, and at the same time render possible the subdivision of gymnosperms into the major groups. If such characters can be found then we can indicate the gymnosperms from which they have been inherited by A.

As already stated above (p. 135), in the course of evolution of various gymnosperm classes the most stable character is the seed type – platyspermic, radiospermic and secondary platyspermic. Platyspermic seeds are always non-cupular, and when accompanied by a cupule-like cladosperm (as the *Ginkgo* collar), they retain the major features of platyspermy. All the secondary platyspermic seeds of Pinopsida, from the Carboniferous cordaitanthaleans to the living conifers, invariably have an integument developing from two primordia and vascularized by two bundles in less advanced forms (supplementary bundles, if present, extend out from the main ones and are arranged in one plane with them; in more advanced forms the vascularization disappears). In Cycadopsida the radiospermy and the cupule are persistent, whereas in more advanced forms the outer integument is of cupular origin.

Considering the remarkable stability of these seed characters in the phylogenetic lineages, there is reason to suppose that they also continued persistent during the transition from gymnosperms to A. All Pinopsida can be excluded from the probable ancestors of A., inasmuch as their phylogenetic independence is clearly apparent from a wide syndrome of characters, both of the reproductive organs and vegetative parts. If we admit that all other gymnosperms are allied to Ginkgoopsida or Cycadopsida (but not to some as yet undefined class) then it is necessary to choose between these two classes. For this purpose analysis should be made of the A. seed type.

In A., unlike gymnosperms, the number of integuments varies considerably even within one genus. If a single integument is present then nothing definite can be said as to which organ it corresponds to – either to the only integument in Ginkgoopsida, the inner integument (vascularized nucellus) in Cycadopsida, or to the cupule fused with the integument. In bitegmic ovules of A. the vascularization of the outer integument is very diverse or absent. The vascularization of the inner integument is rare, but if more than one bundle passes through the inner integument then they are arranged in a ring, i.e. as in radiospermic forms. It is thereby conceivable that radiospermy is typical of A. as a whole. Accordingly, the A. ancestors should be sought among radiospermic gymnosperms (Meyen, 1982a, 1984a). Then all representatives of Ginkgoopsida, including Callistophytales, Arberiales, Peltaspermales, Caytoniales and Leptostrobales, cannot be regarded as probable A. ancestors. This is consistent with the previously mentioned fact of the abaxial attachment of the seeds in Ginkgoopsida. Angiosperms with leaf-like carpels are characterized by adaxial or marginal, but not abaxial placentation.

If we admit the hypothesis postulating the origin of A. from radiospermic gymnosperms, then we should direct our attention to the Mesozoic taxa of the latter (Lagenostomales and Trigonocarpales seemingly were exterminated before the end of the Palaeozoic), among which only Bennettitales and Cycadales are known. Bennettitales have been frequently discussed in connection with the origin of A. However, the possibility of establishing new phylogenetic linkages between these groups is still not exhausted. In the morphological section of Chapter 8 we shall discuss the possibility of deriving the carpel from the bennettitalean fructification with the aid of gamoheterotopy (transfer of characters of one sex to another). Such derivation may abandon a major objection against the bennettitalean ancestry of angiosperms. Otherwise bennettites reveal numerous similarities to the supposedly primitive angiosperms, such as homoxylary wood, paracytic stomata, non-lamellar nexine, presence of organs resembling nectaries, variously organized involucre members, tendency to the bisexuality of fructifications, etc. With the admission of the gamoheterotopic origin of the carpel, the Bennettitales become the most suitable gymnosperm order to derive primitive flowering plants. It seems very suggestive, that bennettitalean stems of the *Sahnioxylon* (= *Homoxylon*) type were repeatedly placed among angiosperms or their putative ancestors, and that the mid-Cretaceous angiosperm fructifications recently described as *Lesqueria* (Crane and Dilcher, 1984) were previously ascribed to *Wil-*

liamsonia. Female fructifications of cycads can easily be modified into a carpel, even more so since in cycads, in addition to a circinate leaf vernation, the straight vernation type is also known. Both marginal and adaxial placentation are known in cycads. Other characters common to cycads and A. can be noted. However, the structure of the reproductive organs in the Mesozoic cycads has as yet been poorly studied. We are acquainted with only a few genera.

The A. ancestors might also have belonged to a still unknown group of Mesozoic radiospermic gymnosperms. The probable existence of such a group is evidenced by the absence in the geological record of the roots of radiospermic orders Gnetales and Welwitschiales. The Gnetales are represented today by a single genus *Gnetum*, which shows affinities in most characters to A., not to gymnosperms. It is quite possible that the living *Gnetum* is only the last survivor of a rather diverse group, the other members of which were more closely related to primitive A. According to the structure of their leaves, vessels and many other characters, *Gnetum* display more affinities to A. than any other gymnosperms (Muhammad and Sattler, 1981). It should be noted that we still know nothing concerning the linkages between the Palaeozoic and Mesozoic radiospermic gymnosperms. Judging by the palynological data A. probably appeared first of all in the Equatorial Zone. Localities of the lowermost Cretaceous megafossils have not been found in these places. The Jurassic and Lower Cretaceous equatorial floras are known chiefly from miospore assemblages. All this is indicative of large gaps in our knowledge of Mesozoic radiospermic gymnosperms. It is no wonder then, that we cannot as yet name the order of radiospermic gymnosperms that gave rise to the Early Cretaceous A.

The oldest angiosperms

In determining the systematic affinities of the oldest fossils to A. we have got to proceed not on the basis of comparison with definite orders and families of extant A., but to rely on a syndrome of characters known among living higher plants only in A. So far we have dealt only with dispersed parts of the oldest A. – pollen, leaves, fruits – for which the organic connections are for the most part unknown. In the preAlbian deposits we must content ourselves with finds of detached leaves and dispersed pollen; the A. pollen appears in the stratigraphic section earlier than the leaves, namely in the Barremian. No indisputable A. have been recovered from earlier deposits.

Under discussion in the following chapter are the Cretaceous and only those younger A. that are most essential to the understanding of the A. evolution, as a whole.

Leaves

Among the A. leaves described in the literature are Triassic *Furcula* (belonging more readily to Peltaspermales; Fig. 63, *q*, *r*) and *Sanmiguelia* (a plant of uncertain systematic affinity, probably allied to articulates) and certain other pre-Cretaceous remains. Their A. affinities are now refuted. Among the oldest indisputable A. leaves are *Dicotylophyllum pusillum* (Fig. 76, *e* (17); Vakhrameev and Kotova, 1977) from the Lower Cretaceous (Barremian or Aptian) of the Transbaikal Region. This is a small lanceolate leaf with a distinct midrib extending nearly to the apex. The lateral veins (7–8 pairs) are slightly curved upwards and do not reach the margin. Tertiary veins are inconspicuous.

More complete data are available on the Aptian and Albian A. leaves, which have been traced through the stratigraphic succession along the Atlantic coast of the USA (Doyle and Hickey, 1976; Hickey and Doyle, 1977). This sequence, supplemented with data from other regions, is shown in Fig. 76, *e*. It has been reliably assigned to the Aptian, Albian and Cenomanian stages. In the Aptian–Lower Albian, five leaf types are restricted to the lower part of the sequence (Zone I). The leaf laminae are small, entire, the venation reticulate with disorganized secondary veins. The meshes formed by the secondary veins and their branches are of irregular configurations and sizes. The secondary veins can be poorly distinguished from those connecting them. Only in the genus

Quercophyllum (Fig. 76, *e* (14)) is the margin denticulate, but the teeth are irregularly spaced. At the tooth apex is a wide glandular area. In *Quercophyllum*, *Rogersia* (Fig. 76, *e* (12)) and *Ficophyllum* (Fig. 76, *e* (15)) the midrib is distinct. In *Proteaephyllum* (Fig. 76, *e* (16)) it is less obvious and a tendency is apparent towards palmate venation, whereas the leaf lamina is reniform. These leaves display affinities to dicots in their venation (particularly to Magnoliidae), while *Acaciaephyllum* (Fig. 76, *e* (13)) with a long attenuated stem embracing the base may be compared to monocots. Its secondary veins, emerging at acute angles from the lower part of the midrib, are bent inwards and extend to the apex where they converge. These leaves are comparable to those of the subclass Liliidae (commonly Alismatidae is regarded as the most primitive subclass).

In the upper part of this stratigraphic interval among *Ficophyllum* larger leaves appear (up to 20 cm in length) with more regular venation. Leaves have been encountered with more complex dissection of the margin into lobes (*Vitiphyllum*; Fig. 76, *e* (9)) and irregular reticulate venation. Leaves occur with parallelodromous venation (*Plantaginopsis*; Fig. 76, *e* (11)). In leaves of this interval the petiole is ill defined from the leaf lamina. Probably these were simple leaves, rather than leaflets of compound leaves. Leaves of *Acaciaephyllum* were found helically attached to the shoot.

In the higher Albian sections certain types continue from the Zone I and certain new ones appear. One of the latter includes leaves with palmate venation, from oval-cordate to pedate. The main radial veins dichotomize and form symmetrical meshes of several orders. These are of the genera *Populophyllum* and *Menispermites*, which are putatively affiliated with *Proteaephyllum* of the preceding assemblage and can be compared to the living aquatic Nymphaeales. They have a long petiole and the leaf base varies from flat to funnel-shaped, as in floating and aerial lotus (*Nelumbo*) leaves. Simple small and medium-sized leaves with pinnate venation, a smooth or bidentate margin persist. *Alismaphyllum* with arrow-shaped leaves seemingly belong to monocots. Pinnate leaves (*Sapindopsis*) are of first appearance. At the outset the leaflets are distinctly separated from the rachis only in the lower part of the leaf, whereas higher they are decurrent on to a winged rachis. Upwards in the section typical compound-pinnate leaves appear with basally asymmetrical leaflets, comparable to Rosidae or Hamamelidae, and somewhat earlier, palmately dissected platanoid leaves (*Araliaephyllum*) with two lateral lobes and occasionally additional lobes in the lower part of the large ones. The secondary veins are not quite regular; the tertiary veins are weak and irregular. In different leaf types the venation becomes increasingly more regular upwards of the sequence. Higher up (in the uppermost Upper Albian and Cenomanian) the platanoid leaves become increasingly diverse (Fig. 76, *e* (7)), leaves appear with a notched apex (*Liriophyllum*; Fig. 76, *e* (3)), deeply dissected into lobes (*Dewalquea*) and other types. Among magnoliid leaves of the Upper Albian–Lower Cenomanian, Upchurch and Dilcher (1984) recognized several types comparable to Winteraceae, Chloranthaceae, Laurales, Illicales, etc. Some leaf species cannot be allied with extant families and orders, although their attribution to the Magnoliidae is clear. These and other observations document the existence of numerous extinct groups of the Magnoliidae and suggest that most extant families did not originate until later in the Cretaceous.

Assemblages of the same age, occurring in other regions, are quite comparable to those described above. In the Middle Albian floras of Western Kazakhstan (Vakhrameev, 1952; Vakhrameev and Krassilov, 1979) leaves were found with irregular venation, asymmetrically lobed, pedate with the main veins disorderly flexed in different directions. Higher in the section compound pinnate leaves appear of the *Sapindopsis* type.

In the Upper Cretaceous the diversity in the leaves continues to increase. Practically all the types characteristic of A. as a whole appear. From the Senonian onwards palm leaves occur. Notwithstanding this high diversity in the leaf forms the epidermal characters are remarkably uniform (Upchurch, 1984). The anticlinal cells are straight, hairs when present are simple consisting of a few cells, glands are common, at times peltate; hydathodes occur. The stomata are mainly anomocytic, although paracytic ones also occur. Epidermal investigations commonly do not support the affiliation of the Upper Cretaceous leaves with any extant genera. In the Palaeogene the identifications of extant genera are frequently confirmed by associated fruits and pollen. Most of the Oligocene and Miocene leaves are usually ascribed to the living genera. However, certain specific characters of the Neogene leaves are occasionally depreciated. For example, Ferguson (1971), having performed scrupulous analytical studies of epidermal characters in Miocene dicots from the Kreuzau flora (FGR), concluded that only 29 of the 65 identified species can be validly affiliated with extant families and only 17 with extant genera. The same author (Ferguson, 1974) demonstrated that although over 80 fossil species of *Laurus* have been described in the literature, only one indisputably belongs to this genus. These conclusions deserve special attention, though they should be verified on materials of other floras.

Dispersed pollen

For the distinction of the most primitive A. pollen from that of Mesozoic gymnosperms of importance is a knowledge of the fine exine structure as observed under the electron microscope. In gymnosperms the nexine shows distinct lamination (Fig. 42 (22, 25, 27); Fig. 58, *m*, *q*; Fig. 76, *a* (left)), whereas in A. (Fig. 76, *a* (right), *b*, *c*) it is solid, i.e. consists merely of one foot layer, or is two-layered (where the upper is the foot layer, and the lower corresponds to the non-laminated endexine). The A. endexine is fine grained. Sometimes in the cross-section of the foot layer a white line appears – a very thin electron-transparent lamina, which often lies near the boundary between the foot layer and endexine. From the foot layer numerous columellae extend outwards forming a columellar layer covered at the top by a tectum. The latter is imperforate (without any holes) or perforated (see details: Walker and Doyle, 1975). The pollen with the columellar layer and tectum is termed tectate (Fig. 76, *b*). It occurs in very many A. The Triassic and Jurassic tectate pollen belongs to cheirolepidians (*Classopollis*; Fig. 72, *h*, *i*). Its nexine is lamellar, as in other gymnosperms.

The tricolpate pollen of *Eucommiidites*, known since the Jurassic, was referred to A. Under its tectum lies a coarse-grained layer, underlain by a lamellar nexine. This pollen was found in microstrobili (of cycads?) and in the micropyle of isolated seeds, i.e. it could not have belonged to A. According to the sexine structure this pollen was affiliated to chlamydosperms (Gnetales, etc.). Rare monocolpate grains with reticulate-columellate exine were recorded in the Upper Triassic in the USA, but these grains might be rather affiliated with *Eucommiidites*, considering that the nexine is thick and seemingly lamellar, and additional colpi may be present. The discovery of an A. pollen was reported even from the Carboniferous (*Tetraporina*, *Triporina*, etc.) but this was found to be algal walls. The affiliation of other pre-Cretaceous microfossils with A. has been repudiated (*Poroplanites*, *Trifossapollenites*, etc.; for details see: Doyle *et al.*, 1975; Hughes, 1976; Pant and Kidwai, 1977).

So far there are no data on the origin of the tectate pollen with a non-laminated nexine. Such pollen appear in small quantities in the Barremian. The oldest A. pollen has been studied in most detail in the stratigraphic section along both coasts of the Atlantic (Gabon, England, Brazil, USA; Doyle *et al.*, 1977, 1982). The diversity of the pollen and leaves are of parallel increase upwards of the sequence (Fig. 76, *d*, *e*). This once again convinces us that both types of fossil remains actually belong to A., although the organic connections between the pollen and leaf types are still unknown. In this respect of interest is the association of the previously mentioned *Dicotylophyllum pusillum* with the pollen of *Asteropollis* in beds in the Transbaikal region, where no other A. pollen was found. This pollen has a branching colpus with 3–5 rays. Judging by the observations under a light and scanning electron microscope, the exine in *Asteropollis* is tectate with large perforations in the tectum.

In the Barremian, only a monocolpate pollen is known, which is assigned to several genera. At first

Asteropollis and *Clavatipollenites* (of most wide occurrence) appear and soon after *Retimonocolpites*, *Liliacidites* and *Stellatopollis* appear. Best studied among them is the genus *Clavatipollenites* (Fig. 76, *d* (1)). They are oval or round grains with a distinct distal colpus. The nexine is thick, non-lamellar. The well-developed columellar layer is covered by a tectum with perforations of uniform size occurring over the entire surface outside of the colpus. The ridges (muri) dividing the holes in the tectum, have a lumpy surface. In the inaperturate part the nexine is seemingly composed only of a non-lamellar foot layer. The nexine in the colpus is covered by indistinct verrucae of irregular configurations. The foot layer here is lamellar and beneath it lies a non-lamellar endexine with inner sculpturing. Chlonova (1984) and D. W. Walker consider that *Clavatipollenites* and *Asteropollis* are closely similar in their exine structure to the living genera *Ascarina*, and *Hedyosmum* of the family Chloranthaceae. Their common features are irregularly stellate sulcus (in *Asteropollis* and *Hedyosmum*), tectal reticulum and lumpy or spinose supratectal sculpture. This gives certain support to the viewpoint advocated by Leroy (1983) that *Hedyosmum* is the most primitive living A. Kuprianova (1981) also maintains that the Chloranthaceae belong to the most primitive living angiosperms.

The genus *Retimonocolpites* (Fig. 76, *d* (2)) has much larger holes in the tectum, which is transformed into a fine reticulum with wide meshes. Muri bear sculpture of transverse ridges, or spines. The columellar layer is practically reduced, so that the tectum separates from the foot layer. The foot layer shows the white line splitting near the colpus. The endexine is present only near the colpus and beneath it and has a granular structure. Such an exine structure with a spinose tectum and the absence of the columellar layer is unknown in living A. and gymnosperms. According to the ultrastructural exine characters observed this genus displays affinities to monocots.

Round grains with a wide circular equatorial colpus belong to the genus *Afropollis* (Doyle *et al.*, 1982). Outside the colpus the tectum is pierced by holes forming a reticulum. The columellar layer is either distinct, or absent; the nexine is rather thick.

The genus *Stellatopollis* (Fig. 76, *d* (4)) involves elliptical to nearly round grains with a reticulate tectum. Muri bear clavate projections (from 4 to 8 per mesh) with triangular, more seldom round heads expanded apically. The heads, arranged around a hole, are closely spaced and form a stellate figure. Such an exine is termed stellate or crotonoid. The ultrathin sections show that the tectum and its projections are non-lamellar, the columellae are short, the foot layer is non-lamellar (probably lamellar beneath the colpus), the endexine is very thin or absent everywhere except around the colpus; the sculpture of the colpus is verrucose. The stellate exine is known in living dicotyledons (Euphorbiaceae, Buxaceae, Thymeleaceae), and something like it in monocotyledons (Liliaceae, Atherospermataceae); however, these groups can hardly be considered phylogenetically linked.

In the Barremian, but somewhat later, appears the genus *Liliacidites* (Fig. 76, *d* (3)), which is affiliated with monocots and differs from *Clavatipollenites* in that the size of the tectal perforations strongly decreases on the narrowed ends of the grains. Such an exine differentiation is known today only in monocots of the subclasses Liliidae (Liliaceae, Amaryllidaceae, Bromeliaceae) and Alismatidae (genus *Butomus*).

The above-described genera continue into the Aptian where new morphological types additionally appear. The major innovation is the appearance of tricolpate grains of the genus *Tricolpites* (Fig. 76, *d* (5)). These are elongate with a rather distinct columellar layer and reticulate exine. The exine structure is very similar to that of *Clavatipollenites*, but differs in the presence of a 'white line' and well-developed fine-grained endexine. It is believed that *Tricolpites* is phylogenetically linked to *Clavatipollenites*, i.e. the tricolpate pollen can be derived from the monocolpate, but did not evolve independently from the inaperturate pollen, as has sometimes been conjectured. Other Aptian tricolpate grains (*Striatopollis*) have a striate reticulate structure. This association of pollen forms continues into the Lower Albian.

In the Middle Albian tricolpoporate forms

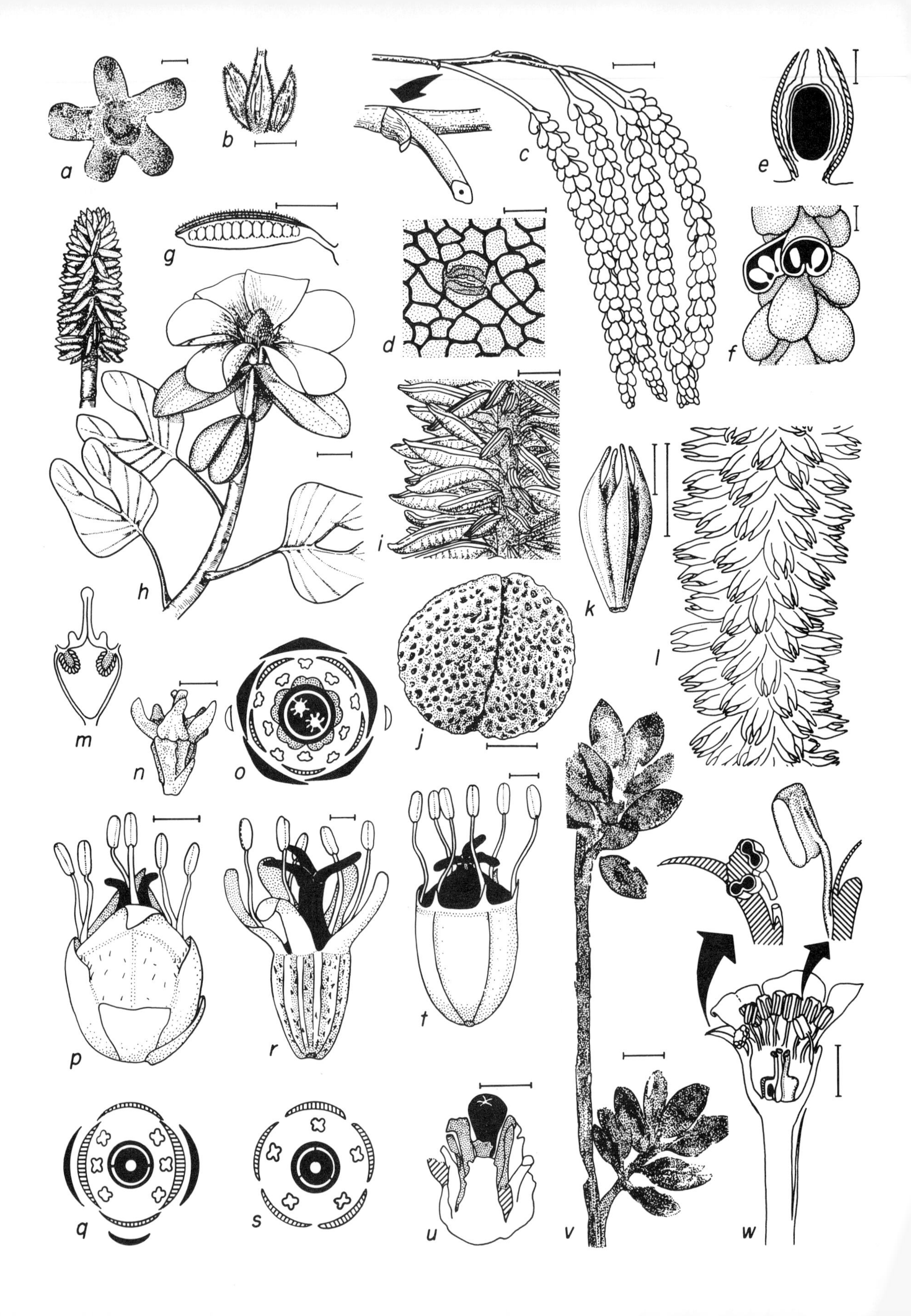
a
b
c
e
g
d
f
h
i
k
l
j
m
n
o
p
r
t
q
s
u
v
w

appear, which increase in number in the Upper Albian (genus *Tricolpoporopollenites*, *Tricolporoidites*). The details concerning the structure of the colpi, pores, also of the exine, as a whole, have not been studied in these forms. The pollen is of triangular or subtriangular outlines. In the Upper Albian polyporate pollen appear. The same set of pollen types persists in the lowermost Cenomanian.

Higher in the section (at the beginning of the Middle Cenomanian) triangular forms with angular pores appear. This type is represented by the genera *Complexiopollis* and *Atlantopollis*, assigned to the group *Normapolles* (Fig. 76, *d* (7)), which is of wide occurrence in the early Caenophyte (Upper Cretaceous) palynological assemblages. Recently the pollen *Normapolles* was attributed to flowers of dicots, closely related to the order Juglandales (Friis, 1983; Friis and Skarby, 1982). Practically all the major structural elements typical of the pollen of living A. appear in the Upper Cretaceous, but little is known as yet about the systematic affinity of the Upper Cretaceous dispersed pollen. The pollen of monocots can be validly distinguished from those of dicots. The extant families and genera can be clearly identified in the Palaeogene dispersed pollen, but even here the number of such taxa is not large (Muller, 1981). Some palynologists maintain that even the Miocene pollen belong in most part to extinct genera or differ so conspicuously from pollen of extant genera that the names of the latter are not applicable to them.

Fructifications

Certain Cretaceous and in part Palaeogene flowers, inflorescences, fruits and infructescences which have been studied in more detail are under discussion below. The materials are selected so as to provide an insight into the level of advancement attained by A. during definite epochs.

The Lower Cretaceous fructifications have as yet been inadequately studied (Dilcher, 1979; Stewart, 1983) and their kinship to A. is not always obvious. The genus *Onoana* (Hauterivian–Aptian) has been described from round fruits (?) with a thick wall, pierced by canals. Of similar aspect are the fruits of the family Icacinaceae (subclass Rosidae), in which *Onoana* was included. The Cenomanian–Turonian fructifications, also poorly studied, have been assigned to the genus *Phytocrene* of the same family. The genus *Kenella* (Albian; Fig. 77, *b*) was described from spindle-shaped fruits (?) with longitudinal ribs and numerous setae on the surface. Together with leaves of *Nelumbites*, fructifications have been recovered from the Albian that are externally similar to the fruits of *Nelumbo*.

The Lower Albian of the USA yielded an infructescence (Dilcher, 1979, Figs 28, 29) which consists of an axis terminated by a receptacle-like area with 6–8 conduplicate (folded lengthwise) transversely striated carpels, dehiscent with an adaxial suture. Below the carpels are scars that probably correspond to abscissed carpels, parts of the perianth or to the androecium. Detached car-

Fig. 77. Fructifications of angiosperms; Albian (*b*), Cenomanian (*a*, *c–l*), Senonian (*m–u*), Maestrichtian (*v*), Eocene (*w*); North America (*a*, *c–l*, *u*, *w*), Kolyma River basin (*b*), Sweden (*m–t*), Amur River basin (*v*); *a*, *Calycites parvus* Newb., calyx-like organ; *b*, fruits *Kenella harrisiana* Samyl., *c–f*, *Prisca reynoldsii* Ret. et Dilch., branch with infructescence in bract axils (*c*), stoma at infructescence stalk (*d*), ovule, outer integument shaded, megaspore in black (*e*), part of infructescence (*f*); *g–i*, *Archaeanthus linnenbergeri* Dilcher et Crane, scheme of follicle, seeds shown by dotted line (*g*), blooming shoot and mature infructescence (*h*), part of infructescence (*i*); *j–l*, catkin-like aggregation of synangia (*l*), its pollen (*j*) and synangium (*k*); *m*, placentation in *Scandianthus*, section view; *o*, *S. costatus* Friis et Skarby, flower with strongly protruding sepals and one bracteole at bottom; *o*, *Scandianthus*, flower diagram; *p*, *q*, reconstruction (*p*) and diagram (*q*) of *Caryanthus* flower; *r*, *s*, reconstruction (*r*) and diagram (*s*) of flower *Manningia crassa* Friis; *t*, *Antiquocarya*, reconstruction of flower; *u*, flower incertae sedis; *v*, infructescence *Nyssidium arcticum* (Heer) Iljinsk.; *w*, flower *Paleorosa similkameensis* Bas. in section, above – details of stamen structure. Scale bar = 3 cm (*h*), 1 cm (*b*, *c*, *g*, *i*, *l*, *v*), 1 mm (*a*, *f*, *k*, *u*, *w*), 0.5 mm (*n*), 200 μm (*e*, *p*, *r*, *t*), 20 μm (*d*), 5 μm (*j*). Modified from: Dilcher, 1979 (*a*, *g*, *i–l*); Samylina, 1968 (*b*); Retallack and Dilcher, 1981c (*c–f*); Dilcher and Crane, 1984–1985 (*h*); Friis and Skarby, 1982 (*m–o*); Friis, 1983 (*p–t*); Tiffney, 1977 (*u*); Krassilov, 1973b (*v*); Basinger, 1976 (*w*).

pels of the same type have been described under the unsuitable name *Carpolithus curvatus*. Another infructescence from the same locality (Dilcher, 1979, Fig. 30) consists of three open carpels attached on three sides to a common stalk and probably basally fused. Together with this specimens were recovered with 3–8 basally fused carpels. Beginning from the Albian, together with platanoid leaves, reproductive axes appear that bear a large number of heads, similar to the inflorescences of *Platanus*. Their structure has not been studied in detail.

The genus *Caspiocarpus* (Middle Albian; Vakhrameev and Krassilov, 1979) refers to apical paniculate infructescence, crowning canaliculate stem with petiolate palmately dissected leaves (of the *Cissites* type). The basal branches of the infructescence occur in the shape of compound racemes, other branches occur as short racemes with spirally arranged fruits (about 10), follicles, dehiscent along a ventral suture, and, in the distal part, also along a dorsal suture. In the fruits are up to four (arranged in two rows) anatropous ovules with a longitudinal suture. The integument is thick, double. The outer integument is longer than the inner one. The micropyle is short, chalasa – wide. According to the fruits and ovules C. is most closely allied to Ranunculales.

The genus *Archaeanthus* (Cenomanian; Fig. 73, *g–i*; Dilcher and Crane, 1984, 1984–1985) comprises infructescence consisting of a long axis with helically arranged carpels with a pedicel. The carpels dehisced along the dorsal suture, accompanied by a crest with hairs. The seeds are small and, judging by their position, the placentation was marginal. Fruits of this type have been interpreted as follicles. Below the carpels two levels of tightly spaced scars appear on the axis. The scars of the upper level probably belong to stamens, those of the lower level to perianth members. Associated leaves (*Liriophyllum*) are large, entire-margined bilobed, with a thick midrib and petiole. Secondary veins extend pinnately from the midrib; two apical veins extend along the margins of the axial notch in its lower part.

Fructifications of the genus *Prisca* (Fig. 73, *c–f*) represent a raceme consisting of elongate multifollicles attached in alternating order. Each multifollicle consists of a thin axis, the lower part of which serves as the stalk, whereas the upper part serves as a receptaculum for numerous tightly spaced follicles of helical arrangement. The carpels are conduplicate with submarginal placentation. The ovules are orthotropous with two thin integuments. Judging by the sizes of the mature and immature follicles, their growth terminated at the moment of pollination, which is typical of many fossil gymnosperms of the classes Ginkgoopsida and Cycadopsida, but not of A. There are no signs of the specialization of the stigmatic surface even near the adaxial suture. Only on the inner surface of the follicles near the ovules have rudimentary papillae been found. Each multifollicle was subtended by a short bract with a stem-embracing base, acuminate apex and reticulate venation. In immature specimens indistinct traces of a perianth were found on the axis. Associated leaves (*Magnoliaephyllum*; Fig. 76, *e* (1)) are simple, entire-margined, narrow, oval to lanceolate, with brochidodromous venation.

The Cenomanian yielded also detached organs, comparable with the perianth and described as *Calycites* (Fig. 77, *a*). They are radially symmetrical, consisting of 4–6 lobes basally fused, at times bipartite.

From the Cenomanian of the USA a catkin-like male fructification has been recovered (Fig. 77, *j–l*). The axis was densely covered by stamen-like organs with four anthers in each. The pollen is monocolpate, of the *Retimonocolpites* type. In their overall habit the fructifications rather resemble gymnosperm microsporoclads. In the same deposits small actinomorphic flowers were found. Five petals with five sepals. Five stamens are situated opposite the petals; the stamen filament is basally attached to the petal. The gynoecium is composed of carpels closely spaced in a ring, and is similar to fruits described from the Cretaceous and Palaeogene under the name *Nordenskioeldia*. D. F. Basinger and D. Dilcher, who discovered these flowers, suggested that they might have belonged to the ancestors of Hamamelidales and Rosales (see also Dilcher and Crane, 1984). Friis and Skarby (1982; Friis, 1983, 1984–1985) described several genera of isolated flowers from the Upper Cretaceous (Senonian) of Sweden. The genus *Scandianthus* (Fig. 77, *m–o*) displays affinities to the order Saxifragales. These are small actinomorphic flowers subtended by two bracteoles. The perianth is composed of five sepals. Ten stamens are arranged in two cycles. The gynoecium is formed by two fused carpels. The ovary is lower and bears two apical lateral placentae with numerous ovules. The genera *Caryanthus* (Fig. 77, *p*, *q*), *Manningia* (Fig. 77, *r*, *s*) and *Antiquocarya* (Fig. 77, *t*) display affinities to the living Juglandales in their general organization, and vary in the number of members and symmetry of the flowers. All the flowers are epigynous and bisexual. In the anthers a pollen of the *Normapolles* type was found. From the Campanian of the USA another type of bisexual flower has been described, with a pentamerous carpel and gynoecium of five fused carpels (Fig. 77, *u*).

Unisexual flowers coming from the same deposits in two areas belong to platanaceous inflorescences (Friis, 1984). The male globose heads consist of flowers composed of 5–6 stamens surrounded by thick or membranaceous perianth parts. The connective is enlarged apically to form a peltate or almost sagittate cap. The pollen is tricolpate, reticulate with closely spaced verrucae in the colpi. The globose female heads produce sessile wedge-shaped ovaries surrounded by small or membranaceous perianth parts.

In the Upper Cretaceous and Palaeocene of wide occurrence is the genus *Nyssidium* (K_2–Pg; Fig. 77, *v*; Crane, 1984) affiliated to the Hamamelidae. It is represented by a 'paniculate' infructescence with a long axis bearing alternating lateral racemes, each consisting of 8–14 spirally arranged fruits. One of these crowns the raceme. The fruits are elliptical, single or paired; the ventral suture is accompanied by crests. Associated leaves belong to *Trochodendroides*.

Head-like male inflorescences of the genus *Tricolpopollianthus* (Danian) consist of staminate flowers with a quite massive receptaculum. The distal part of the stamens is expanded and forms a shield with a leaf-like apex. The anthers were attached to two ribs of the connective. The pollen is spherical, tricolpate, tripartite in the polar view, elliptical in the equatorial view. These inflorescences associate with platanoid leaves and resemble staminate heads of sycamores, but display affinities in their pollen structure to Hamamelidaceae. The Danian also yielded female inflorescences, closely similar to platanaceous. Of the same age are the most ancient fossil remains of sedge fruits encased in typical sacs.

The mode of pollination of Cretaceous flowers is not known. From the small sizes of the pollen and exine structure the inference was drawn that the Early Cretaceous A. were already entomophilous. However, it is more likely that the Early Cretaceous unisexual flowers, clustered in racemose, paniculate and catkin-like inflorescences, were anemophilous. In general no one-to-one correspondence exists between the modes of pollination, pollen sizes and the exine structure.

The Eocene is the next stratigraphic interval from which a profusion of A. fructifications are known. Nearly all of them belong to extinct genera, and a certain portion belong to extinct families. For instance, only 30–35% of the seeds and fruits in the Eocene 'London clay' can be affiliated with extant genera. Only occasionally can the Eocene flowers be indisputably assigned to living genera (e.g. *Cinnamomum* and *Quercus* in the Baltic amber belonging to the uppermost Eocene or lowermost Oligocene). *Trianthera*, *Adenanthemum*, *Sahnianthus*, *Sahnipushpam*, *Eokachyra*, *Combretanthites*, *Eomimosoidea*, etc. can only be assigned to extinct genera. Flower petrifactions of the genus *Paleorosa* (Middle Eocene; Fig. 77, *w*) have been well studied.

Eocene flowers (the same as for leaves, pollen, fruits and seeds) can be commonly affiliated with certain living families. Occasionally the structure of the flower combines characters of closely related families of one order. For example, *Sahnianthus* combines characters of the families Lythraceae and Sonneratiaceae of one order Myrtales. Hence, we are once again confronted here (as in conifers) with synthetic types. Among the Eocene flowers the bisexual condition prevails, with a well-developed, usually double perianth. Nearly all the flowers are distinctly actinomorphic, less frequently bilaterally symmetrical, which is reflected in a slight difference in the petal sizes. Even more scarce are distinctly zygomorphic flowers. The perianth has a small number of members (3–6), the petals are mostly free or slightly fused at the base. Flowers with a corolla fused in the lower part into a tube are also known. The narrowest part of the corolla is directly above the ovary. In some flowers the androecium is reduced to two stamens. Catkin-like inflorescences have also been encountered. Judging by the morphology of the flowers, the Eocene A. were probably both anemophilous and entomophilous. Their pollinators could probably have been coleoptera, diptera, hymenoptera, lepidoptera. In the Eocene, flower types associating with other pollinators are not known (Crepet, 1979). No flowers are known with traces of high specialization, typical, for instance, of Orchidaceae and Compositae (the assignment of the Eocene genus *Protorchis* to orchids is debatable). Compositae were of first appearance only in the Oligocene.

In the Cretaceous and Palaeogene no fructifications have been found that give confirmation to the primitiveness of the entire syndrome of the 'ranalian complex' (Dilcher, 1979; Dilcher and Crane, 1984–1985; Stewart, 1983). From the 'primitivity code', relying on the ranalian complex, the palaeobotanical record confirms the primitivity of the monocolpate pollen and the successive development from this pollen to a tricolpate one, and further to a tricolpoporate, triporate and multiporate pollen (Fig. 78, *m*). It also supports the primitiveness of the bitegmic ovules as compared with the unitegmic ovules, of the follicles relative to other fruit types, of actinomorphic flowers relative to zygomorphic ones, etc. (Dilcher and Crane, 1984–1985). The formation of bisexual flowers and of the syncarpous gynoecium was seemingly accomplished by the end of the Early Cretaceous, that of the zygomorphic flowers in the Eocene, and of polypetalous in the Oligocene.

Synthetic types, combining characters of different genera of one family, have been described even from the Oligocene. For instance, the genus *Fagopsis* (family Fagaceae) in the structure of the stamens and pollen shows affinities to *Fagus*, *Quercus* and *Trigonobalanus*, according to the staminate inflorescences affinities to *Fagus*, whereas in the structure of the female dichasia, cupule and nutlet affinities to *Castanea*, *Castanopsis* and *Lithocarpus* (Manchester and Crane, 1983). Because of this caution is warranted concerning the assignment to extant genera (in part also families) of detached organs (e.g. only leaves, or only fruits), not verified by finds of other parts of the same plants. If we take into consideration the association of various organs encountered in burials, then it is conceivable that certain families appeared during the Late Cretaceous. The proportion of living genera of A. in the Eocene hardly exceeded one third of the total generic composition. The extant genera prevail since the Miocene. Tracing the pollen types characteristic of living families and species we arrive at nearly the same conclusions (Muller, 1981).

Chapter 4

Palaeopalynology

All kinds of microfossils recovered from sediments by palynological methods (maceration in acids and alkalis, separation by heavy liquids) are termed palynomorphs. This refers to spores and pollen, acritarchs, dinoflagellates, tiny fruiting bodies of fungi, etc. The discipline dealing with palynomorphs is called palaeopalynology or merely palynology. Some palynomorphs have been under consideration in other chapters. This chapter treats those sections of palynology which are most closely associated with investigations of the morphology, systematics and evolution of higher plants. Other information on palynology can be gleaned from the literature reviews (Chaloner, 1968b; Chaloner and Muir, 1968; Faegri and Iversen, 1975; Kremp, 1965; Paleopalinologia, 1966; R. Potonié, 1956–1970; 1973; R. Potonié and Kremp, 1970; Shaw, 1970; Taylor, 1981).

CERTAIN CONCEPTIONS AND TERMS

In the following section attention is focused on certain specific morphological features of spores and pollen, the taxonomic significance of which has been recognized only recently; owing to this, descriptions of them cannot be found in many special manuals. Some concepts and terms need a fuller explanation, considering that they are variously treated in the literature.

Miospores

It is not always possible to distinguish spores from pollen and prepollen, microspores from megaspores and isospores. Because of this of wide usage in palaeopalynology is the term 'miospores', which corresponds to isospores of homosporous plants, microspores of heterosporous plants, small megaspores (when microspores corresponding to them are unknown), prepollen and pollen. It is arbitrarily taken that miospores are less than 200 μm in size. This size has no special biological meaning and has been chosen for technical reasons. It is determined by the size of the meshes in standard sieves and by the specimen preparation techniques.

Prepollen

The transition from pteridophytes to gymnosperms was concomitant with the change in the place and time of the germination of spores. In pteridophytes the germination of spores may occur in sporangia (in zygopteroid *Biscalitheca* up to 10 cells were formed – Mamay, 1957), but more often outside them; the germination is proximal, i.e. on the side of the dehiscence slit, facing inwards to the tetrad. Only in rare instances do pteridophytes reveal bipolar (on the proximal and distal sides) germination (certain ferns of the families Osmundaceae, Loxogrammaceae, Polypodiaceae and Ophioglossaceae). In ferns equatorial apertures are absent. Spores develop into a free-living gametophyte. The reduced gametophyte in the more advanced gymnosperms is already formed in the microsporangium and is retained within the intact microspore wall. Intracellular gametophytes have been observed in pollen of callistophytans (Fig. 47, *l*). Frequently the folds of the exine (in Cordaithantales and Bennettitales) have been mistaken for gametophyte cells (prothallial). Only in the ovule does the microspore germinate and form a pollen tube that conveys the sperm to the archegonia. Microspores with encased gametophytes and a primary distal aperture for germination are termed pollen. In angiosperms germination does not necessarily take place on the distal side, but may occur through any aperture (colpus, colpus-pore, pore) that, however, is thought to have been derived from the distal aperture (colpus). The equatorial swellings in some angiosperms (for example, Upper Cretaceous *Pemphixipollenites*) seemingly preceded the formation of the equatorial pores.

The process of acquiring gymnospermy was not synchronic with the transition from typical spores to typical pollen. In lagenostomans and certain other primitive gymnosperms no traces can be found of a distal aperture, whereas a proximal monolete or trilete slit is present (Fig. 57, *a*, *e*, *r*; Fig. 58, *p*), in just the same way as in ferns. It is probable that the germination was proximal. This combination of proximal germination (as in spores) and the penetration of the microspores into the ovules (as in pollen) is taken as the conceptual premise defining the prepollen (Chaloner, 1970a). Morphological characters of the wall, that render possible reliable distinctions between spores and spore-like prepollen among dispersed miospores are not known. In this case one can only take for reference certain pteridophytes or primitive gymnosperms yielding the same miospores.

In cycads and *Ginkgo* the pollen germinates from two sides (Fig. 78, *a* (C)). From the distal side a pollen tube is issued, but it is has only a gaustorial function, i.e. it protrudes into the nucellar wall and adsorbs water and nutrients. The spermatozoids are released from the opposite (proximal) side, where there is no aperture. It is possible that this was also characteristic of many fossil gymnosperms, where the microspores resemble externally a true pollen. Indirect indication of this is found in the ultrathin sections of the exine. Until they were being employed the trigonocarpalean prepollen *Monoletes* was described as pollen of indeterminate polarity. Investigations with a transmission electron microscope (TEM) served to indicate that the exine is composed of a lamellar nexine and alveolar sexine. In *Dolerotheca* and *Aulacotheca* during the earlier stages of ontogenesis along the proximal slit the nexine of such pollen was strongly thickened in the shape of a triangular ridge facing the surface, whereas the sexine practically wedged out in this place. In the course of ontogenesis the nexine then underwent resorption beneath the ridge and a narrow slit was formed (Fig. 58, *m*). Two furrows are present on the distal side (Fig. 58, *k*, *l*), along which the sexine thinned out, but no attenuation of the nexine is apparent in this place. In *Codonotheca* and *Rhetinotheca* (?) a proximal slit is lacking, in the structure both of the nexine and the sexine. The distal side bears two colpi that are marked by the thinning out of the nexine. These genera probably demonstrate the transition from proximal to distal germination. It is not known on which side the pollen tube was formed (if it was actually produced), what the gametes were like (spermatozoids of spermia), and how they were vectorized to the archegonia.

The transition from the proximal aperture to the distal has been described in the Carboniferous Cordaitanthales. The change from grains with a distal colpus and occasionally with a proximal slit (Middle Carboniferous) to grains lacking a proximal slit (Upper Carboniferous) can be observed in

Callistophytales. The pollen tube issued from the distal side (Fig. 47, *n*).

Several types of prepollen are known. Radially symmetrical round grains were produced by the Lagenostomales (*Feraxotheca*, *Telangiopsis*), some Trigonocarpales (*Potoniea*) and other primitive forms. The second type includes bilaterally symmetrical pollen of the *Monoletes* type (see p. 171). The third type involves cavate (pseudosaccate) prepollen of primitive gymnosperms (*Pacalathiops*, *Simplotheca*, etc.) and certain Trigonocarpales (*Parasporotheca*). Grains of the fourth type are saccate and quasisaccate with a functional proximal slit and no traces of a distal aperture. Other types can be noted. The change in the polarity of germination was of independent ocurrence in various gymnosperm groups (Millay and Taylor, 1974, 1976) and in each case no clear-cut distinction between pollen and prepollen can be drawn.

Stratification of the miospore wall

In fossil material the entire wall is not preserved. Only in exceptional cases is the inner layer preserved – the intine, consisting chiefly of cellulose. The other layers comprise the exine and are of good preservation, being composed of sporopollenin – an organic polymer, resistant to oxidation and leaching. The chemical composition of various layers varies in different plants, which is reflected in the differences in their state of preservation, optical and electron-optical density (Brooks *et al.*, 1971; Brooks and Shaw, 1973; Shaw, 1970). Miospore walls of some plants are of very poor preservation (for example, larch, aspen, poplar, etc.). At times the layers may be of differential preservation and undergo separation during burial (e.g. the body may fall out of the saccate grain and be buried separately from the saccus).

Many schemes have been proposed for the stratification of miospore walls (Kremp, 1965; Walker and Doyle, 1975). The schemes adopted below are most widely acknowledged in the current literature, especially in studies of miospores employing electron microscopy techniques. The comparison of the layers in different plants is based on the density of the walls, their relative positions, inner structure, chemical features. For convenience we will begin our discussion with the wall stratification in pollen of gymnosperms and angiosperms. In electron microphotographs of transverse cuts the exine can be seen above the intine. Its lower layer is the endexine. It is non-lamellar or of lamellar texture. Higher up occurs a three-layered ectexine. Its lower layer (foot layer) is non-lamellar, the same as the upper layer (tectum or tegillum), at times pierced by pores. The middle layer (endosexine) may be of three major types: (1) granular, consisting of numerous adhered grains; (2) alveolar; the meshes may have walls and luminae, varying in relative sizes and orientation; the meshes may be connected by pores; (3) columellar, consisting of columns that connect the foot layer with the tectum. In angiosperms the expansion of the pores can result in a reticulate structure of the tectum (semitectate exine), and then in its reduction (intectate exine) with the exposure of the columellar layer. Granular and alveolar types intergrade. In those cases where the endexine and foot layer cannot be separated under the light microscope, these two layers are collectively termed the nexine, whereas the columellar layer and tectum are termed the sexine. When the endexine is absent, the term nexine refers only to the foot layer. If the nexine is absent then beneath the tectum only one layer may be present, for example, the granular layer. In many gymnosperms and angiosperms the exine may bear sculpture (Fig. 78, *b*, *c*), which should be distinguished from tiny grains, at times spinulose, that were deposited on the exine surface after its formation, and can be assigned to the perine. These grains (orbicules or Ubisch bodies) sometimes coalesce and cover large portions of the exine. In places where they are aggregated into stacks they are associated with fine membranes, which, in the same way as the orbicules, are obviously of tapetal origin (Fig. 58, *m*).

The similarity in the position and fine structure of the exine layers does not imply the homology of the layers of different plants, provided that the criteria of homology are the succession in the formation of the layers, their chemical composition

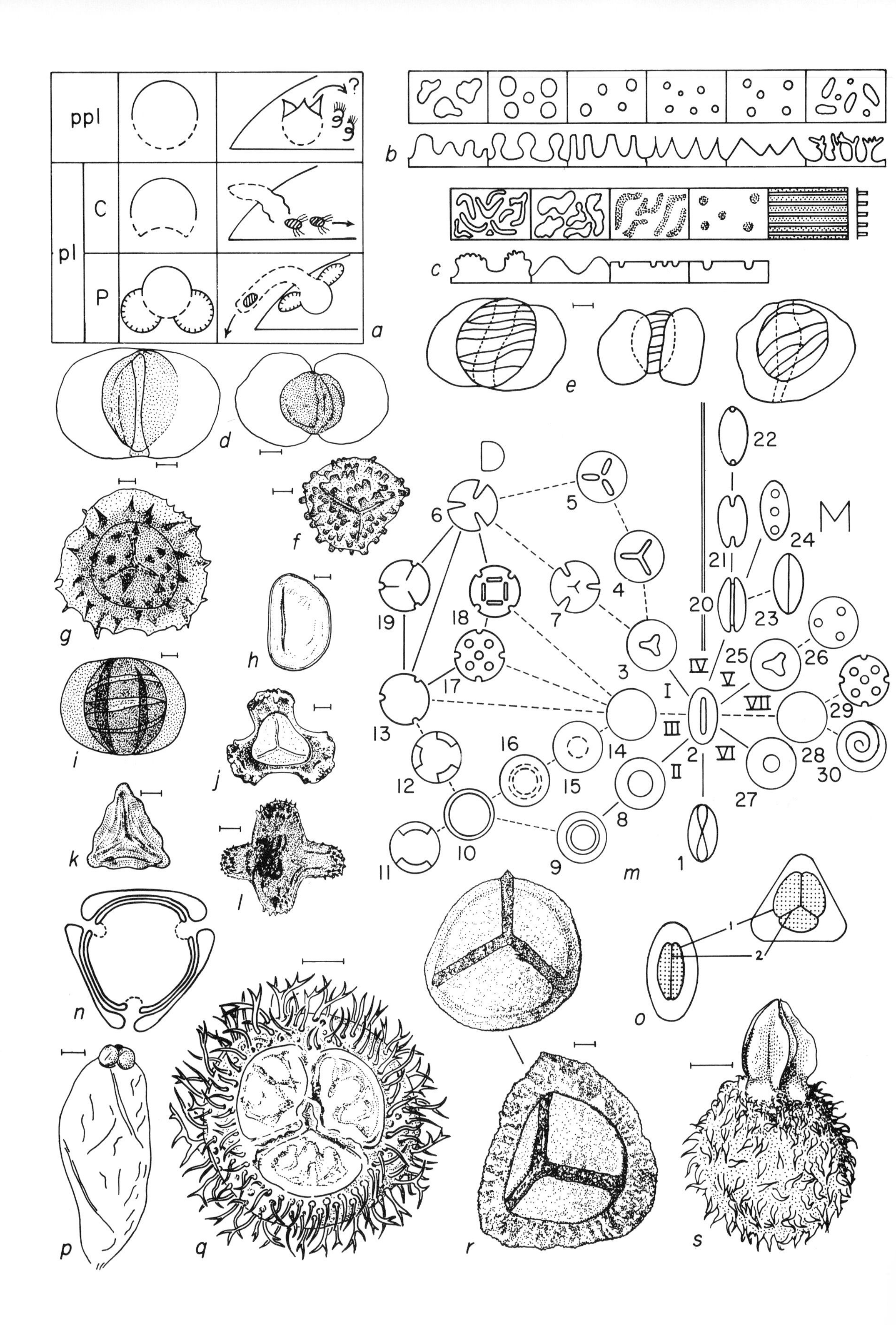

ppl
pl
C
P
?
a
b
c
d
e
f
g
h
i
j
k
l
m
n
o
p
q
r
s
D
M
1
2
3
4
5
6
7
8
9
10
11
12
13
14
15
16
17
18
19
20
21
22
23
24
25
26
27
28
29
30
I
II
III
IV
V
VI
VII

and sporopollenin source (intracellular relative to the gametophyte, or tapetal relative to the sporophyte). In some extant seed plants (*Lilium*, *Ceratozamia*) the nexine is formed during the later stages of the exine development. In *Classopollis* (Cheirolepidiaceae) the laminated nexine was of quite late formation. However, in *Potoniea* and *Monoletes* (Trigonocarpales) the lamellar nexine was formed first, whereas the alveolar exine layers were located outside it; the orbicules and external membranes were evidently deposited from tapetal secretion. The inner reticulum on the *Pinus* sacci, which corresponds in position to the alveolar layer in Trigonocarpales, is deposited without the aid of the tapetum. The problems bearing on the homologization of the exine layers in extant and fossil seed plants have been discussed in more detail in the book edited by Ferguson and Muller (1976), and in the paper by Taylor and Alvin (1984).

Spores of pteridophytes differ considerably in the inner structure of their walls (Lugardon, 1972, 1976); however, in palaeobotanical systematics little attention is given to these differences. The conceptions and terms used to denote the sculptural elements of spores in pteridophytes are the same as those adopted for the pollen. In many pteridophytes the perisporium forms a continuous layer and bears a sculpture of ribs, folds, processes, etc. The exine proper may be unilayered and homogeneous, or consist of several layers of different structures (lamellar, granular, homogeneous; Lugardon, 1972, 1976).

Cavum, quasisaccus, saccus

In many miospores the wall is exfoliated and forms a cavity. In pteridophytes the layers between which the cavity appears are denoted by different terms. It is commonly observed that the cavity divides the endexine and ectexine. Occasionally these are referred to by the terms nexine and

Fig. 78. Miospores and megaspores; *a*, distinctions between prepollen (ppl) and pollen (pl) of *Cycas*-type (C) and *Pinus*-type (P), proximal side shown by solid line, distal side by dashed line (to left, grain structure; to right, germination of grain in ovule); *b*, *c*, exine relief in surface view and in section; *b* (from left to right), verrucate, pilate, baculate, echinate, with coni and capilli; *c* (from left to right), cristate, rugate, vermiculate, foveolate, costate; *d*, different quasisaccate pollen from microsporangium of *Permotheca vesicasporoides* Esaul., Gom. et S.M. (Cardiolepidaceae), body stippled, normal (to left) and aberrant (to right) grains, Upper Permian, Russian platform; *e*, variation of striated quasisaccate pollen in microsporangium of *Arberiella vulgaris* (Arberiales), Upper Permian, India; *f*, *Raistrickia grovensis* J. M. Schopf, Carboniferous, USA; *g*, *Kraeuselisporites cuspidatus* Balme, Lower Triassic, Australia; *h*, *Laevigatosporites*, Carboniferous, Euramerian area; *i*, *Taeniaesporites*, Lower Triassic, Australia; *j*, *Tripartites incisotrilobus* (Naum.) Pot. et Kr.; *k*, *Gleicheniidites*, Lower Cretaceous, England; *l*, *Aquilapollenites attenuatus* Funk., Maestrichtian, USA; *m*, putative transformations of pollen apertures in dicots (D) and monocots (M), Roman numerals mark major trends of transformation of initial type, (1) ancestral monosulcate type, (2) initial monosulcate type, (3–30) derived aperture types (3, 4, trichotomosulcate; 5, distally tricolpate; 6, tricolpate; 7, trichotomosulcate + tricolpate; 8, distally monoporate; 9, distally operculate; 10, equatorially zonocolpate; 11, equatorially dicolpate; 12, equatorially tricolpate; 13, equatorially triporate; 14, inaperturate; 15, proximally monoporate; 16, proximally operculate; 17, periporate; 18, pericolpate; 19, tricolporate; 20, monosulcate; 21, dicolpate; 22, diporate; 23, meridionally zonocolpate; 24, distally meridionally triporate; 25, trichotomosulcate; 26, distally subequatorially triporate; 27, distally monoporate; 28, inaperturate; 29, periporate; 30, spiraperturate); *n*, one of the pollen types of *Normapolles* group (genus *Trudopollis*); *o*, common features of monolete and trilete spores (1, curvature; 2, laesura, or dehiscence slit; contact areas stippled); *p*–*s*, megaspores; *p*, *Cystosporites*, at apex – three aborted megaspores of the tetrad, Middle Carboniferous, USA; *q*, *Setosisporites*, contact area devoid of trichomes, small gula in place of rays junction, Middle Carboniferous, Ruhr; *r*, *Triletes brassertii* Stach et Zerndt, intact specimen (below) and body devoid of flange, Carboniferous, Western Europe; *s*, *Lagenicula crassiaculeata* Zerndt with large gula, Carboniferous, Western Europe. Scale bar = 100 μm (*p*–*r*), 10 μm (*d*–*l*, *s*). Modified from: Chaloner, 1970b (*a*); Kremp, 1965, after Couper and Grebe (*b*, *c*, *o*); Pant and Bhatnagar, 1971–1973 (*e*); Schopf J. M. *et al.*, 1944 (*f*); Chaloner (*g*, *i*), Kosanke (*h*, *j*), Hughes (*k*), Penny (*l*) and Tschudy (*n*) in Tschudy and Scott, 1969; Muller, 1970 (*m*); R. Potonié, 1973 (*p*, *q*); Kremp, 1965, after Dijkstra (*r*) and R. Potonié and Kremp (*s*).

sexine, intexine and exoexine, etc., accordingly. Spores with cavities belonging to pteridophytes and primitive gymnosperms are most frequently termed cavate, and the cavity itself a cavum (Fig. 42 (1)). The cavum cannot always be easily distinguished from the equatorial disc-like flange lacking an inner cavity, that was formed by the perine or exine. In the latter case the flange is denoted by different terms, dependent upon the thickness and type of cross-section. A wide membranaceous flange is termed a zone, a narrower and thicker flange a cingulum, etc. (Kremp, 1965). The cavum evolved independently in different pteridophytes. It is described in various lycopsids (*Selaginella*, *Spencerites*, *Endosporites*; Fig. 17, *v–x*) and progymnosperms (Protopteridiales, Protopityales; Fig. 38, *j*, *k*). In the latter the cavum is attached to the body only proximally above the equator.

Until recently gymnosperm microspores were most frequently termed pollen and subdivided into two major types – asaccate and saccate. It was believed that the structure of the saccate pollen was very similar to that in pines. Here the exine splitting occurs between the nexine and sexine, so that the alveolar layer remains on the inner surface of the saccus, which acquires an intrareticulate structure, and the foot layer composes the surface of the body (Fig. 42 (26); in cavate forms the surface of the body is composed of the endexine). On the basis of general phylogenetic conceptions the saccate pollen could be derived from cavate spores of progymnosperms. Studies of the ultrathin sections of fossil pollen by TEM have shown that many genera of dispersed miospores formerly considered saccate, have, in effect, no sacci in the form of hollows (Fig. 42, 23). This was first demonstrated for the genera *Lueckisporites*, *Strotersporites*, *Inferpollenites*, *Parillinites*, *Ovalipollis*, *Polarisaccites*, *Triadispora*, *Jugasporites*, *Podosporites*, *Parvisaccites* and *Guttulapollenites* from the Permian and Triassic of Western Europe (Scheuring, 1974), and then for the genera *Cannanoropollis*, *Plicatipollenites*, *Platysaccus*, *Protohaploxypinus* and *Striatopodocarpites* from the Upper Palaeozoic of Australia (Foster, 1979). The structure which under the light microscope looks like a saccus (Fig. 49 (1, 4–9, 11); Fig. 52, *u*; Fig. 53, *n*; Fig. 55, *k*; Fig. 64, *w*; Fig. 65, *l*; Fig. 69, *f*, *n*, *u*, *x*; Fig. 73, *g*; Fig. 78, *d*, *e*, *i*) and was accordingly described as such by palynologists, is composed of a spongy exine mass with numerous small cavities, irregular in form, either of random or radial orientation. This spongy mass, comparable with the alveolar layer of the sexine, covers the body on all sides, thinning on the proximal and distal sides. If ribs are borne on the proximal side (Fig. 38 (22)), then they also consist of the spongy mass that thins out in the grooves between the ribs. The spongy mass thins out or disappears in the area of the distal aperture. This picture can frequently be seen under the light microscope. Instead of distinct meshes, typical of saccate pollen and clearly restricted to the saccus wall, the light microscope reveals an intricate intertwining of the luminae and interconnections between them; an inner cavity, corresponding to a saccus, cannot be seen even in that case when the pollen grains are slightly flattened. Under fluorescence microscopy this spongy mass shines brightly and numerous inner luminae of various sizes can be well detected in it. A true saccus looks different under the luminescent microscope. The luminosity of its walls is less than that of the pollen body, whereas the inner reticulum exhibits a much brighter luminous flux clearly visible against the darker outer wall of the saccus.

The spongy masses simulating the saccus were termed protosaccus, and the corresponding miospores were termed protosaccate (Scheuring, 1974). These terms are etymologically irrelevant, inasmuch as they convey the implication that the protosaccate structure precedes the saccate, which is not always so. Because of this the terms 'quasisaccus' and 'quasisaccate' are more appropriate. It can easily be understood how the alveolae undergo resorption inside the quasisaccus, resulting in the formation of a true saccus. In *Taeniaesporites* (al. *Lunatisporites*) *noviaulensis* (Triassic) a thick alveolar layer comprises the wide proximal ribs of the sexine, just as in the quasisaccate Permian genus *Lueckisporites*, but large cavities are present in both sacci. The outer walls of the saccus are still quite thick and within a large part resembles more the spongy mass of a quasisaccus, than the thin-walled alveoles in extant conifers. The same type of thick spongy wall around a large cavity was observed in the pollen *Nuskoisporites* that was produced by Upper Permian conifers *Ortiseia* (Clement-Westerhoff, 1984), also in the pollen *Potonieisporites* that was produced by '*Lebachia*' *lockardii* from the uppermost Carboniferous (Mapes and Rothwell, 1984). In *Felixipollenites macroreticulatus* from the Middle Carboniferous the thin-walled alveolar layer closes the proximal and distal sides, but the cavity in the saccus is large. Inside the saccus on the body a thin alveolar layer is retained. The quasisaccate organization is typical of extant *Dacrydium* (Podocarpaceae) and some species of *Tsuga* (Pinaceae), although other members of both families have true sacci. Evidently, the transition from the quasisaccate to the saccate organization proceeded in different groups, but the opposite process could have occurred. A true saccus with thin-walled alveoles on its inner surface is known in the Middle Carboniferous Cordaitanthales (*Cordaitanthus concinnus* and *C. schuleri*; dispersed miospores *Sullisaccites* and *Felixipollenites*; Fig. 64, *p–v*) and Callistophytales (*Idanothekion*; Fig. 47, *k–m*; some dispersed miospores of the genus *Vesicaspora*). However, in the descendants of both plant types we encounter quasisaccate miospores. The exine structure in older putatively saccate miospores has not been studied by TEM. As was previously stated, cavate spores can be observed in progymnosperms. The structure of cavate prepollen of gymnosperms (*Paracalathiops*, *Simplotheca*, etc.; Fig. 57, *a*) and of cavate spores of homosporous progymnosperms (Fig. 38, *j*, *k*) was compared with the trigonocarpalean saccate prepollen of the genus *Parasporotheca*, where two semicircular sacci are formed as a result of splitting of the sexine, of which one layer remains on the body, so that its surface looks granular (Millay and Taylor, 1979). In microspores of the most advanced progymnosperms (Archaeopteridales) the inner exine layer is lamellar, whereas the outer layer is composed of a dense spongy granular mass with small luminae. These two layers are treated as the endexine and ectexine (the foot layer and tectum are absent). From these spores the quasisaccate and then the saccate pollen of gymnosperms can be derived.

TAXONOMY OF DISPERSED MIOSPORES

The case history of the taxonomy of dispersed miospores was discussed at length by Potonié and Grebe (1974). Only during the 30s of the present century were the first attempts made to construct a full classification system of dispersed miospores. The first system was proposed by A. C. Ibrahim in 1933. Almost immediately afterwards, activities were directed along three lines, which can be arbitrarily designated as formal, congregational and actualistic. In the first line of approach the chosen taxonomic characters were assigned a definite weight. In strict accordance with the weight (rank) of the characters the ranks of miospore parataxa were distinguished. No provision was made for the change in the weight of the characters upon transition from one parataxon to another. The system suggested by Naumova (1937–1939) was constructed in this manner. In her system all miospores (at that time this term was not used) were subdivided into two types – spores and pollen, corresponding to pteridophytes and seed plants. The classes Rimales (with a definite dehiscence slit) and Irrimales (without it) were identified among the spores. The class Rimales was subdivided according to the slit structure into the groups Triletes (trilete slit) and Monoletes (monolete slit). The group Triletes involved two parataxa of no certain rank – Azonotriletes (without a flange) and Zonotriletes (with a flange). Nine subgroups were distinguished in the first, and ten in the second. The distinctions of the subgroups of spores without flanges were based on their exine relief (Leiotriletes – smooth, Trachytriletes – shagreen, Acanthotriletes – spinose, Lophotriletes – tuberculate, Dictyotriletes – reticulate, Chomotriletes – girdled by concentric ridges, etc.). The subgroups of the flanged spores were recognized according to the structure of the flange (Euryzonotriletes – a dense wide flange, Hymenozonotriletes – membranaceous flange, Stenozonotriletes – narrow flange, etc.). The classification of monolete miospores and pollen was based on the same principle. Pollen was classified according to the presence of pores and sacci, their number and other characters. This resulted in the construction of a fixed system of 49 subgroups including all possible types of miospores. The subgroups were subsequently assigned to the status of genera.

Owing to the simplicity of its application Naumova's system was widely used in the USSR. However, the subgroups (genera) were found to be very broadly defined and of small stratigraphic significance. For detailed stratigraphic subdivision it was necessary to rely on species, the number of which in some genera was very high, so that orientation among them was difficult.

The congregational approach also began to develop during the 1930s, but was of further development following the publication of the studies contributed by Potonié (1956–1970), Potonié and Kremp (1954, 1970), J. M. Schopf, Wilson and Bentall (1944), and others. These authors also distinguished suprageneric taxa according to the fixed weight of characters, but the genera were derived by means of grouping of similar species, and the key characters were allowed to change from one genus to another, as was the weight of these characters. The number of the genera was not fixed by the structure of the system, but depended upon the particular material under consideration. The discovery of miospores with a new combination of characters or with characters that were not taken into account earlier, could lead to the establishment of a new genus. Some genera were classified on the basis of a combination of simple characters, while others on specialized characters of rare occurrence. For example, the genus *Leiotriletes* included smooth subtriangular trilete spores. Spores of the genus *Reinschospora* have triangular equatorial contours with concave sides, infilled with numerous setae. The genus *Cristatisporites* was identified on the basis of the denticulation of the flange. The variability in the characters within a genus became evident only in the course of investigations, and no limits for the variability were established beforehand. For instance, if within a group of species, associated by transitions and united into a single genus the sculpture changed from granular to microtuberculate, this did not require that the respective species should be divided into different genera, only because in other cases the distinction of the genera was based on the same difference in the sculpture. In the taxonomy of extant organisms this method of establishing the taxa by successive grouping of specimens into species, species into genera, etc. (i.e. in 'ascending order' from taxa of inferior rank to taxa of increasingly higher rank) was termed by Smirnov (1925) congregational (see Chapter 2). It should be remembered that in botany the transition to congregational systematics and the rejection of the Linnaean artificial system of suprageneric taxa furnished the basis for the modern system adopted today for the classification of plants. The congregational system is frequently termed a natural system. The advantage of this system is that we are not restricted to any premeditated fixed weight of the characters. Such fixation of the weight leads inevitably to the artificial grouping of objects into a taxon, or the separation of closely related objects.

Congregational systematics, employed by most palynologists, is more difficult in application than formal classifications. But the genera congregations derived in this manner, being much narrower in circumscription, as compared with Naumova's generic subgroups, were found to be incommensurably greater in significance for stratigraphic purposes. This was demonstrated first on miospore assemblages from the Carboniferous coal measures of Western Europe and North America, and then on materials from other deposits.

Both the formal and congregational approaches do not link the miospore system with taxa established from entire plants (eutaxa). Because of this both approaches were contested, giving rise to adverse criticism by the palynologists, who argued against the distinction of miospore parataxa and advocated the establishment of eutaxa of entire plants. The supporters of this viewpoint, proceeding on the basis of actualistic observations, affiliated miospores to eutaxa of extant (and rarely, extinct) plants. The 'actualistic' approach resulted in many names derived from extant genera appearing in the lists of miospores – not only of the Cainozoic and Mesozoic, but even of the Paleozoic. Sometimes such genera names were changed in order to emphasize certain peculiarities of fossil miospores. In this manner the genera *Protopicea*, *Equisetosporites*, etc. appeared. Unfortunately, miospores of different plants may be practically identical (at least under the light microscope), and the variability in the spores and pollen within the bounds of the eutaxa, even of generic rank, may be extremely large. Because of this the inferences derived on the basis of the 'actualistic' approach resulted in serious errors concerning the stratigraphic and geographic distribution of genera and families. Bennettites were 'discovered' in the Carboniferous, the genus *Podocarpus* and the order Welwitschiales were discovered in the Permian, etc. These invalid data penetrated into the reference literature (Osnovy paleontologii, 1963a, 1963b; Paleopalinologia, 1966, etc.).

Hence, congregational systematics has proved to be the most practical. Its suprageneric taxa are (in ascending order) subinfraturma, infraturma, subturma, turma and anteturma. They are distinguished in the same manner as the taxa in Naumova's system, but on the basis of a wider syndrome of characters. The major intention of these taxa is to group the genera in such a manner as to make the system workable. It does not presuppose the association with eutaxa. As an example can be taken the congregational system constructed for monosaccate (including quasisaccate) miospores. They are assigned to the anteturma Variegerminantes (germination may be both distal and proximal), the turma Saccites (saccate miospores), subturma Monosaccites (one saccus). The latter is divided into infraturmas Sphaerosacciti (the body is freely inserted inside the saccus). Monpolsacciti (the saccus envelops the body from the proximal or distal side), Dipolsacciti (the saccus is attached near the equator, leaving the poles free). Subinfraturmas can be distinguished on the basis of other characters. This system can be further improved by taking account of the distinctions between the saccus and quasisaccus, the presence of an intrareticuloid in certain miospores (*Cladaitina*), etc.

The same principle has been adopted for the systematics of all palynomorphs (see also the section on Acritarchs (p. 54)). The classification of disaccate and quasidisaccate miospores relies on the presence of ribs, the structure of the distal colpus, the form and nature of attachment of the saccus (quasisaccus). Distinctions are made of haploxypinoid grains, in which the saccus (quasisaccus) has a uniform oval contour, the contour of the body included (Fig. 78, *e* (left)), and diploxypinoid grains, where the sacci (quasisacci) narrow to the place of their attachment to the body, so that the general grain contour becomes distinctly lobed (Fig. 78, *d* (right), *e* (in centre)). Spores and spore-like prepollen are classified according to the general symmetry, the presence of a perisporium, structure of the dehiscence slit, sculpture and structure of the exine. The classification of the angiosperm pollen is based on the number, structure and arrangement of the apertures, sculpture and structure of the exine, etc.

The application of the congregational principle deprives the investigator once and for all of the possibility of governing by the fixed weights of the characters, and urges the need to define again every time the weight of each character. Such an approach affords a more natural grouping of miospores of greater stratigraphic and phytogeographic significance. A disadvantage of the congregational approach is the large dependency of the taxa upon subjective judgements of individual palynologists. Because of this attempts were made to bring the congregational approach in line with the formal approach. For instance, it was suggested that distinctions of the genera should be based on the miospore organization, and of the species on the details of the exine sculpture and structure. On the other hand, attempts were made to weigh the major characters of large miospore groups, and to define such characters that most closely establish the affinities of the miospores to the suprageneric eutaxa (Bharadwaj, 1974). In this case three different lines of approach are brought together in the systematics of miospores; in the ideal the miospore system is thought to be brought closer to the eutaxa system.

As will be shown below, between the miospore types and the eutaxa of entire plants a one-to-one correspondence does not exist. This means that the construction of a miospore system, isomorphous to a eutaxa system, is, in principle, impossible, and there is no sense in attempting to achieve this. Instead, we should (1) bring some of the miospore taxa closer in line with the eutaxa, if possible, (2) develop such a general miospore system that would merely serve an information-retrieval function, disclaiming pretensions to replacing the eutaxa system for entire plants.

The first task can be solved mainly by means of monotopic sampling and congregational analysis (see Chapter 2). In this case we should be prepared to find that any character may be variable within the congregation, defined as a genus. A good example of this is the Cretaceous genus *Afropollis*, that has been assigned to angiosperms (Doyle *et al.*, 1982). Due to its typical tectal muri this genus can readily be identified among dispersed miospores. Notwithstanding the presence of this specific stable character, the genus reveals a variable character – the absence or presence of columellae, which is commonly of paramount importance in the distinction of pollen of gymnosperms and angiosperms. In *Afropollis* this character forms a basis for delimiting only species, at best.

With regard to the second task, of most importance is the convenience in the application of the information-retrieval system. Because of this we should not be appalled by the fact that certain miospore genera may occur simultaneously in different places of the system.

CORRELATIONS BETWEEN MIOSPORE PARATAXA AND EUTAXA

In the systematization of dispersed miospores we should take account of the variability of the spores and pollen perceived in eutaxa, preservation factors and the variations observed in the given sample. All this increases the naturalness of parataxa, bringing them more in line with the eutaxa.

Certain miospore characters are sufficiently distinct to allow for their affiliation with definite eutaxa. For example, Equisetaceae and Calamostachyaceae have very typical elaters, that so far have not been noted in other higher plants. Microspores *Aratrisporites* have so far been encountered only in lycopsids allied to the genus *Pleuromeia* and probably belonging to the family Pleuromeiaceae. Certain opposite examples may be offered where similar miospores are linked with quite different eutaxa. Mention has already been made above of the pollen form-genus *Stellatopollis*, which is characterized by a very peculiar crotonoid sculpture. Among the extant angiosperms such a sculpture has been observed only in single genera of six families, assigned to two subclasses, Dilleniidae and Caryophyllidae. Disaccate or quasisaccate miospores with typical proximal ribbing of the body are known in Peltaspermales, Arberiales and Pinales.

No less difficulties arise with regard to the opposite phenomena – the polymorphism of spores, not only within the bounds of a eutaxon, but even within one sporangium. In miospore systematics much weight is assigned to the number of rays of the proximal slit, the presence, number and shape of the sacci, colpi and pores, the perisporium structure, exine sculpture, etc. All these characters may be at times of very high intraspecific variability, at least within certain species and genera.

The first monolete miospores appear later than the trilete, namely in the Middle Devonian, and in conspicuous amounts from the Middle Carboniferous. Among the Carboniferous miospores such forms occur that are closely similar in sizes and sculpture, but differ in their dehiscence slit – mono- or trilete. In many extant plants spores of both types have been found in one sporangium, although one type is most dominant. The same has been noted for the Middle Carboniferous plants. For example, in the marrattiaceous *Scolecopteris minor* and *S. fragilis*, besides monolete spores, grains have been found with an underdeveloped third ray; these grains predominate in *S. latifolia*, where the transition can be perceived both to typical monolete, and typical trilete spores (Millay,

1979). The genus *Scolecopteris* includes species with the predominance of either monolete, or trilete spores. The same grouping of species is known in other pteridophytes. The Carboniferous genus *Spencerites* (lycopsids) is characterized by trilete spores and only one species – *S. majusculus* – has monolete spores (Leisman and Stidd, 1967). The same holds true for the genus *Bowmanites*. Sometimes this difference is a significant generic character. For example, among Trigonocarpales a trilete prepollen is known only in the genus *Potoniea*. The variability in the proximal slit is generally typical of gymnosperms. In the operculate pollen (*Jugasporites*) from the sporangium of *Ullmannia frumentaria* the transition was observed from the monolete slit via the dilete to the trilete (R. Potonié and Schweitzer, 1960). Saccate (and quasisaccate) miospores exhibit a great variability in the sizes, configurations, number and angle of inclination of the sacci (quasisacci), which is clearly apparent in the pollen from sporangia of extant conifers and fossil gymnosperms (Fig. 78, *d*, *e*). Grains encountered in one sporangium may fit different genera of dispersed miospores. Mention has already been made above of the variability of the pollen *Cladaitina*, found in the sporangia of *Cladostrobus* (Gomankov and Meyen, 1980). All stages of the exfoliation of the saccus from the body, beginning from the grains that in general seem asaccate, have been preserved here. All these stages, as well as the detached sacci and bodies are often folded and boat-shaped, simulating a monocolpate pollen. This variation corresponds to several genera established by palynologists for dispersed miospores. Probably to this pollen belongs the type-material of the genus *Ginkgocycadophytus*, which was mistaken for pollen of ginkgos and cycadophytes.

Many similar examples may be offered. The range of intraspecific variability in many characters is at times comparable to the diversity observed in several genera. As a whole the range of variability increases when we take the eutaxa of ever higher rank. Frequently certain species of the genus, or singular genera of the family exhibit deviations according to a definite character of the spores. For example, most of the Trigonocarpales have pollen of the *Monoletes* type (Fig. 58, *k*, *l*), but the pollen in *Potoniea* is trilete (of the *Punctatisporites* type; Fig. 58, *p*), whereas in *Parasporotheca* it is saccate (Fig. 42 (20)). The family Pinaceae is characterized by a saccate pollen, predominantly disaccate, but in the genera *Larix* and *Pseudotsuga* it is asaccate, and in one *Tsuga* section it is quasisaccate. The extant genus *Dacrydium* yields a quasisaccate pollen, *Saxegothaea*, which is asaccate, whereas in all other Podocarpaceae, two to three, seldom more, true sacci occur. All this causes difficulties for the palaeobotanist. For instance, the Mesozoic genera *Dictyophyllum*, *Clathropteris*, *Thaumatopteris* and *Hausmannia*, included in Dipteridaceae, have trilete spores, whereas in the only suriving genus of the family – *Dipteris* – the spores are monolete. These Mesozoic genera have been affiliated with the extant genus *Cheiropleuria* (family Polypodiaceae) that has trilete spores.

Sometimes it happens that the range in the known diversity of the spores in the taxon increases in the course of investigations. At first it was considered that the Peltaspermaceae are characterized by an asaccate boat-shaped pollen. However, among them have been identified striated and quasisaccate pollen types (Meyen, 1984a), which the palynologists affiliated with Pinales, Welwitschiales and Arberiales. The Carboniferous ferns *Botryopteris* exhibit a high diversity in their exine ultrastructural characters (Millay and Taylor, 1982).

Not only miospores similar in type, but also similar transformations linking various forms can be observed in different eutaxa. This is evident from the examples cited above. The transition from the trilete spores to monolete develops in the same manner in different eutaxa, namely through the reduction of one of the rays and straightening of the remaining into one line. The transitions from the monosaccate (quasimonosaccate) pollen to the disaccate (quasidisaccate) in different gymnosperms develops through the appearance of bilaterality, narrowing of the saccus (quasisaccus) on opposite sides and the disappearance of interconnections (Lele, 1974, p. 245). The parallelism in the transformations of the saccate and quasisaccate grains is remarkable. The very formation of the quasisaccate forms accompanied by similar variability proceeded independently in the Arberiales, Peltaspermales and Pinales. The transition from the quasisaccate to the saccate forms was independent through similar transformations in the Cordaitanthales and Pinales. A striking parallelism is observed in the change in the polarity of germination in different gymnosperms. The transition from proximal to distal germination, from the prepollen to the pollen occurred independently in all three classes of gymnosperms (Chaloner, 1970a; Meyen, 1984a; Millay and Taylor, 1976; Taylor, 1981). The parallelism of many trends can be observed in the pollen structure of monocots and dicots (Fig. 78, *m*; Zavada, 1983).

Some of these parallelisms can be treated as evolutionary transformations along similar lines, beginning from a certain initial point. This interpretation is pertinent with respect to the change in the germination polarity, the transition from prepollen to pollen, from a monocolpate pollen to a porate and tricolpate. In other cases the polymorphism in a definite character persists in different groups, including those of a single phylogenetic lineage. There is no justification with this for the belief that within each group the evolution proceeded along parallel lines of forms in a definite direction. This phenomenon is termed transitive polymorphism (see Chapter 8; Meyen, 1978b; Meyen, 1984a). It refers, for instance, to the sexual dimorphism manifested in the distinctions between microspores and megaspores. In angiosperms, particularly in the case of heterostyly, the intraspecific differentiation of the pollen into two and even three distinct types was noted. Dimorphism is known in megaspores (e.g. *Isoetes dixitei*). Of more wide occurrence is persistent orderly polymorphism with gradual transition between various modifications. This can be best illustrated in the repeatedly mentioned transition, linking the mono- and trilete proximal slits. Although the monolete slit evolved originally from the trilete, the belief that the evolutionary changes developed in the same direction in each group having both slit types is not warranted. The probability is strong that in each taxon the variability in this character persists, but one or another type may be of predominant occurrence.

Both the evolutionary parallelism and transitive polymorphism are reflected in all characters of the miospore structure, from its overall organization to the fine details of the exine. Certain similarities cannot as yet be cogently interpreted. For example, in certain ferns and angiosperms the exine is uniform, not divided into layers, and is overlain by an interrupted layer of orbicules of tapetal origin. In the Upper Carboniferous microspores *Densosporites* (lycopsids) the exine is two-layered. The inner layer is thin, the outer layer spongy with irregular lumina, resembling the alveolar layer in certain gymnosperms. For the interpretation of such similarities detailed systematic TEM studies should be made of spores of the propteridophytes and progymnosperms. In propteridophytes the sporoderm subdivision into a smooth body and sculptured outer layer is known, and in progymnosperms into a cavum, but the fine structure of these spores has not been studied well enough.

The classification of miospores of simple organization entails particularly serious difficulties. This concerns the round miospores with a trilete or monolete slit and smooth or thin-sculptured exine (form-genera *Calamospora*, *Leiotriletes*, *Laevigatosporites*, *Punctatisporites*, etc.); their classification relies on the sizes of the spores, thickness of the exine, length of the rays, slight variations in the sculpture. The value of such shifts is not high. This is exemplified in the genus *Calamospora* (Good, 1977). Such spores have been encountered in the strobili of *Calamostachys*, *Palaeostachya* and *Pendulostachys* (also in Noeggerathiales). Characters judged to be of species significance were found to be associated with intraspecific variations or with the state of preservation of miospores. Spores, extracted from sporangia, sometimes differ in their maturity. For example, in the sporangia *Senftenbergia plumosa* (Botryopteridales) immature spores are smooth, while mature ones are spinose of the *Raistrickia* type (Grauvogel-Stamm and Doubinger, 1975). Changes in the sculpture in the course of maturity have been noted in spores of extant *Anemia* (Markova, 1964) and many other plants.

Thus, no distinct correspondence is apparent between the miospore parataxa and eutaxa. Each large eutaxon yields an orderly diversity of spores with the dominance of one or a few types. A similar diversity can commonly be perceived in other taxa, although other types may predominate. This is not a fortuitous phenomenon. It is due in part to the fact that the spore organization is only of a limited number of types, differing in the arrangement of the apertures, exine exfoliation, perisporium structure, grain configuration, as well as in finer characters (ornamentation, ultrastructure of the walls). Although these characters occur in multiple combinations, a definite order can be observed in the modifications of the characters, as well as certain prohibitions (e.g. the sacci never exhibit any prominent sculpturing). Because of this the repetition of certain similar types in different eutaxa is inevitable. It can be stated that between the miospore system and eutaxa system a mutual multiple (polymorphic) correspondence exists. An important implication from this is the necessity of constructing an independent taxonomy and nomenclature for miospores. Attempts to link the miospore taxa strictly with eutaxa leads to confusion.

This holds true particularly for miospores of insufficiently good preservation. We have already mentioned above the genus *Cladaitina*, where detached sacci and bodies simulate a monocolpate asaccate pollen. In spores of the Calamostachyaceae and Equisetaceae elaters are commonly detached. The loss or poor preservation of single elements is conducive to errors in identification of the specimens. It has become customary to introduce different generic names for plant megafossils that differ essentially in their state and kind of preservation. A similar approach is necessary in the systematics of miospores.

Considering the foregoing, there seems no sense in ignoring the possibility of more natural grouping of miospores into parataxa. It is possible to take account of the variability of spores in eutaxa. This can be accomplished on the basis of repeated associations linked by transitions in the samples. Much can be gained from data on spores in sporangia of plants, known from the same deposits. Examples showing the variability of spores in sporangia were cited above. It is frequently possible to recognize similar variation series among dispersed miospores recovered from the same localities. The variation series can be constructed irrespective of the data on the polymorphism of spores in sporangia. Some such series when they are of multiple recurrence in different samples, may be interpreted as representing intraspecific variability. It has been suggested that the term 'morphon' should be introduced for the designation of groups of palynological taxa, united by continuous changes in the morphological characters (Van der Zwan, 1979). Suggestions have been advanced that in studies of the variability of fossil miospores the binomial nomenclature should be rejected, and that instead 'biorecords' should be used, consisting of a formalized system of symbols (Hughes, 1970). The concept 'palynodemes' was suggested with reference to miospore groups of one layer that in detailed studies differ somewhat but are linked by transitions (Visscher, 1971). These concepts have not as yet received due recognition.

In establishing the rank of variability in miospores of serious help can be studies of miospores in the micropyle of ovules, or of those that are adhered to the cuticle of leaves of a definite type. Little attention is still given to coprolites of herbivorous animals. In them mass accumulations of monotypic miospores occur (Gomankov and Meyen, 1979; Meyen, 1981). Often in rock deposits with a profusion of macroremains of one species, one type of miospores is dominant, its variability being interpreted as intraspecific. For example, in palynological assemblages of paper coals, composed of compressions of *Phylladoderma* leaves (family Cardiolepidaceae), the miospores are dominated by the *Vesicaspora* type. The latter, dependent upon the direction of flattening in the process of burial, may resemble either a monosaccate or disaccate one, so that the palynologists identified among them different genera. Taking into consideration the fact that they underwent burial together, also the diversity of forms in the associated microsporangia *Permotheca* (Fig. 78, *d*) and micropyle of the seeds, these forms may be treated in terms of intraspecies variability (Gomankov and Meyen, 1980).

MORPHOLOGICAL EVOLUTION OF MIOSPORES

The diversity of miospores is of orderly increase upwards in the geochronological scale, reaching its maximum in recent deposits. Many miospores are comparable to definite eutaxa. Sometimes the morphological types of miospores exhibit a distinct geographical restriction. Amazing parallelisms are evident in the evolution of the miospores, not only in the similarity in the trends of morphological transformations, but also in the high synchronism in their development in various taxa and biotas. Some examples of this are cited below.

In this aspect the most ancient palynomorphs of the Pre-Cambrian, Cambrian and Ordovician have not as yet been analysed. Because of this we will commence directly with the miospores of higher plants and closely similar to them, as for instance, spongiophytaceous algae, that display affinities to higher plants in the major features of miospores. Miospores with a wall resistant to maceration and an indisputable trilete slit appear in the uppermost Ordovician, and are of gradual increase in number and diversity to the end of the Silurian. The oldest spores occur mostly in tetrads, with isolated spores dominating from the Llandoverian upwards. If the genera are treated in a narrow sense, then to the end of the Silurian their number attained about 20. These genera have been classified on the basis of rather minor features. Silurian miospores are round or roundish-triangular, of simple organization. They are smooth or with a low relief in the form of small tubercles, pits, spines or radial folds. In the Silurian, curvatures appear – flexed ribs connecting the ends of the rays. At the same time miospores appear with an equatorial thickening, somewhat later they appear with a thin flange (perisporium), and still later with a proximal thickening of the exine at the place of junction of the rays. Miospores with such thickenings and curvatures are termed retusoid (after the genus *Retusotriletes*). They are typical of many propteridophytes (later they also associated with algae of the *Protosalvinia* type). The above-cited characters occur in various combinations, which furnishes the basis for distinctions of the genera.

In the Lower Devonian likewise of predominant occurrence are round or roundish-triangular miospores. They exhibit a more contrasting relief, and a more intricate structure of their outgrowths (for details see Alpern and Streel, 1972). The end of the Lower Devonian is marked by the incoming of miospores with forked and anchor-shaped exine outgrowths. In the lowermost Devonian miospores appear with small papillae at the place of convergence of the rays, or with a distinct triangular contour of the proximal thickening. Of practically simultaneous appearance is the reticulate relief of the exine (*Dictyotriletes*) and cavate organization. From the end of the Lower Devonian many spores exhibit differences in their relief on the proximal and distal sides, so that smooth contact areas are formed along which the miospores came into contact in the tetrads. Miospores with a cavate organization and cingulum appear, as well as rare monolete spores.

During the Lower Devonian the miospores gradually increase in size. The average size of the miospores is about 20 μm at the base of the Siegenian, and about 40 μm at its top (Alpern and Streel, 1972). In the middle of the Silurian miospores appear that are already over 50 μm in size, whereas from the middle of the Lower Devonian they are over 100 μm, and by the end of the Lower Devonian over 200 μm. The rise of the heterospory can be assigned to the end of the Early–beginning of the Middle Devonian. In the uppermost Middle Devonian megaspores have been encountered that reach 1600 μm in size (Stockmans and Streel, 1969). In the Middle Devonian an increase in the number and diversity of the cavate (pseudocavate) miospores can be observed, some of which (*Rhabdosporites*) belong to progymnosperms. In the Upper Devonian, in addition to a rather abundant amount of megaspores, miospores *Archaeoperisaccus* appear that have been treated as the saccate pollen with a distal colpus. Later they were judged to be monolete spores, belonging to lycopsids. Meyer-Melikyan and Raskatova (1984) studied the exine structure in such spores and assigned them to conifers, notwithstanding the fact that according to their lamellar nexine and alveolar-granular sexine they can more readily be paralleled with microspores of *Archaeopteris*. Inasmuch as the Upper Devonian marks the incoming of the first seeds, it is conceivable that the Upper Devonian miospores belong in part to prepollen.

In the Lower Carboniferous the amount and diversity of the cavate prepollen increase conspicuously. Other prepollen, morphologically similar to pteridophyte spores, cannot be identified among the miospores. The diversity of other miospores also increases. Sometimes the successive increase in the complexity of a particular miospore type can be traced. For example, in the Devonian triangular spores with rounded angles already appear. In the lowermost Carboniferous such spores exhibit concave interradial portions of the equatorial contour. Higher in the sequence the concave sectors become deeper (*Waltzispora*), and then on the equatorial elevations located along the extension of the rays, additional sculptural ornamentation appears (*Mooreisporites*, *Murospora*, *Ahrensisporites*, *Tripartites*; Fig. 78, *j*). Different modifications occur in other triangular miospores. Spines of various sizes appear, so that the spores become of round external contour (*Diatomozonotriletes*). Some specific miospore types have a narrow stratigraphic range. This applies to trilete miospores with a membranous perforated flange (*Retispora*; Fig. 20, *z*), that occur characteristically in the Devonian–Carboniferous boundary beds.

Monolete spores (mainly marattialean) increase in number in the Middle Carboniferous. This is concomitant with a change in the polarity of germination in certain gymnosperms. The number of quasisaccate and saccate forms increases. Quasimonosaccate forms occur prior to the quasidisaccate, but with a small gap in time. At the end of the Lower Carboniferous of first appearance are quasimonosaccate forms, and soon after – saccate forms. Practically of concurrent incoming among them is the bilateral quasisaccus (saccus) that evolved through its narrowing on opposite sides of the equator. During the first half of the Middle Carboniferous the narrowing of the bridges leads to the division of the quasisaccus into two. During the second half of the Middle Carboniferous a proximal striation appears in the quasisaccate miospores. At the end of the Middle Carboniferous of first appearance are asaccate miospores of the *Vittatina* type with a striated body. In most of the miospores, considered above, including those with a striated body, the monolete, dilete and trilete slit are retained. It can often be observed likewise in *Vittatina*.

This succession of the miospores has been traced in the Carboniferous Equatorial Belt. Recently new data have been obtained (Peterson, 1983), indicating that the mono- and disaccate (quasisaccate), including the striated miospores, as well as miospores of the *Vittatina* type, are of first appearance in the lower Middle Carboniferous of Middle Siberia. The same miospore types are known in the Upper Palaeozoic of Gondwana, but owing to the uncertainty in dating the deposits from which they come, inferences concerning their synchronous or diachronous appearance, relative to those of the Equatorial Belt, are not warranted. However, a remarkable fact is that the general successive order in the miospores of Gondwana is the same; at first quasimonosaccate miospores appear, then quasidisaccate with a smooth body, and soon after with a striate body, and eventually, forms of the *Vittatina* type. Inferences concerning the systematic affinities of these miospores of different phytochorias can be derived only on the basis of indirect evidence. The quasimonosaccate and monosaccate pollen in the Equatorial Belt and Siberia belong to the two orders Cordaitanthales and Pinales, whereas in the Gondwana localities they belong to Arberiales, to which most of the Gondwana quasidisaccate, striate or non-striate miospores belong, as well as some of the miospores of the *Vittatina* type. The complete absence of arberians in the Equatorial floras and Carboniferous–Early Permian assemblages of Siberia suggests that similar changes in the miospore types were of parallel development in different taxonomic groups. Here we are again faced with evolutionary parallelism, other examples of which (the appearance of cavate spores and heterospory in various groups of Devonian and Lower Carboniferous plants) have been previously illustrated (p. 256).

As a whole, the geological sequence reveals in ascending order an increase in the diversity of characters in the miospores, although certain specific characters disappear (e.g. the forked and anchor-shaped outgrowths, as in *Hystricosporites* and *Ancyrospora*, disappear near the Devonian–Carboniferous boundary). Evidently, a prominent change in the morphological types of gymnosperm miospores occurred in the Permian. At this time miospores with a proximal slit were replaced in several gymnosperm lineages by alete miospores. There is reason to believe that gymnosperms in the main had already acquired during the Permian a pollen instead of a prepollen. Lists of Permian miospore assemblages contain a large percentage of asaccate monocolpate miospores of the *Cycadopites* type. In most cases the 'boat-shaped' monosaccate and protomonosaccate miospores, and their detached sacci and bodies, were mistaken by the palynologists for monocolpates. The true proportion of asaccate monocolpate miospores in the Permian assemblages remains unclear. In general the Permian miospores are represented in equal measure by (1) pteridophyte spores, and (2) quasisaccate and saccate pollen (and prepollen?). These types vary in proportions from one phytochorion to another, and within a definite phytochorion, depending upon the facies types of the enclosing rocks.

Triassic miospores maintain to a large extent the Permian habit. At the same time new miospores appear, particularly in the Upper Triassic, that become dominant in the Jurassic and Lower Cretaceous. The quasimonosaccate forms decrease in numbers and diversity, the striated quasisaccates become gradually of less diversity (they disappear in the Upper Triassic), instead of *Vittatina* other asaccate striated miospores appear (*Gnetaceaepollenites*, *Ephedripites*, *Weylandites*). Among the quasisaccate miospores those forms persist that have a proximal slit (e.g. *Triadispora*), but alete forms play a major role. Of considerable increase are asaccate monocolpate miospores belonging to ginkgos, cycads, bennettites, peltasperms and leptostrobans. In the Upper Triassic the composition of the miospores becomes similar to the Jurassic. Miospores of the genus *Classopollis* appear (Fig. 72, *h*, *i*), that determine the aspect of Jurassic and Cretaceous miospore assemblages in many regions. Triassic spores of pteridophytes are of transitional aspect and represented in part by morphological types, transitional from the Permian, and in part by forms typical of the Jurassic and Cretaceous. The Triassic is characterized by cavate miospores of *Densoisporites* (with a trilete slit) and *Aratrisporites* (with a monolete slit), that are allied with the Pleuromeiaceae and are of nearly global dispersal.

Jurassic and Cretaceous pteridophyte spores correspond, as a whole, to the extant forms according to the general diversity in the type of their organization. It is no wonder that many palynologists affiliated spores of this age with families and genera of extant plants. Studies of spores from sporangia and comparison of the miospore assemblages with the megafossils revealed many errors in such identifications. The same holds true for the gymnosperm pollen, not taking into account the aforementioned *Classopollis* pollen. Besides this pollen, the Jurassic and Cretaceous yield the saccate (chiefly disaccate) and quasisaccate forms that belong to Pinales, Caytoniales, Peltaspermales, and asaccate striate forms of uncertain affinities, also monocolpate asaccate forms produced by Ginkgoales, Leptostrobales, Bennettitales and Cycadales, tricolpate forms of the *Eucommiidites* type (see p. 236), and asaccate pollen, comparable to the pollen of Araucariaceae, Taxaceae and other conifer families (Bharadwaj, 1974). Of these types only *Classopollis* and *Eucommiidites* are unknown among extant gymnosperms.

During the Cretaceous pollen of *Classopollis* and monocolpate pollen of the *Cycadopites* decrease gradually in numbers. This decrease in the role of gymnosperm pollen is concomitant with the increase in the number and diversity of angiosperm pollen beginning from the Middle Cretaceous (see above, p. 236). Cainozoic miospores are represented chiefly by extant morphological types, although the systematic affinities of many genera are still uncertain.

The data presented above show the general

succession of miospores in the geological time scale. It is more fully expressed in the Equatorial assemblages. In the extraequatorial phytochoria the change from one to another miospore type may be less distinct, occasionally lagging behind somewhat, the miospore types are of less diversity (e.g. monolete pteridophyte spores are not characteristic of the Palaeozoic and Mesozoic extra-equatorial areas), at times other types, different from those in the Equatorial region, predominate, few endemic types occur. Nevertheless, the observed succession, if taken as the general tendency, is of global significance. Kuprianova (1969) suggested that three major types of miospores should be recognized. The first type has a proximal ray aperture, corresponds to pteridophytes, and prevails in the Palaeozoic. The second type has one distal aperture in the shape of a colpus; it corresponds to gymnosperms and early angiosperms, and was dominant in the Mesozoic. The third type has apertures equatorial or scattered over the entire grain (colpi, pores, complex apertures) and was most dominant in the Cainozoic. This scheme has been suggested as the general trend, revealing multiple deviations and complications. As an important addition to this scheme should be noted the earlier appearance and wider distribution of some miospore types. This refers particularly to the prepollen with proximal and bipolar apertures, which was probably more widespread, especially in the Palaeozoic. This prepollen prevailed in the Carboniferous and Permian gymnosperms and had not lost its significance during the Mesozoic. It occupies a transitional position between the first and second types.

Problems bearing on the evolutionary transformations in miospores have been discussed in detail by R. Potonié (1973), Bharadwaj (1974), Muller (1970, 1981) and Meyen (1984a).

DISPERSED MEGASPORES

Heterospory is a characteristic feature of certain bryophytes and many pteridophytes – progymnosperms, Protopityales and Archaeopteridales, barinophytes, articulates (some Calamostachyales, Echinostachyaceae, Bowmanitales), lycopsids (Selaginellales, Isoetales), Noeggerathiales, zygopterids (Stauropteridaceae), leptosporangiate ferns (family Platyzomaceae) and 'aquatic ferns' (Marsiliales, Salviniales). All seed plants are heterosporous. No matter how reconstructions are made of the phylogenetic links between the taxa of higher plants, an inevitable conclusion derived from them is that the heterospory evolved independently in different groups. Its independent occurrence in the Middle–Upper Devonian has been established for Lycopodiopsida, Progymnospermopsida and Barinophytopsida. Microspores and megaspores are commonly found in different sporangia, but in *Barinophyton* they occur together in one sporangium. In *Chaleuria* (p. 124) spores of one sporangium belong to two size-classes, that differ also in morphological characters. The co-occurrence of micro- and megaspores in sporangia was observed in *Archaeopteris* (Medyanik, 1982).

Valid distinctions between microspores and megaspores in dispersed fossils are not always possible, inasmuch as small (less than 200 μm) megaspores and very large (over 600 μm) microspores are known. In most cases megaspores can be easily distinguished from microspores by their thicker wall and larger sizes (over 200 μm). The largest megaspores have been recorded from the Carboniferous, and reach 9 mm in diameter. The proximal slit in megaspores is trilete, even when the microspores are monolete. Occasionally micro- and megaspores can be distinguished only when the organic association in the strobili are known. Megaspores of *Calamostachys americana* are practically the same as microspores; the smallest megaspores (140 μm) are close in size to the biggest microspores (120 μm), although distinctions are clearly apparent in the average sizes of both (230 and 90 μm).

In lycopsids, articulates and zygopterids the megaspores in the sporangia are reduced to one tetrad and further (in *Lepidocarpon*, *Achlamydocarpon*, *Caudatocorpus*, *Calamocarpon*) to one functional megaspore, which reaches a large size and bears strongly underdeveloped (aborted) megaspores on the proximal side (Fig. 18, *u*; Fig. 78, *p*). The same process occurred in the formation of gymnosperm seeds. In the oldest seeds of *Archaeosperma* (order Lagenostomales), in addition to a functional megaspore three aborted ones are retained (Fig. 43, *f*).

Megaspores are frequently morphologically similar to microspores. In some *Archaeopteris*, certain Calamostachyaceae and other groups, the megaspores look like enlarged microspores. The exine is divided into the same layers, the exine ultrastructure is also identical (e.g. in *Archaeopteris*), but more often distinctions are obvious (*Selaginella*, *Marsilia*, etc.). The same classification scheme is applied to the structure of the layers. They may have a perine, tapetal membranes and orbicules. All these data are essential for the understanding of the evolution of heterospory, homologization of the seed parts, taxonomic relationships, etc. (Pettitt, 1966; Taylor, 1981; Taylor and Brack-Hanes, 1976; Zimmermann and Taylor, 1970–1971).

Dispersed megaspores are extracted from sediments by various palynological methods, by sifting through screens, or they can simply be liberated from the matrix by a needle. They are studied in dry conditions or macerated. The sizes and sometimes the relief of the megaspores may differ considerably, depending upon the techniques employed and the medium in which they are embedded. For example, macerated megaspores and megaspores embedded in glycerol-jelly are greater in size and show a more contrasting relief. Because of this megaspores should be described in dry conditions prior to chemical treatments. Sometimes we come across detached megaspore membranes of gymnosperm seeds. Megaspore sections are commonly encountered in thin sections of coals, particularly of the Carboniferous.

For stratigraphic purposes Carboniferous, Permian and Triassic megaspores of Isoetales and Selaginellales are more often employed, as well as those of the Upper Cretaceous and Cainozoic Salviniaceae. Less studied are megaspores of the Devonian, Permian and Jurassic, among which lycopsids predominate.

The classification of dispersed megaspores is based on the same principles as that of the miospores, but it has not been so amply elaborated. The Cretaceous and Cainozoic megaspores of Salviniaceae and sporocarps of Marsiliaceae have usually been described within the genera *Azolla*, *Salvinia*, *Marsilia*, or extinct genera. Older dispersed megaspores are sometimes also affiliated with genera of entire plants (eutaxa), e.g. with the Triassic genus *Pleuromeia* (the same megaspores have been described as *Talchirella daciae* or *Triletes polonica*). Dispersed megaspores should preferably be treated as separate form-genera independent of eutaxa, barring those cases when a mass of similar megaspores are accompanied in monodominant burials by a specific heterosporous plant (e.g. masses of megaspores occur together with lycopsids of *Viatcheslavia* and Pleuromeiaceae).

The taxonomic categories and nomenclature used for the suprageneric taxa are the same for megaspores and miospores. These taxa are distinguished according to the presence of flanges (Fig. 78, *r*), the relationship between the sculpture on the proximal and distal sides, number of envelopes (in macerated remains), etc. The genera are based on the combination of a larger number of characters, whereas the species are based on the degree of development of one or another character, the size of the megaspores and their elements, e.g. the sculpture (Fig. 78, *q*). In general, of most importance in the systematics of megaspores are the following characters: (1) Overall outline (round, subtriangular, etc.). (2) Structure of the contact area, i.e. of the proximal area which was in contact with other megaspores of the tetrad. Inside this area a trilete slit occurs (monolete megaspores do not exist), which is elevated at times. The ends of the slit are sometimes united by arcs (curvatures). The relief of the contact area may differ from the rest of the surface relief. (3) The sculpture and its arrangement on the surface, which may be smooth, covered by spines, swellings, simple or branching processes, of reticulate pattern, pitted, etc. Frequently the sculpture differs essentially on the distal and proximal sides. Sometimes the distal side is smooth, whereas the proximal is sculptured, but usually the reverse situation is observed. Even more frequently the sculpture on the distal side is more contrasting. (4) The proximal side may bear a specific outgrowth, either entire or divided into lobes (Fig. 78, *s*). This outgrowth ('gula') is typical of the Carboniferous lepidophytes. The proximal side may be covered by an additional sporopollenin layer. (5) In macerated megaspores various numbers of envelopes can be recognized. The inner envelopes bear at times specific round thickenings near the ray junction or along the rays. (6) The inner structure of the envelopes (entire, without visible structure, lamellar, granular, alveolar, filamentous, consisting of intertwined filaments of various thickness).

Certain types of dispersed megaspores are depicted in Fig. 78, *p–s*.

Chapter 5

Epidermal–cuticular studies

Without epidermal–cuticular studies (ECS) distinctions between species, genera and suprageneric taxa (up to and including the rank of class) of fossil plants would in many instances be impossible. This refers especially to Peltaspermales, Ginkgoales, Leptostrobales, Cordaitanthales, Pinales, Cycadales, Bennettitales and many angiosperms. Because of ECS the systematics of many groups of plants as well as current notions on their stratigraphic and geographic distribution underwent a radical change. ECS have a paramount role in deciphering plant morphology, particularly in the reconstruction of the original, life-time connections between various parts according to their epidermal structure. For instance, by means of ECS organic connections have been established between the foliage of *Czekanowskia* and female capsules of *Leptostrobus*, also between leaves of *Phylladoderma* and fructifications of *Cardiolepsis*. Dispersed cuticles can also be used to characterize rock deposits. Their classification is discussed below. Detailed information on ECS can be gleaned from reviews: Dilcher, 1974; Edwards *et al.*, 1982; Kidwai, 1981; Krassilov, 1968, 1978a; Mehra and Soni, 1983; Meyen, 1965; Pant, 1965; Rasmussen, 1981; Roselt and Schneider, 1969; Samylina, 1980; Sinclair and Sharma, 1971; Stace, 1965; Tomlinson, 1974; Van Cotthem, 1970.

PERTINENT CHARACTERS USED IN ECS

The characters used in ECS include those related both to the epidermis proper and the cuticle covering the epidermis. The cuticle bears imprints of the epidermal cells and reflects their shape and arrangement. The characters pertinent to the cuticle are various projections, inclusions, the thickness, colour, etc. Distinctions are usually drawn between the walls of epidermal cells impregnated by cutin (cutinized), and the outer cell cover consisting of the cuticle, i.e. that underwent cuticularization but not cutinization. However, in palaeobotany the term cuticle applies to that membrane which consists of cutin and is left over after maceration of the compressions. In recent years studies of the cuticle relief are conducted with the aid of scanning electron microscopy (SEM). Although the observations and descriptions of some features became of higher precision

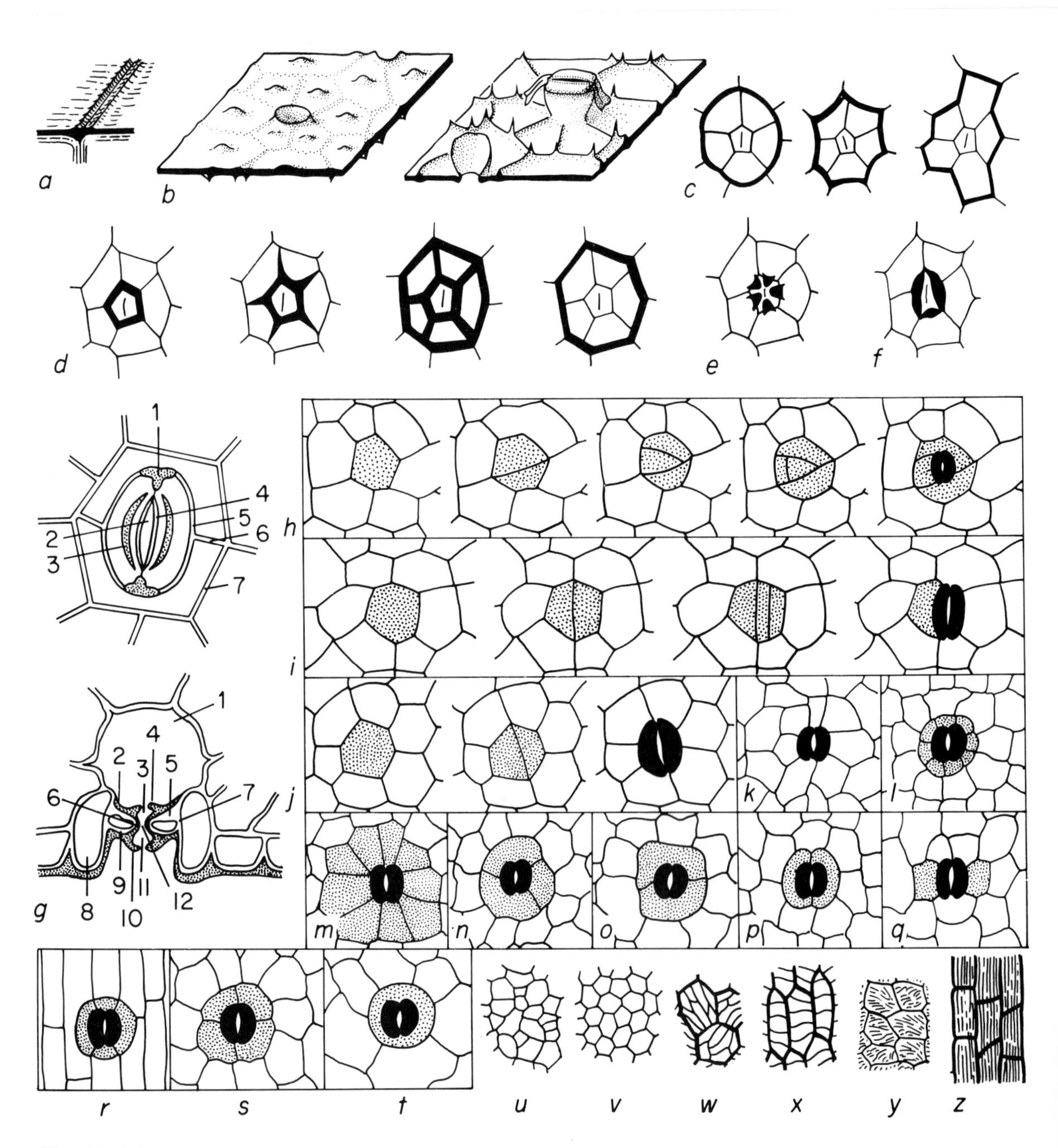

Fig. 79. Epidermal–cuticular structures of higher plants: *a*, cutin suture above radial wall; *b*, reflection of cellular structure on cuticle, outer view (to left) and view from inside of leaf (to right), two stomata are shown in one of which guard cells are abolished; *c*, various outlines of stomata; *d*, different types of cutinization of subsidiary cells (proximal, star-shaped, wheel-shaped, distal); *e*, *f*, proximal papillae (*e*) and lips (*f*) on subsidiary cells; *g*, stoma in surface and section view; numerals in upper figure show T-shaped thickening of poles of guard cells (1), aperture (2), outer ledge (3), poral (4) and epidermal (5) walls of guard cells, radial (6) and tangential (7) walls of subsidiary cells; numerals in lower figure show substomatal cavity (1), stomatal flap (2), back cavity (3), inner ledge (4), dorsal (5), poral (6) and epidermal (7) walls of guard cells, subsidiary cell (8), outer wall of guard cell with outer ledge (9, 10), front cavity (11), ▶

the systematic characters have remained the same in scope as in studies with the light microscope. The microstructure of the compression or impression surface may also be studied without having to isolate the cuticle. For this purpose reflected light is employed, also replicas, latex casts and SEM. Natural maceration renders possible film pulls and transfer preparations and ECS can then be accomplished without further chemical treatments.

The observed structural elements may differ, depending on the state of preservation of the material studied and on the techniques employed, but the terms used for their designation may be the same. For instance, with reference to the radial walls of the cells one should bear in mind the preserved wall itself, as is frequently the case in non-macerated compressions of fern leaves. In cuticle preparations anticlinal flanges of cuticle can be observed but not the walls themselves. Such a flange (Naht, in the German literature) is a cutin rib that penetrated between the cell luminae (Fig. 79, *a–c*). Its thickness and structure are linked with the wall structure, but reflect only certain of its characters. Because of this the usage of the term 'wall' in descriptions of the cuticle is a simplification. The same applies to the usage of the term 'guard cells' with respect to the cuticular cover, corresponding to only part of the walls of the guard cells. The same can be said of the expression 'epidermal characters', which at times designate only cuticular features.

Epidermal characters can be conditionally divided into topographical and structural. The first refers to the arrangement and orientation of the non-stomatal cells, stomata and trichomes. The cells are arrayed irregularly, in files, as alternating bands of cells varying in shape and orientation, or otherwise. Cell files have been frequently observed only along the veins or on one side of the leaf. According to the arrangement of the stomata, hypostomatic, epistomatic and amphistomatic leaves can be distinguished. In the first instance the stomata are gathered on the lower side, in the second on the upper side, and in the third on both sides. The orientation of the stomata (determined by the orientation of the guard cells) relative to the leaf axis, may be longitudinal, oblique or transverse. The stomata may be arranged in files, bands, or groups. Bands may consist of files or of irregularly arranged stomata. Adjacent stomata often have common subsidiary cells.

Variation in the topographical characters (their total syndrome may be termed the architectonics of the epidermis) are manifest principally in the change in number and mutual arrangement of the elements, but not in the characters of individual elements. Because of this the relevant transformations may be described in terms of oligomerization, polymerization and heterotopy. ECS provide good examples of these modes of evolutionary morphological transformations, and may serve as models in their studies, which will be discussed in more detail in Chapter 8.

The term structural characters is used to designate the characters of the non-stomatal cells, stomata and trichomes (i.e. the tectology of the epidermis). The characteristics of the cells include the following distinctions: thickness and configuration of the radial walls (i.e. the walls between the epidermal cells), relief of the periclinal walls (that compose the leaf surface). Radial walls may be straight, flexed, minutely- or largely sinuous, etc. The periclinal walls may be smooth, striate (Fig. 79, *y*), ribbed (Fig. 79, *z*), with cuticular papillae (ocells; Fig. 79, *b*), or epidermal papillae (when the cell cavity enters the papillae).

The term stoma is frequently used to denote only the guard cells with an aperture between them, whereas the entire structure which includes

---

aperture (12); *h–j*, mesogenous (*h*), mesoperigenous (*i*) and perigenous (*j*) ontogenetic types of stomata; *k–t*, some stomatal types established irrespective developmental patterns, guard cells in black, subsidiaries stippled (*k*, anomocytic; *l* cyclocytic; *m*, actinocytic; *n*, anisocytic; *o*, diacytic; *p*, paracytic; *q*, brachyparacytic; *r*, tetracytic; *s*, staurocytic; *t*, pericytic); *u*, *v*, anisogonal (*u*) and isogonal (*v*) patterns of non-stomatal epidermal cells; *w*, cellular packets; *x*, septate cells; *y*, epicuticular striation; *z*, epicuticular ribbing. Modified from: Doubinger *et al.*, 1964 (*a*); Stace, 1965 (*g*); Dilcher, 1974 (*h–t*).

also the neighbouring specialized (subsidiary and encircling) cells is termed the stomatal apparatus. However, more often the term stoma refers to the entire structure. If the guard cells come into contact with ordinary non-stomatal cells, then such stomata are called non-specialized, lacking subsidiary cells. The term subsidiary cells refers to epidermal cells that are contiguous to the guard cells and can be distinguished from the surrounding epidermal cells by the degree of cutinization or other characters. At times such modifications are evident in cells further distant from the guard cells; these cells are termed encircling cells. Originally the term encircling cells was used only in the ontogenetic sense. Those cells that were formed during the division of the parent subsidiary cells by walls not contiguous to the guard cells were defined as encircling cells. The subsidiary and encircling cells taken together are termed neighbouring.

Various detailed classifications for stomata have been proposed (Dilcher, 1974; Kidwai, 1981; Rasmussen, 1981). They are based not only on the structure of mature stomata, but also on their ontogenesis. It is well known that according to the mature stomata it is impossible to judge the type of their ontogenesis. Nevertheless, of wide current usage in palaeobotany was the classification of stomata into haplocheilic and syndetocheilic types (Florin, 1931; Osnovy paleontologii, 1963b). The former type involves stomata in which only the guard cells are formed from the mother cell, whereas those that occur on the sides of the mother cell serve as the neighbouring cells. In syndetocheilic stomata the mother cell gives rise not only to the guard cells, but also to the lateral subsidiary cells (occurring laterally on either side of the guard cells). In this case the ordinary non-stomatal cells, frequently non-specialized, serve as the polar cells (situated on the extension of the stomatal aperture). The polar subsidiary cells are then said to be absent.

Later another ontogenetic classification suggested by Pant (Fig. 79, *h–j*; Pant, 1965) became widely recognized, in which the stomata were divided into perigenous, mesogenous, mesoperigenous. The terms perigenous (Fig. 79, *h*) and haplocheilic stomata coincide in meaning. The term mesogenous stoma (Fig. 79, *j*) applies to those stomata where all the subsidiary cells (both the polar and lateral) are derived from the same mother cell as the guard cells. In mesoperigenous stomata (Fig. 79, *i*) some of the subsidiary cells are developed according to the perigenous types, and others are developed according to the mesogenous type. Syndetocheilic stomata may be mesogenous and mesoperigenous.

In some cases the arrangement of the subsidiary cells allows judgement of the ontogenesis of the stomata, but the inferences cannot be verified in fossil forms unless underdeveloped or teratological stomata are found (Krassilov, 1976–1978). Because of this of more wide application is the classification based on the mature stomata, without consideration of their ontogenesis. The distinctions of the types rely on the number, shape and arrangement of the subsidiary cells. Most frequently the following types are recognized: (1) anomocytic – the guard cells are surrounded by non-specialized cells (Fig. 79, *k*); (2) anisocytic – the guard cells are surrounded by three cells, one of which is much smaller than the others (Fig. 79, *n*); (3) paracytic – on either side of the guard cells are situated one or more parallel subsidiary (and encircling) cells (Fig. 79, *p*); (4) diacytic – the guard cells are surrounded by a pair of subsidiary cells, where the common wall between them is perpendicular to the stomatal aperture (Fig. 75, *o*); (5) actinocytic – the subsidiary cells spread out radially from the guard cells (Fig. 79, *m*). Some other types are depicted in Fig. 75, *k–t*. The stomata are called papillocytic when papillae are overhanging the stomatal pit (Fig. 79, *e*), and jugatocytic when the pit bears a cutin rib along its edge (Fig. 79, *d*, *f*).

The guard cells may be superficial (lying on the leaf surface) or sunken. Together with the guard cells the subsidiary and encircling cells may also be sunken. In some Lycopodiopsida, Cordaitanthales, Dicranophyllales, Bennettitales, Pinales and other plants the entire stomatal band is sunken, so that stomatiferous furrows are formed (Figs. 18, *k*; Fig. 65, *b–e*; Fig. 66, *b*, *d*, *f–l*). The guard cells of the stomata in the furrows may also be sunken. Plants are known with stomata elevated above the leaf surface.

Trichomes are represented by uni- and multicellular hairs, at times by complex multicellular structures (stellate or peltate trichomes in ferns and angiosperms). Trichomes may become abscissed or mechanically separated in the process of fossilization or compression recovery, so that only the base of the trichome may be left on the leaf.

Of other characters taken into consideration in ECS we shall mention only those of the structure of the guard cells. These are currently being studied in detail with the aid of SEM. In lycopsids, ferns and many angiosperms the guard cells are devoid of any striking and persistent characters. In Bennettitales (Fig. 62, *c*, *d*, *g*, *l*, *p*) and various Ginkgoopsida (Fig. 52, *l*, *m*, *o*, *p*; Fig. 53, *b*, *f*) the guard cells bear peculiar cutin thickenings on the dorsal side, i.e. on the side facing the leaf surface (p. 179), resembling a butterfly with outspread wings. In most cycads and conifers similar in outline thickenings, when present, are composed of lignin, not of cutin. Because of this they are easily destroyed during maceration. Other structures may be borne on the guard cells (folds, striations, etc.).

SYSTEMATIC SIGNIFICANCE OF THE CHARACTERS

The first attempts to apply ECS to palaeobotanical systematics date back to the last century (J. Bornemann, R. Zeiller, etc.). Systematic investigations in this direction commenced at the beginning of this century (H. H. Thomas, N. Bancroft, A. G. Nathorst, etc.), from which it soon became apparent that the generic and species systematics of vegetative shoots of certain plant groups require treatment of ECS. This was demonstrated on conifers, ginkgos, leptostrobans, cycads, bennettites, pteridosperms and angiosperms. Nowadays ECS are used in studies of practically all groups of higher plants. The data available on extant plants were widely used by the palaeobotanists.

The time-tested experience of ECS convinces us that these investigations are very fruitful in certain cases, whereas in other cases they are practically useless. The same holds true for extant plants. However, the following is evident. If the systematics is based on the above-cited topographic characters, dominant type of stomata, structure of the guard cells and trichomes, and comparison is made between relevant organs (e.g. normal foliage leaves) and their portions (e.g. middle parts of the leaf), then the defined epidermal–cuticular types characterize one particular species or a group of species of one genus. More rarely all the species of a genus, or all genera of a family, reveal a uniform epidermal structure. No examples are known where the syndrome of epidermal characters is highly variable within the species or even within the genus (given that all the above-cited conditions are observed). Naturally, the epidermal characters may differ considerably in different organs. But such differences in the epidermal–cuticular characters in fossil remains of homologous parts are inconsistent with the affiliation of these remains to any particular species.

In ECS we are, of course, confronted with the variability and inconsistency of the weight of the characters. Those characters that are of significant taxonomic weight within one taxon may vary in another taxon. In this respect the epidermal–cuticular characters do not differ from any others. We can never know beforehand what character is persistent and which is variable or useless in systematics. For example, in Cordaitanthales the hypostomatic leaves persist within the entire vast genus *Rufloria* (probably corresponding to several natural genera), but genera may be found among the conifers where the arrangement of the stomata on the lower and upper sides of the leaf depend upon the degree of its appression to the stem. In Podocarpaceae and Pinaceae the guard cells are always of longitudinal orientation whereas in Taxodiaceae the stomatal orientation is variable.

In many cases ECS essentially improve the systematics of fossil plants, but, on the other hand, they have led to considerable and not always justifiable splitting of the taxa. Frequently the species and genera are keyed out on the basis of characters the taxonomic significance of which is easier to postulate than prove in each case. Ordinarily the palaeobotanists limited their sampling to singular preparations for each species, and the variability of the epidermal characters was not studied in large monotopic sampling. The application of ECS to monotopic sampling has revealed the great variability of characters formerly regarded as stable. This phenomenon was recorded by Harris and coauthors (1974) in their studies of the Middle Jurassic ginkgos and lepidostrobans, by Dilcher (1973) in relation to Eocene angiosperms, and by Gomankov and Meyen (1980 and in press) in studies of the Upper Permian genus *Tatarina*. The present author had to cope with the same phenomenon in studies of the Upper Palaeozoic cordaitanthaleans, conifers (*Quadrocladus*, *Kungurodendron*), *Phylladoderma* (Cardiolepidaceae) and other plants. Leaves coming from a single bed exhibit gradual transitions between somewhat different epidermal types. In this case the affiliation of the extreme forms of the monotopic sampling to one species becomes questionable. The extreme forms of resulting series prove to be similar to the typical forms of another series coming from another bed, and according to the adopted canons, seemingly belong to a different species. Eventually, most of the specimens coming from one locality have to be affiliated with one quite polymorphous species, and the number of specimens belonging to the same species, but recovered from other localities, becomes very small. Even with regard to the latter doubt arises concerning their specific affinities.

Of importance is the fact that the variation concerns those characters that are customarily used for the distinction of species or genera (distinctness of stomatal files, degree and type of cutinization of the subsidiary cells, the presence and size of the papillae on non-stomatal and stomatal walls, etc.). At the same time certain characters, which were earlier regarded as of minor significance (e.g. the thin epicuticular striation) may exhibit high stability.

The foregoing does not discredit the significance of ECS for systematics, but merely serves to indicate that in the application of epidermal–cuticular characters, as of any others, studies should not be limited to single specimens, and the attribution of undue taxonomic weight to any one of the characters is not warranted.

EVOLUTION OF EPIDERMAL–CUTICULAR CHARACTERS

The epidermal characters of many taxa are outlined in Chapter 3. A summary is presented here of the data on the trends in the epidermal–cuticular characters in the geochronological scale. Regrettably, neither the origin of the cuticle, nor that of the stomata is known. A cuticle-like covering is known in *Protosalvinia*, spongiophytaceous, and closely related to them algae, but only in those Devonian representatives that co-existed with higher plants. On the basis of ontogenetic observations Pant (1965) set forth a hypothesis, postulating that the stomata evolved from conceptacles of brown algae. However, many phylogenists derive the higher plants from green algae. Pant stressed the similarity in the ontogenesis of stomata and sporangia. It is possible that the ancestral structures of the stomata were of a reproductive function.

The stomata in propteridophytes are already well developed (Edwards *et al.*, 1982). In *Rhynia* they are composed of two superficial bean-shaped guard cells with a distinct wall between them and an aperture (Fig. 13, *i*). The walls facing the aperture are slightly thickened. In *Sawdonia* (Chaloner *et al.*, 1978; Edwards *et al.*, 1982; Rayner, 1983) the guard cells are sunken and the cuticle covers that part that is exposed in the stomatal pit. The walls between the polar parts of the guard cells are ill defined. In *Zosterophyllum* that has similar stomata, this wall is not evident (Fig. 14, *g*). Because of this it is thought that *Zosterophyllum* actually had one 'guard' cell with an aperture in its centre. Somewhat similar stomata are known in moss sporophytes. Until studies are performed on the permineralized remains it is difficult to decipher the structure of the stomata in zosterophylls. In many higher plants the boundary between the guard cells near the poles is generally poorly perceptible on the cuticle. This can be explained by the strongly adpressed position of the guard cells, owing to which a cutin flange is not formed. In propteridophytes specialized subsidiary cells are lacking or can scarcely be detected. But uni- and multicellular hairs and emergences are already present. The epidermal cells spread out radially at the hair bases.

In the Early Devonian *Drepanophycus spinaeformis* (Lycopodiopsida) stomata have been found that have distinct subsidiary and encircling cells (Fig. 17, *g*). They have been described as paracytic, but this is probably a misinterpretation; some stomata were actinocytic. Lycopsids evolved rapidly in relation to their epidermal structure. In the Early Carboniferous leaves appeared in this group with two dorsal furrows (in the lower surface), at first shallow and then deep (Fig. 18, *k*).

Dorsal furrows evolved independently in gymnosperms, first in *Dicranophyllum* and the Angara genera *Rufloria* and *Angaropteridium* (in part), and then in other groups. Other examples can be cited that illustrate the parallelism in the evolution of the epidermal–cuticular structures. The most striking examples are provided by repeated stomatal types in ferns and angiosperms, although the common ancestors of these taxa – the Devonian progymnosperms or allied to them trimerophytalean propteridophytes – had weakly specialized stomata of the same type as in propteridophytes. Stomatal files and bands appear in the Carboniferous and from this time onwards are typical of Lycopodiopsida, Cordaitanthales and conifers. Later they can be encountered in different cycadophytes and some ginkgoopsids as well as in angiosperms.

In the Palaeozoic higher plants anomocytic and actinocytic stomata prevail. Paracytic stomata, in addition to *Drepanophycus*, have been noted in the Middle Carboniferous trigonocarpalean *Alethopteris sullivantii* and conifers *Swillingtonia* (= *Lebachia*?), but these observations are doubtful. Paracytic stomata are typical of bennettites. They are known in Mesozoic–Cainozoic ferns and many angiosperms. The complete set of stomatal types known in higher plants can be observed only since the Cainozoic. In the Permian a butterfly-shaped cutinization of the guard cells appears (in Peltaspermales), that is characteristic also of the Mesozoic bennettites (see above).

The early (Cretaceous) angiosperms are relatively uniform in the syndrome of epidermal characters (see p. 235). Their present epidermal–cuticular diversity developed only later in the Cainozoic.

CLASSIFICATION OF DISPERSED CUTICLES

The term dispersed cuticle refers to leaf scraps with preserved epidermal structure and naturally macerated cuticular fragments. Dispersed cuticles are frequently found in sediments and are of practical application in view of their palaeobotanical characterization. A large amount of material is available to the palynologists as a result of maceration of specimens. For special studies of the dispersed cuticles bulk-maceration techniques are employed; the material is disaggregated and then sieved.

Dispersed cuticles cannot usually be allied with any definite genera or species established from complete remains. In this case for the designation of specific types of dispersed cuticles the investigators used a system of formalized symbols, which is inconvenient in routine studies of these materials. Because of this a binary nomenclature was introduced for dispersed cuticles. The first scheme for the genera of dispersed cuticles was originally introduced by Meyen (1965), and was similar in construction to that proposed by Naumova for dispersed miospores (see p. 249). The choice fell to a few characters, the combination of which limited the number of genera beforehand. Later the suggestion was advanced that this principle should be applied to the distinction of suprageneric groupings (anteturmas, turmas, subturmas), and that the genera should be narrowed in bulk, whereas their number should not be fixed beforehand (Roselt and Schneider, 1969; see also; Kovach and Dilcher, 1984). Neither the first nor the second system is of wide current use in practical classifications of dispersed cuticles. As before, of more common usage is an arbitrary set of formalized symbols or the names of corresponding eutaxa. This is partly due to the fact that dispersed cuticles are still being incidentally described in the process of studies of other plant remains, and only occasionally are they used for stratigraphic purposes.

Chapter 6

Plant palaeoecology

A comprehensive study bearing on plant palaeoecology has been contributed by Krassilov (1972b, 1975). Because of this only a brief outline is presented here, in which attention is called to certain of the most important aspects of palaeoecological investigations in palaeobotany. A short but very informative account of palaeoecological studies was compiled by Scott and Collinson (1983).

Ecology developed initially as a science concerned with the environment and mode of life of organisms, whereas now it is practically being converted into a general study bearing on ecosystems. From the earliest times ecology has been subdivided into autecology, which deals with the life and environmental conditions of representatives of individual taxa, and synecology, which deals with the natural associations of organisms – from those of elementary communities to the biosphere, inclusive. Synecology of plants is synonymous with geobotany ('plant sociology').

Palaeoecology of plants is concerned with the same topics: the reconstruction of aut- and synecology of fossil plants based on studies of fossil remains and their incorporation in rock. Studies of plant incorporation in sediment form an independent branch of science – taphonomy, which constitutes a part of palaeoecology. Studies of the pure productivity of extant communities may be regarded to a certain extent as the analogue of taphonomy in ecology. The ultimate aim of plant palaeoecology is the reconstruction of plant components of the biosphere for past geological periods.

The monograph submitted by Krassilov on plant palaeoecology consists of three subdivisions that deal with (1) taphonomy, (2) life forms, (3) vegetation. A different structure has been adopted below for plant palaeoecology, including: (1) the study of incorporation (taphonomy); (2) morpho-functional analysis – the reconstruction of the ecology of representatives of any definite taxon according to morphological–anatomical characters; (3) palaeogeobotany, i.e. the reconstruction of the life forms, of plant communities, and of their abiotic environments; (4) the interactions between plants and other living components of the ecosystem (fungi, insects, tetrapods, etc.).

Taphonomy deals with (1) the preservation of fossil organisms (taphonomy of individuals, or it may be called autotaphonomy; for recent developments in this field, including experimental studies, see Rex, 1983, 1985; Rex and Chaloner, 1983) and (2) the structure, classification and the process of incorporation (syntaphonomy). The term burial refers to the 'bed or part of the bed containing plant or animal remains. The burial is the basic unit in taphonomic classifications. The totality of plant remains (or other organic remains – S.M.), occurring in the burial is termed taphocoenosis' (Krassilov, 1972b, p. 70). In studies of the structure of burials the palaeobotanist records their configuration (layer, lens, concretion, etc.), amount, distribution and orientation of the plant remains, state of preservation, thickness and composition of the plant-bearing beds (Fig. 80). The burials are classified according to those characters that can be directly observed and characters that can be reconstructed, such as the palaeogeographical environments to which they are restricted (paralic, limnic, deltaic, lacustrine, fluvial, flood-plain,

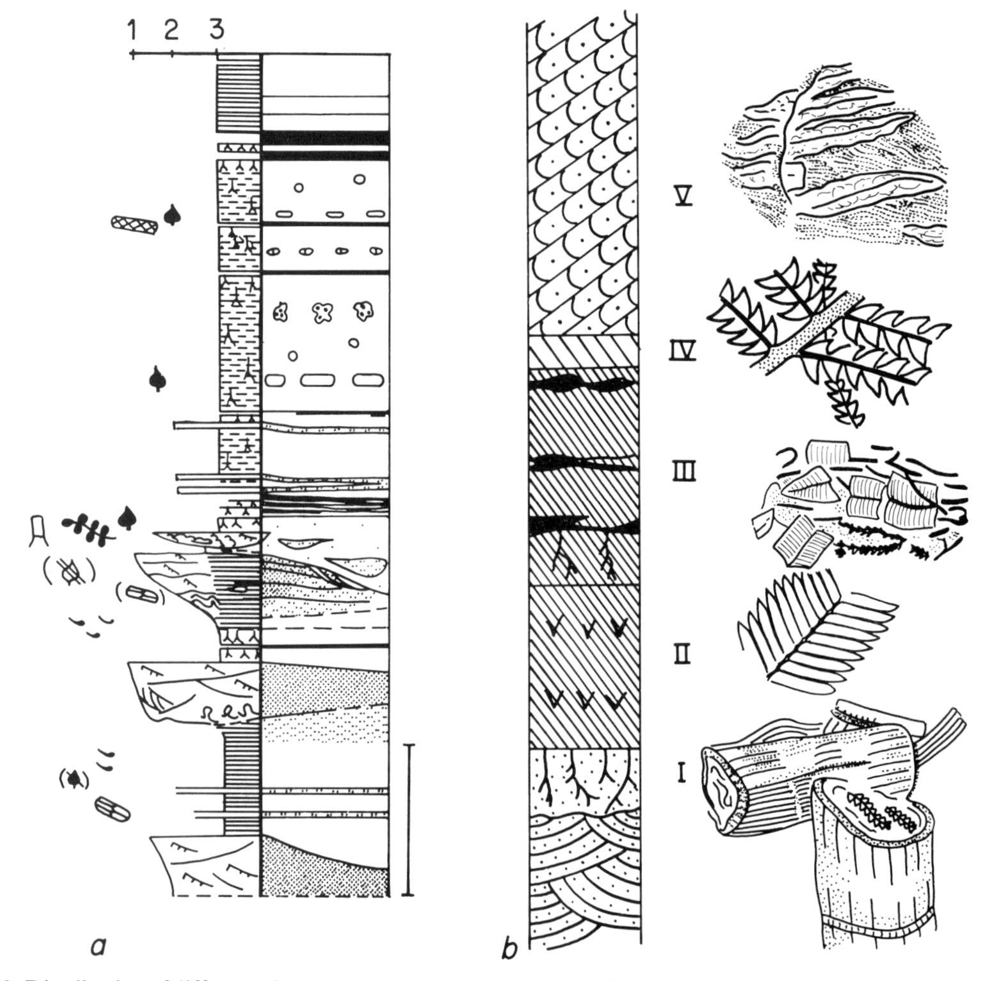

Fig. 80. Distribution of different plant remains in burials, and types of burials; *a*, Middle Carboniferous (Westphalian B) of Yorkshire, below are lake and deltafill sediments, above are floodplain and swamp sediments, various kinds of plant remains are shown at left side, (1) sandstones, (2) siltstones, (3) shales, scale bare = 5 m; *b*, Jurassic supradeltaic sediments in Donbass, feruginous sandstone with *Equisetites* (I), gypsiferous clay with *Ptilophyllum* (II), coaly clay with *Pityophyllum*, *Elatides* and *Taeniopteris* (III), white clay with *Todites* (IV), cross-bedded sandy silt with *Dictyophyllum* (V). Modified from: Scott, 1978 (*a*); Krassilov, 1975 (*b*).

channel, oxbow, etc.). Taphocoenoses are also variously classified. The most pertinent characters available from observations are the dimensions of the remains (mega-, meso- and microfossils), their affiliation with definite organs (miospores, megaspores, seeds, leaves, stems, etc.) or with definite taxa, the kind of fossilization, degree of fragmentation. A taphocoenosis dominated by one definite species is termed monodominant, by several species oligodominant, and by many such species polydominant.

According to the size, degree of fragmentation, orientation and systematic composition of the remains, taking into consideration the lithological–genetical characters of the enclosing rocks, it is possible to determine whether or not the fossils underwent transportation prior to burial. Autochthonous inclusions and taphocoenoses are those where the plants underwent burial *in situ* in their life position. This refers to remains of rhizomes, rhizophores and roots (e.g. stigmarian soils), fossil stands of trees (Davitashvili and Zakharieva-Kovacheva, 1975), certain plants (e.g. bryophytes) entombed in travertines, etc. If certain plant parts underwent insignificant transport, the plant-bed and taphocoenosis are termed hypautochthonous. These assemblages occur in peat beds or at the bottom of water bodies near the sites of parent communities. Autochthonous and hypautochthonous taphocoenoses reflect, although not fully, the primary composition of the plant communities. In allochthonous taphocoenoses the plant remains underwent considerable transport, so that the primary plant communities are reflected either in impoverished or mixed state. Taphocoenoses composed of reworked plant remains (particularly miospores) stand apart from the above-cited types. At times the entire palynological assemblage proves to be reworked.

Both the composition of the taphocoenoses and of the process and site of incorporation in sediment, show changes in the course of plant evolution. The gradual invasion of the plants onto the land resulted in a decrease in the rate of erosion processes, which was conducive to the formation of stable river valleys, the areal expansion of stable continental sedimentation. The fossil occurrences in the Silurian–beginning of the Devonian are for the most part restricted to coastal-marine facies that probably accumulated in marshlands. However, in the Silurian and later in the Early Devonian of increasingly wider occurrence are intracontinental assemblages in alluvial deposits. In the uppermost Devonian and in the Lower Carboniferous, first paralic and then limnic burials appear, the latter being associated with coal-bearing formations. Of simultaneous appearance are mangrove-like burials. In the Carboniferous with the development of the seasonal leaf-fall of wide occurrence are mass burials of leaves, termed leaf-roofings by Krassilov. In general, beginning from the Carboniferous the diversity in the types of burials and taphocoenoses does not exhibit an essential increase. However, the representativeness of the taphocoenoses could have changed, considering that the preservation ability varied in different plant groups (for a review of the data on recent accumulations of plant remains see Collinson, 1983). For instance, experimental investigations established that angiosperms, as a whole, are of more rapid decomposition in water, compared with gymnosperms (Kalugina, 1974), and, in consequence, the probability of their leaving recognizable fossils in sediments is smaller. Uncalcified algal thalli undergo more rapid decay compared with higher plants.

Morphofunctional analysis aims to reconstruct the physiology and autecology of the plants. It is clear that we can judge the functions of fossil plants only by comparison with extant forms (the actuality principle). Unfortunately, the structural types and functional types cannot be arranged in a general one-to-one correspondence. Each type of organ, tissue, cell may correspond to more than one functional type. This phenomenon, termed multifunctionality, causes serious difficulties in defining the functional meaning of morphological structures. However, at times their meaning is apparent. This refers to the decrease in the number and disappearance of the stomata on the submerged parts of aquatic and semiaquatic plants. But even such seemingly obvious morphofunctional interpretations become inapt in the course of time. For example, the presence of large air chambers in roots, stems and leaves is typical of plants in moist and aquatic habitats. However, the same type of air chambers occur in plants not resistant to excessive moisture. Different ecological factors frequently cause the same effects in plants. For instance, the strong development of the cuticle, submergence of stomata, reduction of the leaf lamina, its dense pubescence, i.e. such characters as are termed xeromorphic, results not only from dryness of the environment, but also from an increase in the light intensity and deficiency of nitrogenous nutrients. These factors, as well as the physiological drought (unfavourable chemical composition of the water) determine the xeromorphic habit of many bog plants.

No definite functional interpretation can be offered for most of the specific characters observed in plants and used in systematics. Although the leaf is known to perform a photosynthetic function, in the physiological and ecological sense the tremendous diversity of the leaf blades is inexplicable. An increase in the degree of dissection of the leaf lamina, as a whole, can be correlated with an increase in the mean annual temperatures and moisture, but this general tendency is complicated by a great number of exceptions and the difficulties incurred in differentiation of one factor from another (Dolph and Dilcher, 1979). At present the percentage of plants with entire margined leaves is much larger in tropical rain forests, while in forests where a temperate climate prevails the leaf margin is more frequently denticulate. In the Late Palaeozoic gymnosperms an opposite relationship is evident.

Only a limited number of characters can be cited that reveal a distinct ecological restriction. Among them are stems without growth rings as observed in trees that exist in climates with weak seasonal fluctuations of the temperature and humidity (Creber and Chaloner, 1984). Perennial stems with the parenchyma-rich wood called 'manoxylic' cannot survive under temperatures below zero, and are characteristic of plants of a frostless climate. Although the above-cited xeromorphic characters are typical of many plants existing in moist habitats, occasionally most of the plants of a definite flora reveal such characters, and then an inference can be drawn concerning the aridity or semiaridity of the climate. Contrariwise, the absence or paucity of xeromorphic characters gives evidence to the high moisture content of the habitat, or of the humidity of the climate of the territory.

When the ecological requirements of a definite taxon are well known in its extant members, the palaeobotanists extrapolate these requirements over the geological past. As a consequence, having established that the genera sequence in the stratigraphic section of Iceland, from the Miocene to the Quaternary, is comparable to that which occurs from south to north in the modern vegetation cover, the inference was drawn that the mean annual temperatures underwent a decrease (Akhmetyev *et al.*, 1978). Such inferences are even more risky when we deal with older deposits. At times only narrowly specialized species of a taxon survive to the present day. Moreover, in the course of evolution the ecological preferences of plants of one taxon may be subjected to changes. This is evident, for instance, in the phenomenon of extra-equatorial persistence (see p. 345). Even the Neogene genera, survivors of which are known to date, could have had other ecological preferences in the past. For example, *Liquidambar*, *Platanus* and *Liriodendron* live today in broad-leaved forests, whereas during the Neogene these plants were typical of coniferous forests (Wolfe, 1981).

The ecological range of each taxon is limited, but we still have no means of judging the limitations other than by simple plotting the present-day area of the taxon against the patterns of the areal distribution of various ecological factors. Any attempt to determine only on the basis of known plant characters the requirements for the plants in quantitative parameters are doomed to fail from the outset. From studies of the habit of extant species of *Selaginella* it would obviously be impossible to predict that some species of this genus populate rain forests, while others populate deserts. From studies of the morphological–anatomical characters of reeds (*Phragmites australis*) it is impossible to derive the inference that these plants are capable of living in a vast area extending from the mouth of the Northern Dvina River to the mouth of the

Ganges River. The modern genus *Taxodium* populates both the paludal lowlands and highland forests, which can by no means be inferred from its characters.

The palaeobotanists frequently neglect the above-cited difficulties of morphofunctional analysis and attempt direct interpretations of one or another character, heedless of their multifunctionality. For instance, having observed a thick cuticle in a plant and deeply sunken stomata, the palaeobotanists conclude that the plant is a xerophyte. Having found hook-like processes on the stem they infer that they have met with a climbing plant. Having established that the prepollen in certain trigonocarpaleans do not correspond in size to the pollen of wind-pollinated plants, a conclusion is drawn that these plants were entomophilous.

Such conclusions are not always wrong, but they are poorly founded, because the multifunctionality of the characters is disregarded. The latter should be taken into consideration, not merely by citing other probable functions of one or another character, but by involving in the analysis all the data on other plants of the same taphocoenosis, along with the full complex of taphonomic and other palaeoecological observations. No matter how paradoxical it may seem, it is much easier to obtain a true reconstruction of the entire palaeoecosystem, verified independently by different observations, than to reconstruct the ecological requirements of any single plant taken by itself. This implies that special attention should be focused on the underlying principles for the reconstruction of palaeoecosystems in their palaeobotanical constituent.

That part of ecology (as the science bearing on ecosystems) that treats of the plant constituent of ecosystems corresponds to geobotany. The main objectives of geobotany are any plant associations up to the vegetation cover of the Earth, in the whole. Accordingly, the objectives of palaeogeobotany involve taphocoenoses and their assemblages, on the one hand, and those primary plant communities which gave rise to these taphocoenoses, on the other hand. Palaeogeobotany aims ultimately to reconstruct the history of the Earth's vegetation – the structure and changes in the plant communities, their regularities and specific history.

The basic notions of geobotany are the life form, plant community and vegetation. The conception of life forms has given rise to the most animated debates (see review: Serebryakova, 1972). It can be more easily understood when considered in comparison with other general conceptions. In this respect it is convenient to proceed from the animal kingdom, not from the plant kingdom. A rather complete mental image of the animal population of a site can be gleaned from the lists of taxa, the number of individuals in each, and their areal distribution. Additional information on the habit of the members of the taxa is usually unnecessary, except in those instances when they are represented by oppressed forms. In plants the same species, not mentioning the taxa of higher rank, may be represented by different modifications repeatedly occurring in other taxa. These modifications may be more important in characterizing the vegetation cover, than a knowledge of the systematic affinities of the plants. For instance, if our description is limited to an indication that the vegetation cover is dominated by the Siberian pine (*Pinus sibirica*) then it is not known whether this pertains to a high-stem tree or to a low growing one. Such a coincidence in one taxon of different modifications of the external habit is typical of many plants and each modification, occurring repeatedly in different taxa and associated with certain particular ecological factors, is termed a life form. A majority of the species are characterized by a single life form, but within the genus or families several life forms can frequently be identified.

The classification of plants on the basis of life forms cuts across the limits of the units (species, genera, families, etc.) in ordinary taxonomic systems of plants. On the other hand, this set of life forms determines the aspect of the vegetation cover of the territory (forest, meadow, shrub thickets, etc.) and the structure of the plant assemblages. This brings us to the concept of vegetation; this term implies the distribution of the taxa, life forms and communities within a given territory. As a result of the parallelism of life forms in different taxa, a change in the taxa in communities may occur without any essential changes in the structure of the community and in the aspect of the vegetation cover.

In palaeobotany there are three major ways of reconstructing life forms. First of all, by taking into consideration the association of various dispersed parts in the burials. If we find in the burial one particular type of leaf, female and male fructifications of such a habit that tends to indicate their possible organic connections, whereas other plant remains are absent from the burial, then it may be postulated that they are parts of the same plants. T. M. Harris formulated a general empirical rule, which states that if we find in the burial any fructifications (they are commonly of more scarce occurrence than vegetative remains) then among the vegetative parts in the burial should be found those with which the fructifications had been associated. This rule turned out be heuristic. Of paramount importance is the stable recurrence of associated parts in the section and area. For example, microstrobili *Cladostrobus* always associate with leaves of *Rufloria*, and this fact served as the starting point in establishing their organic connections. After having established the organic connections between dispersed parts it is often possible to form an opinion concerning the aspect of the reassembled plants.

Secondly, we can mentally envisage the connections of detached parts encountered in the burial on the basis of markers that serve to indicate that these plants belong to a single plant type. The major impediment in such reconstructions is the absence of valid markers. Examples of successful reconstructions were illustrated in Chapter 3 (for *Pseudosporochnus*, *Archaeopteris*, *Caytonia*, etc.). Various characters were used as markers – epidermal–cuticular, secondary wood structure, the presence of specific sclereid nests in the bark, etc. At times it is possible to find different parts, e.g. fructifications and vegetative organs, in full organic connection.

It is seldom possible to reconstruct fully the overall habit of the plants according to the markers. Usually the connections are only established between individual parts. For instance, we are aware of the fact that the polysperms *Beania* associate with microstrobili of *Androstrobus* and leaves of *Nilssonia*. Reconstruction of the leptocaulic shoots (with slender branches) was suggested, but the overall habit of the plant is still unknown. It happens that for the reconstructions data are used that characterize not one but several plant species together. This relates to the most widespread reconstructions of *Lepidodendron*. In some species connections were observed between the rhizophores and stems with leafy cushions of the *Lepidodendron* type but not of any particular species of the genus. Leafy shoots were found connected with strobili of the *Lepidostrobus* type but from this it was not possible to identify the *Lepidodendron* species on which the strobili of a certain *Lepidostrobus* species were borne. Hence, a reconstruction of the genus *Lepidodendron* was suggested in a general form, as if all the species of the genus should have belonged to a single arborescent life form. In reality it is known that liana-like forms existed among the *Lepidodendron* species. Only recently DiMichele (1981) and DiMichele and Phillips (1985) found it possible to propose a well-founded reconstruction for individual species of these plants (Fig. 18, *c–e*).

The third way of reconstruction of the life forms is based on the application of typological extrapolations. For this purpose data are taken that are available on the life forms of other plants, which are comparable in a certain chosen manner to those under study, and the reconstructions are modelled after these plants. For instance, the reconstructions performed during the last century for cordaites, according to which they are depicted as shapely trees with tall trunks, were modelled after certain modern conifers, although they might have been just as well reconstructed after the model of other conifers or, let us say, ginkgos, i.e. plants with a different kind of crown. After the discovery of roots of cordaites with air chambers, the hypothesis was set forth that certain cordaites were mangrove plants (Cridland, 1964). The extant genera with aerial roots typical of mangrove plants was then taken as an actualistic model (Fig. 64, *f*). Many fossil cycadophytes were depicted after the extant cycadaleans as thick-stemmed (pachycaulic) trees with an apical crown of leaves. Only later did it come to be realized that among the fossil cycadophytes, leptocaulic forms are more prevalent.

The necessity of appealing to models, including actualistic ones, is dictated by the fact that extant and well-studied fossil plants allow for a sufficiently complete classification of the life forms. Proceeding on the basis of such a classification it is possible to trace the recurrent sets of life forms in different taxa, as well as that combination of characters that is typical of any particular life form in different taxa. This line of approach ensures a more valid reconstruction of the life forms in the extinct taxa. It has been noted that pendant fructifications are more characteristic of trees than of shrubs. Because of this the inference was drawn that the above-mentioned polysperms *Beania* were suspended from branches that belonged to trees. It has been indicated that trees with deciduous leaves do not usually exceed 20–30 m in height, since in a leafless state a deficiency is inevitable in the water supply to the upper branches, due to the absence of the leaves which serve as pumps. Because of this there is reason to believe that deciduous trees of the Palaeozoic and Mesozoic also did not exceed 30 m in height. In general, the larger the taxon, the greater is the probability of a larger life-form diversity in it. For instance, the extant articulates are represented by a single genus *Equisetum*, that includes only herbaceous forms. In the Carboniferous the articulates were much more diverse and involved various life forms including arborescent, which could not be inferred from the actualistic model (genus *Equisetum*). From a knowledge of the modern conifers it was impossible to conclude that certain fossil ones could be herbaceous plants. Moreover, the very possibility of herbaceous forms in gymnosperms was refuted. However, Grauvogel-Stamm (1978) established that the Triassic conifers *Aethophyllum* (Voltziaceae; Fig. 71, *j*) were herbaceous.

There is one more method of reconstruction of life forms that differs from those described above and relies on the general conception of the structure of the vegetative cover derived from indirect characters (e.g. from the structure of palaeosoil profiles; see below), or from certain analogies to extant types of vegetation. Proceeding on this basis we can judge the dominant (paramount) role of a certain group acccording to the abundance of the remains (including miospores), whereby this group may be attributed to a definite life form. It is obvious that in such investigations we proceed not from a reconstruction of the life forms to the reconstruction of the vegetation, but the other way round.

An extensive literature is devoted to the classification of life forms. Most widely acknowledged is the classification proposed by K. Raunkiaer, which is based on the position of the reproduction buds. In this classification the following types are distinguished: epiphytes (lacking roots in the soil), phanerophytes (trees, shrubs, caulescent succulents, herbaceous caulescents and lianas), chamaephytes (the reproduction buds occur at the soil surface), hemicryptophytes (turf-forming; the reproduction buds occur in the soil surface or immediately below the surface), cryptophytes or geophytes (winter buds hidden in the soil), therophytes (annual plants). Some classifications also involve life forms of algae and fungi. In palaeogeobotany, due to the paucity of data, the application of detailed classifications is impossible, as a rule. Because of this we need to content ourselves with such fuzzy terms as tree, shrub, grasses, and speak of the arborescent, shrubby or herbaceous habit. No reliable criteria are available for the identification of therophytes, epiphytes and lianas; frequently we do not know the root-system structure.

The terms tree, shrub and herbaceous plant have no definite meaning in palaeobotany, and only reflect a notion of the size of the aerial part of the plants, at times the presence of secondary wood in the axes. Any plants with stems exceeding 7–10 cm in diameter are termed arborescent. Plants with more slender axes and secondary wood in them are treated as shrubs. Plants with slender axes without secondary wood are termed herbaceous. Obviously such as approach is unsatisfactory. It does not take into consideration the complexity of the concepts such as tree, shrub and herb, the multiplex and variable combinations of criteria that are used for the distinctions between these concepts (Gatzuk, 1976). As in the systematics of fossil plants, where besides eutaxa parataxa are used, it is desirable to have a classification of the life forms of fossil plants that takes into account first of all those characters that are readily available to the palaeobotanist. In this respect the classification proposed by E. Schmid (1963; quoted from Serebryakova, 1972; pp. 134–6), which relies on the degree of lignification of the axes and the arrangement of the growth points, deserves attention.

The evolution of life forms, particularly of angiosperms, has for long been a debatable problem. Many botanists are convinced that the evolution of angiosperms was accompanied by the somatic reduction from the tree to the shrub and further to herbaceous plants. A similar phenomenon was conjectured for certain ferns, articulates and lycopsids. These problems cannot be solved easily because of the gaps in our knowledge bearing both on the life forms of extinct plants and their phylogenetic links. For instance, phylogenetic continuity was believed to exist between the Upper Palaeozoic sigillarias, Triassic pleuromeias, Cretaceous *Nathorstiana* and extant quillworts. This lineage served to illustrate the somatic reduction in certain lycopsids. However, all these constructions were upset by the discovery of plants with life forms such as those in the Triassic *Isoetes* and peculiar life forms such as the Carboniferous chalonerias.

Nevertheless, there is reason to believe that all higher plants evolved ultimately from very tiny herbaceous propteridophytes. In the Early Devonian among trimerophytes forms appeared that reached to one metre in height and exhibited a very intricate branching and distinct main axis. The Middle Devonian progymnosperms (with secondary wood) and ferns externally similar to them looked like shrubs or small trees. Evidence indicating an increase in the sizes of plants during the Devonian is found in the preserved fragments. According to measurements made by Chaloner and Sheerin (1979), during the Silurian the thickest axes were 0.3 cm in diameter, during the Lower Devonian they were 7 cm, and during the Upper Devonian they were about 1.5 m. The records gave indications of the presence of succulents with short axes in the Lower Devonian, but then it was found that these plants (*Mosellophyton*) do not belong to higher plants, but are algae of the *Prototaxites* type. In the Early Devonian forms appeared that had fairly thick prostrate axes, that trailed along the ground, and from which vertical axes extended. Practically all the major types of life forms distinguished in Raunkiaer's classification had probably evolved already by the Early Carboniferous. From this time onwards the diversity in the life forms of such large taxa as the lycopsids, articulates and ferns long persisted in their evolution (transitive polymorphism). The somatic reduction phenomenon has so far been traced for palaeobotanical material only within single lineages (e.g. within the lineage from the arborescent progymnosperms to the Triassic conifers).

Poor knowledge of the life forms of fossil plants is a serious impediment in the reconstruction of communities. By analysing Cainozoic assemblages that contain taxa surviving to date, the palaeobotanist can judge the life forms from those extant representatives of these taxa. Accordingly, reconstructions are made of the communities and of the aspect of the vegetation. When we descend further down in the geochronological scale, we lose the possibility of such direct application of the actuality principle. It is true that certain communities are very stable. The community *Taxodium–Nyssa* persists since the Eocene. The modern forests of New Zealand are characterized by a combination of *Agathis* and *Lygodium*, known also in the Jurassic taphocoenoses (see other examples in Krassilov, 1975). However, reverse cases are known (see above on *Liquidambar*, *Platanus* and *Liriodendron*). From comparison of the extant communities and fossil communities similar to them, and from actual observations on the fossilization process, inferences may be drawn concerning the loss of information in the process of burial (Collinson, 1983, and references cited in this paper). For instance, in studies of one of the lakes it was demonstrated that in its central part in the recent deposits plant remains occur exclusively of coastal plants and among them only 25% of the species are present (McQueen, 1969; quoted from Scott, 1979).

Reconstructions of the communities and plant vegetation of the Mesozoic, and even more so of the Palaeozoic, may be performed in the following manner. At first it is necessary to establish the associations of taxa of repeated occurrence in different beds, and the affiliation of the respective taphocoenosis to a definite facies type. Studies are then made of the structure of the taphocoenosis, the abundance of the representatives of the major taxa (for this purpose the usual geobotanical methods of observations and estimations are employed), areal and stratigraphic distribution of the types of taphocoenoses. Various classifications were proposed for the taphocoenoses (which have already been discussed above). As a consequence, various classifications were suggested for the vegetation. Some of them, advanced by Fissunenko (1965, 1973), Shchegolev (1965, 1979), Havlena (1971), Oshurkova, (1978, 1981), Drägert (1964), etc., were based on the moisture content of the biotope and hydrophilicity of the vegetation. The following types of vegetation have been distinguished: hygrophilous – in coastal swamp lowlands and river flood-plains; meso-hygrophilous – not swamped aggradational plains, mesophilous – uplands and less moist areas of sedimentation including river valleys; xerophilous vegetation – on well-drained slopes. Other classifications (e.g. Scott, 1977, 1978, 1979; Scott and Collinson, 1983) were based on reconstructed geomorphological elements. They distinguished different vegetation types: in swamps (oligotrophic and eutrophic, after Havlena, 1971), in different parts of river and lake valleys, also in uplands contiguous to the sedimentary basins (Figs 81, 82). At times distinctions were made only of 'coal-forming' and 'not coal-forming' vegetation. The first type was equated to the taphocoenoses accompanying coal seams, the second to taphocoenoses in the barren intervals of the sequence. However, it was shown that the taphocoenoses within the coal seam and at its roof differ essentially. For instance, in the Middle Carboniferous coal seams of the Euramerian area remains of lycopsids prevail, whereas in the sediments above the coal remains of fern-like leaves prevail. The term 'coal-forming vegetation (community)' is therefore irrelevant for the taphocoenosis of the immediately overlying beds. It has been suggested that the association of the burial with coal seams should be defined by the terms 'anthracophilic' and 'anthracophobic' communities, respectively (Meyen, 1969a), without any further implications regarding the relationship between the vegetation and coal formation.

Very complete reconstructions were accomplished by Phillips and collaborators (DiMichele and Phillips, 1985; DiMichele *et al.*, 1985; Phillips, 1981; Phillips and DiMichele, 1981; Phillips *et al.*, 1985 Phillips and Peppers, 1984, etc.), through studies of the Pennsylvanian swamp vegetation of Illinois. For this purpose plant remains in coal balls were studied (with quantification of taxa according to the geobotanical rules), and vast palynological and coal-petrographical investigations were performed. Lycopsids, ferns, trigonocarpaleans and articulates, more seldom cordaitanthaleans, were found to be the major components of the peat-forming vegetation. Arborescent forms, encountered in different groups, were the major components involved in coal formation and produced the major mass of miospores. The assemblages and those organs that comprised the peat mass exhibited changes, dependent upon the water table of the habitat. For instance, arborescent lepidophytes (*Lepidodendron*, *Lepidophlois*) populated wet freshwater swampy areas; cordaitanthaleans, trigonocarpaleans and arborescent ferns revealed better adaptation to physiological drought and decreasing water table; temporary marshlands were populated by chalonerias. Of most diversity were the communities with *Sigillaria* and *Paralycopodites*, that populated the dryest parts of the swamps. The change in the type of the assemblage was demonstrated in the geological section.

Interesting data on the life conditions in coal-ball peats may be obtained from studies of the shoot–root ratios in the plant remains (Phillips *et al.*, 1985; Raymond and Miller, 1984). In the Recent mangrove peats the ratios are low (0/100–20/80), whereas in freshwater peats they are higher (40/60–60/40). These ratios probably provide evidence of the salinity of the coal-ball peats in which unusually high ratios (70/30–100/0) may be observed. In most cases the shoot–root ratios of a taxon within a coal-ball generally matches that of the coal ball, with the exception of *Medullosa*, which had high shoot–root ratios regardless of that in the coal ball as a whole.

At present little can be said of the changes in the

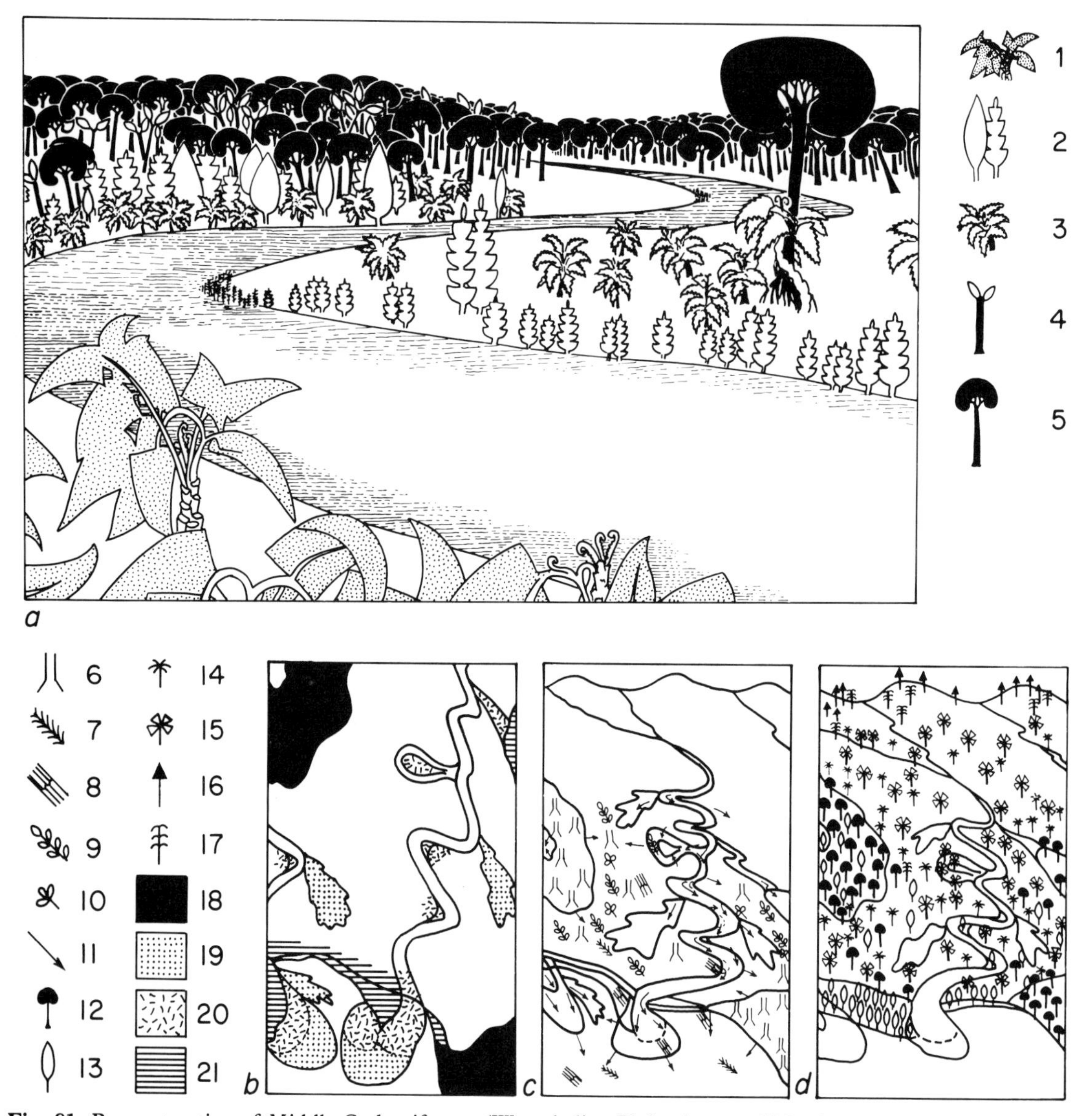

Fig. 81. Reconstruction of Middle Carboniferous (Westphalian B) landscape of Northern England and Southern Scotland; *a*, river and flood plain landscape; *b–d*, various components of reconstruction (*b*, depositional environments; *c*, distribution of fossils showing possible directions of derivation; *d*, original plant communities reconstructed); 1, pteridosperms; 2, calamostachyans; 3, ferns; 4, sigillarias; 5, 6, lepidodendrons; 7, leafy shoots of lepidodendrons; 8, calamostachyans; 9, pteridosperms; 10, ferns; 11, source of remains; 12, lepidodendrons; 13, calamostachyans; 14, ferns; 15, pteridosperms; 16, conifers; 17, cordiatanthaleans; 18, swamp sediments; 19, silt; 20, sand; 21, calamostachyalean ('reed') swamp. Modified from Scott, 1979.

structure of the communities and diversity of their components in the history of terrestrial vegetation. In order to solve this problem it is not enough to gather data only on localities of different ages. The large and most abundant assemblages of the Carboniferous and later periods contain hundreds, but not thousands, of plant species. If we take single layers and the corresponding local floras of the Palaeozoic and Mesozoic, then their diversity amounts to tens of species. At the same time, Recent local floras are much richer. For example, in the Moscow district over 1300 species have been registered (excluding bryophytes), of which about 40 fall to cryptogamic and gymnospermous plants. The floral diversity is incomparably greater in the southern regions. The relatively lower diversity of the Palaeozoic and Mesozoic floras (and, accordingly, the communities) can probably be explained by their impoverished composition, rather than by mere loss of information. Evidently the formation of rich and diverse assemblages became possible after the incoming of angiosperms.

On the basis of the taphocoenoses sequence it is possible to trace the succession of communities in any definite locality or area. Although these successions may be associated with specific local environmental factors (particularly with regard to plant assemblages in volcanics, where their accumulation may be practically instantaneous, thereby violating the successional series; Burnham and Spicer, 1984; Spicer, 1984), comparison of the successions in many geological sections of different areas allows for the rearrangement of palaeoecosystems on a regional or even global scale. Analysis of the successive series is highly effectual in local, regional and global stratigraphic studies (Krassilov, 1977a).

In taphocoenoses catenas can be traced – the successive series of assemblages from the water's edge upwards along the slope. Of the vegetation on slopes and even mountains we can judge from the *in situ* burials (in travertines and volcanics; at times river valleys of the uplands are preserved), or from indirect data. For instance, Chaloner and Muir

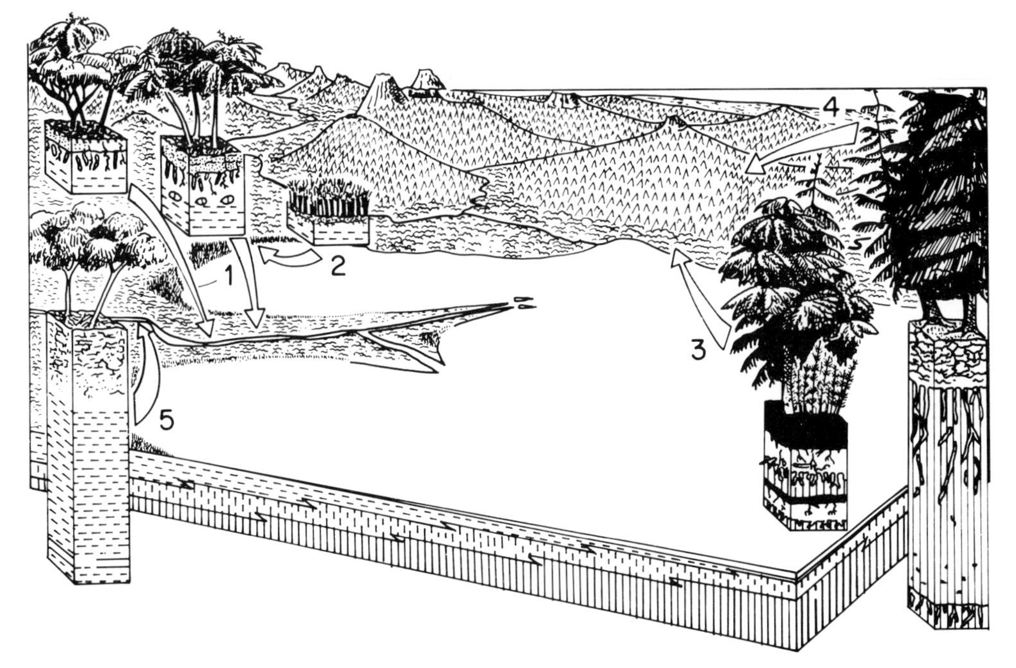

Fig. 82. Reconstruction of vegetation, soil profiles and relief near Sydney in late Early Triassic (lower part of Newport Formation); 1, *Dicroidium* tickets; 2, meadows with Pleuromeiaceae; 3, swampy forest with *Dicroidium* and *Voltziopsis*; 4, coniferalean forest with *Voltziopsis* and conifers producing *Brachyphyllum*-like leaves; 5, shrubs with *Taeniopteris*-like leaves. Modified from Retallack, 1977.

(Chaloner, 1968b; Chaloner and Muir, 1968) demonstrated that during transgressions of the sea over the coastal lowland, the palynological assemblages in marine sediments begin to be dominated by miospores of plants that populated the slopes. This phenomenon (Neves' rule) was established for the Palaeozoic, Mesozoic and Cainozoic deposits. The slope vegetation may be reflected by plant remains which are encountered in the lowland sediments, but reveal distinct traces of transportation.

Climatic changes may cause shifts in certain successions of the slope vegetation, which is reflected in the change of the taphocoenoses. For instance, during the cooling of the climate the slope vegetation descends to the lowlands. In a geological section this can be reflected in that certain plants pass from allochthonous accumulation into autochthonous (Krassilov, 1975, 1977a).

Due to the parallelism in the life forms in different taxa, the structure of the community may remain the same, despite the changes in the taxa. For instance, nowadays the coastal marshes and mangroves are composed of angiosperms, whereas in their place during the Mesozoic were horse-tails, ferns and gymnosperms, and during the Palaeozoic other articulates, ferns and gymnosperms, also lycopsids. In the Permian the herbaceous vegetation cover of Angaraland consisted of ferns and articulates (Gomankov and Meyen, 1980; Meyen, 1982b). Naturally, in the geological past certain peculiar 'extinct' types of communities could have existed, and the overall habit of the vegetation cover underwent radical changes, particularly during the Devonian.

In the geological record are represented only those types of plant communities, the components of which could be buried. Certain communities can be judged only from dispersed miospores and cuticles, wood fragments and other remains. In assemblages chiefly of autochthonous and hypautochthonous composition, allochthonous components can be distinguished with difficulty. Because of this we need to seek indirect indications of the evolution of the communities through analysis of extant communities. The possibility of such an approach was convincingly demonstrated by Plotnikov (1979), who attempted to reconstruct the evolution of communities, taking into account the predominant occurrence in past periods of definite taxonomic groups, relying on the present-day structure of the communities formed by these groups. In this manner he established the approximate productivity of communities of the terrestrial biota in the whole during different geological periods. The productivity of the Early Devonian land area, inhabited by propteridophytes, was found to be 20 times less than that of nowadays.

For the reconstruction of the vegetation distributed over different landscape elements, valuable data can be obtained through studies of buried soils (palaeosols). Complete soil profiles are rarely preserved, and the most ancient soils may differ from the modern types. Nevertheless, studies of the Palaeozoic (beginning from the Lower Devonian) and younger palaeosols serve to prove that according to their features it is possible to judge the life-forms of the dominant plants, of the vegetation structure and its distribution relative to the sedimentary conditions (Fig. 82); also, inferences can be drawn concerning the depth of occurrence and chemical composition of subsurface waters, the palaeoclimatic and palaeogeographic conditions, evolution of the interrelations between the vegetation and relief, time of formation of the forest, steppe and other vegetation types, etc. (for details see: Remy, 1980; Retallack, 1981, 1983a, 1983b, 1985). By tracing palaeosols containing remains of subsurface parts of plants, it is possible to correlate with definite precision different stratigraphic sections and to define the levels of regression in marine sequences, which may be overlooked in ordinary lithological investigations (Remy, 1980).

There are other means of reconstruction of the communities and ecology of individual taxa, that are not based on palaeobotanical material. Valuable information on the life of plants and their communities can be obtained from studies of associated animals. The composition of aquatic faunas may witness the degree of eutrophication of the water body in which the plants lived. Kalugina (1974) demonstrated that during the Early and

Middle Jurassic, horsetails of the Baikal region populated oligotrophic water basins. This is evidenced by maggots of Plecoptera buried together with them. The same author advanced interesting considerations suggesting that the invasion of angiosperms into the water bodies, resulting in the input of a large amount of easily decaying tissues, would tend to cause eutrophication.

Finally, attention should be focused on the interactions between plants and other organisms in the palaeoecosystems. In the literature bearing on this theme only general considerations are usually outlined, concerning, for instance, the impact of herbivorous animals on plants and their communities, the interactions between entomophilous angiosperms and pollinators, etc. Publications on the results of concrete investigations of the interactions between plants and other organisms, are more scarce, although of greater interest (Crepet, 1979; Hughes and Smart, 1967; Krassilov and Rasnitsyn, 1982; Scott and Taylor 1983; Scott *et al.*, 1985, etc.). Studies of coprolites of herbivorous animals, studies of their gut content, traces of animal activities on plants (burrows, tracks), fungal remains (fruiting bodies, hyphae) on the surface and inside of the plants, comparison of the types of flowers with the insects known from the same deposits are all very informative. These studies provide valuable information that at times sheds light on the interactions between the plants and other organisms of the ecosystem, and on other aspects of plant ecology. For instance, Scott and Taylor (1983) through investigations on plant remains in coprolites established that the assemblages of soil arthropods (chiefly myriapods, millipedes and centipedes), which are of particular importance in soil ecosystems, were already formed during the Carboniferous. Coprolites packed with remains of peltaspermaceous *Tatarina* (Meyen, 1981) and occurring in deposits abundant in aquatic animals, clearly indicate that the shores of the water body were densely covered by thickets of *Tatarina* shrubs and were probably frequently flooded.

The interactions between plants and other components of the ecosystems are in need of more extensive special investigations. The probability is strong that the change from stromatolite to algal bioherms at the Pre-Cambrian–Phanerozoic boundary is due in part to the impact of herbivorous animals on the build-up of stromatolitic mats, but this is pure speculation. Little as yet is known of the historic interactions between zooxanthellae or other algae and the animal population of the reef-building communities. Judging from the footprints, tetrapods probably appeared in the Middle Devonian, and during this time should already have exercised an influence on the coastal vegetation of water bodies. But very little can be said of this influence, not only during the Devonian, but in later periods of the Palaeozoic and even Mesozoic. More data are available on the interactions between plants and insects, and insects and mammals, but this problem awaits more extensive and specially organized investigations. In the more distant future special studies should be made of the evolution of the ecosystem types. Interesting conceptions advanced in relation to this problem (McArthur, 1972; Pianka, 1978; Ponomarenko, 1984; Stebbins, 1974, etc.) are in need of more detailed examination and concrete definition.

Chapter 7

Palaeofloristics

As a simplification, it may be said that floristics treats of the geographical distribution of taxa, the plotting of the plant system against the geographical background. In palaeofloristics we are concerned with both the spatial distribution and the chronological ranges of the taxa. Palaeofloristics may be defined as the study of the distribution of the taxa – in time and space. By plotting the taxa areas on maps we observe the concentration of the boundaries in certain areas of the Earth. This implies that the taxa form natural associations, each of which is confined to a definite territory. It can also be seen that the associations are inlain one in another, forming a system of subordination. Each association corresponds to a territorial floristic unit – the phytochorion. Phytochoria may be of different ranks. In descending order they are as follows: kingdom, area, province, district, region. Each is denoted by a specific geographical name. The taxa list of a phytochorion or its part is termed flora.

In the identification of phytochoria we are compelled to take into account the frequency distribution of one or another taxon. Because of this the boundaries between phytochoria are not reflected as lines along which the association of taxa change completely, but as bands within which more or less abrupt changes occur in the frequency distribution of the taxa. The boundaries between the phytochoria when they pass on land are always blurred to a greater or lesser degree. In ecology and biogeography such blurred boundaries are termed ecotones.

In the chapter Plant palaeoecology (Chapter 6) it was stated that no direct relationship exists between the structure and taxonomic composition of communities. The same taxon may occur in different communities, either retaining or not retaining its life forms. On the other hand, the replacement of the taxa in the community may occur without changes in its structure (to be more exact, without changes in the place of this community in the general classification of the communities). Notwithstanding this, the distribution of the taxa, as well as the structure and distribution of the communities, are controlled by the same factors: orographic, edaphic, climatic, etc. Hence, a close correlation is apparent between the floristic (according to the distribution of the taxa) and geobotanical (according to the distribution of communities) zonality. Many boundaries are at the same time floristical and geobotanical. It has been suggested that the territories delimited by such complex boundaries, should be termed vegetation divisions (Schmithüsen, 1961). Reciprocal links between floristic and geobotanical zonation are also inevitable, because in modern phytocoenology the classification of the communities themselves is based on the floristic principle (the Braun-Blanquet method).

In any zonation certain characters should be substituted by others. It is in principle impossible to trace directly the distribution of taxa over vast territories, because in this case all individuals of a definite taxon should be plotted on the map. Therefore, even in pure floristic zonations we must rely on the distribution of communities, if they can be visually traced. On the other hand, by the distribution of an easily traced taxon we can judge the distribution of the community. Both cases involve the application of the data on physical–geographical features that attend the changes in the vegetation cover of the territory. For instance, there is no need to perform detailed floristic and geobotanical investigations along the entire boundary between a swamped plain and the adjacent upland. It is sufficient to study certain sites along this boundary, and if they are uniform, according to the characters of interest to us, then the boundary may be drawn simply with the aid of the hypsometric map. Thus, in zonation of the territory (as in any other typological investigation) important characters that are difficult to trace are substituted by others, associated with them, but more readily available from observations. The latter are discarded in the end results, so that one gets the impression that these characters were not involved in the study.

In studies of the modern vegetation cover the floristic characteristics may be separated from the geobotanical (although they are often perfunctorily combined), whereas in the general case this is impossible when we turn to palaeobotanical materials. First of all, owing to the peculiarities of the palaeobotanical records, only certain vegetation types may be represented, while others may be missing (e.g. the vegetation of high-mountainous areas is commonly not represented in macrofossils). As a consequence, the palaeofloristic zonality maps, even when they are plotted by means of superposition of the areas of different taxa, are as it were inserted in the distribution pattern of certain specific vegetation types. Secondly, the palaeobotanical taxonomy sometimes more readily reflects the classification of life forms of a given taxon, than its division into genera and species according to all-inclusive characters. For example, the genus *Calamites* involves pith casts of nearly all the Carboniferous and Permian Calamostachyaceae that have quite thick stems and a well-developed pith cavity. By tracing the spatial distribution of the genus *Stigmaria*, we obtain, in effect, the distribution pattern of some generalized taxon of indeterminate rank (since such rhizophores were characteristic of many lycopsids), and, what is most essential, the distribution of thickets of strongly flooded habitats (including those of the mangrove type). Thirdly, the tracing of fossil taxa relies, to a greater extent than the tracing of the living taxa, on indirect characters, such as the distribution of sea and land areas, the distribution of a given sedimentary formation, or simply of a definite plant-bearing bed, etc. The palaeofloristic map is thereby based on a wide complex of ecosystemic features, some of which cannot be easily discarded in characteristics of the phytochoria.

Because of this it is not fortuitous that the palaeofloristic zonality maps, as well as the characteristics of the phytochoria, are loaded with extraneous characters, primarily palaeoclimatic details. Often these maps show the distribution of arid belts and glaciations. At times palaeoclimatic terms are introduced into the phytochorial nomenclature. The phytochoria acquire the nature of 'vegetation divisions', reflecting certain generalized types of palaeoclimates.

It stands to reason that palaeofloristic zonality maps cannot be treated as palaeoclimatic maps, considering that the boundaries between the phytochoria may not be associated with the climatic divides, but rather with the hindrances along the migration routes. Remote phytochoria with similar climatic characteristics may be composed of different taxa sets, having been formed from different parent floras. Moreover, some boundaries on the maps may result from secondary tectonic connections or isolation of territories (if we admit the mobilistic concept). As a consequence, the palaeofloristic maps, no matter what geobotanical and palaeoclimatic features are used for their compilation, can serve only as an important component of palaeoclimatic maps, but cannot substitute for them.

Many palaeobotanists have been disconcerted by the fact that the term flora has been used in reference to the geological past, although assemblages of plant remains and corresponding taxa lists cannot be regarded as floras of past times. Essential corrections should be made on the incompleteness of the data, mixture of the remains in deposition (resulting also from redeposition), lack of correspondence between the parataxa and eutaxa. Owing to this the terms taphoflora, stratoflora, geoflora, etc. have been introduced. Not contending the need of these terms, which may be desirable in discussing certain specific problems, we consider it more reasonable to retain the general term flora, for the sake of brevity, as well as for the unity of the ensuing historic picture. In describing the history of the plant world we attempt as much as possible to divert our attention from the defects in the geological record, from the detached state of the remains, and, as a matter of fact, to outline a certain reconstructed distribution pattern, namely of floras of the geological past. It is another matter that in the process of reconstruction we need to delimit strictly the observed plant assemblages of fossil remains from the conjectured composition of the parent floras or vegetation types. In descriptions of reconstructed history only the term geoflora may be preferable, since it implies a long existing flora within a vast territory that corresponds to a phytochoria of high rank (kingdom, area).

Palaeofloristic investigations culminate in the reconstruction of the florogeny. The generalized branch in botany dealing with this can be termed florogenetics. In the same way as phylogenetics aims to reconstruct the phylogeny of organisms and for this purpose all available data are incorporated, florogenetics aims to reconstruct the origin of floras. The historical continuity of the floras may be represented in the form of a florogenetic tree (Meyen, 1969a; Chaloner and Meyen, 1973) of which the proposed version (Fig. 83) will be discussed at the end of this chapter.

For the determination of the origin of the floras it is necessary to establish the origin and distribution pattern of the floras comprising taxa. Hence, florogenetics involves plant phylogeny with all its constituents and plots of the phylogenetic lineages projected on to the palaeogeographical and palaeofloristic background. The plant dispersal can be obtained only from knowledge of the aut- and synecology of the taxa. Therefore florogenetics also includes plant palaeoecology. In this case it is of particular importance that florogenetics does not merely use the data provided by systematics, phylogenetics palaeofloristics and palaeoecology of plants, but synthesizes all the above-cited disciplines. Hence, florogenetics is a much wider discipline that has been termed the historical geography of plants (Wulff, 1943), phytochorionomy (Takhtajan, 1978), phylogenetic geography of plants (Engler, 1899, quoted from Takhtajan, 1978), etc.

In describing the history of floras today it is impossible to leave unheeded the mobilistic concept. It is evident that the positional relations between the continents were different in past geological times. Unfortunately the available mobilistic reassemblies of the continents are convincing only beginning from the Jurassic. For the Palaeozoic valid reassemblies are available only for Gondwana (although many uncertainties occur here). The positions of Cathaysia and Kazakhstan, the palaeogeographic interrelations between the north-east USSR and north-west of North America, the palaeogeographical situation of Middle Asia, etc. are still much debated. It is difficult to assess the validity of various reassemblies with the aid of Palaeozoic floras. Because of this in discussing the history of the Palaeozoic and Triassic floras only in several instances is mention made of the positional relations between the continents.

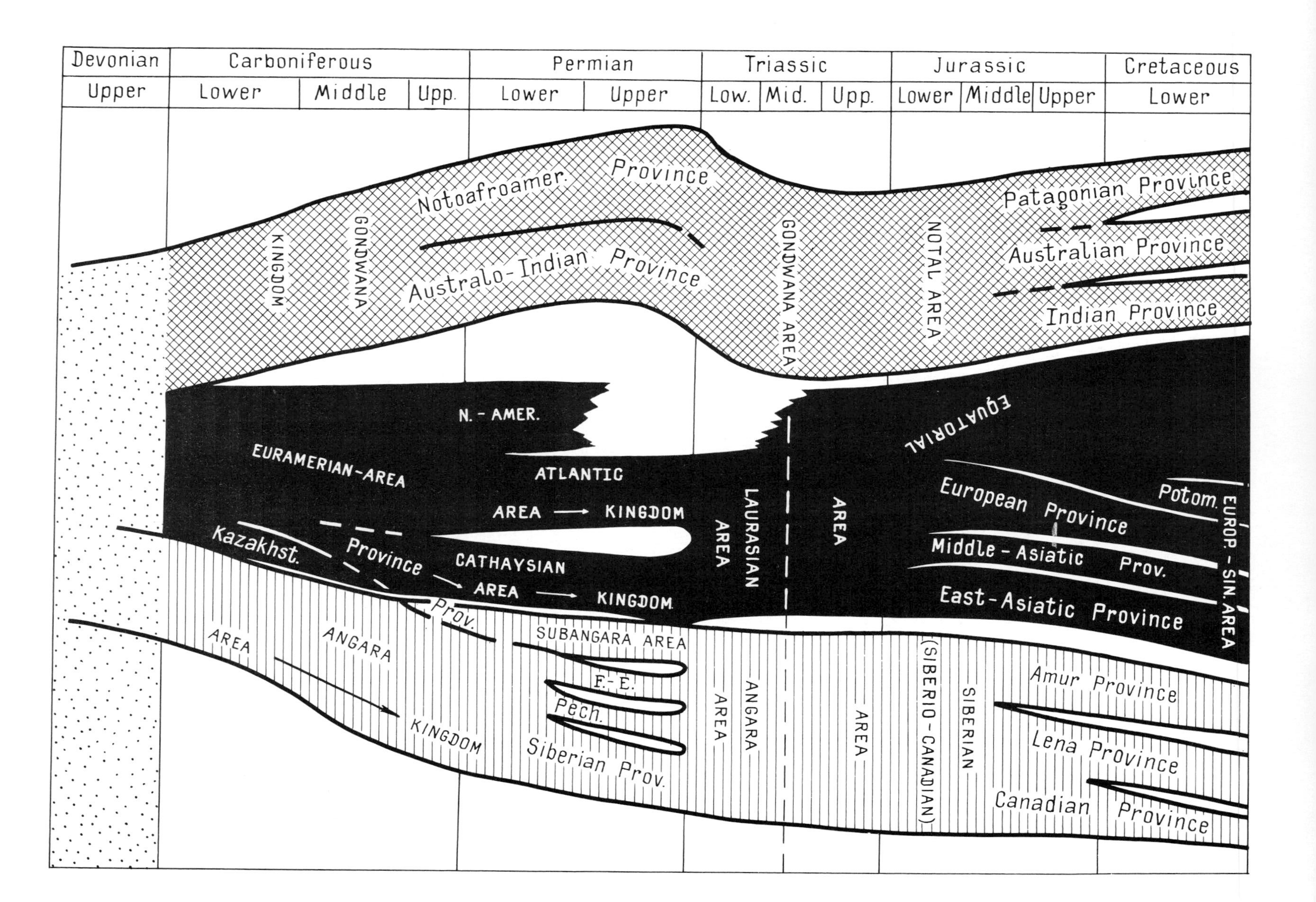
Devonian
Upper
Carboniferous
Lower
Middle
Upp.
Permian
Lower
Upper
Triassic
Low.
Mid.
Upp.
Jurassic
Lower
Middle
Upper
Cretaceous
Lower
Notoafroamer. Province
Australo-Indian Province
GONDWANA
KINGDOM
GONDWANA AREA
NOTAL AREA
Patagonian Province
Australian Province
Indian Province
N.-AMER.
EURAMERIAN-AREA
ATLANTIC
AREA → KINGDOM
Kazakhst.
Province
CATHAYSIAN
AREA → KINGDOM
Prov.
LAURASIAN
AREA
AREA
EQUATORIAL
European Province
Potom.
Middle-Asiatic Prov.
East-Asiatic Province
EUROP.-SIN. AREA
AREA
ANGARA
KINGDOM
SUBANGARA AREA
F.-E.
Pech.
Siberian Prov.
ANGARA
AREA
AREA
(SIBERIO-CANADIAN)
SIBERIAN
Amur Province
Lena Province
Canadian Province

HISTORY OF FLORAS

In this section an outline is given only of floras composed of higher plants. The Pre-Cambrian biota of algae and acritarchs was briefly described in the section on systematics (for details see J. W. Schopf, 1983). Algal assemblages of the Phanerozoic can be termed floras (algofloras), but an outline of their history involves a wide array of material more closely related to the palaeozoology of marine invertebrates, rather than to palaeobotany. In particular, attention should be turned to reef-building algae and, consequently, to the biogeography of reef-building organisms in general. The distribution of planktonic algae (nannoplankton, diatoms, etc.) can be shown only against the background of the biogeography of seas and oceans. Because of this only a very brief outline of the biogeography and stratigraphic distribution of algae is given in the discussion on their systematics (for details see Tappan, 1980).

No illustrations and characteristics of the taxa, with rare exceptions, are presented in this chapter, since the more important taxa have already been discussed in Chapter 3, where any particular taxon can easily be found by using the index.

In discussing the history of land vegetation it seems convenient to proceed on the basis of natural periods. Such periods can be distinctly recognized in the history of regions taken separately. On a global scale the situation becomes more complicated. Easily recognizable are very large periods, that are suitable only for the organization of a portion of the outlined material, namely the early (pre-Carboniferous) history of terrestrial vegetation, prior to the formation of distinct phytochoria. It should be noted that the systemic boundaries of the standard geochronological scale at times only artificially divide the history of individual phytochoria.

Kryshtofovich (1957) recognized seven great floras, successively replacing each other on the Earth, and corresponding to eras in the evolution of plants. The pre-Devonian floras he termed phycomycophytic, and the corresponding era he termed Thalassophytic. These were followed by psilophytic (Early–Middle Devonian) and anthracophytic (Upper Devonian–Early Permian) floras, belonging to the Palaeophytic era. The next Mesophytic era is composed of palaeomesophytic (Late Permian–Triassic) and neomesophytic (Jurassic–Early Cretaceous) floras. The Cenophytic era is composed of palaeocenophytic (Late Cretaceous–Neogene) and neocenophytic (Quaternary) floras. Other schemes for the periodization of the vegetation cover have been suggested (Vakhrameev *et al.*, 1970, 1978). Regardless of how natural such planetary floras and eras may seem, we consider that the terms palaeophyte, mesophyte and cenophyte are convenient in application to the terrestrial plant world and hereafter they are used in our descriptions.

Some of the forthcoming floras are composed of hundreds of genera and species, which exhibit a sophisticated distribution in time and space. Because of this more attention will be paid to those taxa that determine the aspect of the flora or are of particular interest from the point of view of links between the major floras. Detailed characteristics of the phytochoria can be found in the literature (J. M. Anderson and H. M. Anderson, 1985; Archangelsky, 1970; Beck, 1976b; Dilcher and Taylor, 1980; Gu and Zhi, 1974; Hughes, 1973; Meyen, 1982b; Paleopalinologia, 1966; Plumstead, 1962; W. Remy and R. Remy, 1977; Tschudy and Scott, 1969; Vakhrameev *et al.*, 1970, 1978), and in the references cited in the text.

Fig. 83. Evolution of phytochoria (florogeny) during Late Palaeozoic and Mesozoic; N.-Amer., North American area; Pech., Pechora province; F.-E., Far Eastern province; Euramerian and Angara areas comprise Arctocarbonic kingdom; North American, Atlantic and Cathaysian areas of Early Permian comprise Amerosinian kingdom; Far Eastern, Pechora and Siberian provinces enter into Angara area which, together with Subangara area, constitutes Angara kingdom; Potomac, European, Middle Asiatic and East Adiatic provinces of Jurassic and Cretaceous are united into European–Sinian (Europ.-Sin.) area. Compiled by S. V. Meyen and V. A. Vakhrameev.

THE RISE OF LAND VEGETATION

The population of land by organisms of any kind in the Pre-Cambrian and Early Palaeozoic, up to the Silurian, inclusive, remains a matter of speculation. Hypotheses have been offered, postulating that the Pre-Cambrian and Lower Palaeozoic land was inhabited by algae and lichens. This hypothesis has found a substantial support in the finding of animal burrows in a late Ordovician terrestrial palaeosol (Retallack, 1985). The presence of the burrows suggests a plant cover sustaining the soil fauna. Retallack also noted that no traces of underground parts are observed in the Ordovician palaeosol. The absence of a dense and well-developed vegetation cover should have had an impact on the weathering conditions, erosion and sedimentation processes on land, since there were no factors regulating the surface run-off. The relief would have been subjected to rapid peneplanation, the landscape being characterized by vast shallow water basins of unstable configurations, that were populated first by procaryotes, and then by eucaryotes. Thereby, the clastic material (excluding pelitic) from the depths of continents did not reach the marine basins, and, hence, it would have been supplied chiefly by coastal abrasion. Accordingly, the thicknesses of terrigenous marine sediments, accumulating in unit time, would have been immeasurably smaller than during periods of developed terrestrial vegetation. An interesting hypothesis was put forward by Ponomarenko (1984), according to which the appearance of propteridophytic terrestrial vegetation would have facilitated the formation of intracontinental and coastal settling basins, evoking the expansion of the euphotic zone in marine basins and amplification of the oxygen supply to the deep water layers.

Finds of microfossils (palynomorphs), externally similar to spores of higher plants, and confined to the Pre-Cambrian and Lower Palaeozoic, gave rise to hypotheses on land colonization long before the appearance of indisputable plant remains in the geological records. From time to time macrofossils, resembling higher plants (Fig. 84, *a–e*) have been recovered from

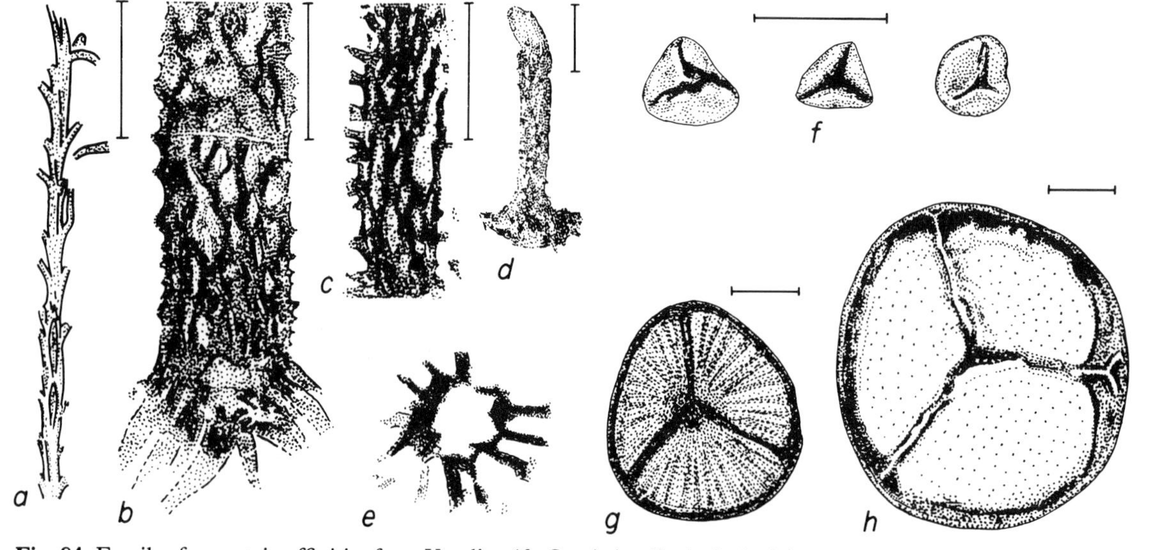

Fig. 84. Fossils of uncertain affinities from Vendian (*f*), Cambrian (*b–e*), Ordovician (*a*), Silurian (*h*) and Downtonian (*g*); Western Europe (*a*), Aldan River basin (*b–e*), Leningrad area (*f*), Libya (*g*, *h*); *a*, *Boiophyton pragense* Obrh.; *b–e*, *Aldanophyton antiquissimum* Krysht.; *f*, *Ambiguaspora parvula* Volk.; *g*, *Emphanisporites rotatus* McGreg.; *h*, *Ambitisporites avitus* Hoffm. Scale bar = 1 cm (*a–d*), 10 μm (*f–h*). Modified from: Obrhel, 1959 (*a*); Ananiev in Osnovy paleontologii, 1963a (*d*, *e*); Volkova, 1976 (*f*); Richardson and Ioannides, 1973 (*g*, *h*).

deposits beginning from the Upper Pre-Cambrian. Among these the best known is the genus *Aldanophyton* in the Cambrian of Siberia (Fig. 84, *b–e*). They are spiny hollow axes with expanded lower parts, bearing strap-shaped rhizoid-like appendages. Despite their good preservation, no anatomical structures are observed. Owing to the restriction of occurrence of these remains to bituminous limestones, that were deposited far from the shoreline, serious obstacles arise concerning their attribution to higher land plants.

In the uppermost Pre-Cambrian very tiny spores with a distinct trilete fold have been found (Fig. 84, *f*), but so far there is no evidence to witness that they belonged to higher plants. It is probable that they might more readily belong to problematical algae (Vendotaenides), occurring in the same beds. Pre-Cambrian, Cambrian and Lower Ordovician spores with indisputable trilete slits have not as yet been found. Notifications of such findings were regarded as doubtful or have proved to be erroneous. Crush folds of the walls were mistaken for the trilete slit. All these fossils are now being circumscribed among acritarchs.

Chaloner (1970b) formulated criteria for the assignment of the oldest remains to higher plants. They involve: spores with a maceration-resistant wall and a slit of dehiscence, tracheids and a cuticle with stomata. If, proceeding on this basis, we admit that the complete set of characters cannot be observed in any one plant-remains, then the incoming of higher plants can be inferred from the stratigraphic level, where at least various plants exhibit the complete set of characters.

The first spores with a typical dehiscence slit and a maceration-resistant wall (genus *Ambitisporites* in the Llandoverian) are of first appearance at the top of the Ordovician (Fig. 84, *h*; Fig. 85). The spores (Gray, 1985) are characterized by their arrangement in diads and tetrads, occasionally encased in a common envelope. They may be algal cysts. The number of trilete spores increases considerably in the Ludlow, where they are restricted chiefly to coastal sediments (Aldridge *et al.*, 1979). No

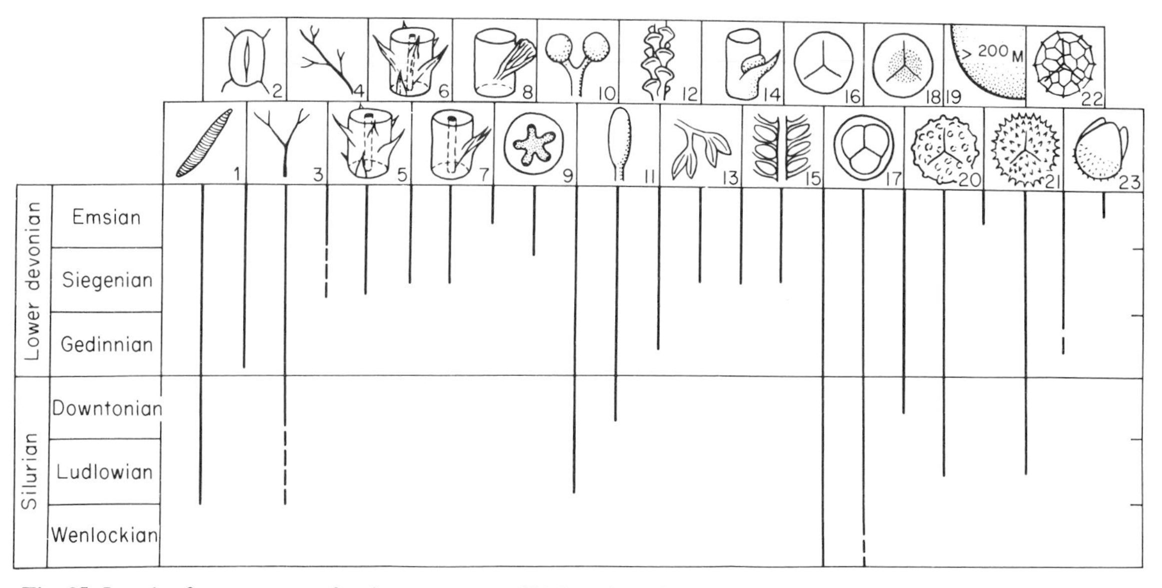

Fig. 85. Levels of appearances of major structures of higher plants in geochronological scale: 1, tracheids; 2, stomata; 3, dichotomizing axes; 4, shoots with main axes; 5, non-vascularized appendages (emergences); 6, vascularized appendages; 7, forked leaf-like organs; 8, planated leaves; 9, actinostele; 10, globose terminal sporangia; 11, elliptical sporangia; 12, sporangial aggregations; 13, longitudinally dehisced sporangia; 14, axillary sporangia; 15, sterile appendages in sporangial aggregations; 16, smooth trilete spores; 17–23, spores of different ornamentation and sizes. Modified from Chaloner and Sheerin, 1979.

orderly geographical distribution of these fossils has been noted, although associated acritarchs reveal a definite provincial restriction (Richardson *et al.*, 1981). Spores in sporangia and axes with tracheids are known from the Ludlow. The most ancient axes yielding cuticles with stomata come from the lowermost Devonian (Gedinnian). As a matter of fact, not only spores have been encountered throughout the entire Silurian sequence, but also dispersed fragments of cuticles with a cellular structure (but lacking stomata); also tubular microfossils with spiral wall thickenings (Al-Ameri, 1983–1984; Banks, 1975; Gray and Boucot, 1977; Pratt *et al.*, 1978). These cuticular fragments and tubes have been interpreted as higher plant parts, but, more likely, they belong to algae, allied to *Nematothallus*, *Nematoplexus* and *Prototaxites* (Edwards and Rose, 1984; Niklas and Smocovitis, 1983), and, probably, to spongiophytes. Among the Silurian dispersed fragments Al-Ameri (1983–1984) noted tissues that resembled higher plant remains. These remains tend to increase in number in the direction of the shoreline. The systematic affinities of the Silurian and oldest Devonian spores (Fig. 84, *g*, *h*) are uncertain. In their structure they resemble spores from sporangia of the Lower Devonian propteridophytes (spores of *Ambitisporites* have been found in sporangia of *Cooksonia pertonii*). However, some of their types are known in sporangia, not only of propteridophytes, but also of algae; other types have not as yet been encountered *in situ* (Fig. 84, *g*).

The appearance of higher plant megafossils in the Wenlockian of the Silurian, also their increase in number in the Přidolian and further to the base of the Devonian is concomitant with an increase in the diversity of dispersed spores. Both these phenomena are commonly regarded as reflections of the advent of higher plants, as well as their parallel invasion of the land (Chaloner and Lawson, 1985). However, it seems much more probable that the oldest higher plant remains belong to aquatic and semiaquatic, rather than to teɩrestrial forms. They were small herbaceous plants and from the subsequent history of the vegetation cover it is evident that herbaceous land plants (never submerged in water) have never been encountered incorporated in sediment as megafossils. Nearly all the known herbaceous fossil plants were aquatic, boggy or epiphytic. All the Lower and Middle Devonian palaeosols containing underground plant parts, are characteristically subaquatic and hydromorphic types, i.e. they either underwent inundation by water, or were of constant high moisture. From comparison of the taphonomic and morphological–anatomical observations (absence of stomata in the lower portions of *Zosterophyllum* shoots; trabecular structure and wide intercellular spaces in the *Asteroxylon* cortex, etc.) it can be inferred that the appearance and increase in the number and diversity of higher plant megafossils in the uppermost Silurian–lowermost Devonian (Fig. 85) reflect, not so much the advent of terrestrial plants, as their transition from land to water (Meyen, 1978e, 1981)). The rise of higher plants should be assigned to earlier Silurian epochs, and the problem concerning the environment where this process developed remains open. One cannot exclude that it might have occurred not during the invasion of the land by algae, but within the terrestrial algal population. The spores encountered in the lower parts of the Silurian probably belonged to these algae, adapted to land conditions, or to the earliest higher plant types that evolved from them ('rhyniophytoids' – Pratt *et al.*, 1978). It is not fortuitous that the number of these spores increases towards the shoreline. In the beginning of the Devonian, plants began to populate land areas further distant from the coast, and already during the Gedinnian higher plant megafossils (*Cooksonia*) underwent burial in typical alluvial deposits.

SILURIAN–DEVONIAN FLORAS

On the basis of the distribution of megafossils in stratigraphic sequences Banks (1980a) proposed subdividing the interval from the Přidolian to the Famennian into seven floristic zones. Each zone is denoted by Roman numerals and names of the characteristic genera, which do not necessarily correspond in range to the entire zone. Since no clear-cut phytochoria can be identified in the

Devonian, the characteristics of these zones are given disregarding localities or their groups. For the Devonian McGregor (1979) and Richardson (1974) established regional and global palynological zones that do not match the zones of megafossils. More detailed palaeobotanical characteristics of different regions are available in the works of Ananiev (1959), Andreeva *et al.* (1962), Banks (1980a), Edwards (1973, 1980), Iurina (1969), Li and Cai (1978), Petrosyan (1968) and Stepanov (1975).

Zone I (*Cooksonia*)

According to the latest finds (Edwards and Fanning, 1985; Edwards *et al.*, 1983), this seemingly begins in the uppermost Wenlockian and continues further into the Ludlowian and Downtonian (Přidolian). Floras of Zone I are known only from a few localities (Podolia, Czechoslovakia, England, Ireland, New York State). Higher plants (*Cooksonia*), represented by tiny specimens with slender axes (up to 3 mm in width), associate with algae *Parka*, *Pachytheca* and *Prototaxites*, also with aquatic animals. This is probably a marshland assemblage. In Podolia thalloid forms (*Praehepaticites*) are known, that probably belong to algae or propteridophytic gametophytes.

The Zone I may also comprise the plant assemblages of Australia containing *Baragwanathia*, *Salopella*, zosterophyllophytes and other plants (Garratt *et al.*, 1984; Tims and Chambers, 1984). The Silurian (Wenlockian–Ludlowian) age of the Yea Formation yielding these plants has been much debated in the literature, but the latest data on the Silurian age of graptolites coming from this formation (Garratt and Rickards, 1984) appear more convincing than data previously published. Accordingly the stratigraphic ranges of relevant taxa as shown in Figs. 11 and 16 probably need extension down to the Silurian.

Zone II (*Zosterophyllum*), Gedinnian (*Dittonian*) – Lower Siegenian

This flora is known in the same areas as that of Zone I (also in Kazakhstan and Kuzbass), and is similar in composition and overall habit, though different in the appearance and wide occurrence of *Zosterophyllum* and in places of *Sporogonites*. In the upper part of the Zone *Parka* disappears. The Gedinnian of Kuzbass (Stepanov, 1975) yielded peculiar genera *Stolophyton* (axes with lateral elliptical stalked sporangia), *Juliphyton* (this genus may probably be allied to *Zosterophyllum*) and *Uscunajphyton* (a forked axis bearing on the outer upper side of both branches upward flexed appendages with solitary sporangia).

Zone III (*Psilophyton*), Upper Siegenian–Emsian

This flora is implied in speaking of the Lower Devonian flora, which is sometimes termed 'psilophytic'. This flora and those of the former zones we will term hereafter 'propteridophytic'. In Zone III about 30 genera have been identified (the richest localities yield up to 15–20 genera), that belong to propteridophytes, the earliest lycopsids (*Drepanophycus*, *Protolepidodendron*, *Baragwanathia*) including ligulate (*Leclercqia*), barinophytes (*Barinophyton*, *Protobarinophyton*). Propteridophytes are represented by all three classes, including trimerophytes of first appearance (*Psilophyton*, *Trimerophyton*, *Pertica*). Affiliated to the Rhyniales are *Cooksonia*, and, probably, some *Taeniocrada*, and to Zosterophyllales are *Zosterophyllum*, *Gosslingia*, *Renalia*, *Rebuchia*, *Sawdonia* and *Crenaticaulis*. Floras of Zone III are known from Siberia, Kazakhstan, Donbass, West Europe, North America, China, South America, Australia. Hueber (1983) believes that the floras containing *Baragwanathia* in Australia, which for long have been regarded as Silurian, may also belong to Zone III. *Baragwanathia* has been found by him in the Emsian of Canada.

Attempts were made to distinguish palaeofloristic provinces in the Early Devonian. It has been suggested, for example, that the Kazakhstan and Siberian floras should be distinguished as independent provinces. The specific features of Gondwana floras were stressed. Although the probable existence of Early Devonian phytochoria cannot be denied, still they cannot be fully established and mapped either by megafossil, or by miospore data. Many Early Devonian megafossil genera are known only from one locality. On the other hand, some genera are of wide distribution (*Zosterophyllum*, *Sawdonia*, *Drepanophycus*, *Taeniocrada*). So far practically no orderly changes in space can be observed in the generic composition, even more so in the suprageneric groups. It has been noted that in some places (West Germany, Kazakhstan, South America) the role of lycopsids increases. Zakharova (1985) reported the absence of *Psilophyton* and related trimerophytes in Siberia, whereas *Margophyton*, which is of wide occurrence in Eurasia, is absent in North America. The impossibility of defining the Early Devonian phytochoria is evidently due to subjective reasons (poorly studied plant remains in the majority of the localities), as well as to objective factors. All or nearly all known Early Devonian plants were annual or had annual aerial shoots. Therefore their distribution might have been little dependent upon the climatic zonality, especially if it was not very contrasting. Besides, many (if not the majority) Lower Devonian genera were aquatic or semiaquatic, which was also favourable for their distribution in different climatic belts. According to Schweitzer's (1983b) reconstructions, the Early Devonian landscapes (Rhine District) appeared as vast marshlands with very weakly dissected topography. These areas intermittently underwent inundation by the sea, which resulted in the formation of multiple biotopes, intermediate between marine and continental types. Many Early Devonian plants (*Sawdonia*, *Renalia*, *Zosterophyllum*, etc.) formed pure stands with smaller interposed clumps of other plant taxa (Gensel and Andrews, 1984).

Zones IV (*Hyenia*) and V (*Svalbardia*)

These encompass the entire Middle Devonian. The boundary between them passes somewhat higher than the Eifelian–Givetian boundary. The Middle Devonian flora is termed *Hyenia* or *Protopteridium* flora, after the characteristic genera (*Protopteridium* = *Rellimia*). The incoming of these genera does actually mark a new stage in the evolution of higher plants. *Hyenia* evidently has the same anatomical structure as *Calamophyton*, that appears in the Middle Eifelian and is affiliated to cladoxylalean ferns. The typical cladoxylalean anatomical structure with numerous xylem bundles running along the stem, appears in the uppermost part of Zone III. *Rellimia* is one of the oldest still quite primitive representatives of progymnosperms. Supporting evidence to the appearance of *Rellimia* at the base of the Middle Devonian is provided by the simultaneous appearance of spores *Rhabdosporites langii*, associated with this genus.

In the Middle Devonian flora, including over 50 genera, many genera are retained of propteridophytic floras, belonging to the Zosterophyllales (*Sawdonia*, *Hicklingia*) and Trimerophytales (*Pertica*, *Psilophyton*). A remarkable genus is *Chaleuria* with incipient heterospory. Its relationships with trimerophytes and progymnosperms are unclear. Barinophytes continue in this zone. Besides the above-mentioned *Hyenia* and *Calamophyton*, ferns are represented by the genera *Pseudosporochnus*, *Langoxylon* and probably *Protocephalopteris*. Apart from *Rellimia*, the genera *Aneurophyton*, *Tetraxylopteris* and *Triloboxylon* belong to protopteridialean progymnosperms. The boundary between Zones IV and V is defined by the appearance of the archaeopteridalean progymnosperms *Svalbardia*. In the Middle Devonian the first dispersed megaspores have been found, that might have belonged either to progymnosperms or to lycopsids. The latter are of wide distribution in both Middle Devonian zones. They are represented by *Drepanophycus*, *Protolepidodendron* and *Leclercqia*, that are continuous from the Lower Devonian, and by *Colpodexylon* (Eifelian), *Archaeosigillaria* and *Lepidodendropsis* (since the Givetian), which are of first appearance here. The Middle Devonian lycopsids are represented by protostelic forms with an exarch or slightly mesarch xylem. Their stele is in the shape of a denticulate wheel in cross-section. The reports on the Middle Devonian *Lycopodites*, comparable to the living *Lycopodium*, need to be checked. Of first appearance in the Givetian are Ibykales (*Ibyka*, *Iridopteris*) – the probable ancestors of articulates. The genus *Honseleria* from the Givetian of West Germany is thought to belong to articulates, but, most likely, it is a member of the Ibykales. Among the Middle Devonian plants of uncertain systematic affinities very typical is the genus *Barrandeina*. The affinity of the heterosporous genus *Enigmophyton*, producing fan-shaped leaves with numerous bifurcating veins, also remains uncertain. Many localities of Zone V yield spongiophytaceous and algae (brown?) closely related to them, belonging to the genera *Protosalvinia*, *Spongiophyton*, *Orestovia*, *Bitelaria*. In Kuzbass compressions of *Orestovia* together with certain other so far poorly studied genera form coal seams (sapromyxite or barsassite coals). The habitats of these coal-forming communities are as yet unknown. Thin coal seams, composed of *Orestovia*, *Bitelaria* and related plants are known in the Voronezh region.

The Middle Devonian localities are distributed in the same regions as the propteridophytic flora, but most of them are of Givetian age. Up to 30 species have been identified in the richest localities. Eifelian flora is known from a few localities and is similar in aspect to the propteridophytic flora. References to Middle Devonian floras in the literature imply first of all Givetian floras of Europe, Siberia, Kazakhstan and North America. Of common occurrence in most localities are remains of *Rellimia*, *Pseudosporochnus* (stems of some specimens attain 20 cm in diameter) and *Protolepidodendron*; however, only in a few localities have they been validly identified. In some localities (Central Kazakhstan, Spitsbergen, North Africa, New York State, China) lycopsids with rather thick stems (up to 8 cm in diameter) have a paramount role. In some of them from Kazakhstan, morphologically resembling *Tomiodendron*, Senkevich (1984) found a large ligular pit in the leaf axil. In the Middle Devonian of Kazakhstan of common occurrence are lepidophytes that have been referred to *Lepidodendropsis* according to the leaf cushion structure and configuration. The stem anatomy has been studied in detail. Iurina (1985) has shown that these lepidophytes produced *Flemingites*-like strobili. It is still unclear whether the localities, rich in lepidophytes in the aforementioned regions, can be united into an independent phytochorion, or whether the occurrence here of lepidophytes with rather thick stems was controlled by local facies conditions. It has been noted that the Middle Devonian Gondwana floras (poorer in composition) differed from coeval floras of northern continents. *Hyenia* and *Calamophyton*, of common occurrence in Europe, so far have not been found in China. No clearly defined phytochoria can be distinguished according to miospores; there are literally cosmopolitan genera and even species (*Rhabdosporites langii*, *Grandispora naumovii*, etc.), but some taxa are of restricted geographical distribution (McGregor, 1979).

Zone VI (*Archaeopteris*), Frasnian–Famennian

The flora of Zones VI and VII is known in the literature as the *Archaeopteris*-flora. Following Banks, this term is used only in reference to the floras of Zone VI. The *Archaeopteris* flora is of wide occurrence, the same as the *Protopteridium* flora, and comparable in its diversity (over 50 genera). The genus *Archaeopteris* proper is of wide occurrence throughout all the Northern Hemisphere and Australia. This, coupled with the fact that *Archaeopteris* is a dominant plant, emphasizes the cosmopolitan nature of the Upper Devonian flora.

The incoming of *Archaeopteris* in the beginning of the Late Devonian played a paramount role in the evolution of vegetation cover. They were real trees with thick trunks (up to 1.5 m in diameter), composed of secondary wood. According to observations in the Upper Devonian of the USA, *Archaeopteris* trees formed forests along the banks of rivers that drained into an inland sea. Judging by the abundant wood fragments of *Callixylon* in many localities these forests covered vast areas during the Late Devonian. Trees that had trunks with wood of the *Dadoxylon* and *Palaeoxylon* types also existed during the Late Devonian. Continuous from the previous zones were the protopteridialean progymnosperms *Aneurophyton*. It is probable that some of the thicker trunks (up to a metre in diameter) belonged to the latter (*Eospermatopteris*).

In the Upper Devonian we still find a number of propteridophytes (*Taeniocrada*, *Sawdonia*, *Thursophyton*) and barinophytes; in addition to these, cladoxylaleans (*Pseudosporochnus*, etc.) occur, also genera of uncertain affinities (*Tortkophyton*, *Platyphyllum*) and algae *Protosalvinia* and *Prototaxites*. In many Upper Devonian localities lycopsids play an important role. Among them are surviving primitive types (*Drepanophycus*), genera that appeared in the Middle Devonian (*Archaeosigillaria*, *Lepidosigillaria*, *Colpodexylon*, *Lepidodendropsis*, *Lycopodites*) and new genera (*Protolepidodendropsis*, *Leptophloeum*, *Pseudolepidodendropsis*). Some lycopsids had thick trunks (in *Lepidiosigillaria* – up to 40 cm in diameter).

With regard to the phytochoria of Zone VI the same can be said as was stated above. Practically no distinct geographical distribution tendencies can be established. The genus *Leptophloeum* is absent in Europe and Siberia, being very common in Gondwana, China, Japan, Kazakhstan and North America, where it occurs in assemblages with dominant lycopsids (the latter are very scarce in Siberia, and, contrariwise, are abundant in the Equatorial floras and in Gondwana). In the previous zones certain differences were noted between the Gondwana and Northern localities; in Zone VI this cannot be perceived, probably because the Gondwana floras have been poorly studied. Some genera are of very wide distribution, whereas others reveal a geographical restriction. For instance, megaspores of the *Nikitinsporites* type and corresponding to them monolete microspores *Archaeoperisaccus*, occur only northward of the conjectured Frasnian Palaeoequator.

Zone VII (*Rhacophyton–Cyclostigma*)

This corresponds to the Upper Famennian and probably in places to the lowermost Tournaisian. Correlations between plant-bearing beds and marine sequences are at times not amply warranted. It is necessary to take into account that in the Devonian–Carboniferous boundary beds the aspect of the plant assemblages may depend upon the facies environment, controlling the proportions of plants more typical of the Devonian or Carboniferous. In the USSR the systematic boundary was long taken at the base of the *Wocklumeria* zone, whereas in other countries it was taken at the base of the *Gattendorfia* zone. At present most of the stratigraphers are inclined to place the boundary along the conodont zone of *Siphonodella sulcata*, which from the palaeobotanical viewpoint differs only slightly from the former level at the base of the *Gattendorfia* zone. This level has been taken in the present work for the boundary in defining the age of plants and floral assemblages. As a result of the elevation of this boundary in the USSR to this level, certain floras that traditionally were regarded as Lower Carboniferous (Meyen, 1982b), should now be regarded as Upper Devonian. Accordingly, the phytogeographical situation for the Upper Devonian, in the whole, should be differently treated.

Zone VII, when established by Banks, was identified only by the genus *Rhacophyton*, to which now can be added the widespread genus *Cyclostigma*. The flora of Zone VII is more impoverished compared with that of the preceding zone (less than 30 genera in reliably dated localities). In Western Europe, on the Bear Island and in the USA the assemblages are dominated by lepidophytes (*Cyclostigma*, *Protolepidodendropsis*, *Archaeosigillaria*), progymnosperms (*Archaeopteris*, *Callixylon*) and fern-like plants (*Rhacophyton*, *Cladoxylon*, *Cephalopteris*). Indisputable articulates first appear (*Pseudobornia*, *Sphenophyllum* and *Eviostachya*), also the first gymnosperms represented by ovules, either detached, or encased in cupules (*Archaeosperma*, *Spermolithus*, etc.). Plants of uncertain affinities include *Sphenopteris*, *Sphenopteridium*, *Platyphyllum*, *Dadoxylon*. According to Scheckler (1982b) the genus *Rhacophyton* could have undergone adaptation to paludal environments in the coastal lowlands and river valleys. Accumulations of its remains comprised thin coal seams. In coeval peat bogs it is probable that no other plants existed. Lycopsids penetrated into the bogs at a later period.

Zone VII seemingly marks the beginning of the phytogeographical differentiation of the Late Palaeozoic type. Most likely the flora of the Bystryanskaya suite of the Minusa Basin belongs to this zone. In this suite *Pseudolepidodendropsis*, *Cyclostigma* and *Sphenophyllum* were noted. To *Pseudolepidodendropsis* have been assigned lepidophytes that belong to another, probably endemic genus (Meyen, 1982b). A single specimen of *Archaeopteris* had also been recorded in this suite. Considering the abundance of *Archaeopteris* in the underlying Tubinskaya suite (Upper Devonian) there is reason to believe that the paucity of this genus in the Bystryanskaya flora, as well as the extremely limited amount of taxa in the latter (notwithstanding the large amount of plant remains) is not fortuitous. From this time onwards, up to the beginning of the Middle Carboniferous, plants with thick pycnoxylic stems were lacking in Siberia. Their disappearance may have been associated with the climate becoming drier. The isolation in Siberia of the independent Angara area is clearly reflected in the Tournaisian flora that replaced the Bystryanskaya.

In order to define the increasing floristic differentiation more profound attention should be given to the palynological data. Peculiar spores *Retispora lepidophyta* with a perforated flange (Fig. 20, *z*; Fig. 86) probably belonging to cyclostigmas, are of wide occurrence practically all over the world in the uppermost Devonian. The palynologists maintain that the miospore assemblages at the end of the Devonian display significant generic unity throughout Europe, Kazakhstan, Siberia, North America and Australia. At the same time, comparison of the palynological assemblages of the lower part of the Tournaisian shows that their species and, in part, their generic composition vary quite considerably from one region to another. The appearance of spore assemblage with *Retispora* is

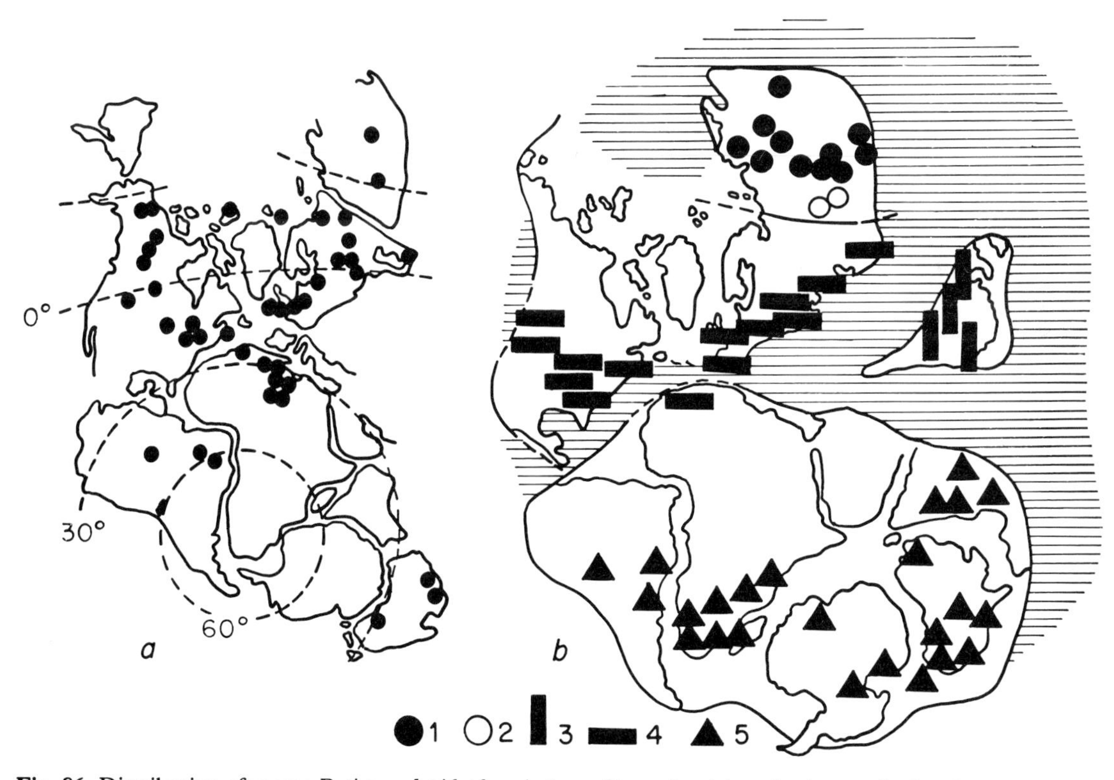

Fig. 86. Distribution of spores *Retispora lepidophyta* in latest Devonian (*a*), and scheme of palaeofloristic zonation of late Carboniferous–early Permian (*b*); 1–5, localities of: Angara area (1), Kazakhstan province (2), Cathaysian (3) and Euramerian (4) areas, Gondwana kingdom (5). Modified from: McGregor, 1979 (*a*); Novik and Fissunenko, 1979 (*b*).

probably due to the warming and stabilization of the climate. In South America this assemblage has been found in beds overlying marine-glacial deposits. On the other hand, the regionalization of the miospore assemblages at the beginning of the Tournaisian witnesses the increasing climatic differentiation.

CARBONIFEROUS AND PERMIAN FLORAS

Up until Zone VII, inclusive, it has been possible to discuss the evolution of the Earth's vegetational cover in its entirety. For younger periods, individual phytochoria should be treated separately. The number of phytochoria and the differences between them increase from the Early Carboniferous to the end of the Permian (Fig. 83). In the Early Carboniferous the Angara and Euramerian areas may already be regarded as independent. Judging by palynological assemblages, the same time marks the separation of the Gondwana area (Clayton, 1985), where the Tournaisian and in part Visean miospores belong to the endemic *Granulatisporites frustulentus* mioflora (Australia). Beginning from the Middle Carboniferous the Cathaysian area gradually separates from the Euramerian. During the Carboniferous and Permian the phytochoria gradually increase both in numbers and rank, up to and including kingdom. The division of the vegetation cover into phytochoria is closely associated with the climate. The Euramerian and Cathaysian floras are restricted to the Equatorial Belt, where a tropical and subtropical climate prevailed; the Angara flora is extratropical (Boreal), and only in the Permian extended in some places into the Equatorial Belt. The Gondwana floras correspond to the Southern (Notal) Extratropical Zone.

Euramerian flora

This flora occurs in the Carboniferous and Lower Permian throughout North America, North Africa, Europe, the Caucasus, Asia Minor and Middle Asia (Fig. 86*b*). In China and South-Eastern Asia only the Early and Middle Carboniferous floras may be assigned to the Euramerian area; higher in the sequence they give way to floras of the Cathaysian area (a province in the Middle Carboniferous).

It is difficult to compile a single list of Euramerian plants, because some taxa have been established from impressions–compressions, whereas others (independent of the first) have been established from petrifactions in coal balls, etc. Correlations between these taxa are often unknown. Even more difficulties arise with regard to the systematic affinities of dispersed miospores. The Euramerian miospore assemblages occurring successively from the uppermost Devonian to the lowermost Permian have been subdivided into a large number of palynological zones (Clayton *et al.*, 1977; Clayton, 1985). The Euramerian floras exhibit gradual changes at the boundary between the Devonian and Carboniferous. Some assemblages of the lowermost Permian maintain a Carboniferous aspect. Because of this the determination of both Carboniferous boundaries in plant-bearing beds is fraught with difficulties.

Beginning from the Tournaisian persistent plant associations can be recognized. One of these Tournaisian assemblages consists chiefly of lycopsids that differ from those in the Devonian. Typical genera are *Lepidodendropsis* and *Sublepidodendron*; the first species of *Lepidodendron* appear. Vegetative shoot remains occur, associated with rhizophores (*Stigmaria*, *Protostigmaria*) and specialized strobili (*Lepidostrobus*, *Flemingites*), which are unknown from the Devonian. Lepidophyte spores comprise a large percentage of the miospore assemblages. Lycopsids invaded the paludal ecosystems, becoming dominant. Other assemblages partly associated with deltaic lowlands are dominated by ferns, Lagenostomales, Calamopityales and, perhaps, progymnosperms. The Tournaisian ferns, apparently, did not as yet have developed leaf laminae (pinnules). Most of the fern-like fronds probably belonged to Calamopityales and Lagenostomales. When dealing with impressions–compressions it is difficult to recognize the foliage of each group. A single list is usually given, including the genera *Cardiopteridium*, *Fryopsis*, *Rhodeopteridium*, *Adiantites*,

Anisopteris, *Sphenopteridium*, *Triphyllopteris*, etc. In calciferous sandstones and tufaceous rock deposits the same plants occur as petrifactions (see the account of this material in Scott *et al.*, 1984). In this case distinctions are possible between the genera of ferns (in most part with clepsidroid petioles), calamopityans (*Calamopitys*) and lagenostomans (*Lyginorachis*, etc.). In the Tournaisian, seeds of the *Hydrasperma* type are known. Fructifications (seeds, sporangia), both of gymnosperms and ferns, were borne on leafless axes. Among the Tournaisian articulates, in addition to sphenophylls, plants appear with pith casts of *Archaeocalamites*.

Evidently, in the Tournaisian the equatorial sea shores were already populated by thickets of the mangrove type, among which lycopsids dominated. From the Tournaisian upwards of the Euramerian Carboniferous sequence stigmarian soils are of frequent occurrence. The first coal-bearing series of intracontinental (limnic) origin are of Tournaisian age. The significance of these phenomena is under discussion in Chapter 8. The spatial differentiation of Euramerian floras had probably already occurred in the Tournaisian. For instance, differences become evident between the floras of the Urals and China and those of western Europe and North America.

The Visean flora is chiefly similar in aspect to the Tournaisian. The gymnosperm genera, established from leaf impressions, persist. Visean ferns belong to the Zygopteridales and Botryopteridales (Cladoxylales disappear), among which no forms have as yet been established with fully developed pinnules. Of first appearance are Trigonocarpales (seeds *Trigonocarpus* and *Holcospermum*, with *Neuropteris* foliage). Among the lepidophytes the number and diversity of *Lepidodendron* increase; a typical genus is *Eskdalia*; the first sigillarias and *Lepidophloios* appear. Of common occurrence is *Archaeocalamites*. The genus *Cheirostrobus* is known only in the Visean. Two major plant assemblages can be distinguished. One, associated with coal seams (anthracophilic), is represented chiefly by lepidophytes. The other (anthracophobic) consists of plants with fern-like foliage.

Floras of the Serpukhovian (Namurian A) are closely similar to Visean, although during this age certain Early Carboniferous species disappeared (*Sublepidodendron*, *Eskdalia*, etc.), and, by contrast, some genera, typical of the Middle Carboniferous appeared (*Mariopteris*, *Alethopteris*, *Pecopteris*, *Cordaites*). Instead of *Archaeocalamites*, *Mesocalamites* plays an increasingly greater role, and later it is replaced by *Calamites*.

During the Visean–Serpukhovian the geographical distinctions between the Euramerian assemblages, both of miospores (Clayton, 1985; Sullivan, 1967) and of megafossils, increase considerably. Reliable distinctions are apparent in the Kazakhstan province (Meyen, in Vakhrameev *et al.*, 1970, 1978; Oshurkova, 1984; Radchenko, 1985), which is characterized by impoverished floral assemblages (*Sphenophyllum*, *Lepidophloios*, *Bothrodendron*, *Knoxisporites*, *Rotaspora*, *Tripartites*, *Grandispora*, etc., are absent from the assemblage) with a small number of endemic forms, as reflected by the published plant lists. However, there is reason to believe that the endemism of the Kazakhstan floras would be much greater if we take into consideration the assemblage-genera and assemblage-species.

At about the Lower and Middle Carboniferous boundary (between the Lower and Middle Namurian of the western European time-scale) the floras undergo a basic change. The Culmian (European Lower Carboniferous) floras give way to the Westphalian type. The stratigraphic level of this 'floral break' (Florensprung), its relation to the breaks in sedimentation, and its abruptness, have been discussed in the literature. Naturally, no instantaneous changes occurred in the floral assemblages, but within the range of the Carboniferous the changes were, at any event, quite rapid. It is important that this change was concomitant with the abrupt reduction (at times two-fold) of the species composition. The 'floral break' probably corresponded in time to the cooling of the climate in the Angara area ('Ostrogian episode'; Meyen, 1982b). The Middle Carboniferous flora is most completely represented in the Late Bashkirian–Early Moscovian where it is of peak diversity (several hundred species). Near the Bashkirian–

Moscovian boundary (in the Westphalian B) the Euramerian coal-swamp flora of Illinois shows the 'first drier interval' (DiMichele *et al.*, 1985; Phillips *et al.*, 1985).

The Euramerian Middle Carboniferous flora has been extensively studied. This flora embodies traditional views on the Carboniferous flora in general, as they are given in textbooks and in the popular literature, where frequently no heed is given to the fact that the luxurious forests, depicted in the reconstructions, are restricted exclusively to the Equatorial Belt. The Middle Carboniferous floras are represented mainly by diverse arborescent lepidophytes (*Lepidodendron*, *Lepidophloios*, *Chaloneria*, *Sigillaria*), Zygopteridales (*Corynepteris*), Botryopteridales (*Botryopteris*, *Senftenbergia*, *Oligocarpia*, *Discopteris*) and Marattiales (*Scolecopteris*, *Psaronius*), Calamostachyales (*Calamites*, *Calamostachys*, *Annularia*, *Asterophyllites*), Bowmanitales (*Sphenophyllum*, *Bowmanites*), Trigonocarpales (*Neuropteris*, *Paripteris*, *Alethopteris*, *Pachytesta*, *Linopteris*, *Medullosa*), Lagenostomales (*Lyginopteris*, *Lagenostoma*, some *Sphenopteris*), Callistophytales, Cordaitanthales (*Cordaites*, *Cordaitanthus*). Among the gymnosperms of uncertain affinities worthy of mention are *Odontopteris* and *Lonchopteris*. The Middle Carboniferous have yielded the Noeggerathiales, herbaceous lycopsids (*Paurodendron*), Dicranophyllales. Reconstructions of Middle Carboniferous landscapes have been discussed in Chapter 6. In addition to this we can note that the swampy areas were dominated by lycopsids (*Lepidophloios*, *Lepidodendron*), the fossil remains of which comprised up to 65% of the peat-bulk, marattiaceous ferns (*Psaronius–Scolecopteris*) and trigonocarpaleans (*Medullosa*); in places cordaitanthaleans predominated. The diversity of the floras in swamps is evident from the following example. The Herrin coal seam, best studied for the coal balls (uppermost Middle Carboniferous, Illinois, USA), contains 68 species (59 genera) and, taking into account the assemblage-taxa, 44 species (29 genera). At the beginning of the Middle Carboniferous pollen of the lebachiaceous type appear, and in the middle of this epoch, the first allochthonous megafossils of shoots of the *Lebachia* type (*Swillingtonia*; Scott and Chaloner, 1983). Probably at first conifers populated the uplands surrounding the aggradational plains. Floras of the habitats beyond the limits of the coal-source swamps have been studied in most detail in Illinois, where the plant remains have been recovered from infilled erosional downcuts. This flora of 'upland' palaeoenvironments is composed of *Lesleya*, *Megalopteris*, *Lacoea*, *Lepidodendron*, *Mesocalamites*, *Asterophyllites*, *Calamostachys*, *Cordaites*, etc. (Leary, 1981).

In the Upper Carboniferous (Stephanian) lycopsids (particularly Lepidocarpaceae) decrease markedly in number, whereas the number of ferns increases (see Chapter 8; Fig. 94). This change is reflected in all types of plant remains: impressions–compressions, petrifications, in coal balls, miospores (the 'second drier interval'; Phillips *et al.*, 1985). The synchronism in the changes of the dominant plants in Donbass, western Europe and North America suggests that a widespread climatic change occurred at this time. The decrease in the coal content in the Upper Carboniferous sequence, and the diminishment of the area of paralic coal-accumulation can probably be explained by the progressive drying of the climate in the Equatorial Belt, which evidently corresponded in time to the Gondwana glaciation. The increase in dryness of the climate probably first affected the plants that inhabited areas outside the coal-producing swamps. The changes in the assemblages, as reflected in the plant impressions in the barren layers, occur somewhat lower in the section (in the middle of the Westphalian D), than in the plant assemblages in coal seams (Pfefferkorn and Thomson, 1982).

The Upper Carboniferous assemblages are dominated by marattialean ferns (*Asterotheca*, *Pecopteris*, *Psaronius*), Cordaitanthales, Calamostachyales, Bowmanitales, Trigonocarpales (*Neuropteris*, *Alethopteris*), pteridosperms (*Callipteridium*, *Odontopteris*, *Taeniopteris*). In the Upper Carboniferous assemblages occur with dominating conifers (*Lebachia*, *Ernestiodendron*). Lepidophytes are represented mainly by sigillarias (subsigillarias). The major forest-forming plants were tree ferns and cordaitanthaleans. In the Late

Carboniferous the vegetation of drylands and sites had already acquired the Permian (Autunian) aspect.

In the Middle Carboniferous the Kazakhstan province was clearly distinguished, revealing a much impoverished flora, compared with the rest of the Euramerian area (*Sigillaria*, *Lepidophloios*, *Mariopteris*, *Alethopteris*, *Cordaites* and many other plants are absent). Of the endemic genera mention should be made of *Caenodendron*. Otherwise the endemism of the Kazakhstan floras, dominated by *Lepidodendron*, *Neuropteris*, *Mesocalamites* and *Sphenopteris*, is manifest only at the species level. The independence of the Kazakhstan floras supports the view that during the Early and Middle Carboniferous Kazakhstan was separated from the rest of the Euramerian area by an oceanic basin. According to certain mobilistic concepts both landmasses came into contact during the second half of the Middle Carboniferous. During this time the flora of the Kazakhstan province became of Angara type.

Upwards of the Carboniferous sequence peculiarities increase in the floras of China (Wagner *et al.*, 1983). Although the Middle Carboniferous flora, as a whole, is of the Euramerian aspect here (similar pteridosperms, calamites, cordaitanthaleans, etc.), certain differences are apparent, being reflected in the large percentage of endemic species, certain endemic genera and practically complete absence of many plants, typical of Europe and North America (e.g. sigillarias). Neuropterids are typical of many localities. From this time onwards the Cathaysian province is of independent existence, and already in the Late Carboniferous may be assigned to a specific area (see below). In the west of the USA (Cordilleran Province) in the Middle–Upper Carboniferous certain peculiarities of the flora can be outlined, particularly the early appearance of conifers. Other local peculiarities are apparent in the Middle and Upper Carboniferous Euramerian assemblages, that are reflected in the distribution of individual genera. According to these features certain palaeobotanists (Novik and Fissunenko, 1979; Pfefferkorn and Gillespie, 1980; etc.) distinguish independent provinces, which, however, are essentially less distinct than the Permian provinces. Because of this we shall not enlarge upon them here.

At about the Carboniferous–Permian boundary the Euramerian floras exhibit an increase in the number of conifers (*Lebachia*, *Ernestiodendron*) and peltaspermaceous *Callipteris*, a decrease in the number of lepidophytes. The fern-like fronds in gymnosperms become strongly reduced in size. Otherwise, the Lower Permian Euramerian (Autunian) flora is similar to the Upper Carboniferous one, so that at times the Stephanian assemblages cannot be distinguished from the Autunian. Typical Autunian assemblages dominated by conifers, *Callipteris*, pollen of *Potoniesporites* and *Vittatina* occasionally appear in the barren facies from the middle of the Upper Carboniferous (Doubinger, 1983; Doubinger and Langiaux, 1982; Winston, 1983). In coal-free areas assemblages of the Permian aspect, containing numerous conifers, were probably of earlier appearance. Indirect evidence of this is found in the palynological data available on the Middle–Upper Carboniferous of the Urals (Chuvashov and Dyupina, 1973; Chuvashov *et al.*, 1984). Plant megafossil assemblages of the Permian aspect have been reported from the lowermost Upper Carboniferous and even from the uppermost Middle Carboniferous of Middle Asia (Sixtel *et al.*, 1975, 1981; Savitskaya and Iskandarkhodzhaev, 1984); but the systematic composition and stratigraphic position of these assemblages are in need of precision. Certain macrofossil conifer assemblages of typical Permian aspect associate in Kazakhstan with marine Middle Carboniferous faunas. It is thereby probable that the flora that, at the end of the Carboniferous–beginning of the Permian, began to replace anthracophilic floras in the west of the Euramerian area, was of much earlier appearance in the ecotone phytochoria in the east of the area, migrating from here westwards. However, it should be borne in mind that the oldest firmly established conifers are known in the Westphalian B in the Euramerian area, where they probably originated in the uplands around the aggradational lowlands.

In the Autunian the coal content decreases significantly in the section; and the fossil plant assemblages, reflecting swampy lowland communities, decrease in proportion. Judging by the palynological data the Autunian coal-forming vegetation of swamps maintained the Stephanian aspect for a long time. The vegetation of the flood-plains and slopes of river valleys and lake shores underwent more pronounced changes. Evidently, beginning with the Permian coniferalean and conifer-fern type forests were of wide distribution and these produced masses of quasimonosaccate pollen. The latter show a drastic increase in percentage near the Carboniferous–Permian boundary. At the same level, striate asaccate pollen (of the *Vittatina* type) and quasi-disaccate pollen (of the *Protohaploxypinus* type) increased in proportion. All these changes in the Euramerian floras, occurring at the end of the Carboniferous–beginning of the Permian, are commonly associated with the aridization of the climate. However, the available data, including those on palaeosols, tend to indicate that the changes in the plant communities were provoked to a great extent by the dissection of the relief and the disappearance of swampy lowlands, rather than by a decrease in precipitation (Remy, 1980).

Knowledge of the Euramerian floras from other parts of the Early Permian and beginning of the Late Permian is meagre. In this stratigraphic range poor assemblages have been encountered, where strongly impoverished Autunian plant assemblages occur together with conifers (*Ortiseia*), that display more affinities with the Upper Permian (Clement-Westerhoff, 1984). During this time interval coal-formation terminated completely. In the Euramerian area volcanic activities are widely evident.

Zechstein flora

The Upper Permian of Western Europe traditionally comprises Zechstein deposits, represented by marine carbonate-evaporite facies, reflecting an arid climate. In the Permian stratigraphic scale adopted in the USSR only the upper part of the Upper Permian corresponds to the Zechstein. The Zechstein Sea invaded Europe from the north and continued from Ireland in the west to the Baltic region, in the east. Another sea basin, contiguous to Europe at the south, was connected with the Tethys Ocean. The shores of both basins were covered by similar vegetation. The Zechstein flora has been studied in most detail. It is not rich (about 20 species) and is composed chiefly of conifers (*Ullmannia*, *Pseudovoltzia*, at times *Quadrocladus*) with a small admixture of other plants (*Pseudoctenis*, *Sphenobaiera*, *Taeniopteris*, peltaspermaceous *Lepidopteris*). A specific feature is the presence of a small number of articulates (*Neocalamites*, *Paracalamites*) and practically complete absence of ferns (the species of *Pecopteris* and *Sphenopteris*, described here, more readily belong to Peltaspermales) and cordaites (single leaves of *Cordaites*). In the palynological assemblages most predominant are the monosaccate *Nuskoisporites* and quasidisaccate *Lueckisporites*, *Taeniaesporites*, *Jugasporites*, *Protohaploxypinus*; *Vittatina* also occurs. Miospore assemblages of closely similar composition have been recovered from the Upper Permian in Canada. Flora similar to the Zechstein type populated areas extending far westward. According to the miospores and scarce finds of plant megafossils, it extends also into the Northern Caucasus and Precaspian. In the Pamirs (Rabnou locality) numerous *Lepidopteris* have been encountered (this genus is relatively scarce in the Zechstein) and rare conifers of the *Ullmannia* type, associated with Zechstein miospores (*Lueckisporites*, etc.; V. I. Davydov's collections, identified by A. V. Gomankov and S. V. Meyen). A flora of the Zechstein type with *Ullmannia*, *Pseudovoltzia*, *Lepidopteris* and *Taeniopteris* has been discovered in North China in the lower part of the Shischienfeng group (Wang, 1985), overlying beds with North Cathaysian floras.

North American flora

Lower Permian localities are known in the east of the USA, where they are similar in composition to the Autunian (Chaloner and Meyen, 1973; Read and Mamay, 1964). They are restricted to the Dunkard Formation. Younger Early Permian flora has been found only in the south-western USA, where it is restricted to the Artinskian–Kungurian stratigraphic equivalents. This flora was long associated with the Cathaysian, and, as a consequence, the inference was derived concerning the direct connection between East Asia and North America, via the Pacific. This was based on numerous finds of plants attributed to the Cathaysian genera *Gigantopteris* and *Tingia*. Later the distinctions between America and Cathaysian gigantopterids became evident, and the plants, identified as *Tingia*, were referred to the endemic genus *Russellites*, which shows affinities to the Euramerian–Cathaysian genus *Plagiozamites*. Even if we disregard these plants, very great differences can be observed between North American and Cathaysian floras. The former reflects a semiarid environment, being restricted to variegated beds, and contains many conifers (they are extremely rare in Cathaysia) and specific gymnosperms, affiliated with the endemic genera *Supaia*, *Glenopteris* and *Tinsleya*. Also present are cycad-like *Archaeocycas* and *Phasmatocycas* and fronds, closely related to the Angara *Compsopteris* and *Callipteris*. Conifers are in part similar to the Zechstein *Pseudovoltzia*. Considerable amounts of ferns occur here (*Pecopteris*).

An idea of the Upper Permian floras of North America can be gleaned only from the miospore assemblages. Quasisaccate pollen (striate and non-striate) and *Vittatina* predominate. These floras have much in common with the Upper Permian assemblages of Western Europe and the Ural Region.

Cathaysian flora (Li and Yao, 1979, 1980, 1982; Wagner *et al.*, 1983; Wang, 1985)

In the Late Carboniferous the former Cathaysian Province became an independent area. Although many Euramerian genera (*Lepidodendron*, *Neuropteris*, *Odontopteris*, *Cordaites*) and even species (*Sphenophyllum*, *Annularia*, *Pecopteris*) continue to exist here, the overall habit of the flora is to a great extent determined by Cathaysian plants – endemic genera (*Tingia*, *Cathaysiodendron*) and species (of *Lepidodendron*, *Bothrodendron*, *Ulodendron*, *Sphenopteris*, *Pecopteris*, *Rhacopteris*, *Callipteridium*). It should be remembered that at the Middle–Upper Carboniferous boundary *Lepidodendron* practically disappears from the Euramerian flora. The Upper Carboniferous *Emplectopteris* and *Emplectopteridium* of China might have been ancestors of the Permian gigantopterids (Gigantonomiales). The aridization of the climate, gradually involving the Equatorial Belt, which was manifested in the Euramerian area, is much less apparent in China and Southeastern Asia, where coal-bearing deposits are of wide development almost to the end of the Permian. These climatic distinctions between both phytochoria, already reflected in the Late Carboniferous, become more apparent in the Permian, when we can refer to Cathaysia as an independent plant kingdom. The major portion of the Permian of Cathaysia practically lacks conifers (dominant in the Euramerian area); *Callipteris*, cordaites are rare and sigillarias are absent. Physiognomically the Cathaysian Lower Permian floras, restricted to coal-bearing series, more readily resemble the Upper Carboniferous than the Lower Permian Euramerian flora.

Localities of the Cathaysian Upper Carboniferous floras occur mostly in China, and those of the Permian floras are known also in Korea, Japan, Vietnam, Laos, Thailand, Indonesia and Malaysia. Certain distinctions can be traced in the composition of the assemblages in the northern and southern localities, so that two Cathaysian phytochoria can be defined, the boundary between them extending in a latitudinal direction through the upper course of the Huang He to the lower course of the Yangtze River. The differences between them are reflected in the distribution of individual species and genera, including gigantopterids (*Gigantopteris* is of very rare occurrence in the northern phytochorion, but quite common in the southern; *Gigantonoclea* has been encountered only in the south). The Lower Permian Cathaysian floras, as a whole, are characterized by the gradual increase in the sequence, in ascending order, of the role of typical Cathaysian plants – *Lobatannularia*, *Tingia*, *Plagiozamites*, *Emplectopteris*, *Cathaysiopteris*, gigantopterids, endemic species of *Sphenopteris*, *Pecopteris*, *Taeniopteris*. Mesophyte elements appear (leaves, morphologically similar to *Pterophyllum* and *Nilssonia*; several species of *Cladophlebis*). Species of *Lepidodendron*, *Cathaysiodendron*, *Pecopteris*, *Sphenopteris*, *Cordaites*, of first appearance in the Carboniferous, continue to the Upper Permian. The number of Cathaysian endemic forms increases in the Upper Permian. Representatives appear (or are of more frequent occurrence) of endemic forms of *Schizoneura*, *Sphenophyllum*, *Fascipteris*, *Gigantopteris*, *Gigantonoclea*, *Rhipidopsis*, in some places – conifers. *Tingia* continues to exist. Some genera occur that are common to the Angara flora (*Compsopteris*, *Comia*, *Rhipidopsis*). The distinctions of the Cathaysian flora from the Euramerian, North American and Zechstein floras are clearly apparent in miospore assemblages, among which endemic forms occur (Ouyang, 1982), whereas the percentage of quasisaccate miospores is small, and exhibits a marked increase only in the northern phytochorion during the second half of the Upper Permian (Yao and Ouyang, 1980). The relationship between the Cathaysian and coeval floras at the suprageneric level is not quite clear, because many Cathaysian forms are of uncertain systematic affinities.

During the Late Permian the Cathaysian flora spread out widely westward. Its localities are known in Turkey, Iraq (Fig. 87), Saudi Arabia and Syria (Čtyroky, 1973; El-Khayal *et al.*, 1980; Hill and El-Khayal, 1983; Lemoigne, 1981; Wagner, 1962; Wang, 1985; O. P. Yaroschenko and Meyen,

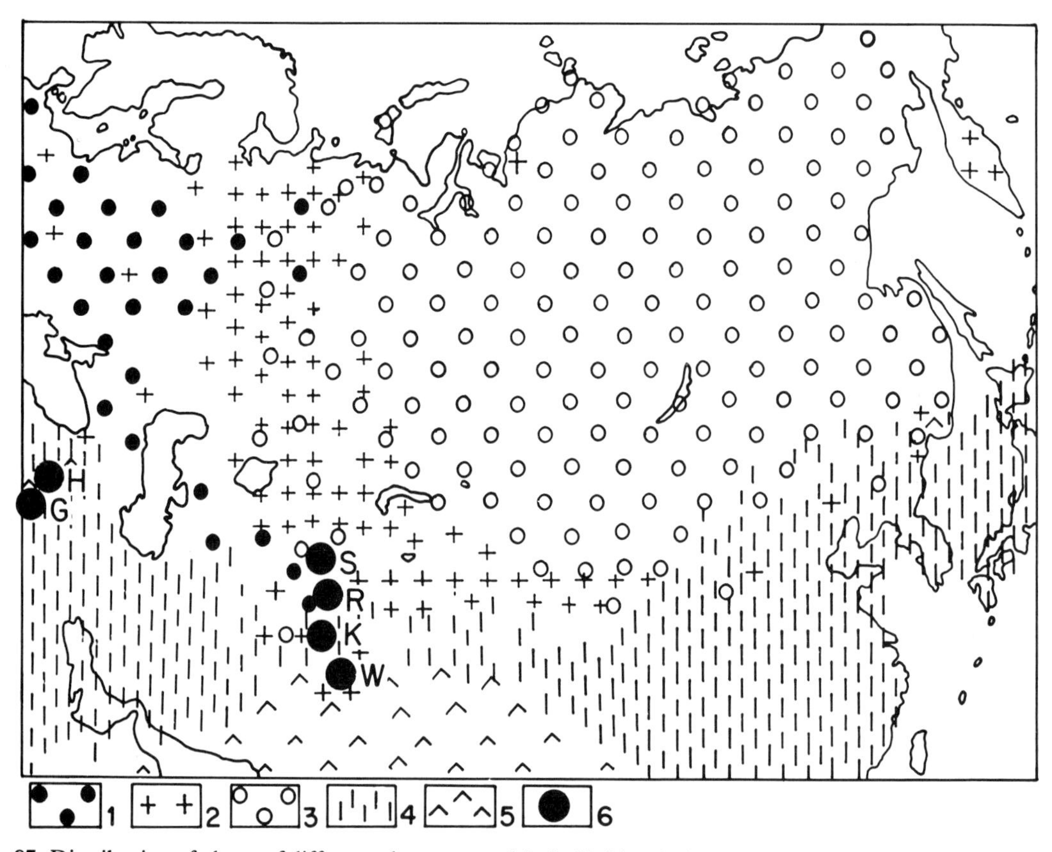

Fig. 87. Distribution of plants of different phytogeographical affinities during late Early Permian and Late Permian; plants of Atlantic kingdom (1), Subangara area (2), Angara area (3), Cathaysian kingdom (4), Gondwana kingdom (5); 6: localities Hazro (H), Ga'ara (G), Sarytaipan (S), Kabul (K) and Wargal (W) yielding mixed plant assemblages. Modified from Meyen, 1982b.

own observations). This suggests the existence of a belt of a thermophilic and hydrophilous flora that populated the Tethys coasts. Contiguous to this belt to the south was the territory inhabited by the Gondwana floras. It is probable that from here *Glossopteris* penetrated into Turkey (Archangelsky and Wagner, 1983). Worthy of note is the absence of mixed assemblages of Cathaysian and Gondwana plants at the south and east of Asia, whereby the present geographical proximity of the localities of both floral types in Tibet (Li *et al.*, 1982a, 1982b), the Himalayas and in Primorye may be interpreted as having been engendered by the horizontal displacements of continental blocks.

Angara flora

A special review (Meyen, 1982b) is devoted to the systematic composition, distribution and history of the Angara flora. The independent status of the Angara phytochorion from the beginning of the Carboniferous is disputed by some palaeobotanists. At least, the species, if not the generic composition, of the Siberian Tournaisian flora is endemic, but more important is the fact that the Siberian assemblages are extremely impoverished and composed exclusively of lepidophytes and sphenophylls. Despite vast collecting, many genera of lepidophytes, ferns and gymnosperms, typical of Euramerian floras, have not been found here. Stigmarias, *Lepidostrobus* and seeds are utterly absent.

So far no reliable data can be advanced concerning the position of the boundaries between the Devonian and Carboniferous, Tournaisian, Visean and Serpukhovian in the plant-bearing beds of Siberia. The four lepidophyte assemblages, recognized here, have been tentatively assigned to the following stratigraphic levels: the lepidophyte assemblage I (Bystryanskaya) to the Famennian, II to the Tournaisian, III to the Visean, IV to the Serpukhovian. Assemblage I, encountered in the Bystryanskaya and Altaiskaya Suites of the Minusa Basin, has been discussed above. Assemblage II is richer, dominated by lepidophytes (*Eskdalia*; of first appearance is the endemic genus *Ursodendron*; *Lepidodendropsis* was noted here), and contains a small amount of plants with fern-like foliage (of the *Aneimites*, *Adiantites* and *Triphyllopteris* types) and *Sphenophyllum*. Of common occurrence are problematical plants, arbitrarily named *Caulopteris ogurensis*. In assemblage III, besides *Ursodendron* and *C. ogurensis*, the first *Angarophloios* appear. Localities of assemblages II and III are known in other parts of Siberia, also in Mongolia (Durante, 1976; Durante and Pavlova, 1983), although their composition may vary somewhat; besides lepidophytes, fern-like plants of the *Chacassopteris* type may occur here. In the south of Mongolia in the lowermost Carboniferous lepidophyte assemblages have been recovered that contain *Lepidostrobus* and *Stigmaria*, probably belonging to the Euramerian floras.

The richest assemblage IV is restricted to the Ostrogsky horizon (Kuzbass) and its stratigraphic equivalents elsewhere in Northern Asia. In some assemblages, chiefly in sandstones, lycopsids predominate (several species of *Tomiodendron*, *Lophiodendron*, and *Angarophloios*), whereas in fine-grained rock types, articulates and fern-like plants (*Angaropteridium*, *Abacanidium*, *Chacassopteris*) predominate.

Beginning from assemblage III, i.e. approximately from the Visean, practically nothing in common remains between the Angara and Euramerian floras, if we compare coeval assemblages. However, certain affinities can be observed between the lepidophytes found in III and IV assemblages, and the older Euramerian and Upper Devonian lepidophytes. The Angara lepidophytes preserve the archaic stelar organization (the protostele is rounded or stellate in cross-section), the absence of true leaf scars, strongly developed infrafoliar bladder, axillary position of the ligular pit. Judging by the absence of strobili (*Lepidostrobus*, *Lepidocarpon*, etc.) in the Angara burials, sporophylls were not clustered in strobili, the same as in the primitive lepidophytes of the Equatorial Belt. The archaic nature is evident in the plants with fern-like foliage, as well. The Serpukhovian *Chacassopteris* is similar to the Devonian–Tournaisian ferns; up until the end of the Lower Carboniferous most prevalent are plants with rounded cardiopteroid pinnules (in the

Euramerian flora they are abundant only in the Tournaisian and Visean, where other ferns occur together with them). Plants with pycnoxylic stems are absent from the Angara flora. Such plants were present in the Devonian and appeared again only in the Middle Carboniferous. From the Early Carboniferous we are thus faced with the survival of ancient morphological types, characteristic of Angaraland, i.e. with the phenomenon of extra-equatorial persistence (see Chapter 8).

The paucity and peculiarities of Angara flora of the Early Carboniferous can be attributed to the aridity of the climate (the plant-bearing rocks are often variegated, contain evaporites and lack traces of coal-formation). The lithological features give evidence for a decrease in the aridity of the climate during the Early Carboniferous (Grizer and Ul'masvai, 1984). Thick perennial manoxylic stems of lepidophytes (up to 30 cm in diameter) provide evidence for a frost-free climate, proof of which is also found in the lithologies. Stems up to 10 cm in diameter are known in the North-East USSR, near the North Pole of that period.

At about the Early–Middle Carboniferous boundary the Angara flora undergoes a complete changeover. The number of lepidophytes decreases markedly, and the assemblages are dominated by *Angaropteridium* and *Rhodeopteridium*, tentatively affiliated with pteridopserms. This assemblage, termed Pteridospermalean, also includes the first leaves of cordaites and articulate stems (*Mesocalamites*, *Paracalamites*). This floral turnover can probably be attributed to the cooling of the climate (the 'Ostrogian episode'), that involved Angaraland in its entirety, and evoked a response in the Euramerian floras (the 'floral break'). Simultaneously, in the southern parts of Siberia the first indications of coal formation appear. The accumulation of the Siberian coal measures proper commenced when the pteridospermalean flora was replaced by the Mazurovskaya flora containing numerous cordaitanthaleans (*Cordaites*, *Rufloria*), pteridosperms *Angaropteridium* and *Angaridium*, callistophytalean *Paragondwanidium*, articulates *Paracalamites*. Euramerian migrants *Neuropteris* and *Dicranophyllum* appear. Lepidophytes (*Angarodendron*, *Angarophloios*, *Caenodendron*) are mostly confined to individual layers, where their remains may dominate; in places coal seams consist completely of accumulations of their stem cuticles (Kuznetsk, Minusa and Tunguska Basins). The origin of the major components of the Mazurovskaya flora is unknown. Considering that the oldest cordaitanthaleans are known in the Serpukhovian of Europe, it is conceivable that the Angara cordaitanthaleans are descendants of the immigrants from the equatorial phytochoria. It is probable that *Paragondwanidium* and plants with foliage of *Angaropteridium*, allied to the Early Carboniferous Equatorial *Cardiopteridium*, are of the same origin. Later the evolution of these groups in Angaraland was autochthonous; their Euramerian allies did not appear any more in Angaraland.

The succeeding Alykaevskaya flora is closely similar to the Mazurovskaya, but richer and more diverse. A greater number of Euramerian migrants appear. They include Bowmanitales (*Sphenophyllum*), Calamostachyales (*Calamostachys*, Fig. 26, *k*; *Annularia*), Dicranophyllales, and certain types of seed. At the same time, *Neuropteris* becomes more diverse and ferns with small fronds and tiny pinnules (*Sphenopteris*, *Pecopteris*) increase in number. A characteristic feature is the presence of fructifications *Krylovia*, *Gaussia*, etc., associated with *Cordaites* and *Rufloria*. The Mazurovskaya and Alykaevskaya assemblages include over 100 species. This enrichment of the flora, particularly owing to the incoming of Euramerian genera, together with intensive coal accumulation, indicates the warming and humidification of the climate. But the distinct growth rings in the pycnoxylic stems of the gymnosperms, suggest that the climate was seasonal. Floral diversity and coal formation decrease towards the Northeast USSR and towards Eastern Kazakhstan, which can be accounted for by the prevalence of a colder climate in the northern areas (where glacial-marine beds were noted in the Middle Carboniferous) and a more arid climate in the southern areas, defined as the Kazakhstan province (until this time the latter constituted a part of the Euramerian area).

The Alykaevskaya flora was replaced by the Promezhutochnaya (Intermediate), so termed after the same-named Kuzbass horizon. In the Verkhoyansk region it is associated with marine faunas of the lowermost Permian. Most of the Euramerian elements are absent, and predominant are leaves of cordaites (*Rufloria*, *Cordaites*, scaly *Crassinervia*, *Nephropsis*), which are associated with female fructifications *Vojnovskya*, *Gaussia* and *Bardocarpus* and profusely branching microsporoclads resembling *Kuznetskia* and *Telangiopsis*. Of common occurrence are surviving Alykaevskaya genera *Angaropteridium*, *Angaridium*, *Neuropteris*, *Paragondwanidium*. Abundant articulates (*Phyllotheca*, *Paracalamites*, *Phyllopitys*, *Tchernovia*, *Annulina*, *Annularia*) and gymnosperms *Zamiopteris* are present; also ferns occur. Higher in the section – in the Ishanovskaya flora – Carboniferous elements disappear and the species composition exhibits certain changes. Practically the same flora continues further into the uppermost Lower–lowermost Upper Permian (Kemerovskaya and Usyatskaya floras). *Sylvella* seeds and *Salairia* mosses appear here. These four floral assemblages (from the Promezhutochnaya to the Usyatskaya) are often united into one Upper Balakhonskaya flora.

The Upper Balakhonskaya flora is superseded by the Kolchuginskaya, consisting of a sequence of several assemblages. It is dominated by cordaites (*Cordaites*, *Rufloria*, scaly *Crassinervia*, *Lepeophyllum*, *Nephropsis*) with fructifications *Cladostrobus* and *Kuznetskia*, and seeds *Tungussocarpus*, *Sylvella*, etc., and the same articulates as in the former assemblages (but without *Phyllopitys* and *Annulina*). Quite abundant are ferns (*Prynadaeopteris*, *Pecopteris*), leafy mosses (*Polyssaievia*, *Protosphagnum*), plants of uncertain affinities (*Glottophyllum*, *Psygmophyllum*). In the upper part of the Upper Permian the genus *Rufloria* – one of the most widespread in the Angara flora – disappears. Somewhat earlier small-leaved cordaites appear ('sulcial') with thin furrows along the veins on the upper leaf side and numerous false veins. Typical of the Kolchuginskaya flora are callipterids (*Callipteris*, *Comia*) and genera of mesophytic habit (*Rhipidopsis*, *Tomia*, *Ginkgoites*, *Yavorskyia*), that increase in number upwards in the section.

The foregoing description of the Upper Balakhonskaya and Kolchuginskaya floras is typical of Kuzbass. The Upper Balakhonskaya flora maintains its habit throughout the entire Asian part of Angaraland, becoming somewhat impoverished in the Northeast USSR and in Kazakhstan. The Kolchuginskaya flora exhibits a greater spatial differentiation. In the Tunguska Basin, Northern Mongolia and Northeast USSR callipterids are absent, ferns are less abundant, *Psygmophyllum* has not been found. In Western Taimyr *Neuropteris* was recovered from the Upper Permian. In Primorye and in the south of Mongolia Cathaysian plants occur (ferns, *Taeniopteris*, etc.). In these areas, also in the Kuznetsk and Tunguska Basins single localities are known of leaves of the *Glossopteris* type (and in Primorye also of *Gangamopteris*).

Closely similar to the Siberian floras is that of the Pechora province (North Fore-Urals), occurring in the stratigraphic range from the Kungurian to the end of the Permian. If differs in having a much greater amount of ferns, in its endemic mosses (*Intia*, *Vorcutannularia*), the lepidophytes *Viatcheslavia* and *Paikhoia*, the genus *Psygmophyllum*, and abundant *Rhipidopsis*, *Callipteris* and *Comia* (in the Kazanian–Tatarian). A characteristic feature is the presence of plants known in the Carboniferous–Lower Permian of the Euramerian area (*Danaeites*, *Oligocarpia*, *Sphenophyllum thonii*). Typical of the Pechora Upper Permian are cardiolepids, the leaves of which (*Phylladoderma*) comprise coal seams, also the genus *Compsopteris*. The floras of the Pechora province and Primorye display certain affinities. Permian floras comparable to those of Pechora and Primorye, are known in Alaska (S. H. Mamay; see Meyen, 1979c), at the extreme north of Greenland (Wagner *et al.*, 1982) and in North China (Huang, 1977, 1983).

Subangara flora

The Angara area during the Permian was surrounded by a territory with floras of an intermediate type (ecotone), or of the Subangara area (Meyen, 1981, 1982b), which included the previously defined Ural–Kazakhstan and East-European areas (the latter are now discarded as independent areas). The Subangara flora is best represented on the Russian Platform and in the Fore-Ural region. It is characterized by the presence of typical Angara plants with an admixture of endemics and of genera of adjacent phytochoria. The relationship between the Subangara flora and adjacent floras is not sufficiently understood, due to the paucity of the localities and inadequate studies. Because of this it is premature to delimit the area and to suggest any subdivision for it.

The Subangara area was already of incipient formation during the Carboniferous. In the Middle–Upper Carboniferous of the Middle Urals palynological assemblages are more closely related to the Permian Subangara, than to the Carboniferous Euramerian and Angara assemblages (Chuvashov and Dyupina, 1973; Chuvashov *et al.*, 1984). They are characterized by the dominance of quasisaccate pollen, including quasidisaccates with a striate body (of the *Protohaploxypinus* type), and the incoming in the Upper Carboniferous of *Vittatina*. The Permian habit of the Uralian miospore assemblages that appears to be controlled by the development of coal-free formations and the prevalence of an arid and semiarid climate, have led to stratigraphic errors. In Euramerian assemblages single miospores of the *Protohaploxypinus* type occur in the Westphalian, whereas rare *Vittatina* appears from the beginning of the Upper Carboniferous. In the Tunguska Basin miospores of the same type also appear in deposits of about the same age, but once again only as single grains (Peterson, 1983). In the Urals these miospores determine the aspect of the assemblages. At the beginning of the Lower Permian the Ural assemblages become similar to the Euramerian. There is reason to believe that the Euramerian area was populated by plants that were formerly more typical of the Urals. There is a high probability that the flora, typical of the Ural Middle–Upper Carboniferous, also populated Kazakhstan and Middle Asia; however, miospores and macrofossils from these areas have so far been poorly studied.

From the Artinskian (Vladimirovich, 1981) in the Middle Fore-Urals megafossil assemblages appear that at first are of poor composition, combining Angara (*Rufloria*, *Paracalamites*) Euramerian (conifers, *Callipteris*), and endemic plants (*Psygmophyllum*). In the Kungurian the floras are already more abundant and diverse. They also include Angara genera (*Rufloria*, *Nephropsis*, *Sylvella*, *Bardocarpus*, *Phyllotheca*, *Paracalamites*, *Annulina*), as well as plants known in the Pechora province and absent in Siberia (*Psygmophyllum*), also endemic genera (*Peltaspermum*, *Mauerites*, *Biamopteris*, *Entsovia*). There are also many sphenophylls, various seeds and other genera, as yet poorly studied. The Kungurian of the Fore-Urals contains abundant conifers (*Kungurodendron*, etc.) and peltasperms of the *Comia*, *Callipteris*, *Sylvopteris* types, also ferns, etc. Similar pteridosperms are known in North America. The miospore assemblages of both regions also show affinities. The Artinskian–Kungurian flora is frequently termed 'Bardian', and is wrongly thought to extend to the Pechora Fore-Urals. Certain Bardian plants occur in South-East Kazakhstan. In general in the Early Permian Kazakhstan was populated by typical Angara, Bardian, Euramerian (?), as well as endemic plants (Salmenova, 1978, 1979, 1982, 1984). The composition and interrelations between the Kazakhstan assemblages are both in need of further explication.

The Ufimian flora of the Fore-Urals (Vladimirovich, 1982) is very poor. It is dominated by heterosporous lepidophytes *Viatcheslavia*. Evidently they grew in dense thickets along the coasts of the sea that advanced on to the Fore-Urals from the north. In Tataria numerous leaves of *Phylladoderma* have been found from the Ufimian. Identification of the Ufimian deposits in other regions is fraught with difficulties. Probably of Ufimian age is the North Afghanistan flora with cordaites, Subangara *Phylladoderma* and *Pursongia*, Equatorial fern types, Euramerian conifers, Cathaysian *Lobatannularia* and *Taeniopteris* (Meyen, 1981, 1982b).

In the Kazanian (Esaulova, 1984; Meyen, 1979b; Vladimirovich, 1984) and Early Tatarian the Subangara flora was dominated by conifers (*Timanostrobus*, *Quadrocladus*, etc.), peltasperms (*Rhaphidopteris*, *Phylladoderma*); there were also osmundaceous ferns (*Thamnopteris*), articulates, *Psygmophyllum*, dicranophylls (*Entsovia, Mostotchkia*), *Signacularia* (closely allied to *Viatcheslavia*). Cordaites occur on the Russian Platform only in the lowermost Kazanian. In the Southern Fore-Urals they rise to the Upper Tatarian substage. In the Upper Tatarian flora of the Russian Platform (Gomankov and Meyen, in press) most predominant are Peltaspermaceae (*Tatarina*) and Cardiolepidaceae (*Phylladoderma* subgen. *Aequistomia*); of common occurrence are conifers (*Sashinia, Quadrocladus*), protosphagnalean mosses, occasionally *Rhaphidopteris* and *Glossophyllum* (with associated cladosperms of *Stiphorus*). The present writer suggested that this flora should be termed *Tatarina*-flora, after the dominant *Tatarina*. This flora has been found only in a few localities, and is restricted to single grey-coloured lenses of lacustrine origin, intercalated in redbeds. These plants seemingly grew along lake shores. Judging by the palynological data, the *Tatarina*-flora was widespread, extending far northwards. Associated miospore assemblages with *Scutasporites, Vittatina, Protohaploxypinus*, etc., have been recorded by Balme (1979) from the north of Greenland (Cape Stosch). Plant assemblages of the *Tatarina*-flora have been found in the Pechora Fore-Urals, where this flora probably replaced the flora of the Pechora province. The Late Tatarian (?) flora of Nan-Shan (North China; Durante, 1980) contains very similar peltasperms and conifers to those on the Russian Platform. Callipterids, ferns, *Rhipidopsis* and other plants also occur here, that give an indication for connections with the Angara floras of Siberia and the Far East. It is probable that the Subangara flora grew on Kamchatka, where redeposited Permian miospore assemblages of the Subangara type (*Protohaploxypinus, Vittatina*, etc.) have been found in the Mesozoic (Dyufur *et al.*, 1977).

The Subangara area underwent changes in its boundaries. For instance, in Tien Shan the pure Euramerian floras of the lowermost Permian are replaced by Subangara floras of the Kungurian–Ufimian with *Rufloria, Equisetinostachys, Nephropsis, Zamiopteris*, conifers, etc. Further eastward (in Northern China) the Subangara flora replaces the Cathaysian, but this happens much later, namely during the Tatarian (Durante, 1980). In general, the Subangara flora spread out to the south and east from areas proximal to the Urals. Its composition differs, depending upon the neighbouring floras. In Western Angaraland it contains Pechora, West European and North American plants, whereas in the east it contains Siberian and Far East plants.

Gondwana flora

In the literature the term Gondwana flora refers to assemblages, dominated by leaves of *Glossopteris* and *Gangamopteris*. It is also called the *Glossopteris*-flora. It has been implied that older Gondwana floras are closely related to Euramerian. The endemic nature of the older Gondwana floras has now been firmly established. Unfortunately, the Lower Carboniferous floras are poorly studied. They are known in Argentina, East Australia and Kashmir. The major components are slender-stemmed lepidophytes of the *Lepidodendropsis, Cyclostigma, Archaeosigillaria* types, fern-like plants (probably in part progymnosperms and gymnosperms) *Rhacopteris* (?), *Rhodeopteridium, Charnelia, Cardiopteridium* and the articulate *Archaeocalamites*. Since the fructifications of these plants are either unknown, or poorly studied, it is impossible to judge how these plants were related to those of the Euramerian and Angara areas. Evidence of the distinctions between the Lower Carboniferous Gondwana and Euramerian floras is provided by the palynological data (Kemp *et al.*, 1977). Tournaisian miospore assemblages of Australia differ considerably from the Euramerian ones. The younger assemblages are even more distinct.

Later assemblages of the pre-*Glossopteris* floras are best known from Argentina, Brazil (Archangelsky *et al.*, 1978–1980) and Australia (Retallack, 1980). They are composed of *Fedekurtzia, Nothoracopteris, Botrychiopsis* (= *Gondwanidium*), probably allied to progymnosperms, also *Sphenopteridium, Bergiopteris* (simple pinnate fronds of this genus resemble *Angaropteridium* of Angaraland), lepidophyte *Brasilodendron*, articulates *Paracalamites*; conifers (?) and *Cordaites* also occur.

Floras, closely similar in composition but dominated by the genus *Botrychiopsis*, occurring together with articulates and lepidophytes, populated Gondwanaland during the glaciation that began in the Late Carboniferous (or at the end of the Middle Carboniferous). A hypothesis has been put forward that the periglacial plains were covered by '*Botrychiopsis*-tundra' (Retallack, 1980). During the glaciation plants appeared that were already typical of the *Glossopteris* flora; among them was *Gangamopteris*. The Middle–Upper Carboniferous of Argentina yielded shoots referred to *Ginkgophyllum* (Archangelsky and Leguizamón, 1980), but more likely belonged to *Dicranophyllum* – a genus, known also in the Euramerian and Angara floras of this age.

The first appearance of the *Glossopteris* flora is not reliably dated. Most likely it was in the Late Carboniferous. Deposits containing *Glossopteris* flora are frequently termed Lower Gondwana. They are known within all the Gondwana continents, including Antarctica. Archangelsky and Arrondo (1975) divided the Gondwana area into the western (Notoafroamerican) and eastern (Australoindian) provinces. In both provinces glossopterids (order Arberiales) played a major role; higher in the section *Glossopteris* gradually replaces *Gangamopteris*. Judging by the recent data, the species unity among *Glossopteris* of different Gondwana continents is not significant (Chandra and Surange, 1979). In the Notoafroamerican province, located nearer to the Equatorial Belt, forms occur that are of the Euramerian aspect or belong directly to Euramerian genera and species (Archangelsky and Cúneo, 1984). This refers chiefly to ferns, including marattialean (*Asterotheca*, *Dizeugotheca*, *Psaronius*), and articulates (*Sphenophyllum*, *Annularia*). In some places most predominant are conifers (*Paranocladus*) and lepidophytes. Fossil lepidophytalean trunks, up to 58 cm in diameter, were described from the uppermost Carboniferous–lowermost Permian of Patagonia and La Rioja (Cesari, 1982; Cúneo and Andreis, 1983); they suggest a frost free climate in this area. The warming of the climate is witnessed by the Euramerian plants in coeval deposits of different parts of Argentina and Brazil. A remarkable feature is the absence of stigmarias in these lepidophytes, the same as in the Angara lepidophytes. In the Australoindian flora admixtures of Euramerian genera, as well as lepidophytes (besides mega- and microspores) are practically lacking. Apart from glossopterids, the Gondwana flora contains conifers *Buriadia* and *Walkomiella*, articulates (*Sphenophyllum*, *Phyllotheca*, *Paracalamites*, *Schizoneura*), osmundalean and other ferns. Leaves are known that are similar to *Cordaites* (commonly wrongly included in the genus *Noeggerathiopsis*). In the older Lower Gondwana assemblages they associate with *Botrychiopsis* and *Gangamopteris*. Certain Arberiales of Gondwana had foliage of the cycadophytes type (*Rhabdotaenia*, *Pteronilssonia*). Macrofossils of lepidophytes are scarce in Australoindian floras, but a profusion of diverse megaspores and microspores occur, that are referred to lepidophytes allied to selaginellas.

From the foregoing description it becomes immediately apparent that the *Glossopteris* flora exhibits striking differences from other Upper Palaeozoic floras, although the opinion long prevailed that it was closely related to the Angara flora. As a matter of fact, common genera are of very rare occurrence in these floras. Other common genera have been established from the vegetative shoots, whereas the associated fructifications differ considerably. For instance, vegetative shoots of *Phyllotheca*, associated with fertile shoots of *Tchernovia* (in Angaraland) and *Gondwanostachys* (in Gondwana), belong to different families.

On the other hand, Lower Gondwana miospore assemblages are very closely similar externally to the Permian of the Russian Platform, Fore-Urals, Western Europe and North America. This applies to certain spores (like *Granulatisporites*, *Calamospora*, *Horridisporites*, *Apiculatisporites*, etc.) and, particularly, to pollen. In Gondwana post-glacial deposits, where leaves of *Gangamopteris* are dominant, quasimonosaccate pollen prevails, whereas in those dominated by *Glossopteris* quasidisaccate pollen, with a striate or non-striate body, prevails. Pollen of the *Vittatina* type also occur. The same pollen types are known in the northern floras, although the form-genera may differ. The sequence of successive occurrence of the major types is also similar: at first quasimonosaccate pollen appear, then quasidisaccate, non-striate, later – the same but striate, and last of all – *Vittatina*. Discussions arose in the literature concerning the question whether this similarity in the miospores can be regarded as proof of the connection between the Gondwana and northern phytochoria. It is probable that the major factor that determined this similarity was parallelism in pollen evolution and not common eutaxa (Meyen, 1981, 1982b). For instance, it has been established that the Gondwana quasidisaccate striate pollen belongs – at least in part – to Arberiales, whereas the same pollen in the Subangara area belongs to Peltaspermales.

TRANSITION FROM THE PALAEOPHYTE TO MESOPHYTE

Concepts concerning the transition from the Palaeophyte to the Mesophyte were strongly influenced by the European material, on which they are based. In this area the Upper Permian (Zechstein) flora displays more affinity with the Triassic than with the Lower Permian flora. At the same time the Zechstein marine fauna exhibits a distinct Palaeozoic aspect. Since Cenophyte floras also already appear in Europe in the middle of the Cretaceous and not at the Mesozoic–Cainozoic boundary, the postulation was offered that the evolution of floras proceeds in advance of the faunal evolution (the 'law of evolutionary disconformity', R. Potonié, 1952). Of continuous appearance to date in the literature are schemes showing the periodization of the plant kingdom, according to which the Palaeophyte–Mesophyte boundary coincides with the Lower–Upper Permian boundary. A more scrupulous analysis of the factual data revealed that this opinion is fallacious. This does not mean that the boundary should be shifted to a different stratigraphic level. Comparison of the data on different phytochoria shows that the Palaeophyte–Mesophyte boundary is not a distinct world-wide changeover, but rather represents a transitional period of long duration over many millions of years, corresponding to several stages and several boundaries (Meyen, 1973b).

Detailed reconstructions of this transition are impossible, because of the inadequate stratigraphic data, and, particularly, due to the variance in correlations and large breaks in the stratigraphic section. For instance, certain Siberian assemblages are regarded by some palaeobotanists as Upper Permian, whereas they are regarded by others as Anisian. According to Gomankov (1983), the hiatus between the Tatarian and Triassic of the Russian Platform, corresponds to a part of the Dzulfian, the entire Dorashamian and lowermost Induan of Tethys. Notwithstanding the uncertainty in stratigraphic correlations, certain transitional assemblages may be nominated, some of which in all events occur in the Permian, while others occur in the Triassic. For instance, it is inconceivable how the Bunter sandstone flora of Western Europe could have passed on into the Permian from the Triassic, and the cordaite flora of the coal measures of Siberia from the Permian into the Triassic.

Mesophytic elements, i.e. members of those groups, which became dominant in the Mesozoic floras (to be more exact, from the middle of the Triassic to the middle of the Cretaceous), appear in various Permian floras. In the Zechstein of Western Europe voltzialean conifers and peltaspermaceous *Lepidopteris* appear. *Sphenopteris* species, described from here, are allied to *Rhaphidopteris* – a genus occurring commonly in the Jurassic, and closely related to the Jurassic genus *Pachypteris*. In the North American Lower Permian floras voltzialean conifers are present together with plants similar to cycads. The genera *Supaia* and *Glenopteris* probably belong to the Peltaspermales. Mesophytic elements are particularly abundant in the Subangara flora. They are Peltaspermales, Voltziaceae, Osmundaceae, etc. Lepidophytes *Viatcheslavia* and *Signacularia* might have been ancestors of the Triassic Pleuromeiaceae. Cathaysian floras are very rich in cycadophytes; some leaves have been directly assigned to the genera *Nilssonia* and *Pterophyllum*, albeit their epidermal structure has not been studied. In the Gondwana flora the Osmundaceae have been encountered. The Arberiales might have been the ancestors of the Mesozoic Pentoxylales. The above-cited Mesophytic elements can only in part be clearly linked with earlier plant types. For instance, peltasperms of the *Mauerites* and *Rhaphidopteris* types evidently evolved from older callipterids. Mesozoic conifers can be derived from the Permian Voltziaceae. The Mesozoic Ginkgoales and Leptostrobales probably evolved from the Subangara Peltaspermales.

The transition from the Palaeophytic assemblages to Mesophytic occurs in different areas at different times. In some places this transition is quite abrupt, whereas in others it is outstretched in the stratigraphic sequence, being of single-, or multistaged occurrence. Abrupt changes are accompanied by breaks in sedimentation or by unfossiliferous intervals. In western Europe a considerable proportion of the Zechstein miospore taxa continues up into the lowermost Triassic, but in most West European sequences the lowermost Triassic lacks palaeobotanical characteristics. On the Russian Platform and in the Fore-Urals many Mesophytic elements are noted from the Kungurian, but the flora acquires a distinct Mesophytic aspect in the upper part of the Tatarian, i.e. long before the end of the Permian (provided that the correlation of the Tatarian and Middle Permian of Tethys is correct). In Cathaysia many Mesophytic elements already appear in the Lower Permian. The change from Palaeophyte to Mesophyte is well pronounced in the miospore assemblages, according to which this change occurs in the northern province within the Upper Permian, whereas in the Southern province it already occurs during the Lower Triassic (Yao and Oyang, 1980). Obviously the oldest Mesophytic miospore assemblages of the northern province correspond to plant megafossil assemblages of the Zechstein type (Wang, 1985). In Gondwana the *Glossopteris* flora reaches beds commonly assigned to the Triassic (Panchet Series), although their Permian age cannot be excluded (Foster, 1979). The overlying deposits (Parsora Series), firmly belonging to the Triassic, contain certain elements of the *Glossopteris* flora. In Gondwana, as a whole, the transition between the floras is gradual.

The situation is more intricate in Siberia. Here on indisputable Permian coal measures with cordaite flora lies a thick sedimentary–volcanogenic series with a rich and specific Korvunchanskaya flora (this term is commonly used to define only the flora of the Tunguska Basin, but it is convenient to apply the epithet 'Korvunchanskaya' to the entire Siberian flora that replaces the cordaite flora. Hence, the term Korvunchanskaya flora may also be used in referring to the flora of the Maltsevskaya Series in Kuzbass). The Korvunchanskaya flora is usually assigned to the Triassic. However, if we rely upon the associated fauna of ostracods and phyllopods, then we cannot exclude the possibility that the lower part of the deposits containing this flora may correspond to the uppermost Tatarian, or to the hiatus between the Permian and Triassic in the sections of the Russian Platform (Sadovnikov, 1981).

The Korvunchanskaya flora is distinctly Mesophytic. Ferns predominate (*Cladophlebis*; in the lower part of the section Osmundales and Marattiales are more abundant; higher up Matoniaceae appear). There are many peltasperms (*Lepidopteris*, *Scytophyllum*, *Madygenopteris*, *Tatarina*, *Peltaspermum*, etc.), *Taeniopteris* is also present, and, of Palaeophytic elements, *Sphenophyllum* and *Yavorskyia*. The Late Korvunchanskaya (Putoranskaya, Maltsevskaya) flora is dominated by conifers *Quadrocladus* (associated with pollen of *Lueckisporites*), and, in places, Pleuromeiaceae (*Tomiostrobus*, *Pleuromeia*, spores *Aratrisporites*). Certain plants are typical of the Triassic (*Pleuromeia*, *Aratrisporites*, *Scytophyllum*, *Madygenopteris*), others, on the other hand, are typical of the Permian (*Sphenophyllum*, *Tatarina*, *Quadroclaus*, *Lueckisporites*). In either case the 'Permian' genera of the Korvunchanskaya flora (e.g. *Tatarina* and *Quadrocladus*) appear in Siberia much later than in the neighbouring Subangara area. However, in the Verkhoyansk region and in Taimyr the genus *Pleuromeia* was found that is of very wide occurrence in the Triassic. In India *Lepidopteris* appears only in the Triassic, whereas in the northern floras it is known from the Upper Permian. These facts and others suggest that the transition from Palaeophyte to Mesophyte was concomitant with the migration of plants. Evidently the barriers that formerly divided the phytochoria were breached and the climatic conditions became more uniform on a global scale. This implies that the transition from Palaeophyte to Mesophyte is a consequence of the general successive evolutionary changes in the Earth's physical condition (Meyen, 1973b, 1981).

TRIASSIC FLORAS

For a long time the succession in Triassic floristic assemblages was represented as reflected in European material. A general concept of the development of the plant kingdom during the Triassic, for the Earth as a whole, did not exist. A number of comprehensive reviews on Triassic floras (including miospores) of all major regions are available at present (J. M. Anderson and H. M. Anderson, 1983; J. M. and H. M. Anderson, 1985; Archangelsky, 1986b; Chaloner, in Tschudy and Scott, 1969; Dobruskina, 1980, 1982; Kimura, 1984; Lele, 1974–1976; Maheshwari, 1976; Saksena, 1974; Retallack, 1977).

In the development of the Triassic geochronological scale, particularly the series and stages, the palaeobotanical data were not taken into consideration. The evolutionary stages in the development of the vegetation cover do not match the units of the geochronological scale. In fact no such stages can be defined for the Earth as a whole. In Gondwana the *Glossopteris* flora was gradually replaced by the *Dicroidium* flora that existed to the end of the Triassic, becoming gradually more abundant in northern plant types. In the northern floras basic changes occurred in the middle of the Middle Triassic; thereafter the floras acquired a general aspect typical also of the Lower–Middle Jurassic assemblages. Because of this the Triassic floras should be discussed in terms of regional developmental stages. Dobruskina (1982) proposed that the major floras should be named after the typical genera, even when the pertinent genus may not be present in all floral localities, and may reveal a wide stratigraphic distribution. For instance, the flora of the Norian and Rhaetic is termed *Lepidopteris* flora, although the genus *Lepidopteris* already appears in the Permian, and is absent in the Norian localities of South-Eastern Asia.

The Triassic in most areas of the world is coal free. It is difficult to obtain continuous (in area and in vertical range) palaeobotanical information on the deposits even with the aid of palynological investigations. Besides, the relationships between the majority of Triassic miospore genera and the families and orders of higher plants are unknown.

As was noted in the previous section, the degree of phytogeographic differentiation at the beginning of the Triassic undergoes a sharp drop compared with the Late Permian. This is best reflected in the distribution of the family Pleuromeiaceae. The genus *Pleuromeia* or associated trilete cavate spores of the *Densoisporites* type are widespread over the area from Western Europe to China, Taimyr and Primorye. Seemingly, closely affiliated to *Pleuromeia* may be the genera *Tomiostrobus* (Kuzbass, Taimyr, Pechora, Fore-Urals, Verkhoyansk region), *Cylostrobus* (Australia, South America) and *Skilliostrobus* (Australia) that produced monolete cavate microspores of the *Aratrisporites* type. The latter occur nearly all over the world and are known in higher sections of the Triassic, where they associate with other lepidophytes. The pleuromeiaceous genera usually occur in assemblages, en masse; other macrofossil plant types are either lacking or rare. These layers include many aquatic invertebrates, and, in places, amphibia. In different localities this fauna may be freshwater, brackish or marine. It is believed that pleuromeiaceous plants comprised dense thickets on flat and frequently inundated coasts. Pleuromeiaceous micro- and megaspores are frequently the most predominant miospores in palynological assemblages. Certain other miospores are of wide occurrence in the Triassic and in transitional Permian–Triassic beds. This applies first of all to the quasidisaccate pollen of the *Taeniaesporites* type with wide ribs on the distal side of the body. This pollen could have been produced by some kind of voltzialean conifers.

Although some Early Triassic genera are of very

wide occurrence, a certain regionalism can be perceived in their distribution pattern. The zonation of Triassic floras outlined below is given according to Dobruskina (1982) with certain amendments. As a whole, the major phytochoria of the Late Permian continue to exist during the Early Triassic and Anisian, although the distinctions between the floras are less contrasting. The Gondwana and Angara phytochoria are preserved in the rank of areas. Between them is the Laurasian area, to which we can assign the '*Voltzia*' flora of the Bunter sandstone of Western Europe (Lower Triassic–Anisian). It includes, besides *Pleuromeia*, numerous conifers (*Voltzia*, *Aethophyllum*, *Yuccites*), articulates (*Schizoneura*, *Echinostachys*, *Equisetites*, *Equisetostachys*), ferns (*Neuropteridium*, *Crematopteris*) and cycadophytes (Grauvogel-Stamm, 1978). A miospore assemblage similar to the European has been found in China (Wang, 1985; Yao and Oyang, 1980). We can judge the Lower Triassic flora of North America only from the miospore assemblages of Greenland (Balme, 1979) and Canada, which are similar to the European. These assemblages contain quasisaccate striate (*Protohaploxypinus*, *Taeniaesporites*), non-striate (*Klausipollenites*), asaccate striated (*Gnetaceaepollenites*) and non-striate (*Cycadopites*) pollen of gymnosperms, and spores, a great proportion of which belong to lycopsids (*Kraeuselisporites*, *Densoisporites*, etc.; Dobruskina and Yaroshenko, 1983).

The *Pleuromeia* and *Voltzia* floras disappeared at about the middle of the Middle Triassic. Later Triassic floras became ever more similar in composition to the Jurassic. At times their burials show a greater resemblance to those of the Jurassic, rather than to the Upper Permian and Lower Triassic. Because of this beginning from the Middle Triassic, the same major phytochoria as in the Jurassic may be adopted for the Extra-Gondwana floras. Two major floras can be outlined in the Equatorial area. One corresponds to the Ladinian and Carnian and is termed the *Scytophyllum* flora. The other, Norian–Rhaetic, is termed the *Lepidopteris* flora. Their localities are known throughout the Equatorial area. The richest localities are restricted to the upper Middle- and Upper Triassic in Western Europe (German Basin), i.e. to the regional stratigraphic subdivision 'Keuper'. Frequently all the Extra-Gondwana floras of this age are referred to the Keuper flora. Here we encounter practically all the major Mesozoic plant groups, appearing at different stratigraphic levels. They include Bennettitales (*Pterophyllum*, *Sturiella*), Caytoniales (*Sagenopteris*) and Leptostrobales (*Czekanowskia*) firmly unknown in older deposits. Very abundant are Ginkgoales (*Ginkgoites*, *Sphenobaiera*), Cycadales (*Nilssonia*, *Ctenis*), ferns of Mesozoic families and genera (*Cladophlebis*, *Clathropteris*, *Camptopteris*, *Phlebopteris*, *Danaeopsis*), horsetails (*Equisetites*, *Neocalamites*). The conifers belong to more advanced Voltziaceae (*Cycadocarpidium*, *Tricranolepis*, *Swedenborgia*, *Podozamites*). The first Cheirolepidiaceae occur (*Brachyphyllum*, pollen of *Classopollis*); probably also present are Pinaceae and Podocarpaceae. Of the plant groups inherited from the Permian Marattiales (*Asterotheca*) have been encountered; widespread are Peltaspermaceae (*Scytophyllum*, *Lepidopteris*, *Peltaspermum*, *Glossophyllum*). The Equatorial area extends from Western Europe through all of Eurasia to China, Japan and South-Eastern Asia. The distinctions between the floral assemblages in various parts of this territory are immeasurably less than in the Permian, and somewhat less than in the Jurassic. The flora exhibits conspicuous changes in a west–east direction, which renders possible the distinction of provinces (sectors, after Dobruskina), as shown in Fig. 88. For example, in the east of the area the Marattiales are absent. In Donbass and further east conifers of the *Podozamites* and *Cycadocarpidium* types are more abundant.

The Madygen flora of Fergana belongs to the Equatorial subarea. This flora became widely known due to the errors in its datings. Although it is composed of plants of the *Scytophyllum* flora and displays many features in common with other Ladinian–Carnian floras of Eurasia (endemics also occur), it has been assigned to the uppermost Permian–lowermost Triassic; Permian Cathaysian, Angara, Gondwana and Zechstein genera have been wrongly recorded among this flora. This led to phytogeographical and florogenetic misconceptions. The Upper Triassic floras of East Asia were divided by Chinese palaeobotanists (Li and Zhou, 1979, quoted from Kimura, 1984) into the *Danaeopsis–Bernoullia* (North China) and *Dictyophyllum–Clathropteris* (South China) floristic provinces. The boundary between them roughly corresponds to that between the Siberian and European–Sinian areas of Dobruskina.

The palaeobotanical findings for the upper part of the Carnian and lower part of the Norian in Eurasia are not reliable. The Upper Norian–Rhaetic ('*Lepidopteris*') flora contains besides Peltaspermaceae (*Lepidopteris*), abundant bennettites (*Pterophyllum*, *Anomozamites*, *Wielandiella*), cycads (*Ctenis*, *Pseudoctenis*, *Nilssonia*), conifers (*Cycadocarpidium*, *Elatocladus*, etc.), ginkgos (*Ginkgoites*, *Baiera*, *Sphenobaiera*), leptostrobans (*Czekanowskia*, *Hartzia*, *Phoenicopsis*). There are many ferns belonging to the Marattiales (*Marattiopsis*, *Danaeopsis*), Osmundales (*Todites*), Dipteridaceae (*Dictyophyllum*, *Clathropteris*), Matoniaceae (*Phlebopteris*). Nearly all these plants continue up into the Jurassic. The Rhaetic and Lower Liassic floras compose a uniform assemblage. During the Norian–Rhaetic the latitudinal zonality in the Equatorial area is more clearly reflected, whereas the distinctions between the west and east become smoothened. Nevertheless,

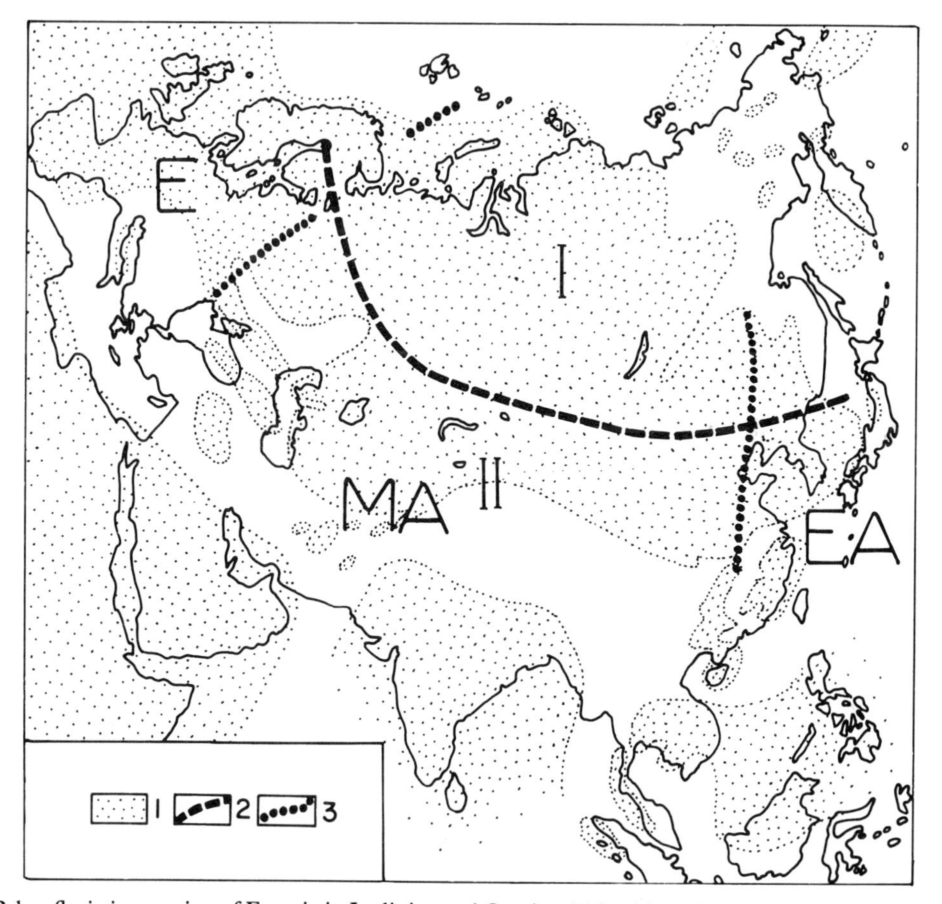

Fig. 88. Palaeofloristic zonation of Eurasia in Ladinian and Carnian (Triassic); I, Siberian area; II, European–Sinian area; European (E), Middle Asiatic (MA) and East Asiatic (EA) provinces (sectors); 1, land; 2, boundary between areas; 3, boundaries between provinces. Modified from map provided by Dobruskina in 1984.

many florules contain endemic genera and species of local occurrence.

The Middle–Upper Triassic flora of North America (Atlantic Coast and Southwestern USA) may belong to a separate subarea or province of the Equatorial area (Dobruskina and Yaroshenko, 1983). Genera occur here that are common to Eurasia (*Selaginellites*, *Neocalamites*, *Equisetites*, *Todites*, *Phlebopteris*, *Clathropteris*, *Pseudoctenis*, *Zamites*, *Otozamites*, *Brachyphyllum*, *Voltzia*), but some genera are endemic and some are quite peculiar (*Dinophyton*, *Marcouia*, *Eoginkgoites*). A striking fact is the absence of indisputable peltasperms. In the southwest of the USA conifers and *Dinophyton* predominate; ferns and cycadophytes are less abundant. In northeastern USA cycadophytes predominate, ferns are less abundant; even more scarce are conifers and articulates; ginkgos and leptostrobans are completely absent.

The Siberian–Canadian area includes Siberia, the Pechora Fore-Urals, the eastern slope of the Urals, Spitsbergen, most of Mongolia and Primorye. These regions are dominated by horsetails, ferns (*Cladophlebis*, *Todites*), peltasperms (*Lepidopteris*, *Scytophyllum*, *Glossophyllum*), and ginkgos (*Sphenobaiera*). Leptostrobans and caytonias (*Sagenopteris*) are also known here. Cycadophytes and Dipteridaceae are either absent or very rare.

The flora of the Gondwana area (J. M. Anderson and H. M. Anderson, 1983; J. M. and H. M. Anderson, 1985; Retallack, 1977, 1980), known from all Gondwana continents, is termed the *Dicroidium*-flora after the dominant genus *Dicroidium* (Umkomasiaceae), which is practically omnipresent. The genera *Johnstonia* and *Xylopteris* are closely related to this genus; associated fructifications belong to *Umkomasia*, *Pilophorosperma* and *Pteruchus*, and stems to *Rhexoxylon*. It is probable that this family is of northern (Subangara?) origin. Also present here are conifers *Voltziopsis* and *Rissikia*, as well as putative northern immigrants *Lepidopteris* and *Peltaspermum*. In some places numerous articulates (*Neocalamites*) and ferns (*Todites*, *Cladophlebis*, *Asterotheca*) occur; cycadophytes of northern origin (*Taeniopteris*, *Otozamites*, *Pseudoctenis*) and ginkgos (*Ginkgoites*, *Baiera*, *Stachyopitys*) have been encountered. There are also genera of uncertain affinities (*Chiropteris*, *Yabeiella*, *Linguifolium*) and problematical gymnosperms *Heidiphyllum* (which resemble *Podozamites*). This flora is most prolific from the second half of the Lower Triassic up to the Carnian; it probably existed to the end of the Triassic. According to the repeated occurrence of associations different types of the *Dicroidium* vegetation have been recognized, that correspond to coastal-marine lowlands, river valleys, broad-leaved forests, etc. The uppermost Triassic coastal-marine sediments of Australia yield associations with numerous remains of *Pachypteris* (= *Pachydermophyllum*), i.e. the genus typical of the European Jurassic. Retallack (1977) maintains that this association belongs to thickets of the mangrove type. Such thickets are commonly developed under conditions of a frost-free climate. The indistinct geographical differentiation of the Middle–Late Triassic floras of the Equatorial, Siberian and Gondwana areas, as well as the wide occurrence of certain genera, suggest that the climatic zonality was not pronounced. It is probable that a frostless climate prevailed all over the Earth.

JURASSIC AND LOWER CRETACEOUS FLORAS

Considering the inheritance of floras, it is convenient to discuss the Lower Cretaceous and Jurassic floras together. In the evolution of the vegetation cover within the time span from the beginning of the Jurassic to the middle of the Cretaceous two developmental stages can be recognized, that reflect the major climatic changes. In the Early–Middle Jurassic a warm and humid climate prevailed. In the Late Jurassic the Equatorial area became much drier, this did not particularly influence the Siberian area, but it evoked the northward shifting of the areal boundaries. The pronounced distinctions between the Gondwana and northern floras disappear in the Jurassic.

The Jurassic and Early Cretaceous are usually referred to as the period of gymnosperm supremacy. As a matter of fact, in the Equatorial area and contiguous to it at the north and east ecotone regions, the dominant plants were cycads, bennettites, conifers, and, in places, peltasperms. In the Siberian area ginkgos and leptostrobans were predominant, but ferns were not inferior to them. It is remarkable that the Jurassic and Lower Cretaceous floras can be so succinctly characterized. This can be explained by the extensive information available on the composition of the floras, as well as by the distinct distribution of the dominant and subordinate groups.

The Jurassic and Cretaceous areas differ not only in the prevalence of definite major groups, but also in their generic and species composition. Nevertheless, floras of different areas exhibit many common genera and even species. Because of this, in the Jurassic and Cretaceous, as in the Triassic, areas – not kingdoms (as in the Late Palaeozoic) – are taken as the major phytochoria, despite the fact that the taxonomic affinities between the areas are greater in the Jurassic and Cretaceous than during the Late Palaeozoic. The nomenclature and phytochoria characteristics are given below after Vakhrameev (1964, 1978, 1981, 1984). Krassilov (1972c, 1985) suggested other names for the major phytochoria. Detailed descriptions of the Mesozoic flora of Eastern Asia have been presented by Kimura (1984). Data on the Jurassic palynofloras of Siberia were summarized by Ilyina (1985).

Lower–Middle Jurassic floras

The Lower and Middle Jurassic floral zonation coincide (Fig. 89). The Siberian area extends over north-eastern Europe, the Urals (with the exception of its southern extension), most of Kazakhstan, all of Siberia and the greater northern part of China. The climate in this area was warm and humid, favouring coal formation, that was particularly intensive during the Middle Jurassic. The Jurassic vegetation of the Siberian area was formerly termed the 'Ginkgoalean taiga'. It was later established that some of the plants, assigned to the Ginkgoales (*Phoenicopsis*, *Czekanowskia*), belong to the Leptostrobales. The Ginkgoales are represented here by *Ginkgo*, *Baiera*, *Sphenobaiera*. The Lower–Middle Jurassic assemblages include many ferns belonging to *Cladophlebis*, *Coniopteris*, *Raphaelia*, etc. Conifers are represented by the genera *Schizolepis*, *Pityophyllum* (Pinaceae) and *Podozamites*. Other plants are less abundant. They include singular bennettites (*Pterophyllum*, *Anomozamites*), cycads (*Nilssonia*), caytonias, marattiaceous, matoniaceous and dipteridaceous ferns. Of rare occurrence or altogether absent are Cheirolepidiaceae, which can be established chiefly by the presence of pollen of *Classopollis*. Only in certain places of Siberia the Toarcian have yielded a large amount (up to 20%) of this pollen and certain spores typical of more southern areas. It is probable that, as reflected in the palaeotemperature measurements, the climate became considerably warmer during the Toarcian. In the Middle Jurassic, in contrast to the Lower Jurassic, ferns decrease in diversity, bennettites and cycads increase in number, *Sagenopteris* appears.

The Equatorial area extended through nearly all of Western Europe, the Caucasus, Middle Asia, Southern China and South-Eastern Asia. In America localities are known in Greenland (upper assemblage of the Rhaetic–Liassic flora of the Scoresby Sound), at the south of Alaska, Vancouver Peninsula, Mexico and Cuba. Most prolific in this area are Bennettitales (*Anomozamites*, *Pterophyllum*, *Ptilophyllum*, *Otozamites*, *Nilssoniopteris*, *Zamites*, *Williamsonia*, *Williamsoniella*), Cycadales (*Nilssonia*, *Beania*, *Ctenis*, *Pseudoctenis*), Matoniaceae (*Phlebopteris*), Marattiaceae (*Marattiopsis*) and Dipteridaceae (*Clathropteris*, *Dictyophyllum*, *Thaumapteris*). In the Middle Jurassic, besides these, of abundant occurrence are *Coniopteris*; typical elements are *Klukia*, *Gleichenia* and *Stachypteris*; the dipteridaceous ferns decrease in number. The coal-free coastal-marine deposits contain numerous Peltaspermales (*Pachypteris*, *Rhaphidopteris*) and Cheirolepidiaceae (*Brachyphllum*, *Pagiophyllym*, *Classopollis*). Conifers are represented by Pinaceae (*Schizolepis*), Taxodiaceae (*Elatides*) and *Podozamites*. Also present are Ginkgoales (*Ginkgo*, *Baiera*, *Sphenobaiera*) and Leptostrobales (*Czekanowskia*). In some places the localities are dominated by articulates (*Equisetum*, *Neo-*

calamites). The Equatorial flora, as a whole, is much richer than the Siberian–Canadian. For instance, in the Lower Jurassic of the Siberian–Canadian area about 70 genera and over 120 species have been established, whereas in the coeval Equatorial flora over 200 genera and 500 species have been established.

In the Equatorial area several provinces have been recognized (so far only in Eurasia), according to the proportions of the above-cited groups. In the European province (Europe, the Caucasus) all the above-cited groups are present, but ginkgos and leptostrobans are sparse and restricted to coal-bearing deposits. The Yorkshire flora belongs to this province, and is one of the best studied floras of the world. The Middle Asiatic province differs from the previous in the more abundant ginkgos, leptostrobans, nilssonias and *Cladophlebis*; on the other hand, the abundance of conifers and pteridosperms is less. In the East Asiatic province (South China) the Cheirolepidiaceae are rare and the Leptostrobales are practically absent.

In the Middle Jurassic burials different plant associations can be distinguished, that correspond to different types of vegetation (Krassilov, 1975). Among them are horse-tails, the remains of which

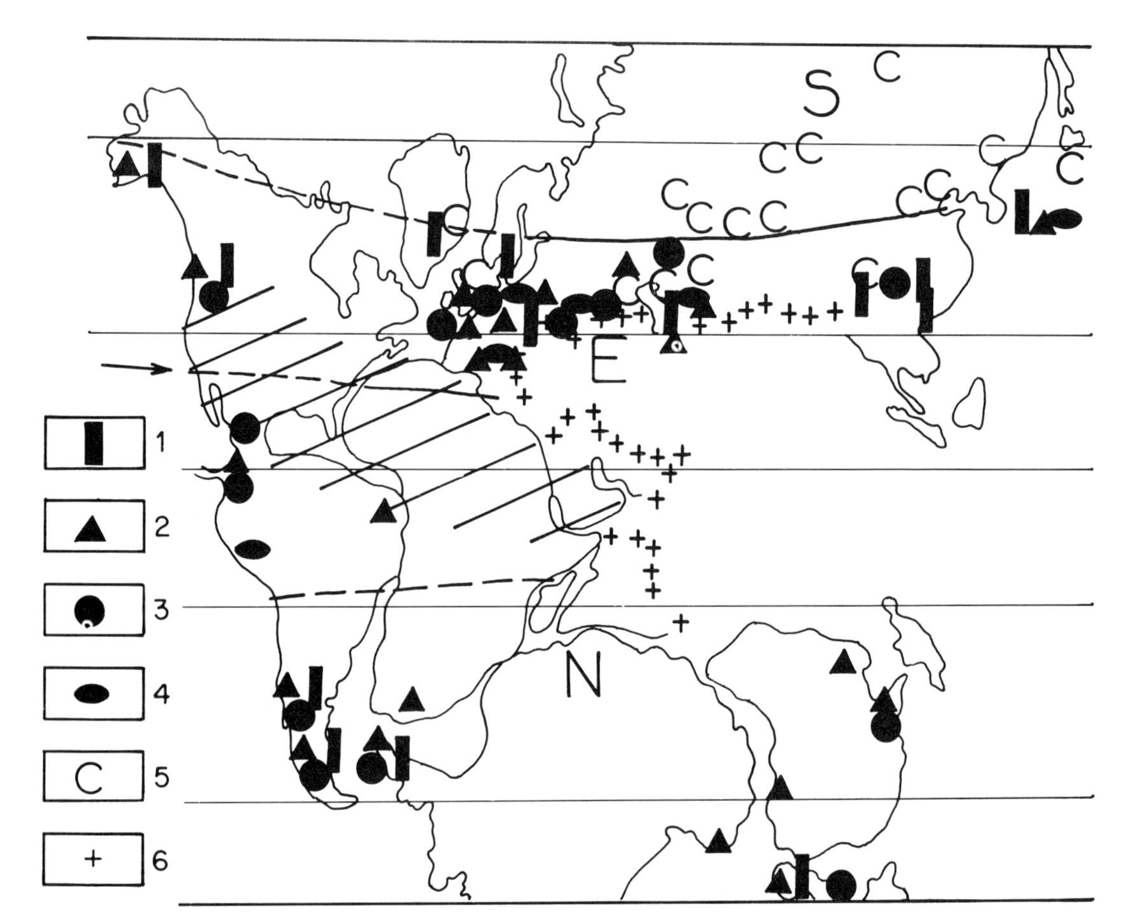

Fig. 89. Palaeofloristic zonation of Early and Middle Jurassic, and distribution of some characteristic plants; S, Siberian area; E, Equatorial area, shading shows arid belt with a putative palaeofloristic boundary within (at arrow); N, Notal area; 1, *Dictyophyllum*, *Thaumatopteris*, *Phlebopteris*, 2, *Otozamites*, *Dictyozamites*, *Zamites*, *Sphenozamites*; 3, *Ptilophyllum*; 4, *Klukia*; 5, czekanowskians (Leptostrobales); 6, shore line of Tethys ocean. Modified from map provided by Vakhrameev in 1984.

at times comprise coal seams. The plant association with numerous *Ptilophyllum* is restricted to coastal-marine, at times gypsiferous facies, and probably occupied the ecological niche of the modern mangrove vegetation. Other plant associations, including forest types with different dominants, have been recognized. The Equatorial area in the present understanding corresponds to the European–Sinian and Equatorial areas of Vakhrameev (1985) together. Vakhrameev draws the boundary between these areas within the arid belt (Fig. 89, at arrow).

The Lower–Middle Jurassic floras of the Gondwana continents have been much more poorly studied and their localities are scanty. They are known in Brazil, Columbia, Argentina, Australia, and New Zealand. After the decline of the *Dicroidium* flora the Gondwana continents were populated by plants known in the Equatorial area. Among the dominant groups, as in the Equatorial flora, are bennettites (*Otozamites*, *Zamites*, *Dictyozamites*, *Ptilophyllum*), dipteridaceous and matoniaceous ferns, *Coniopteris*, conifers *Brachyphyllum* and *Pagiophyllum*. In some places (e.g. in the lowermost Jurassic of Australia) a large percentage of *Classopollis* is observed in the miospore assemblages. However, in contrast to the Equatorial area, Gondwana is poor in Ginkgoales, and Leptostrobales are absent, *Nilssonia* and *Podozamites* are scarce, and Pinaceae are absent. There are Podocarpaceae (*Mataia*). Pollen of Podocarpaceae is of common occurrence in the palynological assemblages.

Although the Gondwana flora displays less distinctions from the Equatorial flora than the Siberian–Canadian, the territory where it occurs can be identified as a separate area. This area was formerly termed 'Australic', which is seemingly etymologically felicitous, but its translation into other languages is fraught with difficulties (it is usually confused with the adjective 'Australian'). Because of this it is more convenient to adopt the term Notal area.

Upper Jurassic and Lower Cretaceous floras

The Upper Jurassic is discussed below in the range accepted in the USSR, i.e. including the Callovian, which in Western Europe is assigned to the Middle Jurassic. From the palaeobotanical and palaeoclimatic viewpoints the Callovian is more closely related to the Late Jurassic (Doludenko, 1984). This epoch marks the beginning of a strong warming, which is reflected in the shifting towards the north of the boundaries between the areas, and in the development of the Equatorial arid belt. In some areas (Middle Asia, south of Western Siberia), the aridization commenced at the end of the Middle Jurassic.

Both the composition and interrelations between the floras that were established during the Late Jurassic, in general features persisted during the Early Cretaceous (Fig. 90). Because of this the floras of both epochs are discussed together. This is also convenient considering that the datings for certain assemblages are not warranted. This refers particularly to the Indian flora of the Rajmahal Group. It has been assigned to different epochs (more often to the Middle Jurassic), but now some palynologists refer it to the Early Cretaceous.

In the Late Jurassic and Early Cretaceous four

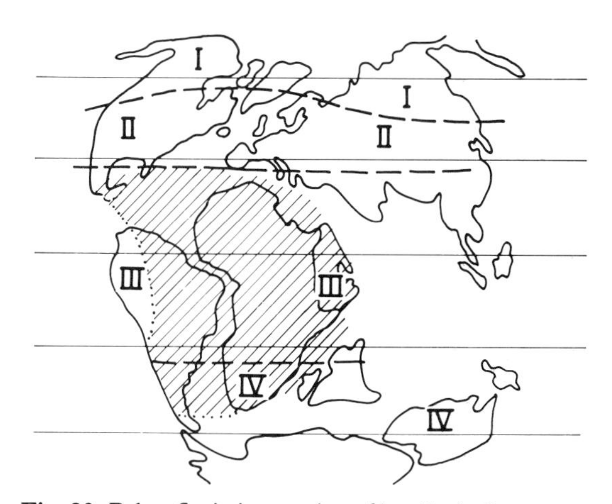

Fig. 90. Palaeofloristic zonation of late Early Cretaceous (Albian); Siberio–Canadian (I), European-Sinian (II), Equatorial (III) and Notal (IV) areas; shading shows arid and semiarid environments. From Vakhrameev, 1984.

major areas have been recognized, corresponding to different palaeobotanical climatic belts. They include from north to south: the Canadian–Siberian, European–Sinian, Equatorial (North Gondwana) and Notal (South Gondwana) areas. The distribution of these areas is shown in Fig. 90. The first phytochorion is termed Siberio–Canadian, but not Siberian, since there is reason to believe that this flora extended into Canada, at least during the Early Cretaceous. The independent status of the European–Sinian and Equatorial areas (in the Early–Middle Jurassic this territory constituted a single Equatorial area with several provinces) is determined by the formation of the arid belt, which, as was mentioned above, corresponds approximately to the Equatorial area. According to its palaeofloristic and palaeoclimatic features the Notal area corresponds to the European–Sinian. It is probable that a southern analogue of the Siberian–Canadian area did not exist during the Late Jurassic and Early Cretaceous.

The Siberio–Canadian area during the Late Jurassic and Early Cretaceous occupied the same area as earlier, but its southern boundary was shifted considerably northwards (in Eurasia). The climate of the area was humid and, for the most part, frostfree. In many places, and especially in Siberia, coal accumulation continued. The generic composition of ginkos and leptostrobans was of general persistence. Both groups have been studied in most detail in the Upper Jurassic–Lower Cretaceous of the Bureja Basin, where (besides *Ginkgo*, *Baiera* and *Sphenobaiera*), *Eretmophyllum*, *Pseudotorellia* and some other genera were also recorded. Among the conifers most predominant throughout the area are *Pityophyllum* and *Podozamites* – more seldom, *Elatides* and *Athrotaxites* – and – even more rare – Cheirolepidiaceae. Cycadophytes increase in number (*Nilssonia*, *Ctenis*, *Pterophyllum*, *Butefia*, *Heilungia* = *Sinophyllum*), and on the basis of their abundance the Amur province is distinguished. Ferns are of abundant occurrence (they are particularly diverse in the Amur province). *Phlebopteris* and *Clathropteris* disappear, and various genera of Osmundaceae and Gleicheniaceae appear; as before there are many *Cladophlebis*, *Coniopteris* and *Raphaelia*. Of wide occurrence are horsetails (*Equisetum*); leafy mosses and liverworts have been encountered. Besides the Amur province, the Lena province is distinguished. Both provinces persisted in the Neocomian.

The Lower Cretaceous flora of the Siberian area shows many features in common with the Upper Jurassic, but distinctions are also apparent. The flora is more diverse, due to the increase in the number of species, but of particular importance is the incoming (from the Albian) of angiosperms (*Trochodendroides*, *Menispermites*, *Celastrophyllum*, *Cissites*, etc.). In the Albian–Cenomanian of Western Siberia (Kya River) a pollen appears, that has been ascribed to *Clavatipollenites incisus* Chlon. (Chlonova, 1984), but differs from typical representatives of this genus in the absence of a tectum and the arrangement of columellae in the form of a more or less regular reticulum.

The European–Sinian area (earlier Vakhrameev considered it a subarea of the Indo–European area) is subdivided into several provinces, and extends from Europe into Middle Asia and further on to the east of Asia. Unlike the Siberio–Canadian flora, that of the European–Sinian area exhibits a lesser degree of inheritance from older floras of the same territory. Coal accumulation ceases practically everywhere, and arid and semiarid climatic conditions prevail. This is reflected in the taxonomic composition of the floras (a pronounced decrease in the role of ferns and horse-tails), and determined the xeromorphic habit of many plants (leathery leaves, thick cuticles with deeply sunken stomata). Xeromorphism is particularly typical of many Bennettitales (*Pterophyllum*, *Ptilophyllum*, *Otozamites*, *Cycadeoidea*, etc.), Cycadales, Peltaspermales (*Pachypteris*) and Cheirolepidiaceae (*Brachyphyllum*, *Pagiophyllum*, *Frenelopsis*). The latter are typical elements of the entire area. They produced large amounts of *Classopollis* pollen, which dominates the miospore assemblages, in some places completely comprising them. Different opinions exist concerning the habitats of the Cheirolepidiaceae. It has been suggested that they comprised mangrove thickets or grew along the sea coasts. Studies of the burials of cheirolepidiaceous logs from the Upper Jurassic of England (Francis,

1983) serve to indicate that the climate was markedly seasonal (warm and humid winter, hot and dry summer); similar extant forests of *Callitris* are known along the Rottnest Island coast at the south of Western Australia. Analysis of the distribution pattern of the remains (including pollen) of the Cheirolepidiaceae shows that these plants could have existed in a wide range of habitats, growing at times far from the shoreline. Of other plants mention should be made of ferns (*Nathorstia*, *Stachypteris*, *Weichselia*), sometimes with thick xeromorphic pinnules, the Caytoniales, Pinaceae, Araucariaceae and Podocarpaceae. The Ginkgoales and particularly Leptostrobales are rare and in the Upper Cretaceous leptostrobans disappear completely, as do ferns of the genus *Coniopteris*. In the west of the area in the beginning of the Cretaceous, the humidity of the climate increased, and along with this a considerable increase is observed in the number of schizaeaceous (*Ruffordia*, *Pelletixia*, *Cicatricosisporites*), gleicheniaceous (*Gleichenia*, *Gleicheniidites*) and matoniaceous (*Matonidium*, *Phlebopteris*) ferns. The aridity was also less pronounced at times in other places and then the Cheirolepidiaceae were replaced by other conifers, primarily by Pinaceae. In the Barremian of first appearance is the pollen of angiosperms, whereas in the Aptian, and particularly in the Albian, their leaves, at times fructifications and wood fragments appear first. The successive appearance of angiosperms has been discussed in Chapter 3. Here it should be noted that the most complete angiosperm assemblages are restricted to the Aptian–Albian of the USA Atlantic Coast (Potomac province) and Western Kazakhstan. In the Albian the dispersal of the tricolpate pollen matches that of peltate leaves (probably aquatic ancestors of the subclass Rosidae) and palmately dissected leaves (allied with the ancestors of Hamamelidales).

The foregoing description refers chiefly to the European province. The Middle Asiatic province differs somewhat in the proportions of certain genera. The East Asiatic province was located outside the arid belt (it thins out eastwards). Its Upper Jurassic flora has been poorly studied. During the Early Cretaceous the East Asiatic flora was of variegated composition. It contained numerous ferns (*Ruffordia*, *Gleichenia*, *Nathorstia*, *Onychiopsis*, *Polypodites*), cycads and bennettites (*Nilssonia*, *Nilssoniopteris*, *Pterophyllum*, *Zamites*), conifers (*Athrotaxites*, *Brachyphyllum*, *Pagiophyllum*, *Araucariodendron*, *Ussuriocladus*, *Nageiopsis*); there are also leaves and fruits of angiosperms (in the Albian). In some places in East Asia plant assemblages show affinities to the European Lower Cretaceous assemblages containing *Weichselia* and Cheirolepidiaceae.

The Equatorial area includes localities in South America (excluding Patagonia and South Chile) and most of Africa (excluding the southern end). At present it is difficult to judge the floristic distribution of this area in relation to other places. The flora of this area has been reconstructed chiefly on the basis of palynological data (Brenner, 1976; Doyle *et al.*, 1977, 1982; Herngreen and Chlonova, 1981). Most typical is the presence of abundant *Classopollis* pollen, co-occurring with pollen of other gymnosperms, namely of *Araucariacites*, asaccate *Dicheiropollis* (probably conifers allied to the Cheirolepidiaceae), striated pollen *Ephedripites*, also with the genera *Elaterosporites*, *Elaterocolpites*, *Galeacornea*. In the Barremian angiosperm pollen appears, that is represented at first by forms related to *Clavatipollenites*, also *Liliacidites*, *Stellatopollis* and *Asteropollis*. For the Aptian and Albian the genus *Afropollis* is particularly characteristic (Doyle *et al.*, 1982). At times it penetrated to the south of the European–Sinian area. The further succession of the angiosperm pollen has been discussed in Chapter 3. The Equatorial area is located mainly in the arid belt, which is evidenced also by the lithological criteria. Only around the margins of this area was the climate more humid.

Plant megafossils are known only from a few localities. These are chiefly ferns *Weichselia* (associated spores belong to *Dictyophyllidites*) and *Piazopteris*, also gymnosperm woods. More complete plant assemblages have been described from Columbia (Pons, 1982). They contain *Cupressinocladus*, *Weichselia* and leaf fragments of angiosperms affiliated with Hamamelidae, Dilleniidae and Magnoliidae. The assemblage with platanoid leaves already belongs to the Upper Albian or Lower Cenomanian.

The Notal area includes the southern parts of Africa and South America, India, Australia and Antarctica. Unfortunately, the rich flora of the Rajmahal Hills in India (Surange *et al.*, 1974) is not reliably dated. This flora consists of numerous permineralized remains of Bennettitales (including fructifications), Pentoxylales (the major source of our knowledge of these plants) and other plants. The Rajmahal flora was earlier regarded as Middle Jurassic or somewhat younger, but according to the palynological data it is conceded to be more readily of the lowermost Cretaceous (*Cicatricosisporites* occur here, but angiosperm pollen does not as yet appear). The Notal area is characterized by abundant Bennettitales (*Otozamites*, *Dictyozamites*, *Ptilophyllum*, *Pterophyllum*, etc.), Cheirolepidiaceae (*Brachyphyllum*, *Pagiophyllum*, *Classopollis*) and other conifers, including Araucariaceae and Podocarpaceae. Various ferns occur (*Cladophlebis*, *Onychiopsis*, *Hausmannia*, *Gleichenites*), some Ginkgoales (*Ginkgoites*, *Karkenia*), Caytoniales, horse-tails. The Pentoxylales are known from the Upper Jurassic and Lower Cretaceous of India, Australia and New Zealand. This group was probably quite widespread in the Notal area (Drinnan and Chambers, 1984). The first angiosperm pollen appears in the Notal area later than in the Equatorial area. In Australia, from where fuller data are available, *Clavatipollenites* appears in the Lower Albian, and, at about the same level, *Asteropollis*. In Patagonia the Lower Cretaceous yielded many peculiar gymnosperms (*Mesosingeria*, *Almargemia*) and plants with fern-like foliage that can be affiliated with Cycadales (*Ruflorinia*, *Ticoa*, etc.). Typical pollen of *Dacrycarpus* (Podocarpaceae) appears in the Upper Jurassic of New Zealand. Pentoxylales have been found here. The entire area shows the absence of Leptostrobales and Pinaceae (saccate pollen similar to pinaceous probably belong to the Podocarpaceae). Outstanding among these floras is the Madagascar flora, which Appert (1973) assigned to the Upper Jurassic. This flora contains chiefly ferns – schizaeaceous (*Mohriopsis*), matoniaceous (*Phlebopteris*, *Matonia*, *Piazopteris*), *Coniopteris*, *Dictyophyllum*, etc. Together with them horsetails have been found. Vakhrameev considers the Notal area to be the opposite, not of the Siberio–Canadian area, but of the European–Sinian area.

The Upper Jurassic–Lower Cretaceous floras are of particular interest because angiosperms appeared first in them. According to the palynological data they appear first of all (in the Barremian) on both sides of the modern Atlantic – in Equatorial Africa, Brazil, England, i.e. in the Equatorial area and at the south of the European–Sinian area. The first pollen of angiosperms is associated with a great abundance of *Classopollis* pollen, and is restricted to sediments that accumulated under arid conditions. Most likely angiosperms appeared under arid climatic conditions on the sea coasts of low latitudes, although they were capable of populating habitats of higher moisture. Their further dispersal coincides with the drift episodes and moving apart of the Gondwana continents. A progressive migration of angiosperms inland of the continents has been noted. Evidently the current opinion is correct that the first angiosperms had considerable advantages over gymnosperms in being capable of populating disturbed and other unstable habitats (Retallack and Dilcher, 1981a). They composed pioneer associations and during all kinds of palaeogeographical rearrangements supplanted the gymnosperms. By the end of the Aptian and in the Albian they spread world-wide. The palaeobotanical data do not support the popular hypothesis of the first appearance of angiosperms in Southeastern Asia. Judging by palynological observations, angiosperms appeared here at the end of the Aptian–Albian, the same as in most areas of the Earth.

The Late Jurassic–Early Cretaceous floras evolved variously in different phytochoria, but certain changes were evidently synchronous and associated with global climatic changes – warming and cooling episodes. The latter are established according to the percentage of cycadophytes in the assemblages, the dissection of the leaf laminae, through reconstructions of the altitudinal vegetation belts, and their migration. The above-mentioned Late Jurassic warming of the climate gave way to a short-term Valangian–Barremian cooling span, which was followed again by the Aptian warming episode and then by the cooling of the climate at the end of the Albian.

This outline of the Early Cretaceous history of floras was based on plant megafossils, whereas palynological material was of subordinate significance. However, there are also palaeofloristic zonation schemes for the Early Cretaceous, based chiefly on miospore data. The most complete and detailed schemes of such a type have been worked out by Herngreen and Chlonova (1981). The palynological zonation is in general agreement with that of plant megafossils, but certain distinctions are evident. For instance, in the Siberio–Canadian area a phytochorion has been distinguished at the north and northeast of Siberia that is termed the Boreal–Arctic subprovince, and is characterized by its impoverished miospore composition and complete absence of *Classopollis*. On miospore zonation maps the boundary between the Siberio–Canadian and European–Sinian areas is not reflected. This can probably be explained by the fact that on the basis of pollen data it is impossible to distinguish phytochoria that differ in the proportions of ginkgos, leptostrobans, cycads and bennettites.

TRANSITION FROM MESOPHYTE TO CENOPHYTE. UPPER CRETACEOUS FLORAS

In many places of the Earth, beginning from the Upper Cretaceous the general habit of floras changed, owing to the dominance of angiosperms. Angiosperms also occur elsewhere in associations dominated by conifers. In the Late Cretaceous many plants that determined the aspect of earlier floras disappeared. Among the first plants that disappear are the Peltaspermales; Caytoniales are only of rare occurrence. Areas of the Bennettitales, Cycadales (excepting nilssonias), Leptostrobales and Ginkgoales became smaller. The Leptostrobales and Bennettitales disappeared at the end of the Cretaceous. Upper Cretaceous conifers, ferns and lycopsids are represented by families living today.

Upper Cretaceous floras are frequently termed Cenophytic because of the dominance of angiosperms in many deposits. The fact that the change from Mesozoic to Cainozoic faunas happens at the Cretaceous–Palaeogene boundary gave rise to the idea that the floral evolution develops in advance of the faunal evolution, in the same way as occurs at the Mesozoic–Palaeozoic boundary. However, special analysis of the continental biota shows that its evolution during the Late Cretaceous was unduly simplified (for details see Krassilov, 1985; Vakhrameev, 1977).

The beginning of the Upper Cretaceous is marked by the appearance of a triangular triporate pollen of the group *Normapolles* in Europe and North America. The Upper Cretaceous angiosperms differ markedly from those of the Cainozoic, and the change from the older to the extant groups occurs at the end of the Cretaceous–beginning of the Palaeogene. Special calculations have shown that the impression concerning the dominance of angiosperms in many Upper Cretaceous localities (Far East, Canada) is fallacious; conifers are most predominant there. In Siberia and New Zealand pollen of angiosperms become dominant only in the Senonian. In the Turonian of New Zealand over 70% of the assemblage is composed of cryptogams, cycads and conifers. The amount of angiosperms depends largely upon the facies conditions. For instance, in the Cenomanian of certain sites in France single pollen grains of angiosperms are present. It should be remembered that living conifers belong to the major forest components that cover vast areas.

As a whole, the transition from the Mesozoic organic world to the Cainozoic was very long, occurring in some groups of organisms earlier, in others later, but in general was asynchronous in different areas. This is of great importance in palaeofloristics. Analysis of the transition from the Mesophyte to Cenophyte should be made on the basis of the existing phytochoria and precise datings of the local floristic changes. The concept of the concerted advent of angiosperms at the beginning of the Cretaceous was based on materials available from Europe and the USA, where assemblages dominated by angiosperms are of common occurrence from the Cenomanian. Postulating that this was typical world-wide the investigators developed hypotheses on the cause of the 'sudden and ubiquitous' advent of angiosperms on land at the Early–Late Cretaceous boundary.

The successive appearance of angiosperms in the megafossil and miospore assemblages has already been discussed in the sections devoted to the systematics of angiosperms, palynology and Early Cretaceous floras. It was noted that difficulties arise in the classification of Cretaceous angiosperm fossils, primarily in the determination of the systematic affinities of the pollen and leaf remains. The opinion is still current that Cretaceous angiosperms were closely related to extant genera. However, other implications have been derived from detailed investigations of Cretaceous fructifications and pollen. The fructifications were discussed in Chapter 3. Only a small quantity of pollen of extant genera has been recorded from the Senonian, belonging to a few (about 10) genera (*Nypa*, *Ilex*, *Nothofagus*, *Alnus*, *Symplocos*, etc.). The number of extant families in these deposits is only slightly more. Most of the pollen types cannot be compared with extant genera and families.

In the palaeofloristic zonation of Late Cretaceous continents it is necessary to take into account the data both on leaves and miospores, owing to the absence of megafossils in many regions. As yet no uniform zonation covering both types of remains has been proposed. The suggested schemes, based on megafossils (Vakhrameev, Krassilov, Takhtajan), include only certain areas (the Northern Hemisphere, Eurasia, or the USSR). Global schemes have been proposed on the basis of miospores (Zaklinskaya, Samoilovich, Chlonova, Herngreen, Srivastava; Fig. 91). However, marked contradictions are evident between these schemes. Krassilov (1985) suggested a palaeofloristic map compiled on the basis of both plant megafossils and miospores. This map also differs essentially from the others. Nevertheless, it can be maintained that the modern distribution pattern of the major phytochoria was already established during the Late Cretaceous. At the north were the phytochoria corresponding to the present-day Holarctic kingdom; to the south of this were phytochoria that can be tentatively compared with the Palaeotropical and Neotropical kingdoms. The southern analogue of the Holarctic kingdom is represented today by the Cape, Holantarctic and Australian kingdoms.

Correspondence between the Late Cretaceous and extant phytochoria does not imply climatic equivalence. During the Late Cretaceous the modern type of climatic differentiation did not exist. There is no evidence for glaciations during that time. It is probable that a frostfree climate prevailed on the Earth, which can be inferred from the palaeotemperature measurements and atmospheric circulation models. Indirect indications are also provided by palaeobotanical data. However, their palaeoclimatic interpretation is not reliable. Climatic characteristics for Late Cretaceous floras usually rely on the presence of extant genera and families. It is assumed that they invariably associate with certain climatic zones. This assumption, even given that the identifications are reliable, is hardly tenable. It seems more probable that many genera underwent in the course of time marked changes in their climatic restrictions. This is

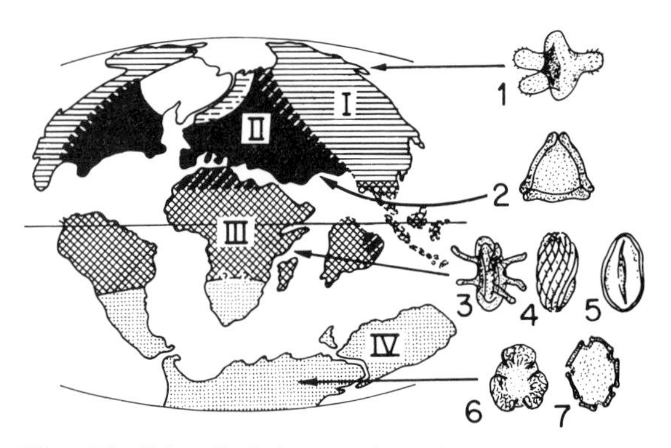

Fig. 91. Palaeofloristic zonation of Late Cretaceous (mostly Senonian) by palynological data; areas with *Aquilapollenites* (I), *Normapolles* (II), Equatorial (with palms; III), Notal (with *Nothofagidites*; IV); 1, *Aquilapollenites*; 2, *Trudopollis* (from *Normapolles* group); 3, *Elaterosporites*; 4, *Ephedripites*; 5, *Psilamonocolpites* (palm-like pollen); 6, *Microcachrydites* (Podocarpaceae); 7, *Nothofagidites*. Modified from Herngreen and Chlonova, 1981.

evident from the fact that genera associated in a single fossil assemblage may exist at present in different climatic zones. The possibility of extra-equatorial persistence should also be considered (see Chapter 8).

Reliable correlations between the Late Cretaceous phytochoria and modern climatic zones, showing preference to any of the suggested viewpoints, are still impossible. This pertains specifically to the distinction of Late Cretaceous analogues of the extant tropical flora. Current opinions concerning this are divergent. Some authors hold that the Late Cretaceous Equatorial Belt was covered by subtropical laurophyllous plants, whereas plant formations comparable to the extant tropical vegetation were non-existent. According to another viewpoint the Equatorial zone with tropical floras not only originated during the Cretaceous, but also by the end of the Cretaceous already underwent subdivision into analogues of the modern kingdoms (Palaeotropics and Neotropics).

Two aspects of this problem should be distinguished: (1) the presence of a tropical zone during the Late Cretaceous (i.e. with mean temperatures of the coldest month of the year above a definite level, e.g. +16 or +18°C), and (2) the presence of plant associations, physiognomically and taxonomically similar to the modern tropical ones. A tropical climatic zone undoubtedly existed at that time, but what type of plant association populated these sites and to which extant associations they can be compared, still remains unclear.

The modern pattern of the major phytochoria was formed during an essentially different arrangement of the continents. The Tethys Ocean was not yet closed during the Late Cretaceous, whereas the proto-Atlantic Ocean did not prevent biogeographical connections. Western Gondwana (Africa and South America) was already separated from Eastern Gondwana. The separation of Africa from South America commenced, so that the distance between them amounted to several hundred kilometres. Seas were created between other Gondwana continents. Nevertheless, floristic links between these parts of Gondwana still persisted.

Vakhrameev (Vakhrameev *et al.*, 1970, 1978), on the basis of megafossils, distinguished in the Northern Hemisphere the Siberio–Canadian and Euro–Turanian areas. The boundary between them in Eurasia passed in a submeridional direction not far from the Urals, crossed Kazakhstan and China. The position of the eastern branch of this boundary is unclear. The flora of the east of North America and Greenland is close to the Euro–Turanian. The transition zone between both areas was distinguished as the Turkmeno–Kazakhstan province. On the basis of megafossils it is difficult to establish any clear geographical trends, that may be reflected in the distribution of the major taxa, or distinct morphological leaf types. Most clearly opposed are the Late Cretaceous floras of the USA and Western Europe and Senonian floras of Southern Kazakhstan, on the one hand, and the Siberian floras on the other. The first contain many angiosperms with relatively narrow, more often entire-margined leaves (or leaflets of compound leaves). They are represented by *Dewalquea* (= *Debeya*), *Manihotites*, *Anisophyllum*, *Magnoliaephyllum*, etc. Genera with platanoid leaves play a less important role (such leaves are sometimes directly referred to *Platanus*, particularly if they associate with suitable fructifications; *Credneria* belongs to this group), and leaves similar to *Ficus* (e.g. *Diospyrophyllum*). Palms are known here. Conifers (*Sequoia*, *Araucaria*) are common in the deposits, whereas the Ginkgoales are practically absent. The Leptostrobales are absent. The association of *Sequoia* with laurophyllous angiosperms is characteristic. The habit of the flora was much influenced by the arid climatic belt which covered South Europe, the Near East, Middle Asia and China.

The Siberio–Canadian flora differs in its much greater amount of platanoid leaves (*Credneria*, *Platanus*, *Pseudoprotophyllum*, *Paraprotophyllum*, *Grewiopsis*) and forms of the *Trochodendroides* (particularly characteristic of the entire area) and *Rhamnites* types. Leaves also occur that are more widespread in the southern floras. The Cenomanian floras are frequently dominated by small leaves, similar to *Viburnum*, and those affiliated with *Menispermites*. Platanoid leaves are prolific since the Senonian. In a north–south direction conifers decrease in amount, while the proportion of entire-margined leaves of angiosperms and remains of cycads increases. A typical genus is *Quereuxia*, that probably belonged to aquatic plants. Leaves of *Trochodendroides* are associated with the infructescence *Nyssidium*. Palms are completely lacking. Besides angiosperms, the Siberio–Canadian flora contains abundant ferns, conifers (*Parataxodium*), ginkgos (*Ginkgoites*); persisting are cycads (*Nilssonia*), and at the east, leptostrobans (up to the Senonian, *Phoenicopsis*). The absence of *Nilssonia* at the extreme north of Alaska and Chukotka (?) furnished the premise for the distinction of the Beringian area. The Siberio–Canadian flora thrived under moderately warm and humid climatic conditions. Coal accumulation continued in the east of the USSR and western USA and Canada. At the beginning of the Late Cretaceous a cooling phase is known (the Amkian episode, after E. L. Lebedev), which was followed by a warming episode, then another cooling episode at the end of the Cretaceous. During the cooling episodes the role of conifers in the floras increased.

The Greenland flora is subject to much controversy. This flora is united either with the Siberio–Canadian, or Boreal Late Cretaceous area (Takhtajan), or with the Euro–Turanian area (Vakhrameev), or even with the Sakhalin flora, that is treated separately from that of the Siberio–Canadian area (Krassilov). By the presence of *Phoenicopsis* and *Ginkgoites* the Greenland flora shows affinities to the Siberio–Canadian; but in the presence of *Dewalquea* (= *Debeya*) and other southern forms shows affinities to the Euro–Turanian flora. It is composed of laurophyllous genera in combination with platanoid genera, and conifers are represented by the combination of *Sequoia* and *Parataxodium*. In other words, the Greenland flora is composed of elements from different areas. A similar combination of elements was found by Krassilov (1979) in Sakhalin.

For the solution of this controversy further detailed studies are needed of the floral composition and their stratigraphic interrelations. Comparisons are frequently made between Upper Cretaceous floras of various standard stages.

On the basis of megafossils no clear idea can be gleaned at present concerning the Upper Cretaceous floras of the Equatorial and Notal areas. Data are available so far on isolated localities. Furthermore, most of the determinations are in need of revision. Therefore, all that can be said is that in these areas the major components were angiosperms, conifers and ferns.

The palynological zonation of the Upper Cretaceous (Fig. 91) was performed heedless of the distribution of megafossils. Attention was devoted to the distribution of the dominant groups of *Normapolles* (Fig. 78, *n*; this is a pollen of the ancient Juglandes) and *Aquilapollenites* (Fig. 78, *l*; the affinity of this pollen within angiosperms is unknown). The pollen *Normapolles* is typical of the east of North America and Europe, i.e. of the Atlantic Sector of the Earth. The corresponding phytochorion was variously named, e.g. also the Euro–Turanian area. The pollen *Aquilapollenites* is typical of northeastern North America and all of Siberia that was also differently termed, e.g. the Siberian–Canadian area. Transitional assemblages have been recorded in Western Siberia, Kazakhstan, the Fore-Urals and southwest USA, and individual provinces have been distinguished. The boundaries between the phytochoria, established by both pollen groups, at first appeared meridional, passing from northeastern Kazakhstan through Western Siberia up to the modern Arctic Ocean. Later on, miospore assemblages, dominated by *Aquilapollenites*, were shown to occur in northwest Europe. These data together with those on other key-genera indicate that the boundary between these major floras was latitudinal with an elbow-like flexion (which has been taken for the meridional boundary) in Western Siberia–Kazakhstan.

The palynological data allow tracing of the xerophytic vegetation in the south of the Euro–Turanian area, which corresponds to the arid climatic belt. Here *Classopollis* and striate gymnosperm pollen (*Gnetaceaepollenites*, *Ephedripites*) persist. In the Upper Cretaceous of Kazakhstan, particularly of Southern Kazakhstan, the *Classopollis* pollen reach 60% of the total. This pollen is abundant in the south of the Siberio–

Canadian area, e.g. it dominates the Santonian of Japan.

South of the Euro–Turanian area the Equatorial floras are traced in the palynological assemblages of Northern Equatorial Africa and north of South America. These areas are united into a phytochorion, to which different names were attached, e.g. the Afro–South American area. As in the Early Cretaceous, it seems best to use the name Equatorial area. Remarkable features of this area are the absence of saccate pollen of conifers, the presence of a large number of spores with elaters, striated pollen, also pollen of palms and other tropical plants, as well as Proteaceae. In the Upper Cretaceous of Egypt the pollen of *Normapolles* is absent. The presence of some common genera (*Buttinia*, *Proxapertites*, etc.), both in Africa and South America, serves as evidence that the Atlantic Ocean during the Late Cretaceous did not hinder the plant migration. The Malaysian province with localities on the islands of Southeastern Asia, belongs to the tropics. Proteaceous pollen is absent there.

Data on the Upper Cretaceous miospore assemblages of India are scarce. The *Aquilapollenites* and *Normapolles* were both recorded here, but the age- and geographical relationships between these pollen assemblages are not quite clear. Further to the south is the Australo–Antarctic province (area) with localities in Australia, New Zealand, in the south of South America and Antarctica. Of particular interest is the predominance of pollen of *Nothofagus* (it appears in the Upper Santonian) – a typical Cainozoic plant of the southern continents. Pollen of Proteaceae and Podocarpaceae also occurs. As in the Early Cretaceous, various types of angiosperm pollen are of later appearance in Australia than in the Equatorial and Euro-Turanian areas. For instance, tricolpate pollen first appears only at the beginning of the Turonian, whereas polyporate (*Australopollis*) appears at the end of the Turonian.

The foregoing description outlines the general features of the Upper Cretaceous phytochoria. The cooling and warming of the climate evoked essential changes in plant associations. For instance, as a result of the Santonian–Campanian warming episode in the Siberio–Canadian area plant associations appeared with a high content of laurophyllous angiosperms, whereas the coniferalean broad-leaved associations were driven further northwards. At the same time in Japan the amount of pollen *Classopollis* increased, which gives an indication of the aridization of the climate. Such climatic episodes exercised a strong impact on the floral aspect, while the absence of marked temperature differences between the phytochoria was conducive to the shifting of floristic associations over long distances. Because of these factors the floristic zonation of the Earth for the Upper Cretaceous, as a whole, entails difficulties. In this case, when the floristic boundaries are subjected to considerable shifting, the zonation of the territory should be performed for narrower stratigraphic units. For this purpose present knowledge is too meagre; information should be available on a greater number of localities and the stratigraphy of fossiliferous beds should be studied in more detail.

The terminal Cretaceous event (extinction of dinosaurs, drastic changes in the marine biota) has been widely discussed in the literature. Its explanation as an asteroid or cometary impact has become increasingly popular. The iridium anomaly recorded at the Cretaceous–Tertiary boundary in many parts of the world is considered to be conclusive evidence in favour of this explanation. Hickey (1984) and Tschudy (1984) have surveyed the available data on plant megafossils and miospores of the assemblages bracketing the boundary. They have concluded that the change in the assemblages at this boundary do not differ essentially from those at boundaries where no catastrophes have ever been invoked (e.g. between the Palaeocene and Eocene). However, there are data showing that at least in some places the iridium anomaly coincides with significant changes in miospore assemblages of continental beds (Smit and van der Kaars, 1984). The overall floral changes at the Cretaceous–Tertiary boundary is more pronounced in northern biotas where by the Palaeocene the *Aquilapollenites* province virtually lost its identity (Tschudy, 1984).

CAINOZOIC FLORAS
(M. A. Akhmetyev)

The major geological and climatic events of the Cainozoic had a profound influence on the evolution of the vegetation cover of the Earth. It came to be realized that, during the Cainozoic, control on global climatic changes was provided not so much by external factors, in relation to the planet, as by changes in the oceanic and atmospheric circulation. For instance, the circumambient equatorial circulation of ocean waters that was established at the beginning of the Eocene, led to a prominent warming of the climate even in high latitudes, and to the expansion of the Equatorial and Subtropical zones. The ice-sheet in Antarctica originated as a result of the formation of circumpolar currents, evoked by the separation of the continent from neighbouring landmasses. The closing of the Isthmus of Panama during the Pliocene resulted in the isolation of the Pacific Equatorial waters from those of the Atlantic, and the formation of stable meridional currents (Gulf Stream, Labrador currents), which can be considered one of the reasons for the development of the glaciation in the Northern Hemisphere.

In the Northern Hemisphere the major events that occurred during the Palaeocene can be associated with the opening of the Northeastern Atlantic. The first violation of the land connections that existed between Eurasia and North America occurred about 50 million years ago, as a result of the separation of Spitsbergen from Greenland. Only at the end of the Palaeocene, after the liquidation of the Greenland–Iceland–Faroes Rapids, was America completely isolated from Eurasia. The conjunction of the Indostan and Xizang plates at the end of the Eocene resulted at first in a more active biota interchange between India and Extratropical Asia. This applies to the beginning of the Oligocene, when the Alpine Mountain ranges were not so highly elevated (Geng and Tao, 1982). The Alpine folding during its first developmental stages was concomitant with the division of the Tethys into the western and eastern parts, as a result of the conjoining of the African and Arabian plates at the beginning of the Neogene. This caused the isolation of marine biotas of the Indo–Pacific and Atlantic–Mediterranean areas. Following the formation of the Alpine mountain chains an arid climatic belt was formed to the north of them, extending from Western Europe to Southeast Asia. The continuous meridional circulation – from the Equator over Eurasia – ceased. The end of the Miocene (Messinian) was characterized in Western Tethys by the isolation of the Mediterranean from the Atlantic, and the subsequent expansion of the arid belt in Southeast Europe and the Near East.

The most crucial events in the Cainozoic history of Eastern Asia are associated with the formation and subsequent liquidation of the 'Bering Bridge', the development of volcanic belts extending parallel to the eastern borders of the continent, and the development of the inland Japan Sea, Sea of Okhotsk, etc. Up to the Miocene, inclusive, the Japanese insular biota was similar to that of the mainland, evidence for which can be found in the preceding history of this part of East Asia. The Yamata Uplift in the central part of the modern Japan Sea remained a landmass up to the end of the Palaeogene. Only during the Nishikurosawa transgression at the Early–Middle boundary did this structure undergo submergence below the sea surface.

The most significant events within the American continent occurred on the Pacific Coast. The origin of the Laramian volcanic belt at the Cretaceous–Palaeogene boundary caused the isolation of the inner regions of North America from the Pacific Ocean, and the formation of a vast arid area during the Eocene (Mexico and central areas of the USA). Another important event that had already occurred at the Miocene–Pliocene boundary was the uplift of the Isthmus of Panama and liquidation of the strait dividing North America from South America, which had existed from the middle of the Mesozoic. This led to a renewed interchange between the biotas of the American continents ('The Great American Faunal Interchange'). The closing of the Isthmus of Panama, which led to the formation of the Gulf Stream and Labrador currents, caused a cooling trend and deterioration of the climate within the entire northeastern part of North America and Greenland (Berggren, 1981).

In the Southern Hemisphere the most decisive event was the separation of Australia from Antarctica at the beginning of the Eocene. It is probable that at the end of the Eocene the uplift in the Tasmania Sea was submerged and the long-existing biotic connections between New Zealand via Lord Howe with New Caledonia were violated. This resulted in the subsequent high endemism of the biotas of these islands, reflected at the generic- and family level, that persists to date. The beginning of the complete thermal isolation of Antarctica falls to the Eocene–Oligocene boundary. The opening of the Drake Strait led to the liquidation of the land connections with South America. With the formation of the continuous circumpolar streams the growth of the ice sheet commenced (Oligocene). The change in the main directions of oceanic circulation from the latitudinal to the meridional at the end of the Palaeogene had a profound influence on the development of the glaciation in the Southern Hemisphere. This happened after the submergence of the long-existing uplifts in South America (Rio Grande, etc.). The formation of the Antarctic ice sheet led to the first Palaeogene lowering of the World ocean level (about 30 million years ago). The immense Cainozoic regression was, apparently, accompanied by the restoration and expansion of land connections. The formation of the Andean Belt in South America and its elevation at the beginning of the Neogene evoked the redistribution of atmospheric precipitations. This was conducive to stronger latitudinal differentiation of the vegetative cover at the south of the equator, whereas in the 'rain shadow', east of the Andes, the formation of steppes and semi-deserts began. The breakup of Gondwana led to the further isolation of Madagascar and other islands in the western part of the Indian Ocean. This probably explains the high endemism of the modern Malgache biota.

From the foregoing description a general idea can be gleaned of the palaeogeographical setting for the evolution of the Cainozoic phytochoria. In outlining their history it seems convenient to proceed from the modern floristic zonation, which is given below after Takhtajan (1978).

SALIENT FEATURES OF CAINOZOIC PALAEOFLORISTICS. MAJOR PHYTOCHORIA

A distinctive feature of the Cainozoic palaeofloristics is that flowering plants become the dominant plant groups that determine the major features of individual phytochoria and their subordination. At the end of the Cainozoic the process of differentiation of the floras was of maximal expression, which resulted in the intricate mosaic picture of the modern phytochoria. Takhtajan (1978) distinguished the following six modern kingdoms: Holarctic, Palaeotropic, Neotropic, Cape, Australian and Antarctic.

For the Cainozoic, as a whole, even on the basis of the actualistic model, it is very difficult to draw a comprehensive picture of the development of the phytochoria. Because of this an attempt is made to consider only the major and most persistent phytochorion (Figs 92, 93). In the extratropical part of the Northern Hemisphere during the entire Cainozoic the Holarctic kingdom remained a phytochoria of the first rank. Several areas and provinces have been recognized here, each exhibiting a notable evolution (Engler, 1879). The entire equatorial zone has been united into the Tropical kingdom, which, in turn, is subdivided into three major areas – Neotropic, African and Indo–Malesian. In the extratropical part of the Southern Hemisphere at the beginning of the Cainozoic only a single Notal kingdom existed (Engler, 1882). During the Palaeogene the Australian kingdom came into existence. Owing to the absence of palaeobotanical data it is impossible to trace the history of formation of other modern superior-rank phytochoria in the Southern Hemisphere – the Cape kingdom, New Caledonian and Malgache subkingdoms.

Holarctic kingdom

During the Cainozoic the extratropical part of the Northern Hemisphere exhibited a distinct climatic and floristic differentiation into the temperate and subtropical zones. At the beginning of the Palaeogene the Boreal and Tethyan areas existed here. In the process of subsequent differentiation at the end of the Palaeogene and during the Neogene both areas underwent transformation and were divided into independent phytochoria of the different ranks.

Boreal area

The Cainozoic Boreal area was inherited from the Cretaceous. Extending throughout the entire Arctic Belt, it was contiguous at the south with the Tethyan area. In Europe its boundaries extended across England, the Baltic Region, Belorussia, and further from the Middle Urals gradually passed southward to the Altais and Mongolia and emerged at the Pacific coast in USSR–Primorye and North Japan. In North America its boundaries nearly matched the modern border between USA and Canada. During the Palaeocene and at the beginning of the Eocene the Boreal area was subdivided into the Beringian, Sakhalin–Primorye and Tulean provinces (Budantsev, 1983). At an early stage of development the Boreal flora maintained the aspect of the Cretaceous flora, particularly at the generic level. It remained of temperate affinities, mesophilous, with the dominance of Taxodiaceae, broad-leaved trees and shrubs. A characteristic feature, particularly of the Arctic regions, was the presence of broad leaves that can be attributed to the specific light-conditions in the high latitudes together with the humidity of the climate, especially during vegetation growth periods. The major forest elements were *Ginkgo*, *Metasequoia*, *Glyptostrobus*, *Magnolia*, *Trochodendroides*, *Platanus*, also *Osmunda*, *Thuja*, *Alnus*, *Betula*, *Quercus*, *Juglans*, *Populus*, *Grewiopsis*, *Acer*, *Aesculus*, *Vitis*, etc. Angiosperms of uncertain affinities that became extinct during the Palaeogene (*Trochodendroides*, *Grewiopsis*, *Macclintockia*, *Protophyllum*,

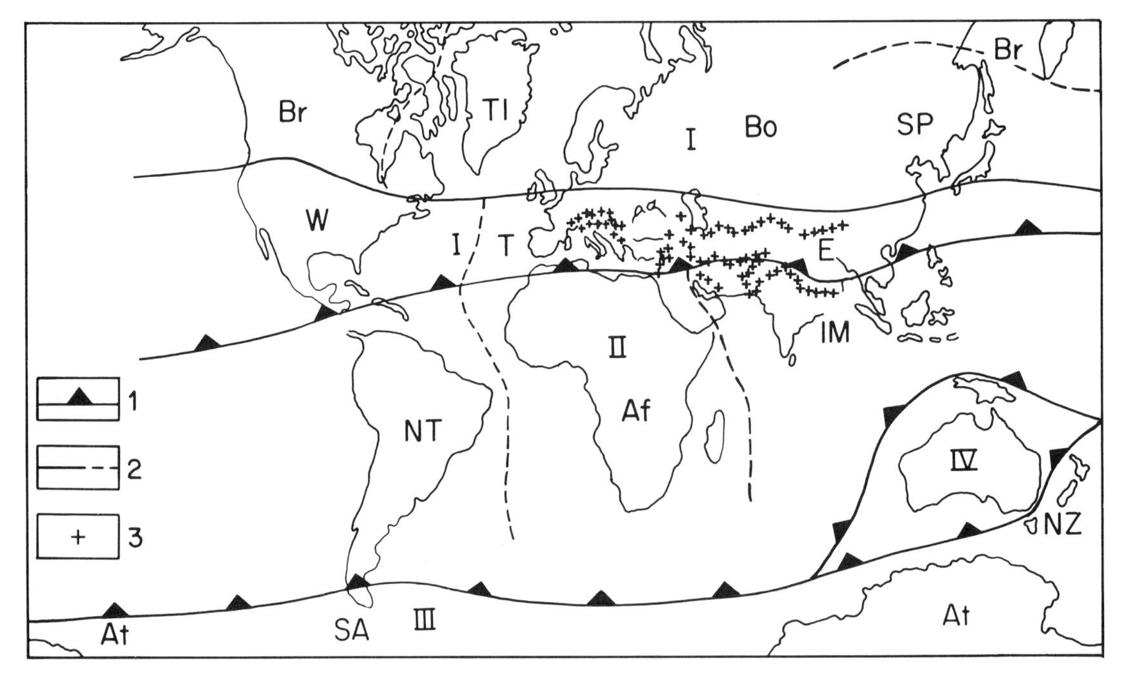

Fig. 92. Palaeofloristic zonation of Early–Middle Eocene; Holarctic kingdom (I), Boreal area (Bo) with Beringian (Br), Tulean (Tl) and Sakhalin–Primorskian (SP) provinces, Tethyan area (T) with Western (W) and Eastern (E) provinces; Tropical kingdom (II), Neotropical (NT), African (Af) and Indo–Malesian (IM) areas; Notal kingdom (III), Antarctic (At), New Zealand (NZ) and South American (SA) areas; Australian kingdom (IV); 1, 2, boundaries between kingdoms (1) and areas and provinces (2); 3, shore line of Tethys ocean. Courtesy of M. A. Akhmetyev, 1984.

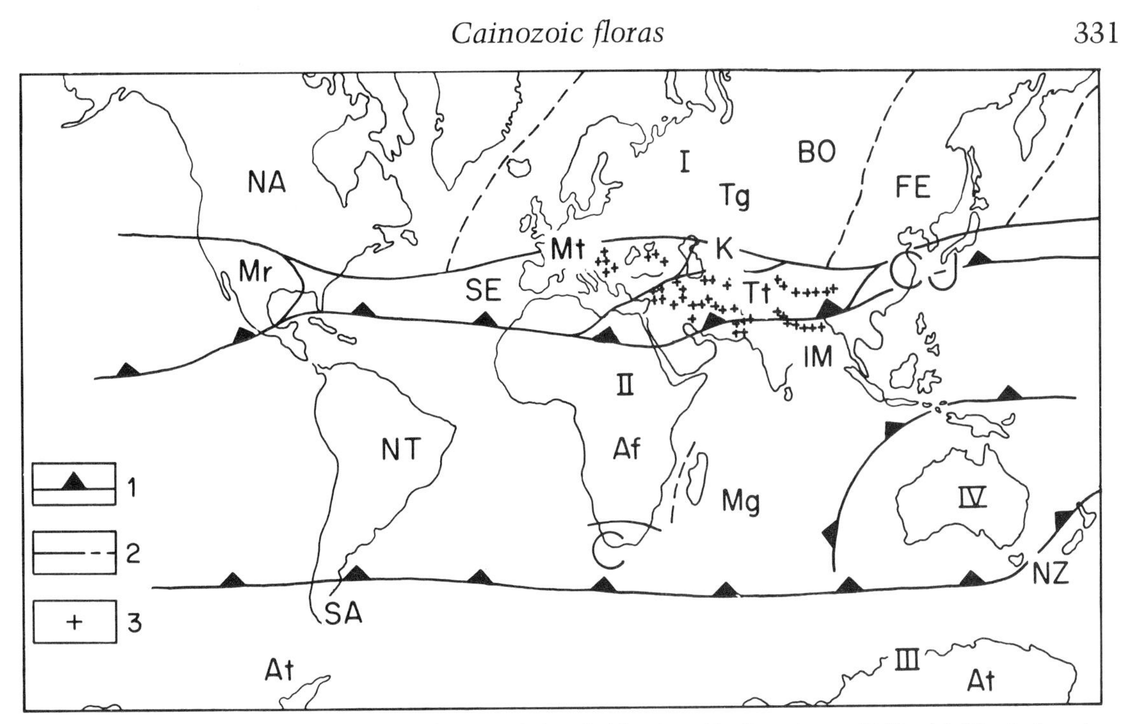

Fig. 93. Palaeofloristic zonation of Early Miocene; Holarctic kingdom (I), Boreal area (BO) with North American (NA), Turgaian (Tg) and Far-Eastern (FE) provinces, Madrean area (Mr), Mediterranean area (Mt) with South European (SE), Kazakhstan (K), Turkestan (Tt) and Chinese–Japanese (CJ) provinces; Tropical kingdom (II), Neotropical area (NT), African area (Af) with Cape (C) and Malagasian (Mg) subareas (?), Indo–Malesian area (IM); Notal kingdom (III), Antarctic (At), New Zealand (NZ) and South American (SA) areas; Australian kingdom (IV); 1, 2, boundaries between kingdoms (1) and areas and provinces (2); 3, shore line of Tethys ocean. Courtesy of M. A. Akhmetyev, 1984.

etc.) played an appreciable role. The Boreal zone was covered at the extreme south (Zaissan Depression, depressions of the Gobian part of Mongolia) by the *Taxodium–Trochodendroides* forests, probably occurring along river valleys.

The Beringian province included Northeast Asia (including Kamchatka), Alaska and West Canada. The major types of vegetation here consisted of coniferous broad-leaved forests with gigantic leaves of the major forest components. Besides panboreal elements, represented by *Equisetum arcticum*, *Metasequoia occidentalis*, *Trochodendroides arctica*, '*Acer*' *arctica*, *Viburnum asperum*, etc., certain local species of the same genera were present. The Sakhalin–Primorye province occupied part of East Siberia and the South of the USSR Far East. It differs from the Beringian province in the species composition, also in the presence of more thermophilous angiosperms (Hamamelidaceae – *Liquidambar*, *Corylus*; *Alangium*, *Celtis*, *Dryophyllum*, etc.). The gigantic size of the leaves is typical of only part of the plants (*Alangium*, Platanaceae). The Tulean province involved the eastern part of Canada, Greenland, the Faroe Islands and, probably, the north of the European continent. The floral composition was formed on the basis of the rapid break-up of the previous Greenland flora at the Cretaceous–Palaeogene boundary. During this period, after the meridional straight at the west of Canada became dry land, a large wave of Pacific migrants appeared here (*Metasequoia*, *Trochodendroides*, *Ulmus*, *Alnus*, *Betula*, *Tilia*, *Grewiopsis*, *Acer*, etc.). The generic composition of the floras was similar to that of the Beringian province, but apart from certain dominant panboreal species, the species composition displayed conspicuous distinctions.

The Early Palaeogene flora of the Boreal area underwent the first conspicuous transformation during the Eocene, when the global warming of the climate brought about its enrichment in migrants from the Tethyan area. Their invasion, also the more widespread distribution of south-boreal elements, is most distinctly reflected in Kamchatka and Alaska, where palm finds are known from the 60°N. Lat. (Budantsev, 1983; Wolfe, 1978, 1980). By the middle of the Eocene the southern boundary of the Boreal area shifted everywhere 5–10° northwards. At the end of the Eocene due to the cooling trend, the broad-leaved flora of the Early Palaeogene disappeared, giving way to another boreal flora – the Turgaian (*s.l.*). The sites of maximum plant concentrations, that constituted the core of the future Turgaian flora within this area, were the southern mountain regions around the border of the Tethys area. Leaves and fructifications of amentiferae (Juglandaceae, Salicaceae, Betulaceae, Fagaceae) are known from certain transitional floras of the Cretaceous and Palaeogene, e.g. the Takhoba flora of Sikhote Alin'. These floras, spreading out to the north and south along the mountain systems, under conditions of a practically frostfree climate for riparian formations, demonstrate a high plasticity in adjustment to diurnal and seasonal temperature fluctuations in mountain regions. Because of this these plants exhibited a greater viability to exist during the lowering of the annual temperatures (by 5–8°) at the Eocene–Oligocene boundary. Many genera of arborescent and shrubby deciduous dicots, together with Pinaceae and Taxodiaceae, formed a new boreal flora, that Kryshtofovich (1928, 1955, 1958) termed the 'Turgaian'. There is probably justification for the term 'Turgaian' flora in defining a specific floral type, i.e. for the designation of an assemblage of genera of arborescent gymnosperms and deciduous angiosperms, that were adapted to a temperate warm and humid climate. Regarding the term Turgaian area, or, to better express it, province, this can be used only in a narrow regional sense. It can be used to denote a restricted territory; for example, only the W. Kazakhstan at the beginning of the Oligocene, or most of Kazakhstan at the beginning of the Oligocene, or most of Kazakhstan and the south of Western Siberia at the Oligocene–Miocene boundary (Kornilova, 1963; Zhilin, 1984).

The Turgaian floral type includes *Osmunda*, *Ginkgo*, Pinaceae (commonly in formations of slope series), *Sequoia*, *Metasequoia*, *Taxodium*, *Glyptostrobus*, *Populus*, *Myrica*, *Carya*, *Pterocarya*, *Alnus*, *Betula*, *Fagus*, *Quercus*, *Castanea*, *Ulmus*, *Zelkova*, *Celtis*, *Magnolia*, *Rhus*, *Cercidiphyllum*, *Liquidambar*, *Platanus*, *Acer*, *Vitis*, *Tilia*, etc. (Iljinskaya, 1962). A great majority of the Turgaian genera are living today, although the areal distribution of certain representatives has dwindled considerably (*Metasequoia*, *Glyptostrobus*, *Ginkgo*, *Parrotia*, etc.). Within most of Kazakhstan the temperate Turgaian flora replaced the subtropical Poltavian at the boundary between the Eocene and Oligocene. Individual derivatives of the Poltavian flora (narrow-leaved poplars, wax-myrtles, sclerophyllous oaks and some laurels) persisted in the riparian assemblages up to the Middle Oligocene. The Turgaian flora from the southern parts of the Boreal area of the Eurasian and American continents at the end of the Oligocene and during the Miocene spread out to the south and southwest (in Europe), as the subtropical flora receded. During the Oligocene the Turgaian flora, considerably impoverished due to the absence of the most thermophilous genera, completely dominated the higher latitudes, and as the climate underwent further deterioration, it was rapidly replaced by another flora, of even more temperate climate. The heritage of past Turgaian floras can be recognized today in refugia (Balkan, Carpathian, Black Sea Region, Talysh, East Asian, Appalachian, etc.). During the maximal flourishing of the Turgaian flora (Oligocene and Miocene) the Boreal area was subdivided into three provinces: North American (Alaska, Canada, Oregon, Washington, and other northern states of the USA with the typical Kenaian groups of Alaska, John Day, Maskall, etc. – Columbian Plateau; Wolfe, 1978, Chaney, 1959), Turgaian (temperate climatic zone of Europe and Kazakhstan, South and Middle Siberia) and the Far Eastern (entire Pacific sector of the Boreal area in Asia). The floras of these provinces exhibited distinctions in their species composition and in the proportions of palaeoendemics. The warming of the climate, that occurred at the Early–Middle Miocene boundary, was conducive to a temporary expansion of the ecotonal zone between the Boreal and Mediterranean areas. This is reflected most distinctly in Central Europe and in the Far East.

At the end of the Miocene, with the further cooling-off and differentiation of the climate, the Turgaian flora disappeared from most of the Boreal area, giving way to a new Boreal flora type. The latter included cold-resistant conifers and small-leaved arborescent types, various herbaceous plants, that furnished the basis for the floras in the modern Euro–Siberian and Atlantic–North American areas. Common plants of this new floral type are *Picea*, *Pinus*, *Abies*, *Larix*, *Tsuga*, *Populus*, *Salix*, *Betula*, *Alnus*, *Quercus*, *Acer*, *Fraxinus*, *Tilia*, *Ulmus*, etc. Among the shrubs most prolific are Rosales (*Spirea*, *Rosa*, *Sorbus* ex sect. *Eo-Sorbus*), *Corylus*, *Betula* ex sect. *Fructicosae*, *Lonicera*, etc. The Turgaian floral type persisted nearly to the end of the Neogene at the south of Central and Eastern Europe, Northeast China, Korea, Japan and the Appalachians.

From the middle of the Miocene as the cooling of the climate became more severe, the latitudinal differentiation of the Boreal area increased and a host a new zonal types of vegetation evolved (taiga, tundra). In Iceland the formation of taiga vegetation, as a result of impoverished coniferous broad-leaved and beech forests of the Turgaian type, occurred during the Late Miocene (Akhmetyev *et al.*, 1978). The degradation of the taiga and formation of the tundra occurred about 4 million years ago. It is probable that the taiga and tundra elsewhere in the Boreal area are of the same age.

The rise of the steppe vegetation in Europe began in the Late Miocene (8–9 million years ago). Simultaneously steppes appeared in Kazakhstan, North Mongolia, central North America and other regions of the Boreal area.

Tethyan area

The Tethyan area existed as an independent phytochorion only during the first half of the Palaeogene. It included the subtropical floral belt that extended practically throughout the entire world northwards of the Tethys coast. In America it included the territory occupied today by the USA, in Eurasia it included Europe, the Near East, most of Kazakhstan, Middle Asia, China and Japan. Very little palaebotanical information is available on the Palaeogene of the Central and Southeastern areas of China. However, judging by the presence in the Palaeocene and Eocene of red-beds, variegated deposits and evaporites, a dry subtropical climate prevailed in this part of Asia. The natural boundary dividing the Early Palaeogene Boreal and Tethyan areas in China was the latitudinal system of mountain ranges, one of the components of which was Tsin Lin (Hsü Jen, 1983). With the beginning of the disappearance of the Tethys as a uniform latitudinal ocean basin, engendered by the early Alpine orogenic phases, the climatic conditions in different parts of the formerly undifferentiated monsoon subtropical zone underwent a conspicuous change. The earlier existing western (American) and eastern (Eurasian) provinces of the Tethyan area were divided into four independent phytochoria. Two of them during the Neogene constituted parts of the Holarctic Kingdom (Chinese–Japanese area and Atlantic Coast province and Mexican Lowland province of the Atlantic–North American area). In that part of the Tethys area where the climatic changes were more contrastive during the Palaeogene, and a humid subtropical climate gave way to the seasonal climate (Mediterranean type), the Mediterranean and Madrean areas were formed. The differentiation of the Tethyan area during the middle of the Palaeogene controlled the remarkable similarity (existing until recently) reflected on the generic level between the floras in the dry subtropics of the American continent and in the Mediterranean, also between the humid temperate warm and subtropical floras of the Appalachians, China and Japan (Graham, 1972; Tanai, 1972).

In the subtropical forests of the Tethyan area evergreen trees and shrubs prevailed. The herbaceous vegetation was more impoverished compared with that in forests of the temperate zone, and consisted chiefly of ferns. These plants

(*Lygodium*, etc.) together with certain dicots (Vitaceae, etc.) composed the rich groups of lianas and eiphytes. The major forest components were Fagaceae (most diverse in Eurasia) and Lauraceae. Palaeocene and Eocene floras of America, judging by the richest localities in California, Texas, Nebraska, Colorado and New Mexico (Elsinore, Wilcox, etc.), included tree-like ferns, palms, swamp cypresses, many arborescent angiosperms (*Celastrum*, *Parathesis*, *Persea*, *Sapindus*). The form-genera *Dryophyllum* ('synthetic' type – progenitor of the modern narrow-leaved oaks and chestnuts) and *Debeya* have been found. A zonal vegetation type was represented by lauraceous–oak–palm forests with *Bumelia*, *Clethra*, *Colubrina*, *Dodonaea*, *Pithecolobium*, *Platanus*, *Ficus*, *Persea*, *Quercus*, *Sabal*, etc. The Tethyan flora of Eurasia, typified by the Gelinden flora, was characterized by the occurrence of *Cinnamomum*, *Litsea*, *Neolitsea*, *Persea*, *Laurus*, *Quercus* (including *Cyclobalanopsis*), *Castanopsis*, *Lithocarpus*, palms (*Nypa*, *Trachycarpus*, *Sabal*, *Livistona*). At the junction of the Tethys with the Atlantic Ocean the subtropical flora moved still further towards the north, reaching to Great Britain (floras of the London and Hampshire basins). Due to the direct proximity of the Tethyan area to the Tropical and Boreal, and the absence of any serious orographic barriers between them, ecotone zones and insular patches of alien intrazonal vegetation existed (e.g. impoverished mangroves in the Eocene subtropics of South England).

In the Eurasian part of the Tethyan area the composition of the form-genera, even among the forest components, was wider than in America. Besides *Debeya* and *Dryophyllum*, it included *Ushia*, *Protoacerophyllum*, genera showing affinities to Myrtaceae, Apocynaceae, Araliaceae, and probably also plants producing pollen of the *Normapolles* type.

The Mediterranean and Madrean areas are discussed below. We can judge the subtropical floras in East Asia in part by the transformations in the Late Palaeogene and Neogene floras of Japan and Korea. In China, outside the regions of occurrence of red-beds, Upper Palaeogene and Neogene continental deposits are practically absent.

Mediterranean area

The Mediterranean area was formed out of the Tethyan, in its eastern part. Its initial boundaries differ considerably from its modern boundaries. In past times it included at the east the interfluvial area between the Yangtze and Huang He rivers, extending nearly to the Pacific Coast. However, it was much narrower in a latitudinal direction, and was initially limited by regions located only to the north of the Tethys Coast. The major transformations during the Eocene are reflected in the replacement of monsoon subtropical floras by 'summer-seasonal' and semi-arid subtropical floras, occurring against the differentiation of the Tethys, as a result of the uplifts in the Eurasian Alpine Zone. The monsoon subtropical climate in the Middle of the Eocene gave way to a hot climate with drought seasons of long duration.

Migrants from the Tethys area, that earlier underwent adaptation to dry habitats, furnished the basis of the Mediterranean (Poltavian) floras. They are represented primarily by Lauraceae and Fagaceae. The palaeobotanical data do not provide supporting evidence for the concept (Popov, 1963) advocating that the ancient Mediterranean flora is a result of interactions between '*Welwitschia*' and '*Ginkgo*' floras. The earliest provincial differentiation of the Mediterranean area was already determined by the degree of aridization of the climate. At the west and north of the area the South European (Poltavian) and Kazakhstan provinces were distinguished here. The Turkestan province, contiguous to the latter at the south, extended eastwards into China. The East European and Kazakhstan provinces, where the aridization was of weaker manifestation, differ one from another in the relative content of tropical and Boreal elements (*Quercus*, *Castanopsis*, *Dryophyllum*) and Lauraceae (*Laurus*, *Cinnamomum*); the major components of sclerophyllous forests in these provinces were *Sequoia*, *Pinus*, *Magnolia*, *Andromeda*, *Apocynophyllum*, *Arbutus*, *Myrica*, *Cassia*, and other Fabaceae, *Rhus*, palms *Nypa* (Kornilova, 1963; Makulbekov, 1982). The flora of the Turkmenian province, judging by the type-localities in the USSR (Badkhyz) and China (Tchansha) was composed exclusively of small-leaved sclerophyllous trees and shrubs: *Chamaecyparis*, *Cinnamomum*, *Dryandra*, *Palibinia*, *Rhus*, *Zizyphus*, *Amygdalis*, Fabaceae. Pollen of desert plants have been recorded here (*Nytraria*, *Zygophyllum*, *Calligonum*, *Frankenia*, *Ephedra*, etc.). The Chinese–Japanese province (existed from the late Eocene to Pliocene) with palms, Lauraceae, *Terminalia*, *Camelia*, etc., corresponds to the humid subtropics in the Mediterranean area.

During the Neogene in the process of differentiation and activation of the uplifts in the Alpine Fold System, the area underwent further differentiation. The cooling at the Eocene–Oligocene boundary, and subsequent humidization of the climate in the northern parts of the Mediterranean area, resulted in the replacement of the subtropical flora by a moderately warm and temperate flora, which occupied most of the territory of Kazakhstan, Central and Eastern Europe. These areas passed from the Mediterranean into the Boreal area. As a consequence the former is delimited at the west by the neighbouring edge of the Mediterranean basin, and the boundary of the area was shifted southwards into the marginal regions of Africa. Judging by the floral composition in Spain, South France, the Balkans and Fore-Asia (the Near East), during the Neogene in the west of the area oak–laurel forests persisted, and maquis communities were more widespread over the territory. During the Miocene most of the Caucasus constituted a part of the ecotone zone between the Mediterranean and Boreal areas, where floral elements of subtropical and temperate affinities underwent mixture. At the east of the Mediterranean area by the end of the Neogene several new provinces were formed on the basis of the Turkmenian province. The flora of each province differed in the endemic genera of dry shrubs and herbaceous plants (Fabaceae, Asteraceae, Chenopodiaceae, Rosaceae, Rubiaceae, Zygophyllaceae, Brassicaceae, etc.).

The heritage of the earlier floras is apparent in the modern Mediterranean floras in a re-worked state in South Europe, the Near East, Western Himalayas and several other regions (forests with *Argania* and *Tetraclinis* in Morocco; forests with *Pinus halepensis*, *Quercus ilex* and *Q. suber* in E. Spain and the French Maritime Alps; forests with *Pinus brutia* in the Near East, etc.).

Madrean area

From the moment of its rise in the end of the Eocene this area evolved autochthonously within the same boundaries (Southwestern USA, Mexican Highlands). This area, having originally acquired individuality in the subtropical zone of seasonal precipitation, in the process of increasing drying and cooling of the climate experienced conspicuous transformations due to the intake of a considerable amount of boreal elements adapted to deficient moisture conditions. Owing to the investigations of Axelrod (1958), the evolution of Madrean floras has been traced on the basis of reliable dating of the major events.

As was already noted, the initial zonal type of vegetation in the Madrean area was represented by subtropical forests with Lauraceae and Fagaceae. Under conditions of increasingly more pronounced seasonal fluctuations, maintaining originally the high mean annual temperatures, during the Eocene the floras of the Green River Formation (second half of the Middle Eocene), and even to a greater extent the Florissant floras (beginning of the Oligocene) were already dominated by plant types that earlier composed forests on dry slopes. They include pines, Lauraceae, Fagaceae, Fabaceae. The composition of the newly formed dry subtropical forests was immeasurably wider and included *Rhus*, *Astronium*, *Ilex*, *Mahonia*, *Arbutus*, *Arctostaphyllos*, *Eugenia*, *Geonothus*, *Rhamnus*, *Dodonaea*, etc. A typical feature is the occurrence of all the major components of the Miocene chaparral among the xerophilous and microphyllous elements of the understory shrubs. As the mean annual temperatures and amount of seasonal precipitations decrease, from the Middle of the Miocene, the incoming in increasing numbers of boreal elements is apparent in the Madrean area, these elements becoming dominant by the end of the Miocene. They are represented by the Aceraceae, Hydrangeaceae, Juglandaceae, Platanaceae, Rosaceae, Salicaceae, Ulmaceae and other families that composed the slope series, as well as the riparian formations. By this time herbaceous plants increased notably in diversity.

Further differentiation of the Madrean area was triggered by active uplifts in the Cascade Range, Sierra Nevada, and Coast Range at the end of the Pliocene. In the 'rain shadow' the formation of semi-desert and desert vegetation began (Great Plains, Mojave and Sonora Deserts, etc.).

Tropical kingdom

During the Cainozoic the Tropical kingdom involved three vast floristic areas: the African, Indo–Malesian and Neotropic. The dispersal of angiosperms within the Equatorial Zone was mainly terminated by the beginning of the Palaeogene. This determined the presence of a large number of genera among the forest components of the tropical rain forests of the past.

African area

Due to the limited number of localities of plant remains (chiefly represented by woods), it is very difficult to draw even a general picture of the Cainozoic evolution of floras and the vegetation of Africa, and to define the boundaries of the area and their changes through time. Africa was already separated from other parts of Gondwana during the Cretaceous. In the Albian its land connections with America were obstructed, and by the middle of the Late Cretaceous its land connections with Antarctica. The fact that communications existed in the past between America and Africa is evident from the disjunct area patterns for several tropical families (Bromeliaceae, Rapateaceae, Velloziaceae, etc.). Proof of the Gondwana connections is found in certain plants persistent to date at the south of Africa and Madagascar. They exhibit disjunct areas with Australia at the family and generic levels (Proteaceae, Restionaceae, Winteraceae, Dilleniaceae, Sterculiaceae, Cunoniaceae, Ericaceae, etc.). During the Palaeocene and at the beginning of the Eocene most of Africa was covered by tropical rain forests. Their composition can be reconstructed only for the north of the continent, in the zone immediately adjacent to the coast of the ancient Tethys Ocean. The genera that occur here are highly diverse and belong to tropical families Arecaceae, Ebenaceae, Fabaceae, Lauraceae, Monimiaceae, Rutaceae, Sterculiaceae, Ternstroemiaceae. Insular intrazonal areas of tropical vegetation spread out along the Tethys coast far to the north, invading the subtropics of the Holarctic kingdom.

Seemingly, during the second half of the Eocene–beginning of the Oligocene, when the seasonal climate became more distinctly expressed, the tropical forests north of the Equator began to be replaced by open woodlands of the savanna forest type with Fabaceae, Annonaceae, Combretaceae, Ebenaceae, Euphorbiaceae, Sterculiaceae etc. (van der Hammen, 1983). This type of forest vegetation extended from Senegal and Mauritania, in the west, to the areas of the Big African Horn, in the east. Nearer to the Tethys coast these forests were replaced by Lauraceae with admixtures of Fabaceae, palms and other arborescent plant types that underwent adaptation to hot seasonal climatic conditions. Lauraceous forests with a preponderance of plants of tropical origin formed a kind of ecotone zone between Europe and Africa. In contrast to the Miocene lauraceous forests of the Balkans and other areas of the Northern Mediterranean, they lack conifers and beeches. The latter appeared in Africa during the Neogene when the entire north of the continent became part of the Mediterranean area. This transition was gradual. At the beginning of the Pliocene the Mediterranean and subtropical elements were represented in about equal amounts. By the end of the Pliocene, with an increase in the aridization of the climate, new xerophytes and halophytes (*Artemisia*, Chenopodiaceae, *Ephedra*, Tamaricaceae) began to actively invade the area, populating semi-arid and arid landscapes. This resulted in the extinction or pushing aside of plants of African origin. A refugial type of the ancient North African laurophyllous forests are the modern forests on the Canary Islands, that contain *Laurus canariensis*, *Dracaena*, *Sideroxylon*, *Appolinias*, and other relict trees and shrubs. On the Abyssinian Plateau the rain forests (climatically more temperate than elsewhere in Africa north of the equator) persisted nearly to the end of the Neogene. The aridization of the climate was more weakly manifested here and led to the replacement of rain forests by open woodlands with *Acacia*. South of the equator within the territory occupied today by Namib and Kalahari Deserts, tropical rain forests were preserved until the Miocene. In their stead sclerophyllous forests appeared, containing elements of the neighbouring Southern Cape area. Desert landscapes did not appear here before the end of the Pliocene.

The final isolation of the Cape and Malgache floras from the African had probably terminated by the Miocene. Tertiary floral finds are practically unknown in these regions.

Indo–Malesian area

During the entire Cainozoic the Hindustan peninsula, Indochina and Malaysia constituted the Indo–Malesian area of the Palaeotropic kingdom. It was here that the modern Palaeotropic flora was formed. From these places during optimal climatic phases, the flora migrated to the northeast as far as Japan, and also to the west and northwest. The Palaeogene floras of the Intertrappean Beds in India, which are not younger in age than the Eocene, contain over 200 genera. Conifers are represented here by Araucariaceae and Podocarpaceae. Palms are of remarkable diversity (more than 30 species according only to the wood remains). Some of them are allied to the extant *Cocos*, *Nypa*, *Amomocarpum*, *Triccocotes*, *Viracarpon*, etc. At the beginning of the Oligocene wood remains of evergreen angiosperms already included representatives of most of the tropical families of the Old World (Flacourtiaceae, Guttiferae, Tiliaceae, Elaeocarpaceae, Simaroubaceae, Burseraceae, Sapindaceae, Anacardiaceae, Fabaceae, Combretaceae, Sonneratiaceae, Euphorbiaceae, etc.). The spectrum of palaeotropic families, established according to palynological data, in the Early Palaeogene of Assam is no less in range. It also includes Nymphaeaceae, Rhizophoraceae, Meliaceae, Olacaceae, Myrtaceae, Araliaceae, etc. Genera occur that are widespread in the Southern Hemisphere (*Podocarpus*, *Araucaria*, Proteaceae, *Casuarina*, *Nothofagus*, etc.). Along the sea shores, around the periphery of the Deccan Trappean Plateau omnipresent were mangrove vegetation with *Nypa*, *Rhizophora*, *Sonneratia*, *Brownlowia*. In the inner parts of the Plateau rain forests were gradually replaced by evergreen subtropical forests.

The most important tropical family of the Old World is Dipterocarpaceae. The centre of origin was probably Malaysia. From here Dipterocarpaceae spread out to the east (the Philippines) and to the west (India, Burma). Dipterocarps appeared in India at the beginning of the Miocene. In Lakhanpal's belief (1970), the migration of the Dipterocarpaceae (in the same way as the Fabaceae) could have occurred via two land migration routes (the possible distribution of Dipterocarpaceae by water routes is excluded because they lose their germinating capabilities in sea waters). The first route extended along the Indian Ocean coast. Having appeared in East India at the beginning of the Miocene, the Dipterocarps rapidly reached the southern end of the Hindustan Peninsula and Ceylon, which was not separated from the mainland. Evidence for this can be found in the fossil remains of Dipterocarpaceae in the Cuddalore Series (first half of the Miocene). It is characteristic that the Dipterocarpaceae are at present most diverse in South India and Ceylon. The discovery of the Dipterocarpaceae in the Siwalik Formation (Neogene) of the Himalayan Foredeep indicates that the second migration route passed around the north of the Hindustan Peninsula. It is probable that along this route, via Arabia, the Dipterocarpaceae reached Africa at the end of the Neogene. Previous records of Dipterocarpaceae in the African Eocene require confirmation.

The formation of subtropical and temperate forests in the Himalayas began to a large extent on the basis of tropical floras (at an early stage) at the end of the Eocene. The active uplifts that occurred during the middle of the Miocene and beginning of the Pleistocene resulted in an even greater differentiation of the vegetation cover, associated with the altitudinal zonality. Conifers play an increasingly greater role in the composition of mountain forests, which is evident from the successive increase in the pollen of Pinaceae (particularly *Pinus*) in the Sivalik Formation. The flora and vegetation of Malaysia, Indochina and India remained tropical, humid, up to the end of the Neogene. However, beginning from the Miocene some regions reflect various degrees of aridization. In the Himalayan Foredeep this is apparent from the change upwards of the Sivalik sequence from large-leaved evergreen plants, restricted to the lowermost beds, to small-leaved sclerophyllous forms. Along the Kutch Bay (Northwest India) at the boundary with the African area the tropical rain forests of the Palaeogene were replaced during the Neogene by impoverished deciduous forests with the dominance of Fabaceae.

Due to the paucity of the palaeobotanical data available on Southeast Asia, no definite conclusions can be drawn concerning the degree of differentiation during the Neogene in the vast modern Indo–Malesian subkingdom. Because of this it is assumed that up to the Pleistocene this region was a single floristic area with certain provincial distinctions in its western (Hindustan) and eastern (Indonesian–Fijian) parts. The south of China and mountain regions of Indochina at the end of the Neogene passed from the Tropical into the Holarctic kingdom. Information on the vegetation of New Guinea and adjacent areas of Micronesia (Guam, Marshall Islands) and Fiji is available only for the Miocene. In New Guinea in the coastal zone mangroves occurred with Rhizophoraceae, Sonneratiaceae, Palmae, which as the distance increases away from the shore are replaced by rain forests. Higher up extended the savanna belt with *Casuarina* and *Eucalyptus*. The landform levels located above the savanna belt were occupied by 'cloud forests'. In Micronesia of wide occurrence were Neogene mangroves and forests with *Pisonia* (Nyctaginaceae).

Neotropic area

As a higher rank phytochorion the Neotropic area was formed by the beginning of the Palaeogene. Throughout the Cainozoic it involved most of South and Central America together with the Caribbean Sea basin. The latter two regions in the northern part constituted the ecotone zone with the combination of North American (Boreal), as well as newly formed Madrean and South American (Neotropic) elements. The boundaries of this area, especially at the south, coincided with the boundaries of the tropical and subtropical zones of the Southern Hemisphere, shifting in time relative to the climatic changes. The zonal vegetation type of the Neotropic area was represented during the Palaeogene by rain forests of polydominant composition. The generic composition was very large, including over a hundred angiosperm genera that continue to live today in the tropics of the New and Old World. This supports the viewpoint that most of the angiosperm genera had already occurred and spread through the tropics during the Cretaceous, when America and Asia were closer to each other. The richest Neotropic flora of the Eocene – the Rio Pichileufu (Rio Negru, Argentina, 41°S. Lat., after Berry) – is composed of over 100 species, including arborescent ferns, *Zamia tertiaria*, *Ginkgo patagonica*, *Araucaria*, *Fitzroya*, *Libocedrus*, *Podocarpus* and angiosperm families Myricaceae, Ulmaceae, Moraceae (*Ficus*), Proteaceae (*Lomatia*, *Embothrium*), Polygonaceae, Nyctaginaceae, Winteraceae (*Drimys*), Myristicaceae, Monimiaceae, Mimosaceae (*Inga*), Caesalpiniaceae (*Cassia*), Fabaceae (*Dalbergia*), Rutaceae, Burseraceae, Meliaceae (*Trichilia*, *Cedrela*), Malpighiaceae, Euphorbiaceae, Buxaceae, Anacardiaceae (*Schinus*, *Astronium*), Celastraceae, Hyppocrateaceae, Icacinaceae, Sapindaceae (*Paullinia*, *Sapindus*, *Cupania*), Tiliaceae, Sterculiaceae (*Buettneria*, *Sterculia*), Dilleniaceae (*Tetracera*), Flacourtiaceae, Cochlospermaceae, Lauraceae (*Nectandra*, *Phoebe*, etc.), Myrtaceae (*Myrcia*, etc.), Araliaceae (*Oreopanax*), Myrsinaceae, Sapotaceae, Styracaceae, Symplocaceae, Apocynaceae (*Allamanda*, *Plumeria*), Asclepiadaceae, Bignoniaceae, Rubiaceae (*Coprosma*), etc. A characteristic feature of the South American Eocene floras is the presence of representatives of native families, which at the present day are endemic in the Neotropic kingdom (Cochlospermaceae, Heliconiaceae, etc.).

The climatic changes during the Cainozoic are reflected not so much in the composition of the Neotropic floras, as in the shifting of their northern and especially southern boundaries. During the maximal warming episode of the Eocene and first half of the Miocene, the boundary was shifted southwards not less than 10–15°, reaching to the island of Terra del Fuego.

The differentiation of the vegetation cover of South America increased conspicuously during the Miocene, as a consequence of the concerted uplifts in the Andes, which by this time was completely formed as a mountain structure. Under conditions of prevailing westerly winds in the 'rain shadow', the tropical forests were replaced first by savannas, and then, at the boundary between the Pliocene and Pleistocene, by steppe and desert vegetation (Peruvian Desert, Patagonia). Rich tropical Miocene floras have been found on the Bolivian Plateau at heights up to 4 km, from which it can be inferred that during the last 5–7 million years the Andes underwent elevation of at least 2 km. The Potosi and Pislipampa floras include most of the tropical genera (*Annona*, *Bauchinia*, *Cassia*, *Dalbergia*, *Drepanocarpus*, *Euphorbia*, *Inga*, *Heliconia*, *Jacaranda*, *Pithecolobium*, *Protium*, *Terminalia*, *Sideroxylon*, etc.). These plants could not have existed above 2000 m in mountainous areas.

The major plant migration routes did not remain constant during the Cainozoic. In the Eocene, when the low-mountain relief and plain landscapes prevailed, the plants migrated mainly from south to north. It is believed that several tens of Tertiary and Recent genera of the North American subtropics are of southern origin (Balanophoraceae, Begoniaceae, Bromeliaceae, Cannaceae, Chloranthaceae, Cochlospermaceae, Combretaceae, Cunoniaceae, Elaeocarpaceae, Flacourtiaceae, Gesneriaceae, Myrsinaceae, Proteaceae, etc.). In the Pliocene the direction of migration changed to the opposite. Plants of North American floras penetrate from the north via the mountain chains to Colombia and Venezuela (*Juglans*, *Alnus*, *Quercus*, Berberidaceae, Caprifoliaceae, Myricaceae, Rosaceae, etc.). By the end of the Pliocene, with an increase in the floral differentiation, the Neotropic area was transformed into a kingdom.

Mangrove vegetation during the Cainozoic

Mangroves are one of the leading, although facultative, tropical plant formations of the Old World. Very little information is available on the Cretaceous mangrove vegetation and there are no means of judging its role in the tropical zone. Fructifications of *Nypa* palms have been reported from the Upper Cretaceous of Indonesia and the Guinean coast of Western Africa. The distribution of mangroves during the entire Cainozoic was not, however, limited to the Equatorial Zone. Besides other coastal communities, frequently together with coral reefs, they penetrated into the subtropics, occupying an intrazonal position (Plaziat *et al.*, 1983). The species composition of the mangroves in this case underwent impoverishment. The major indicator plants of mangroves are *Nypa*, *Sonneratia*, *Rhizophora*, *Paleobruguiera*, *Ceriops*, *Kandelia*, *Avicennia*, etc. Their pollen, fruits, seeds and wood remains frequently occur in marine-coastal sediments. Most abundant are finds of well-preserved fruits of the palm *Nypa*. During the Palaeogene mangroves were probably weakly differentiated and only during the Neogene, with the closing of the Isthmus of Panama, did they become divided into Atlantic and Indo–Pacific mangroves. The former were dominated by *Rhizophora*, the latter by *Avicennia* and *Kandelia*. During the Eocene mangroves were of maximum distribution to the north and south. They migrated along the Tethys coasts, reaching Great Britain and the Ukraine. On the Pacific coast of North America individual mangrove components (*Kandelia*) reached beyond the 60°N. Lat.; in Japan they reached to Hokkaido. In the Southern Hemisphere traces of mangrove vegetation have been reported from the Eocene of S. Australia, New Zealand and South America. In Australia during the Middle and Late Eocene mangroves were of wide occurrence along the modern southwestern coast. The sea level at that period was 300 m higher than at present, and the archipelago was very favourable for the expansion of mangroves over vast areas. The major components of Australian mangroves were *Nypa*, *Sonneratia*, *Avicennia*, Rhizophoraceae.

The second mangrove expansion to the north and south from the equator occurred at the Early–Middle Miocene boundary. Pollen of *Avicennia* has been recovered from the Bourdigalian of Languedoc (France) and Sardinia, and on the southeastern coast of Honshu, between Osaka and Hiroshima, coeval sediments yielded pollen of *Rhizophora*, *Sonneratia*, *Avicennia*.

History of palms

During the Cainozoic palms continued to be the most conspicuous plants of the Tropical kingdom (Trivedi, 1982). As today, their area did not extend beyond the Equatorial and Subtropical Zones of the Earth, being delimited by the 44°N. and 44°S. Lat. Palms reached their peak diversity during the Recent period, and have been subdivided (Moore) into 15 groups (over 200 genera and about 2700 species). It is difficult to correlate fossil finds of palms with the extant genera. According to their wood, fructifications and other parts, form-genera have been recognized, among which are *Palmoxylon*, *Rhizopalmoxylon*, *Palmocarpon*, *Palmorachis*, *Palmophyllum*; the pollen is represented by at least 20 genera (*Monocolpites*, *Palmopollenites*, *Palmidites*, etc.). It is thought that the ancient palms were massive branching plants, up to 3–5 m in height, with compound rosette leaves. They comprised thickets along river banks and spread out along sea coasts. The representatives of various palm groups indicate that in most cases the provenance which furnished the plants was in the area where they are concentrated today, and only as a result of climatic changes and palaeogeographic reconstructions they migrated from these centres of origin. The palaeoecological data suggest that at the beginning of the Late Cretaceous palms were of greater diversity and covered vast areas (Brazil, Western Africa, India). The fossil remains delineate three probable palm centres of origin: Antarctic, Malesian and South American–West African. The notable decrease in the diversity of palms in Africa during the Recent period is associated with the Pleistocene cooling of the climate.

Notal kingdom

In the Southern Hemisphere at the beginning of the Cainozoic the Notal kingdom persisted in existence. It involved the entire Antarctic continent, but was already then divided into the western and eastern parts. Close to this continent were Australia, New Zealand and South America. Africa, already separated from Antarctica during the middle of the Cretaceous, by the beginning of the Cainozoic was floristically quite independent of it, although not so isolated as today. The available palynological data, though meagre, indicate that at the beginning of the Palaeogene the extreme south of Africa, which at present is included in the Cape kingdom, constituted an independent area of the Notal kingdom located at the boundary between the temperate and subtropical zones. Evidence for the ancient connections between the African and other Notal biotas is found in the Palaeogene palynological assemblages of Africa which exhibit the presence of *Araucaria*, nowadays unknown on this continent, and *Podocarpus*. Probably the existence of past connections is also reflected in three groups of extant African Proteaceae (*Brabeium*, *Dilobeia* and 13 of the 19 genera of Proteeae tribes). As to the disjunct areas existing today of some other angiosperms of Africa, Madagascar and other continents of the Southern Hemisphere, they may be partly of ancient origin, emphasizing the Early Palaeogene and even Cretaceous connections (for Restionaceae, Chloranthaceae, Winteraceae, Sterculiaceae, *Hibbertia*, Dilleniaceae, *Cunonia*, *Adansonia*, etc.). At the same time, for most of the herbaceous plants, especially Asteraceae, that are particularly capable of dispersal over long distances, these connections are newly formed, having been established at the end of the Neogene, probably already during the Quaternary. So far in the Cainozoic of Africa no remains have been found of *Nothofagus*, which throughout the Cainozoic was the most significant plant type of the temperate zone of the Southern Hemisphere.

The Palaeogene flora of the Notal kingdom during the Cainozoic was more homogeneous, compared with the present day. Its analogue is the extant New Zealand flora with various conifers

(*Agathis*, *Podocarpus*, *Dacridium*, *Phyllocladus*), *Nothofagus*, Proteaceae, Myrtaceae, Araliaceae, Winteraceae and representatives of other angiosperm families known in the Southern Hemisphere from the beginning of the Cainozoic. The differentiation of the Notal kingdom into areas followed immediately after the isolation of different parts of Antarctica. The separation of Australia from Antarctica during the first half of the Palaeogene, already in the Eocene led to the isolation of the independent Australian area within the Notal kingdom. By the end of the Palaeogene, as Australia drifted to the north, this area underwent transformation into an independent kingdom, losing all earlier existing connections with the major forest components of the temperate zone. After this the floristic interchange between New Zealand and South America was accomplished via Western Antarctica. The transformation of these two regions into individual regions of the rank of areas occurred at the beginning of the Oligocene, after the formation of the Drake Strait. At the beginning of the Neogene all communications completely ceased, due to the formation of the glacier sheet of Antarctica.

Antarctic area

The information available on the Cainozoic flora and vegetation of Antarctica is scanty. Macro- and microfossil remains have been recovered from several localities on the Antarctic Peninsula and surrounding islands (South Shetland, Seymour, Alexander, etc.; Barton, 1964); palynological assemblages have been recorded from detached moraine blocks on Blake Island in the McMurdo Sound, also from drill cuts taken in the Ross Glacier. During the first half of the Palaeogene Antarctica was covered by coniferous small-leaved forests. The major forest constituents were conifers (*Araucaria*, *Podocarpus*, *Phyllocladus*, *Dacrydium*, *Microcachrys*) in combination with southern beeches of all three groups – *Nothofagus menziesii*, *N. fusca*, *N. brassii*. Proteaceae were represented by *Grevillea*, *Banksia*, *Beauprea*. Winteraceae and other angiosperms were present. Some ferns were arborescent (*Cyathea*). The climate during the first half of the Palaeogene was probably similar to the modern climate in the coastal areas of Chile between 37° and 49°S. Lat. (the mean annual temperature 5°C; average summer temperature 8–10°C; the mean annual fallout 500 mm). By the end of the Oligocene, with the development of mountain valley glaciers, the forest composition experienced significant impoverishment, and, probably at the beginning of the Neogene, underwent complete degradation. By the middle of the Miocene, when a stable ice sheet was formed, the vegetation, mainly shrubby–herbaceous, was preserved only along the ocean coast and on adjacent islands.

New Zealand area

New Zealand is perhaps the only region in the world where the ancient core of the floras is preserved to date, having been subjected to the least alterations through time. To a great extent this can be explained first of all by the long prevalence of uniform climatic conditions (the influence of the surrounding ocean, its situation at the junction of the subtropical and temperate climatic zones, etc.). It is probable that the isolation of the island, which was of different degree through time, also had a paramount role. The degree of isolation gradually increased to the Quaternary, and at least by the beginning of the Neogene New Zealand had already become an independent floristic area within the Notal kingdom. With a great degree of confidence it might be said that the bridges that connected New Zealand with South America and New Caledonia persisted until the end of the Palaeogene. The Tasmania Sea up to this time also was not an impediment to the migration from Australia of plants, capable of dispersal over long distances. For example, pollen similar to that of the extant *Acacia*, known in Australia since the beginning of the Miocene, occurs only in the Pliocene of New Zealand. In New Zealand, much later than in Australia, appeared the pollen of *Triporopollenites bellus*, *Bombacacidites bombaxoides*, *Polypodites utimulatus*, *Milforsia homeopunctata*, etc. At the same time, it is suggestive that plants so important in Australia, such as *Eucalyptus* and *Xanthorrhoea* are absent to date in New Zealand. Analysis of the extant New Zealand flora shows that during various stages of the Cainozoic history to its ancient floristic core were added cosmopolitan, Australian, New Caledonian, Antarctic, Palaeotropic and other elements (McQueen *et al.*, 1968; Mildenhall, 1980).

The Palaeogene forests of New Zealand, as in Antarctica, were of mixed composition, but not rich. Among the conifers the major forest constituents were *Agathis*, *Araucaria*, *Podocarpus*, *Libocedrus*, *Dacrydium*. Among angiosperms the major role belonged to *Nothofagus*, Proteaceae, Myrtaceae (*Leptospermum* and *Metrosideros*), Rubiaceae, Araliaceae, Winteraceae, etc. During the Neogene, with an increase in the degree of isolation of the island, several forest components disappeared from the vegetation cover (*Araucaria*, *Athrotaxis*, *Casuarina*, *Nothofagus* ex gr. *brassii*, many Proteaceae). In their stead migrants of neoendemics appeared. In the opinion of most specialists the late differentiation of several taxa was facilitated by the high degree of hybridization. This was particularly reflected in Asteraceae, also in *Carmichaelia*, *Hebe*, *Coprosma*, *Acaena*, *Aciphylla*, *Epilobium*, *Myosotis*, *Ranunculus*. The present high degree of endemism of the New Zealand flora is determined by an intricate combination of palaeo- and neoendemics, that occur in practically all the major vegetation divisions.

Due to the relative stability of the climatic conditions, the Palaeogene and Neogene forests were of quite persistent composition in New Zealand. Changes of cooling and warming episodes resulted mainly in redistribution of the roles of the major forest components. During the cooling of the climate in the *Nothofagus* the role of the '*menziesii*' group increased, especially that of '*fusca*', whereas the role of '*brassii*' decreased. Of notable increase was the content of conifers, particularly of Podocarpaceae. Warming of the climate led to an increase in the role of *Nothofagus brassii*. In the palynological assemblages of conspicuous increase was the content of pollen of *Bombacacidites*, *Cupaniedites*, *Anacolosidites*, of palms *Rhopalostylis*. On the Western Island, especially during the maximum warming period of the Eocene and first half of the Miocene, traces of mangrove vegetation are recorded. The formation of open stretches at the south of the Island at the end of the Neogene and during the Pleistocene, can be attributed to the conspicuous cooling of the climate and an increase in the role of herbaceous plants and shrubs. The Pleistocene cooling led to the reduction of Proteaceae (only *Knightia* and *Persoonia* were preserved); *Microstrobus*, *Microcachrys*, *Acacia*, *Cranwellia*, etc., disappeared.

South American area

Throughout the Cainozoic the southern end of the American continent constituted a part of the Notal kingdom. The northern boundary of the South American area, which was already formed during the middle of the Palaeogene, underwent shifting through time, reaching the 40°S. Lat. during the cooling span of Oligocene and the end of Neogene, and descending to Terra del Fuego (Menendez, 1969, 1971) during the warming of the climate (Eocene, first half of Miocene). The repeated shifting of this boundary evoked the formation of the ecotone zones (within 5° of latitude), where admixtures of Antarctic and Neotropic elements can be observed among the major forest constituents. The composition of such forests reflects the Oligocene flora of Rio Turbio of Argentina, that includes over 50 genera and about 90 species, some of which are of indisputable Antarctic origin (*Nothofagus*, *Saxegothopsis*, *Embothriophyllum*, *Rhoophyllum*), whereas some are of Neotropic or Tropic cosmopolitan origins (*Annona*, *Bignonia*, *Buettneria*, *Cupania*, *Nectrandra*, *Phoebe*). In the flora of Terra del Fuego throughout the entire Palaeogene and Neogene such components as *Araucaria*, Podocarpaceae, *Nothofagus*, Proteaceae continued to exist. Due to the uplift of the Andes during the Neogene temperate forests of the south of the American continent were preserved up to the Pliocene only in its western mountainous part, while east of the mountain range steppe and semi-desert formations of the temperate zone developed, which are similar to those that can presently be observed in Patagonia.

Australian kingdom

The Australian kingdom separated from the Notal only at the end of the Eocene, as a result of the increasing splitting of the continents in the Southern Hemisphere, and drifting of Australia into the low latitudes.

One distinctive feature of the Palaeocene and Eocene flora of Australia during the period when it constituted a part of the Notal kingdom was the combination of plant types that commonly existed under different climatic conditions. Besides typical Early Palaeogene elements of the Notal kingdom (*Araucaria*, *Podocarpus*, *Dacrydium*, *Microcachrys*, *Nothofagus*), important forest constituents were cycads (*Lepidozamia*, *Bowenia*, *Pterostome*) that partly became extinct by the end of the Palaeogene, also other tropical and subtropical plants (*Casuarina*, *Cupania*, *Beauprea*, Myrtaceae, *Santalum*, *Banksia*, etc.). In the coastal zone mangrove associations with *Avicennia*, *Rhizophora*, *Sonneratia*, *Nypa* developed. Such a mixture of different plant types could have occurred only under conditions of a warm and humid climate, without any considerable seasonal and daily temperature differences.

Northern Australia during the Palaeogene was located in the 'critical' latitudes. As a consequence, many scientists consider that during the second half of the Palaeogene the north of the continent was covered by open woodlands, including sclerophyllous arborescent forms and various shrubs. Indirect indication for this is found in the occurrence of the *Adansonia* pollen. In emphasizing the specific features of the Notal flora, stress is frequently laid on the fact that many plants, having already appeared during the Cretaceous, underwent extremely meagre morphological changes and are very closely similar to their Recent analogues (*Banksia*, *Nothofagus*).

The large abundance of palaeoendemics of ancient Australian and New Zealand origin in the Recent flora of New Caledonia and Fiji, serves to suggest that their dispersal occurred long before the opening of the Tasmania Sea and the Lord Howe Uplift. Australia exhibited closer connections with these regions during the Palaeogene, and until the floras of these islands were pervaded by palaeotropical elements and neoendemics became of wide occurrence, as a result of the isolation of the territory, they probably constituted the northeastern part of the Notal kingdom.

During the Neogene, with an increase in the climatic differentiation, the floristic differentiation of Australia became more enhanced (Sluiter and Kershaw, 1982). As the continents of the higher latitudes drifted into the low latitudes during the middle of the Miocene, the rain forests, which until then covered most of the continent were pushed to the northeastern coast. They were replaced in vast areas, giving way to light forests with specific zonal types. The major forest constituents were eucalypts, casuarinas and acacias. Later these began to be pushed out by savannas. As the aridization of the climate increased, its peak falling to the end of the Pliocene–beginning of the Pleistocene, the western and central regions separated into two floristic areas, corresponding approximately to the modern Southwestern Australian and Eremeian areas. The western area, despite the absence of mountain systems, was distinguished, the same as today, by high species diversity and high endemism. This can be explained by the long preservation of landscapes with a combination of marine, edaphic and climatic barriers, that prevented migration. By the end of the Pliocene, here in the zone of maximum precipitation (up to 800 mm and more) eucalypt forests existed, which were replaced within drier areas by light forests composed of acacias. Heaths developed over the lateritic soils, whereas the arid belt (where the precipitations were less than 300 mm) was covered by herbaceous associations with Poaceae.

New Guinea, which at present is included in the Papua area of the Palaeotropic kingdom, at least to the middle of the Neogene constituted a part of the Australian kingdom. Rain forests occurred here that were similar in composition to the rain forests of Australia at the beginning of the Neogene. Of predominant occurrence were cycads, Araucariaceae, Cunoniaceae, Monimiaceae, Magnoliaceae, Myrtaceae, *Nothofagus*. The uplifts in the central part of the island caused an increase in the altitudinal zonation of the vegetation. The invasion of palaeotropic elements, chiefly of Indo–Malesian origin, increased considerably as New Guinea approached the Wallace Line. In contrast to other neighbouring regions, now included in the Palaeotropic kingdom, the cooling of the climate in the Southern Hemisphere at the end of the Pliocene, exercised a profound bearing on the New Guinea vegetation (Khan, 1971). This cooling trend, being concomitant with the drying of the climate, led to the formation of the middle mountain savanna belt, which even now divides the tropical rain forests of the coasts from the high mountain forests, where abundant relics of ancient floras of the Southern Hemisphere are still preserved.

MAJOR FEATURES OF FLOROGENY

Figure 83 shows the florogeny for the time span from the Carboniferous to the Early Cretaceous. The evidence so far available is insufficient to establish definitely the further interrelations between the phytochoria, and to designate the rise of Recent phytochoria. The proposed phylogeny of higher plants (Fig. 11) to a certain extent conveys florogenetic implications. By comparison of the schemes, depicted in the figures (see also Fig. 99), and taking into consideration the foregoing outline of the history of the floras, certain inferences can be drawn concerning the major features of florogeny and the factors and mechanism governing the florogenetic process.

1. Two major cycles of geographical differentiation are clearly apparent in Fig. 83. As already stated above, for the Devonian it is as yet impossible to clearly delimit the phytochoria. A distinct floristic differentiation appeared at the beginning of the Carboniferous and gradually increased further onwards, which is reflected both in the number of phytochoria and their ranks. The maximum differentiation is observed in the Late Permian, when the floras of different kingdoms exhibit distinctions in the dominant plant groups and overall generic composition. The greatest distinctions are apparent between the Gondwana and other floras. Near the Permian–Triassic boundary the affinities between the floras increase and during the Early Triassic the degree of floristic differentiation decreases markedly. Although the major phytochoria persist, their floras reveal many features in common and certain plants are nearly cosmopolitan. Later on, the differentiation process recommenced and continued up to the Quaternary. The destruction of natural landscapes by man, which occurs today, and the artificial spreading of plants, leads to a certain degree of dedifferentiation of the vegetation cover.

It is probable that the described phases of differentiation and dedifferentiation cannot be attributed to any single factor. Most likely they can be accounted for by phenomena resulting from the intricate interference of climatic changes, relationships between the continental plates and water bodies, and evolution of continental relief. It is clear that the rise of endemism in the Gondwana flora may be associated with the influence of Tethys, which hindered meridional migration, and with the specific climatic features of Gondwanaland. An increase in the differentiation of the floras at the end of the Palaeozoic was associated with orogenic processes and the more severe climatic contrasts that then prevailed on the Earth. In defining these floristic changes no individual factors should be sought that might have been responsible for these changes; rather it is necessary to disclose and bring to light the interrelations between various factors, each of which provided a constant control on florogeny.

2. In the evolutionary literature animated discussions proceed concerning the model of punctuated equilibria, describing the speciation (see Chapter 8). The process of evolution of the floras is well described by this model. During the last century the palaeobotanists were already calling attention to the fact that periods of slow change in the floras alternate with brief intervals of rapid change. In most cases climatic changes were the major factors that caused rapid floristic changes. However, although most of the changes were concomitant with definite climatic episodes, the ultimate control was sometimes provided by other factors, which evoked changes in the plant migra-

tion routes. For example, the rapid spread of the Angara flora over Kazakhstan at the middle of the Carboniferous can probably be attributed to the establishment of dry-land connections with Siberia. The appearance of northern plants in India during the Neogene reflects the disappearance of the Tethys as a phytogeographical barrier. The rearrangement of the floras is not necessarily associated with their phytogeographical status. A flora that has been subjected to a rapid and radical rearrangement of its composition may retain its phytogeographic position. For instance, Siberian floras, prior to and after the Ostrogian climatic episode (a cooling span at the Early–Middle Carboniferous boundary), belong to the same phytochorion – the Angara area.

3. The shifting of the boundaries between the phytochoria during the cooling and warming of the climate cannot always be discerned, due to the scarcity of localities in the ecotones. Climatic changes may provoke the migration of individual taxa. Warming episodes are reflected in the appearance of Equatorial plants in extraequatorial floras, as, for example, the Euramerian plants in the Middle–Upper Carboniferous of Southern Angaraland, *Classopollis*, cycads and bennettites in the Mesozoic of Siberia. The opposite process of the penetration of single extraequatorial plants into the Equatorial floras during cooling spans, has not been noted so far for the Palaeozoic and Mesozoic. This asymmetry in the migration is probably associated with the different degree of integrity in the Equatorial and extraequatorial ecosystems, and their different capability to resist the invasion of alien elements.

4. The distribution of phylogenetic lineages within the phytochoria (Fig. 11) clearly shows that the origins of practically all supreme-rank taxa are restricted to the Equatorial Belt, where the taxa and morphological structures exhibit the maximum diversity. Only a small number of supra-generic taxa can be named that originated in the extraequatorial floras (Arberiales and Gondwanostachyaceae in Gondwana, Vojnovkyaceae, Rufloriaceae and Tchernoviaceae in Angaraland). It may thereby be inferred that the main formative processes in higher plants are restricted to the Equatorial and neighbouring ecotone phytochoria. The evolutionary implications of this conclusion are discussed in Chapter 8.

5. Certain taxa that migrated into the extra-equatorial floras continue to exist and persist in them longer than in Equatorial floras, imparting a more ancient aspect to the floras where they occur. For instance, in the Upper Permian flora of the Angara area many plants according to the level of their evolutionary advancement, correspond to the Carboniferous (even in part to the Lower Carboniferous) plants of the Euramerian area. Progymnosperms of a primitive aspect (*Fedekurtzia*) survived in Gondwanaland up to the Middle–Late Carboniferous. This extraequatorial persistence is of interest from the evolutionary point of view, and should be taken into consideration in stratigraphic correlations and palaeoclimatic reconstructions (see Chapter 8).

6. If any taxon is known, both from megafossils and miospores, then the stratigraphic horizon of its first appearance is usually earlier in the miospore assemblages (sometimes by a whole stage). In some cases this difference in the time of first appearance of fossils of various types may be attributed simply to the mass production of miospores. This, seemingly, can account for the incoming of angiosperm pollen at a lower stage (in the Barremian), than leaves of the same plants (in the Aptian). However, in other cases it might be assumed that the origins of the taxa were restricted to biotopes located beyond the limits of the aggradational plains, so that the macrofossil remains of plants of these biotopes were not preserved. It is probable that the first conifers appeared under such conditions (see Chapter 3).

7. It might be interesting to use the general florogenetic data for stratigraphic purposes. In analysing the scheme of florogeny (Fig. 83) and the distribution of the phylogenetic lineages in the geochronological scale (Figs 11, 16, 99), the periods of florogeny and phylogeny should not be placed in a one-to-one correspondence with the stratigraphic units of the standard scale. For stratigraphic purposes of primary importance are not so much the levels of appearance of new plant groups, as the levels of rearrangements of the floras, even

when they preserve their phytogeographical restriction (see Chapter 8).

The formulated considerations that have emerged from this study may be regarded as the general features (general rules) of florogeny. However, they have not been worked out to full satisfaction, and are in need of further precision and more comprehensive explication. This problem may be suggested as one of the most important problems for future investigations. Extensive florogenetic investigations have not as yet received due attention in palaeobotany, notwithstanding the fact that these studies may largely contribute to a synthesis of palaeobotanical knowledge, and the integration of palaeobotany with other disciplines.

Chapter 8

Relationship between palaeobotany and other fields of natural history

In his studies the palaeobotanist uses physical and chemical techniques. When collecting, studying and interpreting data he or she has to turn to various fields of biology, geology and physical geography. Mathematical methods, including computer simulation, are also of use in palaeobotany. Palaeobotany is connected with archaeology in the case of studies of plant remains associated with human artifacts, and with astronomy (in dendrochronology and when reference is made to global events, probably connected with astronomical causes). The present chapter will deal not with all relationships between palaeobotany and other fields, but with the most important applications of palaeobotany to geology and biology.

STRATIGRAPHY

In their stratigraphic uses fossil plants do not generally differ from other groups of organisms. The principal way to use palaeobotanical data in stratigraphy is in comparative analysis of the taxa ranges in geological sections. Two somewhat different tasks are approached in this way. One task is the correlation of actual sections by the horizons of appearance and/or dominance of certain taxa. This requires the repetition of this sequence in each section. Such similarity in the succession of horizons is known as homotaxis and the succession itself as a homotaxial succession. (There is some misunderstanding of homotaxis in the literature, when this is understood only in the sense of similarity in the taxonomic composition of floras or faunas in different sections.) Once homotaxis is established, it is possible to correlate the sections under study.

The second task is the further use of the homotaxial succession once it is established. It may be assigned the status of a local or regional (up to global) biostratigraphical scale that summarizes the data from all measured sections. This scale may be used for constructing a standard stratigraphical scale as well as for determining the stratigraphical position of such assemblages of organisms each of which is found in its region outside the general sequence. For example, in some sections of the Upper Palaeozoic of Kuzbass the homotaxial succession of plant genera and species has been established by assemblages of which the regional phytostratigraphical units are outlined. In the Tunguska basin it is difficult to compose sufficiently complete succession due to weak outcropping, tectonic fractures and other reasons. Therefore, the different outcrops are sometimes correlated by correlating each of them separately with the Kuzbass, the generalized section of which is the key one.

The following examples demonstrate how efficient such phytostratigraphical studies may be. From plant megafossils only (without data on miospores), seven successive floristic assemblages are indicated for the coal-bearing Middle Carboniferous of West Europe; if the sampling is more complete, it is possible to differentiate some additional assemblages. The corresponding subdivision of the section made the basis for the regional stratigraphic scale and has been successfully used during more than half a century. For subdivision and correlation of the Devonian–Carboniferous boundary beds miospore assemblages are widely used. The zones of wide geographical distribution established by them are not less detailed than those established by the conodonts and foraminifers and have the advantage of being distinguishable in continental and in littoral marine deposits.

It is obvious that both tasks cannot be solved without well-elaborated plant systematics, without exact diagnostics of taxa forming homotaxial succession. Another point is no less important. In each of the partial sections, working from the data on which the homotaxial succession of taxa is based and from it the corresponding phytostratigraphic scale, it is necessary to reveal the levels of the appearance and disappearance of each of the taxa used. As both tasks are labour consuming an attempt is often made to use approximate systematics and to calculate the quantitative participation of genera and even of taxa of a higher rank instead of the exact diagnostics of the taxa. In a similar manner a summary characterization is given for whole members, subformations and even formations instead of the detailed characteristics of the section by layers. The results of phytostratigraphic studies are often presented in exactly this way and sometimes the original material is collected similarly. Thus, for many regions of the USSR the palynological characteristics of the strata are available only as lists of genera or suprageneric groups of uncertain volume ('conifer-like pollen') with indications of percentages in assemblages. Similar palynological characteristics, in the best case, serve the aim of the local stratigraphy and are always fraught with stratigraphic mistakes.

In the above text phytostratigraphical studies under conditions of persistent homotaxial succession along the strike were dealt with, i.e. within the limits of a certain phytochorion, most often within a province or district. In the case of interprovincial and even more in inter-area correlations serious difficulties arise, as the homotaxis in the taxa distribution is violated. Palaeobotanists often tried to overcome these difficulties by using for long-distance correlation (not only between areas but also between kingdoms) single common taxa. Thus the Carboniferous–Permian boundary in Kuzbass (Angara kingdom) was drawn from the appearance of the genus *Callipteris* by analogy with West Europe (Amerosinian kingdom). In these correlations the use of single taxa is seen to be very risky. To do it it is necessary first to make sure that a given taxon (a genus or species) is natural enough and does not comprise representatives of different eutaxa. Secondly it is necessary to incorporate some general reasons justifying the assumption of a synchronous appearance of this taxon in different phytochoria. In the opposite case serious mistakes are possible. For instance, in the above-mentioned example of *Callipteris* a mistake, now clearly evident, was committed in the correlation of sections. It is quite understandable. Though the Angara and European *Callipteris* species are closely related and classified with the same family Peltaspermaceae, it is doubtful whether they belong to one natural genus. At present it is customary to compare the levels of appearance of *Callipteris* in the Kuznetsk and Pechora basins. It is not yet possible to judge the relations of the Kuznetsk and Pechora species; however, in this case there are some common foundations for correlation of sections. *Callipteris* is known only from the periphery of the Angara area and it is possible to believe that it moved into this area due to climatic changes (warming up and some aridization). One may then assume that the appearance of *Callipteris* in the south-west and in the north-west of this area was approximately simultaneous.

The degree of naturalness of genera which are used for stratigraphic correlations may be evaluated only after the life-time connection of different organs, commonly attributed to different genera, is established. It is known, for example, that the European species *Callipteris conferta* associated with microsporoclads *Pterispermostrobus gimmianus*, pollen *Vesicaspora* and seeds *Cyclocarpus*. The Pechora and Far-Eastern *Callipteris* possessed peltoids of the *Peltaspermum* type. The sets of taxa corresponding to natural genera and species may be called assemblage-genera and assemblage-species (see Chapter 2). It is clear that the stratigraphic correlation should be based as far as possible on assemblage-genera and assemblage-species, rather than on their separate components. Hence, the reconstruction of the initial association of the detached parts is of immediate practical significance.

General foundations necessary for stratigraphic comparison between different phytochorias have been mentioned above. In a general case ecosystemic notions of different degree of concreteness and generality are dealt with. Realization and wide use of these notions permits a much deeper understanding of stratigraphical problems, and their solution can be found in cases where the analysis of distribution of taxa in sections fails. It is exemplified best of all by comprehension of local and regional events, fixed in the stratigraphical boundaries, in a wide interregional (up to global) context. This approach provides us with an instrument for interregional and even global correlations. During recent years an increasing number of data appear indicating that many phytostratigraphical boundaries strikingly dissimilar in different regions by their expression are traces of one global climatic episode, cooling off or warming up. In the previous chapter it was demonstrated that with the climatic episodes are connected directly or indirectly the changes of the plant assemblages at the boundaries between the Devonian and the Carboniferous, of the Lower and Middle Carboniferous (Pennsylvanian and Mississippian), of the Middle and Upper Carboniferous, the Carboniferous and Permian, the Permian and Triassic, the Middle and Upper Jurassic, the Albian and Cenomanian, of the Cretaceous and Paleogene, the Eocene and Oligocene, as well as at the boundaries between the other standard stratigraphic units. Plant communities react sensitively to climatic changes (Figs 94–97) which are now revealed to be mostly sudden and, moreover, synchronous in different parts of the world.

In this way palaeobotany proves to be intimately connected with the solution of the most important stratigraphic task, namely with perfecting the international stratigraphic scale by choosing the most natural boundaries between its subdivisions. These boundaries must be clear cut, easily observable over long distances and maximally isochronous over their strike. The climatogenic boundaries best fit these requirements and it is exactly the study of plant remains that makes it

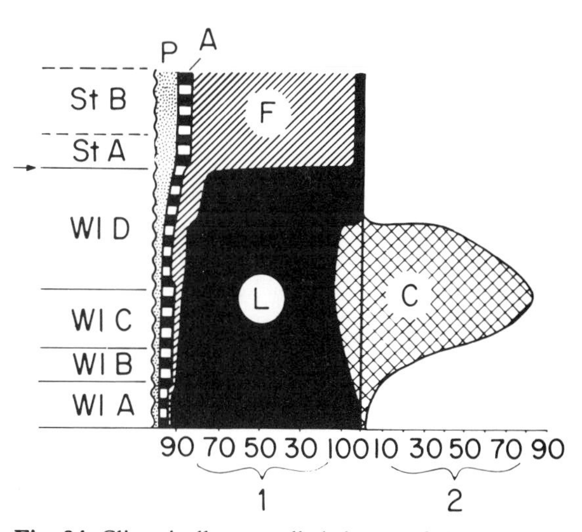

Fig. 94. Climatically controlled change of plant associations in peats of Euramerian area at the Middle–Upper Carboniferous boundary (at arrow); Wl A, Wl B, Wl C, Wl D, St A, St B – Westphalian A, B, C, D, and Stephanian A and B; P, pteridosperms; A, articulates; F, ferns; L, lycopsids; C, Cordaitanthales; horizontal scale – percentage of plant groups; 1, general pattern; 2, regional variations in content of Cordaitanthales. After Phillips, 1981.

easiest to reveal and to follow such boundaries. By integrating data on plant megafossils and miospores the relevant boundaries may be traced not only in the continental but also in the near-shore marine strata.

There are many other possibilities in using palaeobotanical material for stratigraphy. Included are the synchronization of ore seams (including coal seams) by miospore assemblages and by the characteristic taphocoenoses of macrofossils, the recognition of soil horizons in predominantly marine deposits and the use of such horizons as markers. Using acritarchs and remains of marine algae it is possible, conversely, to reveal traces of marine transgressions in continental strata, and also to use them as markers, etc.

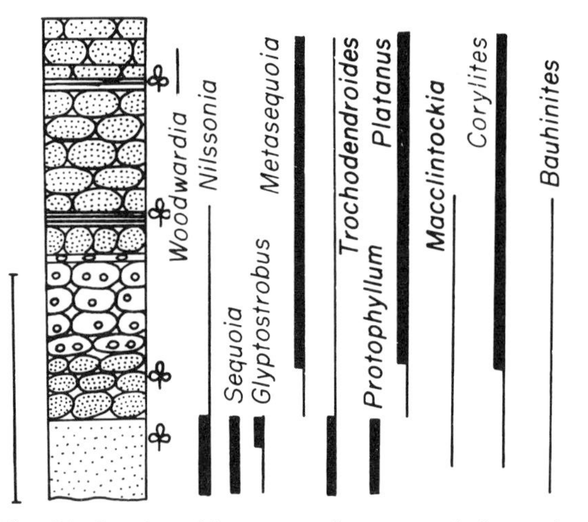

Fig. 96. Stratigraphic ranges of genera and change in dominance at the Maestrichtian–Danian boundary in South Sakhalin. From Krassilov, 1977a.

LITHOLOGY

Connections between palaeobotanical and lithological studies were established in the last century when geologists began to reconstruct the mechanism of formation of the continental, especially the coal-bearing strata. Three main ways of utilization of palaeobotanical materials for the consideration of lithological problems may be outlined.

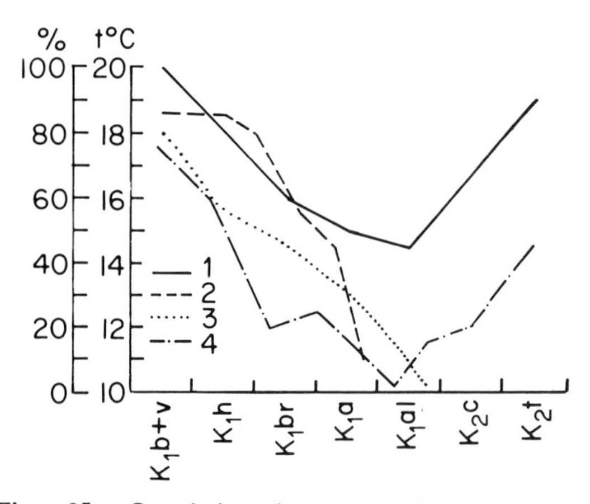

Fig. 95. Correlation between palaeotemperature measurements and *Classopollis* pollen content in Cretaceous of southern regions of the USSR; 1, palaeotemperatures of Middle Asiatic seas according to N. A. Yasamanov's data; 2, *Classopollis* content in Turkmenia and adjacent parts of Uzbekistan according to N. A. Fokina's data; 3, same in SE Caucasus, according to data by S. B. Kuvaeva, M. M. Aliev and R. A. Aliev; 4, same in Crimea, according to data by S. B. Kuvaeva and B. T. Yanin; horizontal scale – Cretaceous stages from Berriasian to Turonian. From Vakhrameev, 1978.

1. The indicator value of fossil plants, when by their taxonomic composition or by the structure of taphocoenoses inferences are drawn on sedimentatory environments. As was mentioned in the previous section, marine interbeds and soil horizons may be revealed from plant remains, and this sometimes leads to a complete revision of the accepted ideas of the sedimentatory environments over vast intervals of the geological section. For example, a series of soil horizons was revealed by W. Remy (1980) from characteristic plant remains in the Devonian of the Rhine–Ardennes area. The deposits comprising these horizons were supposed to be marine throughout and this was reflected in palaeographical charts. The multiplicity of soil horizons obviously indicates repeated regressions, not thought of before. The whole sedimentatory regime of the deposits appeared in a new light. Palaeobotanical evidence may indicate climatic conditions of sedimentation (see below). The area analysis of taphocoenoses yields most important information for the reconstruction of palaeolandscapes (see Chapter 6; Figs 81, 82).

2. The study of the immediate participation of higher plants (algae being omitted here) in the sedimentation when they supply dispersed or concentrated organic material to the sediment. In the latter case this is a source material for sedimentary organic matter, primarily for coal. Links between palaeobotany and coal geology are traditional. The source plant communities which gave rise to coal seams were evaluated at first on the basis of plant remains occurring in the seat earth and roof shales of seams, then on the basis of the permineralized remains in coal balls or coalified remains within seams, and still later by the composition of miospores in a seam. At present the analysis of a source plant material is performed by all the available methods combined, the change of plant remains both in the section and along the strike of the seams being scrutinized. To study the source material of highly metamorphosed coals, the technique of ion-beam etching of a cut face is very efficient (Kizilshtein and Shpitzgluz, 1984). The data obtained are important for the classification of coals and for predicting their technological properties. No less important is the elucidation of the general situation during coal accumulation. This contributes to understanding the changes of coal along the section and strike both of particular seams and of the coal measures as a whole.

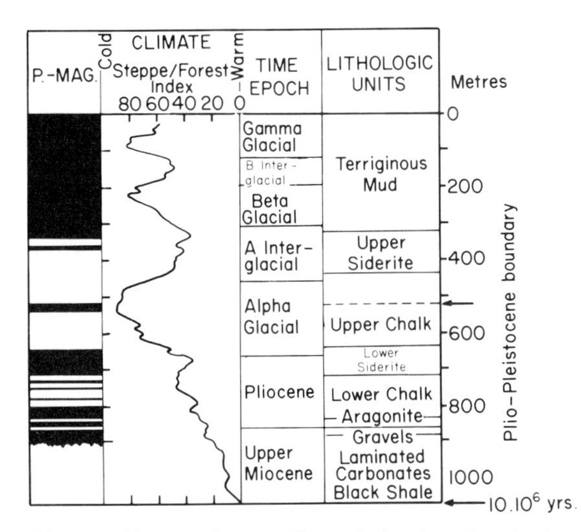

Fig. 97. Change of steppe/forest index in palynological assemblages against palaeomagnetic episodes, lithological and palaeoclimatic units in Neogene and Quaternary deposits of Black Sea. Modified from Traverse, 1982.

3. The study of the significance of terrestrial plants as the most important controlling factor of sedimentation. Unfortunately, in this respect mostly general speculations are available while actual research has been performed on a very limited scale. The dynamics of vegetation of past epochs is litle known, including the dynamics of conquering the land by plants, interrelations between the relief and vegetation in the past. One opinion is that the land has been populated by plants since the Pre-Cambrian. Another opinion has it that coverage of land with vegetation comparable to that of the present day was established only in the Cainozoic after the herbaceous angiosperms had appeared. Before that placores (uplands) in the arid and semi-arid climate are said to have had no vegetation cover. Both viewpoints are speculative and have not been the subject of serious study.

Summarizing palaeobotanical data it is possible to point out certain principal aspects of the relations between vegetation and sedimentation. First of all, the extreme rarity of reliable indications of continental deposits of the Cambrian and the Ordovician is to be noted. This may be related to the peculiarity of the most ancient continental strata, the origin of which may not be evident and to the rarity of such strata. In any case, it may be concluded that terrestrial vegetation during these periods, even if it existed, was not developed adequately to control the formation of the continental strata in a form familiar to us, with well-known alluvial, lacustrine and other lithogenetic rock types. It is important that the early Palaeozoic palaeosols contain no traces of root systems (Retallack, 1981). Continental deposits are known from the Silurian containing plant remains as microfossils (e.g. in Pennsylvannia, USA). In the uppermost Silurian macrofossils begin to occur in near-shore marine deposits. In Chapter 7 it has already been stated that the expansion of plants on to the continents might have led to the return of plants into aquatic habitats at the end of the Silurian. In the Devonian the terrestrial vegetation was widespread and the density of plant cover was high enough judging by the abundance and diversity of miospores in marine deposits. Nevertheless, as already mentioned, the productivity of terrestrial propteridophyte communities was about twenty times less than that of the present-day forest communities. How that influenced solid and liquid drainage is not yet clear.

At the very end of the Devonian, plants indeed seemed able to populate the uplands, including the slopes of valleys, to such a degree as to control the ablation of clastic material and allow low-ash peat to be accumulated. In the same period shrubby and then arboreal vegetation of the intertidal zone appeared, sometimes compared directly to the present-day mangroves. The formation of a belt of this vegetation along the shores in the equatorial zone should have rendered an intensive influence on the marine sedimentation. Possibly the appearance of this belt limiting the shore erosion contributed to the formation of extensive seaside lowlands with their paralic sedimentation.

The lithological analysis of marine, especially carbonate, strata makes wide use of data on marine algae. The corresponding range of problems is disregarded here since it is more connected with the palaeoecology of marine animals and sedimentary–petrographic studies than with palaeobotany in its traditional sense.

PALAEOCLIMATOLOGY

Palaeoclimates are evaluated on the basis of the presence of characteristic plants, the composition of communities, and by morpho-anatomical characters. However, some taxa and morpho-anatomical characters may change their climatic affinity with time. This manifests itself in the form of extraequatorial persistence, which was first traced in the late Palaeozoic of the northern hemisphere. Many morphological types characteristic of the early–middle Carboniferous of the Equatorial belt occur in the Angara kingdom when the conditions warmed up. There they persisted until the late Permian. The possibility in principle of such phenomena motivates the approach to palaeoclimatic reconstructions of analysing the whole pattern of floras for large time intervals rather than isolated floras. One has not so much to give independent palaeoclimatic characteristics to separate floras as to compare floras sequentially replacing one another in space and time, drawing conclusions on trends in climatic changes.

It is difficult for a palaeobotanist to say how warm or cold, dry or humid the climate was in quantitative terms, or to indicate annual averages for temperature and precipitation, but it is possible to state, and sometimes very confidently, the trends of climatic changes, periods of warming up or cooling off, aridization and humidification, increase or decrease in seasonality of climate (Figs 95, 97). Some features are indicative of a frostfree climate with a high mean annual temperature.

Though palaeoclimatology needs the ‘absolute’ (metric) characteristics of temperature and humidity, the relative (non-metric) characterization given by palaeobotany is no less important. Palaeoclimatology may be compared to geochronology where palaeontological data do not indicate the duration of chronological subdivisions but indicate their sequence, the asynchroneity of geological events. Just as on geological maps the rock of the same time interval may be outlined using plant remains, so the representation of areas with floras of the same composition and similar climatic affinity (isofloras) on maps demonstrates the distribution of palaeoclimates. Differentiation of the present-day climatic zones makes wide use of the data on vegetation cover. In contradistinction to metric parameters issued by meteorological stations and reflecting the climatic situation for short time intervals, the vegetation cover reflects the averaged secular characteristics of a climate. It indicates major climatic boundaries which can be rationally taken as a basis for climatic zoning. The same is true of the geological past.

In evaluating palaeoclimates it is impossible to rely only on palaeobotanical data without the support of other evidence. The fact is that similar changes in morpho-anatomical characters and in the systematic composition of plants may be connected with changes in different climatic parameters (see Chapter 6). For instance, in the Middle–Upper Carboniferous of the Angaraland an impoverishment of the assemblages of plant remains may be noted from the inner areas of the continent (south of Middle Siberia) to the north-east (to the Verkhoyansk region), south (to Mongolia), and south-west (to Kazakhstan). Predominantly leaves of cordaites and *Angaropteridium* remain in the assemblages. However, the causes of impoverishment seem to be different: a cooling of the climate in the north-east direction and aridization in the other directions. This may be seen in the lithological characters of rocks.

TECTONICS AND PLATE MOVEMENT

The common character of the Upper Palaeozoic floras of the Gondwana continents formerly motivated the idea of the past existence of Gondwana and was then widely used as a basis for the concept of continental drift. At present mobilistic and fixist (steady state) concepts are also often checked against palaeobotanical evidence. The following main implications of palaeobotanical data in tectonics may be named. The similarity of floras of different continents may indicate direct terrestrial connections between continents. However, the neighbourhood of significantly different floras testifies to a former disjunction of territories that suffered secondary tectonic convergence. A palaeogeographical reconstruction envisaging the distribution of very similar floras on different shores of a wide water area is open to doubt. The situation of the climatic zones established by fossil plants may indicate relative changes in the position of continents. The plant belts reveal a connection with the orography and orogeny of the region. These are special applications of palaeobotany to separate tectonic considerations. In the more general aspect palaeobotanical facts should be incorporated along with all others into a factual foundation of global historical geological reconstructions.

The similarity of the fossil floras on both sides of the recent ocean is often straightforwardly interpreted as evidence of the moving apart of continents. In reality this similarity points only to land bridges between continents, which might have remained in their places. It is different when fossil plants indicate such a shift of continents in relation to climatic zones which cannot be co-ordinated with the stability of the position of continents. It has been shown that the Upper Palaeozoic belt of frostless climate is located almost entirely in one hemisphere, if the present relative position of the continents is assumed. The northern border of this belt shifts within a one or two thousand kilometre wide band over a period from the early Carboniferous to the middle Cretaceous, this shift being random. The southern analogue of this boundary shifts several thousand kilometres to the south over the same time span, also if the stability of continents is assumed. It is known that the general situation of climatic boundaries of such a high rank on the Earth is controlled not by local palaeogeographical factors, but by general astronomic factors (change in the specific quantity of solar radiation depending on the curvature of the Earth's surface and creating a general latitudinal climatic zonality). It is impossible then to explain the different behaviour of both boundaries under consideration. It must be assumed that a shift not of the southern boundary relative to the stable continents, but of the continents themselves relative to a comparatively stable boundary had been taking place (for more details see Meyen, 1973c).

On certain mobilistic reconstructions the territories belonging to the same phytochoria (areas, provinces) are divided by a large area of water or cross all the latitudinal zones. Thus in some reconstructions a wide Tethys Ocean is shown with the Subangara flora of Afghanistan (the Upper Permian) on the southern shore while other occurrences of the same flora remain on the opposite shore. In other reconstructions the Cathaysian continent is placed in a meridional position so that its south lies in the equatorial zone and its north in high latitudes. Though the Cathaysian floras are somewhat differentiated in space, it is impossible to fit their distribution to the proposed reconstruction. Tectonic reconstructions are sometimes in good agreement with palaeobotanical observations. There is an opinion that Kazakhstan was a separate continent until the middle of the Middle Carboniferous and was situated in low latitudes not far from a continent populated with Angara flora. In the Middle Carboniferous both continents fused. Before this hypothesis was formulated an independent Kazakhstan province had been recognized, populated with Angara plants exactly in the Middle Carboniferous. Comparison of floras on both sides of the Atlantic ocean in the interval from the late Palaeozoic to the Cainozoic is in very good agreement with the views of tectonists as to the time of initiation and development of the Atlantic Ocean.

Some palaeofloristic boundaries are so sharp that their tectonic origin is to be assumed. Over a short distance the most abrupt change of plant assemblages may occur e.g. along the north-east border of Gondwana in the south-east of Eurasia in the late Palaeozoic. The possibility is also to be taken into consideration of sharp boundaries created by orographic factors (similar to the northern border of the present-day Crimean subtropics). A tectonic explanation for palaeofloristic boundaries should therefore be supported by analysis of the whole palaeogeography of a region. For instance, in the Jurassic and the Cretaceous of the Far East and the Japanese islands the boundary between floras of Siberian and equatorial type was situated longitudinally and was probably controlled not by the general zonality, but by the location of orographic barriers (Kimura, 1980). The history of the Tibetan flora is very suggestive (Palaeontology of Xizang, 1982). A Cathaysian flora is known for the north of Tibet in the Permian, whereas a Gondwana flora is known for the south. Until the early Eocene the flora of Tibet was the same as equatorial floras of Asia situated to the north and populated lowland habitats. Then an affinity of the flora with the Himalayas was established and it is obvious that the Himalayas did not prevent penetration of the monsoons of the Indian ocean far into the north. Only in the Quaternary did the Tibetan flora become montane and acquire the present-day composition. The interpretation of the Tibetan flora directly influenced tectonic concepts (Xiao and Gao, 1984). The affinity of the Permian flora of the southern part of Tibet to Gondwana has entailed corresponding palaeogeographic and palaeotectonic reconstructions. It proved that the structural border of Gondwana and Laurasia is not marked by tectonic congestion. The Himalayan structures to the south of Tibet should be attributed to Gondwana; they cannot be considered to be a direct result of the collision of Gondwana and Laurasia.

On the whole it may be noted that incompetence in tectonics deprives a palaeobotanist of the possibility of fully understanding the nature of phytochorias, and comprehending their arrangement and evolution. Equally, a tectonist discussing the fates of continents and forgetting the available palaeobotanical evidence may at best miss an excellent opportunity for independent testing of his conclusion, and at worst commit mistakes bringing the whole work to nothing.

PALAEOBOTANY AND THE THEORY OF EVOLUTION

The contribution of palaeobotany to the rise and development of the theory of evolution is small. In *The Origin of Species* Darwin scarcely referred to palaeobotanical materials. Among palaeobotanists there were almost no evolutionists who suggested influential and original evolutionary concepts. True, O. Heer suggested a theory of Überprägung which in modified form was then supported by O. Schindewolf and somewhat resembles the modern concept of punctuated equilibria (punctualism). If palaeobotanists turned to the evolutionary implications they commonly operated with already available theoretical concepts worked out on the basis of recent or palaeozoological material. The evolutionary morphology of plants owes much to palaeobotanists H. Potonié, O. Lignier, H. H. Thomas and W. Zimmermann, but actually it was only the extension to plants of concepts considered earlier for animals in more detail and more completely. The same can be seen in modern palaeobotanical literature when it refers to evolutionary problems.

Meanwhile palaeobotany possesses such factual material which can make evolutionary discussions more constructive, be a source of new evolutionary ideas and point to the limitations of concepts suggested as universal. Below only those items are considered that can be made specially fruitful if discussed using palaeobotanical data.

Punctualism (hypothesis of punctuated equilibria) vs gradualism

This is the most popular subject for modern discussions on evolution (Gould, 1982b; Gould and Eldredge, 1977; Hecht, 1983; Lister, 1984; Stebbins, 1982; etc.). According to the concept of punctualism, species formation is not a continual (gradualistic) but a structuralized process with phases of prolonged stasis followed by phases of rapid genetical and, correspondingly, morphophysiological reorganizations. Discussions on punctualism take several trends. Most of the discussion participants attempt to analyse their material in terms of punctualism or of gradualism and then pass sentence in favour of one or the other. In so doing a peculiar taxonomic reductionism is very common: the conclusion on a particular group of organisms is universalized. True, more cautious participants of the discussion admit the possibility of combining gradualistic and punctualistic components in the evolutionary process.

In current discussions punctualism is in an unfavourable position as discreteness in the form sequence observed in a geological section may always be referred to the imperfection of the geological record. Mass and rapid transformations of palaeopopulations may be interpreted as the modification variability governed by an external factor and not concerning the gene pool of the population. Thus gradualism may always find refuge in an 'asylum ignorantiae'.

To make the discussion of the concept of punctualism in a palaeobotanical context more productive, it is useful to differentiate certain aspects of plant evolution (the same is true of other organisms). First, morphological and population–taxonomic aspects of transformations should be distinguished. A morphological transformation may be supposed which will be saltational on the line ancestor–descendant (during one generation) but will gradually spread in the population and still more gradually within the limits of the whole species. From the morphological viewpoint such a new formation will then be punctualistic and from the population–taxonomic viewpoint will be gradualistic. Second, it is necessary to distinguish between the reality (the actual realization) of punctualistic transformations and their actual overall role in the evolution. Third, it is necessary to keep in mind that the shortage in facts can easily be supplemented with hypothetic assumptions, and it is not difficult to make them plausible. It is useful therefore to differentiate between our ability to describe something in terms of a certain concept (and with rich semantics it is possible to describe anything in the language of a concept) and our ability to prove the adequacy of the description to fit what is really going on in nature. In the first respect our ability is very high, and on this basis Popper came to the conclusion that there is a need to limit it using the falsification principle.

Now let us turn to concrete palaeobotanical data.

If punctualism is connected only with speciation, palaeobotany should not participate in the discussion. It is not possible yet to point out a single example when palaeobotanists could reliably demonstrate the transformation of one species into another. At most it has been possible to demonstrate the sequence of form species in section outlined by the detached organs and not of assemblage-species (for treatment of that concept, see Chapter 2). However, there are some data on the distribution of some assemblage-species. Assemblage-species of the arborescent lycopsids in the Carboniferous of the USA are most fully reconstructed and followed most thoroughly in section (DiMichele, 1981). For example, the assemblage species *Lepidodendron varium* (Baxter) Taylor emend. DiMichele–*Achlamydocarpon varium* (Baxter) Taylor et Brack-Hanes proved to be distributed over almost all the middle Pennsylvanian. This very completely reconstructed species has thus existed for a very long period, at least 20 million years. The intervals of distribution of other lycopsid assemblage-species found by DiMichele are approximately the same. This is the time interval which is quite comparable to the duration of all previous evolution of the lycopsids with *Lepidodendron* trunks and strobili of the *Flemingites–Achlamydocarpon–Lepidocarpon–Lepidostrobus* type. Taking into consideration the insignificance and non-directional nature of morphological variations in lycopsid assemblage-species throughout their existence it is impossible to assume that the whole evolution of this group of the lycopsids went on at the same rate. The earlier stages of evolution obviously went on much more rapidly and were more channelled.

This example of the lycopsids is significant for all discussion on punctualism. It has been pointed out (e.g. by T. Schopf, 1983) that a prolonged stasis of forms may represent only an artifact of the material preservation, since the transformations of populations might not involve the characters of shells or other solid and easily preserved parts available to palaeozoologists. Transformations might occur in the structure of soft tissues. In the case of the arborescent lycopsids it was possible to observe the stability both of external morphology and cellular structure. Of course, the characters not yet studied may now be referred to (caryotype, protein polymorphism, etc.) but such arguments would mean only an attempt to save the gradualistic hypothesis at any price, even at the price of an 'asylum ignorantiae'.

Other cases of the very wide stratigraphical range of certain species may be cited. However, such facts cannot be included in the discussion while species established by detached organs are not reduced to assemblage-species. Otherwise a serious mistake may be committed, its probability being seen from the following example. The species *Dicksonites pluckenetii* (Schl.) Sterz. is distributed from the Westphalian D to the Lower Permian. Within this interval vegetative fronds show no perceptible variation (to be more correct, their variability is not arranged in any stratigraphic sequence). However, phyllosperms change so much (Fig. 47, *i*, *j*) that Westphalian and Stephanian specimens will probably be attributed to different genera in the future. Thus, from the Westphalian D to the Lower Permian only a certain morphological type of foliage persisted rather than a species comparable to a recent species. This exemplifies the diachronism of evolutionary transformations of different organs.

Examples of the gradual directed transformation of organs in a section may also be pointed out. In the Westphalian A–D of the Rhine–Ruhr district quite a gradual transition from *Neuropteris obliqua* (Brongn.) Zeill. (pinnules with open venation) via *Neuropteris semireticulata* Jost. (pinnules with sparse anastomoses) to typical *Reitculopteris muensteri* (Eichw.) Goth. with reticulate venation was observed by Josten (1962). A gradual change of forms of *Lueckisporites* pollen is described by Visscher (1971) in the Zechstein of West Europe. A gradual transformation of *Afropollis* pollen is demonstrated by Doyle *et al.* (1982) in the Lower Cretaceous of Africa. These and similar observations may be interpreted as proof of the gradualistic model. However, it is more correct to consider them as evidence for non-correspondence of the semophylogeny of certain organs to the punctualistic model. This only means that the punctualistic model is limited and does not fit certain situations, but not erroneous.

On the other hand, such taxonomically significant transformations of plants may be pointed out which demonstrate the limitations of the gradualistic model as well. For instance, in many conifers leaves are epistomatic while the ancestors of conifers (either the Cordaitanthales or Dicranophyllales) had hypostomatic leaves (occasional stomata occur on the upper epidermis). Among ancient conifers there are hypostomatic forms (*Swillingtonia*), amphistomatic forms (*Ernestiodendron*, *Lebachia*, *Quadrocladius*), and epistomatic forms (*Kungurodendron*). The transition from hypostomatic to amphi- and epistomatic leaves may have occurred gradually enough, i.e. the total number of stomata on a certain side of the leaf may have changed gradualistically. But the appearance of stomata, even rare ones, on a side of the leaf previously void of stomata cannot in itself be gradualistic. It is important that quite formed normal stomata appear in a new place. The appearance of stomata in a new place is a 'microsaltation'. Morphogenetically the appearance of a stoma is in no way simpler than that of any other organ. The ontogenetic programme, which has already been perfected, acts in a new place and produces the same organ here.

Medullation of the protostele is produced in the Devonian lycopsids and progymnosperms by exactly the same 'microsaltations'. In this case no gradual penetration of parenchyma cells from the bark through the metaxylem layer to the centre of the stele occurs. Different events happen; some cells along the central axis of the xylem cylinder are differentiated to parenchyma, not to tracheids. A change of ontogenetic programme takes place by saltation.

There was a multitude of similar 'microsaltations' in the evolution of plants (though microgradualistic processes also occur, e.g. in the transformation of the cuticular ornamentation of cells). Each 'microsaltation' possesses no taxonomic significance, but the reality of its manifestation is more important for us than any taxonomic evaluation of this phenomenon. Such peculiarities in the position of organs may be found in Recent plants that could not in principle appear gradualistically. For example, there is the obligate appearance of plantlets with roots on the edges of leaves of *Bryophyllum* and related plants. It was recalled long ago by Leavitt (1909) that in compound leaves of angiosperms the abscission layer may be present simultaneously in the bases both of the whole leaf and of each leaflet. Study of the Cretaceous angiosperms convinces us that compound leaves appeared later than simple leaves. It is an absurd idea that during the transition of a simple leaf into a compound leaf the abscission layer divided and the daughter layers slowly went to the bases of leaflets. There is not much more sense in the assumption that abscission layers in the bases of leaflets developed slowly. It is known that if a plant already has an ontogenetic programme for a certain organ this programme in most cases will function in a new place by the principle of 'all or nothing'.

The origins of the compound leaf and of abscission layers at the bases of leaflets produce taxonomically significant characters. The first character probably originated rather gradually (Hickey and Doyle, 1977), and the second character, as mentioned above, by saltation. This example convinces us that we do not have to choose between punctualistic (saltational) and gradualistic models, but to consider how these models combine in the real evolutionary process.

In terms of genetics, the micro- and macrosaltations considered above may be described as homoeotic mutations supposed to be connected with mutations in regulatory genes. Very little is known about the nature of regulatory genes in the higher eucaryotes, their function, or how their functioning is rearranged. Moreover, the regulation of form building is performed at the epigenetic level (hence the equifinality of ontogenetic programmes and phenotypic modifications). So it is better for the present to speak about ontogenetic regulations in general without specifying their mechanisms. Accordingly, it is better to speak of homoeotic phenomena rather than of homoeotic mutations. Homoeosis is very widely spread in plants as was obvious as far back as the beginning of this century (Leavitt, 1909). Most often homoeosis gives rise to teratological forms. Such is the appearance of marginal seeds and sporangia on the leaves of *Ginkgo* (Sprecher, 1907). However, in some cases homoeotic transformations may be widely distributed in a population and acquire taxonomic significance. The above-mentioned marginal plantlets on leaves of *Bryophyllum* and of other Crassulaceae are the best example. Such fixed homoeosis acquires the status of heterotopy, i.e. obligate change in the position of an organ (in the pesent case the heterotopy may be called saltational; in other cases heterotopy may develop gradualistically). An attempt should then be made to find out which morphological transformations in plant evolution occurred by saltational heterotopy. Such cases will obviously correspond to the punctualistic model. In Chapter 3 saltational heterotopy was supposed in many plants. It is most probable that sporangia on the pinnules of zygopterids, synangia and seeds on fronds of Callistophytales and of some other gymnosperms appeared in this way. Ancestors of these plants possessed leafless sporoclads and polysperms. No transitional forms have been found between these leafless organs and fertile fronds (sporophylls and phyllosperms). Reduced pinnules sometimes occurring on fertile organs (Fig. 47, *q*; Fig. 57, *d*) can hardly be taken as evidence of a gradual transition.

It is significant that leafless fructifications and fertile leaves occur in closely related plants, e.g. within one genus *Botryopteris*. The genera *Musatea* and *Nemejcopteris* have leafless sporoclads, and in the related genus *Corynepteris* sporangia are situated on pinnules (Fig. 31, *l*; Fig. 32, *c*, *d*; see also *Tedelea* and *Senftenbergia* – Fig. 33, *b*, *f*, *j*, *k*).

If the hypothesis on such a transformation of these fructifications is confirmed (only one objection can be made as yet: no transitional forms are known now, but will be found in the future), then a very important role for saltational heterotopies will have to be recognized both in morphological and taxonomical aspects. In the morphological aspect saltational heterotopy can give rise to a new type of organ. Thus a leaf-like organ bearing sporangia, synangia, seeds or polysperms cannot be named a leaf. This is a microsporophyll, phyllosperm or fertiliger. The aforesaid is true even of microsaltations. For instance, a parenchymal cell emerging in the centre of the axis will be a component of the pith. If it emerges at the margin of a lateral vascular bundle it may become a component of a peripheral loop.

In the taxonomical aspect a saltational heterotopy might have been responsible for the emergence of new taxa of high rank (genera, families and orders of the class Polypodiopsida possessing sporophylls; Callistophytales, Gigantonomiales). The principal characters of these taxa have evolved in accordance with the punctualistic model.

On the other hand, cases may be pointed out where the transformation of the characters most important taxonomically better agrees with the gradualistic model. In these cases gradualistic interpretation is not prevented even by large stratigraphic intervals between the occurrences of separate forms. The loss of leaf-likeness (post-heterotopic transformation) of phyllosperms and microsporophylls in the later Callistophytales and then in Peltaspermales and their descendants proceeded very slowly apparently. If medullation as a whole is considered, rather than microsaltations of separate cells, the process indicated will most likely be gradualistic in primitive gymnosperms.

In the evolution of particular groups and perhaps of particular organs the gradualistic and saltational processes interfered in a complicated way and their relative role changed. For example, in the Arberiales a gradual shift of polysperms on to leaves and the same gradual modification of fertiliger may be supposed. However, the number of epiphyllous polysperms (and microsporoclads) was later recorded to change (Fig. 55, *e*, *g*, *n*). These quantitative changes (such as the number of petals in oligomerous perianths) obviously occurred in a saltational way in accordance with the morphogenetic principle 'all or nothing'.

Of course, the assumption of saltational transformations greatly complicates the reconstruction of the evolutionary process on the basis of palaeobotanical material. It is much more difficult to demonstrate that a certain transformation occurred as a saltation than to construct a mental series of unobserved transitional forms and to describe the same transformation in gradualistic terms. On the other hand, the assumption of saltations, including saltational heterotopies, opens up possibilities of suggesting numerous hypotheses that can hardly be verified (as happened in tectonics after modifications in the position of continents and in the Earth's diameter were recognized as admissible). However, both these considerations will lose sense if it is possible to prove convincingly that saltations (micro- and macrosaltations, with or without heterotopy) compose a real mechanism of plant evolution. Saltations will then have to be systematized and their interaction with each other and with gradualistic processes studied.

The theory of evolution and nomothetics of morphology. The nature of the taxon

Two special articles by the author were dedicated to these themes (Meyen, 1973a, 1978d). The same themes are considered more briefly below but in a wider context. The nomothetics of morphology here implies the following. The numerous phenomena of parallelism in plant variations require that the character space (field, framework) of plants is not an amorphous accumulation of various parameters, but is distinctly structuralized. It is demonstrated best of all by leaf segmentation (Fig. 98). There is a very limited number of rules by which (separately and in various combinations) leaf segmentation occurs or, conversely, the transformation of the dissected leaf into the entire leaf. The major modes of segmentation – dichotomy, pinnation, and palmation – combining with each other and with the secondary modes, yield all the diversity of leaf segmentation of higher plants. A pattern of transformation types (rules) is thus formed and the evolution cannot break away beyond it. This situation may be compared to the movement of a train on rails which do not determine the direction and speed of movement, but channel the routes.

This concept of transformation rules and types is essentially different from the common concepts of parallelism. The simple appearance of separate similar forms within different taxa may be understood as parallelism. For instance, ferns and angiosperms may have twice-pinnate leaves. Attention is often drawn to such a repetition of characters (this very repetition is described by the law of homologous series in hereditary variation; Vavilov, 1922), and cladistic analysis tends to exclude it. Much more important is that both in ferns and angiosperms the transition from simple leaves to once-pinnate leaves and then to bipinnate occurs by the same general geometric rules. In both groups similar regular series (it is better to say sets) of forms are seen. For such sets of forms bound by one rule of transformation the term refrain has been suggested (Meyen, 1978d; in the article of 1973 the term 'repeating polymorphic set' – RPS – was used).

In refrains the unity of two opposite trends –

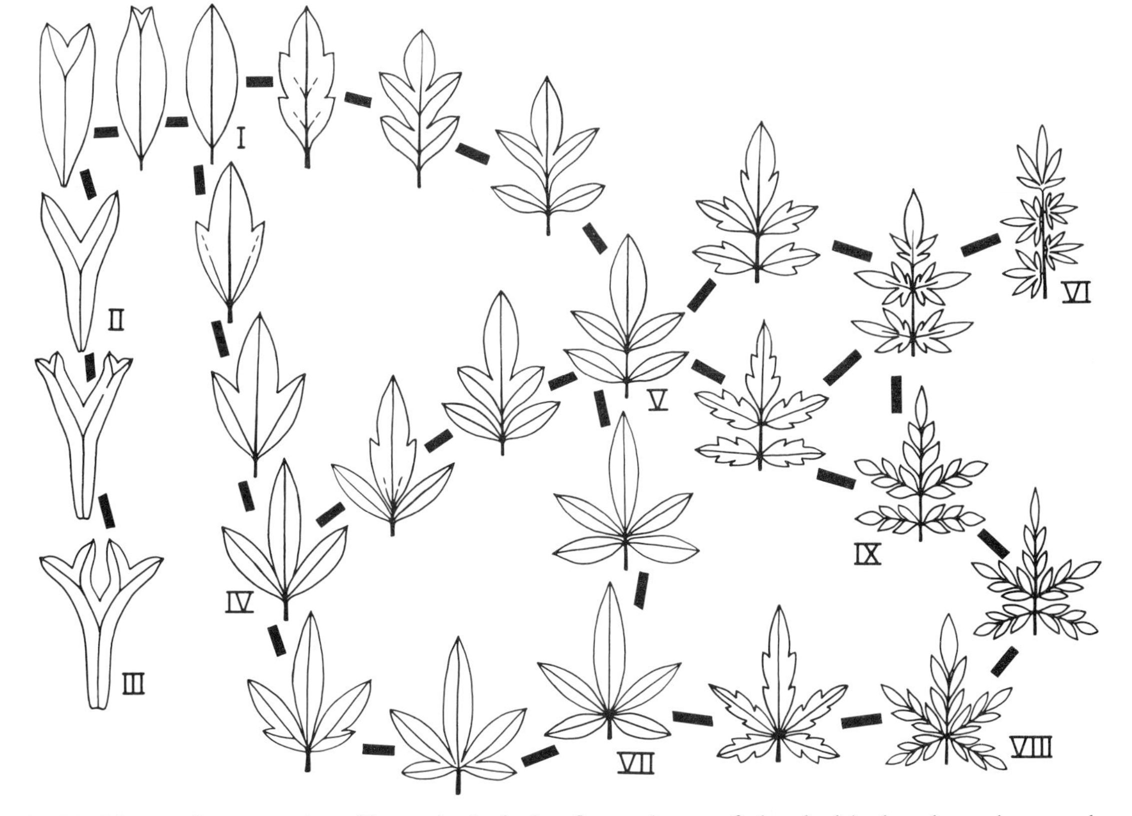

Fig. 98. Scheme of segmentation of leaves (and of other flattened parts of plant body); the scheme does not show segmentation exceeding second order, dissymmetry of parts and combination of furcation with other modi of segmentation. From Meyen, 1973a.

polymorphism and isomorphism – is seen, since the diversity unfolds by invariant rules.

Refrains are outlined more or less easily for all organs of higher plants (see numerous examples in Meyen, 1971a, 1973a, 1978d, 1984a). A creode, a stable ontogenetic track of an organ (Waddington, 1962), may be considered as a special case of refrain. The same track in phylogeny (semophylogeny) is called a phylocreode (Meyen, 1973a).

It was the presence of refrains including phylocreodes, that led to the conclusion about the directed nature of evolution. However, it is only in separate cases that the transformation of organs actually goes in one direction. In most cases evolution follows refrains, changing directions and persisting in former states, and refrains themselves are realized with different degrees of completeness. A typical example of mixing of the two circumstances (the presence of refrains and the orientation of evolution in only one direction) is seen in the studies by Asama (1983) who reduces all evolution to retardation (i.e. to a track reverse in relation to the creode of the ancestral form).

The completeness of refrains, when comparing two taxa, is fuller the higher the variation of the compared characters, while the range of variation depends on the size of a sample. If a sample is very large the tails of a variation curve of a character elongate. In this case, the state of a character is observed as a rare modality which is normally typical of a different taxon. Since very rare modalities of characters are considered to be monstrosites (terats) it turns out that monstrosities in some taxon correspond to the normal characters of another, generally, related taxon. This phenomenon was named 'the law of related deviations' by Krenke (1933–1935) who illustrated it with several hundred examples. Later it was suggested that the same phenomenon should be named Krenke's rule (Meyen, 1973a).

Krenke's rule manifests itself when comparing taxa of any rank and this frequently causes taxonomic problems. For example, a whorled leaf arrangement is generally typical of the articulates as a whole and helical of the lycopsids. However, terats of *Equisetum* are known that possess helicoidal phyllotaxis (Bierhorst, 1971, Fig. 7–1E). Among the lycopsids the typical whorls of leaves are characteristic of *Eleutherophyllum* (Fig. 20, *j–l*) with leaves much reduced to small teeth (as in *Equisetum*). These lycopsids were originally described as *Equisetites*.

The Zosterophyllopsida is characterized by lateral sporangia arranged sparsely as terminal strobili or as fertile zones. *Renalia* does not differ from typical zosterophylls in the structure of sporangia, but it was placed in the Rhyniopsida (Gensel, 1976; Gensel and Andrews, 1984) since the sporangia here are distal and not grouped. However, the distal sporangia may be considered to be a manifestation of Krenke's rule and *Renalia* may be placed in the Zosterophyllopsida.

The larger and more diverse the compared taxa, the better they are studied, and the more perfect their geological record, the better fulfilled is Krenke's rule (as well as refrains). Manifestations of Krenke's rule are not yet systematized but the available material suggests that it is widely applicable, thereby giving rise to important consequences. The main one is the statistical nature of the differences between taxa. Describing and comparing taxa as well as composing identification keys commonly require diverse rare deviations from the norm to be disregarded. For instance, in the characteristics of *Equisetum* rare cases of the formation of fertile zones by proliferation of strobili are not taken into consideration. This permits the Equisetaceae and Tchernoviaceae to be clearly differentiated. Actually the distinction between these families by this character is statistical.

The relation of taxa and characters may be represented in the following analogy. Let us imagine the space of characters as a multidimensional panel with a multitude of lamps, each corresponding to a certain character. If a character is present in a taxon the lamp gives a flash. It will give a flash as often as the character occurs in the taxon. The lamps repeating in different taxa will correspond to convergent forms. A cloud of switched-on lamps corresponds to the habitual concept of a taxon, with some lamps emitting light continuously (obligate characters), some giving flashes with different frequency. If refrains and all teratic characters are introduced into this model, the representation of a taxon will change. First, the lamps will be organized into a complicated system of orderly patterns (refrains). Second, it will be noted that there are no lamps permanently emitting light at all and the impression of permanent light is produced by a very high frequency of flashes. Third, the cloud will considerably expand owing to very rarely flashing lamps.

Now let us consider the relations of taxa in this model. The traditional concept of the transition from one taxon to another will correspond to the shift of the cloud on the panel, partial change of lamps permanently emitting light, as well as partial change and modification of frequency of flashes of other lamps participating in the cloud. Assemblies of lamps in different taxa are fairly distinct. If Krenke's rule is included in the model (a suggested version, not a habitual one), the transition from one taxon to another will acquire quite a different appearance. If the taxa are closely related, the shapes of their clouds will coincide and changes will occur only in the frequencies of flashes. Some lamps formerly emitting almost continuous light will now give very rare flashes, and vice versa. Other frequencies will also change.

Having accepted such a frequency model of a taxon and the transition from one taxon to another, which reflects the observed distribution of characters between taxa more completely, we face new problems. It is customary to believe that the process of speciation involves the formation of new structural genes and changes in general genetic potencies. The frequency model may envisage these processes (a shift of the cloud on the panel), but the change in the frequencies of characters acquires the principal significance. This means not so much that genetic potencies change as that the mode of their realization changes. In this case it becomes very difficult to recognize when a new lamp has joined the cloud and when only an abrupt increase has occurred in the frequency of flashing of the lamp which was formerly present in the cloud, but gave extremely rare flashes. It is also possible that a lamp's circuit was temporarily blocked, i.e. the potency was latent.

The suggested model was formerly (Meyen, 1973a) described in a different language using another analogy (magnetic tape with numerous tracks and a fluctuating reproducer). Essentially, the matter then referred to was later described in the literature as the interaction of structural and regulatory genes. The notions of structural and regulatory genes appeared as a result of thorough genetic and molecular–biological studies. The conclusion was drawn that most of the changes in the evolution of many groups of organisms (especially the more highly organized) had occurred due to rearrangements in the action of regulatory genes rather than to changes in structural genes (Bachmann, 1983; Britten *et al.*, 1982; Davidson, 1982; Gould, 1982a; Templeton, 1981; Valentine, 1980; etc.). It is evident that analysis of the distribution of characters in taxa leads to related concepts, though they are expressed in a different language. This supports both the premises underlying the suggested frequency model, namely a wide distribution (or even universality) of refrains, and Krenke's rule.

If so, we may be guided by these premises and also by the frequency model in further considerations. All this provides a deeper insight into the different modi of plant evolution.

Divergence

Two forms of divergence may be differentiated. Diversification of pollen and leaves of the early Cretaceous angiosperms appears to be a classical form of divergence. All subsequent diversity is derived from one initial type. However, there are complicating circumstances here. Either *Asteropollis* or *Clavatipollenites* may be taken for the initial type of pollen (Fig. 76, *d*; Fig. 78, *m*). However, these genera are quite similar and their general organization coincides. A simple leaf of an angiosperm described from the Lower Cretaceous of Siberia by V. A. Vakhrameev (Fig. 76, *e* (17)) is considered to be the oldest. However, the age of this finding should be defined more accurately. There are no initial stages of divergence of leaves in the more completely studied section of the Atlantic shore of the USA.

The second form of divergence deals with the splitting of synthetic types. The synthetic types were discussed in Chpater 3 when the evolution of conifers and angiosperms was considered. If one proceeds backwards along the geochronological scale from the present to the past the following can be seen in the plant groups mentioned. First extant species occur, then extinct species of Recent genera begin to appear. These extinct species are sometimes distinguished from Recent species by a different combination of characters, such that in one extinct species a combination of characters may be observed which are present separately in Recent species. This is a synthetic type at the species level. In more ancient forms synthetic types at the generic level appear, i.e. the characters of different Recent genera are combined in one extinct genus. Still further back, synthetic types at the family level are found.

It would seem that the synthetic types may be considered to be direct ancestors of the taxa to which the combined characters refer. This is probably sometimes so, but there are such assemblies of synthetic types (among conifers) where the characters of various recent taxa combine differently. It is probably exactly this reason that is responsible for contradictions in the phylogenetic (cladistic) constructions, when 'cycles' which resist cutting appear in cladograms (Doyle *et al.*, 1982; Stein *et al.*, 1984).

All the aforesaid well fits the frequency model of the taxon. It was considered using the example of the relations of only two taxa. If we imagine the historic transformation of several taxa at once having various redistributions of frequencies of different characters from ancestors to descendants, then 'cycles' will exactly result as phylogenetic lineages are restored by the characters of the highest frequency which are considered to be constant.

The phenomenon of synthetic types deserves very thorough analysis. It is quite possible that the distribution of characters between synthetic types of different levels will prove to be an obstacle in the elucidation of reliable phylogenetic relations between taxa, especially between taxa of lower rank. The transition between them is connected with a minimum quantity of novelties in general ontogenetic potencies and is performed mainly due to the change in the regime of the mechanisms regulating the externalization of the potencies (of 'regulatory genes'). The higher the proportion of such 'regulated' characters, the lower the possibility of indicating concrete phylogenetic connections between the taxa. But as each character may prove sooner or later to be a 'regulated' one, all phylogenetic constructions by common morphological characters may turn out to be impossible in principle in relation to taxa specially characterized by 'regulatory' evolution.

Parallelism

In the evolutionary sense parallelism is the independent transit of the same morphological track (or phylocreode) by different taxa. In different phyla we can see the independent transition from a protostele to a siphonostele or to a eustele, from isospory to heterospory and then to the reduction of megaspores in a sporangium to one tetrad or one megaspore. These are the classic and most obvious cases of parallelism. More sophisticated cases are considered below. Though they are not less common, it is more difficult to reveal them, hence their manifestation is less habitual. Transitive polymorphism (iterative form building inclusive) and pseudocycles are meant.

Transitive polymorphism is persistence of diversity (refrain) of a certain organ in the history of the taxon (Meyen, 1978b; Meyen, 1984a). The most trivial example is the persistence of leaf segmentation of different degrees during the phylogeny of many taxa. In the peltasperms, both in the Permian and the Triassic, simple, pinnate, bipinnate and dichotomous leaves are observed (Fig. 49). Here, there are no grounds for believing that a segmentation type was characteristic of particular phylogenetic lines. Permian pinnate fronds are known for part of the species of the genus *Tatarina* (other species have entire or lobed leaves) and for the genera *Compsoptrepis* and *Comia*, which are evidently related to *Callipteris*. Triassic pinnate fronds are included in the genus *Scytophyllum* which is related to *Lepidopteris* so much that the independence of these genera is uncertain. The same peltasperms in the Permian and in the Triassic are observed to have two types of cladosperms – peltoids and palmately veined cladosperms. Again there are no reasons to believe that a cladosperm type persisted in the phylogeny. In the Permian both types associate with leaves of the *Callipteris* type and belong to one natural group. In the Triassic the same types associate with the *Lepidopteris* leaves and form another natural group.

In the Zosterophyllopsida sparse sporangia, fertile zones and terminal strobili are known. The same types of sporangial position manifest themselves independently in different groups of the Lycopodiopsida. A protostele of progymnosperms may be of different configuration. The same configurations are repeated in the protostelic gymnosperms. These examples may easily be continued.

Thus, in the case of transitive polymorphism, refrain (completely or partly) passes from the ancestor to the descendant. It is probable that a new taxon was originated by a species with refrain not wholly represented (but whole in a potential state). In descendants the refrain is then expressed totally. This phenomenon is close to what we have seen in the case of the synthetic types and may obviously be attributed to the 'regulatory' evolution.

In palaeozoology the phenomenon of iterative form building was noted long ago (e.g. the repeated formation of *Pecten*-type shells among bivalves). It may be supposed that iterative form building is a special case of transitive polymorphism.

Transitive polymorphism introduces the same complications into phylogenetic reconstructions as the synthetic types.

The most complicated variety of parallelism is the formation of pseudocycles. The concept of pseudocycles was developed for flowers and inflorescences of the recent angiosperms (Gaussen, 1952). In morphological series of inflorescences repeating transformation are clearly notable with such combination of complication and reduction that a similarity arises between the flower and the inflorescence (Arber, 1950; Troll, 1928), the inflorescence of the first order (e.g. simple umbel) and an inflorescence of a higher rank (e.g. umbels of various levels of complexity). Such a high similarity of organs may result that its origin due to pseudocycles cannot even occur to us. Transformations of seed envelopes (Figs 43–45) can be attributed to pseudocycles. If the suggested concepts are correct then the external similarity of seeds of *Eurystoma* and *Pachytesta* is pseudocyclic (in the first case the external cover of a seed is a true integument, in the second case it is a transformed cupule).

So far there are no explanations for pseudocyclic similarity. Attempts to connect it with vectorized selection are completely unconvincing (what selection factor can lead a structure along a complicated track returning to the initial state?). It is probably more productive here to look for reasons in ontophylogenetic mechanisms again, e.g. in a prolonged unfolding of correlative interactions, perhaps in the inertness and recurrence of morphological transformations.

Convergence

It is customary to understand the recurrence of similarity of organs which were different in ancestral forms as convergence. If the seeds are again referred to (Figs 44, 45), the similarity of seeds of *Ginkgo* and of some conifers is a typical convergence. The similarity of leaves of some conifers (*Podozamites*) and of the Peltaspermales (*Phylladoderma*), of monocolpate pollen of the Ginkgoales and of the Cycadales may also be qualified in the same way. The above cases of convergence were selected (and not the generally known cases like the similarity of *Equisetum* and *Hippuris*) as they had misled morphologists and taxonomists. Nevertheless this is a manifestation of convergence in its classic form. However, as in the case of the previous modes of evolution, less trivial manifestations of convergence will be of interest, namely dedifferentiation of organs and its special cases – gamoheterotopy and ontogenetic introgression. Dedifferentiation of organs was discussed in considering the genus *Sashinia*, in which a striking similarity of vegetative leaves, bracts, sterile scales of the axillary complex and even of seed stalks had certainly recurred. Such dedifferentiation of organs may be qualified as intraorganismic convergence.

Gamoheterotopy is the transfer of characters of one sex to the opposite sex (Meyen, 1984a). The possibility of gamoheterotopy itself in plants remains unproved. This phenomenon is considered to be very common in animals, including birds and mammals (Schmalhausen, 1968). Its manifestation in plants is therefore quite probable. As mentioned in Chapter 3 the similarity of female and male strobiloid fructifications originated in the Cordaitanthales and in the more advanced Rufloriaceae probably as a result of transfer of the general construction of complex polysperms to microsporoclads. One cannot rule out the possibility that the gynoecium of the angiosperms originated as a result of a reverse transfer of characters of male microsporophylls arranged in whorls in the bennettites to female fructifications. If in a whorl of microsporophylls of the bennettites, seeds are transplanted in the place of synangia, quite a suitable image of the ancestral fructification will be obtained. From the latter a whorl of follicles with adaxial laminar placentation can easily be derived. Of course, this is just a guess as yet.

Since gamoheterotopy leads to secondary similarity of previously differing organs it may be considered a special form of convergence, which is intraorganismic in monoecious plants.

The dedifferentiation of organs and gamoheterotopy may occur both as saltations (probably gamoheterotopy in the Cordaitanthales originated in this way) and by a gradual transition, by ontogenetic introgression, i.e. by establishing a gradual transition between ontogenetic programmes that have earlier been dicontinuous, resulting in the recurrence of transitions between different organs. The most convincing example of such ontogenetic introgression is the genus *Kungurodendron* (Fig. 70, *b*) with its gradual transition, of obviously secondary origin, from seed stalks via sterilized seed stalks to sterile scales of the axillary complex. In all more ancient conifers and also in those Cordaitanthales which could be the ancestors of the conifers seed stalks and sterile scales of the axillary complex are discontinuous types of organs.

The recognition of the possibility of secondary origin of transitions between different organs may be important for evolutionary plant morphology (Zimmermann, 1959). The presence of transitions between organs is often treated as an indication of their semophylogenetic connection. For instance, the transition between stamens and petals in the Nymphaeaceae is currently supposed to be a proof of origin of petals from stamens. In other cases transitions between petals and sepals (tepals, elements of involucre), which in turn may be connected by transitions with foliage leaves, are observed. Hence, it is possible to conclude that petals originated in different angiosperms from different organs, i.e. the petal as a semophylogenetic category has a polyphyletic origin. The possibility of ontogenetic introgression makes all these considerations doubtful. Moreover, they are doubtful from the viewpoint of the data available at present on Cretaceous flowers. No Cretaceous flowers possess any hint of transitions between the listed organs: stamens, petals, sepals (tepals) are definitely discontinuous organs.

It is unnecessary to explain that the above convergence types fit in quite well with the same regulatory mechanisms of evolution. In genetic language it means that new structural genes are not necessary for such transformations.

Mosaic structure of plants

All the above-mentioned types of evolutionary modes are illustrated with separate organs and separate aspects of their transformations. If other structures of the same plants were considered, manifestations of other modes might be observed. For such unco-ordinated transformation of different organs and of different aspects of their structure the term mosaic evolution was suggested. The weaker the correlation between the organism's parts the more mosaic the evolutionary process. Independently evolving components of organisms are named blocks (Muzhchinkin, 1978). The presence of enough autonomous blocks enables plants to perform saltational heterotopies without violating the other traits of organization, to change the number of identical organs (oligomerization and polymerization).

In botanical literature the term mosaic evolution is rarely used. More often the term heterobathmy is used instead, which was suggested by A. L. Takhtadjan to designate different evolutionary advancement (degree of specialization) in different parts of one plant. Meanwhile to outline the heterobathmy it is necessary to know the phylogenetic sequence of taxa, to have an unambiguous code of primitivity of characters, to be able to distinguish the primary primitivity of characters from reversions and pseudocyclic formation of character states externally similar to primitive characters, to eliminate those original states which persist due to transitive polymorphism, etc. In short, it is necessary to meet requirements which in practice are too difficult. As these requirements are commonly not fulfilled, many usual examples of heterobathmy provoke suspicions. Sometimes the data of palaeobotany indicate erroneous use of the term heterobathmy. For example, a combination of secondary wood formed only by scalariform tracheids (a supposedly primitive character) with characters which are considered to be advanced in the cycads and in some angiosperms is attributed to heterobathmy. Beck (1970) drew attention to a mistake committed here. In the progymnosperms which are ancestors of all seed plants secondary wood is formed by tracheids with circular and not scalariform bordered pits (scalariform tracheids are present only in the metaxylem). This means that the extension of scalariform tracheids to secondary wood is a character of secondary formation. This is typical heterotopy at the level of cellular structure of the plant. Accordingly, it is not possible to speak of heterobathmy due to secondary formation of this character.

Cope's rule, neoteny

Cope's rule according to which new taxa arise from the least specialized members of the ancestral group have found wide support among zoologists. However, as noted by Vorobyeva (1980), an evolutionary episode of high significance to mankind contradicts this rule, namely the origin of terrestrial vertebrates: amphibia originated from the most specialized fishes – rhipidistia. The ancestors of mammals were also highly specialized reptiles. These examples are already sufficient to reject Cope's rule. Nevertheless, it has penetrated into botany and manifested itself in search of direct ancestors of almost all larger taxa, even if they appeared rather late in the geological record, among the most primitive plants. For instance, ancestors of the angiosperms were looked for among the Palaeozoic pteridosperms, ancestors of the taxads among the early Carboniferous gymnosperms, etc. At present we see that there is no need for all that. With rare exceptions, there are no obvious blind alleys of specialization in plants since they possess efficient ways of despecialization (e.g. dedifferentiation, reactivation of latent potencies, etc.). Neoteny is regarded as one of the methods for overcoming specialization. Palaeontological material has not yet yielded any obvious cases of the manifestation of neoteny. Until now all examples of neoteny have concerned Recent material not supported by reference to concrete fossil plants. True, Stidd (1980) supposed that microsporophylls of the bennettites were of neotenic origin from microsporophylls of the medullosans. The phylogentic connection of both orders is very probable, but so far it has not been supported by findings of transitional taxa. Thus, the role of neoteny in the evolution of higher plants is still completely unclear.

Modes of phylembryogenesis

After studies by Haeckel and especially by Sewertzoff (1939) many evolutionary transformations in animals have been described in terms of coenogenesis, palingenesis, anaboly, archallaxis, deviation, etc. (for more details see: Remane, 1956; Gould, 1977). For the first time these notions were widely introduced into botany by Takhtajan (1954, 1972). However, as in the case of neoteny, these notions can only rarely be involved in the interpretation of palaeobotanical material, as each of them requires simultaneous knowledge both of the phylogeny of a taxon and of the ontogenesis of the organs considered. At the moment we are interested only in the question to what extent the evolutionary innovations are connected with the transformation of early ontogenetic stages. In zoological literature opinion is widespread that the most important evolutionary innovations arise from archallaxis, i.e. modification of early ontogenetic stages. However, in organisms with a high degree of equifinality of the ontogenesis (i.e. even when very significant deviations from the norm are corrected in the course of embryogenesis) manifestations of archallaxis are highly improbable (see the epigenetic concept of evolution: Shishkin, 1981, 1984). In plants such new formations as phyllosperms and microsporophylls of callistophytans must have originated by archallaxis. But generally in plants, with indeterminate growth and feebly determinate general habit the classification of modes of phylembryogenesis in their habitual meaning is unproductive. The main evolutionary and taxonomical significance belongs here to the ontogenetic stages of separate blocks – of leaf, of flower, etc. – rather than the whole plant (e.g. from a seedling to a senile tree).

Modes of evolution and cladistics

Cladistic analysis is discussed briefly in Chapter 2 and in the present chapter (in connection with synthetic types and transitive polymorphism). Cladistics is a peculiar phenomenon in biology (see in more detail: Wiley, 1981). Beginning with the work of Haeckel, who first began to draw phylogenetic trees, phylogenetics was a theoretical pivot for systematics. A procedure for phylogenetic reconstructions was developed, criteria of affinity were outlined, and types of similarity not indicating affinity were grouped. In 1950 Hennig successfully summarized all that had previously been said, but did not actually introduce anything new into phylogenetic studies. The fifties were a period of disappointment in phylogenetics as a foundation for systematics. A tendency to make the system of organisms less biased led to rejuvenation of the views of Adanson, an eighteenth century botanist, who suggested the use of a large number of characters given equal weight a priori, in distinction to the principles of the natural system of A. de Jussieu. The Adanson's method could not be realized in his time as it required the processing of very numerous characters. However, in the fifties progress in computer engineering raised the possibility of doing this. Thus numerical (phenetic) systematics appeared, claiming full objectivity of its dendrograms. Meanwhile in traditional phylogenetic systematics the discussion went on in connection with the concepts of a taxon. The orthodox phylogeneticists (Hennig, Kirjakoff, *et al.*) maintained that real taxa should represent the clades, i.e. the phylogenetic branches. Less radical phylogeneticists admitted the possibility of including in one taxon organisms equally advanced in a certain respect, i.e. standing at a certain evolutionary grade.

In 1966 Hennig's enlarged book was issued in English and unexpectedly caused a strong response. The focus of attention of systematists was shifting rapidly from numerical methods to the cladism of Hennig (actually traditional cladism), but the experience of work with computers was not lost. First, it had trained biologists to deal with more complete calculation of characters and possible connections between taxa. Second, it introduced into phylogenetics new semiotic means such as variance diagrams and matrices which made it possible to present the results of studies explictly. However, as regards theory this modernized cladistics remained at its former level. The only theoretical achievement was the use of a multitude of terms (apomorphy, plesiomorphy, synapomorphy, symplesiomorphy, etc.), but almost no new methods of discrimination of the inherited (plesiomorphic) and newly formed (apomorphic) characters were suggested. The tendency to get rid of anything hypothetical is characteristic of cladistics, obviously reflecting the influence of positivism. Since the incorporation of palaeontological material inevitably introduces much that is hypothetical a tendency appeared ('transformed cladism') to construct a system of organisms predominantly or exclusively on the basis of recent forms. A large role in the plotting or interpretation of cladograms is played by the principle of parsimony which is undoubtedly important as a general regulatory principle in science, but its application as a criterion of selection between concrete statements may lead to ascribing our modes of thinking to nature. However, the main drawback of cladism is that detailed study of the nature of the taxon and of various modes of evolution is not part of its foundation. As noted above, the traditional notion of the taxon should be placed by the 'frequency model', thus leading to essential changes in the understanding of the modes of evolution, including divergence, convergence and parallelism. Since the traditional concept of the taxon dominates in cladism, the three modes indicated are accepted only in their classical form. Thus, cladism is based on very simplified notions of the taxon nature and of the relation of taxa to the space of characters. It is not surprising therefore that increasing use of the cladistic approach has failed to lead to essential perfection of the system of organisms.

As already mentioned, cladism has not yet rendered any influence on palaeobotanical phylogenetics and systematics. However, separate palaeobotanical cladistic studies are already beginning to appear. If research in this line continues to develop, the theoretic neglects of cladistics considered above should be taken into account. It is particularly important to take into consideration transitive polymorphism, the synthetic types and other phenomena connected with the 'frequency model' of the taxon.

Adaptationism

As to the explanation both of the evolutionary process itself and the appearance of each new structure, palaeobotany is true to the ideals of the second half of the last century. Any evolutionary step is considered to be explained if a newly arisen structure is credited with a special function (ecological or physiological) and a selection factor is indicated which governed the transition from the previous state to a new state. This is sometimes called the establishing of the 'biological sense' of innovations. Meanwhile the number of functions and selection factors which can be used for explanations of this kind is incommensurably small compared with the diversity of taxa and morphological structures recorded by systematics and plant morphology. Indeed, to explain the variety in the structure of the spore and pollen exine, the following set of functions is used: (1) a harmomegathic function (volume regulation); (2) protection from drying; (3) transport; (4) provision of efficient germination by pollination (including mechanisms of stigma recognition). The combination of these functions explains the specificity only of some very general or, alternatively, minute specific characters. But on the whole, in spite of the vast literature and special studies performed the functional sense of the great diversity of spores and pollen has to be admitted as unknown. The functional interpretation of leaves is still more demonstrative. Only some specific peculiarities of leaves can be connected with ecological functions (thick cuticle of many xerophytes, isopalisade parenchyma of heliophytes, etc.), but almost nothing can be said on the functional sense of the morphologically most significant characters (segmentation of leaf blades, venation types, edge structure, ontogenetic types of stomata, etc.). The interpretation of fructifications takes into account the efficiency of pollination, fertilization, the distribution of diaspores and of their germination, as well as the protection of spores and embryos. In this case too the diversity of fructifications far surpasses a store of functional explanations. The same is true of most other structural characters of plants.

Certainly, it is always possible to find some wordings which will imitate a functional explanation. Such arbitrary explanations do not satisfy a critical mind and contradict the modern spirit of science. Therefore, efforts have long since been taken to introduce into the functional interpretation more strict evidence, mathematical models, and experiments. Givnish (1978), for example, attempted to find out the functional sense of leaf blade outlines from the viewpoint of their optimum light absorption and of other functions. Niklas (1982b) simulated the growth forms of primitive plants, using a computer to reveal the adaptive advantages of various forms. Using enlarged models of various fructifications, he also (Niklas, 1982c) tried to understand the mechanism of their pollination.

However, functional studies at present, as

before, are characterized by the following fundamental drawbacks.

1. Each organ is represented by a certain plurality of forms. Therefore, the functional analysis should provide an insight into the functions not only of this organ as such (thus, there is no need to look for proofs of the trivial functions, e.g. that leaves are photosynthetic devices, and the phloem serves for transport of metabolites), but also into the functional specificity of various forms. That is where the main problem lies.

2. Many (or all) organs belong to certain refrains. High correlation of particular members of a refrain and certain ecological or physiological factors may be demonstrated. But this does not necessarily offer the functional sense of the rule that binds the components of the refrain. Let us assume, for example, that we have managed to decipher the functional sense of a pinnate, a palmate and an entire leaf. But this will offer no insight into the sequence of primary formation, shifting and reduction of sinuses and lobes, observed in the change of segmentation type of the leaf. This sequence surprises by its orderliness which until now has not been able to find even an approximate functional explanation. The same is true of many other refrains.

3. Morphological tendencies are often counter-directed. In some plants the process goes from A to B, and in others from B to A. It is obvious that the functional explanation is acceptable only if it relates to both tendencies. Otherwise we come to arbitrary statements *ad hoc*. For example, the origin of peltate sporangiophores and adjacent bracts in articulates may be connected with the function of protection of sporangia. If the Equisetaceae is assumed to have originated from the Calamostachyales it is necessary to explain why bracts were reduced in this line so that the sporangiophores became open again (as in primitive forms of the *Pothocites* type). If bracts were to prevent the free distribution of spores, this would contradict the previous explanation. Exactly the same situation is seen with female fructifications of the conifers in which bracts sometimes develop well, sometimes disappear, fusing with the seed scale, but retaining independent vascularization. It may be added that bracts may be leaf-like, membranaceous, woody, with entire or denticulate edge, furcate or needle-shaped, with stomata, or without them, etc. Only a very ingenious mind may relate all this to the fairly simple function of protection. And it is quite incomprehensible why in some cordaitanthales and conifers bracts have almost completely disappeared leaving the axillary complex without any protection.

4. The same structure often receives too many explanations and it is necessary to select some of them. How to do that is generally unknown. For instance, the air sacci of pollen grains were interpreted as (1) a means of transport; (2) outgrowths for protection of the distal furrow and prevention of drying up of the germinal apparatus, (3) floating devices for movement in the necessary direction in micropylar liquid. So which of these explanations is true and in which way may the adequacy of selecting one of them be proved? In addition, the selected explanation should satisfy the previous condition, namely to explain simultaneously the counter-tendency, the reduction of sacci in many conifers (sometimes in some genera of the family). The functional interpretations are often performed by establishing a correlation between a character and a factor of the environment. Stebbins (1950) pointed out that the structure of the perianth is correlated with the quantity of lime in soil. At the same time, he joined the generally accepted explanation of the structure of the perianth by coevolution with pollinators. How can the action of these factors be separated, and the role of each of them demonstrated?

5. If a function of an organ is considered to be an intermediary of selection, it is still necessary to demonstrate that the selection based on this function can actually proceed at an adequate rate. Let us assume that the conclusions drawn by Niklas from models of pollination dynamics are correct. The adaptation to pollination of different fructifications proved to be different. However, this does not suggest yet that the selection factor is indisputable since the selectivity coefficient should be high enough. The efficiency not only of pollination but also of the whole reproductive cycle is known to be

incomparably higher in angiosperms than in gymnosperms. Nevertheless, the angiosperms have not been able to expel the gymnosperms from forest stands for many millions of years. The coefficient of selectivity obviously proved insufficient. Then why should it be considered sufficient in the case of much lesser differences between fructifications of various lagenostomans?

6. The functional interpretation of the origin and evolution of morphological characters is continuously confronted by the paradox of the polymorphic relation of form and function. This is, on the one hand, multifunctionality (performance of different functions by one organ), on the other hand, functional polymorphism (performance of one function by different organs). True, very little can be said about multifunctionality in plants (all these are trivial statements, like the root performing anchoring, absorptive and, sometimes, storing functions). But functional polymorphism is widely spread. And again there is much that is enigmatic here. For example, for what reason were the reduction of leaf blades and the transition of the function of photosynthesis to widened petioles (phyllodization) and to stems (cladodization) necessary? For what reason were complicated pseudocyclic transformations necessary, finally to return to a state differing little from the original state (see above on pseudocycles)? Has the cupular micropyle of the trigonocarpaleans and of their descendants really any advantages over the integumental micropyle of older forms? What selective factors have brought the compound inflorescences (cyathia) of the euphorbias to the imitation of small flowers? Not only are there no answers to these questions, but we do not even know where comprehensible answers are to be sought.

Meditations on these drawbacks of functional interpretations and all adaptionistic ideology bring us to a more general question: is the functional and, consequently, selectionistic explanation the only one possible? This question has occurred to several plant morphologists and the answer has sometimes been negative. The most profound analysis both of this question and the possible answers to it was made by Troll (1937) and Arber (1950) whose work has remained almost unnoticed in palaeobotany. The book by Arber is of particular interest in this respect. Detailed analysis of all aspects of the question formulated above would lead us far away from palaeobotany. So only a few considerations are noted below (see also Meyen, 1973a).

Adaptationism appeared in biology when protoplasm was thought to be a structureless gel and nothing was known of the mechanisms of self-organization and self-regulation of living systems. However, as early as the last century Driesch demonstrated that the phenomenon of equifinality and related phenomena contradict the concept of organisms as passive objects for action of external factors, but Driesch's observations and generalizations were completely at variance with the mental fashion of that time (though animal embryology followed Driesch's path and was little connected with adaptationism). The mental fashion of that time was in accordance with physics, especially thermodynamics. Any spontaneous form production both in living and inanimate nature was rejected as contradictory to the law of increasing entropy.

Since then the situation has changed greatly. The mechanisms of self-assembling at molecular level have been elucidated, the great role of the regulatory systems in organisms has become obvious, the variety of the living world has turned out to be much more orderly than could have been thought of. Application of the approaches of conform geometry revealed the invariance of forms in many such cases where no invariance was observed in terms of Euclidean symmetry transformations (Petukhov, 1981). Finally non-classical thermodynamics (Prigogine, 1980), synergetics, the theory of dissipative structures and related physical concepts demonstrated that in open systems spontaneous self-organization and form-production may occur only on condition of an external energy supply (see also Wiley and Brooks, 1982, 1983; Brooks, 1983). Only a non-specific energy supply is required from the external medium and the open system will be structuralized.

Thus we arrive at the conclusion that biological

systems are highly autonomous and capable of producing structures whose characters depend only very indirectly on the specificity of external factors. The external factors actually resolve characters rather than determine them, i.e. they determine the occurrence but not the mode of development of the characters. The external factors may act as relays switching the available programmes but not creating them. If the external factors affect the programmes directly, the organism reserves the possibility to respond with a whole spectrum of reactions, often non-specific (e.g. a thickness of cuticle may increase in response to an increase in dryness or insolation, to nitrogen deficiency in the soil).

Thus, the former foundation of adaptationism proves to be unstable. Its every stone should be put to the test for reliability.

This is not to suggest that adaptationist studies should be abandoned. The case in question is only that the role of adaptation and function in evolution remains a serious problem. Adaptationism needs much more serious substantiation than it has received up to now. We have to be ready to learn that functionality in principle is unsuitable for explaining the diversity of the living world and will be applicable only to certain peculiarities of organisms. In this case the application of adaptationism in palaeobotany with its fragmentary material will not be fruitful.

Abiotic factors of selection and macroevolution of plants

Zherikhin and Rasnitsyn (Zherikhin, 1979; Zherikhin and Rasnitsyn, 1980) suppose that special reasons accelerating evolution are looked for erroneously. Non-uniformity of evolutionary rate is caused not by accelerating factors but by decelerating factors. Evolution unrestrained by brakes should proceed at a very high speed. This statement seems to find confirmation in palaeobotanical data indicating that evolution is strongly impeded by some abiotic selection factors, first of all temperature. In the section on florogenesis it is noted that almost all suprageneric taxa originated in the equatorial belt and adjacent ecotones (Figs 11, 99). This equatorial cradle seems to be functioning at present (Chernov, 1984). In extraequatorial regions the temperature factor has obviously cut off almost all phylogenetically essential innovations. For the same reason in extraequatorial (especially arctic) floras prolonged survival of archaic forms is observed, having been transformed not higher than the generic level. The relation of selection to innovations is obvious if Krenke's rule is referred to. Van Steenis (1969) noted an important regularity. The forms which appear in plants of temperate climate as teratological forms become normal taxonomic characters in related plants of the tropics (Van Steenis' rule). It was noted long ago that if plants were placed under more favourable conditions the range of their variation increased due to the elongation of the tails of the variation curves (Troll, 1937). Partly due to this the variation spectrum of cultivated plants apparently greatly surpasses that of their wild ancestors.

The environmental temperature probably acts as a limiting factor and not a driving force. In the Earth's history periods of warming up and cooling off have continuously interchanged. If the temperature had been a driving force of evolution, a higher number of newly appearing taxa would have fallen within periods of warming up. Such a relationship has not been recorded. For instance, no taxa of high rank appeared in the middle of the Gzhelian stage (the upper part of the upper Carboniferous) and in the Toarcian stage (the Lower Jurassic). Generally no periodicity of the appearance of supergeneric taxa has yet been noted which could be connected with global fluctuations of climate. Periods very similar in general climatic situation may differ much in the intensity of appearance of taxa of high rank and vice versa. For example, during the Devonian, which cannot be considered to be a period of climatic stasis, evolution progressed from the most primitive propteridophytes to the first gymnosperms, the evolution of the lycopsids greatly advanced, the articulates and primitive ferns appeared. In climatic parameters and palaeogeographic situation the Devonian is close to the Triassic with its much more modest advances of evolution. With such a large climatic change as the late Cainozoic glaciation, only small phylogenetic innovations coincide in time.

General conclusion

The above text briefly reviews the evolutionary problems, for the analysis of which palaeobotanical data are particularly useful. The other aspects of evolution are not referred to. Some general conclusions may be derived from what has been said above.

Each biological discipline contributes to the elucidation of certain aspects of evolution. There is no point in discussing which input is more important. If a universal theory of evolution is to be created, care should be taken that it should describe all kinds of evolutionary processes and be applicable to all groups of organisms. It is evident that if not the most general then certain special kinds of evolutionary processes depend on the properties of organisms undergoing the evolution. Thus, organisms with sexual and asexual reproduction, determinate and indeterminate growth, more or less ontogenetic regulations, stenobiontic and eurybiontic, etc., evolve differently. During geological history the participants and the ecological background of the evolutionary process were changing. It is important therefore to take into consideration the evidence of very different groups of organisms. It is just this that determines the necessity to involve palaeobotanical

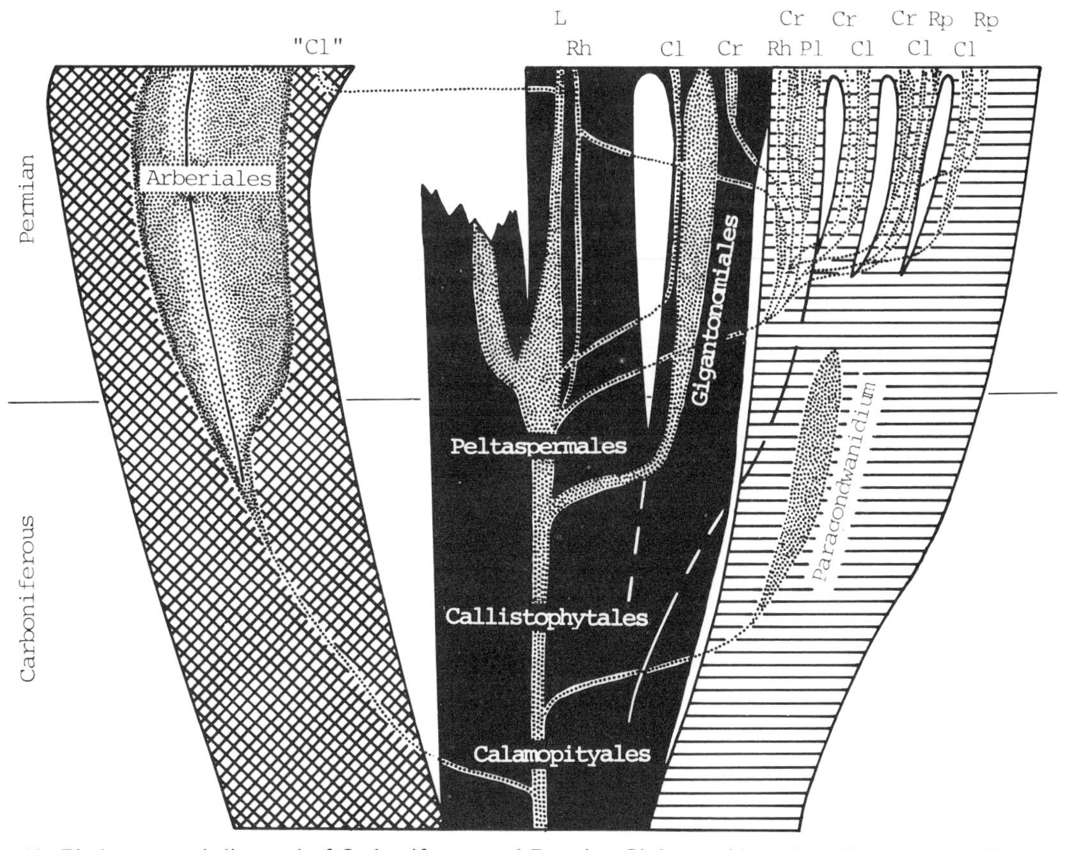

Fig. 99. Phylogeny and dispersal of Carboniferous and Permian Ginkgoopsida against florogeny (see Fig. 83 for nomenclature of phytochoria); 'Cl', *Callipteris*-like plants; L, *Lepidopteris*; Rh, *Rhaphidopteris* and allies; Cl, *Callipteris-Comia-Compsopteris*; Cr, Cardiolepidaceae; Pl, Peltaspermaceae with *Tatarina*-, *Glossophyllum*- and *Lepidopteris*-like leaves; Rp, *Rhipidopsis*.

materials in the general theory of evolution. Palaeobotany has its advantages and drawbacks for the elucidation of evolutionary problems. Among the advantages there is an abundance of plant remains (they are incomparably more numerous than, for example, the remains of vertebrates) which in some cases are more fully preserved than other organisms. The main drawback of palaeobotanical material is that plants are divided into detached organs. For evolutionary studies therefore the most suitable groups of plants should be selected specially and no attempt at evolutionary conclusions should be made from just any material. Palaeobotany needs analogous objects, specially assorted for detailed evolutionary studies, such as *Drosophila* and *Escherichia coli* for geneticists.

Properties of the geological record need special study. It is probable that its perfection at a supra-generic taxonomic level is much underestimated (Meyen, 1984a). It is all too easy to refer the impossibility of reconstruction of phylogenetic lineages to imperfection of the record. Our collections may actually be imperfect as well as our knowledge of the plants already collected. It is demonstrative in this sense that the overwhelming majority of suprageneric taxa established in this century have not been a discovery of plants hitherto quite unknown. Representatives of these new taxa remained in collections for many years with their systematic attribution wrongly determined. For instance the family Callistphytaceae was established in 1970 and the order Callistophytales in 1970. However, undeciphered representatives of the callistophytans were described as ferns by Schlotheim back in 1820 (*Filicites pluckenetii*) and by Renault in 1879 as cordaites (*Poroxylon*). There is every reason to believe that the palaeobotanical record is not so imperfect as to deny wide use of palaeobotanical material in general evolutionary research.

PALAEOBOTANY AND PLANT MORPHOLOGY

Many notions and generalizations concerning plant morphology have already been considered in the previous chapter since they deal equally with the theory of evolution. These are refrains, transitive polymorphism, dedifferentiation of organs, etc. Problems more initimately connected with morphology are treated below. These are (1) the notion of homology and the procedure of homologization in palaeobotany, (2) the telome theory and (3) the evolution of some principal organs of higher plants.

Homology

A vast literature is dedicated to the notion of homology (see reviews: Meeuse, 1966; Remane, 1956; Sattler, 1978, 1984; Stevens, 1984; Tomlinson, 1984; Voigt, 1973). There are many opinions as to what homology is and how to establish it. However, all these opinions in the final analysis are reduced to homologization as a certain classification of parts of organisms. The homologous parts are those which are embraced by one morphological notion being a class of parts (meron). The criteria of attributing parts to one meron may be different. According to the selected criteria, the kinds of homology are differentiated, and, depending on how many criteria are used for homologization, complete and incomplete homology may be distinguished. Each part is absolutely homologous only to itself.

The criteria of homology were considered most thoroughly by Remane (1956). He believes that the basic criteria are (1) the criterion of position, (2) the criterion of special quality and (3) the criterion of transitional forms. The following peculiarities of applying these criteria in palaeobotany may be named.

1. Since the detached parts of plants become buried separately, the criterion of position cannot be fully applied. For example it is unknown how and to what organs the racemose, spicate and umbellate polysperms of the Rufloriaceae were attached (Fig. 65, *f–j*, *q*, *s*). Therefore, no reliable homologization of them with simple axillary or compound polysperms of the Cordaitanthaceae is possible. No life-time position is known for many cataphylls which may belong to vegetative buds, circandr, circasperm or bracts.

2. Special qualities include not only definitive characters but also ontogenetic ones. For instance to distinguish the subsidiary and encircling cells of stomata (after Florin, 1931), the ontogeny of stomatal apparatus should be known. Likewise, to identify certain layers of the exine it is necessary to know the ontogeny of all sporoderm. Sometimes the ontogenetic data may be derived from observations of separate stages, from a judgement about ontogenesis by a definitive structure, or by analogy with other plants, but more often this is impossible.

3. For criteria of position and of special quality, use can be made not only of epimorphological but also of anatomical characters, which may be non-preserved.

For these three reasons palaeobotany needs some morphological terms of its own. The situation here is just the same as in palaeobotanical taxonomy (see Chapter 2) which also uses, in addition to eutaxa, parataxa with special names. Morphology may also deal with mera, eumera and paramera (singular – meron, eumeron, parameron). To name the paramera, neutral terms are used, such as appendage, laterals, fructification, seed-bearing organ, dark body, etc.

Other difficulties in homologization are connected with the specificity not of the palaeobotanical material but of plants themselves possessing lower integration of the individuum than the animals. This leads to violations of all three criteria of homology. Violations of the criterion of position occur in various heterotopies (Sattler, 1975) which probably result in many cases from homoeosis (Leavitt, 1909; Meyen, 1984a). It is essential that the homoeotic transformation may lead to violation of all three criteria of homology and consequently the resulting structure may obtain the status of a new organ *sui generis*. For example in the case of the homoeotic transfer of seeds on to leaves, the new organs are formed, phyllosperms and further cladosperms, since they have new 'special qualities', do not obey the criterion of gradual transition, and finally may occupy a new position in the plant body (in the archetype). Such organs are comparable to hybrids in taxonomy

Violation of the criterion of special quality is met in the dedifferentiation of organs. The Permian conifers *Sashinia* yield the best example (Fig. 69, *s*, *v*) with their secondary similitude of foliage leaves, bracts, sterile scales of the axillary complex, and even seed stalks. And finally, the criterion of transition is violated in cases when gradual transition reconnects organs known to be different and previously discrete. This is exactly the case with the conifers *Kungurodendron* (Fig. 70, *b*) where gradual transition from the sterile scales of the axillary complex to the sterilized seed stalks and further to functioning seed stalks, has undoubtedly recurred. The possibility itself of the reappearance of transitions between previously discrete organs jeopardizes the implications for the origin of organs of one type from another on the basis of transitional forms. The presence of transitions between petals and stamens in the recent nymphaeids, for example, is commonly considered to be a proof of the origin of petals from stamens by sterilization of the latter. Meanwhile in the most ancient flowers with perianth (Basinger and Dilcher, 1984; Dilcher and Crane, 1984) the stamens and petals differ sharply; moreover, the stamens are dorsifixed on the petals. There are no gradual transitions between the petals and stamens in all other Cretaceous flowers. Obviously, these transitions originated in flowering plants as secondary and do not indicate the semophylogenetic connection of these two types of organs.

The telome theory

The telome theory implies that all epimorphological structures of higher plants may be derived from the primary cylindrical axes (telomes) and that transformations of telomes into the other types of organs occur according to a limited number of rules (modes of transformation). In elaborated form the telome theory was suggested by Zimmermann (1930, 1959, 1966), but many of its principles had been proposed earlier (O. Lignier, H. Potonié, etc.). Some general assumptions underlie the telome theory, specifically: (1) the general morphological principle that all morphological diversity of higher plants may be reduced to one original form and to a finite (and small) number of modes of transformation; (2) the conviction that any subsequent (advanced) types of organs may be described in terms of the previous (primitive) organs (i.e. only the most primitive types of organs are interpreted *sui generis*); (3) the notion that all transformations of organs are gradualistic; (4) the conviction that all transformations of plants are of an adaptive nature. Within the scope of the telome theory three criteria of homology listed in the previous section are accepted, actually without limitations.

Discussing the role of the telome theory in plant morphology it is necessary to clearly distinguish (1) the fate of the concrete statements of W. Zimmermann and other 'telomists' in connection with the evolution of particular morphological structures of higher plants, and (2) the penetration of a certain style of thinking into plant morphology. Many concrete conclusions of 'telomists' have suffered severe criticism and have been rejected. Such is, for example, the notion that axes with lateral sporangia originated in all cases from terminal ones by overtopping of dichotomous axes with terminal sporangia. Nevertheless, the style of thinking characteristic of the telome theory has spread widely and dominates modern morphology. Four general assumptions listed above underlie this style of thinking. This finally determines all morphological, and then evolutionary concepts in modern botany and palaeobotany.

The canons of the telome theory have been formed to a considerable degree under the influence of palaeobotanical discoveries concerning presumably Devonian plants. The telomes are quite comparable to axes of propteridophytes of the *Rhynia* type. Studies of other Devonian plants in the twenties and thirties yield enough grounds to suppose that in the morphological evolution from plants of the *Rhynia* type to more advanced Devonian plants the main role belonged to (1) overtopping, (2) planation, (3) incurvation and (4) fusion of cylindrical telome-like axes both sterile and bearing terminal sporangia. (5) Reduction of parts and (6) differentiation of tissues occurred in parallel. Thus these six principal semophyletic modes of higher plants have not resulted from speculative morphological derivation but rested on the empirical material.

It then turned out that in terms of the listed modes, further morphological transformations of higher plants may be described. Since that moment the telome approach has not so much promoted reconstruction of the semophylogenetic linkages between organs as dictated ways of morphological derivation. In cases when palaeobotanical material failed to yield the intermediate forms dictated by the theory, such forms were invented. Numerous representations of various hypothetical intermediate forms have appeared in the morphological literature. Not finding among extinct gymnosperms, for example, such a form as could be used for derivation of the flowers of angiosperms, the morphologists themselves devised the supposed ancestral fructifications. These hypothetical organs were given proper morphological names (anthocorm, gonophyll, etc.).

The construction of missing links in morphological series in itself is quite justifiable, including from a theoretical cognitive viewpoint. Modern heuristics following the ideas of Kant, allocates a leading role to constructive imagination in scientific discoveries. Sometimes hypothetical constructions were confirmed later by discoveries of the relevant real forms. For instance, Lignier (1908) successfully predicted the habit of the Devonian rhyniophytes. But numerous hypothetical forms may be pointed out which have never been found in nature. In this case it is easier to allude to the imperfection of documents rather than to that of hypotheses, as is generally done. Still, as documents are accumulated we are entitled to expect at least partial confirmation of hypotheses. And if all newly found documents rather refute than support a hypothesis then confidence in it decreases. The premises of the hypothesis have then also to be subjected to doubt.

First of all a question arises to what degree the palaeobotanical record is really imperfect in relation to the principal types of morphological structures. The problem of perfection of the record is commonly considered in relation to taxa. Data have been reported (Meyen, 1984a) giving reason to regard the palaeobotanical record as complete for the gymnosperms at family level. The possibility is strong that the same is also true in regard to the main classes of morphological structures. It is not clear whether the discovery of organs of higher plants fundamentally different from those known at present can be expected. The fact is suggestive in this respect that intensive anatomical studies of the fossil pteridophytes and gymnosperms over the last 10–15 years have not discovered essentially new types of stelar organization, structure of wood, epidermis, spores, prepollen and pollen, sporangia, seeds and other organs. There is an impression that knowledge of anatomical diversity in the main traits for these plants is filled up. This knowledge is expanding at present, predominantly due to revision of interpretation of the already known structures rather than to discoveries of completely new plants possessing completely new types of anatomical structure. For instance, the eustelic structure of *Medullosa* and other plants which were earlier regarded as polystelic, was established. But the anatomical structure of *Medullosa* stems has been known for more than a hundred years. The imperfection of the record is not to blame for these stems being treated as polystelic.

Phylogenetic analysis of the gymnosperms has shown (Meyen, 1984a) that for most of their organs the main modes of their transformation are already known at present. This has permitted a real possibility of such transformations to be revealed which earlier were not admitted for consideration at all. Such is the probable formation of, for example, the *Cordaites*-type leaves from petioles of the Lagenostomales (phyllodization). According to the canons of the telome theory a long series of intermediate forms with gradual reduction of the leaf blade segmentation should be arranged between the *Cordaites* leaf and fern-like fronds characteristic of all primitive gymnosperms. Phyllodization does not require such a long series; it is sufficient to admit a delay in development of the whole part of a frond above the furcation point of the main rachis. The other types of non-trivial transformations that probably occurred in the gymnosperms, include gamoheterotopy (the transfer of characters from one sex to another; see below), homoeotic transfer of fructifications, secondary similarity between different organs of one plant, etc. The listed and some other transformations may be described, if necessary, in terms of the telome modes, and each of these transformations may be represented hypothetically as a long series of hypothetic transitional forms. However, the suspicion arises that in so doing we adjust the evidence to preconceived ideas rather than forming notions which are adequate to the evidence.

Telomic modes may be compared to rules of a very simple language which is adequate when it is necessary to describe something equally simple. More complex phenomena require a more complex language for their description. If a simple language is used for describing complex phenomena, the length of the text has to be sharply increased. It is known that any communication may be transmitted using the binary system of 0 and 1 but linguists would hardly insist that in any of our languages there is nothing except 0 and 1.

The supplementation of 'the language' (i.e. of modes) of the telome theory requires a return to the analysis of empirical material. In addition to the palaeobotanical evidence for phylogenetic connections between different organs an analysis is necessary as to what transformations of organs occur in mutations of recent plants, what trends there are of variability of their organs in general. For organs of different types, for different levels of advancement of one organ, a special set of modes may be required and it is not quite obvious that all these modes may be reduced to a very small number of universal transformations.

Taking the aforesaid into consideration, the evolution of different organs of higher plants can be attempted.

Evolution of organs

With the years plant morphology increasingly loses its independence and fuses, on the one hand, with systematics and, on the other hand, with studies in morphogenesis (ontogenesis). This process is particularly obvious in textbooks, though having in their titles the word 'morphology', the contents relate equally to morphology and to systematics (e.g. Bierhorst, 1971; Sporne, 1965, 1974, 1975). As a result comparative analysis throughout (through the whole system) of particular organs (as was the case in works by Goebel, Troll, A. Arber and other representatives of classical morphology) remains somewhat apart. It is probably not occasionally that the literature of the last few years contains no publications describing comparative analysis of the leaf or seed simultaneously on the basis of recent and fossil material (similar to the analysis of the stelar organization of seed plants performed not long ago – Beck *et al.*, 1982–1983).

It is not possible here to present a detailed analysis of a similar kind. Only some palaeobotanical data on particular organs are briefly summarized below. More emphasis will be placed on transformations which generally draw less attention but seem to be important. Moreover, the recurrence of some transformation types at different levels of morphological organization will be of interest (something like pseudocycles) as well as all aspects of the reversibility of transformations. The evolution of many morphological characters is described in Chapters 3–5 and what has been said there will not be repeated.

General habit

The documented evolution of higher plants begins from very small forms of the *Cooksonia* type that are several centimetres long. Chaloner and Sheerin (1979) showed a gradual increase in the size of higher plants during the Devonian, and Niklas (1982b) analysed the strategy of their branching. The increase in size was due to primary tissues, predominantly bark, and then to secondary tissues, both bark (in manoxylic forms) and wood (in pycnoxylic forms). Gensel and Andrews (1984) noticed that branching of the most primitive higher plants (*Cooksonia*, *Zosterophyllum*, *Sawdonia*, etc.) proceeded in one plane. This is also reflected in the elliptic outline of the stele (in the Zosrophyllopsida). The transition to three-dimensional branching probably took place independently in the lineage connecting the Zosterophyllopsida and the Lycopodiopsida, and in the Rhyniales–Trimerophytales. The intensification of three-dimensional branching is observed in different phylogenetical lineages, but a tendency to two-dimensionality remains. The secondarily two-dimensional (plagiotropic) branching of lateral branches is observed in very different groups. Sometimes it involves a whole arborescent individual (part of *Psaronius* and *Lepidodendron*). Independently, the whorled branching originated in different lineages. This was probably preceded by the arrangement of laterals in storeys (as in *Pertica quadrifaria*, Fig. 13, *q*). However, only in the Equisetopsida did the whorled arrangement acquire the significance of an obligate character for a taxon of such high rank.

The intensification of growth caused the appearance of increasingly larger forms. This process was accompanied both by an increase and a decrease in branching. In the first case arborescent forms with a crown appeared (in forests they formed a crown canopy). In the second case stick-like forms appeared with unbranching (as in many lycopsids of the *Tomiodendron* or *Pleuromeia* types, some *Phyllotheca* and the conifers of the *Yuccites* type) or sparsely branching trunks. The same two tendencies are noted for plants with thick false trunks. Apparently in all groups with developed arborescent forms sooner or later a reverse process of transformation of large arborescent forms into small herbaceous forms occurred accompanied by the reduction of secondary tissues. This is noted in the Isoetales, Calamostachyales, Polypodiales, Pinales (*Aethophyllum*), angiosperms.

As the axes differentiated into principal and lateral axes the different types of phyllotaxis (stegotaxis) developed. Regular phyllotaxis appeared independently in different lineages. The appearance of the Fibonacci series has been followed so far only in a lineage from the progymnosperms to the gymnosperms. In the latter the initial type was probably a phyllotaxis 2/5 (Beck *et al.*, 1982–1983).

The origin of roots remains completely unknown and there is little to be said on the evolution of roots. It is probable that the formation of roots was correlated with the development of the main aerial axis. The presence of true roots in the trimerophytans cannot be excluded. It is quite incomprehensible why plants having mesarch and endarch organization of the vascular bundles of stems have exarch metaxylem in roots. The exarch organization of the primary xylem might have been initial for higher plants and might have persisted only in the roots while in the aerial axes there was a transition to the mesarch and endarch organization. The formation of root hairs is not more comprehensible. In *Rhynia* and *Horneophyton* there are non-septate rhizoids. In more advanced plants rhizoids are lacking in sporophytes but are present in gametophytes.

Stelar organization

Beck *et al.* (1982–1983) have recently published a very complete review of the stelar organization of seed plants, and in this article as well as in the review by Schmid (1982–1983) the data are summarized on the evolution of the stelar organization of higher plants in general. Obviously a haplostele (undissected protostele) is the initial type of stele but it is not clear which type of haplostele is more primitive – centrarch or exarch. A stele elliptical in transverse section with obscurely defined peripheral patches of protoxylem is the earliest to occur stratigraphically. A stele round in section with a central protoxylem appears later and probably only when branching has become three-dimensional. The stele is then dissected and soon its medullation takes place. The dissected stele (actinostele) underwent medullation and segmentation, resulting in a multistrand (cladoxyloid) stele and eustele (one ring of endarch or mesarch bundles; some progymnosperms and many seed plants). Polystely was thought to be characteristic of the cladoxylaleans and some gymnosperms (Trigonocarpales, Pentoxylales). As already mentioned in Chapter 3, the Trigonocarpales and Pentoxylales actually have eustelic stems. Perhaps the Cladoxylales also have no polystely. It is not clear yet whether the arthrostele of articulates originated through the fragmentation of an actinostele or through the transformation of a cladoxyloid stele. In a lineage from the trimerophytes to the ferns the medullation of the protostele leads to the formation first of siphonoxylic (medullated protostele) and then siphonostelic (amphiphloic) organization. Still later in this lineage first monocyclic and then polycyclic dictyostele appear with numerous leaf and branch gaps. The stelar organization is not considered here in more detail as this has been done quite comprehensively in the reviews and in books by Stewart (1983) and Taylor (1981).

Leaf

Palaeobotanical data in general agree well with the old notion that leaves of the lycopsids (microphylls) originated differently from leaves of other higher plants (see Richardson, 1984). However, while megaphylls as a morphologically uniform type of organ formed by plantation and fusion of telomes could be contrasted to microphylls, the situation now has become more complicated. The leaf as a certain type of organ in the non-lycopsid plants, descendants of the trimerophytes, proves to be, first, polyphyletic and, second, a result of different processes. Terminal laminar pinnules of the Archaeopteridales probably formed via planation and fusion of the terminal sterile axes observed in the Protopteridiales, but it is quite likely that in the Noeggerathiales in leaf formation (of the *Megalopteris* and *Lesleya* type) the lateral branch of the last order participated. In the zygopterids and botryopterids the formation of laminar pinnules occurred independently and much later (Galtier, 1981). For these two groups of ferns the presence of phyllophores is characteristic besides petioles and stems. If phyllophores are recognized as the organ *sui generis* of an independent morphological status then the problem will arise to what degree phyllophores participated in the formation of fronds of ferns more advanced than the Botryopteridales. The fronds of the primitive gymnosperms are probably quite comparable to fronds of the Archaeopteridales, but it is not quite so obvious that the segmented part of the *Lyginopteris* frond and simple leaf blade of a bennettite *Nilssoniopteris* may be homologized. As mentioned above, cordaitanthalean leaves could have formed due to phyllodization, i.e. transformation of the petioles of the Lagenostomales. Phyllodial origin of the leaves of some Peltaspermales is quite probable (e.g. *Phylladoderma*). Phyllodization is not excluded in descendants of the Trigonocarpales. The terminal pinnules of the cycads and bennettites may correspond not to pinnules, but to petioles of the last order in the Trigonocarpales. Though it is now too early to make such statements for certain, the possibility itself deserves consideration.

In the Lagenostomales and Callistophytales pinnate fronds and scale-like cataphylls at their base may be distinguished (Blanc-Louvel, 1966; Rothwell, 1981). It is not clear whether cataphylls of these gymnosperms may be regarded as leaves. However, sterile scales of the axillary complex of the Cordaitanthales and further foliage leaves of the conifers might have been derived from cataphylls of the lagenostomans (Meyen, 1984a; Rothwell, 1982). If fronds and cataphylls of the lagenostomans are considered to be organs *sui generis*, then leaves of the conifers will prove non-homologous to leaves of the Cordaitanthales.

The morphological status of angiosperm leaves is not clear either. The primitive type in their case is a simple non-segmented leaf with disorganized pinnate venation (see Chapter 3). In any group of gymnosperms taken as a potential ancestor of the angiosperms the venation is well ordered. The possibility cannot be excluded that angiosperm leaves are homologous to some specialized leaf like organs of gymnosperms, e.g. phyllodia.

Thus, the notion of the leaf turns out to be much more complicated than an assembly of microphylls and megaphylls. A complete classification of leaves by classes of homologous organs cannot be provided as yet. At present the following types of leaves may be outlined: microphyllous (of enation origin), planational, cataphyllous (cataphylls themselves, and normal foliage leaves derived from them), phyllodes, cladodes. The degree of external similarity of these leaves that are different in origin may be very high.

Sporangium

The origin of the sporangium remains a puzzle. In the Rhyniales, Trimerophytales, and Horneophytopsida the sporangium wall resembles very closely the external zone of the cortex of vegetative axes. It is probable that sporangia are preceded by the tips of axes having inner zones of their cortex transformed into sporogenous tissue. Persistence of a column of sterile tissue in the *Horneophyton* sporangia suggests that the sporogenous tissue originally surrounded the stele and the adjacent cortical layer. Later a continuous sporangial cavity was formed. Further evolution of sporangia involved mainly the transformation of spores (see Chapters 3 and 4), the wall, a mechanism for dehiscence, the stalk and participation in various aggregates. Two trends may be outlined in the evolution of the wall (Zimmermann, 1959): (1) differentiation of the cells of the outer layer of the wall with the formation of a special mechanism for dehiscence (annulus, stomium, etc.) and transition to the unilayered wall; these transformations occur in propteridophytes and in different groups of pteridophytes; (2) the persistence of a homogeneous wall, the formation of mechanisms for dehiscence on account of the inner layer (endothecium, etc.); this tendency has already been noted in the propteridophytes (an endothecium-like layer in *Rhynia major* – Remy, 1978a) and is more expressed in the gymnosperms and angiosperms.

So far the sporangial stalk seems never to have been an object of special comparative study except in ferns for which a tendency to the formation of a slender stalk of one row of cells was noted long ago. The stalk as an independent organ is already expressed in the Zosterophyllopsida, and in this case a lateral stalk is not quite comparable to terminal fertile branchlets in the genera with apical sporangia (*Renalia*, *Nothia*). In the pteridophytes the sporangial stalk already acquires full morphological independence. The structure of sporangiophores (i.e. stalks) of the articulates, conifers, and some other higher plants acquires important taxonomic significance. At the same time with complication of the stalk in the pteridophytes and gymnosperms its reduction occurs, and in many groups sporangia become sessile again. Vascularization of the stalk also varies. The stalk is devascularized in the ferns; in various lineages of the gymnosperms the vascularization persists or disappears. In the trigonocarpaleans and in part of the callistophytans the vascularization extends not only into the stalk but also into the wall of the sporangium (synangium) and then the vascularization may disappear again.

Fructifications

In the Propteridophytes three main modes of sporangial aggregation are observed: (1) in the fertile zones of the middle part of axes, (2) in the terminal strobiloid structures, (3) in the planated or three-dimensional concentrations of the ramifying axes (sporoclads). To these three types of aggregation a fourth is added later: (4) on the surface of the planated organs (sporophylls). These four types of sporangial aggregation are then repeated in the distribution of the sporangial aggregations themselves, and then in the distribution of seeds, polysperms, flowers and inflorescences. Thus, the four major types of the architectonics of fructifications are dealt with. In the framework of these major types subordinated types repeating in the same degree may be noted (e.g. the position of sporangia and of more advanced fructifications on one side of the axis, on the lower or upper side of the supporting leaf-like organ, etc.). In this case we encounter a phenomenon resembling pseudocycles. On the other hand, in addition to aggregation the disaggregation of fructifications takes place and they again become scattered singly over the plant body.

It is difficult to judge the aggregation processes of sporangia in the oldest higher plants (*Cooksonia*, *Steganotheca*). Fertile zones composed solely of sporangia (without the participation of bracts and sporophylls) are known in the Zosterophyllopsida, Drepanophycales and articulates, and identical purely sporangial terminal strobili are known in the Zosterophyllopsida, Barinophytopsida and articulates. A concentration of spore-bearing branchlets with terminal sporangia appears in the Trimerophytales (*Psilophyton*, *Pertica*) and is characteristic of the Zygopteridales, Botryopteridales, Protopteridiales, Protopityales, Lagenostomales, Calamopityales, Arberiales and other groups. A more advanced stage is the obligate association of sporangia with certain vegetative parts different in various degrees from foliage leaves, the degree of difference being reversible. Such more or less specialized parts may be arbitrarily divided into (1) sporophylls (organ-bearing sessile sporangia or sporangiophores); (2) bracts (planated intersporangial organs); (3) various involucres, including indusia (organs accompanying the whole aggregate of sporangia); (4) sterile appendages similar in structure to sporangiophores but not bearing sporangia (sterilized sporangiophores; e.g. in *Calamophyton*).

Sporophylls are probably formed by heterotopy, i.e. the transfer of sporangia on to foliage leaves. Sporangial aggregates may undergo the same transplantation. The sporophylls are formed independently in different groups. In most cases this process probably proceeds by saltations (by homoeosis). No case of the gradual formation of sporophylls is yet known. In lycopsids the sporangium was supposed to shift from a lateral position on the stem into the leaf axil and only then on to its blade. The axillary sporangia were supposed to be characteristic of *Drepanophycus* (Fig. 17, *h*) but doubts have arisen over the validity of this (see Chapter 3). The axillary position of sporangia in some species of *Lycopodium* is probably of secondary origin. As already mentioned in Chapter 3, the transfer of seeds and synangia to leaves from leafless sporophylls (i.e. the formation of phyllosperms and microsporophylls) in the gymnosperms obviously occurred by saltations. However, the formation of fertiligers in the Arberiales when whole polysperms and microsporoclads were transferred on to the leaf was gradual. Multiplication of the number of fertiligers and microsporoclads on the subtending leaf obviously proceeded by saltations.

After the sporangia have been transferred on to the blades of subtending leaves, the latter undergo transformations and increasingly differ from foliage leaves. The same phenomenon, the postheterotopic transformation of subtending leaves, occurs in the case of the transfer of sporoclads, microsporoclads, seeds and polysperms to leaves. Postheterotopic transformations played a great role in the evolution of fructifications. The formation of specialized seed-like sporophylls in the Lepidocarpaceae, of strobilar fructifications in the Noeggerathiales, of various capsules and other seed-bearing organs of many gymnosperms belong here. It is quite probable that postheterotopic transformation of sporophylls is precisely connected with the appearance of the seed. In *Archaeopteris* the transfer of sporangia on to pinnules is noted, the latter after this event being strongly modified: their leaf blade becomes more dissected (Fig. 35, *p*) and some of the pinnules become saucer-shaped with a strongly dissected margin (Fig. 35, *q*). It is not difficult to imagine how a primary integument is formed from such a saucer-shaped pinnule. Foliage leaves in the neighbourhood of sporangia or other fertile organs also undergo transformations. Thus bracts, involucres and interseminal scales are formed; it is probable that petals, sepals and tepals also belong here. Various vegetative organs attending sporangia, their aggregations, and other fructifications may undergo secondary dedifferentiation, acquire a likeness to foliage leaves and may also disappear (also secondarily). It is probable that this disappearance may be saltational. Thus if the Equisetaceae stem from the Calamostachyales, the secondary disappearance of bracts in strobili may be supposed. No transitional forms demonstrating gradual reduction of bracts are known; it is possible that bracts disappeared saltationally. On the other hand, the postheterotopic transformation of sporophylls, phyllosperms and fertile organs similar to them leads not only to the loss of a leaf-like habit by them, but also to a certain return to an original leafless fertile shoot. It is probable that the strongly expressed dimorphism of tropho- and sporophylls in recent leptosporangiate ferns is of secondary origin, the sporophylls having been transformed into a paniculate organ with highly reduced leaf blades. Such leafless sporophylls are close in general habit to leafless sporoclads of the primitive Botryopteridales and Zygopteridales. Loss of leaf-like habit of male and female fructifications, and a return to the habit of a primarily leafless organ are very demonstratively manifested in microsporophylls in a lineage from the Callistophytales via the Peltaspermales to the Ginkgoales and Caytoniales. At the same time a three-dimensional branching of fructifications characteristic of the most primitive gymnosperms is re-established.

It is commonly very significant taxonomically to which side sporangia, synangia, seeds and polysperms are transplanted by the formation of fertile leaf-like organs (a marginal position occurring incomparably more rarely than an epi- and hypophyllous position). Sporophylls are always epiphyllous in the lycopsids, in the Polypodiopsida they are hypophyllous or rarely marginal, in the Archaeopteridales they are marginal or epiphyllous, in the Noeggerathiales they are epiphyllous; the phyllosperms of the Callistophytales bear seeds on the lower side, the fertiligers of the Arberiales are epiphyllous. In all cases the position of fructifications is very stable within a taxon.

Transformations of fructifications and of vegetative parts largely duplicate each other. This is demonstrated particularly in the cladosperms of the Peltaspermales which as well as leaves may be pinnate, pinnately dichotomous, may have pinnate and palmate venation of the entire blade and may be peltate (Meyen, 1984a). The cladosperms evidently preserve many morphogenetic potencies of the leaf.

Among the modes of transformation of fructifications the peltation, i.e. a formation of shield-like (peltate) structures, stands somewhat aside. Peltation is also characteristic of vegetative parts (the peltate leaves of some lycopsids, ferns and angiosperms), indusia and even trichomes. It is probable that peltation is performed differently in organs of different types and in different plant groups. The peltate sporangiophores of articulates are considered to be formed by fusion of the incurved stalks of sporangiophores arranged in a bunch (Zimmermann, 1959; however, the vascular bundle in the shield of the sporangiophore of *Equisetum* ramifies twice dichotomously – Bierhorst, 1971). The shields of microsporophylls of the conifers are probably formed of the distal lamina by accrescence of the heel. Peltoids of the Peltaspermales most probably originated from bilateral cladosperms by transposition of the cladosperm stalk from a marginal position to the central position.

Soon after heterospory had appeared in higher plants (on this process see Chapter 4) unisexual fructifications were formed (in lycopsids, gymnosperms, later in articulates). The semophylogenetic trends of fructifications of different sex then often coincided. The uniform construction of microsporoclads and ramified polysperms, of microsporophylls and phyllosperms is manifested in the Callistophytales. In the lineage from the Callistophytales via the Peltaspermales to the Ginkgoales the male and female fructifications undergo many similar transformations. In other cases the male and female fructifications evolved along essentially different semophylogenetic trends. In some Lepidocarpaceae, megasporophylls are transformed into seed-like organs but the blades of the microsporophylls have not acquired an integument-like habit. A very strong morphological divergence of trends of male and female fructifications is manifest in the arberians, trigonocarpaleans, bennettites, some cordaitanthaleans and all conifers.

However, this divergence in semophylogenetic trends is not fully irreversible. It may be assumed that at different stages of evolution, and in various plant groups a secondary drawing together of the semophylogenetic programmes of different sexes in relation to certain characters took place. This concerns the phenomenon of character transfer from one sex to another, or gamoheterotopy (Meyen, 1984b). As was already mentioned, the phenomenon of gamoheterotopy is well known and described in detail for various groups of animals (Schmalhausen, 1968). Of importance is that a high integration of the ontogenesis of birds and mammals does not prevent gamoheterotopy. The ontotogenesis of plant individua is much less integrated, as indicated by numerous cases of violation in plants of all three criteria of homology (see above). Thus, in plants manifestations of gamoheterotopy are quite to be expected. Meanwhile the possibility of gamoheterotopy itself has not been considered in plants, therefore in the case of sexual dimorphism the derivation series were made separately for each sex.

Evidently a demonstration of each concrete case of gamoheterotopy on palaeobotanical material is extremely difficult. However, at least one group, the Cordaitanthales (see Chapter 3), reveals such transformations of fructifications which cannot be understood unless gamoheterotopy is admitted. It is essential that gamoheterotopy manifests itself here both in the general construction of fructifications (the Cordaitanthaceae, part of the Rufloriaceae) and in the substitution of microsporangia for seeds (part of the Rufloriaceae). As seeds and microsporangia may act as interchangeable units in the process of evolution, a very intriguing possibility is suggested, namely the possibility of explaining the origin of the carpel of angiosperms.

Considering the origin of the angiosperms, the derivation series are usually constructed separately for male and female generative organs. Correspondingly, the ancestral structure of the carpel is sought among the female fructifications of the gymnosperms. This search has been unsuccessful until now (see Chapter 3). The failure may be explained for two reasons: (1) the wanted female fructification belongs to a yet unknown group; (2) the solution of the problem is sought in the wrong place. The first explanation is quite probable, but also one should not discard the second explanation. If the possibility of gamoheterotopy in the formation of the carpel is admitted it will be necessary to discover in which gymnosperms the microsporangia are located on the bearing organs in the same way as for the seeds of the putative primitive angiosperms. The bennettites are such gymnosperms. It is noteworthy that the synangia in bennettites are situated on the upper surface of the microsporophylls. If the seeds are transplanted to the place of the synangia they will evidently also be on the upper side of the bearing organ. Carpels with such a seed position are considered to be the most primitive among angiosperms. To make it possible to reconstruct the gamoheterotopic transformation, the relationship of the analysed groups should be supported by other characters. Angiosperms and bennettites manifest a similarity in the structure of the stems (it is suggestive that the bennettitalean stems of the genus *Homoxylon* were originally attributed to angiosperms), monocolpate pollen and paracytic stomata. In bennettites fructifications are strobiloid and accompanied by an involucre of sterile scales. Both these characters are to be expected in primitive angiosperms. Finally, the seeds of bennettites may be both uni- and bitegmic, with an embryo developing before the seeds fall off, i.e. again as in various groups of angiosperms.

The described materials on the morphology of various organs of the higher plants demonstrate that these plants possess types of transformations essentially more rich than the modes permitted by the telome theory. Upon reaching a certain level of morphological complexity the higher plants manifest new modes of transformation. The morphological theory which will replace the telome theory should be much less rigid and should permit a wide circle of morphological transformations, many of them at present being almost or altogether beyond the consideration of morphologists and systematists. These modes include:

1. The dedifferentiation of organs, including the formation of secondary gradual transitions between organs that have been discrete before.
2. Heterotopy which may involve organs that are most different, and the postheterotopic transformations of organs.
3. Parallel or divergent development of fructifications of different sex. In the case of divergent development the organs of each sex undergo evolution along different morphological trends. A secondary rapprochement of these trends is possible, i.e. gamoheterotopy.
4. In addition to planation, the reverse process may occur, i.e. the appearance of a three-dimensional instead of a planated structure.
5. Symmetrization, dissymmetrization and asymmetrization of organs. The institution of various regularities instead of disorganized structures (e.g. the Fibonacci series in the phyllo- and stegotaxis). Change in the symmetry type of an organ.
6. The origin of new organs via the superposition of ontogenetic programmes ('hybrid' organs); essentially, such organs cannot be described in terms of organs that have existed previously.
7. Repetition of transformations of similar types in organs of different levels of organization. This includes pseudocycles, manifesting in fructifications, the formation of leaf-like organs from petioles (phyllodia) and stems (cladodia), the recurrence of symmetry transformations in the vegetative organs and fructifications, etc.
8. Oligo- and polymerization of organs.
9. Determination of growth either of separate parts or of the whole individual as well as secondary indetermination.

The listed (and other processes) may proceed differently: (1) saltationally or gradualistically; (2) observing or violating three principles of homology; (3) be directed or kaleidoscopic; (4) irreversibly or reversibly.

References

Akhmetyev, M. A., Bratzeva, G. M., Giterman, R. E. *et al.* (1978) Stratigraphy and flora of the Late Cenozoic of Island. *Trudy Geol. Inst. Akad. Nauk SSSR*, **316**, 188 pp.

Al-Ameri, T. K. (1983–1984) Microstructures of possible early land plants from Tripolitania, North Africa. *Rev. Palaeobot. Palynol.*, **40**, 375–86.

Aldridge, R. J., Dorning, K. J., Hill, R. J. *et al.* (1979) Microfossil distribution in the Silurian of Britain and Ireland. In: *The Caledonides of the British Isles – Reviewed*, Geological Society, London, pp. 433–8.

Allen, K. C. (1980) A review of in situ Late Silurian and Devonian spores. *Rev. Palaeobot. Palynol.*, **29**, 253–70.

Alpern, B. and Streel, M. (1972) Palynologie et stratigraphie du Paléozoïque moyen et supérieur. *Mém. Bur. R. Géol. Min. Fr.*, **77**, 217–40.

Alvin, K. L. (1968) The spore-bearing organs of the Cretaceous fern *Weichselia* Stuhler. *J. Linn. Soc. London (Bot.)*, **61** (384), 87–92.

Alvin, K. L. (1977) The conifers *Frenelopsis* and *Manica* in the Cretaceous of Portugal. *Palaeontology*, **20**, 387–404.

Alvin, K. L. (1982) Cheirolepidiaceae: biology, structure and paleoecology. *Rev. Palaeobot. Palynol.*, **37**, 71–98.

Alvin, K. L. and Hluštík, A. (1979) Modified axillary branching in species of the fossil genus *Frenelopsis*: a new phenomenon among conifers. *Bot. J. Linn. Soc.*, **79**, 231–41.

Alvin, K. L., Spicer, R. A. and Watson, J. (1978) A *Classopollis*-containing male cone associated with *Pseudofrenelopsis*. *Palaeontology*, **21**, 847–56.

Ananiev, A. R. (1957) New fossil plants from the Lower Devonian deposits near village of Torgashino in south-eastern part of Western Siberia. *Bot. Zhurn.*, **42**, 691–702 (in Russian).

Ananiev, A. R. (1959) *Most Important Localities of Devonian Floras in the Sayany–Altai Area*, Tomsk University Press, Tomsk.

Ananiev, A. R. and Stepanov, S. A. (1968) Findings of fructifications in *Psilophyton princeps* Dawson emend. Halle in Lower Devonian of South-Minussa depression (Western Siberia). *Trudy Tomsk Univ.*, **202**, 30–46 (in Russian).

Anderson, H. M. (1976) A review of the Bryophyta from the Upper Triassic Molteno Formation, Karroo basin, South Africa. *Palaeont. Afr.*, **19**, 21–30.

Anderson, J. M. and Anderson, H. M. (1983) *Palaeoflora of Southern Africa. Molteno formation (Triassic)*, Vol. 1, Part 1. Introduction, Part 2 *Dicroidium*, A. A. Balkema, Rotterdam.

Anderson, J. M. and Anderson, H. M. (1985) *Palaeoflora of Southern Africa. Prodromus of South African megafloras Devonian to Lower Cretaceous*. A. A. Balkema, Rotterdam.

Andreeva, E. M., Petrosyan, N. M. and Radczenko, G. P. (1962) New data on phytostratigraphy of the Devonian deposits in Altai–Sayany mountaineous area. *Trudy Vsesoyuzn. Geol. Inst.*, n. ser., **70,** 23–63 (in Russian).

Andrews, H. N. (1961) *Studies in Paleobotany*, Wiley, New York and London.

Andrews, H. N., Arnold, C. A., Boureau, E., *et al.* (1970) Filicophyta. In: *Traité de Paléobotanique* (ed. E. Boureau), T. 4, fasc. 1, Masson et Cie, Paris.

Andrews, H. N. and Mamay, S. (1955) Some recent advances in morphological palaeobotany. *Phytomorphology*, **5,** 372–94.

Andrews, H. N. and Phillips, T. L. (1968) *Rhacophyton* from the Upper Devonian of West Virginia. *J. Linn. Soc. (Bot.)*, **61**(384), 37–64.

Appert, O. (1973) Die Pteridophyten aus dem Oberen Jura des Manamana in Südwest Madagaskar. *Schweiz. Paläontol. Abh.*, **94,** 1–62.

Arber, A. (1950) *The Natural Philosophy of Plant Form*, Cambridge University Press, Cambridge.

Arber, E. A. N. (1921) *Devonian Floras. A Study of the Origin of Cormophyta*, Cambridge University Press, Cambridge.

Archangelsky, S. (1963) A new Mesozoic flora from Ticó, Santa Cruz province, Argentina. *Bull. Brit. Mus. (Nat. Hist.), Geol.*, **8**(2), 47–92.

Archangelsky, S. (1968a) On the genus *Tomaxellia* (Coniferae) from the Lower Cretaceous of Patagonia (Argentina) and its male and female cones. *J. Linn. Soc. (Bot.)*, **61**(384), 153–65.

Archangelsky, S. (1968b) Palaeobotany and palynology in South America: a historical review. *Rev. Palaeobot. Palynol.*, **7,** 249–66.

Archangelsky, S. (1970) Fundamentos de paleobotánica. *Univ. Nac. La Plata, Fac. Cien. Natur. y Mus., Ser. Techn. y Didact.*, **10,** 1–347.

Archangelsky, S. (1981) *Fedekurtzia*, a new Carboniferous frond from Gondwanaland and its fructification. *Amer. J. Bot.*, **68,** 1130–8.

Archangelsky, S. and Arrondo, O. G. (1975) Paleogeografía y plantas fósiles en el Pérmico inferior Austrosudamericano. *Actas I Congr. Arg. de Pal. y Bioestrat. Tucumán, 1974*, **1,** 479–96.

Archangelsky, S., Azcuy, C. L., Pinto, I. D. *et al.* (1978–1980) The Carboniferous and Early Permian of the South American Gondwana area: a summary of biostratigraphic information. *Actas II Congr. Arg. de Pal. y Bioestrat. y I Congr. Latinoamer. de Pal. Buenos Aires*, **4,** 257–69.

Archangelsky, S. and Cúneo, R. (1984) Zonacion del Pérmico continental de Argentina sobre la base de sus plantas fosiles. *Mem. III Congr. Latinoamer. Paleontol., Mexico 1984*, 143–53.

Archangelsky, S. and Leguizamón, R. (1980) El registro de *Ginkgophyllum diazii* en el Carbónico de Sierra de Los Llanos, Provincia de la Rioja. *Bol. Acad. Nac. Ciencias, Cordoba, Argentina*, **53** (3–4), 211–19.

Archangelsky, S. and Wagner, R. H. (1983) *Glossopteris anatolica* sp. nov. from uppermost Permian strata in south-east Turkey. *Bull. Brit. Mus. Nat. Hist. (Geol.)*, **37**(3), 81–91.

Asama, K. (1959) Systematic study of so-called *Gigantopteris*. *Sci. Rep. Tohoku Univ.*, 2nd Ser. (Geol.) **31,** 1–72.

Asama, K. (1962) Evolution of Shansi flora and origin of simple leaf. *Sci. Rep. Tohoku Univ.*, 2nd Ser. (Geol.), spec. vol., No. 5, 247–74.

Asama, K. (1966) Two types of evolution in *Sphenophyllum*. *Bull. Nat. Sci. Mus.*, **9**(4), 577–608.

Asama, K. (1983) Evolution and phylogeny of vascular plants based on the principles of growth retardation. Part 7. *Bull. Nat. Sci. Mus. Tokyo*, Ser. C, **9**(1), 1–21.

Ash, S. (1976) The systematic position of *Eoginkgoites*. *Amer. J. Bot.*, **63,** 1321–31.

Ash, S. (1977) An unusual bennettitalean leaf from the Upper Triassic of the south-western United States. *Palaeontology*, **20,** 641–59.

Ash, S. (1979) *Skilliostrobus* gen. nov., a new lycopsid cone from the Early Triassic of Australia. *Alcheringa*, **3,** 73–89.

Ash, S., Litwin, R. and Traverse, A. (1982) The Upper Triassic fern *Phlebopteris smithii* (Daugherty) Arnold and its spores. *Palynology*, **6,** 203–19.

Awramik, S. M. and Barghoorn, E. S. (1977) The Gunflint microbiota. *Precambrian Res.*, **5,** 121–42.

Axelrod, D. J. (1958) Evolution of the Madro-Tertiary geoflora. *Bot. Rev.*, **24,** 433–509.

Bachmann, K. (1983) Evolutionary genetics and the genetic control of morphogenesis in flowering plants. *Evol. Biol.*, **16,** 157–208.

Baer, K. von. (1959) Differences between artificial and natural classifications of bodies of nature (1819). *Ann. Biol.*, **1,** 367–72 (in Russian).

Balme, B. E. (1979) Palynology of Permian–Triassic boundary beds of Kap Stosch, East Greenland. *Medd. om Grønland*, **200**(6), 46 pp.

Balueva, G. A. (1964) *Asinia paradoxa*, a new species of water ferns in the deposits of Symian suite of the Taz River. *Bot. Zhurn.*, **49,** 1471–3 (in Russian).

Banerjee, M. (1969) *Senotheca murulidihensis*, a new glossopteridian fructification from India associated with *Glossopteris taeniopteroides* Feistmantel. J. Sen Memor. Vol., pp. 359–68. Botanical Society Bengal, Calcutta.

Banks, H. P. (1966) Devonian flora of New York State. Empire State Geogram, a triannual publ. of *Geol. Surv.*, **4**(3), 10–24.

Banks, H. P. (1968) The early history of land plants. In: *Evolution and Environment* (ed. E. T. Drake), Yale University Press, New Haven and London, pp. 73–107.

Banks, H. P. (1975) The oldest vascular land plants: a

note of cautions. *Rev. Palaeobot. Palynol.*, **20,** 13–25.

Banks, H. P. (1980a) Floral assemblages in the Siluro–Devonian. In: *Biostratigraphy of Fossil Plants. Successional and Palaeoecological Analyses* (eds D. L. Dilcher and T. N. Taylor), Dowden *et al.*, Stroudsburg, pp. 1–24.

Banks, H. P. (1980b) The role of *Psilophyton* in the evolution of vascular plants. *Rev. Palaeobot. Palynol.*, **29,** 165–76.

Banks, H. P., Bonamo, P. M. and Grierson, J. D. (1972) *Leclercqia complexa* gen. et sp. nov., a new lycopod from the late Middle Devonian of eastern New York. *Rev. Palaeobot. Palynol.*, **14,** 19–40.

Banks, H. P., Leclercq, S. and Hueber, F. M. (1975) Anatomy and morphology of *Psilophyton dawsonii* sp. n. from the Late Lower Devonian of Quebec (Gaspé), and Ontario, Canada. *Palaeontographica Americana*, **8**(48), 77–127.

Barale, G. (1981) La paléoflore Jurassique du Jura Français étude systématique; aspects stratigraphiques et paléoécologiques. *Docum. Lab. Géol. Lyon*, **81,** 1–467.

Barghoorn, E. S. and Tyler, S. A. (1965) Microorganisms from the Gunflint chert. *Science*, **147,** 563–76.

Barnard, P. D. W. and Long, A. G. (1973) On the structure of a petrified stem and some associated seeds from the Lower Carboniferous rocks of East Lothian, Scotland. *Trans. Roy. Soc. Edinburgh*, **69,** 91–108.

Barthel, M. (1968) *'Pecopteris' feminaeformis* (Schlotheim) Sterzel und *'Araucarites' spiciformis* Andrae in Germar – Coenopterideen des Stephans und Unteren Perm. *Paläontol. Abh.*, Abt. B, **2,** 727–42.

Barthel, M. (1977) Die Gattung *Dicranophyllum* Gr. Eury in den varistischen Innensenken DDR. *Hall. Jb. f. Geowiss.*, **2,** 73–86.

Barthel, M. (1980) Calamiten aus dem Oberkarbon und Rothliegenden des Thüringer Waldes. 100 Jahre Arboretum (1879–1979) Berlin, 237–58.

Barthel, M., Götzelt, V. and Urban, G. (1976) Die Rotliegendflora Sachsens. *Abh. Staatl. Mus. Mineral. Geol.*, **24,** 1–190.

Barton, C. M. (1964) Significance of Tertiary fossil floras of King George Island, South Shetland Islands. In: *Antarctic Geology* (ed. R. J. Adie), Wiley, New York, pp. 603–8.

Basinger, J. F. (1976) *Paleorosa similkameensis*, gen. et sp. nov., permineralized flowers (Rosaceae) from the Eocene of British Columbia. *Can. J. Bot.*, **54,** 2293–305.

Basinger, J. F. and Dilcher, D. L. (1984) Ancient bisexual flowers. *Science*, **224,** 511–13.

Basinger, J. F., Rothwell, G. W. and Stewart, W. N. (1974) Cauline vasculature and leaf trace production in medullosan pteridosperms. *Amer. J. Bot.*, **61,** 1002–15.

Bassoullet, J. P., Bernier, P., Deloffre, R. *et al.* (1977) Classification criteria of fossil Dasycladales. In: *Fossil Algae. Recent Results and Developments* (ed. E. Flügel), Springer, Berlin, pp. 154–66.

Baxter, R. W. (1971) *Carinostrobus foresmani*: a new lycopod cone genus from the Middle Pennsylvanian of Kansas. *Palaeontographica*, B**134,** 124–30.

Baxter, R. W. (1975) Fossil fungi from American Pennsylvanian coal balls. *Univ. Kans. Paleontol. Contribs. Pap.* 77, 1–10.

Baxter, R. W. and Baxendale, R. W. (1976) *Corynepteris involucrata*, sp. nov., a new fertile fern of possible zygopterid affinities from the Pennsylvanian of Kansas. *Univ. Kans. Paleontol. Contribs. Pap.* 85, 1–15.

Baxter, R. W. and Leisman, G. A. (1967) A Pennsylvanian calamitean cone with *Elaterites triferens* spore. *Amer. J. Bot.*, **54,** 748–54.

Beck, Ch.B. (1957) *Tetraxylopteris schmidtii* gen. et sp. nov., a probable pteridosperm precursor from the Devonian of New York. *Amer. J. Bot.*, **44,** 350–67.

Beck, Ch. B. (1960) The identity of *Archaeopteris* and *Callixylon*. *Brittonia*, **12,** 351–68.

Beck, Ch. B. (1962) Reconstruction of *Archaeopteris* and further consideration of its phylogenetic position. *Amer. J. Bot.*, **49,** 373–382.

Beck, Ch. B. (1970) The appearance of gymnosperm structure. *Biol. Revs. Cambridge Phil. Soc.*, **45,** 379–400.

Beck, Ch. B. (1971) On the anatomy and morphology of lateral branch systems of *Archaeopteris*. *Amer. J. Bot.*, **58,** 758–84.

Beck, Ch. B. (1976a) Current status of the Progymnospermopsida. *Rev. Palaeobot. Palynol.*, **21,** 5–23.

Beck, Ch. B. (ed.) (1976b) *Origin and Early Evolution of Angiosperms*. Columbia University Press, New York and London.

Beck, Ch. B. (1981) *Archaeopteris* and its role in vascular plant evolution. In: *Paleobotany, Paleoecology, and Evolution* (ed. K. J. Niklas), Vol. 1, Praeger, New York, pp. 193–230.

Beck, Ch. B., Schmid, R. and Rothwell, G. W. (1982–1983) Stelar morphology and the primary vascular system of seed plants. *Bot. Rev.*, **48,** 681–815.

Berggren, W. A. (1981) Role of ocean gateways in climate changes. *Acta Univ. Stockholm*, **37**(2), 9–20.

Bertrand, P. (1935) Contribution à l'étude des Cladoxylées de Saalfeld. *Palaeontographica*, B**80,** 101–70.

Bharadwaj, D. C. (1974) On the classification of gymnospermous sporae dispersae. In: *Symposium on Structure, Nomenclature and Classification of Pollen and Spores* (ed. D. C. Bharadwaj), B. Sahni,

Institute of Palaeobotany. Spec. Publ. No. 4 Lucknow, pp. 7–52.

Bierhorst, D. W. (1971) *Morphology of Vascular Plants*, Macmillan, New York.

Blanc-Louvel, Ch. (1966) Étude anatomique comparée des tiges et des pétioles d'une Ptéridospermée du Carbonifère du genre *Lyginopteris* Pontonié. *Mém Mus. Nat. Hist. Natur.*, n. ser., Ser. C, Sci. de la Terre, **18**(1), 5–103.

Bocheński, T. (1936) Uber Sporophyllstände (Blüten) einiger Lepidophyten aus dem produktiven Karbon Polens. *Polnisch. Geol. Ges. Jg.*, **12,** 193–240.

Bocheński, T. (1960) Evolution of pinnule venation in the Carboniferous seed ferns Alethopterides (*Alethopteris* and *Lonchopteris*) and the meaning of pinnule venation analysis for diagnosis of species. *Prace Inst. Geol. Warszawa*, **20,** 1–42.

Boersma, M. (1972) The heterogeneity of the form genus *Mariopteris* Zeiller. A comparative morphological study with special reference to the frond composition of the West-European species. Utrecht.

Bonamo, P. M. (1975) The Progymnospermopsida: building a concept. *Taxon*, **24,** 569–79.

Bonamo, P. M. (1977) *Rellimia thomsonii* (Progymnospermopsida) from Middle Devonian of New York State. *Amer. J. Bot.*, **64,** 1272–85.

Bonamo, P. M. (1983) *Rellimia thomsonii* (Dawson) Leclercq and Bonamo (1973): The only correct name for the aneurophytalean progymnosperm. *Taxon*, **32,** 449–53.

Bonamo, P. M. and Banks, H. P. (1967) *Tetraxylopteris schmidtii*: its fertile parts and its relationships within the Aneurophytales. *Amer. J. Bot.*, **54,** 755–68.

Bose, M. N. and Roy, S. K. (1967–1968) On the occurrence of *Pachypteris* in the Jabalpur Series of India. *Palaeobotanist*, **16,** 1–9.

Boureau, E. (ed.) (1964–1975) *Traité de Paléobotanique*. T. II. Bryophyta. Psilophyta. Lycophyta. 1967. T. III. Sphenophyta, Noeggerathiophyta. 1964. T. IV, f. 1. Filicophyta. 1970. T. IV, f. 2. Pteridophylla (première partie). 1975. Masson et Cie, Paris.

Boureau, E. (1971) *Les Sphénophytes. Biologie et Histoire Évolutive*, Librairie Vuibert, Paris.

Boureau, E. and Doubinger, J. (1975) *Traité de Paléobotanique*. T. IV, f. 2. Pteridophylla. Première partie. Masson et Cie, Paris.

Brack, S. D. and Taylor, T. N. (1972) The ultrastructure and organization of *Endosporites*. *Micropaleontology*, **18,** 101–9.

Brack-Hanes, S. D. (1981) On a lycopod cone with winged spores. *Bot. Gaz.*, **142,** 294–304.

Brack-Hanes, S. D. and Thomas, B. A. (1983) A reexamination of *Lepidostrobus* Brongniart. *Bot. J. Linn. Soc.*, **86,** 125–33.

Brack-Hanes, S. D. and Vaughn, J. C. (1978) Evidence of Paleozoic chromosomes from lycopod microgametophytes. *Science*, **200,** 1383–5.

Brauer D. F. (1980) *Barinophyton citrulliforme* (Barinophytales incertae sedis, Barinophytaceae) from the Upper Devonian of Pennsylvania. *Amer. J. Bot.*, **67,** 1186–206.

Brauer D. F. (1984) A permineralized cordaitalean tree from the Middle Pennsylvanian of Iowa, USA. Abstr. of contributed papers and poster session, *2nd Int. Org. Paleobot. Conf. Edmonton.*

Brenner, G. J. (1976) Middle Cretaceous floral provinces and early migration of angiosperms. In: Beck, C. B. (ed.). *Origin and Early Evolution of Angiosperms*. Columbia Univ. Press, New York, pp. 23–47.

Britten, B. J., Davidson, E. H., Dover, G. A. *et al.* (1982) Genomic change and morphological evolution. Group report. *Life Sci. Res. Rept.*, **22,** 19–39.

Brongniart, Ad. (1828–1838) *Histoire des végétaux fossiles ou recherches botaniques et géologiques sur les végétaux renfermés dans les diverses couches du globe*, Dufour et D'Ocagne, Paris.

Brooks, D. R. (1983) What's going on in evolution? A brief guide to some new ideas in evolutionary theory. *Can. J. Zool.*, **61,** 2637–45.

Brooks, J. *et al.* (eds) (1971) *Sporopollenin*, Academic Press, London and New York.

Brooks, J. and Shaw, G. (1973) The role of sporopollenin in palynology. In: *Problems of Palynology* (ed. M. I. Neishtadt), Nauka, Moscow, pp. 80–91.

Brousmiche, C. (1976) Précisions sur les organes de *Nemejcopteris feminaeformis* (Schlotheim). *Ann. Soc. Géol. Nord*, **96,** 233–7.

Brousmiche, C. (1979) *Pecopteris* (*Asterotheca*) *saraei* P. Corsin, 1951, forme fertile de *Sphenopteris damesii* (Stur, 1885). *Géobios*, **12,** 75–97.

Brousmiche, C. (1982) Sur la synonymie de *Crossotheca boulayi* Zeiller, 1886–88 et *Crossotheca bourozii* Danzé, 1956 avec l'espèce-type du genre: *Crossotheca crepinii* Zeiller, 1883. Une nouvelle interprétation de la fructification. *Géobios*, **15,** 679–703.

Bundantsev, L. Yu. (1983) *History of Arctic flora of the Early Cenophyte Epoch*, Nauka, Leningrad (in Russian).

Burago, V. I. (1982) To leaf morphology of the genus *Psygmophyllum*. *Paleontol. Zhurn.*, **2,** 128–36 (in Russian).

Burnham, R. J. and Spicer, R. (1984) Preservation of forest litter in volcaniclastic sediments of el Chichonal, Mexico. Abstr. of contributed papers and poster session, *2nd Int. Org. Paleobot. Conf. Edmonton.*

Cesari, S. N. (1982) Licópsidas arborescentes de la Formación Tupe, Sierra de Maz, Provincia de La Rioja. *Ameghiniana*, **19,** 273–9.

Chaloner, W. G. (1967) Lycophyta. In: *Traité de*

Paléobotanique (ed. E. Boureau). T. 2, Masson et Cie, Paris, pp. 434–802.

Chaloner, W. G. (1968a) The cone of *Cyclostigma kiltorkense* Haughton, from the Upper Devonian of Ireland. *J. Linn. Soc. (Bot.)*, **61**(384), 25–36.

Chaloner, W. G. (1968b) The paleoecology of fossil spores. In: *Evolution and Environment* (ed. E. T. Drake), Yale University Press, New Haven and London, pp. 125–38.

Chaloner, W. G. (1969) Triassic spores and pollen. In: *Aspects of Palynology* (eds R. H. Tschudy and R. A. Scott), Wiley, New York, pp. 291–309.

Chaloner, W. G. (1970a) The evolution of miospore polarity. *Geoscience and Man*, **1**, 47–56.

Chaloner, W. G. (1970b) The rise of the first land plants. *Biol. Revs Cambridge Phil. Soc.*, **45**, 353–77.

Chaloner, W. G. (1984) Evidence of ontogeny of two Late Devonian plants from Kiltorcan, Ireland. Abstr. of contributed papers and poster session. *2nd Int. Org. Paleobot. Conf. Edmonton.*

Chaloner, W. G., Hill, A. J. and Lacey, W. S. (1977) First Devonian platyspermic seed and its implications in gymnosperm evolution. *Nature*, **265**, 233–5.

Chaloner, W. G., Hill, A. and Rogerson, E. C. W. (1978) Early Devonian plant fossils from a Southern England borehole. *Palaeontology*, **21**, 693–707.

Chaloner, W. G. and Lawson, J. D. (eds) (1985) Evolution and environment in the late Silurian and early Devonian. *Phil. Trans. Roy. Soc. London*, Ser. B, *Biol. Sci.*, **309**(1138), 1–342.

Chaloner, W. G. and McDonald, P. (1980) *Plants Invade the Land*, Royal Scottish Museum, Edinburgh.

Chaloner, W. G. and Meyen, S. V. (1973) Carboniferous and Permian floras of the Northern continents. In: *Atlas of palaeobiogeography* (ed. A. Hallam), Elsevier, Amsterdam, pp. 169–86.

Chaloner, W. G. and Muir, M. (1968) Spores and floras. In: *Coal and Coal-bearing Strata* (eds D. Murchison and T. S. Westoll), Oliver and Boyd, Edinburgh, pp. 127–46.

Chaloner, W. G. and Sheerin, A. (1979) Devonian macrofloras. *Spec. Pap. in Palaeontol.*, **23**, 145–61.

Chandra, S. and Surange, K. R. (1979) Revision of the Indian species of *Glossopteris*. *B. Sahni Inst. Palaeobot. Monogr.*, **2**, 1–291.

Chaney, R. W. (1959) Miocene floras of the Columbia Plateau. I. Composition and interpretation. *Carnegie Inst. Wash. Publ.*, **617**, 1–134.

Chaphekar, M. (1965) On the genus *Pothocites* Paterson. *Palaeontology*, **8**, 107–12.

Chaphekar, M. and Alvin, K. L. (1972) On the fertile parts of the coenopterid fern *Metaclepsidropsis duplex* (Williamson). *Rev. Palaeobot. Palynol.*, **14**, 63–76.

Charlton, W. A. and Watson, J. (1982) Patterns of arrangement of lateral appendages on axes of *Stigmaria ficoides* (Sternberg) Brongniart. *Bot. J. Linn. Soc.*, **84**, 209–21.

Chernov, Yu. I. (1984) Biological premises of assimilation of Arctic environment by organisms of different taxa. In: *Faunogenes i Filotsenogenes* (ed. Yu. I. Chernov), Nauka, Moscow, pp. 154–74 (in Russian).

Chlonova, A. F. (1984) Stratigraphic range and distribution of morphological types of the Cretaceous angiosperm pollen. In: *Problemy Sovremennoi Palinologii*, Novosibirsk, pp. 13–17 (in Russian).

Chuvashov, B. I. and Dyupina, G. V. (1973) Upper Palaeozoic terrigenous deposits of western slope of Middle Urals. *Trudy Inst. Geol. i Geokhim., Sverdlovsk*, **105**, 3–208 (in Russian).

Chuvashov, B. I., Ivanova, R. M. and Kolchina, A. N. (1984) *Upper Palaeozoic of eastern slope of Urals. Stratigraphic and geological history*. Ural. Sci. Centre Acad. Sci. USSR, Sverdlovsk (in Russian).

Cichan, M. A. and Taylor, T. N. (1982a) Structurally preserved plants from Southwestern Kentucky: *Stauropteris biseriata* sp. nov. *Amer. J. Bot.*, **69**, 1491–6.

Cichan, M. A. and Taylor, T. N. (1982b) Wood-boring in *Premnoxylon*: plant–animal interaction in the Carboniferous. *Palaeogeogr., Palaeoclimatol., Palaeoecol.*, **39**, 123–7.

Cichan, M. A. and Taylor, T. N. (1983) A systematic and developmental analysis of *Arthropitys deltoides* sp. nov. *Bot. Gaz.*, **144**, 285–94.

Cichan, M. A., Taylor, T. N. and Brauer, D. F. (1984) Ultrastructural studies of *in situ* Devonian spores: *Protobarinophyton pennsylvanicum Brauer*. *Rev. Palaeobot. Palynol.*, **41**, 167–75.

Clayton, G. (1985) Dinantian miospores and intercontinental correlation. *C. r. 6ème Congr. Intern. Stratigr. et Géol. Carbonifère, Madrid, 12–17 Sept. 1983*, vol. 4, pp. 9–23.

Clayton, G., Coquel, R., Doubinger, J. *et al.* (1977) Carboniferous miospores of Western Europe: illustration and zonation. *Meded. Rijks Geol. Dienst*, **29**, 1–71.

Clement-Westerhoff, J. A. (1984) Aspects of Permian palaeobotany and palynology. IV. The conifer *Ortiseia* Florin from the Val Gardena Formation of the Dolomites and the Vicentinian Alps (Italy) with special reference to a revised concept of the Walchiaceae (Göppert) Schimper. *Rev. Palaeobot. Palynol.*, **41**, 51–166.

Collinson, M. E. (1980) A new multiple-floated *Azolla* from the Eocene of Britain with a brief review of the genus. *Palaeontology*, **23**, 213–29.

Collinson, M. E. (1983) Accumulations of fruits and seeds in three small sedimentary environments in

Southern England and their palaeoecological implications. *Ann. Bot.*, **52,** 583–92.

Combourieu, N. and Galtier, J. (1985) Nouvelles observations sur *Polypterospermum*, *Polylophospermum*, *Colpospermum* et *Codonospermum*, ovules de Ptéridospermales du Carbonifère supérieur français. *Palaeontographica*, B**196,** 1–29.

Conkin, J. E. and Conkin, B. M. (1977) North American primitive Paleozoic charophytes and descendants. In: *Geobotany* (ed. R. C. Romans), Plenum Press, New York and London, pp. 173–193.

Cornell, W. C. (1970) The chrysomonad cyst-families Chrysostomataceae and Archaeomonadaceae: their status in paleontology, *Proc. N. Amer. Paleontol. Conv., Chicago, 1969*, Part G, Lawrence, Kansas, pp. 958–65.

Cornet, B., Phillips, T. L. and Andrews, H. N. (1976) The morphology and variation in *Rhacophyton ceratangium* from the Upper Devonian and its bearing on frond evolution. *Palaeontographica*, **B158,** 105–29.

Crane, P. R. (1984) A re-evaluation of *Cercidiphyllum*-like plant fossils from the British early Tertiary. *Bot. J. Linn. Soc.*, **89,** 199–230.

Crane, P. R. and Dilcher, D. L. (1984–1985) *Lesqueria*: an early angiosperm fruiting axis from the mid-Cretaceous. *Ann. Mo. Bot. Gard.*, **71,** 384–402.

Creber, G. T. and Chaloner, W. G. (1984) Influence of environmental factors on the wood structure of living and fossil trees. *Bot. Rev.*, **50,** 357–448.

Crepet, W. L. (1974) Investigation of North American cycadeoids: the reproductive biology of *Cycadeoidea*. *Palaeontographica*, **B148,** 144–69.

Crepet, W. L. (1979) Some aspects of the pollination biology of Middle Eocene angiosperms. *Rev. Palaeobot. Palynol.*, **27,** 213–38.

Cridland, A. A. (1964) *Amyelon* in American coal-balls. *Palaeontology*, **7,** 186–209.

Cridland, A. A. and Morris, J. E. (1960) *Spermopteris*, a new genus of pteridosperms from the Upper Pennsylvanian series of Kansas. *Amer. J. Bot.*, **47,** 855–9.

Čtyroky, P. (1973) Permian flora from the Ga'ara region (Western Iraq). *N. Jahrb. Geol. Paläontol. Monatsh.*, **7,** 383–8.

Cúneo, R. and Andreis, R. R. (1983) Estudio de un bosque de licofitas en la Formación Nueva Lubecka, Pérmico de Chubut. Implicancias paleoclimáticas y paleogeográficas. *Ameghiniana*, **20,** 132–40.

Daber, R. (1968) A *Weichselia–Stiehleria*–Matoniaceae community within the Quedlinburg Estuary of Lower Cretaceous age. *J. Linn. Soc. London (Bot.)*, **61**(384), 75–85.

Daber, R. (1980) Zum Problem der Gabelwedelformen des Karbons und Perms – eine Übersicht. *Schriftenr. Geol. Wiss. Berlin*, **16,** 15–48.

Daber, R. and Helms, J. (1981) *Fossile Schätze*, Fortschritt Erfurt, Leipzig.

Darrah, A., Davies, E. H., Gunter, D. *et al.* (1979) A systematic illustrated guide to fossil organic-walled dinoflagellate genera. *Life Sci. Misc. Publ. Roy. Ontario Mus.*

Davidson, E. H. (1982) Evolutionary change in genomic regulatory organization: speculations on the origins of novel biological structure. *Life Sci. Res. Rept*, **22,** 65–84.

Davitashvili, L. Sh. and Zakharieva-Kovacheva, K. (1975) *Origin of stone forests*, Metsniereba, Tbilisi.

Delevoryas, T. (1955) The Medullosae – structure and relationships. *Palaeontographica*, **B97,** 114–67.

Delevoryas, T. (1962) *Morphology and evolution of fossil plants*, Holt *et al.*, New York.

Delevoryas, T. and Gould, R. E. (1973) Investigations of North American cycadeoids: williamsonian cones from the Jurassic of Oaxaca, Mexico. *Rev. Palaeobot. Palynol.*, **15,** 27–42.

Delevoryas, T. and Hope, R. C. (1981) More evidence for conifer diversity in the Upper Triassic of North Carolina. *Amer. J. Bot.*, **68,** 1003–7.

Delevoryas, T. and Taylor, T. N. (1969) A probable pteridosperm with eremopteroid foliage from the Allegheny Group of Northern Pennsylvania. *Postilla*, **133,** 1–14.

Dennfer, D. V., Ehrendorfer F., Mägdefrau, K. and Ziegler, H. (1978) *Lehrbuch der Botanik für Hochschulen*, G. Fischer, Stuttgart and New York.

Dennis, R. L. (1970) Middle Pennsylvanian basidiomycete mycelium with clamp connections. *Mycologia*, **62,** 578–84.

Dennis, R. L. (1976) *Palaeosclerotium*: a Pennsylvanian age fungus combining features of modern ascomycetes and basidiomycetes. *Science*, **192,** 66–68.

Dilcher, D. L. (1973) A paleoclimatic interpretation of the Eocene floras of south-eastern North America. In: *Vegetation and Vegetational History of Northern Latin America* (ed. A. Graham), Elsevier, Amsterdam, pp. 39–59.

Dilcher, D. L. (1974) Approaches to the identification of angiosperm leaf remains. *Bot. Rev.*, **40,** 1–157.

Dilcher, D. L. (1975) Epiphyllous fungi from Eocene deposits in Western Tennessee, USA *Palaeontographica*, **B116,** 1–54.

Dilcher, D. L. (1979) Early angiosperm reproduction: an introductory report. *Rev. Palaeobot. Palynol.*, **27,** 291–328.

Dilcher, D. L. and Crane, P. R. (1984) In pursuit of the first flower. *Nat. Hist.*, **93**(3), 57–60.

Dilcher, D. L. and Crane, P. R. (1984–1985) *Archaeanthus*: An early angiosperm from the Cenomanian of the Western Interior of North America. *Ann. Mo. Bot. Gard.*, **71**, 351–83.

Dilcher, D. L. and Taylor, T. N. (eds) (1980) *Biostratigraphy of fossil plants. Successional and paleoecological analyses*, Dowden *et al.*, Strousburg.

DiMichele, W. A. (1980) *Paralycopodites* Morey & Morey, from the Carboniferous of Euramerica – a reassessment of generic affinities and evolution of '*Lepidodendron*' *brevifolium* Williamson. *Amer. J. Bot.*, **67**, 1466–76.

DiMichele, W. A. (1981) Arborescent lycopods of Pennsylvanian age coals: *Lepidodendron*, with description of a new species. *Palaeontographica*, **B175**, 85–125.

DiMichele, W. A., Mahaffy, J. F. and Phillips, T. L. (1979) Lycopods of Pennsylvanian age coals: *Polysporia*. *Can. J. Bot.*, **57**, 1740–53.

DiMichele, W. A. and Phillips, T. L. (1977) Monocyclic *Psaronius* from the Lower Pennsylvanian of the Illinois Basin. *Can. J. Bot.*, **55**, 2514–24.

DiMichele, W. A. and Phillips, T. L. (1985) Arborescent lycopod reproduction and paleoecology in a coal-swamp environment of late Middle Pennsylvanian age (Herrin Coal, Illinois, USA). *Rev. Palaeobot. Palynol.*, **44**, 1–26.

DiMichele, W. A., Phillips, T. L. and Peppers, R. A. (1985) The influence of climate and depositional environment on the distribution and evolution of Pennsylvanian coal-swamp plants. In: *Geological factors and the evolution of plants* (ed. B. H. Tiffney), Yale Univ. Press, New Haven, pp. 223–56.

DiMichele, W. A., Rischbieter, M. O., Eggert, D. A. and Gastaldo, R. A. (1984) Stem and leaf cuticle of *Karinopteris*: source of cuticles from the Indiana 'paper' coal. *Amer. J. Bot.*, **7**, 626–37.

Dittrich, H. S., Matten, L. C. and Phillips, T. L. (1983) Anatomy of *Rhacophyton ceratangium* from the Upper Devonian (Famennian) of West Virginia. *Rev. Palaeobot. Palynol.*, **40**, 127–47.

Diver, W. L. and Peat, C. J. (1979) On the interpretation and classification of Precambrian organic-walled microfossils. *Geology*, **7**, 401–4.

Dobruskina, I. A. (1969) Genus *Scytophyllum* (morphology, epidermal structure and systematic position). *Trudy Geol. Inst. Akad. Nauk SSSR*, **190**, 35–58 (in Russian).

Dobruskina, I. A. (1980) Stratigraphic position of plant-bearing beds of the Triassic of Eurasia. *Trudy Geol. Inst. Akad. Nauk SSSR*, **346**, 3–163 (in Russian).

Dobruskina, I. A. (1982) Triassic floras of Eurasia. *Trudy Geol. Inst. Akad. Nauk SSSR*, **365**, 3–196 (in Russian).

Dobruskina, I. A. and Yaroshenko, O. P. (1983) Relations between Triassic floras on both sides of North Atlantics. *Bull. Mosk. Obshch. Ispyt. Prirody, Otd. Geol.*, **58**(3), 83–96 (in Russian).

Dodge, J. D. (1983) Dinoflagellates: investigation and phylogenetic speculation. *Brit. Phycol. J.*, **18**, 335–56.

Dolph, G. E. and Dilcher, D. L. (1979) Foliar physiognomy as an aid in determining paleoclimate. *Palaeontographica*, **B170**, 151–72.

Doludenko, M. P. (1984) Late Jurassic floras of South-Western Eurasia. *Trudy Geol. Inst. Akad Nauk SSSR*, **390**, 3–112 (in Russian).

Doludenko, M. P. and Kostina, E. I. (1985) *Schizolepis fanica* Mesozoic member of the family Pinaceae. *Bot. Zhurn.*, **70**, 464–71 (in Russian).

Doludenko, M. P. and Svanidze, Ts.I. (1969) Late Jurassic flora of Georgia. *Trudy Geol. Inst. Akad. Nauk SSSR*, **178**, 5–116 (in Russian).

Doran, J. B. (1980) A new species of *Psilophyton* from the Lower Devonian of northern New Brunswick, Canada. *Can. J. Bot.*, **58**, 2241–62.

Doran, J. B., Gensel, P. G. and Andrews, H. N. (1978) New occurrences of trimerophytes from the Devonian of eastern Canada. *Can. J. Bot.*, **56**, 3052–68.

Dorofeev, P. I. (1979) On seeds of Cretaceous conifers. *Bot. Zhurn.*, **64**, 305–17 (in Russian).

Doubinger, J. (1983) Études palynologiques dans le bassin stéphanien de Montceau-les-Mines (le couche): aspects stratigraphiques et paléoécologiques. *Mém. Géol. Univ. Dijon*, **8**, 43–50.

Doubinger, J. and Langiaux, J. (1982) Un faux problème: la limite Stéphanien/Autunien. *C. R. Acad. Sci. Paris*, **294**, 395–8.

Doubinger, J., Remy, W. and Gerhardt, H. D. (1964) Entwurf für eine einheitliche diagnostische Beschreibing von Kutikulen. *Fortschr. Geol. Rheinl. Westfalen*, **12**, 11–24.

Downie, C. (1973) Observations on the nature of the acritarchs. *Palaeontology*, **16**, 239–59.

Downie, C., Evitt, W. R. and Sarjeant, W. A. S. (1963) Dinoflagellates, hystrichosphaeres and the classification of the acritarchs. *Stanford Univ. Publ. (Geol. Sci.)*, **7**(3), 3–16.

Doyle, J. A. (1977) Magnoliophyta. In: McGraw Hill Yearbook, *Science and Technology*.

Doyle, J. A. (1978) Origin of angiosperms. *Ann. Rev. Ecol. Syst.*, **9**, 365–92.

Doyle, J. A., Biens, P., Doerenkamp, A. and Jardiné, S. (1977) Angiosperm pollen from the pre-Albian Lower Cretaceous of Equatorial Africa. *Bull. Cent. Rech. Explor.-Prod. Elf-Aquitaine*, **1**(2), 451–73.

Doyle, J. A. and Hickey, L. J. (1976) Pollen and leaves from the Mid-Cretaceous Potomac group and their bearing on early angiosperm evolution. In: *Origin and Early Evolution of Angiosperms* (ed. Ch. B. Beck), Columbia University Press, New York, pp. 139–206.

Doyle, J. A., Jardiné S. and Doerenkamp, A. (1982)

Afropollis, a new genus of early angiosperm pollen, with notes on the Cretaceous palynostratigraphy and paleoenvironments of Northern Gondwana. *Bull. Cent. Rech. Explor.-Prod. Elf-Aquitaine*, **6**(1), 39–117.

Doyle, J. A., Van Campo, M. and Lugardon, B. (1975) Observations on exine structure of *Eucommiidites* and Lower Cretaceous angiosperm pollen. *Pollen et Spores*, **17**, 429–86.

Dràbek, K. (1977) *Polysporia mirabilis* Newberry, 1837 z Nýrăn (Vestfál D). *Cas. Nád. Muz. Odd. Přírodověd.*, **146**, 93–6.

Drägert, K. (1964) *Pflanzensoziologische Untersuchungen in den Mittleren Essener Schichten des nördlichen Ruhrgebietes*, Forschungsber. Landes Nordrhein-Westf. 1363, pp. 1–295.

Drinnan, A. N. and Chambers, T. C. (1984) The Pentoxylales in the Early Cretaceous of Victoria, Australia. Abstr. of contributed papers and poster session. *2nd Int. Org. Paleobot. Conf. Edmonton*.

Durante, M. V. (1976) The Carboniferous and Permian stratigraphy of Mongolia on the basis of palaeobotanical data. *Trudy Sovmestn. Sov.–Mongol. Exped.*, **19**, 5–279 (in Russian).

Durante, M. V. (1980) On relations between the Upper Permian floras of Nan-Shan and coeval Angara floras. *Paleontol. Zhurn.*, **1**, 125–35 (in Russian).

Durante, M. V. and Pavlova, E. E. (1983) The significance of the section of the Deng-Nuru Range for palaeobiogeography. *Trudy Sovmestn. Sov.–Mongol. Exped.*, **21**, 29–32 (in Russian).

Dyufur, M. C., Ereshko, E. M., Lebedev, M. M. *et al.* (1977) On spore–pollen assemblages from metamorphozed deposits of Kamchatka and the age of the embedding beds. *Voprosy Regionaln. Geologii*, **2**, 103–13 (in Russian).

Eaton, G. L. (1980) Nomenclature and homology in peridinialean dinoflagellate plate patterns. *Palaeontology*, **23**, 667–88.

Edwards, D. (1969) Further observations on *Zosterophyllum llanoveranum* from the Lower Devonian of South Wales. *Amer. J. Bot.*, **56**, 201–10.

Edwards, D. (1973) Devonian floras. In: *Atlas of Palaeobiogeography* (ed. A. Hallam), Elsevier, Amsterdam, pp. 105–15.

Edwards, D. (1980) Early land floras. *Syst. Assoc. Spec. Vol.* **15**, 55–85.

Edwards, D., Bassett, M. G. and Rogerson, E. C. W. (1979) The earliest vascular land plants: continuing the search for proof. *Lethaia*, **12**, 313–24.

Edwards, D., Edwards, D. S. and Rayner, R. (1982) The cuticle of early vascular plants and its evolutionary significance. *Linn. Soc. Sympos. Ser.*, **10**, 341–61.

Edwards, D. and Fanning, U. (1985) Evolution and environment in the late Silurian–early Devonian: the rise of the pteridophytes. *Phil. Trans. Roy. Soc. London.*, Ser. B, Biol. Sci., **309**(1138), 147–65.

Edwards, D., Feehan, J. and Smith, D. G. (1983) A late Wenlock flora from Co. Tipperary, Ireland. *Bot. J. Linn. Soc.*, **86**, 19–36.

Edwards, D. and Rogerson, E. C. W. (1979) New records of fertile Rhyniophytina from the late Silurian of Wales. *Geol. Mag.*, **116**, 93–8.

Edwards, D. and Rose, V. (1984) Cuticles of *Nematothallus*: a further enigma. *Bot. J. Linn. Soc.*, **88**, 35–54.

Edwards, D. S. (1980) Evidence for the sporophytic status of the Lower Devonian plant *Rhynia gwynne-vaughanii* Kidston et Lang. *Rev. Palaeobot. Palynol.*, **29**, 177–88.

Eggert, D. A. (1961) The ontogeny of Carboniferous arborescent Lycopsida. *Palaeontographica*, **B108**, 43–92.

Eggert, D. A. (1962) The ontogeny of Carboniferous arborescent Sphenopsida. *Palaeontographica*, **B110**, 99–127.

Eggert, D. A. (1972) Petrified *Stigmaria* of Sigillarian origin from North America. *Rev. Palaeobot. Palynol.*, **14**, 85–99.

Eggert, D. A. (1974) The sporangium of *Horneophyton lignieri* (Rhyniophytina). *Amer. J. Bot.*, **61**, 405–13.

Eggert, D. A. and Gaunt, D. D. (1973) Phloem of *Sphenophyllum*. *Amer. J. Bot.*, **60**, 755–70.

Eggert, D. A. and Kryder, R. W. (1969) A new species of *Aulacotheca* (Pteridospermales) from the Middle Pennsylvanian of Iowa. *Palaeontology*, **12**, 412–19.

Eggert, D. A. and Taylor, T. N. (1971) *Telangiopsis* gen. nov., an Upper Mississippian pollen organ from Arkansas. *Bot. Gaz.*, **132**, 30–7.

Ehret, D. L. and Phillips, T. L. (1977) *Psaronius* root systems – morphology and development. *Palaeontographica*, **B161**, 147–64.

El-Khayal, A. A., Chaloner, W. G. and Hill, C. R. (1980) Palaeozoic plants from Saudi Arabia. *Nature*, **285**, 33–4.

El-Saadawy, W. El-S. and Lacey, W. S. (1979) The sporangia of *Horneophyton lignieri* (Kidston and Lang) Barghoorn and Darrah. *Rev. Palaeobot. Palynol.*, **28**, 137–44.

Engler, A. (1879) *Versuch einer Entwicklungsgeschichte der extratropischen Florengebiete der nördlichen Hemisphäre*. Leipzig.

Engler, A. (1882) *Versuch einer Entwicklungsgeschichte der extratropischen Florengebiete der südlichen Hemisphäre und der tropischen Gebiete*. Leipzig.

Engler, A. (1899) *Die Entwicklung der Pflanzengeographie in den letzten hundert Jahren*, A. V. Humboldt-Centenarschrift, Berlin (not seen; cited after Takhtajan, 1978).

Erwin, D. M. (1984) Growth and development in the

rhizomorph of *Paurodendron fraipontii*. Abstr. of contributed papers and poster session, *2nd Int. Org. Paleobot. Conf. Edmonton.*

Esaulova, N. K. (1984) *Palaeobotanical characteristics of the Kazanian stage in the stratotype sections of the Kama embayment*, Abstract of Thesis, Kazan University.

Evitt, W. R., Lentin, J. K., Millioud, M. E. *et al.* (1977) Dinoflagellate cyst terminology. *Pap. Geol. Surv. Can.*, 76–24, 1–11.

Faegri, K. and Iversen, J. (1975) *Textbook of pollen analysis*, Hafner, New York.

Fairon-Demaret, M. (1979) *Estinnophyton wahnbachense* (Kräusel et Weyland) comb. nov., plante remarquable du Siegenien d'Allemagne. *Rev. Palaeobot. Palynol.*, **28,** 145–60.

Fairon-Demaret, M. (1980) À propos des specimens determinés *Protolepidodendron scharyanum* par Kräusel et Weyland, 1932. *Rev. Palaeobot. Palynol.*, **29,** 201–20.

Fairon-Demaret, M. (1985) Les plantes fossiles de l'Emsien du Sart Tilman, Belgique. I. *Stockmansia langii* (Stockmans) comb. nov. *Rev. Palaeobot. Palynol.*, **44,** 243–60.

Fefilova, L. A. (1978) *Leafy mosses of the Permian of the European North of the USSR*. Nauka, Leningrad (in Russian).

Ferguson, D. K. (1971) The Miocene flora of Kreuzau, Western Germany. 1. The leaf remains. *Verh. K. Ned. Akad. Wet.* (afd. Natuurkunde, Tweede Reeks), **60,** 1–297.

Ferguson, D. K. (1974) On the taxonomy of recent and fossil species of *Laurus* (Lauraceae). *Bot. J. Linn. Soc.*, **68,** 51–72.

Ferguson, I. K. and Muller, J. (eds) (1976) The evolutionary significance of the exine (*Linn. Soc. Symp. Ser.*, No. 1), Academic Press, London and New York.

Filin, V. R. (1978) Division Equisetophyta. In: *Zhizn rastenii* (eds I. V. Grushvitsky and S. G. Zhilin), Vol. 4, Prosveshchenie, Moscow, pp. 131–46.

Fissunenko, O. P. (1965) Regularities in the development of the Carboniferous flora, and phytostratigraphic subdivision of the Donets basin Middle Carboniferous. In: *Geology of Coal Measures and Stratigraphy of the Carboniferous in the USSR* (ed. I. I. Gorsky), Nauka, Moscow, pp. 199–208 (in Russian, English summary).

Fissunenko, O. P. (1973) *Methods and geological significance of ecologic–taphonomic studies as exemplified by the Middle Carboniferous of Donbass*, Abstract of Thesis, Kiev.

Florin, R. (1931) Untersuchungen zur Stammesgeschichte der Coniferales und Cordaiten. I. *Kgl. Sv. Vetensk.-Akad. Handl.*, Ser. 3, **10**(1), 1–588.

Florin, R. (1933) Studien über die Cycadales des Mesozoikums. *Kg. Sv. Vetensk.-Akad. Handl.*, Ser. 3, **12**(5), 1–134.

Florin, R. (1938–1945) Die Koniferen des Oberkarbons und des Unteren Perms. *Palaeontographica* **B85,** H. 1 (1938), 1–62; H.2–4 (1939), 63–242; H.5 (1940), 243–363; H. 6 (1944a), 365–456; H. 7 (1944b), 457–654; H. 8 (1945), 655–729:

Florin, R. (1949) The morphology of *Trichopitys heteromorpha* Saporta, a seed-plant of Palaeozoic age, and the evolution of the female flowers in the Ginkgoinae. *Acta Horti Berg.*, **15,** 80–109.

Florin, R. (1951) Evolution in Cordaites and Conifers. *Acta Horti Berg.*, **15,** 285–388.

Florin, R. (1958) On Jurassic taxads and conifers from north-western Europe and eastern Greenland. *Acta Horti Berg.*, **17,** 257–402.

Florin, R. (1963) The distribution of conifer and taxad genera in time and space. *Acta Horti Berg.*, **20,** 121–312.

Flügel, E. (ed.) (1977) *Fossil Algae. Recent Results and Developments*, Springer, Berlin.

Foster, C. B. (1979) Permian plant microfossils of the Blair Athol Coal Measures, Baralaba Coal Measures, and basal Rewan Formation of Queensland. *Geol. Surv. Queensland, Publ. 372, Palaeontol. Pap.* 45, 1–154.

Foster, C. B. (1983) *Jugasporites* Leschik 1956, a Late Palaeozoic operculate pollen genus. *Mem. Ass. Australas. Palaeontol.*, 1, 327–38.

Fott, B. (1971) *Algenkunde*, G. Fischer, Jena.

Francis, J. E. (1983) The dominant conifer of the Jurassic Purbeck Formation, England. *Palaeontology*, **26,** 277–94.

Friis, E. M. (1983) Upper Cretaceous (Senonian) floral structures of juglandalean affinity containing *Normapolles* pollen. *Rev. Palaeobot. Palynol.*, **39,** 161–88.

Friis, E. M. (1984) Platanaceous inflorescences from the Late Cretaceous of Sweden and Eastern North America. Abstr. of contributed papers and poster session, *2nd Int. Org. Palaeobot. Conf. Edmonton.*

Friis, E. M. (1984–1985) Preliminary report of Upper Cretaceous angiosperm reproductive organs from Sweden and their level of organization. *Ann. Mo. Bot. Gard.*, **71,** 403–18.

Friis, E. M. and Skarby, A. (1982) *Scandianthus* gen. nov. Angiosperm flowers of saxifragalean affinity from the Upper Cretaceous of Southern Sweden. *Ann. Bot.*, **50,** 569–83.

Fryxell, G. (1983) New evolutionary patterns in diatoms. *BioScience*, **33,** 92–8.

Galtier, J. (1970) Recherches sur les végétaux à structure conservée du Carbonifère inférieur Français. *Paléobiol. Continentale*, **1**(4), 1–221 (reprinted in 1976).

Galtier, J. (1974) Sur l'organisation de la fronde des *Calamopitys*, ptéridospermales probables du Carbonifère inférieur. *C. R. Acad. Sci. Paris*, Ser. D, **279**, 975–8.

Galtier, J. (1975) Variabilité anatomique et ramification des tiges de *Calamopitys*. *C. R. Acad. Sci. Paris*, Ser. D, **280**, 1967–70.

Galtier, J. (1981) Structures foliaires de fougères et ptéridospermales du Carbonifère inférieur et leur signification évolutive. *Palaeontographica*, B **180**, 1–38.

Galtier, J. and Grambast, L. (1972) Observations nouvelles sur les structures reproductrices attribuées à *Zygopteris lacattei* (Coenopteridales de l'Autuno-Stéphanien français). *Rev. Palaeobot. Palynol.*, **14**, 101–11.

Galtier, J. and Phillips, T. L. (1977) Morphology and evolution of *Botryopteris*, a Carboniferous age fern. Part 2. Observations on Stephanian species from Grand 'Croix, France. *Palaeontographica*, B**164**, 1–32.

Galtier, J. and Scott, A. C. (1979) Studies of Paleozoic ferns: on the genus *Corynepteris*. A redescription of the type and some other European species. *Palaeontographica*, B**170**, 81–125.

Galtier, J. and Scott, A. C. (1985) Diversification of early ferns. *Proc. Roy. Soc. Edinburgh*, Sect. B, **86**, 289–301.

Garratt, M. J. and Rickards, R. B. (1984) Graptolite biostratigraphy of early land plants from Victoria, Australia. *Proc. Yorkshire Geol. Soc.* **44**(4), No. 27, 377–84.

Garratt, M. J., Tims, J. D., Rickards, R. B., *et al.* (1984) The appearance of *Baragwanathia* (Lycophytina) in the Silurian. *Bot. J. Linn. Soc.*, **89**, 355–8.

Gastaldo, R. A. (1980) Taxonomic considerations for Carboniferous coalified compression equisetalean strobili. *Amer. J. Bot.*, **68**, 1319–24.

Gatzuk, L. E. (1976) Content of the notion 'herbs' and the problem of their evolutionary position. *Trudy Mosk. Obshch. Ispyt. Prirody*, Otd. Biol., Sect. Bot., **42**, 55–130.

Gaussen, H. (1952) L'évolution pseudocyclique. *Ann. Biol.*, **28**, 207–25.

Geng Guocang and Tao Junrong (1982) Tertiary plants from Xizang. In: *Palaeontology of Xizang*, Book 5, Sci. Press, pp. 202–35.

Gensel, P. G. (1976) *Renalia hueberi*, a new plant from the Lower Devonian of Gaspé. *Rev. Palaeobot. Palynol.*, **22**, 19–37.

Gensel, P. G. (1979) Two Psilophyton species from the Lower Devonian of Eastern Canada with a discussion of morphological variation within the genus. *Palaeontographica*, B**168**, 81–99.

Gensel, P. G. (1980) Devonian *in situ* spores: a survey and discussion. *Rev. Palaeobot. Palynol.*, **30**, 101–32.

Gensel, P. G. (1982) *Oricilla*, a new genus referable to the Zosterophyllophytes from the late Early Devonian of Northern New Brunswick. *Rev. Palaeobot. Palynol.*, **37**, 345–59.

Gensel, P. G. (1984) A new Lower Devonian plant and the early evolution of leaves. *Nature*, **309**, 785–7.

Gensel, P. G. and Andrews, H. N. (1984) *Plant Life in the Devonian*, Praeger, New York.

Gensel, P. G., Andrews, H. N. and Forbes, W. H. (1975) A new species of *Sawdonia* with notes on the origin of microphylls and lateral sporangia. *Bot. Gaz.*, **136**, 50–62.

Gensel, P. G., Chaloner, W. G. and Forbes, W. H. (1984) *Spongiophyton* from the Early Devonian (Emsian) of New Brunswick and Gaspé, Canada. Abstr. of contributed papers and poster session, *2nd Int. Org. Paleobot. Conf. Edmonton.*

Gensel, P. G. and White, A. R. (1983) The morphology and ultrastructure of spores of the Early Devonian trimerophyte *Psilophyton* (Dawson) Hueber and Banks. *Palynology*, **7**, 221–33.

Gerasimenko, L. M. and Krylov, I. N. (1983) Postmortal changes of cyanobacteria in algal-bacterial films of thermal springs of Kamchatka. *Doklady Akad. Nauk. SSSR*, **272**, 201–3.

Gillespie, W. H., Rothwell, G. W. and Scheckler, S. E. (1981) The earliest seed. *Nature*, **293**, 462–4.

Givnish, T. J. (1978) Ecological aspects of plant morphology: leaf form in relation to environment. *Acta Biotheoretica*, **27** (suppl.: Folia Biotheoretica, No. 7), 83–142.

Gnilovskaya, M. B. (1984) On the nature of the Vendotaenides. In: *Stratigrafiya i paleontologiya drevneishego fanerozoya*, Nauka, Moscow, pp. 58–64 (in Russian).

Goeppert, H. R. (1864–1865) Die fossile Flora der permischen Formation. *Palaeontographica* **12**, 1–316.

Gollerbakh, M. M. (ed.) (1977) *Zhizn Rastenii (Plant Life)*, Vol. 3. *Vodorosli, Lishainiki (Algae, Lichens)*, Prosveshchenie, Moscow (in Russian).

Golubev, S. N. (1981) *Real Crystals in Skeletons of Coccolithophorids*, Nauka, Moscow (in Russian).

Golubić, S. and Barghoorn, E. S. (1977) Interpretation of microbial fossils with special reference to the Precambrian. In: *Fossil Algae. Recent Results and Developments* (ed. E. Flugel), Springer, Berlin, pp. 1–14.

Gomankov, A. V. (1983) *Palaeobotanical characteristics of the Upper Tatarian substage of the Russian platform.* Abstract of Thesis, Moscow (in Russian).

Gomankov, A. V. and Meyen, S. V. (1979) On representatives of the family Peltaspermaceae from the Permian deposits of the Russian platform. *Paleontol. Zhurn.*, **2**, 124–38 (in Russian).

Gomankov, A. V. and Meyen, S. V. (1980) On relations

between assemblages of plant mega- and microfossils in the Permian of Angaraland. *Palaeontol. Zhurn.*, **4,** 114–22 (in Russian).

Gomankov, A. V. and Meyen, S. V. (in press). *Tatarina*-flora (composition, distribution in Late Permian of Eurasia). *Trudy Geol. Inst. Akad. Nauk SSSR*, **401** (in Russian).

Good, C. W. (1975) Pennsylvanian-age calamitean cones, elater-bearing spores, and associated vegetative organs. *Palaeontographica*, **B153,** 28–99.

Good, C. W. (1977) Taxonomic and stratigraphic significance of the dispersed spore genus *Calamospora*. In: *Geobotany* (ed. R. C. Romans), Plenum Press, New York and London, pp. 43–64.

Good, C. W. (1978) Taxonomic characteristics of sphenophyllalean cones. *Amer. J. Bot.*, **67,** 86–97.

Gothan, W. and Weyland, H. (1973) *Lehrbuch der Paläobotanik*, 3rd edition Akademy-Verlag, Berlin.

Gould, R. E. (1970) *Palaeosmunda*, a new genus of siphonostelic osmundaceous trunks from the Upper Permian of Queensland. *Palaeontology*, **13,** 10–28.

Gould, S. J. (1977) *Ontogeny and Phylogeny*, Belknap, Cambridge, Mass.

Gould, S. J. (1982a) Change in developmental timing as a mechanism of macroevolution. *Life Sci. Res. Rept*, **22,** 333–46.

Gould, S. J. (1982b) Punctuated equilibria – a different way of seeing. *New Sci.*, **1301,** 137–41.

Gould, S. J. and Eldredge, N. (1977) Punctuated equilibria: the tempo and mode of evolution reconsidered. *Paleobiology*, **3,** 115–51.

Graham, A. (1971) The role of Myxomyceta spores in palynology (with a brief note on the morphology of certain algal zygospores). *Rev. Palaeobot. Palynol.*, **11,** 89–99.

Graham, A. (ed.) (1972) *Floristics of Asia and Eastern America*, Elsevier, Amsterdam.

Grambast, L. (1974) Phylogeny of Charophyta. *Taxon* **23,** 463–81.

Grand'Eury, C. (1877) Mémoire sur la flore carbonifère du départment de la Loire et du centre de la France. *Mém. Acad. Sci., Inst. Natur. France*, **24**(1), 1–624.

Grauvogel-Stamm, L. (1978) La flore du grès à *Voltzia* (Buntsandstein supérieur) des Vosges du Nord (France). Morphologie, anatomie, interprétations phylogénique et paléogéographique. *Univ. L. Pasteur de Strasbourg, Inst. Géol., Mém.* **50,** 1–225.

Grauvogel-Stamm, L. and Doubinger, J. (1975) Deux fougères fertiles stéphanien du Massif central (France). *Géobios*, **6,** 409–21.

Grauvogel-Stamm, L. and Duringer, P. (1983) *Annalepis zeilleri* Fliche 1910 emend., un organe reproducteur de Lycophyte de la Lettenkohle de la France. Morphologie, spores *in situ* et paléoécologie. *Geol. Rundsch.*, **72,** 23–51.

Grauvogel-Stamm, L. and Grauvogel, L. (1980) Morphologie et anatomie d'*Anomopteris mougeotii* Brongniart (synonyme: *Pecopteris sulziana* Brongniart), une fougère du Buntsandstein supérieur des Vosges (France). *Sci. Géol. Bull.*, **33,** 53–66.

Grauvogel-Stamm, L. and Schaarschmidt, F. (1979) Zur Morphologie und Taxonomie von *Masculostrobus* Seward und anderen Formgattungen peltater männlicher Koniferenblüten. *Senckenbergiana lethaea*, **60,** 1–37.

Gray, J. (1985) The microfossil record of early land plants: advances in understanding of early terrestrialization, 1970–1984. *Philos. Trans. Roy. Soc. London*, Ser. B, *Biol. Sci*, **309**(1138), 167–195.

Gray, J. and Boucot, A. J. (1977) Early vascular land plants: proof and conjecture. *Lethaia*, **10,** 145–74.

Grierson, J. D. and Bonamo, P. M. (1979) *Leclercqia complexa*: earliest ligulate lycopod (Middle Devonian). *Amer. J. Bot.*, **66,** 474–6.

Grizer, M. I. and Ul'masvai, F. S. (1984) Paleoclimatic and paleogeographic developmental conditions of the East Siberian Upper Paleozoic coal-bearing formation. *27th Int. Geol. Congr.*, Abstr., Vol. 9, part 2 (additional), Moscow, August 4–14. Nauka, Moscow, pp. 89–90.

Grushvitsky, I. V. and Zhilin, S. G. (eds) (1978) *Zhizn Rastenii (Plant Life)*, Vol. 4. *Mosses. Lycopods. Horsetails. Ferns. Gymnospermous plants*, Prosveshchenie, Moscow (in Russian).

Gu and Zhi (Li Zingxue, Deng Longhua, Zhou Zhiyan, Xuren (= Hsu, J.) and Zhu Jianan) (1974) Palaeozoic plants from China. In: *Fossil Plants of China*, Vol. 1, Sci. Press, Peking, pp. 1–226 (in Chinese).

Halle, T. G. (1927) Palaeozoic plants from Central Shansi. *Palaeontol. Sinica*, Ser. A., **2**(1), 1–316.

Hamer, J. J. and Rothwell, G. W. (1983) *Phillipopteris* gen. nov. – anatomically preserved sporangial fructifications from the Upper Pennsylvanian of the Appalachian basin. *Amer. J. Bot.*, **70,** 1378–85.

Hammen, T., van der. (1983) The Palaeocology and palaeogeography of Savannas. In: *Tropical Savannas*, Elsevier, Amsterdam, pp. 19–35.

Haq, B. U. (1978) Silicoflagellates and ebridians. In: *Introduction to Marine Micropaleontology* (eds B. U. Haq and A. Boersma), Elsevier, Amsterdam, pp. 267–75.

Harms, V. Z. and Leisman, G. A. (1961) The anatomy and morphology of certain Cordaites leaves. *J. Palaeontol.*, **35,** 1041–64.

Harris, T. M. (1931–1937) The fossil flora of Scoresby Sound, East Greenland, Parts 1–5. *Medd. om Grön-*

land, **82**(2), 1–102; **85**(3), 1–112; **85**(5), 1–133; **112**(1), 1–176; **112**(2), 1–114.

Harris, T. M. (1961) *The Yorkshire Jurassic flora. I. Thallophyta – Pteridophyta*, British Museum (Nat. Hist.), London.

Harris, T. M. (1964) *The Yorkshire Jurassic flora. II. Caytoniales, Cycadales & Pteridosperms*, Trustees of the British Museum (Nat. Hist.), London.

Harris, T. M. (1969a) Naming a fossil conifer. In: J. Sen Memorial Volume, Botanical Society Bengal, Calcutta, pp. 243–52.

Harris, T. M. (1969b) The Yorkshire Jurassic flora. III. Bennettitales. *Trust. Brit. Mus. (Nat. Hist.) Publ.* 675, 1–186.

Harris, T. M. (1973) Pollen from fossil cones. *Botanique*, **4**, 1–8.

Harris, T. M. (1974) *Williamsoniella lignieri*: its pollen and the compression of spherical pollen grains. *Palaeontology*, **17**, 125–48.

Harris, T. M. (1976) The Mesozoic gymnosperms. *Rev. Palaeobot. Palynol.*, **21**, 119–34.

Harris, T. M. (1979) The Yorkshire Jurassic flora. V. Coniferales. *Trust. Brit. Mus. (Nat. Hist.) Publ.* 803, 1–166.

Harris, T. M., Millington, W. and Miller, J. (1974) The Yorkshire Jurassic flora. IV. Ginkgoales. Czekanowskiales. *Trust. Brit. Mus. (Nat. Hist.) Publ.* 724, 1–150.

Havlena, V. (1971) Die zeitgleichen Floren des europäischen Oberkarbons und die mesophile Flora des Ostrau-Karwiner Steinkohlenreviers. *Rev. Palaeobot. Palynol.*, **12**, 245–70.

Hecht, M. K. (1983) Microevolution, developmental processes, paleontology and the origin of vertebrate higher categories. *Colloq. Int. CNRS*, **330**, 289–94.

Hennig, W. (1950) *Grundzüge einer Theorie der phylogenetischen Systematik*, Dtsch. Zentralverlag, Berlin.

Herak, M., Kochansky-Devidé, V. and Gušić, I. (1977) The development of the dasyclad algae through the ages. In: *Fossil Algae. Recent Results and Developments* (ed. E. Flügel), Springer, Berlin, pp. 143–53.

Herngreen, G. F. W. and Chlonova, A. F. (1981) Cretaceous microfloral provinces. *Pollen et Spores*, **23**, 441–555.

Hickey, L. J. (1973) Classification of the architecture of dicotyledonous leaves. *Amer. J. Bot.*, **60**, 17–33.

Hickey, L. J. and Doyle, J. A. (1977) Early Cretaceous fossil evidence for angiosperm evolution. *Bot. Rev.*, **43**, 3–104.

Hickey, L. (1984) Changes in the angiosperm flora across the Cretaceous–Tertiary boundary. In: *Catastroph in Earth history. The new uniformitarianism* (eds W. H. Berggren, and J. A. Van-Couvering), Princeton Univ. Press, Princeton, pp. 279–313.

Hill, C. R. and El-Khayal, A. A. (1983) Late Permian plants including Charophytes from the Khuff formation of Saudi Arabia. *Bull. Brit. Mus. (Nat. Hist.), Geol.*, **37**(3), 105–12.

Hirmer, M. (1927) *Handbuch der Paläobotanik*, Bd 1, Oldenbourg, Berlin.

Hluštík, A. (1974) Contribution to the systematic and leaf anatomy of the genus *Dammarites* Presl in Sternberg. *Sborn. Narodn. Muz. Praze*, **30B**, 49–70.

Hluštík, A. (1978) Frenelopsid plants (Pinopsida) from the Cretaceous of Czechoslovakia. *Paleontol. Conference '77 – Univ. Karlova, Praha*, pp. 129–41.

Høeg, O. A. (1942) The Downtonian and Devonian flora of Spitsbergen. *Norges Svalbard- og Ishavs-Unders. Skrift.*, **83**, 1–228.

Høeg, O. A. (1967) Psilophyta. In: *Traité de Paléobotanique* (ed. E. Boureau), T. 2, Masson et Cie, Paris, pp. 193–433.

Holmes, J. C. (1979) Further observations on the coenopterid fern genus, *Rhabdoxylon*. *Ann. Bot.*, **44**, 113–19.

Holmes, J. C. (1981) The Carboniferous fern *Psalixochlaena cylindrica* as found in Westphalian A coal balls from England. Part II. The frond and fertile parts. *Palaeontographica*, **B176**, 147–73.

Holmes, W. B. K. (1977) A pinnate leaf with reticulate venation from the Permian of New South Wales. *Proc. Linn Soc. NSW*, **102**, 52–7.

Hörich, O. (1906) *Lyginopteris oldhamia*. In: *Abbildungen u. Beschreibingen foss. Pflanzen-Reste* H. Potonié, Lief. IV, Nr. 69, 1–49.

Hsü Jen. (1983) Late Cretaceous and Cenozoic vegetation in China, emphasizing their connections with North America. *Ann. Mo. Bot., Gard.*, **70**, 279–307.

Huang Benhong. (1977) *Permian flora from the southeastern part of the Xiao Hinggan Lin, Northeastern China*, Geol. Publ. House, Peking (in Chinese).

Huang Benhong. (1983) On late Late Paleozoic palaeophytogeographic regions of eastern Tianshan–Hingan foldbelt and its geological significance. *Contr. Proj. Plate Tect. Northern China*, **1**, 138–55 (in Chinese).

Hueber, F. M. (1983) A new species of *Baragwanathia* from the Sextant Formation (Emsian) Northern Ontario, Canada. *Bot. J. Linn. Soc.*, **86**, 57–79.

Hughes, N. F. (1970) The need for agreed standards of recording in palaeopalynology and palaeobotany. *Paläontol. Abh.*, Abt. B, **3**, 357–64.

Hughes, N. F. (ed.) (1973) Organisms and continents through time. *Palaeontol. Assoc. Spec. Pap.* **12**, 1–334.

Hughes, N. F. (1976) *Palaeobiology of angiosperm origins. Problems of Mesozoic seed-plant evolution*, Cambridge University Press, New York.

Hughes, N. F. and Smart, J. (1967) Plant–insect relationships in Palaeozoic and later time. In: *The Fossil Record* (eds C. H. Harland *et al.*), Geological Society, London, pp. 107–17.

Huth, W. (1912) *Mariopteris muricata*. In: *Abbildungen u. Beschreibungen foss. Pflanzen-Reste* (H. Potonié), Lief. VIII, Nr. 143, 1–14.

Iljinskaya, I. A. (1962) On floral changes in Zaisan depression since the end of the Upper Cretaceous. *Doklady Akad. Nauk SSSR*, **146,** 1408–11 (in Russian).

Ilyina, V. I. (1985) Jurassic palynology of Siberia. *Trudy Inst. Geol. i Geofiz. SO AN SSSR*, **638,** 3–237 (in Russian).

Istchenko, T. A. and Istchenko, A. A. (1981) *Middle Devonian flora of the Voronezh anteclise*, Naukova dumka, Kiev (in Russian).

Istchenko, T. A. and Shlyakov, R. N. (1979) Marchantialean liverworts from the Middle Devonian of Podolia. *Paleontol. Zhurn.*, **3,** 114–25 (in Russian).

Iurina, A. L. (1969) Devonian flora of Central Kazakhstan. *Mater. po geol. Centr. Kazakhst.*, **8,** 3–207 (in Russian).

Iurina, A. L. and Lemoigne, Y. (1975) Anatomical characters of the axes of arborescent lepidophytes of the Devonian, referred to *Lepidodendropsis kazachstanica* Senkevitsch 1961. *Palaeontographica*, **B150,** 162–8.

Iurina, A. L. (1985) *Middle- and Late Devonian floras of North Eurasia*. Abstract of Doctorate Thesis, Moscow Univ., 39 pp. (in Russian).

Jennings, J. R. (1975) *Protostigmaria*, a new plant organ from the Lower Mississippian of Virginia. *Palaeontology*, **18,** 19–24.

Jennings, J. R. (1976) The morphology and relationships of *Rhodea*, *Telangium*, *Telangiopsis*, and *Heterangium*. *Amer. J. Bot.*, **63,** 1119–33.

Johnson, J. H. (1961) *Limestone-building algae and algal limestones*. Colorado School of Mines, Golden, Colo.

Josten, K.-H. (1962) *Neuropteris semireticulata*, eine neue Art als Bindglied zwischen den Gattungen *Neuropteris* and *Reticulopteris*. *Paläontol. Z.*, **36,** 33–45.

Jovet-Ast, S. (1967) Bryophyta. In: *Traité de Paléobotanique* (ed. E. Boureau), T. **2,** Masson et Cie, Paris, pp. 2–186.

Kalugina, N. S. (1974) Changes of subfamily composition of chironomids (Diptera, Chironomidae) as an indicator of possible eutrophication of water bodies in the end of the Mesozoic. *Bull. Mosk. Obshch. Ispyt. Prirody. Otd. Biol.*, **79**(6), 45–56 (in Russian).

Karczewska, J. (1969) Microsporangia and pollen of *Paracalathiops stachei* Remy, 1953 from Chełm I bore-hole, Eastern Poland. *Acta Palaeontol. Polon.*, **14,** 343–9.

Karrfalt, E. (1984) Further observations on *Nathorstiana* (Isoetaceae). *Amer. J. Bot.*, **71,** 1023–30.

Kasper, A. E. and Andrews, H. N. (1972) *Pertica*, a new genus of Devonian plants from Northern Maine. *Amer. J. Bot.*, **59,** 897–911.

Kaźmierczak, J. (1976a) Devonian and modern relatives of the Precambrian *Eosphaera*: possible significance for the early eukaryotes. *Lethaia* **9,** 39–50.

Kaźmierczak, J. (1976b) Oldest organic remains of boring algae from Polish Upper Silurian. *Nature*, **261,** 404–6.

Kemp, E. M., Balme, B. E., Helby, R. J. *et al.* (1977) Carboniferous and Permian palynostratigraphy in Australia and Antarctica: a review. BMR *J. Austral. Geol. Geophys.* 2(3), 177–208.

Kerp, J. H. F. (1981) On the morphology of the genus *Lilpopia* Conert et Schaarschmidt. *Cour. Forsch.-Inst. Senckenberg.*, **50,** 17–21.

Kerp, J. H. F. (1982) Aspects of Permian palaeobotany and palynology. II. On the presence of the ovuliferous organ *Autunia milleryensis* (Renault) Krasser (Peltaspermaceae) in the Lower Permian of the Nahe area (FRG) and its relationship to *Callipteris conferta* (Sternberg) Brongniart. *Acta Bot. Neerl.*, **31,** 417–27.

Khan, A. M. (1971) Palynology of Neogene sediments from Papua (New Guinea). Stratigraphic boundaries. *Pollen et Spores*, **16,** 265–84.

Kidwai, P. (1981) An illustrated glossary of technical terms used in stomatal studies. *Indian J. Forestry*, addit. ser. **1,** 1–35.

Kimura, T. (1980) The present status of the Mesozoic land floras of Japan. In: Professor Saburo Kanno Memorial Volume, pp. 379–413.

Kimura, T. (1984) Mesozoic floras of East and Southeast Asia, with a short note on the Cenozoic floras of Southeast Asia and China. *Geol.* a. *Palaeontol. Southeast Asia*, **25,** 325–50.

Kimura, T. and Sekido, S. (1975) *Nilssoniocladus* n. gen. (Nilssoniaceae n. fam.), newly found from the early Lower Cretaceous of Japan. *Palaeontographica*, **B153,** 111–18.

Kirichkova, A. I. and Samylina, V. A. (1979) On peculiarities of leaves of some Mesozoic ginkgos and czekanowskias. *Bot. Zhurn.*, **64,** 1529–38 (in Russian).

Kizilshtein, L. Ya. and Shpitzgluz, A. L. (1984) Under microscope are cellular structures of Palaeozoic plants. *Priroda*, **6,** 77–85 (in Russian).

Knobloch, E. (1972) Der Gattungsname *Sphenobaiera* Florin ist illegitim. *Taxon*, **21,** 545–6.

Kon'no, E. (1960) *Schizoneura manchuriensis* Kon'no and its fructification (*Manchurostachys* n. gen.) from the *Gigantopteris-nicotianaefolia* bearing formation in Penchihu coal-field, Northeastern China. *Sci.*

Rep. Tohoku Univ., 2nd ser. (geol.), spec. vol., No. 4, pp. 163–88.

Kornilova, V. S. (1963) *Main developmental phases of the Cenozoic floras in Kazakhstan*, Abstract of Thesis, Alma-Ata.

Kovach, W. L. and Dilcher, D. L. (1984) Dispersed cuticles from the Eocene of North America. *Bot. J. Linn. Soc.*, **88,** 63–104.

Kovács-Endrödy, E. (1977–1978) A re-evaluation of the venation structure of *Glossopteris*. *Ann. Geol. Surv.*, **12,** 107–41.

Krassilov, V. A. (1967a) Complex maceration – a far-reaching method of palaeobotanical studies. *Doklady Akad. Nauk SSSR*, **174,** 1191–4 (in Russian).

Krassilov, V. A. (1967b) *Lower Cretaceous Flora of South Primorie and its Significance for Stratigraphy*, Nauka, Moscow (in Russian).

Krassilov, V. A. (1968) On classification of stomatal apparatuses. *Paleontol. Zhurn.*, **1,** 102–9 (in Russian).

Krassilov, V. A. (1969) On reconstruction of extinct plants. *Paleontol. Zhurn.*, **1,** 3–12 (in Russian).

Krassilov, V. A. (1970) Leafy liverworts from the Jurassic of the Bureja basin. *Paleontol. Zhurn.*, **3,** 131–42 (in Russian).

Krassilov, V. A. (1972a) *Mesozoic Flora of Bureja*, Nauka, Moscow (in Russian).

Krassilov, V. A. (1972b) *Palaeoecology of Land Plants*, Vladivostok (in Russian).

Krassilov, V. A. (1972c) Phytogeographical classification of Mesozoic floras and their bearing on continental drift. *Nature*, **237,** 49–50.

Krassilov, V. A. (1973a) Mesozoic bryophytes from the Bureja Basin, Far East of the USSR. *Palaeontographica*, **B143,** 95–105.

Krassilov, V. A. (1973b) Mesozoic plants and the problem of angiosperm ancestry. *Lethaia*, **6,** 163–78.

Krassilov, V. A. (1975) *Paleoecology of Terrestrial Plants: Basic Principles and Techniques*, Wiley, New York and London.

Krassilov, V. A. (1976–1978) Bennettitalean stomata. *Palaeobotanist*, **25,** 179–84.

Krassilov, V. A. (1977a) *Evolution and Biostratigraphy*, Nauka, Moscow (in Russian).

Krassilov, V. A. (1977b) The origin of angiosperms. *Bot. Rev.*, **43,** 143–76.

Krassilov, V. A. (1978a) Electron microscopy of guard cells of stomata. *Paleontol. Zhurn.*, **3,** 128–30 (in Russian).

Krassilov, V. A. (1978b) Mesozoic lycopods and ferns from the Bureja basin. *Palaeontographica*, **B166,** 16–29.

Krassilov, V. A. (1979) *Cretaceous Flora of Sakhalin*, Nauka, Moscow (in Russian).

Krassilov, V. A. (1981) *Orestovia* and the origin of vascular plants. *Lethaia*, **14,** 235–50.

Krassilov, V. A. (1982) Early Cretaceous flora of Mongolia. *Palaeontographica*, **B181,** 1–43.

Krassilov, V. A. (1985) *Cretaceous period. Evolution of Earth crust and of biosphere*. Nauka, Moscow (in Russian).

Krassilov, V. A. and Rasnitsyn, A. P. (1982) Unique finding: pollen in the intestine of Early Cretaceous Xyelidae. *Paleontol. Zhurn.*, **4,** 83–96 (in Russian).

Krassilov, V. A. and Schuster, R. M. (1984) Paleozoic and Mesozoic fossils. In: *New Manual of Bryology.*, Vol. 2, Hattori Bot. Lab., Nichinan, pp. 1172–1193.

Kräusel, R. (1959) Die Keuperflora von Neuewelt bei Basel. *Schweiz. Palaeontol Abh.*, **77,** 5–19.

Kräusel, R., Maithy, P. K. and Maheshwari, H. K. (1961–1962) Gymnospermous woods with primary structures from Gondwana rocks – a review. *Palaeobotanist*, **10,** 97–107.

Kräusel R. and Weyland, H. (1926) Beiträge zur Kenntnis der Devonflora. II. *Abh. Senckenb. Naturforsch. Ges.*, **40,** 115–55.

Kremp, G. O. W. (1965) *Morphologic encyclopedia of palynology*, University Arizona Press, Tucson.

Krenke, N. P. (1933–1935) Somatische Indikatoren and Faktoren der Formbildung. In: *Phänogenetische Variabilität*, Bd. 1. Abh. Abt. Phytomorphogenese, Timiriaseff Biol. Inst., Moscow, pp. 11–415 (in Russian, German summary).

Krylov, I. N. (1975) Stromatolites of the Riphaean and Phanerozoic of the USSR. *Trudy Geol. Inst. Akad. Nauk SSSR*, **274,** 3–243 (in Russian).

Kryshtofovich, A. N. (1928) Greenland Tertiary flora in the North Urals, and botanico-geographical provinces of the Tertiary period. *Priroda*, **5,** 499–502 (in Russian).

Kryshtofovich, A. N. (1955) Development of botanico-geographical areas of the Northern Hemisphere since the beginning of the Tertiary period. *Voprosy Geologii Azii*, **2,** 825–44.

Kryshtofovich, A. N. (1957) *Palaeobotany*, Gostoptekhizdat, Leningrad (in Russian).

Kryshtofovich, A. N. (1958) Origin of flora of Angaraland. *Materialy po istorii flory i rastit. SSSR*, **3,** 7–14 (in Russian).

Kupriyanova, L. A. (1969) On evolutionary levels in the morphology of pollen and spores. *Bot. Zhurn.*, **54,** 1502–12 (in Russian).

Kupriyanova, L. A. (1981) Palynological data on the family Chlorantaceae, its relationships and history of dispersal. *Bot. Zhurn.*, **66,** 3–15 (in Russian).

Lacey, W. S. (1969) Fossil bryophytes. *Biol. Revs. Cambridge Philos. Soc.*, **44,** 189–205.

Lacey, W. S., van Dijk, D. E. and Gordon-Gray, K. D.

(1975) Fossil plants from the Upper Permian in the Mooi River district of Natal, South Africa. *Ann. Natal Mus.*, **22,** 349–420.

Lakhanpal, R. N. (1970) Tertiary floras of India and their bearing on the historical geology of the region. *Taxon*, **19,** 675–94.

Lange, R. T. (1978) Southern Australian Tertiary epiphyllous fungi, modern equivalents in the Australasian region, and habitat indicator value. *Can. J. Bot.*, **56,** 532–41.

Laubenfels, D. J. de. (1953) The external morphology of coniferous leaves. *Phytomorphology*, **3,** 1–20.

Laveine, J.-P. (1967) Les Neuroptéridées du Nord de la France. *Étud. Géol. Atl. Top. Sout., HBNPC*, 1. Flore fossile, fasc. 5, 1–344.

Laveine, J.-P. (1971–1972) Sporomorphes *in situ* de quelques Parispermées (Neuroptéridées) du Carbonifère. *Ann. Soc. Géol. Nord*, **91,** 155–73.

Laveine, J.-P., Coquel, R. and Loboziak, S. (1977) Phylogénie générale des Callipteridacées (Pteridospermopsida). *Géobios*, **10,** 757–847.

Leary, R. L. (1978) Fossils from 'Forgottonia'. A Geological View, *The Illinois State Mus.*, 8–9.

Leary, R. L. (1981) Early Pennsylvanian geology and paleobotany of the Rock Island County, Illinois, Area. Part 1: Geology. *Ill. State Mus. Repts Investigat.*, **37,** 1–88.

Leavitt, R. G. (1909) A vegetative mutant, and the principle of homoeosis in plants. *Bot. Gaz.*, **47,** 30–68.

Leclercq, S. (1936) À propos du *Sphenophyllum fertile* Scott. *Ann. Soc. Géol. Belg.*, **40,** B170–2.

Leclercq, S. (1957) Étude d'une fructification de Sphenopside à structure conservée du Dévonien supérieur. *Mém. Acad. Roy. Belg. Cl. Sci., Coll. in-4°*, Sér. 2, **14**(3), 3–39.

Leclercq, S. and Andrews, H. N. (1960) *Calamophyton bicephalum*, a new species from the Middle Devonian of Belgium. *Ann. Mo. Bot. Gard.*, **47,** 1–23.

Leclercq, S. and Banks, H. P. (1962) *Pseudosporochnus nodosus* sp. nov., a Middle Devonian plant with cladoxylalean affinities. *Palaeontographica*, **B110,** 1–34.

Leclercq, S. and Bonamo, P. M. (1971) A study of the fructification of *Milleria* (*Protopteridium*) *thomsonii* Lang from the Middle Devonian of Belgium. *Palaeontographica*, **B136,** 83–114.

Leclercq, S. and Lele, K. M. (1968) Further investigation on the vascular system of *Pseudosporochnus nodosus* Leclercq et Banks. *Palaeontographica*, **B123,** 97–112.

Leisman, G. A. (1962) A *Spencerites* sporangium and associated spores from Kansas. *Micropaleontology*, **8,** 396–402.

Leisman, G. A. (1964) *Mesidiophyton paulus* gen. et sp. nov., a new herbaceous sphenophyll. *Palaeontographica*, **B114,** 135–46.

Leisman, G. A. and Graves C. (1964) The structure of the fossil sphenopsid cone, *Peltastrobus reedae*. *Amer. Midl. Natur.*, **72,** 426–37.

Leisman, G. A. and Stidd, B. M. (1967) Further occurrences of *Spencerites* from the Middle Pennsylvanian of Kansas and Illinois. *Amer. J. Bot.*, **54,** 316–23.

Lele, K. M. (1974) Palaeozoic monosaccate miospores. In: *Aspects and Appraisal of Indian Palaeobotany* (eds K. R. Surange, R. N. Lakhanpal and D. C. Bharadwaj), B. Sahni Inst. Palaeobot., Lucknow, pp. 232–52.

Lele, K. M. (1974–1976) Late Palaeozoic and Triassic floras of India and their relation to the floras of Northern and Southern Hemispheres. *Palaeobotanist*, **23,** 89–115.

Lele, K. M. and Walton, J. (1961) Contributions to the knowledge of '*Zosterophyllum myretonianum*' Penhallow from the Lower Old Red Sandstone of Angus. *Trans. Roy. Soc. Edinburgh*, **44,** 469–475.

Lemoigne, Y. (1963) Les appendices radiculaires des *Stigmaria* des lycopodiales arborescentes du Paléozoïque. *Ann. Sci. Natur. Bot. et Biol. Végét.*, Sér. 12, **4,** 751–73.

Lemoigne, Y. (1970) Nouvelles diagnoses du genre *Rhynia* et de l'espèce *Rhynia gwynne-vaughanii*. *Bull. Soc. Bot. France*, **117,** 307–20.

Lemoigne, Y. (1981) Flore mixte au Permien supérieur en Arabie Saoudite. *Géobios*, **14,** 611–35.

Lemoigne, Y., Iurina, A. and Snigirevskaya, N. (1983) Révision du genre *Callixylon* Zalessky 1911 (*Archaeopteris*) du Dévonien. *Palaeontographica*, **B186,** 81–120.

Lemoigne, Y. and Zdebska, D. (1980) Structures problematiques observées dans axes provenant du Chert Dévonien de Rhynie. *Acta Palaeobot.*, **21,** 3–7.

Lepekhina, V. G. (1972) Woods of Palaeozoic pycnoxylic gymnosperms with special reference to North Eurasia representatives. *Palaeontographica*, **B138,** 44–106.

Lepekhina, V. G. and Oleinikov, A. N. (1978) On classification of Palaeozoic woods of gymnosperms. *Trudy Vsesoyuzn. Nauch.-Issled. Geol. Inst.*, **289,** 106–17 (in Russian).

Lepekhina, V. G. and Yatsenko-Khmelevsky, A. A. (1966) Classification and nomenclature of woods of Palaeozoic pycnoxylic plants. *Taxon*, **15,** 66–70.

Leroy, J.-F. (1983) The origin of angiosperms: an unrecognized ancestral dicotyledon, *Hedyosmum* (Chloranthales), with a strobiloid flower is living today. *Taxon*, **32,** 169–75.

Li Cheng-Sen. (1982) *Hsüa robusta* – a new land plant

from the Lower Devonian of Yunnan, China. *Acta Phytotaxon. Sin.*, **20**(3), 331–42.

Li Xing-xue and Cai Chong-yang (1978) Devonian floras of China. Paper for the Int. Symp. on the Devonian System, Nanking, pp. 1–8.

Li Xing-xue and Yao Zhao-qi (1979) Carboniferous and Permian floral provinces in East Asia. Paper for the *9th Int. Congr. of Carboniferous Stratigr. and Geol., Nanjing*, Nanjing Inst. Geol. Palaeontol., pp. 1–11.

Li Xing-xue and Yao Zhao-qi (1980) An outline of recent researches on the Cathaysia flora in Asia. Paper for the *First Conf. Int. Org. Palaeobotany, Nanjing*, Inst. Geol. Palaeontol., pp. 1–15.

Li Xing-xue and Yao Zhao-qi (1982) A review of recent research on the Cathaysia flora in Asia. *Amer. J. Bot.*, **69**, 479–86.

Li Xing-xue and Yao Zhao-qi (1983) Fructifications of gigantopterids from South China. *Palaeontographica*, **B185**, 11–26.

Li Xing-xue, Yao Zhao-qi and Deng Longhua (1982a) *An Early Late Permian Flora from Toba, Qamdo District, Eastern Xizang. Palaeontology of Xizang.* Book 5, Sci. Press, Beijing, pp. 17–40.

Li Xing-xue, Yao Zhao-qi, Zhu Jianan *et al.* (1982b) *Late Permian Plants from Northern Xizang. Palaeontology of Xizang*, Book 5, Sci, Press, Beijing, pp. 1–16.

Lignier, O. (1908) Essai sur l'évolution morphologique du regne végétal. *Bull. Soc. Linn. Norm.*, 6-e sér., **3**, 34–62.

Lipps, J. H. (1970) Ecology and evolution of silicoflagellates, *Proc. N. Amer. Paleontol. Conv., Chicago, 1969*, Part G, Lawrence, Kansas, pp. 965–93.

Lister, A. (1984) Evolutionary case histories from the fossil record. *Nature*, **309**, 114–15.

Long, A. G. (1969) *Eurystoma trigona* sp. nov., a pteridosperm ovule borne on a frond of *Alcicornopteris* Kidston. *Trans. Roy. Soc. Edinburgh*, **68**, 171–82.

Long, A. G. (1977a) Observations on Carboniferous seeds of *Mitrospermum*, *Conostoma* and *Lagenostoma*. *Trans. Roy. Soc. Edinburgh*, **70**, 37–61.

Long, A. G. (1977b) Some Lower Carboniferous pteridosperm cupules bearing ovules and microsporangia. *Trans. Roy. Soc. Edinburgh*, **70**, 1–11.

Long, A. G. (1979) Observations on the Lower Carboniferous genus *Pitus* Witham. *Trans. Roy. Soc. Edinburgh*, **70**, 111–27.

Lugardon, B. (1972) La structure fine l'exospore et de la périspore des filicinées isosporées. I. Généralités. Eusporangiées et Osmundales. *Pollen et Spores*, **14**, 227–61.

Lugardon, B. (1976) Sur la structure fine de l'exospore dans les divers groupes de ptéridophytes actuelles (microspores et isospores). In: *The Evolutionary Significance of the Exine* (eds I. K. Ferguson and J. Muller), Academic Press, London and New York, pp. 231–50.

Mägdefrau, K. (1968) *Paläobiologie der Pflanzen*, G. Fischer, Stuttgart.

Maheshwari, H. K. (1972) Permian wood from Antarctica and revision of some Lower Gondwana wood taxa. *Palaeontographica*, **B138**, 1–43.

Maheshwari, H. K. (1976) Floristics of the Permian and Triassic Gondwanas of India. *Palaeobotanist*, **23**, 145–60.

Maheshwari, H. K. and Meyen, S. V. (1975) *Cladostrobus* and the systematics of cordaitalean leaves. *Lethaia*, **8**, 103–23.

Makulbekov, N. M. (1982) Development of the Eocene floras of North-East Kazakhstan. *Paleontol. Zhurn.*, **2**, 151–3 (in Russian).

Mamay, S. H. (1957) *Biscalitheca*, a new genus of Pennsylvanian coenopterids, based on its fructification. *Amer. J. Bot.*, **44**, 229–39.

Mamay, S. H. (1966) *Tinsleya*, a new genus of seed-bearing callipterid plants from the Permian of North-Central Texas. *Geol. Surv. Profess. Pap.* 523E, 1–23.

Mamay, S. H. (1976) Paleozoic origin of the cycads. *Geol. Surv. Profess. Pap.* 934, 1–48.

Mamet, B. and Roux, A. (1977) Algues rouges dévoniennes et carbonifères de la Téthys occidentale. *Rev. Micropaléontol.*, **19**, 215–66.

Manchester, S. R. and Crane, P. R. (1983) Attached leaves, inflorescences, and fruits of *Fagopsis*, an extinct genus of fagaceous affinity from the Oligocene Florissant flora of Colorado, USA. *Amer. J. Bot.*, **70**, 1147–64.

Mapes, G. and Rothwell, G. W. (1980) *Quaestora amplecta* gen. et sp. n., a structurally simple medullosan stem from the Upper Mississippian of Arkansas. *Amer. J. Bot.*, **67**, 636–47.

Mapes, G. and Rothwell G. W. (1984) Permineralized ovulate cones of *Lebachia* from Late Palaeozoic limestone of Kansas. *Palaeontology*, **27**, 69–94.

Marguerier, J. (1970) Sur les différents types cellulaires ponctués chez les Équisétales fossiles. *92ᵉ Congr. Nat. Soc. Savantes, Strasbourg et Colmar, 1967, Paris*, T. 3, pp. 77–92.

Marguerier, J. (1977) Sur la ponctuation simple dans les champs de croisement de quelques structures paléozoïques et mésozoïques. *C. R. 102ᵉ Congr. Nation. Soc. Savantes, Limoges, 1977*, fasc. 1, 79–97.

Markova, L. G. (1964) Significance of morphological features in description and identification of fossil pollen and spores. In: *Sistematika i metody izucheniya iskopaemykh pyltsy i spor*, Nauka, Moscow, pp. 39–44 (in Russian).

Maslov, V. P. (1956) Fossil calcareous algae of the USSR. *Trudy Inst. Geol. Nauk Akad Nauk SSSR*, **160,** 1–301 (in Russian).

Maslov, V. P. (1960) Stromatolites (their genesis, methods of study, connections with facies and geological significance as exemplified by the Ordovocian of the Siberian platform). *Trudy Geol. Inst. Akad. Nauk SSSR*, **41,** 1–187 (in Russian).

Maslov, V. P. (1962) Fossil red algae of the USSR and their connection with facies. *Trudy Geol. Inst. Akad. Nauk SSSR*, **53,** 1–222 (in Russian).

Maslov, V. P. (1963) Introduction to the study of fossil characeous algae. *Trudy Geol. Inst. Akad. Nauk SSSR*, **82,** 1–104 (in Russian).

Maslov, V. P. (1973) (compilator and author). *Atlas of Rock-building Organisms (Calcareous and Siliceous Organisms)*, Nauka, Moscow (in Russian).

Matten, L. C. (1981) *Svalbardia banksii* sp. nov. from the Upper Devonian (Frasnian) of New York State. *Amer. J. Bot.*, **68,** 1383–92.

Matten, L. C. and Banks, H. P. (1966) *Triloboxylon ashlandicum* gen. and sp. n. from the Upper Devonian of New York. *Amer. J. Bot.*, **53,** 1020–8.

McArthur, R. H. (1972) *Geographical Ecology: Patterns in the Distribution of Species*, Harper and Row, New York.

McGregor, D. C. (1979) Spores in Devonian stratigraphical correlation. *Spec. Pap. Palaeontol.*, **23,** 163–84.

McQueen, D. R. (1969) Macroscopic plant remains in recent lake sediments. *Tuatara*, **17,** 13–19.

McQueen, D. R., Mildenhall, D. C. and Bell, C. J. E. (1968) Palaeobotanical evidence for changes in the Tertiary climates of New Zealand. *Tuatara*, **16,** 49–56.

Medvedeva, A. M. (1960) *Stratigraphic subdivision of lower horizons of the Tunguska Series by the method of the spore-pollen analysis*, Izdat. Akad. Nauk SSSR, Moscow (in Russian).

Medyanik, S. I. (1982) Fructification of the Lower Frasnian *Archaeopteris* from South Timan. *Paleontol. Zhurn.*, **2,** 121–7 (in Russian).

Meeuse, D. D. J. (1966) The homology concept in phytomorphology – some moot points. *Acta Bot. Neerl.*, **15,** 451–76.

Mehra, P. N. and Soni, S. L. (1983) Stomatal patterns in pteridophytes. An evolutionary approach. *Proc. Ind. Nat. Sci. Acad.*, **B49,** 155–203.

Menendez, C. A. (1969) Die fossilen Floren Südamerikas. In: *Biogeography and Ecology in South America*, The Hague, pp. 519–61.

Menendez, C. A. (1971) Floras Tertiarias de la Argentina. *Ameghiniana*, **8,** 357–71.

Meyen, S. V. (1964) On morphology, anatomy and nomenclature of the Angaro-Gondwana genus *Noeggerathiopsis*. In: *Gondwana* (ed. V. A. Vakhrameev), Nauka, Moscow (in Russian, English summary).

Meyen, S. V. (1965) On classification of dispersed cuticles. *Palaeontol. Zhurn.*, **4,** 75–87 (in Russian).

Meyen, S. V. (1966) Cordaiteans of the Upper Palaeozoic of North Eurasia (morphology, epidermal structure, systematics and stratigraphical significance). *Trudy Geol. Inst. Akad. Nauk SSSR*, **150,** 1–184 (in Russian).

Meyen, S. V. (1969a) *Comparative historical analysis of the Carboniferous and Permian floras of Eurasia*, Abstract of Thesis, Moscow (in Russian).

Meyen, S. V. (1969b) New data on relationship between Angara and Gondwana Late Palaeozoic floras. In: *Gondwana Stratigraphy IUGS Symp., Buenos Aires, 1967*, UNESCO, Paris, pp. 141–57.

Meyen, S. V. (1969c) New genera *Entsovia* and *Slivkovia* from the Permian deposits of the Russian platform and Fore-Urals. *Paleontol. Zhurn.*, **4,** 93–100 (in Russian).

Meyen, S. V. (1970) Epidermisuntersuchungen an permischen Landpflanzen des Angaragebietes. *Paläontol. Abh.*, B**3,** 523–52.

Meyen, S. V. (1971a) Parallelism and its significance for the systematics of fossil plants. *Geophytology*, **1,** 34–47.

Meyen, S. V. (1971b) *Phyllotheca*-like plants from the Upper Palaeozoic flora of Angaraland. *Palaeontographica*, B**133,** 1–33.

Meyen, S. V. (1973a) Plant morphology in its nomothetical aspects. *Bot. Rev.*, **39,** 205–60.

Meyen, S. V. (1973b) The Permian–Triassic boundary and its relation to the Palaeophyte–Mesophyte floral boundary. *Can. Soc. Petrol. Geol., Mem.*, **2,** 662–7.

Meyen, S. V. (1973c) Über die Hypothese der Kontinentaldrift unter dem Aspekt der Paläobotanik von Karbon und Perm. *Z. Geol. Wiss. Berlin*, **1,** 415–29.

Meyen, S. V. (1976) Carboniferous and Permian lepidophytes of Angaraland. *Palaeontographica*, **B157,** 112–57.

Meyen, S. V. (1976–1978) Permian conifers of the West Angaraland and new puzzles in the coniferalean phylogeny. *Palaeobotanist*, **25,** 298–313.

Meyen, S. V. (1978a) An attempt at a radical improvement of suprageneric taxonomy of fossil plants. *Phyta*, **1,** 76–86.

Meyen, S. V. (1978b) Main aspects of organism typology. *Zhurn. Obshch. Biol.*, **39,** 495–508 (in Russian, English summary).

Meyen, S. V. (1978c) Morphology of propteridophytes ('psilophytes'). *Bull. Mosk. Obshch. Ispyt. Prirody, Otd. Biol.*, **83**(2), 96–107 (in Russian, English summary).

Meyen, S. V. (1978d) Nomothetical plant morphology

and the nomothetical theory of evolution: the need for cross-pollination. *Acta Biotheoretica*, **27** (suppl.: Folia Biotheoretica 7), 21–36.

Meyen, S. V. (1978e) Systematics, phylogeny and ecology of propteridophytes. *Bull. Mosk. Obshch. Ispyt. Prirody, Otd. Biol.*, **83**(4), 72–84 (in Russian, English summary).

Meyen, S. V. (1979a) Ancestors of higher plants. *Priroda*, **11**, 40–9 (in Russian).

Meyen, S. V. (1979b) Permian predecessors of the Mesozoic pteridosperms in Western Angaraland, USSR. *Rev. Palaeobot. Palynol.*, **28**, 191–201.

Meyen, S. V. (1979c) The North American Permian flora – an Angara palaeobotanist's first impression. *IOP Newsletter*, **10**, 9–10.

Meyen, S. V. (1981) *Traces of Indian Herbs*, Mysl, Moscow (in Russian).

Meyen, S. V. (1982a) Gymnosperm fructifications and their evolution as evidenced by palaeobotany. *Zhurn. Obshch. Biol.*, **43**, 303–23 (in Russian, English summary).

Meyen, S. V. (1982b) The Carboniferous and Permian floras of Angaraland (a synthesis). *Biol. Mem.*, **7**, 1–109.

Meyen, S. V. (1984a) Basic features of gymnosperm systematics and phylogeny as shown by the fossil record. *Bot. Rev.*, **50**, 1–111.

Meyen, S. V. (1984b) Gamoheterotopy – a probable process in morphological evolution of higher plants. *IOP Newsletter*, **25**, 4–5.

Meyen, S. V. (1985a) *Sphenobaiera* is under fire again. *IOP Newsletter*, **26**, 6.

Meyen, S. V. (1985b) Three weeks in palaeobotanical laboratories in France. *IOP Newsletter*, **26**, 8–9.

Meyen, S. V. in press. Permian conifers of Western Angaraland. *Palaeontographica* B.

Meyen, S. V. and Gomankov, A. V. (1980) Peltaspermaceous pteridosperms of the genus *Tatarina*. *Paleontol. Zhurn.*, **2**, 116–32 (in Russian).

Meyen, S. V. and Menshikova, L. V. (1983) Systematics of the Upper Palaeozoic articulates of the family Tchernoviaceae. *Bot. Zhurn.*, **68**, 721–9 (in Russian, English summary).

Meyen, S. V. and Smoller, H. G., in press. The genus *Mostotchkia* Chachlov (Upper Palaeozoic of Angaraland) and its bearing on the characteristics of the order Dicranophyllales (Pinopsida). *Rev. Palaeobot. Palynol.*

Meyen, S. V. and Traverse, A. (1979) Remove 'Form-genus' too! *Taxon*, **28**, 595–8.

Meyer-Melikyan, N. R. and Raskatova, L. G. (1984) Exine structure of *Archaeoperisaccus* Naumova (results of electron microscopic study of the pollen sections). In: *Problemy sovremennoi palinologii*, Nauka, Novosibirsk, pp. 91–5 (in Russian, English summary).

Mickle, J. E. and Rothwell, G. W. (1982) Permineralized *Alethopteris* from the Upper Pennsylvanian of Ohio and Illinois. *J. Paleontol.*, **56**, 392–402.

Mildenhall, D. C. (1980) New Zealand Late Cretaceous and Cenozoic plant biogeography: a contribution. *Palaeogeogr. Palaeoclimatol. Palaeoecol.*, **31**, 197–233.

Millay, M. A. (1979) Studies of Paleozoic marattialeans: a monograph of the American species of *Scolecopteris*. *Palaeontographica*, **B169**, 1–69.

Millay, M. A. and Rothwell, G. W. (1983) Fertile pinnae of *Biscalitheca* (Zygopteridales) from the Upper Pennsylvanian of the Appalachian basin. *Bot. Gaz.*, **144**, 589–99.

Millay, M. A. and Taylor, T. N. (1974) Morphological studies of Paleozoic saccate pollen. *Palaeontographica*, **B147**, 75–99.

Millay, M. A. and Taylor, T. N. (1976) Evolutionary trends in fossil gymnosperm pollen. *Rev. Palaeobot. Palynol.*, **21**, 65–91.

Millay, M. A. and Taylor, T. N. (1978) Chytrid-like fossil of Pennsylvanian age. *Science*, **200**, 1147–9.

Millay, M. A. and Taylor, T. N. (1979) Paleozoic seed fern pollen organs. *Bot. Rev.*, **45**, 301–75.

Millay, M. A. and Taylor, T. N. (1982) The ultrastructure of Paleozoic fern spores: I. *Botryopteris*. *Amer. J. Bot.*, **69**, 1148–55.

Millay, M. A. and Taylor, T. N. (1984) The ultrastructure of Paleozoic fern spores: II. *Scolecopteris* (Marattiales). *Palaeontographica*, **B194**, 1–13.

Miller, C. N. (1982) Current status of Paleozoic and Mesozoic conifers. *Rev. Palaeobot. Palynol.*, **37**, 99–114.

Morgan, J. (1959) The morphology and anatomy of American species of the genus *Psaronius*. *Univ. Ill. Ill. Biol. Monogr.*, **27**, 1–108.

Mosbrugger, V. (1983–1984) Organische Zuzammengehörigkeit zweier Fossil-Taxa als taxonomische Problem am Beispiel der jungpaläozoischen Fernfruktifikationen *Scolecopteris* und *Acitheca*. *Rev. Palaeobot. Palynol.*, **40**, 191–206.

Muhammad, A. F. and Sattler, R. (1982) Vessel structure of *Gnetum* and the origin of angiosperms. *Amer. J. Bot.*, **69**, 1004–21.

Muller, J. (1970) Palynological evidence on early differentiation of angiosperms. *Biol. Revs Cambridge Philos. Soc.*, **45**, 417–50.

Muller, J. (1981) Fossil pollen record of extant angiosperms. *Bot. Rev.*, **47**, 1–142.

Mustafa, H. (1975) Beiträge zur Devonflora. I. *Argumenta Palaeobot.*, **4**, 101–33.

Muzhchinkin, V. F. (1978) Mammalian organism as a construction built of separate blocks (with special reference to the otarids). *Zhurn. Obshch. Biol.*, **39**, 777–82 (in Russian, English summary).

Namboodiri, K. K. and Beck, C. B. (1968) A comparative study of the primary vascular system of conifers. *Amer. J. Bot.*, **55**, 447–72.

Nathorst, A. G. (1902) Zur Oberdevonischen Flora der Bäreninsel. *Kgl. Sv. Vetensk.-Akad. Handl.*, Ser. 3, **36**(3), 1–60.

Naumova, S. N. (1937–1939) Spores and pollen of coals of the USSR. *Trans. 17th Int. Geol. Congr.*, **1**, Gosgeolizdat, Moscow, pp. 355–66, (in Russian).

Němejc, F. (1959) *Paleobotanika. I. Všeobecná část. Systematická cast. Bakterie – Sinice – Bičíkovci – Řasy – Houby*, Nakl. Česk. Akad. Věd., Praha.

Němejc, F. (1963) *Paleobotanika. II. Systematická část. Rostliny mechovité, Psilofytové, akaprădorosty*, Nakl. Česk. Akad. Věd., Praha.

Němejc, F. (1968) *Paleobotanika. III. Systematická část. Rostliny nahosemenné*. Akademia, Praha.

Němejc, F. (1975) *Paleobotanika. IV. Systematická část. Rostliny krytosemenné*, Akademia, Praha.

Neuber, E. (1979) *Parka decipiens* Fleming: Grünalge oder Lebermoos? *N. Jahrb. Geol. Paläontol. Monatsh.*, **11**, 681–9.

Neuburg, M. F. (1948) The Upper Palaeozoic flora of the Kuznetsk Basin. *Paleontol. SSSR*, **12**(3,2), 1–342 (in Russian).

Neuburg, M. F. (1960a) Leafy mosses from the Permian deposits of Angaraland. *Trudy Geol. Inst. Akad. Nauk SSSR*, **19**, 1–104 (in Russian).

Neuburg, M. F. (1960b) *Pleuromeia* Corda from the Lower Triassic deposits of the Russian platform. *Trudy Geol. Inst. Akad. Nauk SSSR*, **43**, 65–92 (in Russian).

Neuburg, M. F. (1960c) Permian flora of the Pechora basin. Part I. Lycopods and ginkgoaleans (Lycopodiales et Ginkgoales). *Trudy Geol. Inst. Akad. Nauk SSSR*, **43**, 3–64 (in Russian).

Neuburg, M. F. (1961) New data on morphology of *Pleuromeia* Corda from the Lower Triassic of the Russian platform. *Doklady Akad. Nauk SSSR*, **136**, 445–8 (in Russian).

Neuburg, M. F. (1964) Permian flora of the Pechora basin. Part II. Articulates (Sphenopsida). *Trudy Geol. Inst. Akad. Nauk SSSR*, **111**, 1–139 (in Russian).

Niklas, K. J. (1975–1976) Morphological and ontogenetic reconstructions of *Parka decipiens* Fleming and *Pachytheca* Hooker from the Lower Old Red Sandstone, Scotland. *Trans. Roy. Soc. Edinburgh*, **69**, 483–99.

Niklas, K. J. (1981) Simulated wind pollination and airflow around ovules of some early seed plants. *Science*, **211**, 275–7.

Niklas, K. J. (1982a) Chemical diversification and evolution of plants as inferred from paleobiochemical studies. In: *Biochemical Aspects of Evolutionary Biology* (ed. M. H. Nitecki), Chicago University Press, Chicago, pp. 29–91.

Niklas, K. J. (1982b) Computer simulations of early land plant branching morphologies: canalization of patterns during evolution? *Paleobiology*, **8**, 196–210.

Niklas, K. J. (1982c) Simulated and empiric wind pollination patterns of conifer ovulate cones (ecology, aerodynamics, wind tunnel, pollen, pine). *Proc. Nat. Acad. Sci. USA*, **79**, 510–14.

Niklas, K. J. (1983) The influence of Paleozoic ovule and cupule morphologies on wind pollination. *Evolution*, **37**, 968–86.

Niklas, K. J. and Brown, R. M. (1981) Some chemophysical factors attending fossilization. *BioScience*, **31**, 148–9.

Niklas, K. J. and Phillips, T. L. (1976) Morphology of *Protosalvinia* from the Upper Devonian of Ohio and Kentucky. *Amer. J. Bot.*, **63**, 9–29.

Niklas, K. J., Phillips, T. L. and Carozzi, A. V. (1976) Morphology and paleoecology of *Protosalvinia* from the Upper Devonian (Famennian) of the Middle Amazon Basin of Brazil. *Palaeontographica*, **B155**, 1–30.

Niklas, K. J. and Smocovitis, V. (1983) Evidence for a conducting strand in Early Silurian (Llandoverian) plants: implications for the evolution of the land plants. *Paleobiology*, **9**, 126–37.

Nitecki, M. H. and Toomey, D. F. (1979) Nature and classification of receptaculitids. *Bull. Cent. Rech. Explor.-Prod. Elf-Aquit.*, **3**, 725–32.

Nitecki, M. H., Zhuravleva, I. T., Myagkova, E. I. and Toomey, D. F. (1981) The similarity of *Soanites bimuralis* to archaeocyatids and receptaculitids. *Paleontol. Zhurn.*, **1**, 5–9 (in Russian).

Norris, G. (1978) Phylogeny and a revised supra-generic classification for Triassic–Quaternary organic-walled dinoflagellate cysts (Pyrrhophyta). Part I. Cyst terminology and assessment of previous classification. *N. Jahrb. Geol. Paläontol. Abh.* **155**, 300–17.

Novik, E. O. and Fissunenko, O. P. (1979) *On the problem of the Carboniferous phytogeography*, Preprint 79–1, Inst. Geol. Nauk Akad. Nauk UkrSSR, pp. 1–53 (in Russian).

Obrhel, J. (1959) Ein Landpflanzenfund im mittelböhmischen Ordovizium. *Geologie*, **8**, 535–41.

Oehler, J. H. (1977) Pyrenoid-like structures in Late Precambrian algae from the Bitter Springs Formation of Australia. *J. Paleontol.*, **51**, 885–901.

Oshurkova, M. V. (1978) Paleophytocoenogenesis as the basis of a detailed stratigraphy with special reference to the Carboniferous of the Karaganda basin. *Rev. Palaeobot. Palynol.*, **25**, 181–7.

Oshurkova, M. V. (1981) *Detailed subdivision of coal-bearing deposits by palaeophytological data. Recom-*

mended methods, Vsesoyuzn. Nauch.-Issled. Geol. Inst., Leningrad (in Russian).

Oshurkova, M. V. (1984) Palynological characteristics of the Kazakhstan province of the Euramerian palaeofloristic area of the Carboniferous. In: *Problemy sovremennoi palinologii*, Nauka, Novosibirsk, pp. 101–5 (in Russian, English summary).

Osnovy paleontologii (Manuals of palaeontology) (1963a) *Algae, bryophytes, psilophytes, lycopsids, articulates, ferns*. Izdat. Akad. Nauk SSR, Moscow (in Russian).

Osnovy paleontologi (1963b) *Gymnosperms and Angiosperms*. Gosgeoltekhizdat, Moscow (in Russian).

Ouyang Shu (1982) Upper Permian and Lower Triassic palynomorphs from Eastern Yunnan, China. *Can. J. Earth Sci.*, **19**, 68–80.

Palaeontology of Xizang, Book 5. *The Series of the Scientific Expedition to the Qinghai-Xizang Plaeau*, Sci. Press, Beijing (in Chinese).

Paleopalinologiya (Palaeopalynology) (1966) Vols I and II. *Trudy Vsesoyuzn. Nauch.-Issled. Geol. Inst.* 141 (in Russian).

Pant, D. D. (1958) The structure of some leaves and fructifications of the *Glossopteris* flora of Tanganyika. *Bull. Brit. Mus. (Nat. Hist.), Geol.*, **3**(4), 127–75.

Pant, D. D. (1965) On the ontogeny of stomata and other homologous structures. *Plant Sci. Ser. Allahabad* **1**, 1–24.

Pant, D. D. (1977) The plant of *Glossopteris*. *J. Ind. Bot. Soc.*, **56**, 1–23.

Pant, D. D. and Bhatnagar, S. (1971–1973) Intraspecific variation in Striatites spores. *Palaeobotanist*, **20**, 318–24.

Pant, D. D. and Gupta, K. L. (1968) Cuticular structure of some Indian Lower Gondwana species of *Glossopteris* Brongn. Part 1. *Palaeontographica*, **B124**, 45–81.

Pant, D. D. and Khare, P. K. (1974) *Damudopteris* gen. nov., a new genus of fern from the Lower Gondwanas of the Raniganj coalfield, India. *Proc. Roy. Soc. London*, Ser. B, **186**, 121–35.

Pant, D. D. and Kidwai, P. F. (1968) On the structure of stems and leaves of *Phyllotheca indica* Bunbury and its affinities. *Palaeontographica*, **B121**, 102–21.

Pant, D. D. and Kidwai, P. F. (1977) The origin and evolution of flowering plants. *J. Ind. Bot. Soc.*, **56**A, 242–74.

Pant, D. D. and Mehra, B. (1963a) On a cycadophyte leaf, *Pteronilssonia gopalii* gen. et sp. nov., from the Lower Gondwanas of India. *Palaeontographica* **B113**, 126–34.

Pant, D. D. and Mehra, B. (1963b) On the epidermal structure of *Sphenophyllum speciosum* (Royle) Zeiller. *Palaeontographica*, **B112**, 51–7.

Pant, D. D. and Nautiyal, D. D. (1967) On the structure of *Buriadia heterophylla* (Feistmantel) Seward & Sahni and its fructification. *Philos. Trans. Roy. Soc. London, Ser.* B., *Biol. Sci.*, **252** (774), 27–48.

Pant, D. D. and Verma, B. R. (1963) On the structure of leaves of *Rhabdotaenia* Pant from the Raniganj coalfield, India. *Palaeontology*, **6**, 301–14.

Peterson, L. N. (1983) Miospores of the Carboniferous of the southern part of the Tunguska Basin. In: *Materialy po geologii Sibiri* (ed. A. R. Ananiev), Tomsk University Press, Tomsk, pp. 110–20 (in Russian).

Petrosyan, N. M. (1968) Stratigraphic importance of the Devonian flora of the USSR. In: *International Symposium on the Devonian System* (ed. D. H. Oswald), Alberta Soc. Petrol. Geol., Calgary, pp. 579–86.

Pettitt, J. M. (1965) Two heterosporous plants from the Upper Devonian of North America. *Bull. Brit. Mus. (Nat. Hist.), Geol.*, **10**(3), 81–92.

Pettitt, J. M. (1966) Exine structures in some fossil and recent spores and pollen as revealed by light and electron microscopy. *Bull. Brit. Mus. (Nat. Hist.), Geol.* **13**(4), 221–57.

Pettitt, J. M. (1970) Heterospory and the origin of the seed habit. *Biol. Revs Cambridge Philos. Soc.*, **45**, 401–15.

Petukhov, S. V. (1981) *Biomechanics, Bionics and Symmetry*, Nauka, Moscow (in Russian).

Pfefferkorn, H. W. (1976) Pennsylvanian tree fern compression *Caulopteris*, *Megaphyton*, and *Artisophyton* gen. nov. in Illinois. *Circ. Ill. State Geol. Surv.*, **492**, 1–32.

Pfefferkorn, H. W. and Gillespie, W. H. (1980) Biostratigraphy and biogeography of plant compression fossils in the Pennsylvanian of North America. In: *Biostratigraphy of Fossil Plants. Successional and Palaeoecological Analyses* (eds D. L. Dilcher and T. N. Taylor), Dowden *et al.*, Stroudsburg, pp. 93–118.

Pfefferkorn, H. W. and Thomson, M. C. (1982) Changes in dominance patterns in Upper Carboniferous plant-fossil assemblages. *Geology*, **10**, 641–4.

Phillips, T. L. (1974) Evolution of vegetative morphology in coenopterid ferns. *Ann. Mo. Bot. Gard.*, **61**, 427–61.

Phillips, T. L. (1979) Reproduction of heterosporous arborescent lycopods in the Mississippian–Pennsylvanian of Euramerica. *Rev. Palaeobot. Palynol.*, **27**, 239–89.

Phillips, T. L. (1981) Stratigraphic occurrences and vegetational patterns of Pennsylvanian pteridosperms in Euramerican coal swamps. *Rev. Palaeobot. Palynol.*, **32**, 5–26.

Phillips, T. L., Andrews, H. N. and Gensel, P. G. (1972) Two heterosporous species of *Archaeopteris* from the Upper Devonian of West Virginia. *Palaeontographica*, **B139,** 47–71.

Phillips, T. L. and DiMichele, W. A. (1981) Paleoecology of Middle Pennsylvanian age coal swamps in Southern Illinois/Herrin coal member at Sahara mine No. 6. In: *Paleobotany, Paleoecology, and Evolution* (ed. K. J. Niklas), Vol. 1, Praeger, New York, pp. 231–85.

Phillips, T. L., Niklas, K. J. and Andrews, H. N. (1972) Morphology and vertical distribution of *Protosalvinia* (*Foerstia*) from the New Albany Shale (Upper Devonian). *Rev. Palaeobot. Palynol.*, **14,** 171–96.

Phillips, T. L. and Peppers, R. A. (1984) Changing patterns of Pennsylvanian coal-swamp vegetation and implications of climatic control on coal occurrence. *Int. J. Coal Geol.*, **3,** 205–55.

Phillips, T. L., Peppers, R. A. and DiMichele, W. A. (1985) Stratigraphic and interregional changes in Pennsylvanian coal-swamp vegetation: environmental inferences. *Int. J. Coal Geology*, **5,** 43–109.

Pianka, E. R. (1978) *Evolutionary ecology*, Hagers, New York.

Pigg, K. B. (1983) The morphology and reproductive biology of the sigillarian cone *Mazocarpon. Bot. Gaz.*, **144,** 600–13.

Pigg, K. B. and Rothwell, G. W. (1983a) *Chaloneria* gen. nov.; heterosporous lycophytes from the Pennsylvanian of North America. *Bot. Gaz.*, **144,** 132–47.

Pigg, K. B. and Rothwell, G. W. (1983b) Megagametophyte development in the Chaloneriaceae fam. nov., permineralized Paleozoic Isoetales (Lycopsida). *Bot. Gaz.*, **144,** 295–302.

Pirozynski, K. A. (1976) Fossil fungi. *Ann. Rev. Phytopathol.*, **14,** 237–46.

Pirozynski, K. A. and Weresub, K. L. (1979) The classification and nomenclature of fossil fungi. In: The whole fungus. The sexual–asexual synthesis (ed. B. Kendrick), *Proc. 2nd. Int. Mycol. Conf. Kananaskis, Canada*, Vol. 2, pp. 653–88.

Plaziat, J.-C., Koeninguer, J.-C. and Baltzer, F. (1983) Des mangroves actuelles aux mangroves anciennes. *Bull. Soc. Géol. France*, **25,** 499–504.

Plotnikov, V. V. (1979) *Evolution of Structure of Plant Communities*, Nauka, Moscow (in Russian).

Plumstead, E. P. (ed.) (1962) *Trans-Antarctic Expedition 1955–1958. Scientific Reports, No. 9, Geology, 2. Fossil Floras of Antarctica*, Trans-Antarctic Expedition Committee, London, pp. 1–132.

Poignant, A.-F. (1977) The Mesozoic red algae: a general survey. In: *Fossil Algae. Recent Results and Developments* (ed. E. Flügel), Springer, Berlin, pp. 177–89.

Ponomarenko, A. G. (1984) Evolution of ecosystems, main events. In: *Paleontologiya* (ed. A. Yu. Rozanov), *27th Int. Geol. Congr. Papers*, Vol. 2, Nauka, Moscow, pp. 71–4 (in Russian).

Pons, D. (1982) Découverte du Crétacé moyen sur le flanc est du Massif de Quetame, Colombie. *C. R. Acad. Sci. Paris*, **294,** 533–6.

Pons, D. and Broutin, J. (1978) Les organes reproducteurs de *Frenelopsis oligostomata* (Crétacé, Portugal). *103e Congr. Nat. Soc. Savantes, Nancy, 1978*, sci., fasc. 2, 139–59.

Popov, M. G. (1963) *Fundamentals of Florogenetics*, Nauka, Moscow (in Russian).

Popov, P. A. (1960) Remains of the Tertiary fungi Microthyriaceae Sacc. in the Fore-Enisei part of the West Siberian lowland. *Doklady Akad. Nauk SSSR*, **131,** 1152–5.

Potonié, H. (1903) *Odontopteris brardii* Brongniart. In: *Abbildungen u. Beschreibiungen foss. Pflanzen-Reste* (H. Potonié), Lief. I, Nr. 14, 1–2.

Potonié, H. (1904) *Linopteris neuropteroides* (Gutbier) Potonié. In: *Abbildungen u. Beschreibungen foss. Pflanzen-Reste* (H. Potonié), Lief. II, Nr. 28, 1–2.

Potonié, R. (1952) Gesichtspunkte zu einer paläobotanischen Gesellschaftsgeschichte (Soziogenese). *Beih. Geol. Jahrb.*, **5,** 1–116.

Potonié, R. (1956) Synopsis der Gattungen der Sporae dispersae. I. Teil: Sporites. *Beih. Geol. Jahrb.*, **23,** 1–103.

Potonié, R. (1958) Synopsis der Gattungen der Sporae dispersae. II. Teil: Sporites (Nachträge), Saccites, Aletes, Praecolpates, Polyplicates, Monocolpates. *Beih. Geol. Jahrb.*, **31,** 1–114.

Potonié, R. (1960) Synopsis der Gattungen der Sporae dispersae. III. Teil: Nachträge Sporites, Fortsetzung Pollenites mit Generalregister zu Teil I–III. *Beih. Geol. Jahrb.*, **39,** 1–189.

Potonié, R. (1962) Synopsis der Sporae in situ. *Beih. Geol. Jahrb.*, **52,** 1–204.

Potonié, R. (1966) Synopsis der Gattungen der Sporae dispersae. IV. Teil. Nachträge zu allen Gruppen (Turmae). *Beih. Geol. Jahrb.*, **72,** 1–274.

Potonié, R. (1970) Synopsis der Gattungen der Sporae dispersae. V. Teil. Nachträge zu allen Gruppen (Turmae). *Beih. Geol. Jahrb.*, **87,** 1–222.

Potonié, R. (1973) Phylogenetische Sporologie. Wandel der Sporengestalt der Höheren Pflanzen im Laufe der Erdgeschichte. *Fortschr. Geol. Rheinl.-Westf.*, **22,** 1–142.

Potonié, R. and Grebe, H. (1974) The history of spore classification, how to avoid errors of the past. In: *Symposium on Structure, Nomenclature and Classification of Pollen and Spores* (ed. D. C. Bharadwaj), B. Sahni Inst. Palaeobot. spec. publ. 4, pp. 1–6.

Potonié, R. and Kremp, G. O. W. (1954) Die Gattungen der Paläozoischen Sporae dispersae und ihre Stratigraphie. *Geol. Jahrb.*, **69,** 111–94.

Potonié, R. and Kremp, G. O. W. (1970) Synopsis der Gattungen der Sporae dispersae. VI. Teil. *Beih. Geol. Jahrb.*, **94**, 1–195.

Potonié, R. and Schweitzer, H.-J. (1960) Der Pollen von *Ullmannia frumentaria*. *Paläontol. Z.*, **34**, 27–39.

Pratt, L. M., Phillips, T. L. and Dennison, J. M. (1978) Evidence of non-vascular land plants from the Early Silurian (Llandoverian) of Virginia, USA. *Rev. Palaeobot. Palynol.*, **25**, 121–49.

Prigogine, I. (1980) From being to becoming: time and complexity in the physical sciences. W. H. Freeman and Co., San Francisco.

Radchenko, M. I. (1985) Atlas (identification key) of Carboniferous flora of Kazakhstan. Nauka, Alma-Ata (in Russian).

Radczenko, G. P. (1936) Some plant remains from the region of the Ostashkiny Gory in the Kuznetsk basin. *Mater. po Geol. Zap.-Sib. Kraya*, **1–3**(35), 1–24 (in Russian).

Radczenko, G. P. (1955) Guide forms of the Upper Palaeozoic flora of Sayany-Altai area. In: *Atlas rukovodyashchikh form iskopaemykh fauny i flory Zapadnoi Sibiri* (ed. L. L. Khalfin). Vol. 2, Gosgeoltekhizdat, Moscow, pp. 42–153 (in Russian).

Radczenko, G. P. and Shvedov, N. A. (1940) Upper Palaeozoic flora of the coal-bearing deposits of the western part of the Lower Tunguska River basin. *Trudy Arktich. Nauch.-Issled. Inst.*, **157**, 1–140 (in Russian).

Radionova, E. P. (1972) Microphytolites and similar structures in the Riphaean and Phanerozoic. *Stratigr. i Paleontol.*, Vol. 3. *Itogi Nauki i Tekhniki VINITI*, 74–93 (in Russian).

Radionova, E. P. (1976) Microphytolites and other problematic forms of the Palaeozoic of some regions of the Russian and Siberian platforms. *Trudy Geol. Inst. Akad. Nauk SSSR*, **294**, 86–164 (in Russian).

Ramanujam, C. G., Rothwell, G. W. and Stewart, W. N. (1974) Probable attachment of the *Dolerotheca* campanulum to a *Myeloxylon-Alethopteris* type frond. *Amer. J. Bot.*, **61**, 1057–66.

Rasmussen, H. (1981) Terminology and classification of stomata and stomatal development – a critical survey. *Bot. J. Linn. Soc.*, **83**, 199–212.

Raymond, A. and Miller, D. J. (1984) Shoot–root ratios in Pennsylvanian coal-ball peats. Abstr. of contributed papers and poster session, *2nd Int. Org. Paleobot. Conf. Edmonton.*

Rayner, R. J. (1983) New observations on *Sawdonia ornata* from Scotland. *Trans. Roy. Soc. Edinburgh, Earth Sci.*, **74**, 79–93.

Rayner, R. J. (1984) New finds of *Drepanophycus spinaeformis* Göppert from the Lower Devonian of Scotland. *Trans. Roy. Soc. Edinburgh, Earth Sci.*, **75**, 353–63.

Rayner, R. J. (1985) The Permian lycopod *Cyclodendron leslii* from South Africa. *Palaeontology*, **28**, 111–20.

Read, C. B. and Mamay, S. H. (1964) Upper Paleozoic floral zones and floral provinces of the United States. *Geol. Surv. Profess. Pap.* 454-K, 1–19.

Remane, A. (1956) *Die Grundlagen des natürlichen Systems, der vergleichende Anatomie und der Phylogenetik*, Geest and Portig K. G., Leipzig.

Remy, R. (1966–1968) Ein Beitrag zur Kenntnis von *Sphenophyllum myriophyllum* Crépin. *Argumenta Palaeobot.*, **1**, 125–30.

Remy, R. and Remy, W. (1960) *Eleutherophyllum drepanophyciforme* n. sp. aus Namur A von Niederschlesien. Senckenberg. *Lethaea*, **41**, 89–100.

Remy, R. and Remy, W. (1975) Zur Ontogenie der Sporangiophore von *Calamostachys spicata* var. *eimeri* n. var. und zur Aufstellung des genus *Schimperia* n. gen. *Argumenta Palaeobot.*, **4**, 83–92.

Remy, W. (1978a) Der Dehiszenzmechanismus der Sporangien von *Rhynia*. *Argumenta Palaeobot.*, **5**, 25–30.

Remy, W. (1978b) Die *Sphenopteris-germanica*-Gruppe in den 'Süplinger-Schichten' (Flechtinger Höhenzug) – ein Beleg für das Autun-Alter. *Argumenta Palaeobot.*, **5**, 161–5.

Remy, W. (1980) Wechselwirkung von Vegetation und Boden im Paläophytikum. In: *Festschrift für Gerhard Keller*, Wenner, Osnabrück, pp. 43–79.

Remy, W. (1982) Lower Devonian gametophytes: relation to the phylogeny of land plants. *Science*, **215**, 1625–7.

Remy, W. and Remy, R. (1975) *Sporangiostrobus puertollanensis* n. sp. und *Puertollania sporangiostrobifera* n. gen., n. sp., aus dem Stephan von Puertollano, Spanien. *Argumenta Palaeobot.*, **4**, 13–29.

Remy, W. and Remy, R. (1977) *Die Floren de Edraltertums. Einführung in Morphologie, Anatomie, Geobotanik und Biostratigraphie der Pflanzen des Paläophytikums*, Glückauf GmbH, Essen.

Remy, W. and Remy, R. (1978) *Calamitopsis* n. gen. und die Nomenclatur und Taxonomie von *Calamites* Brongniart, 1828. *Argumenta Palaeobot.*, **5**, 1–10.

Remy, W. and Remy, R. (1980) *Lyonophyton rhyniensis* nov. gen. et nov. spec., ein Gametophyt aus dem Chert von Rhynie (Unterdevon, Schotland). *Argumenta Palaeobot.*, **6**, 37–72.

Remy, W., Remy, R., Hass, H. *et al.* (1980) *Sciadophyton* Steinmann – ein Gametophyt aus dem Siegen. *Argumenta Palaeobot.*, **6**, 73–94.

Retallack, G. J. (1977) Reconstructing Triassic vegetation of Eastern Australasia: a new approach for the biostratigraphy of Gondwanaland. *Alcheringa*, **1**, 247–78.

Retallack, G. J. (1980) Late Carboniferous to Middle Triassic megafossil floras from the Sydney Basin.

In: *A guide to the Sydney Basin* (eds C. Herbert, and R. J. Helby), *Geol. Surv. NSW Bull.* 26, pp. 384–430.

Retallack, G. J. (1981) Fossil soils: indicators of ancient terrestrial environments. In: *Paleobotany, Paleoecology, and Evolution* (ed. K. J. Niklas), Vol. 1, Praeger, New York, pp. 55–102.

Retallack, G. J. (1983a) A paleopedological approach to the interpretation of terrestrial sedimentary rocks: the mid-Tertiary fossil soils of Badlands National Park, South Dakota. *Geol. Soc. Amer. Bull.*, **94**, 823–40.

Retallack, G. J. (1983b) Paleopedology comes down to Earth. *J. Geol. Educat.*, **31**, 390–2.

Retallack, G. J. (1985) Fossil soils as grounds for interpreting the advent of large plants and animals on land. *Phil. Trans. Roy. Soc. London*, **B309**, 105–42.

Retallack, G. J. and Dilcher, D. L. (1981a) A coastal hypothesis for the dispersal and rise to dominance of flowering plants. In: *Paleobotany, Paleoecology, and Evolution* (ed. K. J. Niklas), Vol. 2 Praeger, New York, pp. 27–77.

Retallack, G. J. and Dilcher, D. L. (1981b) Arguments for a glossopterid ancestry of angiosperms. *Paleobiology*, **7**, 54–67.

Retallack, G. J. and Dilcher, D. L. (1981c) Early angiosperm reproduction: *Prisca reynoldsii*, gen. et sp. nov. from mid-Cretaceous coastal deposits in Kansas, USA. *Palaeontographica*, **B179**, 103–37.

Rex, G. (1983) The compression state of preservation of Carboniferous lepidodendrid leaves. *Rev. Palaeobot. Palynol.*, **39**, 65–85.

Rex, G. (1985) A laboratory flume investigation of the formation of fossil stem fills *Sedimentology*, **32**, 245–55.

Rex, G. and Chaloner, W. G. (1983) The experimental formation of plant compression fossils. *Palaeonotology*, **26**, 231–52.

Reymanóvna, M. (1980) A new ginkgoalean plant from the Jurassic of Kraków, Poland. *Abstr. Int. Paleobot. Conf. Reading*, 47.

Reymanóvna, M. (1984) *Pseudotorellia*, a Mesozoic genus close to *Ginkgo*? Abstr. of contributed papers and poster session, *2nd Int. Org. Paleobot. Conf. Edmonton*.

Richardson, J. B. (1974) The stratigraphic utilization of some Silurian and Devonian miospore species in the Northern Hemisphere: an attempt at a synthesis. In: *Int. Sympos. on Belgian Micropal. limits from Emsian to Visean*, Brussels, pp. 1–13.

Richardson, J. B. (1984) Early evolution of leaves. *Nature*, **309**, 749–50.

Richardson, J. B. and Ioannides, N. (1973) Silurian palynomorphs from the Tanezzuff and Acacus Formations, Tripolitania, North Africa. *Micropaleontol.*, **19**, 257–307.

Richardson, J. B., Rasul, S. M. and Al-Ameri, T. (1981) Acritarchs, miospores and correlation of the Ludlovian–Downtonian and Silurian–Devonian boundaries. *Rev. Palaeobot. Palynol.*, **34**, 209–24.

Riding, R. (1977) Skeletal stromatolites. In: *Fossil Algae. Recent Results and Developments* (ed. E. Flügel), Springer, Berlin, pp. 57–60.

Rigby, J. F. (1970–1972) The Notocalamitaceae, a new family of Upper Palaeozoic equisetaleans. *Palaeobotanist*, **19**, 161–3.

Rigby, J. F. (1972) On *Arberia* White, and some related Lower Gondwana female fructifications. *Palaeontology*, **15**, 108–20.

Riggs, S. D. and Rothwell, G. W. (1985) *Sentistrobus goodii* n. gen and sp., a permineralized sphenophyllalean cone from the Upper Pennsylvanian of the Appalachian basin. *J. Paleontol.*, **59**, 1194–202.

Roselt, G. and Schneider, W. (1969) Cuticulae dispersae, ihre Merkmale, Nomenklatur und Klassification. *Paläontol. Abh.*, Abt. B, **3**, 1–128.

Rothwell, G. W. (1972) Evidence of pollen tubes in Paleozoic pteridosperms. *Science*, **175**, 772–4.

Rothwell, G. W. (1976) Primary vasculature and gymnosperm systematics. *Rev. Palaeobot. Palynol.*, **22**, 193–206.

Rothwell, G. W. (1980) The Callistophytaceae (Pteridospermopsida). II. Reproductive features. *Palaeontographica*, **B173**, 85–106.

Rothwell, G. W. (1981) The Callistophytales (Pteridospermopsida): reproductively sophisticated Palaeozoic gymnosperms. *Rev. Palaeobot. Palynol.*, **32**, 103–21.

Rothwell, G. W. (1982) New interpretation of the earliest conifers. *Rev. Palaeobot. Palynol.*, **37**, 7–28.

Rothwell, G. W. (1984) Apical structure and growth in *Stigmaria*; rhizomorphic rooting organ of Lepidodendrales. Abstr. of contributed papers and poster session, *2nd Int. Org. Paleobot. Conf. Edmonton*.

Rothwell, G. W. and Eggert, D. A. (1982) What is the vascular architecture of complex medullosan pollen organs? *Amer. J. Bot.*, **69**, 641–3.

Rothwell, G. W. and Erwin, D. M. (1984) Homologies and evolution of rhizomorphic lycophyte rooting organs. Abstr. of contributed papers and poster session, *2nd Int. Org. Paleobot. Conf. Edmonton*.

Rothwell, G. W. and Erwin D. M. (1985) The rhizomorph apex of *Paurodendron;* implications for homologies among the rooting organs of lycopsida. *Amer. J. Bot.*, **72**, 86–98.

Rothwell, G. W. and Taylor, T. N. (1972) Carboniferous pteridosperm studies: morphology and anatomy of *Schopfiastrum decussatum*. *Can. J. Bot.*, **50**, 2649–58.

Rothwell, G. W. and Warner, S. (1984) *Cordaixylon*

dumusum n. sp. (Cordaitales). I. Vegetative structures. *Bot. Gaz.*, **145,** 275–91.

Round, F. E. and Crawford, R. M. (1981) The lines of evolution of the Bacillariophyta. I. Origin. *Proc. Roy. Soc. London*, Ser. B, Biol. Sci., **211,** 237–60.

Sadovnikov, G. N. (1981) Regional stratigraphic subdivisions of the Upper Permian and Lower Triassic of the Siberian platform and adjoining regions. *Sovetsk. Geologiya*, **6,** 74–84 (in Russian).

Sadovnikov, G. N. (1983) New data on morphology and anatomy of the genus *Kirjamkenia*. *Paleontol. Zhurn.*, **4,** 76–81 (in Russian).

Sagan, E. (1980) Neuentdeckte morphologische Einzelheiten von *Eleutherophyllum mirabile* Stur, *E. waldenburgense* Zimmermann und *E. drepanophyciforme* R. et W. Remy. *Acta Palaeobot.*, **21,** 9–26.

Saksena, S. D. (1974) Palaeobotanical evidence for the Middle Gondwana. In: *Aspects and Appraisal of Indian Palaeobotany* (eds K. R. Surange, R. N. Lakhanpal and D. C. Bharadwaj), B. Sahni Inst. Palaeobot., Lucknow, pp. 427–446.

Salmenova, K. Z. (1978) Permian flora of North Pribalkhashie. *Paleontol. Zhurn.*, **4,** 122–7 (in Russian).

Salmenova, K. Z. (1979) Peculiarities of the Permian flora of South Kazakhstan and its links with neighbouring floras. *Paleontol. Zhurn.*, **4,** 119–27 (in Russian).

Salmenova, K. Z. (1982) On floristic links between southern and eastern parts of Angaraland in the Early Permian. *Paleontol. Zhurn.*, **1,** 3–9 (in Russian).

Salmenova, K. Z. (1984) Permian flora and questions of the phytogeographical zoning of Kazakhstan. *Izvest. Akad. Nauk KazSSR*, Ser. Geol., **1,** 14–20 (in Russian).

Samylina, V. A. (1968) Early Cretaceous angiosperms of the Soviet Union based on leaf and fruit remains. *J. Linn. Soc. (Bot.)*, **61**(384), 207–18.

Samylina, V. A. (1980) Significance of epidermal–cuticular studies of leaves for the knowledge of the Mesozoic gymnosperms. In: *Sistematica and evolyutsiya vysshikh rastenii* (ed. S. G. Zhilin), Nauka, Leningrad, pp. 31–42 (in Russian).

Sarjeant, W. A. S. (1974) *Fossil and Living Dinoflagellates*. Academic Press, London and New York.

Sattler, R. (1975) Organverschiebungen und Heterotopien bei Blütenpflanzen. *Bot. Jahrb. Syst. Pflanzengesch. u. Pflanzengeogr.*, **95,** 256–66.

Sattler, R. (1978) What is theoretical plant morphology? *Acta Biotheoretcia*, **27** (suppl.: Folia Biotheoretica 7), 5–20.

Sattler, R. (1984) Homology – a continuing challenge. *Syst. Bot.*, **9,** 382–94.

Savitskaya, L. I. and Iskandarkhodzhaev, T. A. (1984) Description of plants and some materials to the history of the vegetation development in the Late Palaeozoic of Middle Asia. In: *Stratigrafiya kamennougolnykh i permskikh kontinentalnykh otlozhenii Vostochnogo Uzbekistana i prilegayushchikh territorii* (ed. V. A. Arapov), FAN, Tashkent, pp. 68–80.

Scheckler, S. E. (1974) Systematic characters of Devonian ferns. *Ann. Mo. Bot. Gard.*, **61,** 462–73.

Scheckler, S. E. (1975) *Rhymokalon*, a new plant with cladoxylalean anatomy from the Upper Devonian of New York State. *Can. J. Bot.*, **53,** 25–38.

Scheckler, S. E. (1976) Ontogeny of progymnosperms. l. Shoots of Upper Devonian Aneurophytales. *Can. J. Bot.*, **54,** 202–19.

Scheckler, S. E. (1982a) Anatomy of the fertile parts of *Tetraxylopteris schmidtii* (Progymnospermopsida). *Abstr. Bot. Soc. Amer. Palaeobot. Sect.*, 64.

Scheckler, S. E. (1982b) Effects of paleotopography on the Late Devonian flora and the origin of coal swamps in Southern Laurussia. *J. Paleontol.*, **56,** Part II, suppl. 2, 29.

Scheckler, S. E. and Beeler, H. E. (1984) Early Carboniferous coal swamp floras from Eastern USA. (Virginia). Abstr. of contributed papers and poster session, *2nd Int. Org. Paleobot. Conf. Edmonton*.

Scheuring, B. W. (1974) 'Protosaccate' Strukturen, ein weitverbreitetes Pollenmerkmal zur frühen und mittleren Gymnospermenzeit. *Geol. Paläontol. Mitt. Insbruck*, **4**(2), 1–30.

Schmalhausen, I. I. (1968) *Factors of Evolution. Theory of Stabilizing Selection*, Nauka, Moscow (in Russian).

Schmid, E. (1963) Die Erfassung der Vegetationseinheiten mit floristischen und epimorphologischen Analysen. *Ber. Schweiz. Bot. Ges.*, **73,** 276–324.

Schmid, R. (1982–1983) The terminology and classification of steles: historical perspective and the outlines of a system. *Bot. Rev.*, **48,** 817–931.

Schmithüsen, J. (1961) *Allgemeine Vegetationsgeographie*, Berlin.

Schopf, J. M. (1975) Modes of fossil preservation. *Rev. Palaeobot. Palynol.*, **20,** 27–53.

Schopf, J. M. (1976) Morphologic interpretation of fertile structures in glosopterid gymnosperms. *Rev. Palaeobot. Palynol.*, **21,** 25–64.

Schopf, J. M. (1978) *Foerstia* and recent interpretations of early, vascular land plants. *Lethaia*, **11,** 139–43.

Schopf, J. M. (1982) Forms and facies of *Vertebraria* in relation to Gondwana coal. Geology of the Central Transantarctic Mountains. *Antarct. Res. Series*, **36**(3), 37–62.

Schopf, J. M., Wilson, L. R. and Bentall, R. (1944) An annotated synopsis of Paleozoic fossil spores and the definition of generic groups. *Ill. Geol. Surv. Rep. Invest.*, **91,** 1–93.

Schopf, J. W. (1968) Microflora of the Bitter Springs Formation, Late Precambrian, Central Australia. *J. Paleontol.*, **42,** 651–88.

Schopf, J. W. (ed.) (1983) *Earth's Earliest Biosphere. Its Origin and Evolution*, Princeton University Press, Princeton.

Schopf, T. J. M. (1983) Summary of a critical assessment of punctuated equilibria. *Colloq. Int. CNRS*, **330,** 51–4.

Schweitzer, H.-J. (1960) Die Makroflora des niederrheinischen Zechsteins. *Fortschr. Geol. Rheinl. Westfal.*, **6,** 1–46.

Schweitzer, H.-J. (1963) Der weibliche Zapfen von *Pseudovoltzia liebeana* und seine Bedeutung für die Phylogenie der Koniferen. *Palaeontographica*, **B113,** 1–29.

Schweitzer, H.-J. (1973) Die Mitteldevon-Flora von Lindlar (Rheinland). 4. Filicinae – *Calamophyton primaevum* Kräusel & Weyland. *Palaeontographica*, **B140,** 117–50.

Schweitzer, H.-J. (1977) Die Räto-Jurassischen Floren des Iran und Afghanistans. 4. Die Rätische Zwitterblütte *Irania hermaphroditica* nov. spec. und ihre Bedeutung für die Phylogenie der Angiospermen. *Palaeontographica*, **B161,** 98–145.

Schweitzer, H.-J. (1983a) Der Generationswechsel der Psilophyten. *Ber. Dtsch. Bot. Ges.*, **96, 483-96.**

Schweitzer, H.-J. (1983b) Die Unterdevonflora des Rheinlandes. I. Teil. *Palaeontographica*, **B189,** 1–138.

Schweitzer, H.-J. and Giesen, P. (1980) Über *Taeniophyton inopinatum*, *Protolycopodites devonicus* und *Cladoxylon scoparium* aus dem Mitteldevon von Wuppertal. *Palaeontographica*, **B173,** 1–25.

Schweitzer, H.-J. and Matten, L. C. (1982) *Aneurophyton germanicum* and *Protopteridium thomsonii* from the Middle Devonian of Germany. *Palaeontographica*, **B184,** 65–106.

Scott, A. C. (1977) A review of the ecology of Upper Carboniferous plant assemblages, with new data from Strathclyde. *Palaeontology*, **20,** 447–73.

Scott, A. C. (1978) Sedimentological and ecological control of Westphalian B plant assemblages from West Yorkshire. *Proc. Yorksh. Geol. Soc.*, **41,** part 4(33), 461–508.

Scott, A. C. (1979) The ecology of Coal Measure floras from Northern Britain. *Proc. Geol. Assoc.*, **90,** 97–116.

Scott, A. C. and Chaloner, W. G. (1983) The earliest fossil conifer from the Westphalian B of Yorkshire. *Proc. Roy. Soc. London*, Ser. B, **220,** 163–82.

Scott, A. C., Chaloner, W. G. and Paterson, S. (1985) Evidence of pteridophyte – arthropod interactions in the fossil record. *Proc. Roy. Soc. Edinburgh*, **86B,** 133–40.

Scott, A. C. and Collinson, M. E. (1983) Investigating fossil plant beds. *Geol. Teaching*, **7**(4), 114–22.

Scott, A. C. and Galtier, J. (1985) Distribution and ecology of early ferns. *Proc. Roy. Soc. Edinburgh*, Sect. B, **86,** 141–9.

Scott, A. C., Galtier, J. and Clayton, G. (1984) Distribution of anatomically-preserved floras in the Lower Carboniferous in Western Europe. *Trans. Roy. Soc. Edinburgh, Earth Sci.*, **75,** 311–40.

Scott, A. C. and Rex, G. (1985) The formation and significance of Carboniferous coal balls. *Phil. Trans. Roy. Soc. London,* B **311,** 123–37.

Scott, A. C. and Taylor, T. N. (1983) Plant/animal interactions during the Upper Carboniferous. *Bot. Rev.*, **49,** 259–307.

Scott, D. H. (1920–1923) *Studies in Fossil Botany* (2 vols), Black, London.

Senkevich, M. A. (1984) Devonian ligulate lepidophytes. *Paleontol. Zhurn.*, **3,** 112–19.

Serebryakova, T. I. (1972) Studies in life forms in modern time. Botanika, Vol. 1, *Itogi Nauki i Tekhniki VINITI*, pp. 84–169 (in Russian).

Serlin, B. S. and Banks, H. P. (1978) Morphology and anatomy of *Aneurophyton*, a progymnosperm from the Late Devonian of New York. *Palaeontographica Americana*, **8**(51), 341–59.

Seward, A. C. (1898–1919) *Fossil Plants* (4 vols) Cambridge University Press, Cambridge.

Sewertzoff, A. N. (1939) *Morphological Regularities of Evolution*, Izdat. Akad. Nauk SSSR, Moscow and Leningrad (in Russian).

Seybold, A. (1927) Untersuchungen über die Formgestaltung der Blätter der Angiospermen. I. *Bibl. genetica* 12.

Shaw, G. (1970) *Sporopollenin in Phytochemical Phylogeny*, Academic Press, London.

Shchegolev, A. K. (1965) Flora at the Carboniferous–Permian boundary in the Donets basin. In: *Geology of Coal Measures and Stratigraphy of the Carboniferous in the USSR* (ed. I. I. Gorsky), Nauka, Moscow, pp. 234–44. (in Russian, English summary).

Shchegolev, A. K. (1979) *The Upper Carboniferous of North Caucasus in the Zelenchuk-Teberda Interfluve*, Naukova Dumka, Kiev (in Russian).

Shishkin, M. A. (1981) The regularities of evolution of ontogeny. *Zhurn. Obshch. Biol.* **42,** 38–54 (in Russian, English summary).

Shishkin, M. A. (1984) Ontogeny and natural selection. *Ontogenez*, **15**(2), 115–36 (in Russian).

Shumenko, S. I. (1968) Some aspects of the ontogenesis, variation and systematics of fossil coccolithophorids on the basis of electron microscopic studies. *Paleontol. Zhurn.*, **4,** 32–7 (in Russian).

Shumenko, S. I. (1984) Coccolithophorids (Coccolithophorales). In: *Spravochnik po sistematike iskopaemykh organizmov (taksony otryadnoi i vyshikh grupp)* (eds L. P. Tatarinov and V. N. Shimansky), Nauka, Moscow, pp. 151–3. (in Russian).

Sinclair, C. B. and Sharma, G. K. (1971) Epidermal and cuticular studies of leaves. *J. Tenn. Acad. Sci.*, **46**, 2–11.

Sixtel, T. A., Iskandarkhodzhaev, T. A., Savitskaya, L. I. and Stankevich, Yu. V. (1981) To the knowledge of the Palaeozoic vegetation of Middle Asia. In: *Paleobotanika Uzbekistana* (ed. T. A. Sixtel), Vol. 2, pp. 90–255 (in Russian).

Sixtel, T. A., Savitskaya, L. I. and Iskandarkhodzhaev, T. A. (1975) Plants of the Middle Carboniferous, Upper Carboniferous and Lower Permian of Ferghana. In: *Biostratigrafiya verkhnego paleozoya gornogo obramleniya Yuzhnoi Fergany* (ed. T. A. Sixtel), FAN, Tashkent, pp. 11–144 (in Russian).

Skog, J. E. (1982) *Pelletixia amelguita*, a new species of fossil fern in the Potomac Group (Lower Cretaceous). *Amer. Fern J.*, **72**, 115–21.

Skog, J. E. and Banks, H. P. (1973) *Ibyka amphikoma*, gen. et sp. n., a new protoarticulate precursor from the late Middle Devonian of New York State. *Amer. J. Bot.*, **60**, 366–80.

Skog, J. E. and Gensel, P. G. (1980) A fertile species of *Triphyllopteris* from the Early Carboniferous (Mississippian) of Southwestern Vriginia. *Amer. J. Bot.*, **67**, 440–51.

Sladkov, A. N. (1967) *Introduction to the Spore-Pollen Analysis*, Nauka, Moscow (in Russian).

Sluiter, I. R. and Kershaw, A. P. (1982) The nature of Late Tertiary vegetation in Australia. *Alcheringa*, **6**, 211–22.

Smirnov, E. (1925) The theory of type and the natural system. *Z. Indukt. Abst. Vererbungslehre*, **37**, 28–66.

Smit J. and van der Kaars, S. (1984) Terminal Cretaceous extinctions in the Hell Creek Area, Montana: compatible with catastrophic extinctions. *Science*, **223**, 1177–9.

Smith, D. L. (1962) Three fructifications from the Scottish Lower Carboniferous. *Palaeontology*, **5**, 225–37.

Smoot, E. L. and Taylor, T. N. (1978) *Etapteris leclercqii* sp. n., a zygopterid fern from the Pennsylvanian of North America. *Amer. J. Bot.*, **65**, 571–6.

Smoot, E. L. and Taylor, T. N. (1984) The fine structure of fossil plant cell walls. *Science*, **225**, 621–3.

Smoot, E. L., Taylor, T. N. and Delevoryas, T. (1985) Structurally preserved fossil plants from Antarctica. I. *Antarcticycas*, gen. nov., a Triassic cycad stem from the Beardmore Glacier area. *Amer. J. Bot.*, **72**, 1410–23.

Smoot, E. L., Taylor, T. N. and Serlin, B. S. (1982) *Archaeocalamites* from the Upper Mississippian of Arkansas. *Rev. Palaebot. Palynol.*, **36**, 325–34.

Snigirevskaya, N. S. (1962a) To morphology and systematics of the genus *Botryopteris*. *Paleontol. Zhurn.*, **2**, 122–32 (in Russian).

Snigirevskaya, N. S. (1962b) Remains of fructifications of Sphenophyllaceae in coal-balls of Donbass. *Bot. Zhurn.*, **2**, 122–32 (in Russian).

Snigirevskaya, N. S. (1980) *Takhtajanodoxa* Snig. – new link in the evolution of lycopsids. In: *Sistematika i evolyutsiya vysshikh rastenii* (ed. S. G. Zhilin), Nauka, Leningrad, pp. 45–53 (in Russian).

Snigirevskaya, N. S. (1984) To methods of collectioning of fossil woods and their bearing to the problem of the reconstruction of archaeopterids. *Bot. Zhurn.*, **69**, 705–10 (in Russian, English summary).

Snigirevskaya, N. S., Fissunenko, O. P., Ikonnikova, I. K. *et al.* (1977) On the first finding of *Litostrobus* in coal-balls of the Carboniferous of the Donets basin. *Geol. Zhurn.*, **37**(3), 140–7 (in Russian).

Spicer, R. A. (1984) Palaeobotanical implications of explosive volcanism: evidence from Mount Saint Helens and El Chichon. Abstr. of contributed papers and poster session, *2nd. Int. Org. Paleobot. Conf. Edmonton.*

Sporne, K. R. (1965) *The Morphology of Gymnosperms: the Structure and Evolution of Primitive Seed-plants*, Hutchinson, London.

Sporne, K. R. (1974) *The Morphology of Angiosperms: the Structure and Evolution of Flowering Plants*, Hutchinson, London.

Sporne, K. R. (1975) *The Morphology of Pteridophytes. The Structure of Ferns and Allied Plants*, Hutchinson, London.

Sprecher, A. (1907) Le *Ginkgo biloba* L. Atar, Genève.

Stace, C. A. (1965) Cuticular studies as an aid to plant taxonomy. *Bull. Brit. Mus. (Nat. Hist.), Bot.*, **4**, 1–78.

Stach, E. and Pickardt, W. (1957) Pilzreste (Sclerotinit) in paläozoischen Steinkohlen. *Paläontol. Z.*, **31**, 139–62.

Stanislavsky, F. A. (1976) *Middle Keuper Flora of the Donets Basin*, Naukova Dumka, Kiev (in Russian).

Stebbins, G. L. (1950) Variation and evolution in plants. *Columbia Biol. Ser.*, **16**, 1–643.

Stebbins, G. L. (1974) *Flowering Plants. Evolution above the Species Level*, Belknap Press of Harvard University Press, Cambridge, Mass.

Stebbins, G. L. (1982) Perspectives in evolutionary theory. *Evolution*, **36**, 1109–18.

Stein, W. E. (1982) *Iridopteris eriensis* from the Middle Devonian of North America, with systematics of apparently related taxa. *Bot. Gaz.*, **143**, 401–16.

Stein, W. E., Wight, D. C. and Beck, C. B. (1983) *Arachnoxylon* from the Middle Devonian of Southwestern Virginia. *Can. J. Bot.*, **61**, 1283–99.

Stein, W. E., Wight, D. C. and Beck, C. B. (1984) Possible alternatives for the origin of Sphenopsida. *Syst. Bot.*, **9**, 102–18.

Stepanov, S. A. (1975) Phytostratigraphy of key sections of the Devonian of marginal regions of Kuzbass.

Trudy Sibirsk. Inst. Geol. Geofiz. i Mineraln. Syrya, **211,** 1–150 (in Russian).

Stevens, P. F. (1984) Homology and phylogeny: morphology and systematics. *Syst. Bot.*, **9,** 395–409.

Stewart, W. N. (1976) Polystely, primary xylem and the Pteropsida. 21st Sir Albert Charles Seward Memorial lecture 1973, B. Sahni Inst. Palaeobot., Lucknow, 3–13.

Stewart, W. N. (1983) *Paleobotany and the Evolution of Plants*, Cambridge University Press, Cambridge.

Stidd, B. M. (1978) An anatomically preserved *Potoniea* with *in situ* spores from the Pennsylvanian of Illinois. *Amer. J. Bot.*, **65,** 677–83.

Stidd, B. M. (1980) The neotenous origin of the pollen organ of the gymnosperm *Cycadeoidea* and implications for the origin of higher taxa. *Paleobiology*, **6,** 161–7.

Stidd, B. M. (1981) The current status of medullosan seed ferns. *Rev. Palaeobot. Palynol.*, **32,** 63–101.

Stidd, B. M. and Cosentino, K. (1975) *Albugo*-like oogonia from the American Carboniferous. *Science*, **190,** 1092–3.

Stidd, B. M., Leisman, G. A. and Phillips, T. L. (1977) *Sullitheca dactylifera* gen. et sp. n.: a new medullosan pollen organ and its evolutionary significance. *Amer. J. Bot.*, **64,** 994–1002.

Stidd, B. M., Rischbieter, M. O. and Phillips, T. L. (1985) A new lyginopterid pollen organ with alveolate pollen exines. *Amer. J. Bot.*, **72,** 501–8.

Stockey, R. A. (1975) Seeds and embryos of *Araucaria mirabilis*. *Amer. J. Bot.*, **62,** 856–68.

Stockey, R. A. (1981) Some comments on the origin and evolution of conifers. *Can. J. Bot.*, **59,** 1932–40.

Stockey, R. A. (1982) The Araucariaceae: an evolutionary perspectives. *Rev. Palaeobot. Palynol.*, **37,** 133–54.

Stockmans, F. and Streel, M. (1969) Une mégaspore de grande taille au sommet du Givétien à Sart Dame Aveline. *Ann. Soc. Géol. Belge*, **92,** 37–40.

Storch, D. (1966) Die Arten der Gattung *Sphenophyllum* Brongniart im Zwickau-Lugau-Oelsnitzer Steinkohlenrevier. *Paläontol. Abh.*, B2, 195–326.

Storch, D. (1980a) *Sphenophyllum*-Arten aus drei intramontanen Karbonbecken – pflanzengeographische Besonderheiten im mitteleuropäischen Karbon. *Schriftenr. Geol. Wiss. Berlin*, **16,** 171–273.

Storch, D. (1980b) *Sphenophyllum tenerrimum* besaß trilete Sporen. *Wiss. Z. Humboldt- Univ. zu Berlin, Math.-Nat. R.*, **29,** 387–8.

Streel, M. (1972) Dispersed spores associated with *Leclercqia complexa* Banks, Bonamo and Grierson from the late Middle Devonian of eastern New York State (USA). *Rev. Palaeobot. Palynol.*, **14,** 205–15.

Stubblefield, S. and Banks, H. P. (1978) The cuticle of *Drepanophycus spinaeformis*, a longranging Devonian lycopod from New York and Eastern Canada. *Amer. J. Bot.*, **65,** 110–18.

Stubblefield, S. and Rothwell, G. W. (1981) Embryogeny and reproductive biology of *Bothrodendrostrobus mundus* (Lycopsida). *Amer. J. Bot.*, **68,** 625–34.

Stubblefield, S. and Taylor, T. N. (1984) Fungal wall ultrastructure of fossil endogonaceous chlamydospores. Abstr. of contributed papers and poster session, *2nd Int. Org. Paleobot. Conf. Edmonton.*

Stubblefield, S., Taylor, T. N., Miller, C. E. and Cole, G. T. (1983) Studies of Carboniferous fungi. II. The structure and organization of *Mycocarpon*, *Sporocarpon*, *Dubiocarpon*, and *Coleocarpon* (Ascomycotina). *Amer. J. Bot.*, **70,** 1482–98.

Sullivan, H. I. (1967) Regional differences in Mississippian spore assemblages. *Rev. Palaeobot. Palynol.*, **1,** 185–92.

Surange, K. R., Lakhanpal, R. N. and Bharadwaj, D. C. (eds) (1974) *Aspects and Appraisal of Indian Palaeobotany*, B. Sahni Inst. Palaeobot., Lucknow.

Takhtajan, A. L. (1954) *Problems of the Evolutionary Plant Morphology*, Leningrad University Press, Leningrad (in Russian).

Takhtajan, A. L. (1956) *Higher Plants. I. From Psilophytes to Conifers*, Izdat. Akad. Nauk SSSR, Moscow and Leningrad (in Russian).

Takhtajan, A. L. (1966) Main phytochoria of the Late Cretaceous and Palaeocene on the territory of the USSR and neighbouring countries. *Bot. Zhurn.*, **51,** 1217–30 (in Russian).

Takhtajan, A. L. (1972) Patterns of ontogenetic alterations in the evolution of higher plants. *Phytomorphology*, **22,** 164–71.

Takhtajan, A. L. (1978) *Floristic Areas of the Earth*, Nauka, Leningrad (in Russian).

Takhtajan, A. L. (1980) Outline of the classification of flowering plants (Magnoliophyta). *Bot. Rev.*, **46,** 225–359.

Tanai, T. (1972) Tertiary history of vegetation in Japan. In: *Floristics and Paleofloristics of Asia and Eastern North America* (ed. A. Graham), Elsevier, Amsterdam, pp. 235–55.

Tappan, H. (1980) *The Paleobiology of Plant Protists*. Freeman, Oxford.

Tatarinov, L. P. (1984) Cladistic analysis and phylogenetics. *Paleontol. Zhurn.*, **3,** 3–16 (in Russian).

Taylor, T. N. (1967) On the structure and phylogenetic relationships of the fern *Radstockia* Kidston. *Palaeontology*, **10,** 43–6.

Taylor, T. N. (1970) *Lasiostrobus* gen. n., a staminate strobilus of gymnospermous affinity from the Pennsylvanian of North America. *Amer. J. Bot.*, **57,** 670–90.

Taylor, T. N. (1973) A consideration of the morphology, ultrastructure and multicellular microgametophyte of *Cycadeoidea dacotensis* pollen. *Rev. Palaeobot. Palynol.*, **16**, 157–64.

Taylor, T. N. (1978) The ultrastructure and reproductive significance of *Monoletes* (Pteridospermales) pollen. *Can. J. Bot.*, **56**, 3105–18.

Taylor, T. N. (1981) *Paleobotany. An Introduction to Fossil Plant Biology*, McGraw-Hill, New York.

Taylor, T. N. (1982) Ultrastructural studies of Paleozoic seed fern pollen: sporoderm development. *Rev. Palaeobot. Palynol.*, **37**, 29–53.

Taylor, T. N. and Alvin, K. J. (1984) Ultrastructure and development of Mesozoic pollen: *Classopollis*. *Amer. J. Bot.*, **71**, 575–87.

Taylor, T. N. and Brack-Hanes, S. (1976) The structure and reproductive significance of the exine in fossil lycopod megaspores. *Scanning Electron Microscopy*, **7**, 513–18.

Taylor, T. N. and Brauer, D. F. (1983) Ultrastructural studies of *in situ* Devonian spores: *Barinophyton citrulliforme*. *Amer. J. Bot.*, **70**, 106–12.

Taylor, T. N., Cichan, M. A. and Baldoni, A. M. (1984) The ultrastructure of Mesozoic pollen: *Pteruchus dubius* (Thomas) Townrow. *Rev. Palaeobot., Palynol.*, **41**, 319–27.

Taylor, T. N. and Millay, M. A. (1969) Application of the scanning electron microscope in paleobotany. *Proc. 2nd Ann. Scann. Electr. Micr. Sympos. Chicago*, 1–9.

Taylor, T. N. and Millay, M. A. (1977a) Structurally preserved fossil cell contents. *Trans. Amer. Micros. Soc.*, **96**, 390–3.

Taylor, T. N. and Millay, M. A. (1977b) The ultrastructure and reproductive significance of *Lasiostrobus* microspores. *Rev. Palaeobot. Palynol.*, **23**, 129–37.

Taylor, T. N. and Millay, M. A. (1981) Morphologic variability of Pennsylvanian lyginopterid seed ferns. *Rev. Palaeobot. Palynol.*, **32**, 27–62.

Taylor, T. N. and Stockey, R. A. (1976) Studies of Paleozoic seed ferns: anatomy and morphology of *Microspermopteris aphyllum*. *Amer. J. Bot.*, **63**, 1302–10.

Taylor, W. A. (1984) Spore wall ultrastructure in the Sphenophyllales. Abstr. of contributed papers and poster session, *2nd Int. Org. Paleobot. Conf. Edmonton*.

Templeton, A. R. (1981) Mechanisms of speciation – a population genetic approach. *Ann. Rev. Ecol. Syst.*, **12**, 23–48.

Thomas, B. A. (1970) A new specimen of *Lepidostrobus binneyanus* from the Westphalian B of Yorkshire. *Pollen et Spores*, **12**, 217–34.

Thomas, B. A. (1978) Carboniferous Lepidodendraceae and Lepidocarpaceae. *Bot. Rev.*, **44**, 321–64.

Thomas, B. A. and Brack-Hanes, S. D. (1984) A new approach to family groupings in the lycophytes. *Taxon*, **33**, 247–55.

Thomas, B. A. and Masarati, D. L. (1982) Cuticular and epidermal studies in fossil and living lycophytes. *Linn. Soc. Symp. Ser.*, **10**, 363–78.

Thomas, B. A. and Meyen, S. V. (1984) A system of form-genera for the Upper Palaeozoic lepidophyte stems represented by compression–impression material. *Rev. Palaeobot. Palynol.*, **41**, 273–82.

Tiffney, B. H. (1977) Dicotyledonous angiosperm flower from the Upper Cretaceous of Martha's Vineyard Massachusetts. *Nature*, **262**, 136–7.

Tiffney, B. H. and Barghoorn, E. S. (1974) The fossil record of the fungi. *Occas. Pap. Farlow Herbar. Cryptogam. Bot. Harvard Univ.*, **7**, 1–42.

Tims, J. D. and Chambers, T. C. (1984) Rhyniophytina and Trimerophytina from the early land flora of Victoria, Australia. *Palaeontology*, **27**, 265–79.

Tomlinson, P. B. (1974) Development of the stomatal complex as a taxonomic character in the monocotylodons. *Taxon*, **23**, 109–28.

Tomlinson, P. B. (1984) Homology: an empirical view. *Syst. Bot.*, **9**, 374–81.

Townrow, J. A. (1955–1956) On some species of *Phyllotheca*. *J. and Proc. Roy. Soc. NSW*, **89**(1), 39–63.

Townrow, J. A. (1957) On *Dicroidium*, probably a pteridospermous leaves and other leaves now removed from this genus. *Trans. Geol. Soc. South Africa*, **60**, 1–36.

Townrow, J. A. (1962) On some bisaccate pollen grains of Permian to Middle Jurassic age. *Grana Palynol.*, **3**(2), 13–44.

Townrow, J. A. (1967) On *Rissikia* and *Mataia* podocarpaceous conifers from the Lower Mesozoic of Southern lands. *Pap. and Proc. Roy. Soc. Tasmania*, **101**, 103–36.

Traverse, A. (1982) Response of world vegetation to Neogene tectonic and climatic events. *Alcheringa*, **6**, 197–209.

Trivedi, B. S. (1982) Palms: the big game of plants. *Geophytology*, **12**, 137–43.

Trivett, M. L. and Rothwell, G. W. (1985) Morphology, systematics and paleoecology of Paleozoic fossil plants: *Mesoxylon priapi*, sp. nov. (Cordaitales). *Systematic Bot.*, **10**, 205–223.

Troll, W. (1928) Organisation und Gestalt im Bereich der Blüte. *Monogr. Wiss. Bot. Berlin*, **1**, 1–413.

Troll, W. (1937–1939) *Vergleichende Morphologie der höheren Pflanzen*. 1. Bd. *Vegetationsorgane*. Teil 1 (1937), Teil 2 (1939). Borntraeger, Berlin.

Tschudy, R. H. (1984) Palynological evidence for change in continental floras at the Cretaceous–Tertiary boundary. In: *Catastroph in Earth History. The new uniformitarianism* (W. H. Berggren, and J. A. Van-Couvering, eds), Princeton University Press, Princeton, pp. 315–37.

Tschudy, R. H. and Scott, R. A. (eds) (1969) *Aspects of Palynology*, Wiley, New York.

Upchurch, G. R. (1984) Cuticle evolution in Early Cretaceous angiosperms from the Potomac Group of Virginia and Maryland. *Ann. Mo. Bot. Gard.*, **71**, 522–50.

Upchurch, G. R. and Dilcher, D. L. (1984) Magnoliid leaves from the Cretaceous Dakota formation of Nebraska and Kansas. Abstr. of contributed papers and poster session, *2nd Int. Org. Paleobot. Conf. Edmonton.*

Vakhrameev, V. A. (1952) The stratigraphy and the fossil flora of the Cretaceous deposits of western Kazakhstan. *Regional. Stratigr. SSSR*, **1**, 1–340 (in Russian).

Vakhrameev, V. A. (1964) The Jurassic and Early Cretaceous floras of Eurasia and palaeofloristic provinces of this time. *Trudy Geol. Inst. Akad. Nauk SSSR*, **102**, 1–263 (in Russian).

Vakhrameev, V. A. (ed.) (1977) *Development of Floras at the Mesozoic–Cenozoic Boundary*, Nauka, Moscow.

Vakhrameev, V. A. (1978) Climates of the Northern Hemisphere in the Cretaceous period, and palaeobotanical data. *Paleontol. Zhurn.*, **2**, 3–17 (in Russian).

Vakhrameev, V. A. (1981) Development of floras in the middle part of the Cretaceous period, and older angiosperms. *Paleontol. Zhurn.*, **2**, 3–14 (in Russian).

Vakhrameev, V. A. (1984) Climatic and phytogeographic zonality of the Earth in the Early Cretaceous. *Ezhegodn. Vsesoyuzn. Paleontol. Obshch.*, **27**, 199–209 (in Russian).

Vakhrameev, V. A. (1985) Phytogeography, palaeoclimates and position of continents in the Mesozoic. *Vestnik Akad. Nauk SSSR*, **8**, 30–42 (in Russian).

Vakhrameev, V. A., Dobruskina, I. A., Meyen, S. V. and Zaklinskaja, E. D. (1978) *Paläozoische und mesozoische Floren Eurasiens und die Phytogeographie dieser Zeit*, G. Fischer, Jena.

Vakhrameev, V. A., Dobruskina, I. A., Zaklinskaya, E. D. and Meyen, S. V. (1970) Palaeozoic and Mesozoic floras of Eurasia and the phytogeography of this time. *Trudy Geol. Inst. Akad. Nauk SSSR*, **208**, 1–426 (in Russian).

Vakhrameev, V. A. and Doludenko, M. P. (1961) Upper Jurassic and Lower Cretaceous flora of Bureja Basin and its significance for stratigraphy. *Trudy Geol. Inst. Akad. Nauk SSSR*, **54**, 1–136 (in Russian).

Vakhrameev, V. A. and Kotova, I. Z. (1977) Older angiosperms and associating plants in the Lower Cretaceous deposits of Transbaikalia. *Paleontol. Zhurn.*, **4**, 101–9 (in Russian).

Vakhrameev, V. A. and Krassilov, V. A. (1979) Reproductive organs of flowering plants from the Albian of Kazakhstan. *Paleontol. Zhurn.*, **1**, 121–8 (in Russian).

Valentine, J. W. (1980) Determinants of diversity in higher taxonomic categories. *Paleobiology*, **6**, 444–50.

Van Cotthem, W. R. J. (1970) A classification of stomatal types. *Bot. J. Linn. Soc.*, **63**, 235–46.

Van der Zwan, C. J. (1979) Aspects of Late Devonian and Early Carboniferous palynology of southern Ireland. I. The *Cyrtospora cristifer* morphon. *Rev. Palaeobot. Palynol.*, **28**, 1–20.

Van Steenis, C. G. G. J. (1969) Plant speciation in Malesia, with special reference to the theory of non-adaptive saltatory evolution. *Biol. J. Linn. Soc.*, **1**, 97–133.

Vavilov, N. I. (1922) The law of homologous series in variation. *J. Genetics*, **12**, 47–89 (amplified Russian edition in: 1967. Selected papers. I. Nauka, Leningrad, pp. 7–61).

Vincent, E. E. and Basinger, J. F. (1984) Fossil cones and seeds of *Cupressinocladus* from Paleocene deposits in Southwestern Saskatchewan, Canada. Abstr. of contributed papers and poster session, *2nd Int. Org. Paleobot. Edmonton.*

Visscher, H. (1971) The Permian and Triassic of the Kingscourt outlier, Ireland. A palynological investigation related to regional stratigraphical problems in the Permian and Triassic of Western Europe. *Geol. Surv. Ireland Spec. Pap.* **1**, 1–114.

Vladimirovich, V. P. (1981) *Artikskian Flora of Urals*, VINITI, 2381–81 Dep., 1–51 (in Russian).

Vladimirovich, V. P. (1982) *Ufimian Type Flora of Urals*, VINITI, 3470–82 Dep., 1–101 (in Russian).

Vladimirovich, V. P. (1984) *Kazanian Type Flora of Prikamie*, VINITI, 4571–84 Dep., 1–91 (in Russian).

Voigt, W. (1973) *Homologie und Typus in der Biologie*, G. Fischer, Jena.

Volkova, N. A. (1965) On the nature and classification of microfossils of plant origin from the Precambrian and Lower Palaeozoic. *Paleontol. Zhurn.*, **1**, 13–25 (in Russian).

Volkova, N. A. (1974) Types of damages of envelopes of the Precambrian and Cambrian acritarchs. *Paleontol. Zhurn.*, **4**, 101–8 (in Russian).

Volkova, N. A. (1976) On finding of Precambrian spores with tetrad scar. In: *Paleontologiya. Morskaya geologiya*, Nauka, Moscow, pp. 14–18 (in Russian, English summary).

Vorobyeva, E. I. (1980) Phylogenetic aspects of palaeontology. *Zhurn. Obshch. Biol.*, **41**, 507–21 (in Russian, English summary).

Vozzhennikova, T. F. (1979) Dinocysts and their stratigraphic significance. *Trudy Inst. Geol. i Geofiz.*, SO AN SSSR **422**, 1–223 (in Russian).

Waddington, C. H. (1962) *New Pattern in Genetics and Development*, Columbia University Press, New York and London.

Wagner, C. A. and Taylor, T. N. (1982) Fungal chlamydospores from the Pennsylvanian of North America. *Rev. Palaeobot. Palynol.*, **37**, 317–28.

Wagner, R. H. (1958) *Pecopteris pseudobucklandi* Andrae and its general affinities. *Meded. Geol. Stichting*, n. ser., **12**, 25–30.

Wagner, R. H. (1958–1959) *Lobatopteris alloiopteroides*, una nueva especíe de Pecopteridea del Estefaniense a Español. *Estud. Geol.*, **14**, 81–107.

Wagner, R. H. (1962) On a mixed Cathaysia and Gondwana flora from SE. Anatolia (Turkey). *C. R. 4-e Congr. Avanc. et Étud. Strat. et Géol, Carbonifère, Heerlen, 1958*, T. 3, Maestricht, pp. 745–52.

Wagner, R. H. (1965) Stephanian B flora from the Ciñera-Matallana coalfield (León) and neighbouring outliers. III. *Callipteridium* and *Alethopteris*. *Notas y communic. Inst. Geol. Min. España*, **78**, 5–69.

Wagner, R. H., Soper, N. J. and Higgins, A. K. (1982) A Late Permian flora of Pechora affinity in North Greenland. *Grønl. Geol. Unders.*, **108**, 5–13.

Wagner, R. H. and Spinner, E. (1976) *Bodeodendron*, tronc associé à *Sporangiostrobus*. *C. R. Acad. Sci. Paris*, **282**, 353–6.

Wagner, R. H., Winkler Prins, C. F. and Granados, L. F. (eds) (1983) The Carboniferous of the World. I. China, Korea, Japan and S. E. Asia. *IUGS Publ. 16*, Inst. Geol. Min. España, Madrid.

Walker, J. W. and Doyle, J. A. (1975) The bases of angiosperm phylogeny: palynology. *Ann. Mo. Bot. Gard.*, **62**, 664–723.

Walter, M. R. (ed.) (1976) *Stromatolites. Developments in Sedimentology* 20. Elsevier, Amsterdam.

Walton, J. (1949) On some Lower Carboniferous Equisetineae from the Clyde Area. *Trans. Roy. Soc. Edinburgh*, **61**, 729–36.

Wang Ziqiang. (1985) Palaeovegetation and plate tectonics: palaeophytogeography of North China during Permian and Triassic times. *Palaeogeogr. Palaeoclimatol. Palaeoecol.*, **49**, 25–45.

Watson, J. (1977) Some Lower Cretaceous conifers of the Cheirolepidiaceae from the USA and England. *Palaeontology*, **20**, 715–49.

Wendel, R. (1980) *Callipteridium pteridium* (Schlotheim) Zeiller im Typusgebiet des Saaletrog. *Schriftenr. Geol. Wiss. Berlin*, **16**, 107–70.

White, D. (1912) The characters of the fossil plant *Gigantopteris* Schenk and its occurrence in North America. *Proc. US Nat. Mus.*, **41**, 493–514.

White, M. E. (1981) Revision of the Talbragar fish bed flora (Jurassic) of New South Wales. *Rec. Austral. Mus.*, **33**, 695–721.

Wiley, E. O. (1981) *Phylogenetics. The Theory and Practice of Phylogenetic Systematics*, Wiley, New York.

Wiley, E. O. and Brooks, D. R. (1982) Victims of history – a nonequilibrium approach to evolution. *Syst. Zool.*, **31**, 1–24.

Wiley, E. O. and Brooks, D. R. (1983) Nonequilibrium thermodynamics and evolution: a response to Løvtrup. *Syst. Zool.*, **32**, 209–19.

Williams, G. L. (1978) Dinoflagellates, acritarchs and tasmanitids. In: *Introduction to Marine Micropaleontology* (eds B. U. Haq and A. Boersma), Elsevier, New York, pp. 293–326.

Wilson, G. J. and Clowes, C. D. (1980) A concise catalogue of organic-walled fossil dinoflagellate genera. *N.Z. Geol. Surv. Lower Hutt Rep.*, **92**, 1–199.

Winston, R. B. (1983) A Late Pennsylvanian upland flora in Kansas: systematics and environmental implications. *Rev. Palaeobot. Palynol.*, **40**, 5–31.

Wolfe, J. A. (1978) A paleobotanical interpretation of Tertiary climates in the Northern Hemisphere. *Amer. Sci.*, **66**, 694–703.

Wolfe, J. A. (1980) Tertiary climates and floristic relationships at high latitudes in the Northern Hemisphere. *Palaeogeogr. Palaeoclimatol. Palaeoecol.*, **30**, 313–23.

Wolfe, J. A. (1981) Paleoclimatic significance of the Oligocene and Neogene floras of the Northwestern United States. In: *Paleobotany, Paleoecology, and Evolution* (ed. K. J. Niklas), Praeger, New York, pp. 79–101.

Wray, J. L. (1977) Late Paleozoic calcareous red algae. In: *Fossil Algae. Recent Results and Developments* (ed. E. Flügel), Springer, Berlin, pp. 167–76.

Wray, J. L. (1978) Calcareous algae. In: *Introduction to Marine Micropaleontology* (eds B. U. Haq and E. Boersma) Elsevier, New York, pp. 171–87.

Wulff, E. V. (1943) Historical plant geography. Chronica Botanica, Waltham, Mass. (English edition not seen, cited after Stebbins, 1974).

Xiao Xuchang and Gao Yanlin (1984) Tectonic evolution of Tethys-Himalayas in China. In: Tektonika Azii (ed. A. L. Yanshin), *27th Int. Geol. Congr. Papers*, Vol. 5, Nauka, Moscow, pp. 150–8. (in Russian).

Yao Zhao-qi and Ouyang Shu (1980) On the Paleophyte–Mesophyte boundary. Paper for the 5th Int. Palynol. Conf. Nanjing Inst. Geol. Palaeontol. Acad. Sinica, 1–9.

Zakharova, T. V. (1985) *The flora of the Byskarsk Series of Minussa depression and its stratigraphic significance.* Abstract of Thesis, Tomsk.

Zalessky, M. D. (1927) Flore permienne des limites ouraliennes de l'Angaride. *Mém. Com. Géol.*, n. sér., **176**, 1–52.

Zalessky, M. D. (1934) Observations sur les végétaux

permien du bassin de la Petchora, I. *Bull. Acad. Sci. URSS, Cl. Sci. Math. et Natur.*, **2–3,** 241–90.

Zalessky, M. D. (1937) Sur la distinction de l'étage Bardien dans le Permien de l'Oural et sur sa flore fossile. *Probl. Paleontol. Moscow Univ.*, 2–3, 37–101.

Zalessky, M. D. and Čirkova, H. F. (1935) Observations sur quelques végétaux fossiles du terrain permien du bassin de Kousnetzk. *Bull. Acad. Sci. URSS, cl. sci. math. et natur.* **8–9,** 1091–16.

Zavada, M. S. (1983) Comparative morphology of monocot pollen and evolutionary trends of apertures and wall structures. *Bot. Rev.*, **49,** 331–79.

Zavarzin, G. A. (1974) *Phenotypic systematics of bacteria. Space of logical possibilities*, Nauka, Moscow (in Russian).

Zdebska, D. (1982) A new zosterophyll from the Lower Devonian of Poland. *Palaeontology*, **25,** 247–63.

Zherikhin, V. V. (1979) Use of palaeontological data in ecological prognoses. In: *Ekologicheskoe prognozirovanie* (ed. N. N. Smirnov), Nauka, Moscow, pp. 113–32 (in Russian).

Zherikhin, V. V. and Rasnitsyn, A. P. (1980) Biocoenotic regulation of macroevolutionary processes. In: *Micro- and macroevolution* (eds K. L. Paaver and T. Sutt), Tartu Riiklik Ülikool, Tartu, pp. 77–81.

Zhilin, S. G. (1984) *Main stages of the formation of the temperate forest flora in the Oligocene–Early Miocene of Kazakhstan.* Leningrad: Nauka (in Russian).

Zhou Zhiyan. (1983) A heterophyllous cheirolepidiaceous conifer from the Cretaceous of East China. *Palaeontology*, **26,** 789–911.

Zhuravleva, Z. A. (1964) Oncolites and catagraphia of the Riphaean and Lower Cambrian and their stratigraphic significance. *Trudy Geol. Inst. Akad. Nauk SSSR*, **114,** 1–99 (in Russian).

Zimmermann, R. P. and Taylor, T. N. (1970–1971) The ultrastructure of Paleozoic megaspores membranes. *Pollen et Spores*, **12,** 451–68.

Zimmermann, W. (1930) *Die Phylogenie der Pflanzen. Ein Überblick über Tatsachen und Probleme*, G. Fischer, Jena.

Zimmermann, W. (1959) *Die Phylogenie der Pflanzen. Ein Überblick über Tatsachen und Problem*, G. Fischer, Stuttgart.

Zimmermann, W. (1966) *Die Telomtheorie*, G. Fischer, Stuttgart.

Zoricheva, A. I. and Sedova, M. A. (1954) Spore-pollen assemblages of the Upper Permian deposits of some regions of the north of the European part of the USSR. In: *Materialy po palinologii i stratigrafii. Sborn. Stat.*, Gosgeoltekhizdat, Moscow, pp. 160–201 (in Russian).

Index

Page numbers in *italic* refer to descriptions; page numbers in **bold** refer to figures.